$$\alpha = 0.05$$

ν_2	ν_1									
	10	12	15	20	24	30	40	60	120	∞
1	241.9	243.9	245.9	248.0	249.1	250.1	251.1	252.2	253.3	254.3
2	19.40	19.41	19.43	19.45	19.45	19.46	19.47	19.48	19.49	19.50
3	8.79	8.74	8.70	8.66	8.64	8.62	8.59	8.57	8.55	8.53
4	5.96	5.91	5.86	5.80	5.77	5.75	5.72	5.69	5.66	5.63
5	4.74	4.68	4.62	4.56	4.53	4.50	4.46	4.43	4.40	4.36
6	4.06	4.00	3.94	3.87	3.84	3.81	3.77	3.74	3.70	3.67
7	3.64	3.57	3.51	3.44	3.41	3.38	3.34	3.30	3.27	3.23
8	3.35	3.28	3.22	3.15	3.12	3.08	3.04	3.01	2.97	2.93
9	3.14	3.07	3.01	2.94	2.90	2.86	2.83	2.79	2.75	2.71
10	2.98	2.91	2.85	2.77	2.74	2.70	2.66	2.62	2.58	2.54
11	2.85	2.79	2.72	2.65	2.61	2.57	2.53	2.49	2.45	2.40
12	2.75	2.69	2.62	2.54	2.51	2.47	2.43	2.38	2.34	2.30
13	2.67	2.60	2.53	2.46	2.42	2.38	2.34	2.30	2.25	2.21
14	2.60	2.53	2.46	2.39	2.35	2.31	2.27	2.22	2.18	2.13
15	2.54	2.48	2.40	2.33	2.29	2.25	2.20	2.16	2.11	2.07
16	2.49	2.42	2.35	2.28	2.24	2.19	2.15	2.11	2.06	2.01
17	2.45	2.38	2.31	2.23	2.19	2.15	2.10	2.06	2.01	1.96
18	2.41	2.34	2.27	2.19	2.15	2.11	2.06	2.02	1.97	1.92
19	2.38	2.31	2.23	2.16	2.11	2.07	2.03	1.98	1.93	1.88
20	2.35	2.28	2.20	2.12	2.08	2.04	1.99	1.95	1.90	1.84
21	2.32	2.25	2.18	2.10	2.05	2.01	1.96	1.92	1.87	1.81
22	2.30	2.23	2.15	2.07	2.03	1.98	1.94	1.89	1.84	1.78
23	2.27	2.20	2.13	2.05	2.01	1.96	1.91	1.86	1.81	1.76
24	2.25	2.18	2.11	2.03	1.98	1.94	1.89	1.84	1.79	1.73
25	2.24	2.16	2.09	2.01	1.96	1.92	1.87	1.82	1.77	1.71
26	2.22	2.15	2.07	1.99	1.95	1.90	1.85	1.80	1.75	1.69
27	2.20	2.13	2.06	1.97	1.93	1.88	1.84	1.79	1.73	1.67
28	2.19	2.12	2.04	1.96	1.91	1.87	1.82	1.77	1.71	1.65
29	2.18	2.10	2.03	1.94	1.90	1.85	1.81	1.75	1.70	1.64
30	2.16	2.09	2.01	1.93	1.89	1.84	1.79	1.74	1.68	1.62
40	2.08	2.00	1.92	1.84	1.79	1.74	1.69	1.64	1.58	1.51
60	1.99	1.92	1.84	1.75	1.70	1.65	1.59	1.53	1.47	1.39
120	1.91	1.83	1.75	1.66	1.61	1.55	1.50	1.43	1.35	1.25
∞	1.83	1.75	1.67	1.57	1.52	1.46	1.39	1.32	1.22	1.00

This figure is used by permission of the Biometrika Trust.

PROBABILITY AND STATISTICS

FOR THE ENGINEERING, COMPUTING, AND PHYSICAL SCIENCES

Edward R. Dougherty

Center for Imaging Science
Rochester Institute of Technology

Prentice Hall, Englewood Cliffs, New Jersey 07632

Library of Congress Cataloging-in-Publication Data

Dougherty, Edward R.
    Probability and statistics for the engineering, computing, and
physical sciences / Edward R. Dougherty.
        p.   cm.
    Bibliography: p.
    Includes index.
    ISBN 0-13-711995-X
    1. Statistics.   2. Probabilities.   I. Title.
QA276.12.D66   1990
519.5--dc20
                                              89-35709
                                                  CIP

Editorial/production supervision: **Kathleen Schiaparelli**
Interior design: **Joan Greenfield**
Cover design: **Photo Plus Art**
Manufacturing buyer: **Margaret Rizzi**

 © 1990 by Prentice-Hall, Inc.
A Division of Simon & Schuster
Englewood Cliffs, New Jersey 07632

Printed in the United States of America
10 9 8 7 6 5 4 3 2 1

ISBN 0-13-711995-X

Prentice-Hall International (UK) Limited, *London*
Prentice-Hall of Australia Pty. Limited, *Sydney*
Prentice-Hall Canada Inc., *Toronto*
Prentice-Hall Hispanoamericana, S.A., *Mexico*
Prentice-Hall of India Private Limited, *New Delhi*
Prentice-Hall of Japan, Inc., *Tokyo*
Simon & Schuster Asia Pte. Ltd., *Singapore*
Editora Prentice-Hall do Brasil, Ltda., *Rio de Janeiro*

TO MY THREE SONS,

RUSSELL MICHAEL

JOHN DOUGLAS

EDWARD SEAN

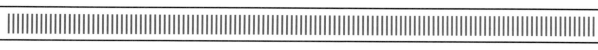

CONTENTS

3 PROBABILITY DISTRIBUTIONS   76

4 IMPORTANT PROBABILITY DISTRIBUTIONS   136

5 MULTIVARIATE DISTRIBUTIONS   199

6 TOPICS IN APPLIED PROBABILITY   265

7 ESTIMATION 309

8 HYPOTHESIS TESTS CONCERNING A SINGLE PARAMETER 374

9 INFERENCES CONCERNING TWO PARAMETERS   461

10 HYPOTHESIS TESTS CONCERNING CATEGORICAL DATA   501

11 ANALYSIS OF VARIANCE   520

12 SIMPLE LINEAR REGRESSION   556

13   MULTIPLE LINEAR REGRESSION   598

14   EXPERIMENTAL DESIGN   636

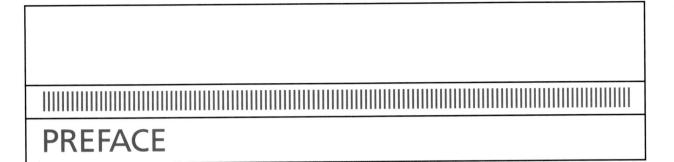

PREFACE

To meet the needs of students in the engineering, computing, and physical sciences, a textbook on probability and statistics should accomplish five basic goals. Specifically, it should focus on contemporary applications, maintain a respectable mathematical level, provide a solid probabilistic framework, give a wide cross section of statistical techniques, and use examples that illustrate models and experimental situations relevant to the audience. The present book aims to accomplish these goals in a text suitable for students with a standard undergraduate calculus background.

Among the key features of the text are

- A consistent blend of theory and examples.
- Approximately 350 carefully worked-out examples.
- Examples that explore special applications and can be employed to tailor a course.
- Over 1000 exercises geared to individual sections.
- Approximately 300 supporting figures.
- Clearly demarcated proofs.
- Chapters that can be skipped without loss of continuity.
- Sections that can be omitted with little or no loss of theoretical continuity.
- In general, the provision of extensive coverage in a setting that allows wide flexibility.

Examples and exercises remain within the disciplines to which the text is geared and illustrate important applications. Topics include the standard areas of manufacturing, quality control, reliability, and materials testing. Also given prominence are modern applications in fields such as computer design, artificial intelligence, robotics, communications, and both image and signal processing.

The text can be used for either a one- or two-semester course. Brief comments on chapter coverage given below will suggest the adaptability to either option. Certainly, full (or almost full) coverage of the material requires two semesters.

Chapter 1 discusses data and measurement. Besides the usual descriptive statistics, it incorporates into the discussion of data the relative-frequency interpretation of probability. Since the sample mean and variance will refer to estimators, this chapter avoids that terminology when referring to their empirical counterparts.

Chapter 2 covers probability spaces. It introduces the usual fundamental properties and gives detailed discussions of counting, conditional probability, and independence. The latter is illustrated by problems on series and parallel systems.

Random variables and expectation are presented in **Chapter 3.** Moment-generating functions, which are used extensively in the text, are discussed in some detail. If desired, Sections 3.3 and 3.4 can be covered lightly.

Chapter 4 discusses a fairly large number of important distributions. The first four sections are dedicated to discrete distributions, while the remainder cover continuous families. Because of its role in computer science, the Poisson distribution receives special emphasis. The normal distribution is discussed extensively, as is the gamma family. Sections 4.2, 4.3, 4.7, and 4.8 can be skipped without undue loss of continuity.

Multivariate distributions, which introductory texts often treat far too lightly, are covered extensively in **Chapter 5.** A good understanding of the multivariate setting is crucial for a deep appreciation of the statistical material in later chapters. It is also important for the type of applications one meets in both engineering and computer science.

Chapter 6 is geared toward applications that are of growing importance to a broad range of disciplines. Although the chapter is independent of the rest of the text, its location is pedagogically appropriate, and, if time permits, it will provide meaningful applications of probability theory. All sections are independent.

Estimation begins the coverage of statistics. Throughout **Chapter 7** and the remainder of the text, care is taken to consistently differentiate between *estimators* and *estimates,* since both the concepts and the language are typically confusing to beginning students. The chapter is structured to provide wide flexibility in coverage. For instance, Sections 7.5 and 7.6 can be covered lightly by omission of some of the detailed examples. Note that the t distribution is introduced in Section 7.8.3. It could have been introduced earlier; however, the student tends to be better motivated when it naturally arises. Should lecture time be available, Section 7.9 can be used to provide an introduction to Bayesian estimation.

A central chapter in any statistics book is the one that introduces hypothesis testing. **Chapter 8** begins with three sections that treat the subject in some generality. The third concentrates on operating characteristic curves, which often receive insufficient attention. Not only are they useful for providing specific numerical results, but they also give an engineer practical insight into both type II error and sample size. Sections 8.4, 8.5, 8.6, and 8.7 give standard applications. Best tests, including the Neymann-Pearson lemma, are discussed in Section 8.8. Although the section can be omitted, there needs to be at least some reference to efficiency. Section 8.9 introduces distribution-free tests, and the chapter closes with an independent section on quality control charts.

Chapter 9 introduces two-parameter inferences. Two points should be noted. First, the design considerations discussed in Section 9.3 on paired observations are important to Chapter 14. Second, the F distribution, so basic to analysis of variance, is introduced in Section 9.6.1, at the point where it naturally arises.

Chapter 10 treats categorical data, the unifying theme being application of the χ^2 test. Goodness-of-fit and testing for independence and homogeneity are considered. The chapter can be skipped in its entirety without loss of continuity.

Analysis of variance is introduced in **Chapter 11.** One-way classification is discussed for both fixed and random effects. The thrust of the chapter is toward understanding the analysis-of-variance approach. Thus, emphasis is placed on the underlying models and the interplay between within-treatment and between-treatment variation. After completing Chapter 11, one can proceed to linear regression or experimental design.

The simple linear regression model is discussed in **Chapter 12.** Least-squares estimation is introduced and the properties of the relevant least-squares estimators are developed. Inferences concerning the model parameters are then discussed. Section 12.6 discusses inferences concerning the correlation coefficient (although this discussion could easily have been put elsewhere).

Chapter 13 concerns the multiple linear regression model. Matrix representation is introduced early in the chapter. Least-squares estimators are found for the general linear model, properties of the estimators are discussed, and inferences pertaining to the model parameters are developed. Polynomial regression is treated as a special case of the general linear model.

Chapter 14 is intended to familiarize students with the principles, mathematical models, and applications of experimental design. Blocking and two-way-classification analysis of variance are followed by an optional section on three-way classification. Detailed sections follow on 2^k factorial design and confounding in the 2^k design.

A year course will ideally cover a good portion of each chapter. It should at least cover the main bodies of Chapters 1 through 5 and Chapters 7, 8, 9, 11, 12, and 14. Such a core should leave ample time to discuss the topics in Chapters 6, 10, and 13 that are most appropriate to a particular student audience. For a one-semester course, there is much less flexibility. One would do well to cover necessary portions of Chapters 1 through 5 (some lightly) and basic topics in Chapters 7, 8, 9, 11, and 12.

COMPUTER USE

In deciding how to include the use of computer software, I was guided by four considerations: (1) providing the student with a general introduction to major competing software packages, (2) providing problems that could be solved by computer in settings where such problems naturally arise, (3) avoiding the sacrifice of statistical coverage to fit in software coverage, (4) avoiding the linking of the text to a particular software package that some might not possess or that might become obsolete. The decision reached was (1) to include an appendix that introduces several commercially available packages and (2) to provide specially marked problems at the end of those chapters where solution by computer is especially appropriate.

ACKNOWLEDGMENTS

Many people contributed to the completion of this text. To all, including the many anonymous reviewers, I offer my sincere appreciation for the advice and criticism that has served to help shape it. I offer specific acknowledgment to Professors Harry Sherman and Francis Sand for providing solutions to the exercises, and to Professor Sand for preparing an appendix on several major computer software packages. Finally, I offer my deepest appreciation to Professor Sand for the numerous conversations that contributed to the final form and content of the text.

Edward R. Dougherty

INTERPRETATION AND DESCRIPTION OF DATA

Statistics involves techniques for making inferences from data in the context of a suitable probability model. This chapter concerns the organization, description, and interpretation of data, and it includes a detailed discussion of the relative-frequency interpretation of probability.

1.1 MEASUREMENT

The twentieth century has witnessed the ever-expanding role of probability and statistics. Even a cursory view of diverse fields such as engineering, psychology, physics, computer science, and economics demonstrates the importance of the statistical approach to scientific understanding. Before entering the main body of this text, we would like to provide some of the fundamental intuitions that drive the probability model and result in the statistical approach to knowledge.

It is the process of measurement, and our understanding of that process, that determines the scientific method. There is nothing wrong with utilizing differential equations to describe behavior; nonetheless, the empirical quantities that give physical relevance to the equations must be determined by some measurement technique. Consequently, the measurement process plays a primary role in forming the kind of knowledge possessed by the empirical sciences.

Consider, for example, weighing a particular object from a collection A of objects. The essential features of the process are (1) the selection of an object w from A, (2) the employment of a physical device (scale) to determine a quantity that is in some manner descriptive of w, and (3) the observation of some metering device that is indicative of the property (weight) being described. Most likely, one is not interested in the specific object w, but rather in learning something about the collection A. In practice, w might simply be the residue of some experiment, and the measurement process might merely be part of some complex procedure (based upon some exper-

1

imental model) which will lead to some understanding of physical behavior. Ultimately, the measurement process will yield knowledge concerning the elements of A. But what sort of knowledge can be obtained in this manner?

To answer the question, let X denote the outcome of the measurement, and suppose we already possess knowledge concerning the behavior under investigation. Even if the same object from A has been weighed previously and the outcome of that process recorded, we cannot expect with certainty to obtain the identical result. Not only has the object possibly changed over time, but there is also variation inherent in the measuring device. The latter, often referred to as "experimental error," is not error in the naive sense; rather, it is intrinsic to the experimental methodology. In sum, our knowledge concerning the measurement X is not deterministic. At best, we perhaps can make confident statements regarding the likelihood of X falling within certain bounds. In short, X is a random quantity and our fundamental knowledge regarding the quality of A under consideration is probabilistic. Even if a set of differential equations is employed to build a mechanical or electrical system, the input to that system, and hence its output, is inherently nondeterministic. Consequently, analysis of the design and quality of the system must start from a probabilistic foundation.

Given a measurement process, with X the random quantity to be measured, we wish to make statements of the form

$$P(a < X < b) = p$$

That is, "The probability of X falling exclusively between a and b is equal to p." The subject matter of probability theory is the development of a systematic mathematical understanding of such (and related) statements.

For a probabilistic analysis to be of scientific value, it must be empirically grounded in the data of experimentation. Although we will often discover practical probability models through reasoning, the test of a model ultimately depends on how it fares relative to actual data. The study of the relationship between probability models and the data of observation belongs to the domain of statistics.

1.2 FREQUENCY DISTRIBUTIONS

Although our main concerns are with the construction of probability models of behavior and with the predictive and decision-making techniques which result, the source of our scientific knowledge lies in *data*. It is therefore appropriate to organize data so as to make them more "comprehensible," rather than to leave them in the **raw** state in which they have been collected.

One way of organizing data is to construct a **frequency distribution** by placing the data into classes and recording the number of data in each class. This latter quantity is the **class frequency.** If there are m classes, then the individual class frequencies will be denoted by $f_1, f_2, \ldots , f_m$.

Suppose the data of Table 1.1 have resulted from testing the lifetimes of 60-watt light bulbs, each datum referring to the number of hours a test bulb survived before burning out. A possible frequency-distribution representation of the data is given in Table 1.2. Several points should be noted. First, eight classes have been chosen in which to place the data. Each class is of **length** 50 (hours), and the extent of the classes is sufficient to include all of the data. Although the selection of equal-length classes is not necessary, it is certainly convenient and quite common. An exception might be the extension of the lower- and upper-end classes in the event

TABLE 1.1 BULB LIFE IN HOURS

963.4	874.8	901.3	822.5	1066.2	939.0	822.9	1023.9
1175.9	1001.7	988.8	950.1	900.0	1092.7	1114.4	1056.2
1074.7	1198.2	1074.1	932.9	1142.2	1132.7	1166.2	1002.4
887.1	1003.4	1109.8	810.1	1152.8	1083.8	1122.8	1187.0
1078.3	1130.2	1087.3	1042.1	1093.1	989.8	1085.1	1023.8
1065.4	1092.7	1114.1	1129.8	1049.9	1021.0	951.9	909.2
1124.8	1143.8	1089.5	995.3	1078.4	1114.1	1001.2	1059.2
1083.8	1021.7	1133.5	1129.8	1021.3	1130.0	956.5	1078.7
882.6	976.1	1072.8	1021.3	1045.0	1121.9	1089.3	1092.1
1055.5	987.2	866.0	1102.1	1123.6	1058.4	1033.2	1066.3
1132.5	1108.4	922.0	970.0	1121.8	1149.9	949.9	984.3
1052.1	1099.3	1121.8	910.0	962.1	1028.1	1043.7	1112.1
1092.2	1075.6	1142.0	1060.1				

there is extreme data at one or the other ends of the scale. However, even in the presence of extreme data, one might very well choose to include classes with zero frequency in order to keep the uniformity of class length.

The center of each class is characterized by a **class mark** x_i. Since the actual raw values are not present in the frequency distribution, the class mark x_i will often be used to represent each datum in the ith class during computations.

Data organized into a frequency distribution can be represented graphically in the form of a **frequency histogram.** This is a bar graph in which each bar is of width equal to the class length and of height equal to the class frequency. The frequency histogram for the distribution of Table 1.2 is depicted in Figure 1.1.

For each class, except the last, the **upper boundary** is equal to the **lower boundary** of the succeeding class. Analogously, except for the first class, the lower boundary is equal to the upper boundary of the preceding class.

When forming a frequency distribution, one or more pieces of data may fall on a class boundary. One way to handle such a problem is to flip a coin to determine the class into which to place the given observation. Such a procedure gives each of the two classes *equal likelihood* (a concept to be rigorously defined later) of containing the boundary datum. For instance, in the bulb-life data, there was a test bulb that lasted 900 hours, and it was placed into the third class by an equal-likelihood technique. A second approach is to avoid boundary data by choosing the boundary values so that they result in decimal expansions that cannot result from the test equipment.

TABLE 1.2 FREQUENCY DISTRIBUTION FOR BULB-LIFE DATA

Class	Class Mark	Frequency
800–850	825	3
850–900	875	4
900–950	925	8
950–1000	975	12
1000–1050	1025	16
1050–1100	1075	28
1100–1150	1125	24
1150–1200	1175	5

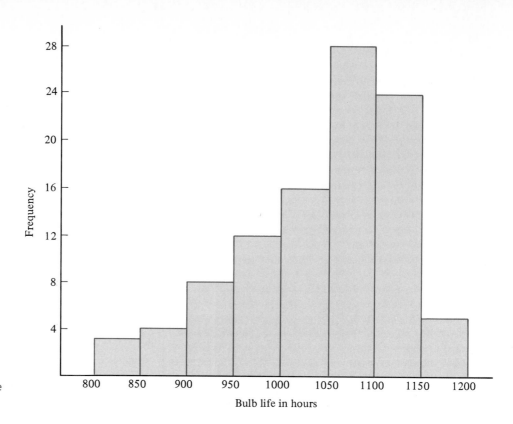

FIGURE 1.1
Histogram for the bulb-life distribution of Table 1.2

EXAMPLE 1.1

A manufacturer of transistors wishes to obtain information regarding the number of defective units in each box of 50. One hundred boxes are randomly selected from the warehouse, and four transistors are randomly selected from each box and tested. The four test units are selected **with replacement.** That is, after each is selected and tested, it is replaced and "mixed" with the remaining 49 in the box before the next selection. The assumption is made that if a particular transistor is tested more than once, later test outcomes are not affected by prior testing.

As a practical matter, each unit in the box might be labeled with a unique integer between 1 and 50, and then a sequence of four random numbers could be generated to select transistors for testing.

The data in this example differ in nature from the bulb-life data considered previously. The bulb-life measurements were **continuous.** That is, each datum could take on any value in a continuous range. In the present example, however, the measurement is **discrete.** It must take on a value from a list of values. In this case, the list consists of values in the set {0, 1, 2, 3, 4}. Because so few values are involved, it is appropriate to choose these values as the class marks. Were there a large number of possible discrete outcomes, then each class would contain more than a single value. The frequency distribution and corresponding histogram are given in Table 1.3 and Figure 1.2, respectively. Notice that we have let the lowest class boundary be -0.5 in order to keep all class lengths equal. The fact that the first class frequency is 40 means that for 40 of the boxes chosen, no defective transistors were among the four selected.

It is sometimes useful to consider the frequency distribution derived by accumulating the individual class frequencies. The resulting compilation is known as the **cumulative frequency distribution.** For each i, the cumulative frequency of the ith class is given by

$$F_i = f_1 + f_2 + \cdots + f_i$$

TABLE 1.3 FREQUENCY DISTRIBUTION
FOR NUMBERS OF DEFECTIVE
TRANSISTORS IN A BOX

Class	Class Mark	Frequency
−0.5–0.5	0	40
0.5–1.5	1	39
1.5–2.5	2	17
2.5–3.5	3	3
3.5–4.5	4	1

For instance, for the frequency distribution given in Table 1.2,

$$F_3 = f_1 + f_2 + f_3 = 3 + 4 + 8 = 15$$

which gives the number of test bulbs whose lives were less than 950 hours. The cumulative frequency distribution for the bulb-life experiment is given in Table 1.4, in which the class marks represent the classes.

Corresponding to the cumulative frequency distribution is the **cumulative frequency histogram.** This is a bar graph like the frequency histogram, except that the cumulative frequencies are used in place of the frequencies. The cumulative frequency histogram for the bulb-life distribution of Table 1.4 is given in Figure 1.3.

Sometimes line graphs are employed instead of bar graphs to depict both frequency and cumulative frequency distributions. To construct the **frequency**

FIGURE 1.2
Histogram for the defective-transistor distribution of Table 1.3

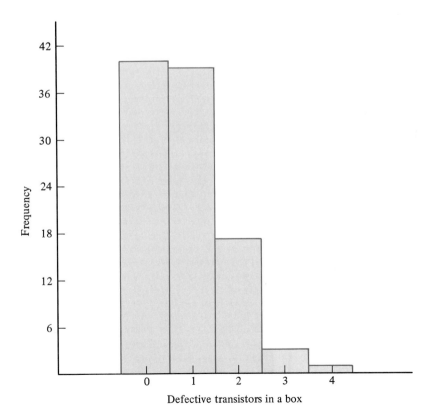

Defective transistors in a box

TABLE 1.4 CUMULATIVE FREQUENCY
DISTRIBUTION FOR BULB-LIFE DATA

Class Mark	Frequency	Cumulative Frequency
825	3	3
875	4	7
925	8	15
975	12	27
1025	16	43
1075	28	71
1125	24	95
1175	5	100

polygon corresponding to the distribution, we first extend the frequency distribution one class in either direction and assign each of the extra classes the frequency 0. We then plot the points (x_i, f_i), $i = 0, 1, 2, \ldots, m, m + 1$, where m is the original number of classes, and connect the pairs of adjacent points by straight lines.

The **cumulative frequency polygon,** or **ogive,** is a piecewise linear representation of the cumulative frequency distribution in which the cumulative frequency of a class is plotted at the upper class boundary. Together with the point $(l_1, 0)$, which represents the cumulative frequency of 0 at the lower boundary of the first class, the points (u_i, F_i), representing the cumulative frequencies at the upper class boundaries for $i = 1, 2, \ldots, m$, are connected by straight-line segments to form the desired line graph.

The frequency polygon and ogive for the frequency distribution of Table 1.2 are illustrated in Figures 1.4 and 1.5, respectively.

FIGURE 1.3
Cumulative frequency histogram for the bulb-life distribution

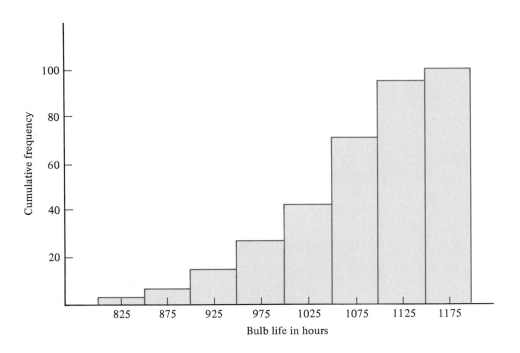

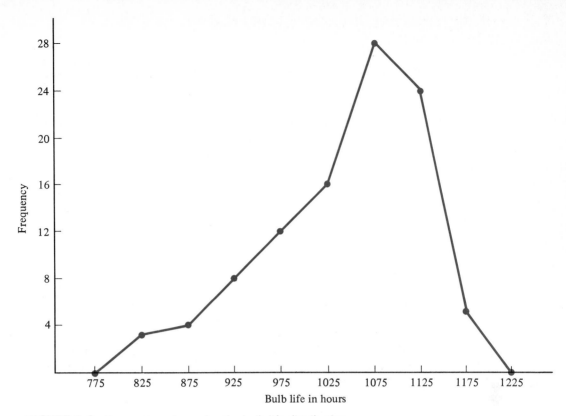

FIGURE 1.4   Frequency polygon for the bulb-life distribution

FIGURE 1.5
Ogive for the bulb-life
distribution

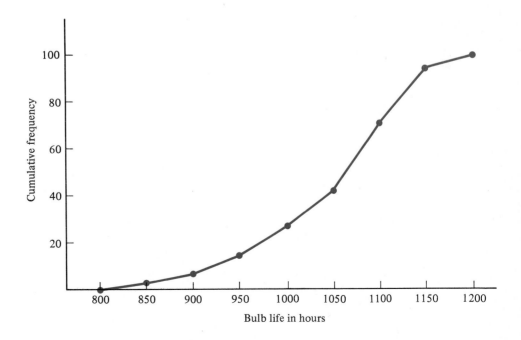

1.3 THE RELATIVE-FREQUENCY INTERPRETATION OF PROBABILITY

In the present section we will intuitively consider the manner in which probability statements of the form $P(a < X < b)$, where X is a random quantity, can be evaluated in terms of observational data that is supposed to be indicative of the overall measurement process from which X derives. Although the discussion will not be mathematically rigorous, it will lay the geometric groundwork upon which much of the subsequent mathematical theory is based.

1.3.1 Relative-Frequency Distributions

Given the class frequencies $f_1, f_2, \ldots, f_m$ of a frequency distribution possessing m classes, we can construct a new distribution utilizing the **relative frequencies**

$$r_i = \frac{f_i}{N}$$

for $i = 1, 2, \ldots, m$, where N is the total frequency of the distribution. The resulting **relative-frequency distribution** gives the proportion of the data belonging to each class. Moreover, a **cumulative relative-frequency distribution** is defined by the **cumulative relative frequencies**

$$R_i = \frac{F_i}{N}$$

for $i = 1, 2, \ldots, m$, where F_i is the cumulative frequency of the ith class. Whereas the sum of the frequencies is N, the sum of the relative frequencies is 1. The relative-frequency and cumulative relative-frequency distributions corresponding to the frequency distribution of Table 1.2 are given in Table 1.5.

We can construct bar graphs representing the newly introduced relative-frequency distributions. These, the **relative-frequency histogram** and the **cumulative relative-frequency histogram,** are identical to the corresponding original histograms except that the scale on the y axis is changed, r_i and R_i replacing f_i and F_i, respectively. A similar rescaling of the y axis in the frequency and cumulative frequency polygons

TABLE 1.5 RELATIVE-FREQUENCY AND CUMULATIVE RELATIVE-FREQUENCY DISTRIBUTIONS FOR BULB-LIFE DATA

Class Mark	Frequency	Relative Frequency	Cumulative Relative Frequency
825	3	0.03	0.03
875	4	0.04	0.07
925	8	0.08	0.15
975	12	0.12	0.27
1025	16	0.16	0.43
1075	28	0.28	0.71
1125	24	0.24	0.95
1175	5	0.05	1.00

results in the relative-frequency and cumulative relative-frequency polygons, respectively.

1.3.2 Relative Frequencies and Probability

The relative-frequency distribution can be given a probabilistic interpretation, albeit one that is full of potential pitfalls. Intuitively, at least, a set of measurements represents a **sample** from the collection of all possible measurements, the latter being known as the **population** of measurements. For instance, the 100 bulb-life measurements of Table 1.1 represent a sample taken from the bulb lives of all bulbs produced by the given manufacturer.

Recalling the discussion of Section 1.1, we wish to view the outcome of a measurement process as a random quantity X and wish to make statements of the form $P(a < X < b)$. If the sample data faithfully reflect the population (a risky assumption) and if the ith class has lower and upper boundaries l_i and u_i, respectively, then the probability that any (future) measurement X lies between the two boundary values is given by

$$P(l_i < X < u_i) = \frac{f_i}{N} = r_i$$

the relative frequency of the ith class. In light of our assumption regarding the faithful representation of the population by the sample, this probability expression appears quite reasonable. The relative frequency r_i represents the proportion of the observed values that are in the ith class, and we are simply extrapolating this proportion to denote the proportion of the population falling within the ith interval.

The problem with the foregoing reasoning is that it involves an inductive leap on a grand scale. The frequency distribution represents merely a sample set of observations, and we have no certainty as to whether or not it yields a faithful representation of the population.

We might, of course, use relative frequencies to *estimate* the values of probability statements regarding the random quantity X. However, this methodology cannot provide a theoretical framework for the development of probability theory. In fact, the very notion of estimation calls for a preexisting theory of probability by which to measure the effectiveness of the estimation process. Moreover, the theory must include a precise formulation of the population–sample relationship.

Given that we understand the intuitive nature of the discussion, let us continue. Suppose we wish to employ relative frequencies to give meaning to the statement $P(a < X < b)$, where a and b are not class boundaries. Assuming the data in each class are *uniformly distributed* throughout the class interval, it seems reasonable to estimate the desired probability by finding the fraction of the data that lies between a and b. Referring to Figure 1.6 and letting c denote the common length of the classes, r_i the relative frequency of the ith class, and l_i and u_i the upper and lower boundaries of the ith class, a straightforward geometrical analysis yields

$$P(a < X < b) = \frac{u_2 - a}{c} r_2 + r_3 + \frac{b - l_4}{c} r_4$$

EXAMPLE 1.2

Let X denote the measurement of bulb life, and suppose it is desired to estimate probabilities regarding X by employing the relative-frequency distribution of Table 1.5. Referring to Figure 1.7, the probability of a measurement falling between 920 and 1060 hours is estimated by

$$P(920 < X < 1060) = 0.6 \times 0.08 + 0.12 + 0.16 + 0.2 \times 0.28 = 0.384$$

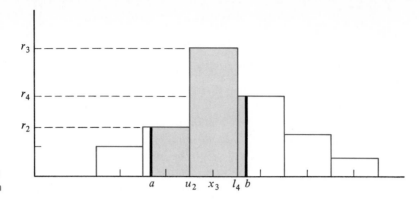

FIGURE 1.6
Estimation of the probability $P(a < X < b)$ by means of a histogram

1.3.3 Probability as an Integral

Again consider the relative-frequency histogram of Figure 1.6, only this time let the y axis be rescaled by dividing each relative frequency by c, the common class length, to obtain the **normalized relative frequencies**

$$p_i = \frac{r_i}{c} = \frac{f_i}{cN}$$

If the tops of the bars of this **normalized relative-frequency histogram** are viewed as portions of a step function $s(x)$, then

FIGURE 1.7
Estimation of $P(920 < X < 1060)$ using the bulb-life histogram

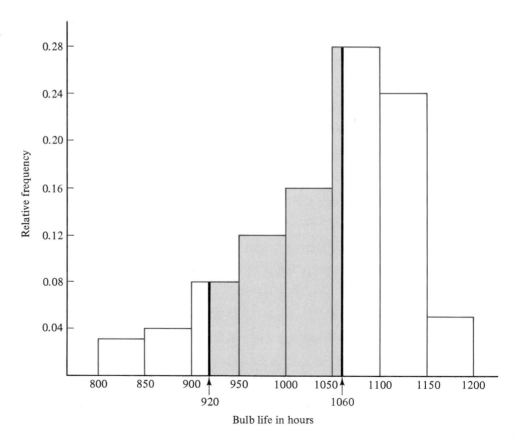

$$\int_a^b s(x)\, dx = (u_2 - a)p_2 + cp_3 + (b - l_4)p_4$$

$$= \frac{(u_2 - a)r_2}{c} + \frac{cr_3}{c} + \frac{(b - l_4)r_4}{c}$$

which is precisely the estimate obtained for $P(a < X < b)$ in Section 1.3.2. That is, the estimate of the probability that X lies between a and b is given by the integral of the normalized relative-frequency histogram over the interval (a, b).

Now let us go one step further. The step function $s(x)$ depends solely on the sample data. Given a new set of observations, it is highly likely that a different histogram, and thus a different step function, would be obtained. Indeed, the result could be a markedly different step function. The problem is that we really would like some function that would, at least in theory, give the actual values of the probability statements, not simply estimates based upon this or that sample. We desire a function $f(x)$ such that

$$P(a < X < b) = \int_a^b f(x)\, dx$$

for all values of a and b, $a < b$.

Figure 1.8 illustrates a function that appears to be reasonable for the histogram of Figure 1.7 (after normalization of the relative frequencies). Of course, the "actual" function that is appropriate for the population might look very different from the step function $s(x)$ derived from a given set of observations. One role of statistical analysis is to provide a methodology for deciding on an appropriate $f(x)$.

EXAMPLE 1.3

Suppose the lengths (in minutes) of 200 phone calls are measured at a particular station and the frequency and relative-frequency distributions are as given in Table 1.6. Since the common class length is $c = 1$, there need be no normalization of the relative-frequency histogram, which, together with the graph of the function

$$f(x) = \begin{cases} \dfrac{1}{2} e^{-x/2}, & \text{if } x \geq 0 \\ 0, & \text{if } x < 0 \end{cases}$$

is illustrated in Figure 1.9. The fact that $f(x)$ appears to be in reasonable agreement with the step function $s(x)$ derived from the particular sample does not necessarily imply that $f(x)$ will serve well as a descriptor of the probabilistic behavior of the random quantity X measuring all possible phone calls from the station. Nonetheless, if we do assume that $f(x)$ describes the behavior of X, then

$$P(a < X < b) = \int_a^b f(x)\, dx = \frac{1}{2} \int_a^b e^{-x/2}\, dx = e^{-a/2} - e^{-b/2}$$

for all a and b such that $0 \leq a < b$. For instance, given an arbitrary phone call, the probability that it lasts for less than 2 minutes is

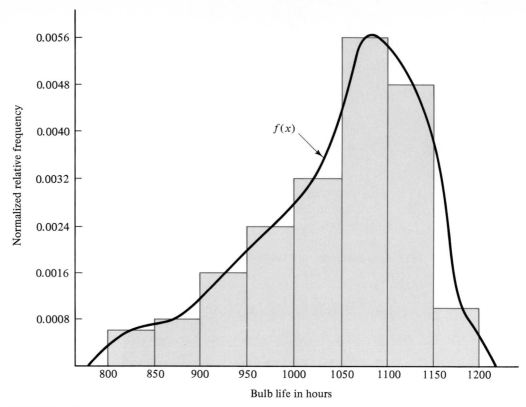

FIGURE 1.8 A function $f(x)$ that appears to fit the bulb-life histogram

$$P(0 < X < 2) = 1 - e^{-1} = 1 - 0.3679 = 0.6321$$

The probability that an arbitrary phone call lasts between 1 and 2.4 minutes is

$$P(1 < X < 2.4) = e^{-0.5} - e^{-1.2} = 0.6065 - 0.3012 = 0.3053$$

That is, assuming that probability statements regarding the length of phone calls can be evaluated by integrating $f(x)$ between the appropriate limits, we can expect that approximately 63% of all calls will last for a duration of less than 2 minutes and that approximately 31% of all calls will last between 1 and 2.4 minutes.

TABLE 1.6 LENGTHS OF PHONE CALLS

Class	Frequency	Relative Frequency
0–1	72	0.36
1–2	54	0.27
2–3	34	0.17
3–4	20	0.10
4–5	6	0.03
5–6	8	0.04
6–7	4	0.02
7–8	2	0.01

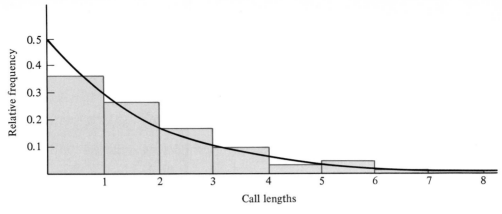

FIGURE 1.9 An exponential function that appears to fit the call-time distribution of Table 1.6

While it is true that the relative-frequency technique and the consequent integration methodology have been developed along intuitive lines without a rigorous mathematical framework, it is one of the paramount functions of probability and statistical theory to give an explicit formulation of the entire matter.

1.4 DESCRIPTIVE MEASURES OF CENTRAL TENDENCY

The frequency distribution provides a coherent organization of data, and the histogram gives a geometrical perspective; however, quantitative description is also needed. The present section introduces measures, called **descriptive statistics,** that are central to the distribution of sample values.

1.4.1 Empirical Mean

The "usual average" of a set of raw data is determined by summing the data and dividing by the total frequency.

DEFINITION 1.1

Empirical Mean (Raw Data). For a data sample

$$\{x_1, x_2, \ldots, x_N\}$$

the empirical mean of the sample is defined by

$$\bar{x} = \frac{1}{N}(x_1 + x_2 + \cdots + x_N) = \frac{1}{N}\sum_{i=1}^{N} x_i$$

The empirical mean is commonly called the **arithmetic mean** and, when the context is clear, we will sometimes simply refer to the *mean* of the sample. The reason for using the adjective "empirical" is that there will be other means introduced subsequently, and the empirical mean is one that has been computed from observed (empirical) data.

TABLE 1.7 MEGAFLOP OBSERVATIONS

3.9	4.7	3.7	5.6	4.3	4.9	5.0	6.1	5.1	4.5
5.3	3.9	4.3	5.0	6.0	4.7	5.1	4.2	4.4	5.8
3.3	4.3	4.1	5.8	4.4	3.8	6.1	4.3	5.3	4.5
4.0	5.4	3.9	4.7	3.3	4.5	4.7	4.2	4.5	4.8

EXAMPLE 1.4

One performance criterion for the speed of a computer's CPU (central processing unit) is the number of floating-point operations that can be performed per second (*flops*). For a super-computer, this rate is measured in *megaflops* (millions of floating-point operations per second). An estimate of this measure can be obtained by averaging the numbers of megaflops achieved when employing some collection of *benchmark* routines. The resulting rates can be interpreted as constituting a sample from some population of rates corresponding to a large class of routines.

Suppose the test rates (in megaflops) given in Table 1.7 are obtained for a particular supercomputer. Then the empirical mean for the sample is

$$\bar{x} = \frac{1}{40} \sum_{i=1}^{40} x_i = \frac{3.9 + 4.7 + \cdots + 4.8}{40} = 4.7$$

One way to recognize the manner in which the empirical mean describes central tendency is to consider the deviations from the mean of the data. For any datum x_i in the sample, define the corresponding **deviation (from the mean)** by

$$d_i = x_i - \bar{x}$$

For instance, if $x_1 = 3.9$ megaflops in Table 1.7, then

$$d_1 = 3.9 - 4.7 = -0.8$$

In general, summing the deviations yields

$$\sum_{i=1}^{N} d_i = \sum_{i=1}^{N} (x_i - \bar{x}) = \sum_{i=1}^{N} x_i - N\bar{x} = 0$$

That is, the empirical mean results in a set of deviations whose sum is 0.

If the raw data have been grouped in a frequency distribution, then the individual observation values can no longer be summed. In such a case, we proceed as though each datum in the class had the same value as the class mark—a reasonable assumption if the data of each class are spread uniformly throughout the class interval. This approach leads to the following definition for grouped data.

DEFINITION 1.2

Empirical Mean (Frequency Distribution). For a frequency distribution with m classes and total frequency N, the empirical mean is defined by

$$\bar{x} = \frac{1}{N} \sum_{i=1}^{m} f_i x_i$$

where f_i and x_i denote the frequency and class mark, respectively, of the ith class.

EXAMPLE 1.5

Table 1.8 gives a frequency distribution for the megaflop data of Table 1.7. Employing the frequency information given in the table yields the empirical mean

$$\bar{x} = \frac{1}{40}(3 \times 3.5 + 8 \times 4.0 + 14 \times 4.5 + 6 \times 5.0 + 4 \times 5.5 + 5 \times 6.0) = 4.6875$$

Bringing the factor $1/N$ inside the summation in the frequency-distribution form of the empirical mean yields an equivalent expression in terms of the relative frequencies $r_i = f_i/N$:

$$\bar{x} = \sum_{i=1}^{m} x_i r_i$$

Basic geometric intuition regarding this formula can be obtained by considering the point-mass graphical representation of the relative-frequency distribution.

Essentially, the point-mass representation treats the distribution in the discrete sense, where the only possible values of the random quantity X are the class marks. The marks are placed along the x axis, and the relative frequencies are plotted against the y axis. The resulting graph consists of the pairs (x_i, r_i). The relative-frequency point-mass representation of the megaflop distribution of Table 1.8 is illustrated in Figure 1.10.

For an arbitrary relative-frequency distribution with class marks $x_1, x_2, \ldots,$ x_m and relative frequencies $r_1, r_2, \ldots, r_m$, it can be seen from the relative-frequency expression for $\bar{x}$ that, in the sense of physics, the empirical mean gives the **center of mass** of the point-mass function (see Figure 1.10).

Going one step further, consider the normalized relative-frequency histogram (with common class length c) of an arbitrary frequency distribution and regard the tops of the bars as a step function $s(x)$. The center of mass (in a continuous sense) is derived by integration and is given by

$$\int_{l_1}^{u_m} x s(x)\, dx = \sum_{i=1}^{m} \int_{l_i}^{u_i} x s(x)\, dx$$

$$= \sum_{i=1}^{m} \frac{r_i}{c} \int_{l_i}^{u_i} x\, dx$$

$$= \sum_{i=1}^{m} \frac{r_i(u_i^2 - l_i^2)}{2c}$$

$$= \sum_{i=1}^{m} \frac{r_i(u_i - l_i)(u_i + l_i)}{2c}$$

$$= \sum_{i=1}^{m} r_i x_i$$

$$= \bar{x}$$

the next-to-last equality following from the relations $u_i - l_i = c$ and $x_i = (u_i + l_i)/2$. Consequently, the center of mass of the normalized relative-frequency histogram is equal to the empirical mean.

As a measure of central tendency, the mean possesses a characteristic that can sometimes render it misleading. Since it weighs each datum equally, if there should exist a small number of extreme observations in the same direction, then these observations will tend to ''pull'' the mean in their direction. Although this is precisely

TABLE 1.8   FREQUENCY AND RELATIVE-FREQUENCY DISTRIBUTIONS
FOR CPU RATES (IN MEGAFLOPS)

Class	Class Mark	Frequency	Relative Frequency
3.25–3.75	3.5	3	0.075
3.75–4.25	4.0	8	0.200
4.25–4.75	4.5	14	0.350
4.75–5.25	5.0	6	0.150
5.25–5.75	5.5	4	0.100
5.75–6.25	6.0	5	0.125

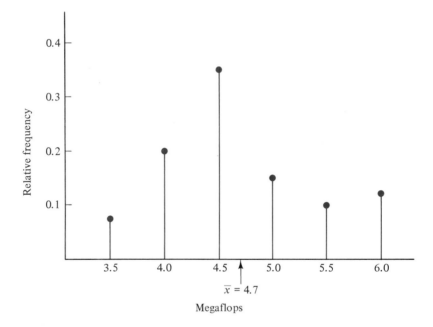

FIGURE 1.10
A center-of-mass interpretation of the empirical mean relative to the megaflop distribution

the behavior expected of the center of gravity, the resulting descriptive statistic might lead to erroneous judgment in the absence of careful scrutiny.

EXAMPLE 1.6

Consider the telephone-call-length frequency distribution given in Table 1.6. The mean for the sample is $\bar{x} = 1.92$. Of the 200 calls monitored, suppose two of them had been different; specifically, suppose the calls in the last class were of durations 76.3 and 84.1 minutes, instead of as originally recorded. Then the class mark 7.5 would be omitted in the sum defining $\bar{x}$, and instead there would be terms for the class marks 76.5 and 84.5, each possessing frequency 1. The resulting empirical mean would be $\bar{x} = 2.65$.

1.4.2 Empirical Median

A descriptive statistic measuring central tendency that is less susceptible to the effect of extreme values is the **median.** It locates the middle of the data.

DEFINITION 1.3

Empirical Median (Raw Data). For a data sample

$$\{x_1, x_2, \ldots . x_N\}$$

the empirical median is defined in one of two ways, depending on whether N is odd or even. Let

$$y_1 \leq y_2 \leq \cdots \leq y_N$$

be a relisting of the data according to increasing magnitude. If N is odd, then the median is defined to be the middle value in the relisting:

$$\tilde{x} = y_{(N+1)/2}$$

If N is even, then the median is defined to be the mean of the two "middle" values:

$$\tilde{x} = \frac{y_{N/2} + y_{(N+2)/2}}{2}$$

EXAMPLE 1.7

Suppose the nine salaries (in thousands of dollars) of the top executives at a certain company are

$$89, \ 80, \ 79, \ 113, \ 68, \ 70, \ 85, \ 87, \ 94$$

Relisting by order of magnitude gives the ordering

$$68, \ 70, \ 79, \ 80, \ 85, \ 87, \ 89, \ 94, \ 113$$

and the median executive salary is $\tilde{x} = \$85,000$. Were the list to include a tenth executive salary, say $81,000$, then the new median salary would be $\tilde{x} = \$83,000$.

In the original collection of nine salaries, if the top executive were to receive an increase of $90,000$, so that his or her new salary was $203,000$, then the new mean would be $\tilde{x} = \$95,000$ (instead of $\$85,000$); however, the median would remain the same.

While at first glance it might appear that the median's resistance to the effect of extreme values makes it superior to the mean, one must be careful not to jump to conclusions. Our ultimate goal is the use of empirical statistics derived from samples to estimate population parameters. In that arena the mean proves to be of superior worth.

To adapt Definition 1.3 to frequency distributions, we find the point on the x axis that divides the area of the histogram in two. Referring to Figure 1.11, suppose

$$F_{k-1} \leq N/2 \leq F_k$$

where N is the total frequency and F_{k-1} and F_k are the cumulative frequencies for classes $k-1$ and k, respectively. Then the point on the x axis that divides the area of the histogram into two equal portions lies within the kth class. This class is called the **median class,** and the median is the point on the x axis that divides the histogram bar of the kth class in such a manner as to split the histogram area into two equal parts. Employing the geometry illustrated in Figure 1.11, $\tilde{x}$ is the median if

$$df_k + cF_{k-1} = cN/2$$

since this would mean that half the area of the histogram ($cN/2$) lies to the left of $\tilde{x}$. Solving for d yields

$$d = \frac{\dfrac{N}{2} - F_{k-1}}{f_k} c$$

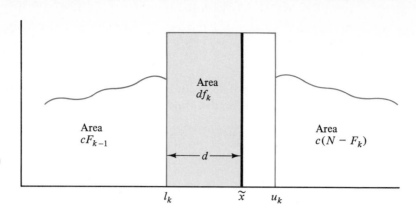

FIGURE 1.11
Graphical determination of the median for a relative-frequency distribution

Since $\tilde{x} = l_k + d$, we are led to the following definition for the median of grouped data.

DEFINITION 1.4

Empirical Median (Frequency Distribution). The empirical median for a frequency distribution is defined to be

$$\tilde{x} = l_k + \frac{\dfrac{N}{2} - F_{k-1}}{f_k} c$$

where N is the total frequency, F_{k-1} is the cumulative frequency for the class immediately prior to the median class, and c, f_k, and l_k are the length, frequency, and lower boundary, respectively, of the median class.

EXAMPLE 1.8

Consider the CPU-rate frequency distribution of Table 1.8. Since $N/2 = 20$, $F_2 = 11$, and $F_3 = 25$, the median class is the third class. Applying Definition 1.4 yields

FIGURE 1.12
Relationship of the empirical mean and the median when the data is skewed to the right

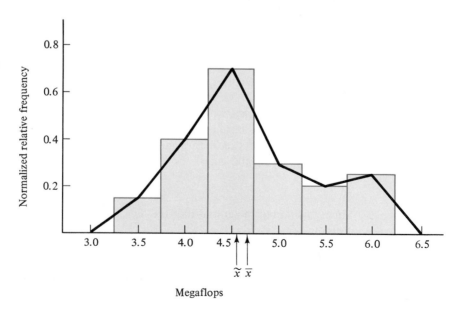

$$\bar{x} = 4.25 + \frac{9}{14} \times \frac{1}{2} = 4.57$$

In Example 1.5, the mean of this frequency distribution was found to be $\bar{x} = 4.6875$, which is to the right of the median. Looking at the normalized relative-frequency histogram for this distribution, which is given in Figure 1.12 (with the normalized relative-frequency polygon superimposed), we see why $\tilde{x} < \bar{x}$: the histogram is **skewed** to the right. That is, relative to the median, there is "extreme" data to the right.

1.5 THE EMPIRICAL VARIANCE—A DESCRIPTIVE MEASURE OF DISPERSION

The last section introduced two descriptive statistics that characterize the central tendency of a collection of data. The present section discusses a descriptive statistic that measures the degree to which the data are dispersed, or spread out.

Consider the two frequency histograms illustrated in Figure 1.13. Although they possess the same mean and median, they differ markedly relative to the degree to which the data are dispersed. In the first, the data are tightly clustered about the mean, whereas in the second, there is only a loose centering about the mean. We desire a descriptive statistic that will reflect these varying degrees of dispersion.

From Figure 1.13 it can be seen that the absolute deviations from the mean,

$$|d_i| = |x_i - \bar{x}|$$

tend to be greater for the data of the second histogram than for the data of the first. A similar comment applies to the squares of the deviations. Thus, the mean of the

FIGURE 1.13
Histograms with the same mean but with differing degrees of dispersion

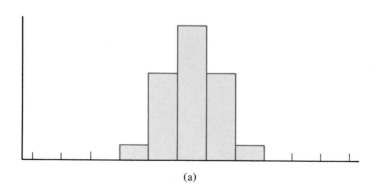

(a)

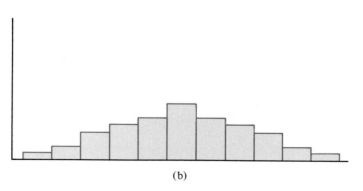

(b)

squared deviations is greater for the second data set than for the first. However, instead of dividing by N to obtain the average of the squares of the deviations, the new descriptive statistic is defined by summing the squares of the deviations and then dividing by $N - 1$. The resulting value is called the **variance** and is the most fundamental empirical measure of dispersion. The reason for dividing by $N - 1$ instead of N will become evident in Section 7.3.

DEFINITION 1.5

Empirical Variance. For the set of raw data

$$\{x_1, x_2, \ldots, x_N\}$$

the empirical variance is defined by

$$s^2 = \frac{1}{N-1} \sum_{i=1}^{N} (x_i - \bar{x})^2$$

In the case of a frequency distribution with m classes,

$$s^2 = \frac{1}{N-1} \sum_{i=1}^{m} f_i (x_i - \bar{x})^2$$

where f_i and x_i are the frequency and class mark, respectively, of the ith class.

EXAMPLE 1.9

Referring to the megaflop data of Table 1.8 and using the rounded-off value $\bar{x} = 4.7$ yields

$$s^2 = \frac{1}{39} (3 \times 1.44 + 8 \times 0.49 + 14 \times 0.04$$
$$+ 6 \times 0.09 + 4 \times 0.64 + 5 \times 1.69) = 0.52$$

Taking the square root of the variance s^2 gives the **empirical standard deviation** s, which for frequency distributions is defined by

$$s = \sqrt{\frac{\sum_{i=1}^{m} f_i(x_i - \bar{x})^2}{N-1}}$$

For instance, in Example 1.9,

$$s = (0.52)^{1/2} = 0.72$$

Sometimes it is difficult to compare the relative variation in two sets of measurements. For instance, suppose the mean weight of twenty 100-lb bags of cement is 99.7 lb, with the standard deviation being 2.4 lb, while the mean weight of thirty 20-lb bags is 20.1 lb, with the standard deviation being 1.5 lb. Then, although the standard deviation of the heavier bags is greater than that of the lighter bags, relatively speaking, it appears that the variation in the lighter bags is more problematic.

To measure relative variation, we can employ the **coefficient of variation,**

$$V = \frac{s}{\bar{x}}$$

the standard deviation divided by the mean. V is expressed as a percentage. Since s is divided by $\bar{x}$, a large standard deviation is offset by a large mean in the computation of V.

EXAMPLE 1.10

Employing the preceding cement-bag information, the coefficient of variation of the sample of 100-lb bags is

$$V = \frac{2.4}{99.7} = 0.024$$

or 2.4%. For the 20-lb bags,

$$V = \frac{1.5}{20.1} = 0.075$$

or 7.5%.

EXERCISES

SECTION 1.2

In Exercises 1.1 through 1.5, do the following:
a) Construct a frequency distribution using the given class marks.
b) Draw the frequency histogram.
c) Find the cumulative frequency distribution.
d) Draw the cumulative frequency histogram.
e) Draw the frequency polygon.
f) Draw the cumulative frequency polygon.

1.1. A steel company transports pipe and other items about its warehouse on a fleet of carts. The tractive resistance gives a measure (in pounds) of the resistive force per ton of combined cart and load weight. A sample of 75 tractive-resistance measurements taken over the company's concrete floors yields the data of Table 1.9. Construct the frequency distribution with class marks 33.8, 35.7, 37.6, 39.5, 41.4, 43.3, 45.2, 47.1, 49.0, and 50.9.

TABLE 1.9 TRACTIVE RESISTANCES

40.4	42.0	43.8	42.2	42.6	44.2	41.7	38.4	47.8	42.9	43.8
37.4	44.7	44.1	43.3	51.8	45.0	44.5	45.3	38.6	38.4	44.2
44.3	42.7	45.3	49.4	44.2	43.8	43.4	43.8	42.0	40.6	42.9
45.4	45.8	45.6	40.7	45.9	44.2	44.0	44.6	49.6	45.3	44.6
45.2	40.6	39.7	38.7	41.0	45.0	42.7	39.7	47.1	43.6	43.4
42.8	44.4	46.4	45.3	43.4	44.8	40.2	41.7	48.6	44.3	45.1
37.0	45.6	41.6	45.1	42.1	32.9	42.8	47.5	41.4		

1.2. A microelectronics company produces a line of specialty computer boards by assembling premanufactured components such as bare boards and chips. Before being hired, prospective employees are tested for manual dexterity. Table 1.10 gives the dexterity scores (on a scale of 1 to 10) for a sample of 80 recent applicants. Construct the frequency distribution using the class marks 5.4, 5.9, 6.4, 6.9, 7.4, 7.9, 8.4, 8.9, and 9.4.

TABLE 1.10 DEXTERITY SCORES

7.6	7.3	7.4	7.1	5.9	7.1	7.4	7.7	8.4	5.2
7.0	8.2	6.7	6.2	8.4	7.5	7.4	7.5	6.4	6.8
7.3	7.7	5.8	8.7	7.5	6.1	7.5	5.9	6.8	7.9
7.0	6.1	6.2	8.3	8.4	5.4	7.6	6.1	9.2	8.9
7.0	8.7	7.7	7.4	7.9	7.3	5.9	6.7	8.0	5.6
9.1	6.0	7.4	8.6	7.2	8.1	7.2	7.2	6.6	8.8
8.1	7.4	8.7	7.4	6.6	7.9	6.1	6.7	7.7	6.4
8.4	8.3	8.5	6.4	7.8	9.6	6.5	6.0	7.2	6.7

1.3. A company that sells used industrial machinery recognizes that the age of a machine is a factor in the expected frequency of repairs and the availability of spare parts. Thus, the company attempts to keep the average age of the machinery in its inventory as low as possible. The age data (in months) of Table 1.11 were obtained from a recent survey of the ages of 70 lathes in the company's inventory. Construct the frequency distribution using the class marks, 5.0, 12.0, 19.0, 26.0, 33.0, 40.0, and 47.0.

TABLE 1.11 LATHE AGES

22	13	4	16	23	38	38	21	33	49	23	35	23	37
27	3	25	38	27	29	39	24	27	24	25	43	26	29
25	33	40	37	5	29	7	24	32	23	18	14	12	18
12	18	7	5	40	21	22	12	35	21	35	27	10	15
32	19	38	16	42	22	32	26	32	20	25	24	11	32

1.4. A large automotive machine shop, as a subcontractor for various automobile and truck manufacturers, fabricates a variety of small parts for engines. The company monitors the overall efficiency of its weekly operations by computing the ratio of the total weekly electric power consumption in kilowatt-hours to the total number of parts produced. Table 1.12 gives the efficiency ratios for the past year of operation. Construct the frequency distribution using the class marks 0.08, 0.17, 0.26, and 0.35.

TABLE 1.12 EFFICIENCY RATIOS

0.24	0.27	0.32	0.20	0.15	0.24	0.34	0.20	0.23	0.32	0.06
0.14	0.27	0.28	0.24	0.27	0.27	0.29	0.27	0.14	0.31	0.29
0.15	0.25	0.26	0.08	0.27	0.17	0.29	0.25	0.15	0.10	0.16
0.21	0.16	0.22	0.30	0.16	0.18	0.18	0.26	0.22	0.19	0.33
0.25	0.11	0.24	0.37	0.30	0.15	0.29	0.25			

1.5. A computer electronics firm maintains quality control by a process in which samples of computer chips are selected randomly from the output conveyor belt on the assembly line. The chips in the sample are then tested and the number of defective chips per hundred is recorded. If any batch of 100 chips from a given assembly line has four or more defective chips, then the line is shut down to make the required adjustments. Table 1.13 gives the recent record of the number of defective chips per 100 produced for assembly lines A, B, C, and D. Treating the data as a single collection, construct the frequency distribution using the class marks 0, 1, 2, 3, 4, and 5.

TABLE 1.13 NUMBER OF DEFECTIVE CHIPS IN A SAMPLE OF 100 CHIPS

0	0	0	1	2	0	2	0	0	1	0	1	1	3	2	0	0	1	3	1	0	1	2	0	0	5	0
1	1	0	1	3	0	0	0	0	1	0	2	0	1	0	1	2	1	0	1	1	3	0	0	0	0	2
2	0	1	0	1	1	1	1	1	0	1	0	1	0	1	0	0	1	1	0	0	1	2	1	1	0	0
1	2	1	0	1	0	2	4	0	2	0	1	0	2	0	0	0	0	0	2	0	0	0	0	3	1	0

SECTION 1.3

In Exercises 1.6 through 1.10, refer to the cited exercise and do parts (a) through (c) relative to the frequency distribution constructed in the cited exercise.

a) *Construct the relative-frequency distribution.*
b) *Construct the cumulative relative-frequency distribution.*
c) *Construct the normalized relative-frequency distribution.*

1.6. Exercise 1.1.
1.7. Exercise 1.2.
1.8. Exercise 1.3.

1.9. Exercise 1.4.
1.10. Exercise 1.5.

In Exercises 1.11 through 1.13, employ the normalized relative-frequency distribution in the cited exercise to estimate the "probabilities" in parts (a) through (d).

1.11. Exercise 1.1
   a) $P(34.75 < X < 40.45)$
   b) $P(32.85 < X < 38.55$ or $46.15 < X < 51.85)$
   c) $P(35.50 < X < 39.50)$
   d) $P(33.80 < X < 37.60$ or $47.10 < X < 50.90)$

1.12. Exercise 1.2
    a) $P(6.65 < X < 8.15)$
    b) $P(5.15 < X < 6.65$ or $8.15 < X < 9.65)$
    c) $P(7.63 < X < 9.00)$
    d) $P(6.00 < X < 7.00$ or $8.50 < X < 9.30)$

1.13. Exercise 1.3
    a) $P(1.5 < X < 22.5)$
    b) $P(1.5 < X < 15.5$ or $36.5 < X < 50.5)$
    c) $P(36.0 < X < 48.0)$
    d) $P(12.0 < X < 18.0$ or $30.0 < X < 36.0)$

SECTION 1.4

In Exercises 1.14 through 1.16, find the empirical mean and empirical median from the raw data of the cited exercise.

1.14. Exercise 1.1.
1.15. Exercise 1.3.
1.16. Exercise 1.4.

In Exercises 1.17 through 1.19, find the empirical mean and empirical median from the frequency distribution of the cited exercise.

1.17. Exercise 1.1.
1.18. Exercise 1.3.
1.19. Exercise 1.4.

SECTION 1.5

In Exercises 1.20 through 1.22, find the empirical variance and the coefficient of variation from the raw data in the cited exercise.

1.20. Exercise 1.1.
1.21. Exercise 1.3.
1.22. Exercise 1.4.

In Exercises 1.23 through 1.25, find the empirical variance and the coefficient of variation from the frequency distribution in the cited exercise.

1.23. Exercise 1.1.
1.24. Exercise 1.3.
1.25. Exercise 1.4.

1.26. Another empirical measure of dispersion that is sometimes employed is the **mean deviation,** which for the raw data $x_1, x_2, \ldots, x_N$ is defined by

$$\text{M.D.} = \frac{1}{N} \sum_{i=1}^{N} |x_i - \bar{x}|$$

Find M.D. for the raw data of Exercise 1.1.

1.27. The definition of the mean deviation can be extended to a frequency distribution. Make the extension and find M.D. for the frequency distribution of Exercise 1.1.

2

PROBABILITY THEORY

Having discussed empirical description, we turn now to the theoretical development of probability. The chapter begins with the underlying notion of a sample space and then introduces upon that space a probability measure. Owing to the importance of finite sample spaces possessing outcomes of equal likelihood, Section 2.3 considers the basic principles of counting. Section 2.4 explains fundamental theoretical properties of probability spaces. Next follows the crucially important topic of conditioning: What is the revised probability of an event in the face of more information? The chapter concludes with a discussion of independence, a topic essential to statistics.

2.1 SAMPLE SPACES AND EVENTS

Our goal is to provide a sound mathematical formulation of probability statements involving random phenomena that closely parallels the normalized relative-frequency histogram interpretation. First, however, we must develop more primitive notions concerning probabilities associated with experimental outcomes.

2.1.1 The Sample Space

As a preliminary to discussing experimental likelihood, we need to fix some universe of discourse that is relevant to the observational data. From a mathematical viewpoint, the universe will be a set, and all questions regarding the experiment must be framed in terms of the elements within this set, which, of practical necessity, must satisfy certain constraints.

DEFINITION 2.1   **Sample Space.** The set S is a sample space for an experiment if every physical outcome of the experiment refers to a unique element of S.

In effect, two requirements are embodied in Definition 2.1. First, every physical outcome of the experiment must refer to some element in the sample space. Second, the uniqueness condition means that each physical outcome must refer to only one element in the sample space. The reasoning behind the two requirements is straightforward. If a particular phenomenon is observed, then that phenomenon must correspond to some entity in the universe of discourse upon which our probabilistic reasoning is based; otherwise, the mathematical model lacks sufficient scope. In addition, should the phenomenon refer to two or more elements within the sample space, then there would be confusion regarding which element to associate with the experimental outcome. From a mathematical perspective, a sample space is simply a set; however, to be of use from a modeling viewpoint, it must satisfy the two requirements embodied in the definition.

EXAMPLE 2.1

Suppose a box contains 100 items of a particular sort, say 100 capacitors, and each capacitor has a unique production number running from 1101 to 1200. If an experiment consists of randomly selecting a single capacitor from the box, then an appropriate sample space would be

$$S_1 = \{1101, 1102, \ldots, 1200\}$$

It would also be appropriate to employ the sample space

$$S_2 = \{1100, 1101, 1102, \ldots, 1200\}$$

One might argue that S_2 is less suitable from a modeling perspective, since no physical observation will correspond to the element 1100; nevertheless, S_2 still satisfies the two necessary modeling requirements. Insofar as probability is concerned, the probability of choosing 1100 will eventually be set to zero. A set that cannot serve as a sample space is

$$S_3 = \{1101, 1102, \ldots, 1199\}$$

since no element in S_3 corresponds to the selection of the capacitor with production number 1200.

Now, suppose the box is believed to contain capacitors having capacitance 0.004 μF (microfarads), but that the production process is suspected of being faulty, so that some of the capacitors might possess capacitance 0.003 μF. Although the original sample space S_1 is still suitable, if our concern is solely with the capacitance of the selected capacitor, another candidate for the sample space is

$$S_4 = \{0.003, 0.004\}$$

No matter which capacitor is selected, observation of the capacitance results in one and only one element in S_4. The set

$$S_5 = \{0.003, 0.004, 1200\}$$

cannot serve as a sample space, because it violates the uniqueness criterion. Specifically, selection of the capacitor having production number 1200 corresponds to two elements of S_5, the two depending on the observed capacitance.

The elements of a sample space are called **outcomes.** An outcome is a logical entity and refers only to the manner in which the phenomena are viewed by the experimenter. For instance, in Example 2.1 the sample space S_4 contains two outcomes. While there might be all sorts of information available regarding the chosen capacitor, once S_4 has been chosen as the sample space, only the measured capacitance is relevant, since only its observation will result in an outcome (relative to S_4).

Many practical sample spaces consist of outcomes that are ordered pairs, ordered triples, or, more generally, ordered n-tuples. Such outcomes belong to one or another product set.

DEFINITION 2.2

Product Set. Given two sets A and B, the product set $A \times B$ is given by

$$A \times B = \{(a, b): a \in A \text{ and } b \in B\}$$

each element of $A \times B$ being an ordered pair. More generally, given n sets A_1, A_2, . . . , A_n, the n-fold product set is

$$A_1 \times A_2 \times \cdots \times A_n = \{(a_1, a_2, \ldots, a_n): a_i \in A_i, \text{ for } i = 1, 2, \ldots, n\}$$

each element of the product set being an n-tuple.

If A and B are finite, then the number of ordered pairs in $A \times B$ is equal to the number of elements in A times the number of elements in B. Indeed, supposing A and B to contain m and k elements, respectively, each of the m elements of A can serve as the first component for any of the k elements of B. Thus, by letting the first component vary over all elements of A, mk ordered pairs are generated. This pairing is demonstrated in the tree diagram of Figure 2.1, where x and y are each paired with the elements 1, 2, and 3. Symbolically,

$$\text{card } (A \times B) = \text{card } (A) \text{ card } (B)$$

where **card** denotes **cardinality** (the number of elements in the set). In words, the cardinality of a product set is the product of the cardinalities. This result generalizes to an n-fold product set and is stated formally in the following theorem.

FIGURE 2.1
Tree diagram illustrating the cardinality of a product set

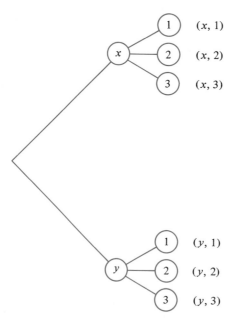

If $A_1, A_2, \ldots, A_n$ are finite sets, then

$$\text{card} (A_1 \times A_2 \times \cdots \times A_n) = \text{card} (A_1) \, \text{card} (A_2) \cdots \text{card} (A_n)$$

Of particular interest is the case where each set in the product is the same—that is, the n-fold product set $A \times A \times \cdots \times A$. In such a case, we write A^n, and it follows at once from Theorem 2.1 that card $(A^n) = [\text{card} (A)]^n$.

EXAMPLE 2.2

Example 1.1 concerned the problem of picking transistors from a box, some of which are known to be defective. If interest is only centered on whether or not a transistor is defective or nondefective, then a suitable sample space for the selection of a single transistor is $S =$

FIGURE 2.2
Tree diagram illustrating the sample space T of Example 2.2

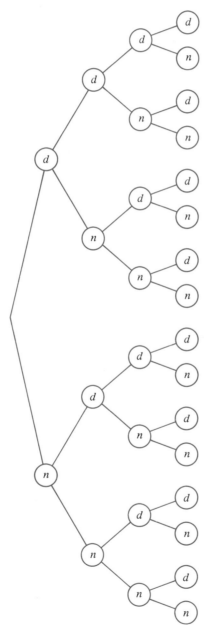

$\{d, n\}$. According to the methodology of Example 1.1, four transistors are to be selected from each box. For a particular box, one way of viewing the problem is that the four selections should be made and the result of each selection recorded. Then, for any box, there would be an ordering of defective and nondefective transistors. A satisfactory sample space for the four selections is the product set

$$
\begin{aligned}
T = S^4 &= S \times S \times S \times S \\
&= \{(d, d, d, d), (d, d, d, n), (d, d, n, d), (d, d, n, n), \\
&\quad (d, n, d, d), (d, n, d, n), (d, n, n, d), (d, n, n, n), \\
&\quad (n, d, d, d), (n, d, d, n), (n, d, n, d), (n, d, n, n), \\
&\quad (n, n, d, d), (n, n, d, n), (n, n, n, d), (n, n, n, n)\}
\end{aligned}
$$

Since S possesses 2 elements, the 4-fold product set contains $2^4 = 16$ 4-tuples.

The construction of T can be seen from the tree diagram in Figure 2.2, where each branch represents a selection of either a d or an n. Reading from left to right generates the outcomes.

EXAMPLE 2.3

The sample spaces discussed in Examples 2.1 and 2.2 were finite. In the present example we consider a computing model that leads to an infinite sample space. Consider a CPU that processes jobs arriving from one of two queues, A and B (see Figure 2.3). Moreover, suppose the next job assigned to the CPU is randomly chosen from among the two jobs at the heads of the respective queues, and neither of the queues is ever empty. Given that job J is at the head of queue B, we desire a sample space that reflects the observations of all job selections until J is chosen. A suitable sample space is

$$
S_1 = \{B, AB, AAB, AAAB, AAAAB, \ldots\}
$$

where A and B denote the choice from queue A or B, respectively, for service.

One might argue that we could stop the list after some number of A's. But after how many? The sample space gives a set containing all logical possibilities, not only those that are likely to occur. If some selection device such as flipping a fair coin is used for the queue selection, then certainly the likelihood of choosing queue A 1000 times before the first selection of a job from queue B is extremely small; nonetheless, such a course of events is still possible.

Another sample space, and one that is equivalent to S_1, is

$$
S_2 = \{0, 1, 2, \ldots\}
$$

which just gives the number of jobs selected from queue A prior to the selection of job J from queue B. Note that for this choice of sample space, the outcomes can be viewed as observed values of a discrete random measurement.

FIGURE 2.3
Queues of jobs arriving for CPU service

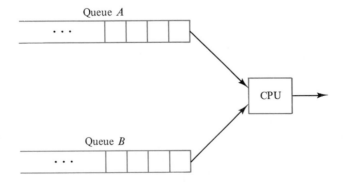

EXAMPLE 2.4

Often, the speed and mass of a projectile must be viewed as random because of **noise** resulting from internal variations of the measuring devices, as well as from unexplained phenomena that are ignored in the physical model. In such a situation, a suitable sample space is the entire Euclidean plane, where a point (x, y) corresponds to a reading of speed x and mass y.

In both Examples 2.3 and 2.4, the sample spaces were infinite; however, there is an important difference. In Example 2.3, the outcomes of the sample space can be put in the form of a list, and the sample space is said to be **denumerable.** In Example 2.4, no listing, either finite or infinite, of the spatial points in the plane is possible. Such a sample space is said to be **nondenumerable.**

2.1.2 Events

In most probability problems, the investigator is interested not merely in the collection of outcomes but in some subset of the sample space. A subset of a sample space is known as an **event.** Two events that do not intersect are said to be **mutually exclusive (disjoint).** More generally, the events $E_1, E_2, \ldots, E_n$ are said to be mutually exclusive if

$$E_i \cap E_j = \emptyset$$

for any $i \neq j$, $\emptyset$ denoting the empty set. If we let E^c denote the set-theoretic complement of E relative to the sample space, then E^c is called the **complementary event** (of E). Finally, if every element of event E lies in event F, then E is said to be a **subevent** of F.

Figure 2.4 employs Venn diagrams to illustrate some of the event relationships. Part (a) of the figure depicts a sample space S and two events E and F. Parts (b), (c), (d), and (e) depict the event formed by intersecting E and F, unioning E and F, taking the set-theoretic subtraction $E - F$, and forming the complementary event of E, respectively. Part (f) illustrates two mutually exclusive events.

EXAMPLE 2.5

Referring to the sample space T of Example 2.2, the events

$$E_1 = \{(d, d, d, d)\}$$

and

$$E_2 = \{(d, n, n, n), (n, d, n, n), (n, n, d, n), (n, n, n, d)\}$$

correspond to the observation of 0 and 3 nondefective transistors, respectively.

EXAMPLE 2.6

Relative to the sample space S_2 of Example 2.3, the event

$$E = \{1, 3, 5, \ldots\}$$

gives a set-theoretic description to the selection of an odd number of jobs from queue A before the selection of the job at the head of queue B.

An event provides the collection of outcomes that corresponds to some classification. Consequently, after an appropriate selection of a sample space, the specification of the relevant event is often the next phase in the solution of a problem. Such specification often proves to be difficult.

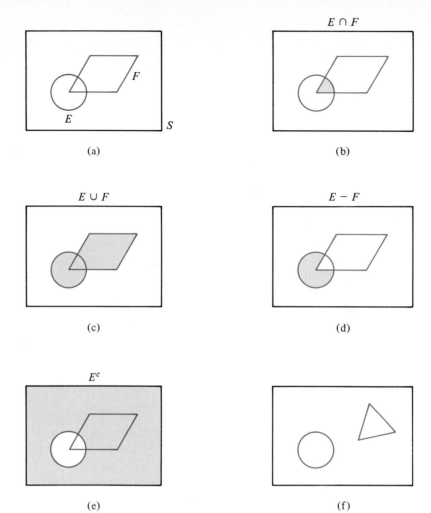

FIGURE 2.4
Venn diagrams
illustrating some event
relationships

2.2 PROBABILITY SPACE

To give mathematically rigorous interpretations to statements of the form $P(a < X < b)$, where X is a random quantity, we are led to consider the probability of an event. For instance, in Example 2.2 it was seen that the selection procedure of Example 1.1 could be phrased in terms of a sample space of ordered 4-tuples. Suppose interest is centered on the likelihood of selecting 3 defective transistors. In terms of the sample space T of Example 2.2, the selection of 3 defective transistors corresponds to the event

$$E = \{(d, d, d, n), (d, d, n, d), (d, n, d, d), (n, d, d, d)\}$$

Probability questions regarding the number of defective transistors are thus thrown back upon the events within the sample space.

2.2.1 Probability Measure

Recalling the relative-frequency interpretation of probability (Section 1.3), certain requirements regarding the eventual rigorous mathematical definition of probability can be conjectured. Consider a relative-frequency histogram with classes $C_1, C_2, \ldots,$ C_m and corresponding relative frequencies $r_1, r_2, \ldots, r_m$. If we let the relative frequency for the class C_i be denoted by $P(C_i)$, the notation P being employed to convey an interpretation of relative frequency as probability, then certain relationships regarding relative frequency are apparent:

1. $P(C_i) = r_i \geq 0$.
2. $P(C_1 \cup C_2 \cup \cdots \cup C_m) = r_1 + r_2 + \cdots + r_m = 1$.
3. For $i \neq j$, $P(C_i \cup C_j) = r_i + r_j = P(C_i) + P(C_j)$.

The first property states that relative frequencies are nonnegative, the second that the probability of all classes taken together is 1, and the third that the relative frequency of the union of two classes is the sum of the relative frequencies. Subject to certain mathematical requirements, the three properties form a model for the axiomatic foundation of probability.

DEFINITION 2.3

Probability Space. Given a sample space S, a **probability measure** P on S is a real-valued function defined on the events of S such that

P1. For any event E, $0 \leq P(E) \leq 1$.
P2. $P(S) = 1$.
P3. If $\{E_1, E_2, \ldots, E_i, \ldots\}$ is any denumerable collection of mutually exclusive events, then

$$P\left(\bigcup_{i=1}^{\infty} E_i\right) = P(E_1) + P(E_2) + \cdots$$

Once S has been endowed with a probability measure, S is called a **probability space.**

The notion of a probability measure is axiomatically defined. To obtain a probability space, one need only begin with a set S and define a function P on the subsets (events) of S that satisfies P1, P2, and P3. Whereas the mathematical theory of probability is concerned with the consequences of the abstract definition, our needs are quite different. To be applicable to the study of physical phenomena, the abstract definition must serve as a model of behavior. Thus, our main thrust will be the application of the model to problems in engineering, computer science, and the physical sciences.

Regarding Definition 2.3 and the aforementioned points concerning relative frequencies, conditions P1 and P2 are direct generalizations, whereas P3, known as the **additivity** property, requires some explanation. A denumerable collection of events can be written in a list, either finite or infinite. Given that the collection is mutually exclusive, meaning that the intersection of any pair of events within the collection is null, the probability of the union is equal to the sum of the probabilities, which is a generalization of the third point concerning relative frequencies. In practice, additivity can prove very useful, since it might be easier to determine the individual probabilities of a disjoint collection of events whose union is the event of interest, rather than to approach the problem directly.

Some examples should help to elucidate the concept of a probability space. The examples demonstrate that while axioms P1, P2, and P3 restrict the choice of functions that may serve as probability measures, we can still work within the restriction to select measures that serve our purposes.

EXAMPLE 2.7

Example 2.1 concerned the selection of a capacitor from a box of 100 capacitors, and the sample space S_1 was constructed by simply listing the production numbers of the capacitors. Assuming the box to be well mixed, so that each capacitor has equal likelihood of being selected, a suitable assignment of probability would be to let the probability of each singleton event be 1/100:

$$P(\{1101\}) = P(\{1102\}) = \cdots = P(\{1200\}) = 1/100$$

More generally, for an arbitrary event E in S_1, let

$$P(E) = \frac{\text{card } (E)}{100}$$

Then $P(E) \geq 0$ and $P(S) = 100/100 = 1$, so the first two requirements of probability are satisfied. Moreover, if E and F are disjoint events, then

$$\text{card } (E \cup F) = \text{card } (E) + \text{card } (F)$$

so that

$$P(E \cup F) = \frac{\text{card } (E \cup F)}{100} = \frac{\text{card } (E)}{100} + \frac{\text{card } (F)}{100} = P(E) + P(F)$$

Extension of additivity in this case to an arbitrary finite union is immediate, hence P3 holds. (Owing to the finiteness of S_1, infinite unions are not required in this example.) Since P1, P2, and P3 are satisfied, P is a legitimate probability measure.

Suppose we were to use the sample space

$$S_4 = \{0.003, 0.004\}$$

where only the microfarad possibilities are listed. From a purely abstract perspective, a probability measure is defined by

$$P(\{0.003\}) = P(\{0.004\}) = 1/2$$

However, the worth of the model is certainly in question. Do we mean by such a probability assignment that the likelihood of picking a capacitor with capacitance 0.003 μF is the same as picking one with capacitance 0.004 μF? If so, are we assuming there are 50 of each type capacitor in the box? In some sense, probability is a measure of belief. The construction of P represents our guess as to the manner in which nondeterministic phenomena are to behave.

For instance, suppose it is known that there are 80 capacitors in the box having capacitance 0.003 μF and 20 having capacitance 0.004 μF. Then it would behoove us to make the probability assignments $P(\{0.003\}) = 0.8$ and $P(\{0.004\}) = 0.2$.

Which sample space is better? While there is no definitive answer, a differentiating criterion can be posited. Simply put, S_1 is more refined than S_4. As such, it can serve as a more detailed model. Specifically, employing S_1, the outcomes of S_4 can be treated as events in S_1, $E_{0.003}$ and $E_{0.004}$ denoting the events consisting of all 0.003-μF and 0.004-μF capacitors, respectively. Assuming the 80 to 20 split described previously,

$$P(E_{0.003}) = \frac{\text{card } (E_{0.003})}{100} = 80/100 = 0.8$$

and

$$P(E_{0.004}) = \frac{\text{card } (E_{0.004})}{100} = 20/100 = 0.2$$

That is, the probabilities that were experimentally appropriate in the case of S_4 are derivable from those appropriate for S_1.

The converse approach does not work. For instance, there is no way to describe the event consisting of all even-numbered capacitors when employing S_4, and clearly this is an event relative to S_1. More simply, we cannot describe the singleton events of S_1 by employing the outcomes of S_4. Generally, unless there is some specific purpose in mind, it is safest to employ the sample space that contains the most information. Moreover, as will be seen shortly, there are benefits to having a sample space (in the finite case) for which each outcome has the same probability when treated as a singleton event.

In Example 2.7 we were careful not to drop the set signs when writing the probabilities of singleton events; that is, the probability of choosing production number 1101 was written as $P(\{1101\})$. Such an approach is technically necessary, since the probability measure P is defined on events, not on elements of the sample space. Nonetheless, to avoid the rigors of cumbersome notation, we will usually write $P(w)$ to denote the probability of the singleton event $\{w\}$, rather than write $P(\{w\})$, and we will refer to $P(w)$ as the **probability of the outcome** w.

EXAMPLE 2.8
This example concerns an approach to probability that has far-reaching impact throughout both probability and statistics. Recall that Example 1.3 involved a frequency distribution resulting from the measurement of phone-call lengths and that we entertained the proposition that the function

$$f(x) = \begin{cases} \dfrac{1}{2} e^{-x/2}, & \text{if } x \geq 0 \\ 0, & \text{if } x < 0 \end{cases}$$

could serve as a descriptor of the probabilistic behavior of the measurement process X, not simply of the frequency distribution. Now consider the sample space R, the real line, and define a probability measure on the events of R by

$$P(B) = \int_B f(x)\, dx$$

for any event (subset of R) B. Mathematically, for P to be a legitimate probability measure, it must satisfy conditions P1 through P3. Since $f(x) \geq 0$, $P(B) \geq 0$ for any event B. Moreover,

$$P(R) = \int_R f(x)\, dx = \int_0^\infty \frac{1}{2} e^{-x/2}\, dx = 1$$

so that P2 is satisfied. Finally, it is a general principle of integration theory that the integral of a function over a union of mutually exclusive subsets of the real line is the sum of the integrals over the individual subsets, thus verifying the validity of P3. In sum, the sample space R together with the probability measure defined by integration with the integrand $f(x)$ constitutes a probability space.

As an illustration of P, the probability of the open interval $(0, 2)$ is

$$P((0, 2)) = \int_0^2 \frac{1}{2} e^{-x/2}\, dx = 1 - e^{-1} = 0.6321$$

Note that this is exactly the result obtained (under an intuitive analysis) in Example 1.3 for the probability that a measurement falls in the interval $(0, 2)$. While at first glance this agreement is obvious from the definition of the probability measure P, it is the approach that is consequential. In fact, once the function $f(x)$ has been determined, the methodology of the present example can be applied, no matter what the source of the frequency distribution, so long as $f(x)$ is nonnegative and possesses total integral 1 over R. We will return to this general question in Section 3.2.

2.2.2 Equiprobable Spaces

In Example 2.7, particular attention was paid to the case where a probability measure was defined on the sample space S_1 in accordance with the heuristic supposition that the selection process was uniform. Such a supposition often is appropriate. Even where it is not, it is sometimes prudent to find a new sample space relative to which the process is uniform. Remember, the mathematical definition of a probability space simply involves a set and a real-valued function on events that satisfies conditions P1, P2, and P3. There is no reason why there cannot be more than a single probability space that can serve as a suitable model for describing the physical phenomena under scrutiny. Should two models be readily available and one be mathematically easier to employ, it is reasonable to choose the simpler of the two.

Given a finite sample space

$$S = \{w_1, w_2, \ldots, w_n\}$$

of cardinality n, the **hypothesis of equal probability** is the assumption that the physical conditions are such that each of the outcomes in S possesses equal probability:

$$P(w_1) = P(w_2) = \cdots = P(w_n) = 1/n$$

In such a case, the probability space is said to be **equiprobable.** There is a subtle but crucial point to be recognized: the hypothesis of equal probability results in a particular assignment of probabilities; however, its use results from the experimenter's belief in the uniform occurrence of the physical phenomena that correspond to the (logical) outcomes constituting the sample space. Whether or not a probability space is equiprobable is simply a question regarding the construction of the probability measure; whether or not such a probability assignment results in a suitable model is another question. The hypothesis of equal probability concerns the modeling assumption.

Given that S is equiprobable, the determination of all probabilities follows at once from the additivity criterion P3. Indeed, if

$$E = \{v_1, v_2, \ldots, v_k\}$$

is any event in S, since

$$E = \bigcup_{i=1}^{k} \{v_i\}$$

additivity implies

$$P(E) = P(\{v_1\}) + P(\{v_2\}) + \cdots + P(\{v_k\}) = \frac{\text{card } (E)}{\text{card } (S)}$$

This is precisely the initial approach taken in Example 2.7. Because of its practical importance, we state it as a theorem.

If the finite sample space S is equiprobable, then

$$P(E) = \frac{card\ (E)}{card\ (S)}$$

for any event E in S.

EXAMPLE 2.9

Consider the sample space T of ordered 4-tuples discussed in Example 2.2, each component of a 4-tuple being a d or an n, depending on whether or not a selection is defective or nondefective. The event E_2 of Example 2.5 corresponds to the selection of 3 nondefective transistors. If T is equiprobable, then each outcome possesses probability 1/16, and

$$P(E_2) = 4/16 = 1/4$$

EXAMPLE 2.10

Consider a box containing two balls, one black and one white. Except for color, the balls are indistinguishable. If the experiment is to reach blindly into the box and choose one ball, it certainly appears that the hypothesis of equal probability is reasonable, since only the sense of vision can be used to distinguish between the balls, and the nature of the experiment eliminates that sense. Thus, the sample space $S = \{b, w\}$ and the probability measure P defined by

$$P(b) = P(w) = 1/2$$

appear to be appropriate.

Now, suppose the subject of the experiment reaches into the box 1000 times, each time recording the color of the selection, replacing the ball, and shaking the box. Moreover, suppose the following frequency distribution results:

Color	Frequency	Relative Frequency
black	648	0.648
white	352	0.352

Speaking intuitively, if the hypothesis of equal probability is justified, relative frequencies nearer to 1/2 might be expected.

Regarding Example 2.10, one may well ask: Is it possible that the use of the hypothesis of equal probability was justified and that the outcomes of the selection process deviated by chance, or is it more likely that the deviation of the relative frequencies from values that the hypothesis seemed to predict indicates that the hypothesis was unwarranted? From the perspectives of theory, application, and philosophy, this question is one of the most intriguing in the entire theory of probability and statistics, and questions like it will be addressed in great depth in future portions of the text.

In an equiprobable space, the computation of probability reduces to counting the number of elements of both the event and the sample space. In practice, certain types of outcomes occur quite often and can be characterized mathematically in standard forms. Consequently, we are led to study some general counting techniques.

2.3 COUNTING

This section treats the problem of counting and is therefore closely related to event probabilities in equiprobable spaces. The treatment involves various **urn models.** In each model, an urn contains a collection of numbered balls and the experiment involves randomly selecting balls according to some protocol. The problem for each protocol is to count the number of ways in which the selection process can be executed. Each process results in a set of elements of a particular sort. The importance of the resulting counting formulas is that many practical problems can be phrased in terms of the selection of balls from an urn.

Given an urn containing n numbered balls, we consider three selection protocols:

1. **Selection with replacement, order counts:** For $k > 0$, k balls are selected one at a time, each chosen ball is returned to the urn (to possibly be reselected), and the numbers of the chosen balls are recorded in the order of selection.
2. **Selection without replacement, order counts:** For $0 < k \le n$, k balls are selected, a chosen ball is not returned to the urn, and the numbers of the chosen balls are recorded in the order of selection.
3. **Selection without replacement, order does not count:** For $0 < k \le n$, k balls are selected, a chosen ball is not returned to the urn, and the numbers of the chosen balls are recorded without respect to the order of selection. Note that this process is equivalent to selecting all k balls at once.

2.3.1 Ordered Selection With Replacement

In selecting k balls from an urn with replacement and recording the order of the selection, the resulting outcomes are k-tuples of integers. If

$$N = \{1, 2, \ldots, n\}$$

gives the numbers of the balls in the urn, then each outcome resulting from the selection process is of the form

$$(n_1, n_2, \ldots, n_k)$$

where n_i is an integer inclusively between 1 and n. Thus, each outcome is an element of the product set N^k. According to Theorem 2.1, the number of possible selections is

$$\text{card}(N^k) = [\text{card}(N)]^k = n^k$$

We saw an application of this formula in Example 2.2.

EXAMPLE 2.11 In deciding the format for a memory word in a new computer, the designer decides on a length of 16 bits. Since each bit can be 0 or 1, the problem of deciding on the number of possible words can be modeled as making 16 selections from an urn containing 2 balls. Thus, there are $2^{16} = 65,536$ possible words.

2.3.2 Ordered Selection Without Replacement— Permutations

As in the case of ordered selection with replacement, selecting without replacement and recording the selection order yields outcomes from N^k. However, since the balls are not being replaced, no component in an outcome k-tuple can be repeated. For instance, if there are seven balls in the urn, numbered 1 through 7, and there are four selections without replacement, then the 4-tuple (5, 2, 1, 6) is a possible outcome, but the 4-tuple (5, 2, 5, 7) is not. Moreover, since the balls are not being replaced, the number of selections must be less than or equal to the number of balls in the urn at the outset. Elements of a product set that contain no repeated components are called **permutations.** Note that we have employed the set N for illustrative purposes, and the notion of a permutation applies to other sets.

DEFINITION 2.4

Permutation. For $0 < k \leq n$, a k-tuple of elements from a set containing n elements for which no component value is repeated is called a **permutation (of n objects taken k at a time).**

In counting the number of possible selection processes without replacement for which order counts, we are counting permutations.

To get an idea of the counting problem, consider the set

$$A = \{1, 2, 3, 4\}$$

The set of permutations of two objects from A is

$$Q = \{(1, 2), (1, 3), (1, 4), (2, 1), (2, 3), (2, 4),$$
$$(3, 1), (3, 2), (3, 4), (4, 1), (4, 2), (4, 3)\}$$

so that there are 12 permutations. Because of the nonreplacement requirement, there are exactly three ordered pairs in Q for each choice of a first component value. Thus, there is a total of $12 = 4 \times 3$ permutations. The next theorem concerns the general case.

THEOREM 2.3

If S is a set containing n elements and $0 < k \leq n$, then there exist

$$n(n - 1)(n - 2) \cdots (n - k + 1)$$

permutations of k elements from S. Letting $P(n, k)$ denote the number of permutations and employing factorials,

$$P(n, k) = \frac{n!}{(n - k)!}$$

Proof: Each permutation is of the form

$$(a_1, a_2, \ldots, a_k)$$

and there are n choices for the first component a_1. Once a_1 is chosen, there remain only $n - 1$ choices for the second component. Consequently, there exist $n(n - 1)$ possible ordered pairs where the second component is different from the first. Once a_2 is chosen, there remain only $n - 2$ choices for the third component, and therefore there are $n(n - 1)(n - 2)$ possible ordered triples where no component is repeated. Continuing, we obtain the result claimed by the theorem.

As for the factorial representation, simply note that

$$\frac{n!}{(n-k)!} = \frac{n(n-1)(n-2) \cdots (n-k+1)(n-k)(n-k-1) \cdots 2 \cdot 1}{(n-k)(n-k-1) \cdots 2 \cdot 1}$$

$$= n(n-1)(n-2) \cdots (n-k+1) \qquad \blacksquare$$

EXAMPLE 2.12

Consider an alphabet consisting of 9 distinct symbols from which strings of length 4 that do not use the same symbol twice are to be formed. Each string is a permutation of 9 objects taken 4 at a time, and thus there are

$$P(9, 4) = 9 \times 8 \times 7 \times 6 = 3024$$

possible words.

Now suppose the 4 symbols are chosen uniformly randomly with replacement. What is the probability that a string will be formed in which no symbol is utilized more than once? If we let S denote the sample space of all 4-symbol words chosen with replacement, let E denote the event consisting of all words with no symbol appearing more than once, and apply the hypothesis of equal probability (which is in accord with the method of uniformly random selection), then the desired probability is

$$P(E) = \frac{\text{card }(E)}{\text{card }(S)} = \frac{P(9, 4)}{9^4} = 0.461$$

2.3.3 Fundamental Principle of Counting

Both the counting of elements in a product set and of permutations are particular instances of a more general principle, namely, counting k-tuples formed according to the following scheme:

1. The first component of the k-tuple can be occupied by any one of r_1 objects.
2. No matter which object has been chosen to occupy the first component, any one of r_2 objects can occupy the second component.
3. Proceeding recursively, no matter which objects have been chosen to occupy the first $j - 1$ components, any one of r_j objects can occupy the jth component.

The **fundamental principle of counting** states that there are $r_1 r_2 \cdots r_k$ possible k-tuples that can result from application of the selection scheme. Equivalently stated, if Q is the set of all possible k-tuples that can result, then

$$\text{card }(Q) = r_1 r_2 \cdots r_k$$

Figure 2.5 gives a graphic illustration of the fundamental principle of counting utilizing a tree diagram. Four possible branches can be chosen for the first selection, 2 for the second, and 2 for the third. As a result, the tree contains $4 \times 2 \times 2 = 16$ final nodes. It is crucial to note that at each of three stages (selections) of the tree, the number of branches emanating from the nodes is the same; otherwise, as illustrated in Figure 2.6, the multiplication technique of the fundamental principle does not apply. The requirement that there be a constant number of emanating branches at each stage corresponds to the condition in the selection protocol that, at each component, the number of possible choices for the component is fixed and does not depend on the particular objects chosen to fill the preceding components.

In the case of a product set $A = A_1 \times A_2 \times \cdots \times A_k$, $r_j = \text{card }(A_j)$ for $j = 1, 2, \ldots, k$. Hence, according to the fundamental principle, the cardinality

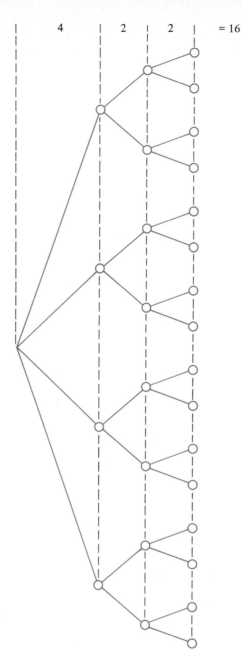

| 4 | 2 | 2 | = 16 |

FIGURE 2.5
Illustration of the
fundamental principle of
counting using a tree
diagram

of the product set is equal to the product of the cardinalities, which is precisely Theorem 2.1.

As for permutations of n objects taken k at a time, the fundamental principle applies with $r_j = n - j + 1$ for $j = 1, 2, \ldots, k$. Theorem 2.3 results.

EXAMPLE 2.13

Suppose a stereo amplifier possesses discrete levels for volume, bass, and treble settings, the number of settings being 12, 8, and 8, respectively. Then, according to the fundamental principle of counting, the total number of settings involving volume, bass, and treble is $12 \times 8 \times 8 = 768$.

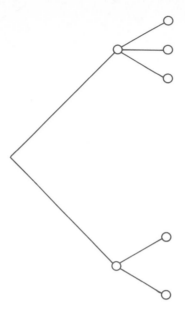

FIGURE 2.6
A tree diagram for which
the fundamental
principle of counting
does not apply

2.3.4 Unordered Selection Without Replacement— Combinations

When selecting balls without replacement from an urn without regard to the order of selection, the result of the procedure consists of a list of nonrepeated elements, the ordering in the list being irrelevant. A list of elements for which the order of the listing can be interchanged without affecting the list is simply a set. Thus, the outcomes are subsets of the set from which the elements are being chosen. Employing the urn-model terminology, each outcome is a subset consisting of k balls from the original collection of n balls in the urn.

Using the set A consisting of the first four integers, the set of all possible subsets that result from selecting two elements according to the unordered, without-replacement protocol is the set of sets

$$\{\{1, 2\}, \{1, 3\}, \{1, 4\}, \{2, 3\}, \{2, 4\}, \{3, 4\}\}$$

Each subset resulting from the unordered, without-replacement protocol is called a **combination (of n objects taken k at a time).** We desire a formula for the number of combinations, which we will denote by $C(n, k)$.

Again consider the set A. It can be seen from Figure 2.7 that there are two 2-tuple permutations for each 2-element combination. The reason is that, according to the permutation theory of Section 2.3.2, given any subset possessing 2 elements, there exist 2! distinct permutations of the subset elements (permutations of 2 elements taken 2 at a time). Thus, each 2-element subset from A yields 2! = 2 permutations. This reasoning generalizes.

THEOREM 2.4

If a set contains n elements, then there exist

$$C(n, k) = \frac{n!}{k!(n - k)!}$$

FIGURE 2.7
An illustration of the two-to-one relationship between combinations of 4 objects taken 2 at a time and permutations of 4 objects taken 2 at a time

combinations (subsets) containing k elements. In particular,

$$P(n, k) = k!C(n, k)$$

Proof: Given any subset possessing k elements, there exist $k!$ distinct permutations of the subset elements (permutations of k elements taken k at a time). Thus, each combination corresponds to $k!$ permutations, and therefore

$$P(n, k) = k!C(n, k)$$

Division by $k!$ yields

$$C(n, k) = \frac{P(n, k)}{k!} = \frac{n!}{k!(n - k)!} \qquad \blacksquare$$

EXAMPLE 2.14

Suppose a committee of 4 engineers is to be selected from a group of 20 working on a particular project. Since the order of selection is irrelevant insofar as the composition of a committee is concerned, there are

$$C(20, 4) = \frac{20!}{4!16!} = \frac{20 \times 19 \times 18 \times 17}{4 \times 3 \times 2 \times 1} = 4845$$

possible committees.

Now suppose that of the 20 engineers, 12 are electrical engineers and 8 are mechanical engineers. Assuming the selection process to be uniformly random, what is the probability that a committee of 4 will consist entirely of electrical engineers? Letting S denote the sample space of all possible committees and E the event comprised of all committees consisting of only electrical engineers, the uniformly random selection process justifies the application of the hypothesis of equal probability; hence

$$P(E) = \frac{\text{card } (E)}{\text{card } (S)} = \frac{C(12, 4)}{C(20, 4)} = \frac{495}{4845} = 0.102$$

EXAMPLE 2.15

Consider the problem of choosing two committees, the first consisting of 3 electrical engineers selected from a force of 14 and the second consisting of 2 computer scientists taken from a group of 12. According to Theorem 2.4, the first committee can be chosen in

$$C(14, 3) = \frac{14!}{3!11!} = 364$$

ways and the second in $C(12, 2) = 66$ ways. In how many ways can the pair of committees be chosen? The answer lies with the fundamental principle of counting. The selection of the first committee can be accomplished in 364 ways, and the selection of the second committee (no matter what the composition of the first committee) in 66 ways. Thus, by the fundamental principle, there are

$$364 \times 66 = 24{,}024$$

possible pairs of committees.

In effect, we are forming ordered pairs of committees, and hence can view the problem as one of product sets, the first set consisting of committees of 3 chosen from among the electrical engineers and the second set consisting of committees of 2 chosen from among the computer scientists.

The binomial theorem (from high school algebra) states that, for any integer $n = 0, 1, 2, \ldots$,

$$(a + b)^n = \sum_{k=0}^{n} \binom{n}{k} a^{n-k} b^k$$

where

$$\binom{n}{k} = \frac{n!}{k!(n-k)!}$$

For instance,

$$(a + b)^4 = a^4 + 4a^3b^1 + 6a^2b^2 + 4a^1b^3 + b^4$$

According to Theorem 2.4, the number of combinations of n objects taken k at a time is equal to the kth binomial coefficient in the expansion for $(a + b)^n$:

$$C(n, k) = \frac{n!}{k!(n-k)!} = \binom{n}{k}$$

As an illustration, letting $a = b = 1$ in the binomial theorem yields

$$2^n = (1 + 1)^n = \sum_{k=0}^{n} \binom{n}{k} 1^{n-k} 1^k$$

$$= \sum_{k=0}^{n} \binom{n}{k} = \sum_{k=0}^{n} C(n, k)$$

Since each binomial coefficient is equal to the number of k-element subsets from a set of n elements, a set with cardinality n possesses 2^n subsets.

The next theorem gives a relation that provides a practical device for the recursive computation of binomial coefficients. Its proof is an exercise in using the factorial recursion equation $m! = m(m - 1)!$.

THEOREM 2.5 For $n \geq 1$ and $1 \leq k \leq n$,

$$\binom{n}{k} = \binom{n-1}{k-1} + \binom{n-1}{k}$$

***Proof*:**

$$\begin{aligned}
\binom{n-1}{k-1} + \binom{n-1}{k} &= \frac{(n-1)!}{(k-1)![n-1-(k-1)]!} + \frac{(n-1)!}{k!(n-1-k)!} \\
&= \frac{k(n-1)!}{k!(n-k)!} + \frac{n(n-k)(n-1)!}{k!n(n-k)(n-1-k)!} \\
&= \frac{kn!}{k!n(n-k)!} + \frac{(n-k)n!}{nk!(n-k)!} \\
&= \frac{k}{n}\binom{n}{k} + \frac{n-k}{n}\binom{n}{k} \\
&= \binom{n}{k}
\end{aligned}$$

■

A graphical way of stating Theorem 2.5 is given in Figure 2.8, known as **Pascal's triangle.** According to the theorem, each coefficient in row n, the first row being row 0, is obtained by adding together the two coefficients that are above it in row $n - 1$. For instance, as illustrated in the figure,

$$10 = \binom{5}{2} = \binom{4}{1} + \binom{4}{2} = 4 + 6$$

FIGURE 2.8
Pascal's triangle

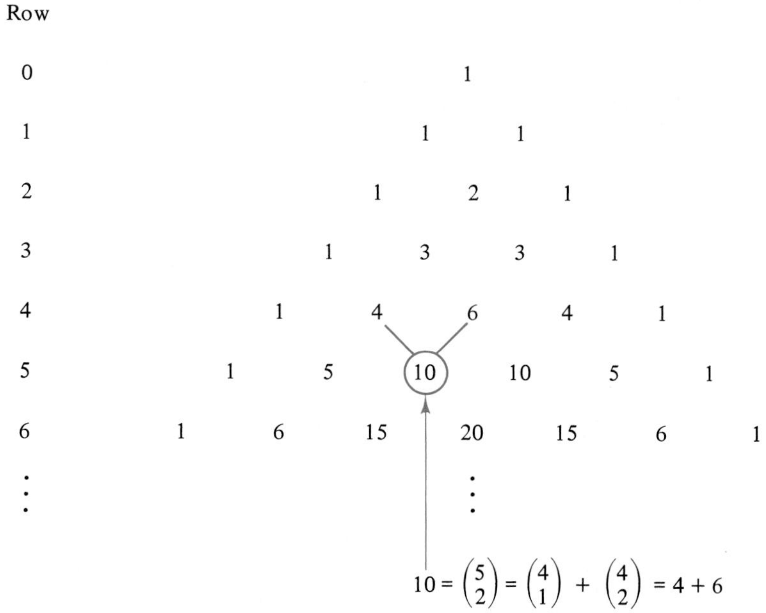
Row
0 1
1 1 1
2 1 2 1
3 1 3 3 1
4 1 4 6 4 1
5 1 5 10 10 5 1
6 1 6 15 20 15 6 1
10 = (5 2) = (4 1) + (4 2) = 4 + 6

which is precisely the relation of Theorem 2.5 applied to $n = 5$ and $k = 2$. Note the symmetry in the rows of Pascal's triangle. This symmetry follows from the identity

$$\binom{n}{k} = \binom{n}{n-k}$$

2.3.5 Placing Objects Into Cells

A close look at the generation of combinations shows that, in effect, two subsets are being formed, one consisting of the objects that are chosen and the other consisting of the remaining objects. Thus, the unordered, without-replacement selection protocol can be viewed as partitioning the original objects into two cells, where the order of objects within the cells is irrelevant. According to Theorem 2.4, if there are n objects, with k to be placed into one cell and $n - k$ to be placed into the other, then there are $C(n, k)$ ways of forming the cells.

As illustrated in Figure 2.9, this partitioning scheme can be extended to an arbitrary number of cells, say r, where we wish to place k_1 objects into the first cell, k_2 objects into the second cell, and so on, with the final k_r objects being placed into the rth cell. Note that

$$k_1 + k_2 + \cdots + k_r = n$$

For the case of two cells, $k_2 = n - k_1$. Our immediate goal is to count the number of ways in which a collection of objects can be partitioned into cells.

To get some understanding of the problem, consider an urn containing balls numbered 1 through 10 and suppose the balls are to be placed into three cells, where the order within each cell is unimportant and where the cells are to contain 2, 3, and 5 balls, respectively. According to Theorem 2.4, the contents of the first cell can be formed in $C(10, 2) = 45$ ways. No matter which balls are placed into the first cell, 8 balls will remain to be selected, and there are $C(8, 3) = 56$ ways of selecting 3 of these balls for the second cell. Finally, the last cell must be filled with the remaining 5 balls, so that no matter how the first two cells are filled, there exists but one way

FIGURE 2.9
Placing n objects into r cells

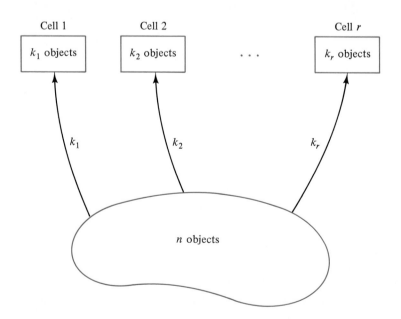

of filling the last cell. In effect, we are forming 3-tuples of subsets of the original set of 10. For instance, one such 3-tuple is

$$(\{2, 4\}, \{1, 10, 5\}, \{3, 6, 7, 8, 9\})$$

Moreover, the number of ways in which each component can be filled is not dependent upon the manner in which the preceding components have been filled. Thus, the fundamental principle of counting applies, and there are

$$45 \times 56 \times 1 = 2520$$

ways to form the cells. The next theorem, stated without proof, provides a general counting formula for forming cells by partitioning. The quantities given in the statement of the theorem are known as **multinomial coefficients.**

THEOREM 2.6

There are

$$\binom{n}{k_1, k_2, \ldots, k_r} = \frac{n!}{k_1! k_2! \cdots k_r!}$$

ways of partitioning n objects into r cells so that k_i objects are in cell i for $i = 1$, $2, \ldots, r$.

EXAMPLE 2.16

Consider partitioning a group of 20 workers into 4 job classifications, requiring 3, 6, 4, and 7 employees, respectively. According to Theorem 2.6, there are

$$\binom{20}{3, 6, 4, 7} = \frac{20!}{3! 6! 4! 7!}$$

ways in which the job classifications can be filled.

2.4 FUNDAMENTAL PROPERTIES OF PROBABILITY SPACES

This section discusses some of the most fundamental theoretical properties concerning probability spaces. In proving the theorems, we must rely solely on the three probability axioms given in Definition 2.3, for these determine the mathematical formulation of probability. Nevertheless, as is often the case in mathematics, an accompanying conceptual model is useful for gaining an intuitive appreciation of the theory.

Consider the Venn diagram of Figure 2.10, which depicts an arbitrary event E in the sample space S, and assume the area of the disk S to be 1. Now imagine a random dart thrower who tosses darts at the disk in such a way that the darts are sprayed uniformly over the disk. Because of the uniform randomness of the darts and the assumption that the area of S is 1, a suitable probability model would be to let the event E possess probability

FIGURE 2.10
An event E as a subset of the sample space S

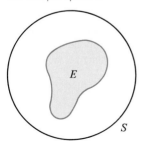

$$P(E) = \text{Area}(E)$$

Such a model certainly satisfies the three requirements of a probability measure. In particular, area is additive.

Our first proposition states that, given the probability of an event E, the probability of the complementary event E^c can be found by subtracting $P(E)$ from 1. Note in Figure 2.10 that

$$\text{Area}(E^c) = 1 - \text{Area}(E)$$

THEOREM 2.7

For any event E,

$$P(E^c) = 1 - P(E)$$

Proof: Since E and E^c are mutually exclusive and $S = E \cup E^c$, additivity implies that

$$P(S) = P(E) + P(E^c)$$

But probability axiom P2 states that $P(S) = 1$. Thus,

$$1 = P(E) + P(E^c)$$ ∎

An immediate corollary of Theorem 2.7 is that the probability of the null event is 0. This is certainly in agreement with our intuitions, since Ø contains no outcomes.

THEOREM 2.8

$P(\emptyset) = 0$.

Proof: Ø is the complement of the sample space S. Hence, according to Theorem 2.7,

$$P(\emptyset) = 1 - P(S) = 1 - 1 = 0$$ ∎

According to Theorem 2.8, the probability of the null event is zero; however, the converse is not true. It is commonplace for nonnull events to have probability zero. For instance, in Example 2.8, any interval (a, b) for which $b \leq 0$ possesses zero probability. In the same example, the event $\{2\}$ also has zero probability. In a similar direction, many events (besides the sample space) in Example 2.8 have probability one. Two examples are the infinite interval $[0, +\infty)$ and the event $R - \{2.7\}$.

EXAMPLE 2.17

We consider the famous *birthday problem*. Suppose M people are selected from a large group and their birthdays recorded. For simplicity, omit leap years. For each selection, there are 365 possible birthdays, which we will label 1, 2, . . . , 365. Since the procedure can be modeled as the selection of M balls with replacement from an urn containing 365 balls, the sample space is the product set

$$S = \{1, 2, 3, . . . , 365\}^M$$

whose outcomes are M-tuples, where each component can be any one of 365 dates. The number of elements in the sample space is 365^M and, employing the hypothesis of equal probability, the probability of any event E is $P(E) = 365^{-M} \text{card}(E)$.

Consider the event described by the statement, "At least two selected birthdays are the same." Then E consists of all outcomes for which there are at least two components occupied by the same birthday. For instance, if $M = 6$, then the following 6-tuples are elements of E:

$$(32, 57, 231, 57, 2, 77)$$

$$(343, 22, 158, 22, 343, 99)$$

$$(55, 55, 249, 192, 3, 55)$$

Clearly, E is quite complicated. However, E^c is easy to describe: it consists of all outcomes possessing no identical components. Thus, E^c consists of all permutations of the 365 dates taken M at a time, and

$$P(E^c) = \frac{\text{card } (E^c)}{\text{card } (S)} = \frac{365 \times 364 \times 363 \times \cdots \times (365 - M + 1)}{365^M}$$

Applying Theorem 2.7 gives

$$P(E) = 1 - \frac{365 \times 364 \times \cdots \times (365 - M + 1)}{365^M}$$

Table 2.1 gives some representative values for $P(E)$. For instance, if there are 23 people selected, then the probability that at least two possess the same birthday is 0.51.

The reason for our interest in the birthday problem is that many people find the resulting probabilities surprising. Indeed, imagine walking into a room of 30 people and asking each person his or her birthday. According to Table 2.1, the probability is 0.71 that there will be at least two duplicate responses. If there were 50 people in the room, then the probability would be 0.97!

Consider the random dart thrower and a situation where E is a subevent of F. Referring to Figure 2.11, we see that

$$\text{Area}(F - E) = \text{Area}(F) - \text{Area}(E)$$

which, according to the model, says that the probability of $F - E$ is equal to the probability of F minus the probability of E. The next theorem gives a statement in terms of general probabilities.

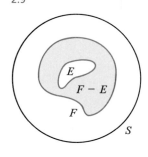

FIGURE 2.11
Illustration for Theorem 2.9

THEOREM 2.9

If E is a subevent of F, then

$$P(F - E) = P(F) - P(E)$$

Proof: E and $F - E$ are mutually exclusive and, since $E \subset F$,

$$F = E \cup (F - E)$$

Additivity yields

$$P(F) = P(E) + P(F - E) \qquad \blacksquare$$

THEOREM 2.10

If E is a subevent of F, then $P(E) \le P(F)$.

Proof: According to Theorem 2.9,

$$P(E) = P(F) - P(F - E) \le P(F)$$

TABLE 2.1 PROBABILITIES FOR THE BIRTHDAY PROBLEM

M	10	20	23	30	40	50
$P(E)$	0.12	0.41	0.51	0.71	0.89	0.97

where the inequality follows from the fact that, according to axiom P1, $P(F - E) \geq 0$. ∎

Whereas additivity provides a method for finding the probability of a union of mutually exclusive events, the next proposition concerns the case of finding the probability of a union of two events that are not necessarily mutually exclusive.

For any events E and F in the same sample space,

$$P(E \cup F) = P(E) + P(F) - P(E \cap F)$$

Proof: The events E and $F - E$ are mutually exclusive, and

$$E \cup F = E \cup (F - E)$$

FIGURE 2.12
Illustration for Theorem 2.11

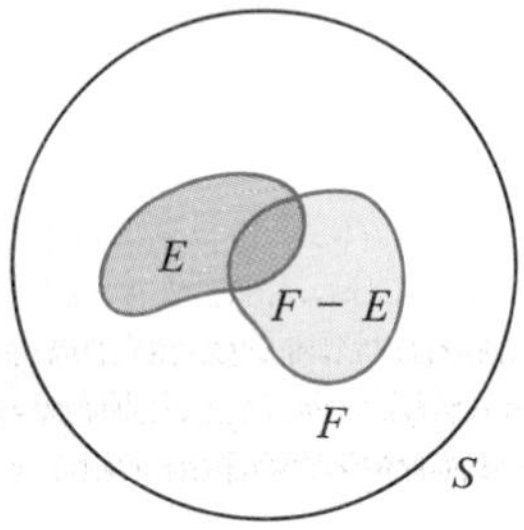

(see Figure 2.12). Therefore, according to additivity,

$$P(E \cup F) = P(E) + P(F - E)$$

But

$$F - E = F - (E \cap F)$$

and $E \cap F$ is a subevent of F. Consequently, by Theorem 2.9,

$$P(F - E) = P(F) - P(E \cap F)$$

Inserting this result into the previous expression for $P(E \cup F)$ yields the result. ∎

Two points regarding Theorem 2.11 should be mentioned. First, if E and F are mutually exclusive, so that $E \cap F = \emptyset$, then the theorem reduces to additivity. Second, the theorem is proved by expressing the event of interest as a union of mutually exclusive events. In practice, one often attempts to reduce a problem to one involving mutually exclusive events so that additivity can be employed.

EXAMPLE 2.18

In the analysis of well water, a number of tests are made for various metallic, chemical, and bacteriological concentrations. Suppose that, in a particular locality, 16% of the tested wells exceed the maximum hardness level of 250 ppm (parts per million), 20% exceed the maximum iron level of 0.30 ppm, and 8% exceed both limits. Given a well selected at random, what is the probability that it exceeds either the maximum hardness or the maximum iron concentration? Letting E and F denote the collections of wells that exceed the maximum hardness and iron concentrations, respectively, we desire $P(E \cup F)$, which, according to Theorem 2.11, is given by

$$P(E \cup F) = P(E) + P(F) - P(E \cap F) = 0.16 + 0.20 - 0.08 = 0.28$$

According to Theorem 2.9, the probability that the chosen well exceeds the hardness limit but not the iron limit is

$$P(E - F) = P(E) - P(E \cap F) = 0.16 - 0.08 = 0.08$$

The next property of a probability measure concerns the decomposition of the sample space and requires a definition.

DEFINITION 2.5

Partition. A partition of a sample space S is a collection of mutually exclusive events $F_1, F_2, \ldots, F_n$ such that

$$S = F_1 \cup F_2 \cup \cdots \cup F_n$$

Figure 2.13 depicts a partition of the random dart thrower's sample space into five events, $F_1, \ldots, F_5$.

THEOREM 2.12

If $F_1, F_2, \ldots, F_n$ form a partition of the sample space S, then, for any event E in S,

$$P(E) = \sum_{i=1}^{n} P(E \cap F_i)$$

FIGURE 2.13
Illustration for Theorem 2.12

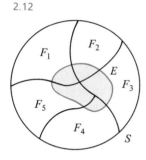

Proof: Because the F_i form a partition of S and $E \subset S$,

$$E = (E \cap F_1) \cup (E \cap F_2) \cup \cdots \cup (E \cap F_n)$$

(see Figure 2.13). Moreover, $E \cap F_1, E \cap F_2, \ldots, E \cap F_n$ are mutually exclusive. Hence, the assertion of the theorem follows by additivity. ∎

2.5 CONDITIONAL PROBABILITY

Intuitively, a probability space models the likelihoods of certain events in the face of uncertainty. Should the degree of uncertainty be narrowed, it should be expected that the probability assignment will be altered so as to reflect the new situation. In effect, the presence of additional information conditions the probability assignment. It should be evident that the probability measure resulting from this conditioning will depend on both the original probability measure and the nature of the additional information.

2.5.1 Conditional Probability Measure

To motivate the definition of conditional probability, we consider an equiprobable space S and events E and F defined by

$$S = \{x_1, \ldots, x_m, y_1, \ldots, y_n, z_1, \ldots, z_p, w_1, \ldots, w_q\}$$

$$E = \{x_1, x_2, \ldots, x_m, y_1, y_2, \ldots, y_n\}$$

$$F = \{y_1, y_2, \ldots, y_n, z_1, z_2, \ldots, z_p\}$$

Owing to the assumption of equiprobability,

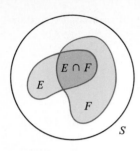

FIGURE 2.14
$E \cap F$ as a subevent of F

$$P(E) = \frac{m + n}{m + n + p + q}$$

$$P(F) = \frac{n + p}{m + n + p + q}$$

Suppose we are given the prior condition that the outcome of the experiment modeled by S will lie in the event F. Then it is natural to remodel the experiment by letting F be the new sample space. Under the given condition, the **conditional probability** that the outcome lies in E refers to the likelihood of the outcome lying in

$$E \cap F = \{y_1, y_2, \ldots, y_n\}$$

since the elements of

$$E - F = \{x_1, x_2, \ldots, x_m\}$$

cannot occur. If the conditional probability of E, given the prior knowledge that F will occur, is denoted by $P(E \mid F)$, then

$$P(E \mid F) = \frac{\text{card } (E \cap F)}{\text{card } (F)} = \frac{n}{n + p} = \frac{P(E \cap F)}{P(F)}$$

Referring to Figure 2.14, in which the area of S is 1, let us once again consider the uniformly random dart thrower. Given that the dart will land in event F, it is clear that F should correspond to a new, conditional sample space. However, since the area of F is less than 1, the conditional probability assignment will have to be normalized so that $P(F \mid F) = 1$, while at the same time keeping the ratios of the probabilities of the events within F invariant. Thus, a reasonable conditional probability assignment is

$$P(E \mid F) = \frac{\text{Area}(E \cap F)}{\text{Area}(F)} = \frac{P(E \cap F)}{P(F)}$$

Both of the preceding analyses yield a similar representation of the conditional probability model. Based upon these, a rigorous mathematical definition of conditional probability can be formulated.

DEFINITION 2.6

Conditional Probability Measure. Let S be a probability space with associated probability measure P. For any event F in S, $P(F) > 0$, the conditional probability measure, relative to F, is defined by

$$P(E \mid F) = \frac{P(E \cap F)}{P(F)}$$

for any event E in S.

Two points should be noted. First, division by $P(F)$ requires that $P(F)$ be nonzero. Second, it must be shown that the conditional probability measure $P(\cdot \mid F)$ is indeed a legitimate probability according to Definition 2.3. This will be done in Theorem 2.13.

EXAMPLE 2.19

At an automobile inspection station, 32% of the cars fail to satisfy the emission standards owing to excess hydrocarbons, 40% fail owing to excess carbon monoxide, and 18% fail on both accounts. Under a frequency interpretation of probability, with E and F denoting the events corresponding to hydrocarbon and carbon monoxide failures, respectively, the probability of failure due to excess carbon monoxide, given failure due to excess hydrocarbons, is given by

$$P(F \mid E) = \frac{P(F \cap E)}{P(E)} = \frac{0.18}{0.32} = 0.56$$

The probability of failure due to high hydrocarbon emission, given failure resulting from excess carbon monoxide, is

$$P(E \mid F) = \frac{P(E \cap F)}{P(F)} = \frac{0.18}{0.40} = 0.45$$

The next theorem shows that, for a fixed event F with nonzero probability, the conditional probability measure $P(\cdot \mid F)$ is a legitimate probability measure on the original sample space.

THEOREM 2.13

Given a probability space S with probability measure P and a fixed event F in S with $P(F) > 0$, $P(\cdot \mid F)$ is a probability measure on S.

Proof: We must demonstrate that P1 through P3 of Definition 2.3 hold for the function $P(E \mid F)$, where E is an event of S. As for P1, for any event E, $P(E \cap F) \geq 0$, and thus both numerator and denominator in the defining relation for $P(E \mid F)$ are nonnegative, assuring that $P(E \mid F)$ itself is nonnegative. Second, since F is a subset of S,

$$P(S \mid F) = \frac{P(S \cap F)}{P(F)} = \frac{P(F)}{P(F)} = 1$$

and therefore P2 is satisfied. With respect to P3, consider a collection of events E_1, E_2, ... Applying the distributive law for sets in conjunction with the additivity of the original probability measure P yields

$$
\begin{aligned}
P(E_1 \cup E_2 \cup \cdots \mid F) &= \frac{P[(E_1 \cup E_2 \cup \cdots) \cap F]}{P(F)} \\
&= \frac{P[(E_1 \cap F) \cup (E_2 \cap F) \cup \cdots]}{P(F)} \\
&= \sum_{k=1}^{\infty} \frac{P(E_k \cap F)}{P(F)} \\
&= \sum_{k=1}^{\infty} P(E_k \mid F)
\end{aligned}
$$

thereby demonstrating that P3 holds for the conditional probability measure. ∎

2.5.2 Multiplication Property

In applications, it is common to utilize knowledge of the conditional probabilities to determine the original probability of certain events. This can be accomplished by cross-multiplication of $P(F)$ in Definition 2.6. Because of its usefulness, the result is stated as a theorem. The assumption that $P(F) > 0$ insures that the conditional probability $P(E \mid F)$ is defined.

THEOREM 2.14

(Multiplication Property). For any events E and F, if $P(F) > 0$, then

$$P(E \cap F) = P(F)P(E \mid F)$$

Due to the commutativity of intersection, so long as $P(E) > 0$, Theorem 2.14 can be rewritten as

$$P(F)P(E \mid F) = P(E \cap F) = P(F \cap E) = P(E)P(F \mid E)$$

EXAMPLE 2.20

In this example, an urn model is employed to demonstrate the manner in which the multiplication property can be fruitfully employed. Consider three urns in which there is a distribution of balls as specified in the following table:

Urn	Red	White
a	3	2
b	3	3
c	4	1

Suppose the following experiment is performed: A fair (uniform) die is tossed. If a 1, 2, or 3 appears on the die, then a ball is randomly selected from urn a; if a 4 or 5 appears, then a ball is randomly selected from urn b; if a 6 appears, then a ball is randomly selected from urn c. A suitable sample space can be obtained by forming the product set

$$\begin{aligned} S &= \{a, b, c\} \times \{r, w\} \\ &= \{(a, r), (a, w), (b, r), (b, w), (c, r), (c, w)\} \end{aligned}$$

The problem is to assign the probability measure.

While it is certainly possible to proceed with a direct approach based on Definition 2.3, it is easier to employ a conditional analysis. First note that there are two events in the product space relating to the choice of a color ball:

$$W = \{(a, w), (b, w), (c, w)\}$$
$$R = \{(a, r), (b, r), (c, r)\}$$

If we let A, B, and C denote the events corresponding to the selections of urns a, b, and c, respectively, then

$$A = \{(a, w), (a, r)\}$$
$$B = \{(b, w), (b, r)\}$$
$$C = \{(c, w), (c, r)\}$$

According to the uniformity of the die outcomes, a satisfactory probability assignment must yield $P(A) = 1/2$, $P(B) = 1/3$, and $P(C) = 1/6$. Moreover, it is evident that there are six conditional probabilities relating to the selection of a particular color ball, given the choice of an urn:

$$P(W\,|\,A) = 2/5, \qquad P(R\,|\,A) = 3/5$$
$$P(W\,|\,B) = 1/2, \qquad P(R\,|\,B) = 1/2$$
$$P(W\,|\,C) = 1/5, \qquad P(R\,|\,C) = 4/5$$

Note that these conditional probabilities have not been derived from the definition; rather, we have discovered them by analyzing the experiment. The point is that if we were to actually go through a probability assignment in S that faithfully reflected the experimental conditions, then these six conditional probabilities would result. Of course, we could go through such an assignment; however, the six conditional probabilities are obtained more easily by consideration of the model, and, in conjunction with the multiplication theorem, they can be used to answer questions regarding the probabilities of selecting a white or red ball. Indeed, since a red ball can be selected in one of three mutually exclusive ways—selecting urn a and picking a red ball, selecting urn b and picking a red ball, or selecting urn c and picking a red ball—then, by additivity and Theorem 2.14,

$$\begin{aligned}
P(R) &= P[(A \cap R) \cup (B \cap R) \cup (C \cap R)] \\
&= P(A \cap R) + P(B \cap R) + P(C \cap R) \\
&= P(A)P(R\,|\,A) + P(B)P(R\,|\,B) + P(C)P(R\,|\,C) \\
&= 1/2 \times 3/5 + 1/3 \times 1/2 + 1/6 \times 4/5 = 3/5
\end{aligned}$$

Similarly, the probability of choosing a white ball is given by

$$\begin{aligned}
P(W) &= P[(A \cap W) \cup (B \cap W) \cup (C \cap W)] \\
&= P(A \cap W) + P(B \cap W) + P(C \cap W) \\
&= P(A)P(W\,|\,A) + P(B)P(W\,|\,B) + P(C)P(W\,|\,C) \\
&= 1/2 \times 2/5 + 1/3 \times 1/2 + 1/6 \times 1/5 = 2/5
\end{aligned}$$

A close examination of Example 2.20 reveals that the explicit representation of the sample space as a product set played only a minor role in the solution methodology. In fact, it was there simply to provide a mathematical framework for

FIGURE 2.15
Tree diagram corresponding to Example 2.20

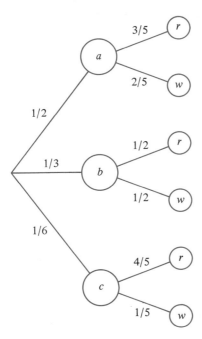

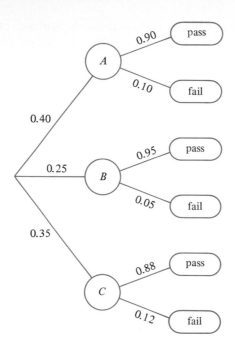

FIGURE 2.16
Tree diagram
corresponding to
Example 2.21

the analysis. From a practical perspective, the solution follows the tree-diagram analysis given in Figure 2.15, where each node of the tree represents a logically possible choice and each branch is labeled with the probability of making that choice. Note that whereas the labels on the three branches constituting the first stage of the tree are straight probabilities, the six labels on the second stage of the tree are conditional probabilities. In effect, to find the probability of choosing a white ball, one simply has to locate the paths that lead to the desired outcome, multiply the probabilities along each appropriate path, and then sum the resulting products. It is just such an approach that makes the multiplication property so easy to use.

Although there was no explicit assignment of probabilities to the individual elements in the sample space of Example 2.20, it should be noted that they are immediately derivable from the events. For instance, since $\{(a, w)\} = A \cap W$,

$$P(\{(a, w)\}) = P(A \cap W) = P(A)P(W \,|\, A) = 1/2 \times 2/5 = 1/5$$

EXAMPLE 2.21

For reasons of availability, a manufacturer of electronic calculators must purchase LSI chips from three suppliers. Suppose the percentages of chips that satisfy minimal endurance specifications are 90%, 95%, and 88% for suppliers A, B, and C, respectively. Moreover, suppose the manufacturer receives 40%, 25%, and 35% of his or her chips from suppliers A, B, and C, respectively. Given an arbitrary chip, what is the probability it will meet minimal endurance requirements? Let E denote the event that the chip passes the endurance test. Then

$$E = (A \cap E) \cup (B \cap E) \cup (C \cap E)$$

By Theorem 2.14,

$$
\begin{aligned}
P(E) &= P(A \cap E) + P(B \cap E) + P(C \cap E) \\
&= P(A)P(E \,|\, A) + P(B)P(E \,|\, B) + P(C)P(E \,|\, C) \\
&= 0.40 \times 0.90 + 0.25 \times 0.95 + 0.35 \times 0.88 = 0.91
\end{aligned}
$$

Figure 2.16 gives a tree-diagram representation of the solution.

Theorem 2.14 can be extended to the intersection of any finite number of events. For instance, in the case of three events E, F, and G, if $P(F \cap G) > 0$, then repeated application of the theorem yields

$$P(E \cap F \cap G) = P(F \cap G)P(E \mid F \cap G) = P(G)P(F \mid G)P(E \mid F \cap G)$$

(Note that $P(F \cap G) > 0$ implies $P(G) > 0$.) Extension to more than three events follows in like manner.

THEOREM 2.15

Given m events E_1, E_2, . . . , E_m for which

$$P(E_1 \cap E_2 \cap \cdots \cap E_{m-1}) > 0$$

the following generalized multiplication property holds:

$$P(E_1 \cap E_2 \cap \cdots \cap E_m)$$
$$= P(E_1)P(E_2 \mid E_1)P(E_3 \mid E_1 \cap E_2) \cdots P(E_m \mid E_1 \cap E_2 \cap \cdots \cap E_{m-1})$$

EXAMPLE 2.22

Consider a group of 30 employees, 20 of whom possess M.S. degrees. What is the probability of selecting without replacement 3 persons who hold M.S. degrees?

One way of solving the problem is to utilize the hypothesis of equal probability and apply Theorem 2.4 to count combinations. Letting H denote the event of interest,

$$P(H) = \frac{\text{card } (H)}{\text{card } (S)} = \frac{\binom{20}{3}}{\binom{30}{3}} = 0.28$$

An alternative approach is to apply the multiplication property for three events. Let E, F, and G denote the events corresponding to selecting a person with an M.S. degree on the third, second, and first picks, respectively. Then

$$P(H) = P(G)P(F \mid G)P(E \mid F \cap G) = \frac{20}{30} \times \frac{19}{29} \times \frac{18}{28} = 0.28$$

The advantage of the latter method is that it applies when the relevant conditional probabilities do not result from counting.

2.5.3 Bayes' Theorem

The intuitive meaning of the multiplication property comes to light when we examine the associated tree diagram, where the somewhat cumbersome sum-of-products expression is replaced by a geometrically transparent tree structure. The tree approach requires that we multiply the branch probabilities along the appropriate paths.

Returning to Example 2.20, suppose a different problem is posed: given a particular outcome of the experiment, namely, that a white ball is chosen, what is the probability that it has come from urn a? At first glance, it might appear that the tree in Figure 2.15 does not contain the necessary information to solve the problem. Such a conclusion would be erroneous. It is the form of the tree that hides the solution.

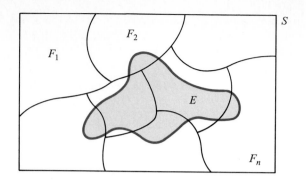

FIGURE 2.17
A partition of S and the induced partition of E

Utilizing the details of Example 2.20, we desire $P(A \mid W)$, which can be found by applying the definition of conditional probability and then the multiplication property:

$$P(A \mid W) = \frac{P(A \cap W)}{P(W)} = \frac{P(W \cap A)}{P(W)} = \frac{P(A)P(W \mid A)}{P(W)} = \frac{1/2 \times 2/5}{2/5} = 1/2$$

From a general perspective, the salient point is that the conditional probability has been reversed: $P(A \mid W)$ has been found by utilizing $P(W \mid A)$. Whereas $P(A \mid W)$ is not observed directly, $P(W \mid A)$ is so observed.

Theoretically, the preceding analysis works because the events A, B, and C partition the sample space of Example 2.20. Specifically, the computation of $P(W)$ in Example 2.20 made use of the partition. In fact, the methodology is quite general and can be extended to a setting where the sample space is partitioned into any finite number of events. Consider the situation depicted in Figure 2.17, where S is a union of the disjoint events $F_1, F_2, \ldots, F_n$ and E is any event. We are given the **a priori** probabilities $P(F_i)$ and $P(E \mid F_i)$, for $i = 1, 2, \ldots, n$, and wish to determine the **a postiori** probabilities $P(F_i \mid E)$. Regarding Example 2.20, the model directly supplied $P(A)$, $P(B)$, $P(C)$, $P(W \mid A)$, $P(W \mid B)$, and $P(W \mid C)$. From these, it is possible to find $P(A \mid W)$, $P(B \mid W)$, and $P(C \mid W)$. The next theorem generalizes the discussion to an arbitrary partition.

THEOREM 2.16

(Bayes' Theorem). Suppose the events $F_1, F_2, \ldots, F_n$ partition the sample space and E is any event. Then, for any k,

$$P(F_k \mid E) = \frac{P(F_k)P(E \mid F_k)}{\displaystyle\sum_{i=1}^{n} P(F_i)P(E \mid F_i)}$$

Proof: Applying the definition of conditional probability in conjunction with Theorem 2.14 yields

$$P(F_k \mid E) = \frac{P(F_k \cap E)}{P(E)} = \frac{P(E \cap F_k)}{P(E)} = \frac{P(F_k)P(E \mid F_k)}{P(E)}$$

Since the F_i form a partition of the sample space, we can apply Theorem 2.12 to the denominator to obtain

$$P(F_k \mid E) = \frac{P(F_k)P(E \mid F_k)}{\displaystyle\sum_{i=1}^{n} P(E \cap F_i)}$$

Applying the multiplication property to each summand in the denominator yields the desired result. ∎

EXAMPLE 2.23

It is common practice to test for the presence of a condition of interest, a positive outcome of the test indicating the condition's presence and a negative outcome its absence. For instance, suppose 40,000 people in a population of 200,000,000 carry a particular virus, and a test has been devised to detect its presence. Moreover, suppose the test results have been shown to obey the following distribution:

	Positive Test	Negative Test
Disease present	99%	1%
Disease not present	2%	98%

On the surface, it appears that a positive test indicates strongly that the patient carries the virus. But let us look a bit deeper. Let T and N indicate a positive and negative test result, respectively, and let V and A denote, respectively, the presence and absence of the virus. Then the information given in the statement of the problem yields the a priori probabilities $P(V) = 0.0002$, $P(A) = 0.9998$, $P(T \mid V) = 0.99$, $P(N \mid V) = 0.01$, $P(T \mid A) = 0.02$, and $P(N \mid A) = 0.98$ (see Figure 2.18). We desire the conditional a postiori probability that, given a positive test result, a person carries the virus. Applying Bayes' theorem to the partition V and A yields

$$P(V \mid T) = \frac{P(V)P(T \mid V)}{P(V)P(T \mid V) + P(A)P(T \mid A)}$$

$$= \frac{0.0002 \times 0.99}{0.0002 \times 0.99 + 0.9998 \times 0.02} = 0.01$$

a very small probability.

An analysis of the denominator in the preceding calculation shows the problem with the test: the **false-positive (false-alarm)** probability $P(T \mid A)$ is too large. Given that so few

FIGURE 2.18
Tree diagram corresponding to Example 2.23

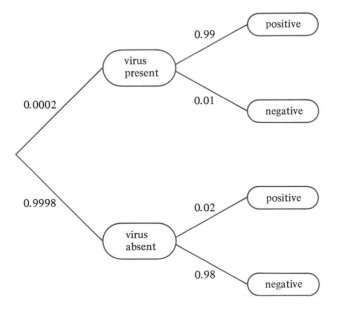

people in the overall population carry the virus, a much smaller false-positive probability is required. Of course, had the virus been more prevalent in the population, a 2% false-alarm rate might have been tolerable.

A very simple interpretation of Bayes' theorem can be given by using the tree diagram resulting from the a priori data. Figure 2.19 shows a tree representation of a partition F_1, F_2, F_3, and F_4 (the first stage of the tree). For each F_i there is one branch leading to the event E. Thus, according to the usual interpretation of the tree, E is the union of the events represented by the four darkened paths, and the probability of E is given by the sum of the products resulting from multiplying the probabilities along the branches of each darkened path:

$$P(E) = \sum_{i=1}^{4} P(F_i)P(E \mid F_i)$$

Now pose the Bayesian question: Given the occurrence of the event E, what is the probability that the path through F_1 was chosen? The answer, $P(F_1 \mid E)$, is provided by Theorem 2.16; however, the same result can be obtained by the following intuitive

FIGURE 2.19

Interpretation of Bayes' theorem by means of a tree diagram

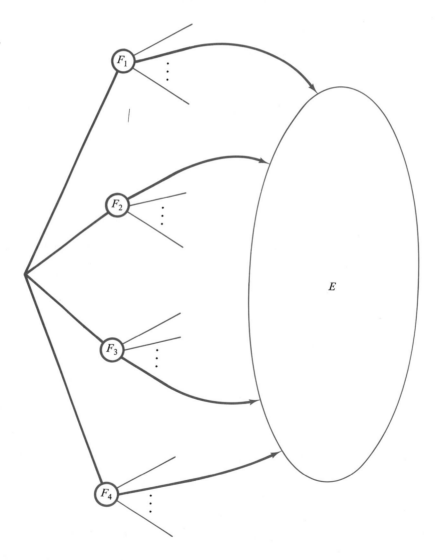

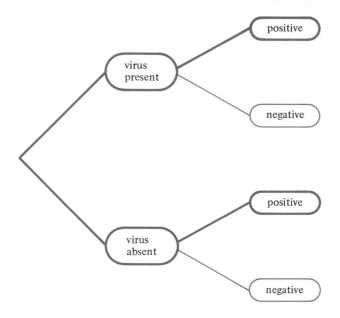

FIGURE 2.20
Depiction of $P(A \mid W)$ in
Example 2.20

(nonrigorous) reasoning. Of the four paths leading to E, only one goes through event F_1 of the partition. Thus, the probability of going through F_1, given that one of the four paths to E has been chosen, is simply the ratio of the probability of going down the path $F_1 \cap E$ to the sum of the probabilities of going down the paths $F_1 \cap E$, $F_2 \cap E$, $F_3 \cap E$, and $F_4 \cap E$. Figure 2.20 depicts the relevant (darkened) branches for the computation of $P(V \mid T)$ in Example 2.23.

2.6 INDEPENDENCE

Intuitively, events are *independent* if the occurrence of one or more of them does not affect the likelihood of occurrence of the remaining ones. For instance, we might like to run the same experiment a number of times and make a succession of observations, none of which is affected by those that preceded it. The present section concerns independence.

2.6.1 Independence of Two Events

The conditional probability of E given F reflects the effect on the likelihood of E due to the stipulation that F must occur. In certain cases, prior knowledge concerning F has no bearing on the likelihood of E; namely, $P(E \mid F) = P(E)$. The next definition formalizes this notion of independence. Note that it takes a slightly different form in order to include the case where $P(F) = 0$, since in such a case the conditional probability $P(E \mid F)$ is undefined.

DEFINITION 2.7

Independent Events. Events E and F in the probability space S are independent if

$$P(E \cap F) = P(E)P(F)$$

If E and F are not independent, then they are said to be **dependent.**

According to Definition 2.7, if $P(F) > 0$, then

$$P(E \mid F) = \frac{P(E \cap F)}{P(F)} = \frac{P(E)P(F)}{P(F)} = P(E)$$

as desired. Analogously, if $P(E) > 0$, then $P(F \mid E) = P(F)$, which demonstrates the symmetry of the independence relation. Moreover, if either $P(E)$ or $P(F)$ is equal to 0, then, by Theorem 2.10, so too is $P(E \cap F)$, and thus E and F are independent. In essence, Definition 2.7 represents a modification of the multiplication property according to the provision $P(E \mid F) = P(E)$.

To see the relationship between independence and selection with replacement, suppose 2 balls are to be selected without replacement, order counting, from an urn containing n balls, r black and $n - r$ white. If F and E represent the selection of a black ball on the first and second picks, respectively, then

$$P(E \mid F) = \frac{r - 1}{n - 1}$$

Should, however, the selection process be with replacement, then

$$P(E \mid F) = \frac{r}{n} = P(E)$$

so that E and F are independent.

EXAMPLE 2.24

Large computers often possess a designated input–output (I/O) processor, called a *channel* processor, to supervise I/O interface. Suppose that observation has shown the CPU to be busy 44% of the time, the channel processor to be busy 70% of the time, and both to be busy 36% of the time. Using the frequency interpretation of probability and letting E and F denote the events that the CPU and channel processor, respectively, are busy at some given instant of time,

$$P(E \cap F) = 0.36 \neq 0.31 = 0.44 \times 0.70 = P(E)P(F)$$

Thus, E and F are not independent. Although the CPU and channel processor work in an asynchronous fashion, their dependence should be expected. If a number of jobs are awaiting the CPU, and a job that has the CPU requires I/O, the operating system gives the CPU to another job while the channel processor provides the necessary I/O interface.

Example 2.24 illustrates the manner in which one might check for independence according to the definition. In practice, however, often the modeling assumption is made that two events are independent. Although the validity of such an assumption might need to be statistically checked, in many situations one can forego the expense of a statistical check because the physical independence of the events is apparent. Given such a model, the modified multiplication property of Definition 2.7 can be directly applied to the solution of a problem.

EXAMPLE 2.25

In a machine shop, engine blocks are automatically routed to one of two stations for cylinder boring. If the probability of a station's failing on a given day is 0.04, what is the probability that both statioins will fail on a given day?

Letting E and F denote the events that one or the other station will suffer a breakdown, it appears that the assumption of independence is warranted and that the solution is given by

$$P(E \cap F) = P(E)P(F) = 0.04 \times 0.04 = 0.0016$$

There is a potential problem here, however. Unless it is assumed that the queue of engine blocks awaiting service is never empty, so that both stations are continuously in operation, E and F might not be independent. Indeed, if there are times when one or the other station is not busy, then a breakdown at one station will increase the burden at the other, thereby creating a situation where one cannot be so confident of independence.

Should events E and F be independent, the question naturally arises as to the independence of the pairs E and F^c, E^c and F, and E^c and F^c. The next theorem asserts that each of these pairs of events is independent whenever E and F are independent.

If E and F are independent, then so are E and F^c, E^c and F, and E^c and F^c.

Proof: In each case we need to check the multiplication property in Definition 2.7. Since, by hypothesis, E and F are independent, $P(E \cap F) = P(E)P(F)$. Since

$$E \cap F^c = E - (E \cap F)$$

Theorems 2.9 and 2.7 yield

$$\begin{aligned} P(E \cap F^c) &= P(E) - P(E \cap F) \\ &= P(E) - P(E)P(F) \\ &= P(E)[1 - P(F)] \\ &= P(E)P(F^c) \end{aligned}$$

so that E and F^c are independent. Owing to the symmetry of the independence relation, a similar argument shows that E^c and F are independent. As for E^c and F^c, their independence follows from

$$\begin{aligned} P(E^c \cap F^c) &= P[(E \cup F)^c] \\ &= 1 - P(E \cup F) \\ &= 1 - [P(E) + P(F) - P(E \cap F)] \\ &= 1 - P(E) - P(F) + P(E)P(F) \\ &= [1 - P(E)][1 - P(F)] \\ &= P(E^c)P(F^c) \end{aligned}$$

where we have applied, in order, De Morgan's law, Theorem 2.7, Theorem 2.11, the independence of E and F, factoring, and Theorem 2.7 once again. ∎

2.6.2 Independence of a Finite Collection of Events

The notion of independence can be extended to more than two events; however, a subtlety appears in the extension to more than two events. Suppose E, F, and G are three events that are **pairwise independent,** meaning that, when taken in pairs, E and F, E and G, or F and G, the pairs are independent:

$$P(E \cap F) = P(E)P(F)$$

$$P(E \cap G) = P(E)P(G)$$

$$P(F \cap G) = P(F)P(G)$$

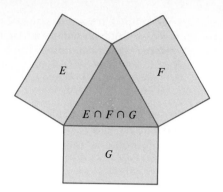

FIGURE 2.21
Pairwise independent
events that are not
independent as a
collection

From pairwise independence, it cannot be concluded that the multiplication rule for the intersection of three events can be reduced to the product of unconditioned probabilities:

$$P(E \cap F \cap G) = P(E)P(F)P(G)$$

But it is precisely such a representation that is required in the extension of independence to three events. The next example demonstrates the difficulty.

EXAMPLE 2.26

Figure 2.21 shows three events, E, F, and G, whose union is the sample space. Their intersection is the triangular region, which also represents the pairwise intersections,

$$E \cap F \cap G = E \cap F = E \cap G = F \cap G$$

Each of the four regions depicted in the figure has probability 1/4. Since

$$P(E \cap F) = 1/4 = 1/2 \times 1/2 = P(E)P(F)$$
$$P(E \cap G) = 1/4 = 1/2 \times 1/2 = P(E)P(G)$$
$$P(F \cap G) = 1/4 = 1/2 \times 1/2 = P(F)P(G)$$

the events are pairwise independent; nonetheless,

$$P(E \cap F \cap G) = 1/4 \neq 1/2 \times 1/2 \times 1/2 = P(E)P(F)P(G)$$

The next definition gives the appropriate generalization of Definition 2.7 to arbitrary finite collections of events in a probability space.

DEFINITION 2.8

Independent Events. A collection of events $E_1, E_2, \ldots, E_n$ is said to be independent if, for any subcollection, $E_{(1)}, E_{(2)}, \ldots, E_{(m)}$, of the events,

$$P(E_{(1)} \cap E_{(2)} \cap \cdots \cap E_{(m)}) = P(E_{(1)})P(E_{(2)}) \cdots P(E_{(m)})$$

In the case of two events, Definition 2.8 reduces to Definition 2.7. In the case of three events, it yields four necessary equations. When applied to the events considered in Example 2.26, it shows they are not independent as a collection. In the case of four events, say E, F, G, and H, Definition 2.8 yields 11 necessary equations. More generally, to check the independence of n events, $2^n - n - 1$

equations must be verified. This follows from counting combinations. Specifically, if there are n events, then there are $C(n, 2)$, $C(n, 3)$, . . . , $C(n, n)$ necessary equations with 2, 3, . . . , n events, respectively, to be checked, and

$$\binom{n}{2} + \binom{n}{3} + \cdots + \binom{n}{n} = 2^n - \binom{n}{1} - \binom{n}{0} = 2^n - n - 1$$

As in the case of two events, practical usage often requires an assumption of independence and the application of the resulting multiplication property, absent of conditional probabilities.

EXAMPLE 2.27

An inspector wishes to check the grade of bolts used in the construction of an amphibious vehicle. Although the specifications called for grade 8 bolts, there is reason to believe that a mix of grades 2, 5, and 8 was used, the suspected mix being 27%, 33%, and 40%, respectively. If the suspicion is correct and if four bolts are chosen randomly with replacement, what is the probability that at least three of them will prove to be of grade 8? If E, F, G, and H denote the selection of a grade 8 bolt on the first, second, third, and fourth picks, respectively, then the event of interest can be expressed as

$$D = (E \cap F \cap G \cap H) \cup (E \cap F \cap G \cap H^c) \cup (E \cap F \cap G^c \cap H)$$
$$\cup (E \cap F^c \cap G \cap H) \cup (E^c \cap F \cap G \cap H)$$

Assuming independence and applying Theorem 2.17, which holds for any collection of independent events,

$$\begin{aligned} P(D) &= P(E \cap F \cap G \cap H) + P(E \cap F \cap G \cap H^c) + P(E \cap F \cap G^c \cap H) \\ &\quad + P(E \cap F^c \cap G \cap H) + P(E^c \cap F \cap G \cap H) \\ &= P(E)P(F)P(G)P(H) + P(E)P(F)P(G)P(H^c) \\ &\quad + P(E)P(F)P(G^c)P(H) + P(E)P(F^c)P(G)P(H) \\ &\quad + P(E^c)P(F)P(G)P(H) \\ &= (0.4)^4 + 4(0.4)^3(0.6) = 0.0256 + 0.1536 = 0.1792 \end{aligned}$$

Figure 2.22 gives a tree-diagram representation of the problem. It is similar to a tree-diagram representation of the multiplication property, except that the probability labels on the branches are not conditional.

EXAMPLE 2.28

Suppose six test missiles are to be fired at a target, each test shot to be rated as either a "hit," if the missile falls within some predetermined distance of the target, or otherwise as a "miss." Given that prior experimentation has shown the probability of a hit to be 0.1, we wish to find the probability of getting at least one hit.

Let E denote the event representing at least a single hit and H_i the event representing a hit on the ith firing, $i = 1, 2, \ldots, 6$. Then

$$E^c = H_1^c \cap H_2^c \cap H_3^c \cap H_4^c \cap H_5^c \cap H_6^c$$

Assuming independence and applying Theorem 2.7 yields

$$\begin{aligned} P(E) &= 1 - P(E^c) \\ &= 1 - P(H_1^c \cap H_2^c \cap H_3^c \cap H_4^c \cap H_5^c \cap H_6^c) \\ &= 1 - P(H_1^c)P(H_2^c)P(H_3^c)P(H_4^c)P(H_5^c)P(H_6^c) \\ &= 1 - (0.9)^6 = 0.47 \end{aligned}$$

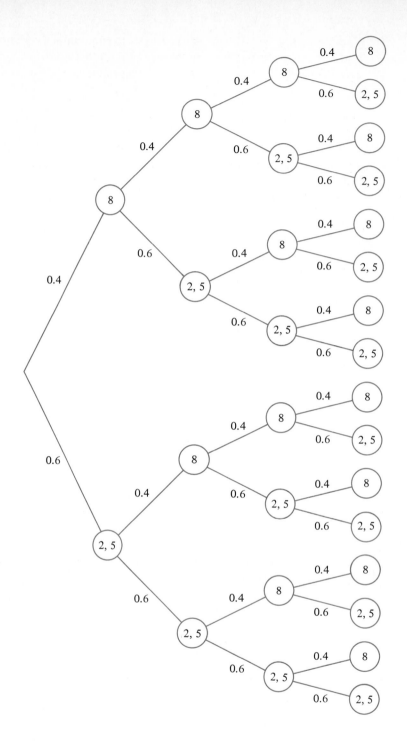

FIGURE 2.22
Tree diagram
corresponding to
Example 2.27

2.6.3 Series and Parallel Systems

In general, a system of components is said to be arranged in **series** if the components $A_1, A_2, \ldots, A_m$ are arranged according to the scheme depicted in Figure 2.23, the intent of the diagram being to represent an arrangement in which the system fails if any component fails. On the other hand, a system is said to be arranged in **parallel** if the components are arranged according to the scheme of Figure 2.24, where the

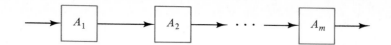

FIGURE 2.23
Serial system

intent is to depict a situation in which the system fails if and only if all components fail.

Suppose the probability of failure of component A_k during some specific interval of time is p_k, $k = 1, 2, \ldots, m$, and that the event representing the failure of A_k is denoted by F_k. If it is assumed that the F_k are independent (i.e., that the failures of the individual components are independent), that F denotes the event that the system will fail during the specified interval, and that the system is arranged in series, then

$$F = F_1 \cup F_2 \cup \cdots \cup F_m$$

By De Morgan's law,

$$F^c = F_1^c \cap F_2^c \cap \cdots \cap F_m^c$$

Thus, by Theorem 2.7, in conjunction with independence,

$$\begin{aligned} P(F) &= 1 - P(F^c) \\ &= 1 - P(F_1^c)P(F_2^c) \cdots P(F_m^c) \\ &= 1 - (1 - p_1)(1 - p_2) \cdots (1 - p_m) \end{aligned}$$

If the system is arranged in parallel, then

$$F = F_1 \cap F_2 \cap \cdots \cap F_m$$

and, due to independence,

$$P(F) = P(F_1)P(F_2) \cdots P(F_m) = p_1 p_2 \cdots p_m$$

FIGURE 2.24
Parallel system

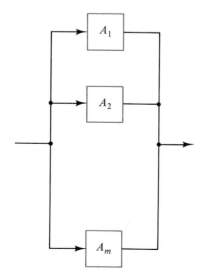

EXAMPLE 2.29

In a *symmetrical multiprocessing system* there are a number of identical processors, any one of which can run a program, access storage, or control I/O. Should any subcollection of the processors be down, the operating system removes the down processors and reconfigures the system so that it will continue to function with the remaining processors. From the standpoint of failure, the system can be considered parallel. In order to apply the foregoing analysis to the system, we will make two assumptions: (1) during a given time interval, each processor possesses the same probability p of failure, and (2) processor failures are independent. The latter assumption is the most problematic. As the overall system degrades due to processor failures, the burden on the remaining processors becomes ever greater. Nevertheless, for the sake of simplicity, independence will be presumed. As a result, the preceding analysis shows the probability of complete breakdown to be

$$P(F) = p^m$$

If $p = 0.1$ and there are four processors, then the probability of total collapse is 0.0001.

EXAMPLE 2.30

One way in which a computer network is classified is according to the arrangement of devices and communication lines. Such an arrangement is referred to as a *network topology*. In a *ring* topology, the computers forming the network are arranged in series in a ring (see Figure 2.25). Although each computer may communicate with any other in the ring, communications must be passed around the ring. Utilizing the ring network of Figure 2.25, assuming that each computer has probability of failure p in a given time interval, and assuming failure independence, we wish to find the probability that communication between computers A and E will break down during the specified interval. Since there are two communication paths from A to E, through A, B, C, D, and E (obviously A and E must function for them to communicate), or through A, G, F, and E, a communication breakdown between A and E will occur if and only if there is a failure in both paths. The difficulty is that the failures of paths A–B–C–D–E and A–G–F–E are not independent because both paths fail should either A or E go down.

To solve the problem, let U and V denote the failures of subpaths B–C–D and G–F, respectively, and let A and E denote the failures of the similarly denoted computers. Then

$$\text{FAILURE} = A \cup E \cup (U \cap V)$$

Using the result of Exercise 2.50, together with the independence of A, E, U, and V, we obtain

FIGURE 2.25
Network topology for
Example 2.30

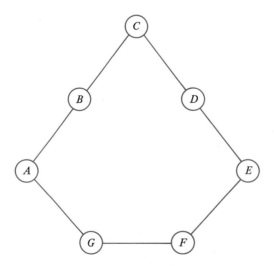

$$\begin{aligned}
P(\text{FAILURE}) &= P[(A \cup E \cup (U \cap V)] \\
&= P(A) + P(E) + P(U \cap V) - P(A \cap E) - P(A \cap U \cap V) \\
&\quad - P(E \cap U \cap V) + P(A \cap E \cap U \cap V) \\
&= P(A) + P(E) + P(U)P(V) - P(A)P(E) - P(A)P(U)P(V) \\
&\quad - P(E)P(U)P(V) + P(A)P(E)P(U)P(V) \\
&= P(A) + P(E) - P(A)P(E) \\
&\quad + [1 - P(A) - P(E) + P(A)P(E)]P(U)P(V) \\
&= 1 + (1 - P(A))(1 - P(E))[P(U)P(V) - 1]
\end{aligned}$$

Applying the previously derived formula for the failure probability of a series system to both U and V gives

$$P(U) = 1 - (1 - p)^3$$
$$P(V) = 1 - (1 - p)^2$$
$$P(\text{FAILURE}) = 1 + (1 - p)^2 [(1 - (1 - p)^3)(1 - (1 - p)^2) - 1]$$
$$= 1 - q^4 (1 + q - q^3)$$

where we have let $q = 1 - p$. For instance, if the probability of any individual computer going down is $p = 0.05$, then $q = 0.95$ and

$$P(\text{FAILURE}) = 1 - 0.89 = 0.11$$

Problems such as the ones presented in Examples 2.29 and 2.30 belong to the theory of **reliability,** a subject to be investigated more fully in Section 6.3.

EXERCISES

SECTION 2.1

2.1. For the experiment of tossing a single die (faces numbered 1 through 6), which of the following are acceptable sample spaces? Explain your answers.
a) $S = \{\text{odd, even}\}$.
b) $S = \{1, 2, 3, 4, 6\}$.
c) $S = \{\text{odd}, 2, 4, 6\}$.
d) $S = \{1, 2, 3, 4, 5, 6\}$.
e) $S = \{\text{prime}, 4, 6\}$.
f) $S = \{2, 3, 5, 1, \text{odd}\}$.

2.2. A card is picked from a deck of 52 cards. List the elements of the following events:
a) Select a heart.
b) Select a picture card.
c) Select a king.

2.3. A robot arm is to select a block from a box containing 30 blocks numbered 1 through 30. Let the sample space be $S = \{1, 2, \ldots, 30\}$ and E, F, G, and H be the events consisting of even numbers, odd numbers, prime numbers, and numbers greater than 20, respectively. Find
a) $E \cup F$ b) $E \cup G$ c) $E \cup H$ d) $F \cup G$
e) $F - G$ f) $H \cap G$ g) $H \cap E$ h) E^c
i) $(E \cap G)^c$ j) $E^c \cup G^c$

2.4. A dart is thrown at the board depicted in Figure 2.26, the board having been segmented into 14 regions—3 white, 4 lightly shaded, 5 mediumly shaded, and 2 darkly shaded. Let the sample space consist of the numbers 1 through 14 corresponding to the regions. Let E, F, G, and H consist of those regions that are lightly or mediumly shaded, nonwhite, mediumly or darkly shaded, and white, respectively. Find
a) $E \cup G$ b) H^c c) $F \cap H$
d) $F \cup H$ e) $G \cup H$ f) $(E \cup G)^c$
g) $E^c \cap G^c$
Describe any subset relations among the given events.

2.5. Two dice are thrown, one red and one white. Let the product set $S = \{1, 2, 3, 4, 5, 6\}^2$ be the sample space. Plot the points of the sample space on a two-dimensional grid and then list the elements of the following events:
a) $\{(x, y): x < y\}$
b) $\{(x, y): x = y\}$
c) $\{(x, y): x^2 + y^2 \leq 4\}$

2.6. The robot arm of Exercise 2.3 is to select a second block from a different box in which the blocks are labeled 1 through 20. Describe the sample space for the two selections. How many elements does it con-

FIGURE 2.26
Dart board for Exercise 2.4

tain? Describe the event consisting of two even-numbered blocks. How many elements are in the event? List the elements in the event determined by the sum of the two blocks being less than 5. How many elements are in the event?

2.7. At the completion of production, lawn mowers are checked to determine whether they meet quality standards. Each tested mower is rated A, B, or C, where A means satisfactory, B means there exists a minor flaw, and C means unusable. For a lot of 20 mowers, what is a suitable sample space?

2.8. There are two urns. Urn A contains 2 white and 3 black balls; urn B contains 3 white, 2 black, and 4 green balls. Describe acceptable sample spaces for the following experiments:

a) An urn is chosen and a ball is picked from the urn.

b) An urn is chosen, a ball is picked from the urn, and a second ball is picked from the other urn.

c) An urn is chosen. If the chosen urn is A, then a ball is selected from A; otherwise, a ball is selected from urn B.

d) An urn is chosen and a ball is selected from urn A no matter which urn has been chosen.

In each of the four experiments, describe the event corresponding to the selection of a green ball.

2.9. A shipment of chemical reagent comes in 40 fifty-gallon drums. Prior to use, each drum must be tested for purity. For a particular reaction, 5 drums are required. Drums are selected from the 40-drum shipment until 5 are determined to be satisfactory or until 2 are found to be unsatisfactory, at which point the shipment is declared unacceptable. Describe a sample space for the possible drum selections.

2.10. In the scenario of Exercise 2.9, suppose drums are selected until 2 are found containing reagent of the desired level of purity. Describe the sample space. Describe the events that exactly 6 drums are tested and that all 40 are tested.

2.11. Use three-circle Venn diagrams to illustrate the following set-theoretic expressions among events.

a) $E \cap (F \cap G)$

b) $(E \cap F) \cap G$

c) $E \cup (F \cup G)$

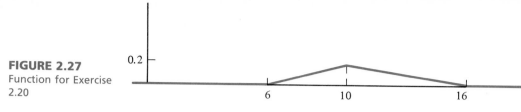

FIGURE 2.27
Function for Exercise
2.20

d) $(E \cup F) \cup G$
e) $E \cap (F \cup G)$
f) $(E \cap F) \cup (E \cap G)$
g) $E \cup (F \cap G)$
h) $(E \cup F) \cap (E \cup G)$

The identities between parts (a) and (b), (c) and (d), (e) and (f), and (g) and (h) represent the associativity of intersection, the associativity of union, the distributivity of intersection over union, and the distributivity of union over intersection.

2.12. Using two-circle Venn diagrams, illustrate the following expressions:
a) $(E \cap F)^c$ b) $E^c \cup F^c$ c) $(E \cup F)^c$
d) $E^c \cap F^c$

The identities between (a) and (b) and between (c) and (d) represent De Morgan's laws.

2.13. Suppose E, F, and G are events in the sample space S. Using intersection, union, and complementation, describe the following events:
a) At least one of the three events occurs.
b) None of the events occurs.
c) All three events occur.
d) E and F occur, but G does not.
e) Exactly one of the three events occurs.
f) Exactly two of the three events occur.
g) At most two of the events occur.
h) Only E occurs.
i) E and F occur.
j) E or F occurs.

SECTION 2.2

2.14. When tossing two fair dice, what is the probability of obtaining
a) A sum of 7.
b) A sum of 13.
c) A sum less than 4.
d) A 4 on the first die and a sum less than 8.
e) A 3 on the first die and a sum less than 8.

2.15. Consider a die for which it is postulated that an odd face is twice as likely to occur as an even face. Based upon this assumption, together with the axioms of probability, find the probabilities of obtaining the following events on a single toss:
a) $\{1\}$
b) $\{2\}$
c) $\{2, 4, 6\}$
d) $\{1, 2, 3\}$

2.16. Referring to Exercise 2.3, assume the robot arm is displaying uniformly random behavior. Assign probabilities to the blocks according to the hypothesis of equal probability, and find the probabilities of the events E, F, G, H, $E \cup G$, $E \cap H$, $E \cap G$, G^c, and $E - H$.

2.17. Assign probabilities to the regions of Figure 2.26 based upon the principle that the landings of the dart will be distributed uniformly over the board (see Exercise 2.4). What are the probabilities of E, F, G, and H? If the dart were tossed 100 times and landed in a darkly shaded region 90 times, what might you be tempted to conclude?

2.18. Assuming the drums of Exercise 2.9 to be drawn randomly, assign probabilities to elements in the sample space in a manner reflecting the fact that 90% of the drums contain reagent of acceptable purity and 10% do not.

2.19. In binary-coded-decimal (BCD) form, each integer, 0 through 9, is given a four-bit code. If the codes are selected uniformly randomly, what is the probability that the code 0110 will not be employed in the coding scheme?

2.20. In observing a machinist grinding valves, it is desired to define a sample space whose elements reflect the time it takes to grind a single valve, the time being measured in minutes. Based upon repeated observation it is determined that the probability of the time being in the interval B is given by

$$P(B) = \int_B f(x)\, dx$$

where $f(x)$ is the function whose graph is depicted in Figure 2.27. Find
a) $P((0, 8])$
b) $P([5, 10])$
c) $P((3, 7) \cup [8, 32))$

SECTION 2.3

2.21. Evaluate
a) $C(9, 4)$
b) $75!/73!$
c) $\begin{pmatrix} 7 \\ 3 \end{pmatrix}$
d) $\begin{pmatrix} 105 \\ 102 \end{pmatrix}$

e) $\begin{pmatrix} 10 \\ 5, 3, 2 \end{pmatrix}$

2.22. Ten technicians are waiting to be interviewed by a personnel manager. If the manager intends to select the interviewees at random but plans to interview only six of them, in how many ways can the interviews proceed?

2.23. A robot must pick up ten items from the floor. In how many ways can the task be performed? If the items are divided into two subcollections, the first containing six items and the second containing four items, and if, once an item from a subcollection is selected, the robot is programmed to pick up the remaining items in that subcollection before proceeding to the other subcollection, in how many ways can the task be performed?

2.24. A driver can choose from among three routes to travel from city A to city B, from among four routes to travel from city B to city C, and from among two routes to travel from city C to city D. In driving from city A to city D, how many routes can he or she take?

2.25. A small generator is assembled in three stages; however, in the first, second, and third stages there are three, four, and six subassemblies, respectively. In how many ways (orderings) can the generator be assembled?

2.26. A computer network consists of ten computers, and the system can remain up so long as at least six computers are functioning. What is the number of computer combinations in which the system will be up?

2.27. A company wishes to build three distribution centers, the centers to be located at different cities. If 12 cities are under consideration, what is the number of possible site combinations? If the 12 cities are divided into geographical regions, East, Midwest, and West, each region contains four cities, and a single center must be in each region, in how many ways can the centers be located?

2.28. How many 8-bit binary words can be formed that possess an even number of 0s and 1s? What is the probability that one of these words chosen at random will possess two 0s?

2.29. In an 8-bit binary code, no more than three consecutive bits may possess the same value (either a 0 or a 1). How many possible codes can be constructed?

2.30. Two 8-bit binary codes are to be *packed* into a single 16-bit word. Assuming the packed codes possess a different number of 1s, how many possible 16-bit words can result from the packing?

2.31. Five letters are to be selected without replacement from the alphabet to form words (possibly nonsense). In how many ways can a word be chosen so that it satisfies the given condition?
a) Begins with an s.
b) Contains no vowels (y not a vowel).

c) Begins and ends with a consonant.
d) Contains only vowels.
e) Has three consecutive vowels somewhere in the word.
f) Begins with three consecutive vowels and ends with two consecutive consonants.
g) Alternates vowels and consonants.

2.32. If the letters of Exercise 2.31 are chosen in a uniformly random manner, what is the probability that a word possessing exactly three consonants will have three consecutive consonants?

2.33. A bin contains 12 thermostats, of which five open at 85° Celsius and seven open at 90° Celsius. If four thermostats are selected at random, what is the probability that they will be evenly divided between the two temperature ratings?

2.34. A dealer has eight tires that appear identical; however, the markings are not on them, and he knows there to be two sets of four. If the dealer randomly selects a combination of four tires from the eight, what is the probability that a set will be chosen?

2.35. From a group of 12 employees, seven must be assigned to assembly, two to painting, two to receiving, and one to shipping. In how many ways can the assignments be made?

2.36. A digitally controlled stereo system has eight settings each for volume, bass, and treble. How many different settings are there for the system as a whole?

2.37. An urn contains five white and four black balls. Three balls are selected at random without replacement. Find the probabilities of the following events:
a) Three white balls are selected.
b) One white and two black balls are selected.
c) At least two white balls are selected.

2.38. Given a collection of 12 electricians, find the following:
a) The number of ways a team of six can be chosen.
b) If four of the electricians are senior grade and eight are junior grade, the number of ways a team of six can be chosen so that there are exactly two senior-grade electricians on the team.
c) Referring to part (b), the number of ways the team can be chosen so that at least two are senior grade.
d) If four electricians are senior grade, five are junior grade, and three are apprentices, the number of ways the team can be chosen so that there are equal representations of the three levels?

2.39. A manager must select two teams of four from among ten workers. In how many ways can this be accomplished? If, among the workers, there are two who will not work together on the same team, how many possible assignments are available to the manager?

2.40. When playing poker with a fair deck, there are $C(52, 5)$ hands a player can be dealt. Find the number of elements in the following events:
a) Flush (all cards of same suit).

b) Full house (three of one kind and two of another—for instance, three kings and two 8s is one possible full house).

c) Four-of-a-kind (four of one kind and one of any other—for instance, four kings and a 5 is one possible four-of-a-kind hand).

What are the probabilities of the events in (a), (b), and (c)?

2.41. Using the binomial theorem, expand the following expressions:
a) $(a + b)^6$ b) $(1 + a)^4$ c) $(1 - a)^4$
d) $(a + 2b)^4$ e) $(a + b/2)^4$

SECTION 2.4

2.42. Four fair coins are tossed. Find the probability of getting at least one tail.

2.43. Based upon relative frequency, it has been determined that 15% of all cars tested emit excessive hydrocarbons, 12% emit excessive CO, and 8% emit excessive amounts of both. Let E and F denote the events that a randomly selected car emits excessive hydrocarbons and CO, respectively. Express the following events in terms of E and F and find the appropriate probabilities.
a) Emissions of both hydrocarbons and CO are excessive.
b) At least one test level is too high.
c) Neither emission is excessive.
d) Hydrocarbon emission is not excessive.
e) Hydrocarbon emission is excessive, but CO emission is not.

2.44. Of 80 employees in a certain company, 26 possess B.S. degrees and 12 possess M.S. degrees. Find the following probabilities concerning an employee selected at random:
a) Possesses both a B.S. and an M.S. degree.
b) Possesses at least one degree.
c) Possesses no degree.

2.45. Suppose $P(E) = 1/2$, $P(F) = 1/4$, and $P(E \cup F) = 5/8$. Find
a) $P(E \cap F)$ b) $P(E^c \cap F)$ c) $P(E - F)$
d) $P((E \cup F)^c)$

2.46. Twelve steel rods are selected at random from a lot of 100 to check for sufficient hardness. Assuming that 94 of the rods in the lot are acceptable, find the probabilities of the following events:
a) All selected rods are sufficiently hard.
b) At least one selected rod is not sufficiently hard.
c) Exactly one selected rod is not sufficiently hard.

2.47. Ten identical computers are subjected to excessively hot operating conditions. Based on past experience, there is a 20% chance of failure for any given computer. Find the probabilities of the following events:
a) None fail.
b) All fail.

c) At least one fails.
d) Exactly one fails.

2.48. A single card is randomly selected from a fair deck of 52. Using propositions demonstrated in Section 2.4, find the probabilities of the following events:
a) Ace or king.
b) Ace or heart.
c) Picture card or 10.
d) A card valued higher than 3.
e) Ace and king.

2.49. Two cards are randomly selected from a fair deck. Find the probabilities of the following events:
a) Two aces.
b) Two kings or two jacks.
c) Two kings or two picture cards.
d) A king and an ace.
e) A king and a nonking.
f) At least one king.

2.50. Theorem 2.11 can be generalized to more than two events. In particular, prove that for three events E, F, and G,

$$P(E \cup F \cup G) = P(E) + P(F) + P(G) - P(E \cap F) \\ - P(E \cap G) - P(F \cap G) \\ + P(E \cap F \cap G)$$

A three-circle Venn diagram can be helpful.

2.51. Of the programmers in a particular department, 20 are fluent in Pascal, 18 in FORTRAN, 9 in COBOL, 15 in both Pascal and FORTRAN, 7 in both Pascal and COBOL, 6 in both FORTRAN and COBOL, and 4 in all three languages. Assuming there are 50 employees in the department, find the probability that an employee chosen at random will possess the following language proficiencies:
a) Proficient in all three languages.
b) Proficient in none of the three languages.
c) Proficient in exactly one of the languages.
d) Proficient in exactly two of the languages.
e) Proficient in at least one of the languages.
f) Proficient in FORTRAN but not in COBOL.
[*Hint*: Draw a Venn diagram and label the various regions.]

2.52. The **symmetric difference** of two events E and F is defined to be the set of all outcomes belonging to either E or F, but not to both. It is denoted by $E \Delta F$. Express $E \Delta F$ in terms of union, intersection, and set subtraction. Show that

$$P(E \Delta F) = P(E) + P(F) - 2P(E \cap F)$$

2.53. A make of tire is tested for durability and braking capability. If 90% of tested tires display the required durability, 80% display the required braking capability, and 78% display both requirements, what is the probability that a tire selected at random will display one but not both of the specified requirements? (Refer to Exercise 2.52.)

2.54. Prove $P(E^c \cap F^c) = 1 - P(F) - P(E) + P(E \cap F)$.

2.55. Prove $P(E \cup F) \geq P(E) + P(F) - 1$.

SECTION 2.5

2.56. Two fair dice are tossed, one red and one green. Find the probabilities of the following events by directly applying the definition of conditional probability:
  a) Sum of 6 given a four on the green die.
  b) Sum of 6 given a six on the green die.
  c) Sum of 7 given a four on the green die.
  d) Sum of 7 given a one on the green die.
  e) Sum of 7 given both dice have values greater than two.

2.57. Four fair coins are tossed. Find the probabilities of the following events:
  a) Exactly two heads appear, given the first coin is a tail.
  b) Exactly two heads appear, given the first coin is a head.
  c) Exactly two heads appear, given at least one tail appears.
  d) Exactly two heads appear, given at least one head appears.
  e) At least one head appears, given the first coin is a tail.
  f) At least one head appears, given the first coin is a head.
  g) At least one head appears, given at least one tail appears.
  h) At least one head appears, given at least one head appears.
  i) At least one head appears, given all tails.

2.58. Three cards are randomly selected from a fair deck. Find the probabilities of the following events by applying the definition of conditional probability:
  a) All hearts, given all cards are red.
  b) All hearts, given one of the cards is a king.
  c) All hearts, given no spades.
  d) All hearts, given two kings.

2.59. Consider the robot arm of Exercises 2.3 and 2.6. Find the probabilities of the following events, assuming the robot is proceeding uniformly randomly:
  a) Two even numbers are selected.
  b) A sum of 40, given the first block is numbered 20.
  c) A sum of at least 32, given the first block is numbered 30.
  d) A sum of at least 32, given the first block is numbered 29 or 30.

2.60. An urn contains four white and six black balls. Two balls are selected without replacement. Find the probabilities of the following events:
  a) Two white balls are selected.
  b) One ball of each color is selected.
  c) At least one white ball is selected.

2.61. During assembly, a product must go through two subassembly stages. After partial assembly at the first station, it is tested. If it is acceptable, then it passes to the second station, where, after final assembly, it is tested again. If the probability of malfunction during the first and second tests is 0.04 and 0.03, respectively, determine the probability that a unit will pass satisfactorily through the line.

2.62. Generalize the result of Exercise 2.61 to an assembly process with n subassembly stations possessing the malfunction probabilities $a_1, a_2, \ldots, a_n$.

2.63. If a driver consistently uses motor oil A, then the probability of a motor repair in excess of \$600 prior to 75,000 miles on the odometer is 0.03; however, for motor oil B, the probability is 0.08. In a test sample, 40 cars have been consistently lubricated with motor oil A and 60 with motor oil B. If a test car is selected at random, what is the probability that it will require \$600 of repair work before being run for 75,000 miles?

2.64. In a box of six fuses, two are defective. If the fuses are selected randomly and without replacement, find the probability that two good fuses have been selected after two, three, and four selections.

2.65. From the urn of Exercise 2.60, three balls are selected without replacement. What is the probability of obtaining exactly three white balls? Employ a conditional probability approach.

2.66. An urn contains m white and n black balls. A ball is picked at random, is replaced, and r balls of the same color are added to the urn. A second ball is picked at random, is replaced, and r balls of its color are added to the urn. The process is repeated once again, and a fourth ball is then selected. Find the probabilities of the following events:
  a) Two white balls are selected on the first two picks.
  b) Three white balls are selected on the first three picks.
  c) Four white balls are selected on the four picks.

2.67. A mailroom utilizes four automatic sorters. The sorters receive 20%, 25%, 35%, and 20% of the incoming mail, respectively, and they possess respective error rates 2%, 3%, 2%, and 1%. What is the error rate of the overall system?

2.68. Referring to Exercise 2.20, find $P([10, 15]|[11, 20])$.

2.69. From the urn of Exercise 2.60, what is the probability that a white ball was selected on the first pick, given that exactly one of the two balls selected is white?

2.70. Referring to Exercise 2.63, given that a car requires motor work in excess of \$600 prior to 75,000 miles, what is the probability it was lubricated with motor oil A?

2.71. For a given raw material used by a chemical company, 96% of all batches possess the desired degree of purity. The test used by the company to test incoming deliveries will correctly identify a batch as acceptable

97% of the time and will correctly identify a batch as unacceptable 95% of the time. If a batch is deemed unacceptable, what is the probability it is actually acceptable? If it is deemed acceptable, what is the probability it is actually unacceptable?

2.72. Referring to Exercise 2.64, if two good fuses are selected in the first three picks, what is the probability that a good fuse was selected on the first pick?

2.73. Referring to Exercise 2.66, what is the probability that a white ball was selected on the first pick, given two white and two black balls have been picked in total?

SECTION 2.6

2.74. For the Venn diagram of Figure 2.28, show that E and F are independent, but E and G are dependent.

2.75. By directly employing the definition, show that obtaining a sum of 7 when tossing two fair dice is independent of getting a two on the first die, but obtaining a sum of 6 is not independent of getting a two on the first die.

2.76. When tossing two fair coins, show that the events corresponding to getting at most one head and exactly one head are dependent.

2.77. Assuming the selection is done with replacement, redo Exercises 2.60 and 2.65.

2.78. A company purchases four identical printers. If the probability of any such printer surviving two years before breakdown (under the expected operating conditions) is 0.7, and failure of the printers is assumed to be independent, then find the probability that at least one printer is used for two years before breakdown.

2.79. In the transmission of an 8-bit word, the probability of an error in a given bit is 0.001. If bit errors are assumed to be independent, find the probabilities of the following events:
a) At most one bit is in error.
b) At least one bit is in error.
c) The first two bits are in error.
d) Two bits are in error.

2.80. A fair coin is tossed five times. What are the probabilities of obtaining exactly one head and of obtaining at least one head?

2.81. Redo Exercise 2.80, assuming the probability of a head is 0.6.

2.82. An urn contains one white and four black balls. If five balls are selected with replacement, what are the probabilities of the following events?
a) At least one white ball is selected.
b) Exactly one black ball is selected.
c) A black ball is selected on the first pick and a white ball is selected on the last pick.
d) A black ball was selected on the first pick, given four white balls have been selected.

2.83. Prove that if $P(E) = 1$, then E and F are independent for any event F.

2.84. In a symmetrical multiprocessing system there are eight identical processors, and the probability of any given processor's failing in a six-month period is 0.05. What is the probability of a total collapse of the system? (Once down, a processor stays down.)

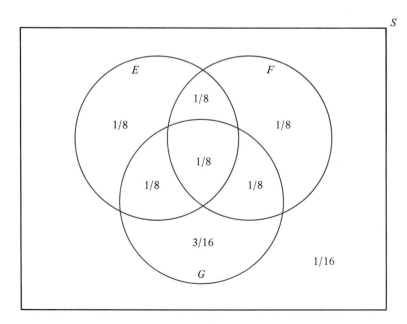

FIGURE 2.28 Venn diagram for Exercise 2.74

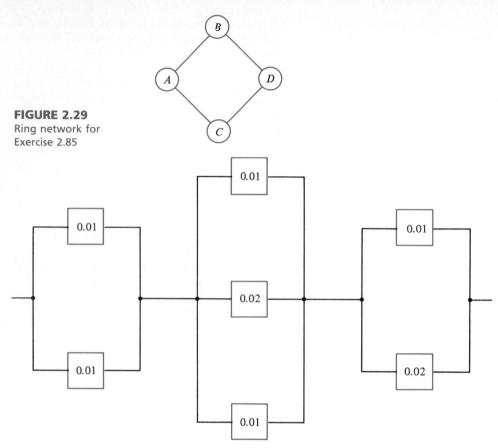

FIGURE 2.29
Ring network for
Exercise 2.85

FIGURE 2.30 Electronic system for Exercise 2.86

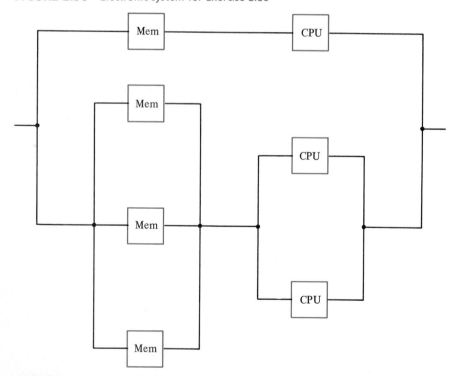

FIGURE 2.31 CPU-memory system for Exercise 2.87

2.85. For the ring network of Figure 2.29, each computer has probability of failure $p = 0.04$ in a given time interval and failures are independent. What is the probability that communication between computers A and D will break down during the specified interval?

2.86. In the electronic system of Figure 2.30, each component is labeled with its probability of failure over some specific time period. Assuming the schematic of the diagram corresponds to a mixed parallel and series design, find the probability of the system's functioning successfully over the given time frame.

2.87. Do the analysis of Exercise 2.86 for the CPU-memory system of Figure 2.31, in which each memory has failure probability 0.005 and each CPU has failure probability 0.01.

3

|||||||||||||||||||||||||||||||||||||||||||||||||||||||||||||||||||||||||||||||||||||||||||||||||||||||||||||||||

PROBABILITY DISTRIBUTIONS

Intuitively, a frequency distribution provides a class representation of the manner in which a collection of sample outcomes is distributed over the real line. More generally, we desire to describe the distribution of a random measurement over the real line. Whereas the frequency distribution results from real data, the description of a random measurement must be described within some abstract mathematical framework. In probability theory, a random variable serves that purpose, and so the present chapter begins with three sections that discuss random variables and their distributions.

Many variables are functions of other variables. Thus, Section 3.4 gives considerable detail on describing the distribution of a function of a random variable that itself possesses a known distribution.

Just as frequency distributions are characterized in part by their means and variances, so too are probability distributions. Sections 3.5 and 3.6 deal with the moments of random variables, with the former concentrating on the mean (expected value) and the latter on the variance. Finally, Section 3.7 introduces a transform technique that plays a key role in the characterization of random variables. Not only does the moment-generating-function transform provide the usual transform benefits (like the Laplace or Fourier), but it also provides a convenient mechanism for finding moments.

3.1 RANDOM VARIABLES

As discussed in Chapter 1, the measurement process leads naturally to a scientific understanding grounded in randomness and variability. The subjects of probability and statistics provide a mathematical framework for this understanding, especially insofar as the observation of random phenomena leads to quantitative description. The concept of a random measurement is manifested in probability theory in the definition of a random variable, and the characterization of random variables and their properties constitutes the main body of probabilistic analysis.

3.1.1 Random Variables as Mappings

In defining a random variable, we take heed of the fact that a measurement is a numerical descriptor of an outcome from some experiment. Since the sample space serves as the logical universe for the collection of outcomes, the following definition makes intuitive sense.

DEFINITION 3.1

Random Variable. A random variable is a real-valued function X defined on a sample space.

Mathematically, a random variable is a mapping

$$X: S \longrightarrow R$$

where the **domain** S is a sample space and R is the set of real numbers. The set of values taken on by X is called the **codomain** and is denoted by Ω_X. Set-theoretically,

$$\Omega_X = \{X(x): x \in S\}$$

EXAMPLE 3.1

In Example 2.2 we considered the problem of selecting, with replacement, four transistors from a box. The sample space was

$$T = \{d, n\}^4$$

which contains 16 elements. If interest centers on the number of defective transistors selected, then the relevant random variable X will provide that number. For instance,

$$X((d, n, n, d)) = 2$$

The codomain is

$$\Omega_X = \{0, 1, 2, 3, 4\}$$

Another random variable of interest on T might be one that marks whether or not the number of defector transistors exceeds some threshold value. For instance, we might consider Y to be 0 if there are 0 or 1 defective transistors and to be 1 otherwise. Then

$$Y((d, n, n, d)) = 1$$
$$Y((n, n, n, n)) = 0$$

and the codomain of Y is

$$\Omega_Y = \{0, 1\}$$

EXAMPLE 3.2

In Example 2.3 we discussed a model that might arise when a CPU must serve two incoming queues of jobs. Using the sample space S_1 discussed therein, let X be the random variable that counts the number of jobs serviced from queue A before the job at the head of queue B is serviced. Then

$$X: S_1 \longrightarrow R$$

with codomain

$$\Omega_X = \{0, 1, 2, \ldots\}$$

For instance, $X(AAAB) = 3$.

We also considered an alternative sample space S_2, constructed so that it would directly represent the number of jobs from queue A chosen prior to the selection of the job at the head of queue B—that is, so that it would reflect the random variable X just defined. Such an approach is basic in the construction of useful probability models. Rigorously, the original codomain Ω_X has been chosen as the new sample space. Thus, the random variable of interest is the identity mapping

$$X: \Omega_X \longrightarrow R$$

defined by $X(x) = x$. The purpose of this approach is transparent: since we are really interested in the numerical values of X, why not take those values as the outcomes in the sample space, rather than employ an intermediate mathematical construct (the original sample space)?

3.1.2   Random Variables and Probability

Given a random variable X, a basic problem is to give meaning to probability statements of the form

$$P(a < X < b) = p$$

Let S be a sample space endowed with the probability measure P and X be a random variable on S. $P(a < X < b)$ is interpreted in the following manner. Let $E_{(a,b)}$ be the event consisting of outcomes x in S for which $a < X(x) < b$:

$$E_{(a,b)} = \{x: a < X(x) < b\}$$

(see Figure 3.1). Since $E_{(a,b)}$ is an event, it possesses a probability $P(E_{(a,b)})$, which measures the likelihood that X lies in the interval (a, b). Thus, we define

$$P(a < X < b) = P(E_{(a,b)})$$

The probabilistic question regarding the random variable X has been thrown back onto the sample space.

More generally, we can define the probability $P(X \in B)$ that X falls in some set

FIGURE 3.1
Illustration of the event
$E_{(a,b)}$

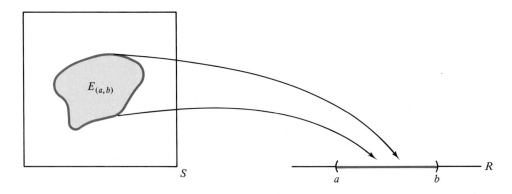

B of real numbers. Consider the event E_B of all outcomes x in S for which $X(x)$ is an element of B,

$$E_B = \{x: X(x) \in B\}$$

The desired probability is defined by

$$P(X \in B) = P(E_B)$$

EXAMPLE 3.3

Referring to Example 3.1, suppose 20% of the transistors are known to be defective and the probability of selecting more than 2 defective transistors is desired. Then

$$P(X > 2) = P(E_{(2,\infty)})$$

where

$$E_{(2,\infty)} = \{(d, d, d, n), (d, d, n, d), (d, n, d, d), (n, d, d, d), (d, d, d, d)\}$$

Using independence, a straightforward tree-diagram analysis yields

$$P(X > 2) = P(E_{(2,\infty)}) = 4(0.2)^3(0.8) + (0.2)^4 = 0.0272$$

(see Figure 3.2, in which the number of defective transistors in each path is listed next to the terminal node of the path).

The probability that the number of selected transistors is greater than or equal to 2 is determined by the event $E_{[2,\infty)}$ consisting of all 4-tuples having a d in at least two components. $E_{[2,\infty)}$ contains 11 outcomes and

$$P(X \geq 2) = P(E_{[2,\infty)}) = (0.2)^4 + 4(0.2)^3(0.8) + 6(0.2)^2(0.8)^2 = 0.1808$$

Of special interest are the probabilities $P(X = x)$, where x is an element of the codomain. These are

$$P(X = 0) = P(E_{\{0\}}) = (0.8)^4 = 0.4096$$

$$P(X = 1) = P(E_{\{1\}}) = 4(0.8)^3(0.2) = 0.4096$$

$$P(X = 2) = P(E_{\{2\}}) = 6(0.8)^2(0.2)^2 = 0.1536$$

$$P(X = 3) = P(E_{\{3\}}) = 4(0.8)(0.2)^3 = 0.0256$$

$$P(X = 4) = P(E_{\{4\}}) = (0.2)^4 = 0.0016$$

Note that the sum of the probabilities over all values in the codomain Ω_X is 1:

$$\sum_{x=0}^{4} P(X = x) = 1$$

This reflects the fact that the probability of the sample space is 1 and

$$\bigcup_{x=0}^{4} E_{\{x\}} = S$$

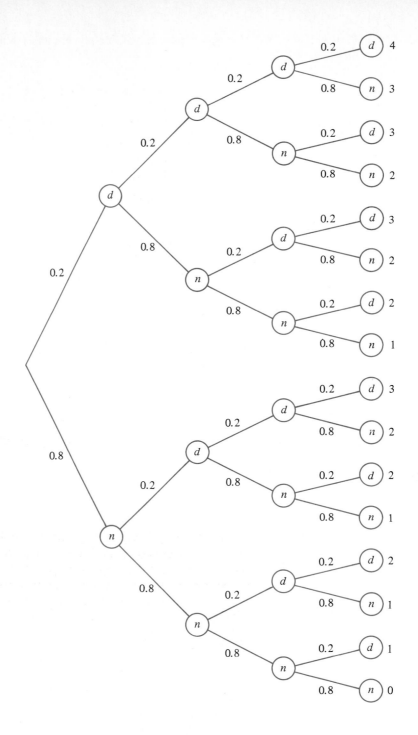

FIGURE 3.2
Tree diagram
corresponding to
Example 3.3

EXAMPLE 3.4

In Example 2.8 we employed the function

$$f(x) = \begin{cases} \dfrac{1}{2} e^{-x/2}, & \text{if } x \geq 0 \\ 0, & \text{if } x < 0 \end{cases}$$

to define the probability measure, the probability of an event in the sample space (the real line) being the integral of $f(x)$ over the event. Recall that the function $f(x)$ was adopted in

Example 1.3 to model the durations of telephone calls at a particular station. Now consider the identity random variable $X(x) = x$. In words, X measures the length of a telephone call. Assuming the model to be appropriate, the probability that a particular phone call is less than two minutes is given by

$$P(X < 2) = P(E_{(-\infty,2)}) = P((0, 2)) = 0.6321$$

as was calculated in Example 2.8. In general,

$$P(a < X < b) = P(E_{(a,b)}) = P((a,b))$$

Moreover, since probabilities are given by integration, and the integral over a point is zero,

$$P(a < X < b) = P(a \leq X \leq b)$$

In particular, for any individual outcome x, $P(X = x) = 0$, which is contrary to the situation in Example 3.3. The random variables of this and the previous example represent two different types of random phenomena.

In employing random variables, one must become adept at handling probability statements, especially insofar as the application of additivity is involved. For instance, since the half-open, half-closed interval $(a, b]$ is the disjoint union of the open interval (a, b) with the endpoint $\{b\}$, additivity yields

$$\begin{aligned}
P(a < X \leq b) &= P(E_{(a,b]}) \\
&= P(E_{(a,b)} \cup E_{\{b\}}) \\
&= P(E_{(a,b)}) + P(E_{\{b\}}) \\
&= P(a < X < b) + P(X = b)
\end{aligned}$$

Many other such relations are possible.

In the future, we will often omit mention of the event E_B when discussing $P(X \in B)$. It is the latter probabilities that are of concern, and the supporting sample-space structure can usually be disregarded, once we possess a convenient manner of obtaining these. For instance, in Example 3.3, once the probabilities $P(X = x)$ are known for all x in the codomain of X, there is no need to refer to the sample space. In Example 3.4, the probabilities $P(X \in B)$ can be ascertained through integration and thus, in a practical sense, the sample space is superfluous. General reasoning supporting these assertions will be given in the next two sections.

For now, we point out that the collection of probability statements $P(X \in B)$ defines a legitimate probability measure on the events in the real line. In other words, no matter what the sample space and random variable, the probabilities of interest, namely those concerning the measurement of random phenomena, are themselves, collectively, a probability measure in the sense of Definition 2.3, and the sample space on which this probability measure is defined is the set of real numbers. The import of this fact, which is proven in Theorem 3.1, cannot be overstated: once a random variable is defined, the "natural" setting for its probabilistic analysis is the real line.

THEOREM 3.1

If X is a random variable on the sample space S endowed with probability measure P, then the probabilities $P(X \in B)$, B a subset of the real line R, define a legitimate probability measure on the events of R.

Proof: Regarding Definition 2.3, certainly P1 is obvious. P2 is demonstrated by

$$P(X \in R) = P(E_R) = P(S) = 1$$

where we have used the fact that $E_R = S$. As for additivity, suppose $B_1, B_2, \ldots$ form a collection of mutually exclusive subsets (events) of R. For $k = 1, 2, \ldots$,

$$E_{B_k} = \{x \in S: X(x) \in B_k\}$$

For any outcome x in S, it cannot be that $X(x) \in B_k$ and $X(x) \in B_j$ if $k \neq j$. Thus, for $k \neq j$,

$$E_{B_k} \cap E_{B_j} = \emptyset$$

By hypothesis, P is a probability measure on S and is therefore additive. Letting

$$B = \bigcup_{k=1}^{\infty} B_k$$

we need to show that $P(X \in B)$ is equal to the sum, from $k = 1$ to ∞, of the $P(X \in B_k)$. In fact,

$$P(X \in B) = P(E_B)$$

$$= P\left(\bigcup_{k=1}^{\infty} E_{B_k} \right)$$

$$= \sum_{k=1}^{\infty} P(E_{B_k})$$

$$= \sum_{k=1}^{\infty} P(X \in B_k) \qquad \blacksquare$$

3.2 PROBABILITY DENSITIES

The manner in which the sample-space structure supports the probabilistic analysis of random measurement processes was revealed in the preceding section, culminating with Theorem 3.1. The probabilities $P(X \in B)$ quantify the manner in which the likelihood of occurrence of X is distributed throughout the real line. In terms of understanding phenomena, it is these probabilities, rather than those of the supporting probability space, that are important. In this and the next section we will develop the general methodology by which probabilities regarding X can be described analytically in terms of functions (such as the one employed in Example 2.8). Ultimately, this leads to an emphasis on models developed directly from observational data. This will bring us full circle back to the frequency distributions of Chapter 1.

3.2.1 Discrete Probability Distributions

The random variables discussed in Examples 3.2 and 3.3 possessed codomains consisting of discrete sets of points in the real line. Such random variables constitute an important class of random variables.

DEFINITION 3.2

Discrete Random Variable. The random variable X is said to be discrete if its codomain consists of a denumerable collection of points, either finite or infinite.

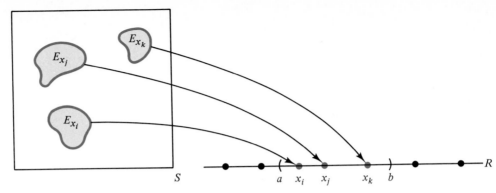

FIGURE 3.3
Illustration of $P(a < X < b)$ for a discrete random variable

If X is discrete, then its codomain can be expressed in the form

$$\Omega_X = \{x_1, x_2, \ldots\}$$

Since the codomain contains the only possible values X can attain, to determine the probability that X lies in the interval (a, b) we need only consider the probabilities of the points in the codomain that lie in (a, b). For instance, if the only points of Ω_X that lie between a and b are x_i, x_j, and x_k, then

$$P(a < X < b) = P(X = x_i) + P(X = x_j) + P(X = x_k)$$

This follows at once from the fact that

$$E_{(a,b)} = E_{\{x_i\}} \cup E_{\{x_j\}} \cup E_{\{x_k\}}$$

(see Figure 3.3). More generally, for any set B of real numbers,

$$P(X \in B) = \sum_{x_k \in B} P(X = x_k)$$

EXAMPLE 3.5

In Example 3.3, the individual probabilities $P(X = x)$ were computed for each x in the codomain

$$\Omega_X = \{0, 1, 2, 3, 4\}$$

Using these probabilities,

$$P(X > 2) = P(X = 3) + P(X = 4) = 0.0256 + 0.0016 = 0.0272$$

which agrees with the value earlier obtained in Example 3.3. In addition,

$$P(1 \le X < 3) = P(X = 1) + P(X = 2) = 0.4096 + 0.1536 = 0.5632$$

and

$$P(1.4 < X \le 2.6) = P(X = 2) = 0.1536$$

In the case of a discrete random variable X, the probabilities $P(X = x)$, x in the codomain of X, provide the desired analytic description of the probability distribution. If, for each x in Ω_X, we plot the probability $P(X = x)$, the resulting graph gives a point-mass distribution of the probabilistic behavior of X over the real

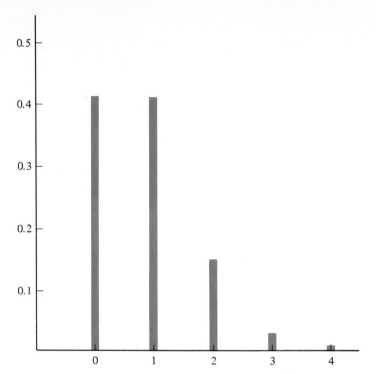

FIGURE 3.4
Probability mass function
for the random variable
of Example 3.3

line. Figure 3.4 gives the point-mass distribution of the probabilities for the random variable of Example 3.3. The following definition formalizes the notion of the point-mass distribution.

DEFINITION 3.3

Discrete Density. If X is a discrete random variable, its density is defined by

$$f_X(x) = \begin{cases} P(X = x), & \text{if } x \in \Omega_X \\ 0, & \text{otherwise} \end{cases}$$

The density of a discrete random variable is also known as its **probability mass function** and its **distribution,** although the latter term is usually employed more loosely rather than as a specific reference to the density. Unless there is possible confusion as to the random variable under consideration, the subscript X is often omitted when referring to the density for X. Moreover, because only nonzero density values are of concern, usually only such values are specified, and it is tacitly recognized that the remaining values are zero.

For the random variable X of Example 3.3, the density can be described in closed form rather than simply as a collection of five values. Indeed,

$$f(x) = \binom{4}{x} (0.2)^x (0.8)^{4-x}$$

for $x = 0, 1, 2, 3, 4$. In Section 4.1, it will be seen that $f(x)$ is a particular instance of a general family of densities whose closed-form expressions are of the form just given.

The next theorem summarizes the discussion pertaining to probabilities that occurred prior to Definition 3.3.

THEOREM 3.2

If X is a discrete random variable with density $f(x)$, then, for any set B of real numbers,

$$P(X \in B) = \sum_{\substack{x \in B \\ f(x) > 0}} f(x)$$

The notation in Theorem 3.2 is meant to imply that the sum is taken over all values $f(x)$ for which $x \in B \cap \Omega_X$—that is, those x in B for which $f(x) > 0$.

EXAMPLE 3.6

In Example 3.2, X counts the number of jobs serviced from queue A before the CPU services the job at the head of queue B, where the choice of which queue to service is made randomly (see Figure 2.3). Suppose the probability of selecting queue A is p, the probability of selecting queue B is $1 - p$, and each selection is independent of the preceding selections. If A_k and B_k denote the events that queue A is chosen on selection k and queue B is chosen on selection k, respectively, $k = 1, 2, \ldots$, then the density for X is given by

$$f(0) = P(X = 0) = P(B_1) = 1 - p$$
$$f(1) = P(X = 1) = P(A_1 \cap B_2) = P(A_1)P(B_2) = p(1 - p)$$

and, in general, for $x = 1, 2, \ldots$,

$$\begin{aligned} f(x) = P(X = x) &= P(A_1 \cap A_2 \cap \cdots \cap A_x \cap B_{x+1}) \\ &= P(A_1)P(A_2) \cdots P(A_x)P(B_{x+1}) \\ &= (1 - p)p^x \end{aligned}$$

Note that the sum of the density values (probabilities) is 1:

$$\sum_{x=0}^{\infty} f(x) = \sum_{x=0}^{\infty} (1 - p)p^x = (1 - p) \sum_{x=0}^{\infty} p^x = 1$$

Table 3.1 provides the first 11 density values for the case $p = 3/4$, and Figure 3.5 illustrates the corresponding density. For $p = 3/4$, the probability that more than two jobs are chosen from queue A prior to a service for queue B is

$$\begin{aligned} P(X > 2) &= 1 - P(X \leq 2) \\ &= 1 - [f(0) + f(1) + f(2)] \\ &= 1 - [1/4 + (1/4)(3/4) + (1/4)(3/4)^2] \\ &= 27/64 \end{aligned}$$

Rather than proceed with an independent selection format, the operating system might be designed to give preference to long-waiting jobs. For instance, given that the initial probability of a job being selected from queue A is p, if a job from queue A is chosen, then, on the next selection, the probability of selecting from queue A might be divided by 2. If another

TABLE 3.1 DENSITY VALUES FOR NONPRIORITY QUEUE SERVICE

x	0	1	2	3	4	5	6	7	8	9	10
$f(x)$	0.2500	0.1875	0.1406	0.1055	0.0791	0.0593	0.0444	0.0334	0.0250	0.0188	0.0140

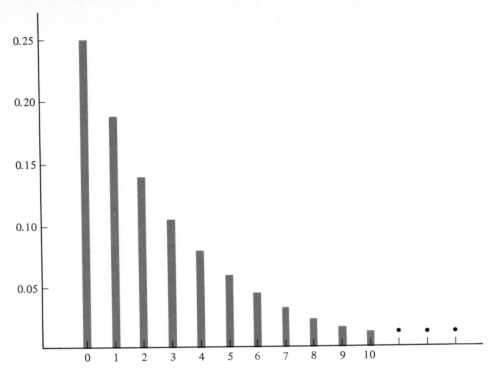

FIGURE 3.5
Probability mass function
for the random variable
of Example 3.6

job from queue A is selected, then the probability of selecting from queue A might again be divided by 2. Continuing in this manner, Theorem 2.15 yields

$$f(0) = P(X = 0) = P(B_1) = 1 - p$$

$$\begin{aligned} f(1) = P(X = 1) &= P(B_2 \cap A_1) \\ &= P(B_2 \,|\, A_1)P(A_1) \\ &= \left(1 - \frac{p}{2}\right) p \end{aligned}$$

$$\begin{aligned} f(2) = P(X = 2) &= P(B_3 \cap A_2 \cap A_1) \\ &= P(B_3 \,|\, A_2 \cap A_1)P(A_2 \,|\, A_1)P(A_1) \\ &= \left(1 - \frac{p}{4}\right)\left(\frac{p}{2}\right) p \end{aligned}$$

and, in general, for $x = 1, 2, \ldots$,

$$f(x) = P(X = x) = \left(1 - \frac{p}{2^x}\right)\left(\frac{p}{2^{x-1}}\right)\left(\frac{p}{2^{x-2}}\right) \cdots \left(\frac{p}{2}\right) p$$

Regardless of the random variable X from which it has been derived, a discrete density $f(x)$ satisfies certain obvious properties. It is nonnegative, positive only on some discrete set of points (the codomain of the random variable), and the sum of its positive values is 1. Of practical interest is that any function $f(x)$ that satisfies these three properties is the probability mass function for some discrete random variable X. In other words, if $f(x)$ satisfies the three properties, then there exists a discrete random variable X having $f(x)$ as its density. This is the import of the next theorem.

The function $f(x)$ is a discrete density if and only if it satisfies the following properties:

(i) $f(x) \geq 0$.

(ii) There exists a denumerable (discrete) set of points Ω, called the **support** of f, such that $f(x) > 0$ if and only if $x \in \Omega$.

(iii) $\sum_{x \in \Omega} f(x) = 1$.

Proof: We need only construct a random variable whose density is $f(x)$. Let Ω be the sample space and define the probability measure P on the events of Ω by

$$P(E) = \sum_{x \in E} f(x)$$

Because of the three stated properties, it is obvious that P is a probability measure in the sense of Definition 2.3. Now define X on Ω as the identity mapping, $X(x) = x$. The fact that $f(x)$ is the probability mass function for X is immediate:

$$f_X(x) = P(X = x) = P(E_{\{x\}}) = P(\{x\}) = f(x)$$

where $E_{\{x\}} = \{x\}$, since X is the identity mapping. ∎

Of practical import is that Theorem 3.3 allows us to merely specify a function satisfying the three properties of the theorem and then proceed in the knowledge that the specification of $f(x)$ determines a random variable whose probability distribution is described by $f(x)$. One might argue that a single mass function might be the density for numerous random variables; however, this nonuniqueness is of no consequence. If both X and Y possess probability mass function $f(x)$, then $P(X \in B) = P(Y \in B)$ for all B. The salient point is that we can begin our probabilistic modeling with the specification of a density and not concern ourselves with the precise specification of a probability space.

3.2.2 Continuous Probability Distributions

A second important class of random variables is comprised of those that are **continuous.** Intuitively, these arise from measurement processes that take on values in an interval of the real line: the measurement varies ''continuously.'' More rigorously, a continuous random variable is one whose probabilities can be derived by integrating some particular function. An example of such a random variable is considered in Example 3.4.

Continuous Random Variable. A random variable X is said to be continuous if there exists a function $f_X(x)$, called its **probability density function,** such that

$$P(X \in B) = \int_B f_X(x)\, dx$$

for any set B of real numbers.

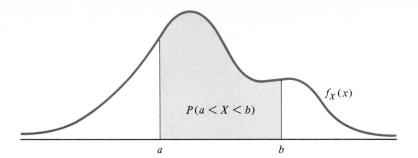

FIGURE 3.6
$P(a < X < b)$ as an
integral of the density
$f_X(x)$ of a continuous
random variable X

Given a density $f_X(x)$ for the continuous random variable X, the probabilities for X are obtained as areas under the curve. For instance, $P(a < X < b)$ is depicted in Figure 3.6. One immediate consequence is that

$$P(a < X < b) = P(a \leq X < b) = P(a < X \leq b) = P(a \leq X \leq b)$$

As in the case of a discrete random variable, the density is sometimes called the **distribution** of the random variable; however, once again, the term ''distribution'' is usually used loosely to refer to the random variable, together with its density.

In the case of a discrete random variable, the probability mass function can be derived by analysis of the probability space; specifically, $f_X(x) = P(X = x)$. There is no analogue for this approach in the continuous case; in fact, since $P(X \in B)$ is defined by an integral over B, $P(X = x) = 0$ for all x.

The question as to whether or not the probabilities resulting from integrating $f_X(x)$ ''actually'' describe the underlying measurement-process distribution is not scientifically meaningful. What is meaningful is whether or not the resulting probabilities are sufficiently in accord with data derived from the process. The degree to which the density satisfies experimental requirements is a concern of statistics and will be addressed later in the text. From the standpoint of probability theory, the density is presumed, and properties of the random variable are described therefrom.

Since probabilities are nonnegative and total probability must be 1, a probability density must yield nonnegative integrals and possess total integral 1. The converse holds: any function satisfying these two properties is the density for some continuous random variable. The next theorem, which is the analogue in the continuous case to Theorem 3.3, and which plays a similar role in applications, gives a precise statement of this fact.

A real-valued function $f(x)$ is the probability density function for a continuous random variable X if and only if

(i) There does not exist a set B of real numbers such that

$$\int_B f(x)\, dx < 0$$

(ii) $\displaystyle\int_{-\infty}^{\infty} f(x)\, dx = 1$

Proof: Suppose $f(x)$ is the density for X. If there were to exist a set B over which the integral of f were negative, then the probability $P(X \in B)$ would be negative, an impossibility. Second, letting S be the sample space on which X is defined,

$$\int_{-\infty}^{\infty} f(x)\, dx = P(X \in R) = P(E_R) = P(S) = 1$$

As for the converse of the theorem, we must specify a probability space and random variable having $f(x)$ as its density. Following the lead of Example 2.8, let R be the sample space, define

$$P(B) = \int_B f(x)\, dx$$

probability on R, since (i) implies P is nonnegative, (ii) implies $P(R) = 1$, and the integral is additive, thus demonstrating P1, P2, and P3 of Definition 2.3. To see that $f(x)$ is the density for X, just note that

$$P(X \in B) = P(E_B) = P(B) = \int_B f(x)\, dx$$

where the second equality follows from the definition of X as the identity mapping.

■

For practical purposes, note that if $f(x)$ is nonnegative, then condition (i) of Theorem 3.4 is satisfied.

As in the discrete case, the random variable guaranteed by Theorem 3.4 is not unique; however, as there, if two random variables possess the same density, then, insofar as their probabilities are concerned, they are indistinguishable.

Before proceeding to the next example, we state a fundamental integral identity:

$$\frac{1}{\sqrt{2\pi}} \int_{-\infty}^{\infty} e^{-x^2/2}\, dx = 1$$

This identity will be proven in Section 4.5.1 and serves as the basis for the normal distribution.

TABLE 3.2 FREQUENCY DISTRIBUTION FOR DE-CIBEL MEASUREMENTS

Class Mark	Frequency	f(Class Mark)
-1.00	4	0.11
-0.75	7	0.26
-0.50	10	0.48
-0.25	18	0.70
0.00	21	0.80
0.25	20	0.70
0.50	9	0.48
0.75	8	0.26
1.00	3	0.11

EXAMPLE 3.7

A manufacturer has designed an instrument to measure noise levels. In order to check the accuracy of the instrument, a number of laboratory tests are done using sounds whose intensities are known. To measure instrument error, the readings on the instrument are subtracted from the known decibel values. The results of 100 tests are grouped in the frequency distribution of Table 3.2 and the normalized relative-frequency histogram is illustrated in Figure 3.7. Also depicted in the figure is a continuous curve that appears to approximate the observational data. It is the graph of the function

$$f(x) = \sqrt{2/\pi} \, e^{-2x^2}$$

which tends to approximate the tops of the histogram bars. (The values of f at the class marks are given in Table 3.2.) Clearly f is nonnegative. Moreover,

$$\int_{-\infty}^{\infty} f(x) \, dx = \sqrt{2/\pi} \int_{-\infty}^{\infty} e^{-2x^2} dx$$

Under the substitution $u = 2x$, $dx = du/2$ and the integral reduces to the one mentioned prior to the example. Consequently, $f(x)$ possesses total integral 1 and satisfies the conditions of Theorem 3.4. Therefore, it is the density of a random variable X.

Applying Definition 3.4, all probabilities relating to X can be found. If $a < b$, then the probability that the measurement error is between a and b decibels is

$$P(a < X < b) = \int_{a}^{b} \sqrt{2/\pi} \, e^{-2x^2} \, dx$$

For instance,

$$P(-1 < X < 1) = \int_{-1}^{1} \sqrt{2/\pi} \, e^{-2x^2} \, dx = 0.9544$$

where the value of the integral has been obtained from Table A.1 (the use of the table is covered in Section 4.5). In terms of the model, 95.44% of all measurements are accurate to within a single decibel.

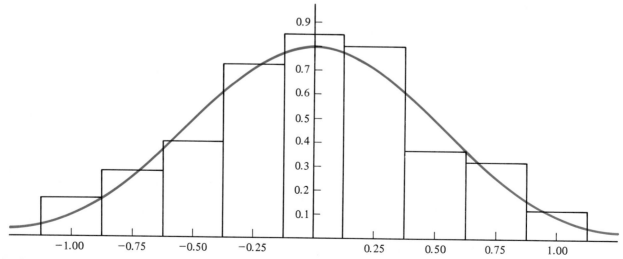

FIGURE 3.7 A continuous density overlaying the histogram for the decibel distribution of Table 3.2

Insofar as the model of Example 3.7 is concerned, f has been chosen to provide the underlying probability distribution for the measurement process at hand. Thus, the identity random variable given in the proof of Theorem 3.4 is appropriate. But what is the measurement process? The question is far from trivial. The original intent of the manufacturer was to analyze the distribution of the instrument error. If this were the case, then the measurement process would involve errors over all possible tests of all possible noise levels. But the actual data came from specific tests on specific noise sources in a specific environment. Can we then interpret the density as defining the probability distribution of all possible measurements? Indeed, can we even interpret it as the error distribution for the sources tested?

These questions cannot be answered simply. Essentially, they depend on the degree to which the model can be statistically verified, which ultimately results in some determination of how well it works. The proof is in the pudding! In addition, even if there were not such obvious conundrums concerning the actual measurement process, we would still be left with the fact that the data represent only a tiny portion of the possible data. The analysis of these profound questions lies in the theory of statistics.

EXAMPLE 3.8

The function

$$g(x) = \begin{cases} xe^{-2x}, & \text{if } x \geq 0 \\ 0, & \text{if } x < 0 \end{cases}$$

does not possess total integral 1. Specifically, integration by parts yields

$$\int_{-\infty}^{\infty} g(x)\, dx = \int_{0}^{\infty} xe^{-2x}\, dx$$

$$= -\frac{x}{2} e^{-2x} \Big|_{0}^{\infty} + \frac{1}{2} \int_{0}^{\infty} e^{-2x}\, dx = \frac{1}{4}$$

Thus, $f(x) = 4g(x)$ is a density according to Theorem 3.4.

If X is a random variable with density $f(x)$, then its probabilities can be obtained by integrating with integrand $f(x)$. For instance, integration by parts shows that

$$P(X < 3) = 4 \int_{0}^{3} xe^{-2x}\, dx = 1 - 7e^{-6}$$

Although the mathematical approaches to the probability density function differ in the discrete and continuous cases, not too much should be made of the distinction. First of all, an integral is a generalized sum (and a sum is a specialized integral), so that the definitions are quite similar. Second, the manner in which discrete densities are discovered might, in practice, be much like the histogram procedure employed in Example 3.7. Whereas, in Example 3.5, the derivation of the probability mass function depended ultimately on a sample-space analysis (the distribution of defective and nondefective items being known beforehand), in practice, the situation is not usually so well defined.

For instance, consider the defective-transistor problem discussed in Example 1.1. Based upon testing, the frequency distribution of Table 1.3 was obtained, and this leads to the relative-frequency distribution in Table 3.3. According to Theorem

TABLE 3.3 RELATIVE-FRE-
QUENCY DISTRIBUTION OF TRAN-
SISTOR DATA

Class Mark	Relative Frequency
0	0.40
1	0.39
2	0.17
3	0.03
4	0.01

3.3, a discrete density is defined by the table, and the corresponding random variable X counts the number of defective transistors selected with replacement from a box. Whether or not it is wise to employ the relative-frequency distribution as a density for the actual distribution of defective transistors over the entire collection of boxes is a profound question that will have to be left open at this point.

3.3 PROBABILITY DISTRIBUTION FUNCTIONS

The previous section described the probabilistic behavior of two types of random variables, discrete and continuous, in terms of a real-valued function. In the present section, we will examine a more general approach to the analytic description of random variable probabilities and then connect this approach to the two techniques introduced in the previous section.

3.3.1 The Probability Distribution Function of a Random Variable

DEFINITION 3.5

Probability Distribution Function. If X is a random variable, then its probability distribution function is defined, for all real x, by

$$F_X(x) = P(X \leq x) = P(E_{(-\infty, x]})$$

In keeping with our custom, we will drop the subscript X in $F_X(x)$ unless it is needed to identify the random variable involved.

EXAMPLE 3.9

For the random variable X of Example 3.4, if $x < 0$, then $F(x) = 0$. If $x \geq 0$, then

$$F(x) = P(X \leq x) = \int_0^x \frac{1}{2} e^{-t/2} \, dt = 1 - e^{-x/2}$$

From the perspective of the phone-call lengths that are modeled by $f(x)$, $F(x)$ is the probability that a call will last less than or equal to x minutes. For instance, the probability that a call will last less than or equal to 8 minutes is

$$F(8) = 1 - e^{-4} = 1 - 0.0183 = 0.9817$$

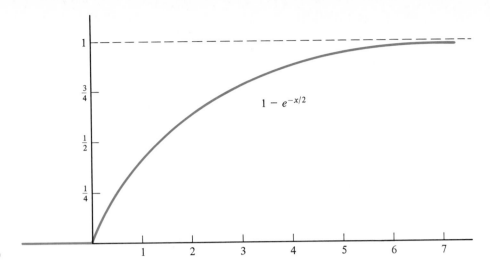

FIGURE 3.8
Probability distribution
function of Example 3.9

Because the probability $P(X \leq x)$ is given by an integral,

$$F(x) = P(X \leq x) = P(X < x)$$

which holds in general for continuous random variables. For instance, the probability that a call lasts less than 8 minutes is also 0.9817.

The graph of $F(x)$ is illustrated in Figure 3.8. Note that $F(x)$ is monotonically increasing and continuous,

$$\lim_{x \to -\infty} F(x) = 0$$

and

$$\lim_{x \to +\infty} F(x) = 1$$

EXAMPLE 3.10

Referring to the selection problem of Example 3.3, $F(x)$ is the probability that less than or equal to x defective transistors are picked. For instance, using the information obtained in Example 3.3,

$$F(2) = P(X \leq 2) = P(X = 0) + P(X = 1) + P(X = 2) = 0.9728$$

On the other hand,

$$P(X < 2) = P(X = 0) + P(X = 1) = 0.8192$$

so that $F(2) \neq P(X < 2)$.

For arbitrary x, if $x < 0$, then $F(x) = 0$. If $0 \leq x < 1$, then

$$F(x) = P(X \leq x) = P(X = 0) = 0.4096$$

If $1 \leq x < 2$, then

$$F(x) = P(X \leq x) = P(X = 0) + P(X = 1) = 0.8192$$

If $2 \leq x < 3$, then

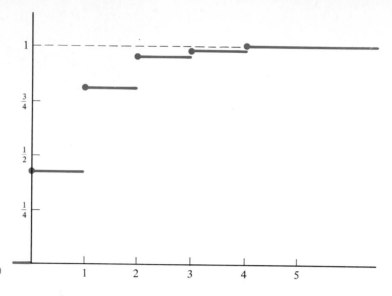

FIGURE 3.9
Probability distribution
function of Example 3.10

$$F(x) = P(X \le x) = P(X = 0) + P(X = 1) + P(X = 2) = 0.9728$$

If $3 \le x < 4$, then

$$F(x) = P(X = 0) + P(X = 1) + P(X = 2) + P(X = 3) = 0.9984$$

Finally, if $x \ge 4$, then $F(x) = 1$. The graph of $F(x)$ is given in Figure 3.9.

   Like the probability distribution function of Example 3.9, $F(x)$ is monotonically increasing; the limit as $x \to -\infty$ is 0 and the limit as $x \to +\infty$ is 1. In the present case, $F(x)$ is not continuous; rather, it is a step function, the jumps occurring at the codomain values of X. Except for those five points, $F(x)$ is continuous. Moreover, at each codomain point, $F(x)$ is continuous from the right; that is,

$$\lim_{x \to a^+} F(x) = F(a)$$

for each a in the codomain Ω_X.

   Because of the manner in which the probability distribution function is defined, various probabilities concerning intervals can be expressed in terms of it. These are summarized in the next theorem.

THEOREM 3.5

If X possesses probability distribution function $F(x)$ and $a < b$, then

(i) $P(a < X \le b) = F(b) - F(a)$
(ii) $P(a \le X \le b) = F(b) - F(a) + P(X = a)$
(iii) $P(a < X < b) = F(b) - F(a) - P(X = b)$
(iv) $P(a \le X < b) = F(b) - F(a) + P(X = a) - P(X = b)$

Proof: The proof is an exercise in the use of additivity. For (i),

$$
\begin{aligned}
P(a < X \le b) &= P(E_{(a,b]}) \\
&= P(E_{(-\infty,b]} - E_{(-\infty,a]}) \\
&= P(E_{(-\infty,b]}) - P(E_{(-\infty,a]}) \\
&= F(b) - F(a)
\end{aligned}
$$

where we have applied, in order, the definition of $P(X \in B)$, set-theoretic subtraction, Theorem 2.9, and the definition of $F(x)$. For (ii),

$$
\begin{aligned}
P(a \leq X \leq b) &= P(E_{[a,b]}) \\
&= P(E_{(a,b]} \cup E_{\{a\}}) \\
&= P(E_{(a,b]}) + P(E_{\{a\}}) \\
&= P(a < X \leq b) + P(X = a) \\
&= F(b) - F(a) + P(X = a)
\end{aligned}
$$

where we have applied, in order, the definition of $P(X \in B)$, set-theoretic union, additivity, the definition of $P(X \in B)$, and part (i). We leave parts (iii) and (iv) as exercises. ∎

3.3.2 The Probability Distribution Function of a Discrete Random Variable

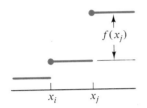

FIGURE 3.10
A jump in a discrete probability distribution function depicted as density value $f(x_j)$

Suppose X is a discrete random variable with codomain $\Omega_X = \{x_1, x_2, \ldots\}$ and probability mass function $f(x)$. Then, according to Theorem 3.2,

$$
F(x) = P(X \leq x) = \sum_{x_k \leq x} f(x_k)
$$

for any x. Consequently, if x_i and x_j are points in the codomain such that $x_i < x_j$ and such that there are no elements in the codomain between them, then $F(x)$ is constant between x_i and x_j:

$$
F(x) = F(x_i)
$$

for $x_i \leq x < x_j$. Figure 3.10 depicts the relevant portion of the graph of $F(x)$. Taken over all x, this reasoning demonstrates that $F(x)$ is a step function with jumps at each point x_k for which $f(x_k) > 0$, the size of the jump being $f(x_k)$. This is exactly the situation that occurred in Example 3.10. The probability distribution function for the

FIGURE 3.11
Probability distribution function for the random variable of Example 3.6

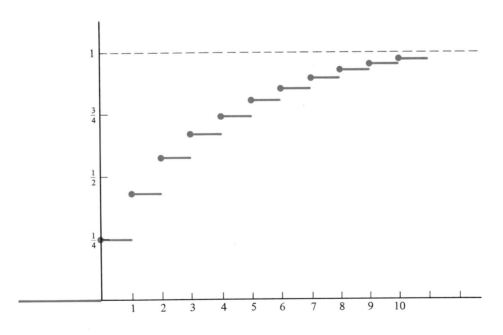

FIGURE 3.12
The probability
distribution function of a
continuous random
variable depicted as an
integral of the density

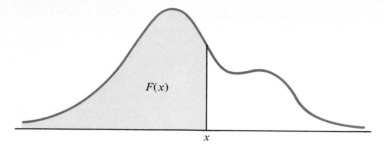

random variable of Example 3.6 (independent selection with $p = 3/4$) is illustrated in Figure 3.11.

Since the density of a discrete random variable is defined by $f(x) = P(X = x)$, the latter three relations of Theorem 3.5 take the form:

(ii) $P(a \le X \le b) = F(b) - F(a) + f(a)$
(iii) $P(a < X < b) = F(b) - F(a) - f(b)$
(iv) $P(a \le X < b) = F(b) - F(a) + f(a) - f(b)$

3.3.3 The Probability Distribution Function of a Continuous Random Variable

According to Definition 3.4, if X is a continuous random variable with density $f(x)$, then its probability distribution function is given by

$$F(x) = P(X \le x) = \int_{-\infty}^{x} f(t)\, dt$$

as depicted in Figure 3.12. Since F is defined by means of integration, it must be a continuous function. Thus, the probability distribution function of a continuous random variable is continuous, as was the case in Example 3.9. Moreover, all the probabilities in Theorem 3.5 reduce to $F(b) - F(a)$.

By differentiating with respect to x and applying the fundamental theorem of calculus, we obtain

$$F'(x) = \frac{d}{dx} F(x) = f(x)$$

at all points x for which f is continuous. For instance, in Example 3.9,

$$F(x) = \begin{cases} 1 - e^{-x/2}, & \text{if } x \ge 0 \\ 0, & \text{if } x < 0 \end{cases}$$

and differentiation retrieves the density utilized in the example.

EXAMPLE 3.11

This example concerns round-off error. Given a real-number measurement q, if the decimal part is less than 0.5, then q is rounded down to the next lowest integer. If the decimal part is greater than 0.5, then q is rounded up to the next highest integer. If the decimal part is exactly 0.5, then a fair coin is tossed to determine whether to round down or up. Let X denote the round-off error. Then X is a continuous random variable, and it seems plausible that the

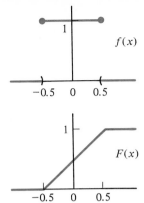

FIGURE 3.13
The density and
probability distribution
function of Example 3.11

values of X will be uniformly distributed between -0.5 and 0.5. Of course, should the input numbers result from some procedure that produces a tendency in the decimal part, then the assumption of uniformity might not be warranted. Be that as it may, if the errors are indeed uniformly distributed over the interval $[-0.5, 0.5]$, then a reasonable density for X would be

$$f(x) = \begin{cases} 1, & \text{if } -0.5 \le x \le 0.5 \\ 0, & \text{otherwise} \end{cases}$$

Integration yields the probability distribution function

$$F(x) = P(X \le x) = \begin{cases} 0, & \text{if } x < -0.5 \\ x + \frac{1}{2}, & \text{if } -0.5 \le x \le 0.5 \\ 1, & \text{if } x > 0.5 \end{cases}$$

Figure 3.13 illustrates both $f(x)$ and $F(x)$. Note that f is continuous at all points except -0.5 and 0.5 and thus F is differentiable everywhere except at those points, with derivative $F'(x) = f(x)$.

Suppose we desire the probability that the round-off error is between -0.13 and 0.22. According to Theorem 3.5,

$$P(-0.13 < X < 0.22) = F(0.22) - F(-0.13) = 0.72 - 0.37 = 0.35$$

The same result could have been obtained by integrating $f(x)$ from -0.13 to 0.22.

The density for the random variable X of Example 3.11 was deduced under the assumption that observations of X would yield outcomes that are spread uniformly over the interval $[-0.5, 0.5]$. As such, we stated loosely that X was *uniformly distributed* over the interval. More generally, if a random variable is to model the uniform distribution of observations across an interval $[a, b]$, a reasonable density would be one that is constant on the interval and zero elsewhere. Of course, it must be normalized so that its total integral is 1.

DEFINITION 3.6

Uniform Distribution. A random variable X is said to possess a uniform distribution and to be **uniformly distributed** over the interval $[a, b]$ if it possesses the density

$$f(x) = \begin{cases} \dfrac{1}{b-a}, & \text{if } a \le x \le b \\ 0, & \text{otherwise} \end{cases}$$

A convenient way to write a uniform density is by employing the **indicator function** of a set. For any set A, its indicator function is defined by

$$I_A(x) = \begin{cases} 1, & \text{if } x \in A \\ 0, & \text{otherwise} \end{cases}$$

If X is uniformly distributed over $[a, b]$, then its density is $f(x) = (b - a)^{-1}I_{[a, b]}(x)$.

A straightforward integration shows that a random variable that is uniformly distributed over the interval $[a, b]$ possesses the probability distribution function

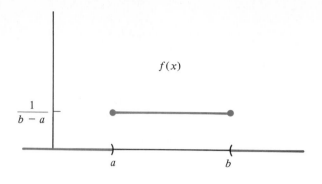

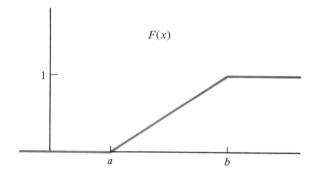

FIGURE 3.14
The density and probability distribution function for the uniform distribution

$$F(x) = \begin{cases} 0, & \text{if } x < a \\ \dfrac{x-a}{b-a}, & \text{if } a \le x \le b \\ 1, & \text{if } x > b \end{cases}$$

(see Figure 3.14).

In fact, there is a whole family of uniform distributions, each one dependent on the choice of endpoints a and b. Although this point will not be emphasized here, in Chapter 4 we will be particularly concerned with families of distributions.

Returning to the main line of thought, the density of a continuous random variable is necessarily equal to the derivative of the probability distribution function only at points where it is continuous. For instance, in Example 3.11, should the density $f(x)$ be redefined to be 0 at $x = -0.5$ and $x = 0.5$, then the same probability distribution function will be obtained. Going further, if any density for a continuous random variable X is changed at a discrete set of points, the probability distribution function will not be affected. Indeed, since the probabilities $P(X \in B)$ are obtained by integration, these are not affected by changing the density at a discrete set of points. Thus, we do not distinguish between continuous random variables whose densities differ only on discrete sets of points. In general, the following definition applies to all random variables, whether continuous, discrete, or otherwise.

DEFINITION 3.7

Identically Distributed Random Variables. The random variables X and Y are said to be identically distributed if they possess the same probability distribution function, namely, $F_X = F_Y$.

We do not distinguish between identically distributed random variables.

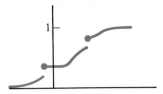

3.3.4 Characterization of Probability Distribution Functions

Certain properties were noted pertaining to the probability distribution functions discussed in Examples 3.9 and 3.10. These had to do with continuity, limits at infinity, and increasing monotonicity. In general, the following theorem, stated without proof, applies to such functions.

THEOREM 3.6

If X is a random variable possessing probability distribution function $F(x)$, then

   (i) $F(x)$ is monotonically increasing.
  (ii) $\lim_{x \to -\infty} F(x) = 0$.
 (iii) $\lim_{x \to +\infty} F(x) = 1$.
 (iv) $F(x)$ is continuous from the right.

Of interest is the fact that any function satisfying the four properties stated in Theorem 3.6 is the probability distribution function for some random variable X. In other words, given a function $F(x)$ satisfying the four properties in the theorem, there can be found some random variable X so that

$$F(x) = F_X(x) = P(X \leq x)$$

for all x. Since the function depicted in Figure 3.15 satisfies the four properties, it must be the probability distribution function for some random variable. Moreover, since the function is neither continuous nor a step function, it is not the probability distribution function of a continuous or discrete random variable. Although this text will focus on continuous and discrete distributions, there exist other kinds. We state the converse of Theorem 3.6 without proof.

THEOREM 3.7

If $F(x)$ satisfies the four properties of Theorem 3.6, then it is a probability distribution function.

3.4 FUNCTIONS OF A RANDOM VARIABLE

Most fundamental relations in science and engineering involve functions in which there is a dependent variable and one or more independent variables. The single-dependent-variable case involves relations of the form $y = g(x)$. Now suppose the independent variable is a random variable X. Random independent variables are common because, in practice, the variable serves as an input to some system whose output is the dependent variable, and the value of the input variable derives from some measurement process. The variability inherent in the measurement process

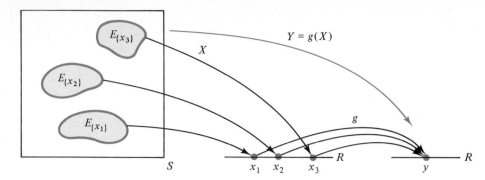

FIGURE 3.16
Illustration of a function
of a discrete random
variable

induces variability into the system model $y = g(x)$, the result being a random output (dependent) variable. Thus, the original equation takes the form $Y = g(X)$, where X and Y are random variables.

3.4.1 Functions of a Single Discrete Random Variable

Consider the discrete input random variable X. If S denotes the sample space on which X is defined, then the function g can be viewed in the manner depicted in Figure 3.16. For any outcome w in S,

$$Y(w) = g[X(w)]$$

Therefore,

$$f_Y(y) = P(Y = y) = P(g(X) = y) = P(\{w \in S: g[X(w)] = y\})$$

and $f_Y(y)$ is determined by the probability of the set of all outcomes w in S for which $g[X(w)] = y$. Because X is discrete and g operates directly on $X(w)$, this set will only be nonempty for a discrete set of y values. Thus, $Y = g(X)$ is also discrete.

In Figure 3.16 there are certain values of x for which $g(x) = y$. The key point is that

$$f_Y(y) = P(Y = y) = \sum_{\{x:g(x)=y\}} P(X = x) = \sum_{\{x:g(x)=y\}} f(x)$$

so that the density for Y is given directly in terms of the density for X.

EXAMPLE 3.12

An important problem in communication is the *compression* of information. In digital image processing, an image is often represented as an array of tiny square picture elements, known as *pixels*. Thus, each pixel corresponds to a spatial location on the image. In a noncolor image, each pixel takes on any one of a number of gray levels. For our purposes, we will assume there are eight possible levels of gray and we will let X be the observed gray level at a particular fixed pixel. Depending on the image captured, X can take on any of eight values, 0, 1, . . . , 7, where 0 denotes white, 7 denotes black, and 1 through 6 denote increasingly darker shades of gray. To transmit the value of X requires at least three bits. To cut the number of bits to be transmitted, an engineer might compress the information. A possible compression can be accomplished by defining the new random variable Y according to Figure 3.17. Y might be interpreted as a coarser representation of the gray scale, 0 being white, 3 being black, and 1 and 2 denoting increasingly darker shades of gray. Should the original gray level at the pixel be $X = 5$, then the level $Y = 2$ is transmitted. The mapping given in the figure determines the system g. The advantage is that Y requires only two bits, whereas X requires three; the

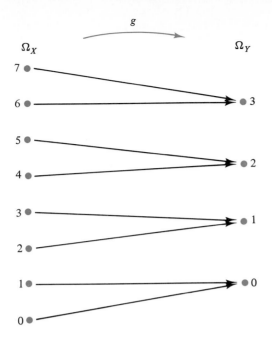

FIGURE 3.17
The compression $Y = g(X)$ of Example 3.12

disadvantage is that a less well-defined picture is transmitted. Should this lack of definition not result in a loss of performance at the other end, for instance, in terms of recognition, then nothing essential is lost.

Now suppose experience has shown X to possess the discrete density defined by the table

x	0	1	2	3	4	5	6	7
$f_X(x)$	0.05	0.10	0.25	0.20	0.19	0.12	0.06	0.03

The density for Y is found as follows:

$$f_Y(0) = P(Y = 0) = f(0) + f(1) = 0.15$$
$$f_Y(1) = P(Y = 1) = f(2) + f(3) = 0.45$$
$$f_Y(2) = P(Y = 2) = f(4) + f(5) = 0.31$$
$$f_Y(3) = P(Y = 3) = f(6) + f(7) = 0.09$$

Although we have considered compression at only a single pixel, in fact, there exists a mapping at each pixel.

In the compression technique of Example 3.12, the operator g is not one-to-one. A function $y = g(x)$ is **one-to-one** if for each value assumed by the dependent variable y, only a single value of x results in y. In Example 3.12, such is not the case. For instance, $g(0) = g(1) = 0$. On the other hand, the linear system $Y = aX$, a a constant, is one-to-one. The salient point concerning one-to-one operators is that there exists an **inverse function** g^{-1} such that $g^{-1}(y) = x$. The inverse system is obtained by solving the original defining relation for x. When considering random variables, we usually write $g^{-1}(Y) = X$ to express the inverse system. For one-to-one operators, the following easy-to-apply theorem relates input and output discrete densities.

Suppose X is a discrete random variable and $Y = g(X)$ is a one-to-one transformation. Then

$$f_Y(y) = f_X[g^{-1}(y)]$$

Proof: Since g is one-to-one,

$$\{x: g(x) = y\} = \{g^{-1}(y)\}$$

a set with only a single element. Hence,

$$f_Y(y) = \sum_{\{x: g(x) = y\}} f_X(x) = f_X[g^{-1}(y)]$$

■

EXAMPLE 3.13

Suppose g is the linear operator

$$Y = g(X) = aX$$

Then, solving for X,

$$X = g^{-1}(Y) = \frac{1}{a}Y$$

According to Theorem 3.8, the density for Y is

$$f_Y(y) = f_X\left(\frac{y}{a}\right)$$

As an illustration, suppose X possesses the density given in the table

x	2	4	6	8
$f_X(x)$	1/8	1/8	1/4	1/2

If $Y = 3X$, then Y possesses the density given by

y	6	12	18	24
$f_Y(y)$	1/8	1/8	1/4	1/2

For instance,

$$f_Y(12) = f_X(12/3) = f_X(4) = 1/8$$

The reason for the simplicity of Theorem 3.8 is that the density of Y is obtained from the density for X by merely changing the scale on the x axis. Figure 3.18 illustrates this similarity of the densities for the random variables of Example 3.13. In general, the scale changing need not be linear.

3.4.2 Functions of a Single Continuous Random Variable

We now turn to the problem of determining the output density for a system in which the input is a single continuous random variable. When the output variable is also

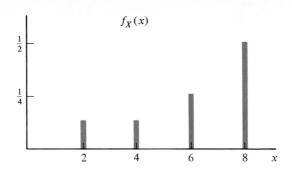

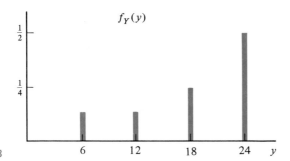

FIGURE 3.18
Densities of Example 3.13

continuous, a key to the approach is that differentiation of a probability distribution function yields the corresponding density. Given the function $Y = g(X)$ and the density f_X, we first find F_Y and then take the derivative.

EXAMPLE 3.14

Consider the *affine* transformation

$$Y = aX + b$$

where $a \neq 0$. For a given value of X, the corresponding value of Y lies on the line $y = ax + b$. First suppose $a > 0$. Then

$$F_Y(y) = P(Y \leq y) = P(aX + b \leq y) = P\left(X \leq \frac{y - b}{a}\right) = F_X\left(\frac{y - b}{a}\right)$$

For $a < 0$,

$$F_Y(y) = P(aX + b \leq y) = P\left(X \geq \frac{y - b}{a}\right)$$

$$= 1 - P\left(X < \frac{y - b}{a}\right) = 1 - F_X\left(\frac{y - b}{a}\right)$$

Note that $F_X(x) = P(X < x)$, since X is a continuous random variable. Differentiating $F_Y(y)$ with respect to y gives the density for Y. For $a > 0$,

$$f_Y(y) = \frac{d}{dy} F_Y(y) = \frac{d}{dy} F_X\left(\frac{y - b}{a}\right) = a^{-1} f_X\left(\frac{y - b}{a}\right)$$

Except for a minus sign in front, we obtain the same expression for $a < 0$. Combining the two results yields

$$f_Y(y) = \frac{1}{|a|} f_X\left(\frac{y - b}{a}\right)$$

Let us apply this result to a specific input distribution. Suppose X is uniformly distributed on the interval $[0, 1]$. Then

$$f_X(x) = \begin{cases} 1, & \text{if } 0 \le x \le 1 \\ 0, & \text{otherwise} \end{cases}$$

For the affine transformation

$$Y = -2X + 4$$

the expression just derived yields

$$f_Y(y) = \frac{1}{2} f_X\left(\frac{y - 4}{-2}\right) = \begin{cases} \dfrac{1}{2}, & \text{if } 0 \le \dfrac{y - 4}{-2} \le 1 \\ 0, & \text{otherwise} \end{cases}$$

$$= \begin{cases} \dfrac{1}{2}, & \text{if } 2 \le y \le 4 \\ 0, & \text{otherwise} \end{cases}$$

EXAMPLE 3.15

In some cases, only the absolute value of a measurement is of interest. For instance, one might be concerned only with absolute error, not signed error. If X measures the signed error, then the relevant transformation is $Y = |X|$. To find the density for Y, first consider $y < 0$. In this case,

$$F_Y(y) = P(Y \le y) = P(|X| \le y) = 0$$

and, differentiating with respect to y, $f_Y(y) = 0$. For $y \ge 0$, Theorem 3.5 yields

$$F_Y(y) = P(|X| \le y) = P(-y \le X \le y) = F_X(y) - F_X(-y)$$

Differentiation gives

$$f_Y(y) = \frac{d}{dy} F_Y(y) = \frac{d}{dy} F_X(y) - \frac{d}{dy} F_X(-y) = f_X(y) + f_X(-y)$$

As an illustration, consider the error-measurement distribution of Example 3.7. If $Y = |X|$, then

$$f_Y(y) = \sqrt{2/\pi}\, e^{-2y^2} + \sqrt{2/\pi}\, e^{-2(-y)^2} = \sqrt{8/\pi}\, e^{-2y^2}$$

for $y \ge 0$ and $f_Y(y) = 0$ for $y < 0$.

In the next example the input random variable is continuous, but the output random variable is discrete. Consequently, we compute the output density directly from the input probability distribution function.

EXAMPLE 3.16

A fundamental operation in the digital processing of images and signals is *quantization*. The readings on sensors are analog (continuous), and these readings must be quantized—that is, they must be converted to a discrete scale to allow digital processing. Here we will consider a very simple case involving a single sensor reading X and a threshold value t. If the reading X is greater than or equal to t, then the stored value is 1; if the reading is less than t, then the stored value is 0. Thus, the output of the system is defined by

$$Y = \begin{cases} 1, & \text{if } X \geq t \\ 0, & \text{if } X < t \end{cases}$$

Note that Y is discrete, having only two values in its codomain. Thus, we will derive its probability mass function directly:

$$f_Y(0) = P(Y = 0) = P(X < t) = F_X(t)$$
$$f_Y(1) = P(Y = 1) = P(X \geq t) = 1 - F_X(t)$$

As an illustration, suppose the reading X possesses the density illustrated in Figure 3.19:

$$f_X(x) = \begin{cases} x, & \text{if } 0 \leq x \leq 1 \\ -x + 2, & \text{if } 1 \leq x \leq 2 \\ 0, & \text{otherwise} \end{cases}$$

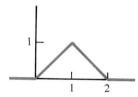

FIGURE 3.19
The density $f_X(x)$ of Example 3.16

For the threshold value $t = \frac{1}{2}$,

$$f_Y(0) = F_X(\tfrac{1}{2}) = \int_0^{1/2} x \, dx = \tfrac{1}{8}$$

and

$$f_Y(1) = 1 - F_X(\tfrac{1}{2}) = \tfrac{7}{8}$$

FIGURE 3.20
Strictly increasing and strictly decreasing functions

At this point we would like to state and prove the analogue to Theorem 3.8 that applies to continuous random variables. There are some subtleties, however. First, as was seen in Example 3.16, a function of a continuous random variable need not be continuous. Second, a one-to-one function can be increasing or decreasing (see Figure 3.20). And third, the differentiability of the function is a concern.

Before stating the theorem, we will review some points from calculus. If $y = g(x)$ is differentiable for all x and possesses a derivative that is either strictly greater than or strictly less than 0, then the derivative of g^{-1} is given by

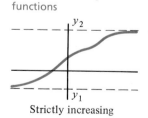

Strictly increasing

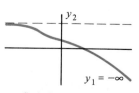

Strictly decreasing

$$\frac{d}{dy} g^{-1}(y) = \frac{1}{g'[g^{-1}(y)]} = \frac{1}{\left.\dfrac{d}{dx} g(x)\right|_{x=g^{-1}(y)}}$$

This derivative is known as the **Jacobian** of the transformation and is denoted by $J_g(y)$. Moreover, there exist values y_1 and y_2, $y_1 < y_2$, either or both possibly infinite, such that for any y between y_1 and y_2, there exists exactly one value x such that $y = g(x)$ (see Figure 3.20).

Suppose X is a continuous random variable, $y = g(x)$ is differentiable for all x, and $g(x)$ is either strictly increasing or strictly decreasing for all x. Then $Y = g(X)$ is a continuous random variable with density

$$f_Y(y) = \begin{cases} f_X[g^{-1}(y)]|J_g(y)|, & \text{if } y_1 < y < y_2 \\ 0, & \text{otherwise} \end{cases}$$

where y_1 and y_2 are the limiting values described prior to the statement of the theorem.

Proof: To begin with, suppose $y_1 < y < y_2$. First consider the case where g is increasing. Then

$$F_Y(y) = P(Y \le y) = P(g(X) \le y) = P(X \le g^{-1}(y)) = F_X[g^{-1}(y)]$$

Because g is increasing, the inequality sign did not flip when applying the inverse across it. Differentiation with respect to y yields

$$f_Y(y) = f_X[g^{-1}(y)]\frac{d}{dy}[g^{-1}(y)] = f_X[g^{-1}(y)]J_g(y)$$

For the case where g is decreasing,

$$F_Y(y) = P(Y \le y) = P(g(X) \le y) = P(X \ge g^{-1}(y))$$
$$= 1 - P(X < g^{-1}(y)) = 1 - F_X[g^{-1}(y)]$$

In this case, the inequality sign is reversed with the application of the inverse operation. Differentiation yields

$$f_Y(y) = -f_X[g^{-1}(y)]\frac{d}{dy}[g^{-1}(y)] = f_X[g^{-1}(y)][-J_g(y)]$$

Since g is decreasing, $g' \le 0$, so that $J_g(y) < 0$ and $-J_g(y) = |J_g(y)|$. Combining the two cases yields the result stated in the theorem when y is between y_1 and y_2.

If $y \le y_1$, then

$$P(Y \le y) = P(g(X) \le y) \le P(g(X) \le y_1) = 0$$

where we have used, in turn, the definition of Y, Theorem 2.10, and the fact that $g(X) > y_1$. The case for $y \ge y_2$ is left as an exercise. ∎

The strict monotonicity of g does not necessarily imply that $g'(x)$ is nonzero for all x. [For instance, $g(x) = x^3$ possesses zero derivative at the origin.] Consequently, the result given in Theorem 3.9 does not apply at all points, the exceptions being those for which the denominator in the Jacobian is zero. However, so long as g' possesses only a finite number of zeros, as will always be the case in this text, the exceptional set is finite. Since the distribution of a continuous random variable is not affected by leaving its density undefined at a finite number of points, the exceptional points have no effect.

EXAMPLE 3.17

The transformation

$$y = g(x) = e^{tx}$$

$t > 0$, satisfies the conditions of Theorem 3.9 with $g'(x) > 0$, $y_1 = 0$, and $y_2 = \infty$. Since the inverse function of the exponential is the logarithm,

$$g^{-1}(y) = \frac{\log y}{t}$$

and

$$J_g(y) = \frac{d}{dy}\left(\frac{\log y}{t}\right) = \frac{1}{ty}$$

Thus,

$$f_Y(y) = \frac{f_X\left(\dfrac{\log y}{t}\right)}{ty}$$

for $y > 0$. For $y < 0$, $f_Y(y) = 0$. For $y = 0$, we can define the density in any way we wish (or leave it undefined).

Now suppose X possesses the density

$$f_X(x) = \begin{cases} be^{-bx}, & \text{if } x \geq 0 \\ 0, & \text{if } x < 0 \end{cases}$$

where b is a positive constant. A straightforward integration shows that f_X possesses total integral 1. In Example 3.4 we considered a special case of this density for $b = \frac{1}{2}$. Let $Y = e^{tx}$. Then, according to the preceding derivation, $f_Y(y) = 0$ for $y < 0$ and

$$f_Y(y) = \begin{cases} \dfrac{be^{-b[(\log y)/t]}}{ty}, & \text{if } \dfrac{\log y}{t} \geq 0 \\ 0, & \text{if } \dfrac{\log y}{t} < 0 \end{cases}$$

for $y > 0$. Simplifying the exponential and recognizing that $\log y \geq 0$ if and only if $y \geq 1$, we obtain

$$f_Y(y) = \begin{cases} \dfrac{by^{-b/t}}{ty}, & \text{if } y \geq 1 \\ 0, & \text{if } y < 1 \end{cases}$$

In the preceding example, the density possessed an unspecified constant b. Thus, as in the case of a uniform distribution, an entire class of densities is actually represented, the particular density of interest resulting from a specific choice of b (which must be experimentally determined). Because this family of densities plays such an important role in applications, it will be studied in its own right in Section 4.6.4. For the time being, we content ourselves with the following definition.

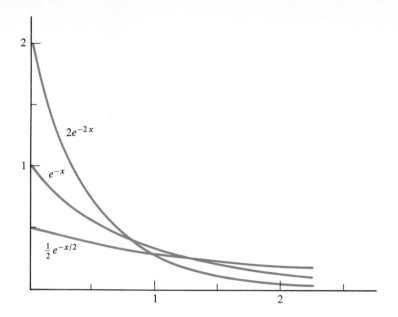

FIGURE 3.21
Various exponential
densities

Exponential Distribution. A random variable X is said to possess an exponential distribution and to be **exponentially distributed** if it possesses the density

$$f(x) = \begin{cases} be^{-bx}, & \text{if } x \geq 0 \\ 0, & \text{if } x < 0 \end{cases}$$

where $b > 0$.

For comparison purposes, the exponential densities for $b = \frac{1}{2}$, $b = 1$, and $b = 2$ are illustrated in Figure 3.21. A straightforward integration shows that, for a given value of b, the probability distribution function is

$$F(x) = \begin{cases} 1 - e^{-bx}, & \text{if } x \geq 0 \\ 0, & \text{if } x < 0 \end{cases}$$

Figure 3.8 illustrates $F(x)$ for the case $b = \frac{1}{2}$.

EXAMPLE 3.18

The integral powers of a random variable play a fundamental role in probability theory. Consider

$$Y = g(X) = X^n$$

where n is a positive odd integer, $n \geq 3$. Then $g(x)$ is increasing, $g(x)$ possesses a continuous derivative, $g'(x) = 0$ if and only if $x = 0$, and Theorem 3.9 applies. Since

$$g^{-1}(y) = y^{1/n}$$

the theorem yields

$$f_Y(y) = f_X(y^{1/n}) \frac{1}{n} y^{(1/n) - 1}$$

Note that $f_Y(y)$ is not defined for $y = 0$ and that

$$J_g(y) = \frac{1}{n} y^{(1/n)-1} > 0$$

for $y \neq 0$. Moreover, since x^n goes from $-\infty$ to $+\infty$, the given expression for $f_Y(y)$ holds for all $y \neq 0$.

As an illustration, let $n = 3$, so that $Y = X^3$, and let X be uniformly distributed on the interval $[0, 2]$ so that

$$f_X(x) = \begin{cases} \dfrac{1}{2}, & \text{if } 0 \leq x \leq 2 \\ 0, & \text{otherwise} \end{cases}$$

Then

$$f_Y(y) = \begin{cases} \dfrac{1}{6} y^{-2/3}, & \text{if } 0 < y^{1/3} \leq 2 \\ 0, & \text{otherwise} \end{cases}$$

$$= \begin{cases} \dfrac{1}{6} y^{-2/3}, & \text{if } 0 < y \leq 8 \\ 0, & \text{otherwise} \end{cases}$$

3.5 EXPECTATION

Whereas Section 1.4 concerned the measurement of central tendency of a data set, the present section concerns the key measure of central tendency relating to discrete and continuous probability distributions.

3.5.1 Expected Value of a Discrete Random Variable

If X is a discrete random variable with probability mass function $f(x)$, then the graphical representation of the distribution consists of a collection of point masses. As in physics, the center of the point masses is found by summing all products of the form $xf(x)$, where x is in the codomain of X and the sum can be finite or infinite. We are led to the following definition.

DEFINITION 3.9

Expected Value (of a Discrete Random Variable). Let X be a discrete random variable possessing discrete density $f(x)$. The expected value of X is given by

$$E[X] = \sum_{f(x)>0} xf(x)$$

where the summation notation means to sum over all x such that $f(x) > 0$.

In Definition 3.9 we make the tacit assumption that the sum is *absolutely convergent*; namely,

$$\sum_{f(x)>0} |x|f(x)$$

converges (is finite). If the codomain happens to be finite, then there is no need to be concerned with convergence.

The "expected value" terminology is meant to connote that $E[X]$ is, in some sense, the "best" guess as to the possible outcome of X. In Section 3.6.3 we will clarify what is meant by "best" in the present context. For now, it is profitable to think of $E[X]$ as representing the center of mass of the probability mass function. Besides being known as the expected value, $E[X]$ is also called the **mean** of X and is often denoted by μ_X.

EXAMPLE 3.19

A minidisk contains nine files, of which four are system files and five are data files. If three files are selected at random without replacement, what is the "expected number" of system files to be selected? The question calls for the expected value of X, where X counts the number of system files selected. Employing the theory of combinations, in conjunction with the hypothesis of equal probability, we can derive the density for X. For $x = 0, 1, 2,$ and 3, consider the number of ways of selecting x system files out of the four available and selecting $3 - x$ data files (see Example 2.15). The number of ways in which this can be done constitutes the number of elements in the event of interest. The number of elements in the sample space is $C(9, 3)$, the number of combinations of nine objects taken three at a time. Thus,

$$f(x) = \frac{\binom{4}{x}\binom{5}{3-x}}{\binom{9}{3}}$$

the values of which are given in the following table.

x	0	1	2	3
$f(x)$	0.12	0.48	0.36	0.05

(Owing to round-off error, the sum of the probabilities in the table is 1.01, rather than 1.00.) The expected value of X is

$$E[X] = 0 \times 0.12 + 1 \times 0.48 + 2 \times 0.36 + 3 \times 0.05 = 1.35$$

Figure 3.22 shows the probability mass function $f(x)$ together with the center of mass $E[X]$.

FIGURE 3.22
The probability mass function and mean for Example 3.19

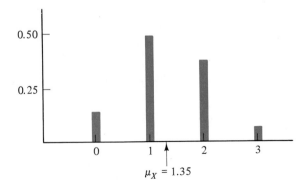

3.5.2 Expected Value of a Continuous Random Variable

The definition of the expected value extends to continuous distributions by means of the center of mass for a region under a curve, in this case the curve being the density.

DEFINITION 3.10

Expected Value (of a Continuous Random Variable). If X is a continuous random variable with density $f(x)$, then the expected value of X is defined by

$$E[X] = \int_{-\infty}^{\infty} xf(x)\, dx$$

As in the case of Definition 3.9, absolute convergence is required in Definition 3.10. Specifically, $E[X]$ is defined if and only if

$$\int_{-\infty}^{\infty} |x|f(x)\, dx < \infty$$

That Definitions 3.9 and 3.10 are consistent can be recognized from the interpretation of an integral as a generalized sum. Once again, $E[X]$ is also called the **mean** of X and is denoted by μ_X.

EXAMPLE 3.20

Let X be exponentially distributed (Definition 3.8). Applying Definition 3.10 in conjunction with integration by parts yields

$$E[X] = \int_{0}^{\infty} bxe^{-bx}\, dx = -xe^{-bx}\Big|_{0}^{\infty} + \int_{0}^{\infty} e^{-bx}\, dx = \frac{1}{b}$$

Referring to Example 3.4, where $b = 1/2$ and X was chosen to model the lengths of telephone calls at a particular station, the expected length of a telephone call is $E[X] = 2$.

EXAMPLE 3.21

In Example 3.7, a density $f(x)$ was utilized to model the errors in a measurement process, the density being illustrated in Figure 3.7. The mean error is given by

$$\mu_X = E[X] = \int_{-\infty}^{\infty} \sqrt{2/\pi}\ xe^{-2x^2}\, dx$$

Although we could proceed with a straightforward integration of the improper integral, it is much easier to recognize that the integrand is an *odd* function. A function $h(x)$ is odd if $h(-x) = -h(x)$, and the integral from $-a$ to a of any odd function is 0. For $a = \infty$, the result holds so long as the integral is absolutely convergent, which in this case it is. Therefore, $E[X] = 0$.

EXAMPLE 3.22

Consider the **Cauchy density**

$$f(x) = \frac{1}{\pi(1 + x^2)}$$

That f constitutes a legitimate density is concluded from

$$\int_{-\infty}^{\infty} f(x)\,dx = \int_{-\infty}^{\infty} \frac{1}{\pi(1+x^2)}\,dx = \frac{\arctan x}{\pi}\bigg|_{-\infty}^{\infty} = 1$$

However, f does not possess an expectation since

$$\int_{-\infty}^{\infty} |x| f(x)\,dx = \int_{-\infty}^{\infty} \frac{|x|}{\pi(1+x^2)}\,dx$$

does not converge when evaluated by improper integration.

3.5.3 Expected Value of a Function of a Random Variable

Often, given a random variable X possessing known distribution $f_X(x)$, we require the expected value of a dependent random variable $Y = g(X)$. From Definitions 3.9 and 3.10, it is evident that $E[Y]$ can be found by first using the techniques of Section 3.4 to find the density $f_Y(y)$ and then either summing or integrating $yf_Y(y)$, depending on whether Y is discrete or continuous. However, one of the most fundamental propositions in probability theory states that the density for $Y = g(X)$ is not required to find $E[Y]$.

THEOREM 3.10

Suppose X is a discrete random variable with discrete density $f_X(x)$ and $y = g(x)$ is any real-valued function. Then

$$E[g(X)] = \sum_{f_X(x)>0} g(x) f_X(x)$$

If, on the other hand, X is a continuous random variable with density $f_X(x)$ and $y = g(x)$ is any piecewise continuous real-valued function, then

$$E[g(X)] = \int_{-\infty}^{\infty} g(x) f_X(x)\,dx$$

Proof: First consider the discrete case. According to the definition of expectation in the discrete case,

$$E[g(X)] = E[Y] = \sum_{f_Y(y)>0} y f_Y(y)$$

For any specific term $y_k f_Y(y_k)$ in the sum, there exists a collection of points x_1, x_2, . . . , x_m such that

$$g(x_1) = g(x_2) = \cdots = g(x_m) = y_k$$

Moreover,

$$f_Y(y_k) = P(Y = y_k) = P(X = x_1) + P(X = x_2) + \cdots + P(X = x_m)$$
$$= f_X(x_1) + f_X(x_2) + \cdots + f_X(x_m)$$

(see Figure 3.16). Hence,

$$
\begin{aligned}
y_k f(y_k) &= y_k[f_X(x_1) + f_X(x_2) + \cdots + f_X(x_m)] \\
&= y_k f_X(x_1) + y_k f_X(x_2) + \cdots + y_k f_X(x_m) \\
&= g(x_1)f_X(x_1) + g(x_2)f_X(x_2) + \cdots + g(x_m)f_X(x_m)
\end{aligned}
$$

If all terms in the defining sum for $E[g(X)]$ are likewise evaluated, the result is precisely the summation in terms of $g(x)f_X(x)$ asserted by the theorem.

For the continuous case, a rigorous proof is beyond the scope of the present text. Nevertheless, we will provide a proof for the special case in which $g(x)$ satisfies the conditions of Theorem 3.9. We will restrict our attention to the case in which $g(x)$ is strictly increasing, the decreasing case being essentially the same. This approach should be instructive and serve to tie together some of the concepts thus far discussed. Now, according to Theorem 3.9, if $g'(x) \geq 0$, then

$$
E[g(X)] = \int_{y_1}^{y_2} y f_X[g^{-1}(y)] \frac{d}{dy} g^{-1}(y) \, dy
$$

For the substitution $y = g(x)$, $dy = g'(x) \, dx$ and $x = g^{-1}(y)$. Moreover, according to the definition of y_1 and y_2, as y varies from y_1 to y_2, x goes from $-\infty$ to ∞. Thus, under the substitution,

$$
E[g(X)] = \int_{-\infty}^{\infty} g(x)f_X(x) \frac{d}{dy} g^{-1}(y)g'(x) \, dx
$$

But

$$
\frac{d}{dy} g^{-1}(y) = \frac{1}{g'[g^{-1}(y)]} = \frac{1}{g'(x)}
$$

Therefore, as claimed in the statement of the theorem,

$$
E[g(X)] = \int_{-\infty}^{\infty} g(x)f_X(x) \, dx
$$

(for the special case considered). ∎

EXAMPLE 3.23

In Example 3.14, we applied the affine transformation $g(X) = -2X + 4$ to the uniformly distributed random variable X possessing the density $f_X(x) = I_{[0,1]}(x)$. According to Theorem 3.10,

$$
E[g(X)] = \int_{-\infty}^{\infty} (-2x + 4)I_{[0,1]}(x) \, dx = \int_{0}^{1} (-2x + 4) \, dx = 3
$$

Directly employing the density for $Y = g(X)$ that was found in Example 3.14 gives

$$
E[g(X)] = \int_{-\infty}^{\infty} y f_Y(y) \, dy = \int_{2}^{4} \frac{y}{2} \, dy = 3
$$

EXAMPLE 3.24

In Example 3.16, we considered the step function

$$g(x) = \begin{cases} 1, & \text{if } x \ge t \\ 0, & \text{if } x < t \end{cases}$$

Since g is piecewise continuous, Theorem 3.10 applies. Letting $f_X(x)$ be the density of any continuous random variable,

$$E[g(X)] = \int_t^\infty f_X(x)\, dx = 1 - \int_{-\infty}^t f_X(x)\, dx = 1 - F_X(t)$$

where F_X is the probability distribution function for X and where we have employed the fact that the total integral of f_X is 1. Looking back at Example 3.16, we see that the density for $Y = g(X)$ is discrete and that, using the density directly,

$$E[g(X)] = 0 \times F_X(t) + 1 \times (1 - F_X(t)) = 1 - F_X(t)$$

precisely the result obtained by use of Theorem 3.10. Note that the theorem applies even when the independent random variable is continuous and the dependent random variable is discrete.

EXAMPLE 3.25

In Example 3.17, we found the density for $y = g(X) = e^{tX}$ when X is exponentially distributed and $t > 0$. Applying Theorem 3.10,

$$E[g(X)] = \int_0^\infty e^{tx} b e^{-bx}\, dx = b \int_0^\infty e^{-(b-t)x}\, dx$$

The integral is finite if and only if $t < b$. Hence, we make this assumption. In such a case, the integral is easily evaluated to yield

$$E[g(X)] = \frac{b}{b - t}$$

whether or not $t > 0$. When $t > 0$, direct application of the density for $Y = g(X)$ found in Example 3.17 yields

$$E[g(X)] = \int_{-\infty}^\infty y f_Y(y)\, dy = \frac{b}{t} \int_1^\infty y^{-b/t}\, dy = \frac{\frac{b}{t}}{1 - \frac{b}{t}} y^{1 - (b/t)} \Bigg|_1^\infty$$

The upper limit is finite if and only if $t < b$, in which case it is 0. Assuming $t < b$, the result agrees with the one obtained by using Theorem 3.10.

3.6 MOMENTS

For any random variable X, $E[X]$, if it exists, is a parameter that carries information regarding the central tendency of the random phenomenon modeled by X. Specifically, $E[X]$ gives the center of mass of the probability distribution. In general, a great deal of information is contained in the density, this information being understood in terms of the manner in which the probability is distributed. The present section describes a collection of parameters known as **moments,** of which $E[X]$ is a particular one. Rather than describe a random variable in full by means of its density, it is often sufficient to give a partial description in terms of the moments of the random variable.

3.6.1 Moments About the Origin

We begin by defining the moments about the origin. These are the expected values of the nonnegative integral powers of the random variable

---

DEFINITION 3.11

Moments about the Origin. For any nonnegative integer k, the kth moment about the origin of the random variable X is

$$\mu_k' = E[X^k]$$

According to Theorem 3.10, if X is a discrete random variable, then

$$\mu_k' = E[X^k] = \sum_{f_X(x)>0} x^k f_X(x)$$

If X is continuous, then

$$\mu_k' = E[X^k] = \int_{-\infty}^{\infty} x^k f_X(x)\, dx$$

In accordance with the covergence requirement regarding expectation, the kth moment exists in the discrete case if and only if the defining sum is absolutely convergent, namely,

$$\sum_{f_X(x)>0} |x|^k f_X(x) < \infty$$

It exists in the continuous case if and only if the defining integral is absolutely convergent. Unless there is a significant difference between the discrete and continuous cases, we will often only make specific reference to the latter. In most situations, the behavior of the discrete case is analogous to that of the continuous case. The terminology "about the origin" is often dropped when discussing the μ_k'.

Note the special cases $\mu_0' = 1$, which holds because the total probability is 1, and $\mu_1' = E[X] = \mu_X$. A particularly important moment is the second moment about the origin,

$$\mu_2' = E[X^2] = \int_{-\infty}^{\infty} x^2 f_X(x)\, dx$$

EXAMPLE 3.26

The random variable X of Example 3.19 was found to have expected value $E[X] = 1.35$. To find the second and third moments, we use the information in the following table.

x	0	1	2	3
x^2	0	1	4	9
x^3	0	1	8	27
$f(x)$	0.12	0.48	0.36	0.05

The second moment about the origin is

$$\mu_2' = E[X^2] = 0 \times 0.12 + 1 \times 0.48 + 4 \times 0.36 + 9 \times 0.05 = 2.37$$

The third moment about the origin is

$$\mu_3' = E[X^3] = 0 \times 0.12 + 1 \times 0.48 + 8 \times 0.36 + 27 \times 0.05 = 4.71$$

EXAMPLE 3.27

If X is exponentially distributed, then $E[X] = 1/b$ (see Example 3.20). The second moment can be found by repeated integration by parts:

$$\mu_2' = E[X^2] = \int_0^\infty x^2 b e^{-bx}\, dx$$

$$= -x^2 e^{-bx}\Big|_0^\infty + \int_0^\infty 2x e^{-bx}\, dx = \int_0^\infty 2x e^{-bx}\, dx$$

$$= -2b^{-1}x e^{-bx}\Big|_0^\infty + 2b^{-1}\int_0^\infty e^{-bx}\, dx = \frac{2}{b^2}$$

EXAMPLE 3.28

Let X be uniformly distributed over the interval $[a, b]$. For $k = 0, 1, 2, \ldots$, the kth moment is

$$\mu_k' = E[X^k] = \frac{1}{b - a}\int_a^b x^k\, dx$$

$$= \frac{1}{b - a}\frac{x^{k+1}}{k + 1}\Big|_a^b$$

$$= \frac{b^{k+1} - a^{k+1}}{(k + 1)(b - a)}$$

In particular, the mean is

$$\mu_X = E[X] = \frac{b^2 - a^2}{2(b - a)} = \frac{b + a}{2}$$

Moreover, the second moment is

$$\mu_2' = E[X^2] = \frac{b^3 - a^3}{3(b - a)} = \frac{b^2 + ab + a^2}{3}$$

In working with moments, the linearity of the integral (summation) plays a useful role. This is illustrated in the next theorem, which provides an easy-to-use mechanism for the algebraic manipulation of moments.

Assuming the first m moments of the random variable X exist, if $g(X)$ is any polymonial function,

$$g(X) = a_m X^m + a_{m-1} X^{m-1} + \cdots + a_1 X + a_0$$

then

$$E[g(X)] = a_m \mu_m' + a_{m-1} \mu_{m-1}' + \cdots + a_1 \mu_1' + a_0$$

Proof: We prove the theorem for the case of a continuous random variable X possessing density $f(x)$. Employing the linearity of the integral, by Theorem 3.10,

$$E[g(X)] = \int_{-\infty}^{\infty} (a_m x^m + a_{m-1} x^{m-1} + \cdots + a_1 x + a_0) f(x)\, dx$$

$$= \sum_{k=0}^{m} a_k \int_{-\infty}^{\infty} x^k f(x)\, dx$$

$$= \sum_{k=0}^{m} a_k \mu_k' \qquad \blacksquare$$

The cases of $m = 1$ and $m = 2$ are especially important. For $m = 1$,

$$E[aX + b] = aE[X] + b = a\mu_X + b$$

Some particular subcases are noteworthy. Setting $b = 0$ shows that multiplying a random variable by a constant has the effect of multiplying the mean by the same constant. Setting $a = 1$ shows that adding a constant to a random variable results in the same constant being added to the mean. Setting $a = 1$ and $b = -\mu_X$ gives

$$E[X - \mu_X] = \mu_X - \mu_X = 0$$

which states that the expected value of the deviation from the mean is zero. Finally, for $m = 2$,

$$E[aX^2 + bX + c] = aE[X^2] + bE[X] + c = a\mu_2' + b\mu_X + c$$

3.6.2 Central Moments

In addition to taking moments about the origin, we also take moments about the mean. These are the central moments. Physically, they represent moments about the center of mass. Of particular importance is the second central moment, which measures the degree to which a distribution is concentrated around its mean.

Central Moments. For any nonnegative integer k, the kth central moment of the random variable X is defined by

$$\mu_k = E[(X - \mu_X)^k]$$

According to Theorem 3.10, in the discrete case,

$$\mu_k = \sum_{f(x)>0} (x - \mu)^k f(x)$$

and, in the continuous case,

$$\mu_k = \int_{-\infty}^{\infty} (x - \mu)^k f(x)\, dx$$

where we have dropped the subscript X in the notation for the mean and simply employed μ. As usual, there is the tacit assumption regarding the absolute convergence of the defining sum and integral. In the latter case this means that

$$\int_{-\infty}^{\infty} |x - \mu|^k f(x)\, dx < \infty$$

Since $E[X - \mu] = 0$, the first central moment is 0.

3.6.3 Variance

The most important central moment is the second. To see why, note that $|X - \mu|$ gives a measure of the distance that the random variable falls from its mean. Thus, as in the case of the empirical variance, the magnitude of $(X - \mu)^2$ is greater or less, depending on the degree to which X is further from or closer to the mean. Since it is itself a random variable, $(X - \mu)^2$ cannot serve as a measure of dispersion. Instead, we employ the mean of $(X - \mu)^2$, which is a numerical quantity giving the "average" value of $(X - \mu)^2$.

DEFINITION 3.13

Variance. The variance of a random variable X is given by the second central moment (assuming it exists):

$$\sigma_X^2 = \mu_2 = E[(X - \mu_X)^2]$$

The variance is also known as the **dispersion.** As in the case of the mean, the subscript X is dropped if the random variable is clear from the context of the discussion. In addition, Var $[X]$ is often used to denote the variance of X. The square root of the variance is called the **standard deviation,** and the latter is customarily denoted by σ.

According to Theorem 3.10, the variance can be written as the integral

$$\sigma^2 = \int_{-\infty}^{\infty} (x - \mu)^2 f(x)\, dx$$

However, rather than compute the variance directly, it is usually easier to employ the following proposition.

The variance of a random variable X is given by

$$\sigma^2 = \mu_2' - \mu^2$$

Proof: By Theorem 3.11,

$$\begin{aligned}
\text{Var}\,[X] &= E[(X - \mu)^2] \\
&= E[X^2 - 2\mu X + \mu^2] \\
&= E[X^2] - 2\mu E[X] + \mu^2 \\
&= \mu_2' - 2\mu^2 + \mu^2 \\
&= \mu_2' - \mu^2
\end{aligned}$$
∎

EXAMPLE 3.29

For the random variable X discussed in Example 3.26, $E[X] = 1.35$ and $E[X^2] = 2.37$. Theorem 3.12 yields the variance

$$\sigma^2 = E[X^2] - E[X]^2 = 2.37 - 1.82 = 0.55$$

The standard deviation of X is $\sigma = 0.74$.

EXAMPLE 3.30

A manufacturer employs two distinct processes in the assembly of automobile transmissions. All components, including the casings, gears, shafts, seals, and bearings are supplied by subcontractors. In measuring the length of time (in hours) a unit stays in production—the time that elapses from the moment the casing enters assembly until the completion of the unit—it is found that processes A and B possess the distributions

$$f_X(x) = \begin{cases} x - 1, & \text{if } 1 \le x \le 2 \\ -x + 3, & \text{if } 2 \le x \le 3 \\ 0, & \text{otherwise} \end{cases}$$

and

$$f_Y(y) = \begin{cases} 4y - 6, & \text{if } 1.5 \le y \le 2 \\ -4y + 10, & \text{if } 2 \le y \le 2.5 \\ 0, & \text{otherwise} \end{cases}$$

respectively (see Figure 3.23). Since both densities are symmetric with respect to the time 2, both possess mean 2. However, their variances are different. The second moment of X is

$$E[X^2] = \int_1^2 (x^3 - x^2)\,dx + \int_2^3 (-x^3 + 3x^2)\,dx = 4.17$$

The second moment of Y is

FIGURE 3.23
Densities $f_Y(y)$ and $f_X(x)$ of Example 3.30

$$E[Y^2] = \int_{1.5}^2 (4y^3 - 6y^2)\,dy + \int_2^{2.5} (-4y^3 + 10y^2)\,dy = 4.04$$

By Theorem 3.12,

$$\sigma_X^2 = 4.17 - 4 = 0.17$$

and

$$\sigma_Y^2 = 4.04 - 4 = 0.04$$

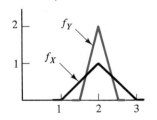

Thus, $\sigma_X = 0.41$ and $\sigma_Y = 0.20$. Although the two production processes result in the same mean time of production, the first exhibits greater variability.

EXAMPLE 3.31

In Examples 3.20 and 3.27, the mean and second moment of an exponentially distributed random variable X were found to be $1/b$ and $2/b^2$, respectively. Theorem 3.12 yields

$$\sigma^2 = \frac{2}{b^2} - \frac{1}{b^2} = \frac{1}{b^2}$$

The standard deviation is $\sigma = 1/b$, which happens to equal the mean. Not only is the mean close to zero for large b, so too is the dispersion. This can be seen in Figure 3.21, where for larger values of b the probability mass is packed more tightly.

EXAMPLE 3.32

In Example 3.28, the mean and second moment for a uniformly distributed random variable with density $(b - a)^{-1}I_{[a,b]}(x)$ were seen to be $(b + a)/2$ and $(b^2 + ab + a^2)/3$, respectively. Thus, by Theorem 3.12, the variance of the distribution is

$$\sigma^2 = \frac{b^2 + ab + a^2}{3} - \frac{(b + a)^2}{4} = \frac{(b - a)^2}{12}$$

An immediate corollary of Theorem 3.11 was that $E[aX + b] = aE[X] + b$. The next theorem provides a corresponding result for the variance.

THEOREM 3.13

For any constants a and b,

$$\text{Var }[aX + b] = a^2 \text{ Var }[X]$$

Proof: By Theorems 3.11 and 3.12,

$$\begin{aligned}
\text{Var }[aX + b] &= E[(aX + b)^2] - E[aX + b]^2 \\
&= E[a^2X^2 + 2abX + b^2] - (aE[X] + b)^2 \\
&= a^2E[X^2] + 2abE[X] + b^2 - (a^2E[X]^2 + 2abE[X] + b^2) \\
&= a^2(E[X^2] - E[X]^2) \\
&= a^2 \text{ Var }[X]
\end{aligned}$$

$\blacksquare$

At the outset of Section 3.5.1 we remarked that the mean is the "best" guess as to the outcome of a random variable. In other words, if one wishes to **predict** the outcome of a random variable X, it is best, in a sense to be given, to predict $E[X]$. In defining a criterion of bestness, it is certainly plausible to define the best prediction as the value that leads to the minimum error; however, for any predicted value c, $|X - c|$ is itself a random variable, and so error cannot be defined by $|X - c|$. As in the case of the variance, we turn to the expected-value operator and define the goodness of the prediction c in terms of the **mean-square error** $E[(X - c)^2]$. In words, the mean-square error gives the expected squared difference between the random variable and the predicted value.

Relative to the mean-square error, $E[X]$ is the best predictor of X: $E[(X - c)^2]$ is minimized if $c = E[X]$. We now demonstrate this minimization property. According to Theorem 3.11,

$$E[(X - c)^2] = E[X^2 - 2cX + c^2] = \mu_2' - 2c\mu + c^2$$

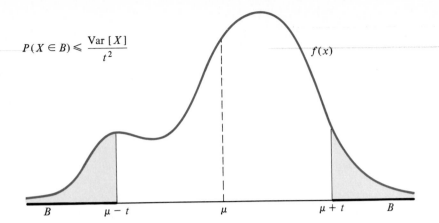

$$P(X \in B) \leq \frac{\text{Var}\,[X]}{t^2}$$

$f(x)$

FIGURE 3.24
Chebyshev's inequality

$B \qquad \mu - t \qquad\qquad \mu \qquad\qquad \mu + t \qquad B$

This expression can be minimized, as a function of c, by applying maximum-minimum theory from calculus. Taking the derivative with respect to c and setting it equal to zero yields

$$-2\mu + 2c = 0$$

Solving for c yields $c = \mu = E[X]$. The second derivative of $E[(X - c)^2]$ with respect to c is 2 and therefore $E[X]$ is a minimum. Since the mean-square error is minimized when $c = \mu$, the minimum mean-square error is Var $[X]$.

3.6.4 Chebyshev's Inequality

The variance provides a measure of the dispersion, or spread, of the distribution. It does so in terms of the absolute deviation from the mean, $|X - \mu|$, which is itself a random variable. The smaller σ^2, the more likely $|X - \mu|$ will be near 0, so that X will be near μ; conversely, the larger σ^2, the more likely $|X - \mu|$ will not be near 0, so that X will not be near μ. The next theorem gives a quantification of the manner in which σ^2 measures the absolute deviation from the mean. It is illustrated in Figure 3.24.

THEOREM 3.14

(Chebyshev's Inequality). If the random variable X possesses mean μ and variance σ^2, then, for any $t > 0$,

$$P(|X - \mu| \geq t) \leq \frac{\sigma^2}{t^2}$$

In particular, if $t = k\sigma$, $k > 0$, then

$$P(|X - \mu| \geq k\sigma) \leq \frac{1}{k^2}$$

Proof: We provide the proof in the continuous case. According to the definition of the variance,

$$\sigma^2 = E[(X - \mu)^2]$$

$$= \int_{-\infty}^{\infty} (x - \mu)^2 f(x)\, dx$$

$$= \int_{-\infty}^{\mu - t} (x - \mu)^2 f(x)\, dx + \int_{\mu - t}^{\mu + t} (x - \mu)^2 f(x)\, dx + \int_{\mu + t}^{\infty} (x - \mu)^2 f(x)\, dx$$

(see Figure 3.24). Since all of the integrals are nonnegative, omitting one yields a smaller quantity. Omitting the second integral yields

$$\sigma^2 \geq \int_{-\infty}^{\mu - t} (x - \mu)^2 f(x)\, dx + \int_{\mu + t}^{\infty} (x - \mu)^2 f(x)\, dx$$

For any x in the interval $(-\infty, \mu - t]$, $|x - \mu| \geq t$, so that $(x - \mu)^2 \geq t^2$. A like statement applies to any x in the interval $[\mu + t, \infty)$. Consequently,

$$\sigma^2 \geq \int_{-\infty}^{\mu - t} t^2 f(x)\, dx + \int_{\mu + t}^{\infty} t^2 f(x)\, dx$$

$$= t^2 \left[\int_{-\infty}^{\mu - t} f(x)\, dx + \int_{\mu + t}^{\infty} f(x)\, dx \right]$$

$$= t^2 [P(X \leq \mu - t) + P(X \geq \mu + t)]$$

$$= t^2 P(|X - \mu| \geq t)$$

Division by t^2 gives the desired inequality. ∎

The second form of the inequality, which arises from letting $t = k\sigma$, is particularly revealing. It states that the probability of X being more than k standard deviations from its mean is less than or equal to $1/k^2$. For instance, no matter what the random variable, the probability of X falling more than 3 standard deviations from the mean is less than or equal to 1/9 (see Figure 3.25).

FIGURE 3.25
Chebyshev's inequality for 3σ

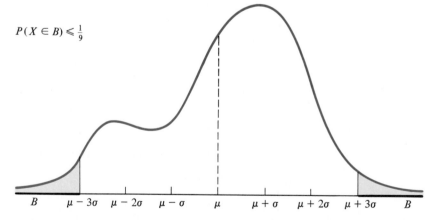

$P(X \in B) \leq \frac{1}{9}$

$B \qquad \mu - 3\sigma \quad \mu - 2\sigma \quad \mu - \sigma \qquad \mu \qquad \mu + \sigma \quad \mu + 2\sigma \quad \mu + 3\sigma \qquad B$

EXAMPLE 3.33

In producing ball bearings, the manufacturing process has shown a radius variation of $\sigma^2 = 0.09$ millimeters. This variance has remained essentially the same regardless of the diameter of the bearings being produced. Let X denote the radius. Given a collection of balls with unknown mean μ, we wish to find the probability that X will fall more than 0.7 millimeters from the mean. Moreover, the density for X is unknown, the only information available being the constancy of the variance over various processes. In the absence of the distribution, the desired probability cannot be found exactly; however, Chebyshev's inequality can be employed to arrive at an upper bound on the desired probability, namely,

$$P(|X - \mu| \geq 0.7) \leq \frac{\sigma^2}{(0.7)^2} = \frac{0.09}{0.49} = 0.18$$

Employing the relative-frequency interpretation of probability, the manufacturer might conclude that no more than 18% of the bearings produced will differ by more than 0.7 millimeters from the mean. In fact, since all that has been discovered is an upper bound on the probability, the actual percentage might be somewhat smaller.

The dilemma pointed out in Example 3.33 is inherent in the use of Chebyshev's inequality. The inequality does not take into account the actual distribution and therefore often yields rather weak conclusions. For instance, applying Theorem 2.7, the inequality can be rewritten as

$$P(|X - \mu| < t) = 1 - P(|X - \mu| \geq t) \geq 1 - (\sigma/t)^2$$

Making the substitution $t = k\sigma$ yields

$$P(|X - \mu| < k\sigma) \geq 1 - k^{-2}$$

which, for $k = 2$, states that the probability of falling within 2 standard deviations of the mean is at least 0.75 (see Figure 3.26). This is not a strong conclusion. We will see in Section 4.5 that, for the distribution of Example 3.7, the actual probability of falling within 2 standard deviations of the mean is over 0.95.

Given the weakness of the Chebyshev inequality, one might ask whether or not it can be improved. As the next example demonstrates, it cannot, unless, of course, the hypothesis is strengthened.

FIGURE 3.26
Chebyshev's inequality for 2σ

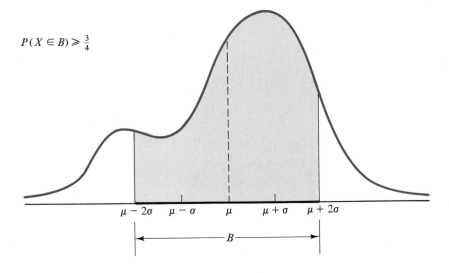

$P(X \in B) \geq \frac{3}{4}$

EXAMPLE 3.34   Consider the discrete random variable with density $f(x)$ given in the following table.

x	-2	0	2
$f(x)$	1/8	3/4	1/8

The mean is $\mu = 0$ and the variance is $\sigma^2 = 1$. Letting $k = 2$ in the standard-deviation form of Chebyshev's inequality gives

$$P(|X - 0| \geq 2) \leq 1/4$$

In fact, based upon the distribution,

$$P(|X| \geq 2) = f(-2) + f(2) = 1/4$$

so that for the present example, at least, the inequality is tight (cannot be improved).

3.7 MOMENT-GENERATING FUNCTIONS

It is common practice in engineering to employ a **transform technique** to move a problem from one environment to another. If, as posed, a problem appears difficult, it might be possible to facilitate the solution by reformulating it in another context. In probability theory, the moment-generating-function transform is useful in attacking certain classes of problems.

3.7.1   The Transform Approach

In general, the transform methodology, as depicted in Figure 3.27, can be summarized in the following manner:

1. Apply the transform $\mathfrak{T}$ to functions and functional relations to yield transformed functions and functional relations.

FIGURE 3.27
Transform methodology

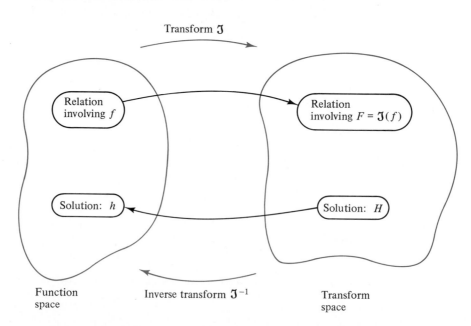

2. Solve the desired problem in the transform environment.

3. Apply inversion to obtain the desired solution.

Step 3 warrants particular attention. Let $\mathcal{T}$ be the transform under consideration, let f and g be distinct functions in the class of functions upon which the transform operates, and let $F = \mathcal{T}(f)$ and $G = \mathcal{T}(g)$ be the transforms of f and g, respectively. Suppose in applying step 2 that we find F to be the solution but that $F = G$. Then inversion cannot proceed, since it cannot be determined whether f or g gives the desired solution to the original problem. In other words, there is no inverse transform $\mathcal{T}^{-1}$. What is required is a **uniqueness** property, namely, that if $f \neq g$, then $F \neq G$, or, equivalently, if $F = G$, then $f = g$. Logically, uniqueness means that the transform is one-to-one and thus there exists a legitimate inverse, $\mathcal{T}^{-1}$. Once a uniqueness property has been established, inversion can be applied.

Regarding inversion, there are two approaches. First, if we solve the problem in the transform world to obtain the solution F and we recognize F as the transform of f, then uniqueness guarantees that f is the solution to the original problem. A second approach is to derive an **inversion formula** to represent $\mathcal{T}^{-1}$. Putting F into the formula yields the desired solution f. In this text we will depend on recognition for inversion.

3.7.2   The Moment-Generating-Function Transform

Given a random variable X and its density $f(x)$, a function $M_X(t)$ of the variable t is defined that is the transformed version of $f(x)$ (or of X, since we do not distinguish between identically distributed random variables). $M_X(t)$ is employed in two ways: (1) It provides an easy way to obtain the moments of X. (2) As a transform of X, it is used to recognize the distribution of X.

DEFINITION 3.14

Moment-Generating Function. The moment-generating function of the random variable X is the function of t defined by

$$M_X(t) = E[e^{tX}]$$

It exists for those values of t for which $E[e^{tX}]$ is finite.

If X is continuous with density $f(x)$, then, according to Theorem 3.10,

$$M_X(t) = \int_{-\infty}^{\infty} e^{tx} f(x)\, dx$$

If X is discrete with probability mass function $f(x)$, then Theorem 3.10 yields

$$M_X(t) = \sum_{f(x) > 0} e^{tx} f(x)$$

EXAMPLE 3.35

Let X be exponentially distributed. According to the results of Example 3.25,

$$E[e^{tX}] = \frac{b}{b - t}$$

Consequently, $M_X(t) = b(b - t)^{-1}$. Recall that in the evaluation of $E[e^{tX}]$ the integral was finite if and only if $t < b$, so that $M_X(t)$ is only defined for $t < b$.

EXAMPLE 3.36

Let X be uniformly distributed over the interval $[a, b]$. Then $f(x) = (b - a)^{-1}I_{[a,b]}(x)$ and

$$M_X(t) = E[e^{tX}] = \frac{1}{b - a} \int_a^b e^{tx} \, dx = \frac{e^{bt} - e^{at}}{(b - a)t}$$

EXAMPLE 3.37

In Section 3.2.1 it was pointed out that the discrete density discussed in Example 3.3 can be written in closed form as

$$f(x) = \binom{4}{x} (0.2)^x (0.8)^{4-x}$$

According to Definition 3.14, the random variable X of Example 3.3 possesses the moment-generating function

$$M_X(t) = \sum_{x=0}^{4} e^{tx} \binom{4}{x} (0.2)^x (0.8)^{4-x}$$

$$= \sum_{x=0}^{4} \binom{4}{x} [(0.2)e^t]^x (0.8)^{4-x}$$

$$= [(0.2)e^t + (0.8)]^4$$

where the last equality is simply the binomial formula for $n = 4$, $a = (0.2)e^t$, and $b = 0.8$.

The next theorem provides a basic property that is useful in the manipulation of moment-generating functions. Note the use of Theorem 3.11 in the third equality in the proof. The theorem applies because e^{bt} is a constant, relative to the expectation, and e^{atX} is a random variable.

THEOREM 3.15

For constants a and b, the moment-generating function of a random variable X satisfies the property

$$M_{aX+b}(t) = e^{bt}M_X(at)$$

Proof: By definition,

$$M_{aX+b}(t) = E[e^{t(aX+b)}] = E[e^{bt}e^{atX}] = e^{bt}E[e^{atX}] = e^{bt}M_X(at) \qquad \blacksquare$$

Two special cases relating to Theorem 3.15 are noteworthy. First, setting $b = 0$ yields

$$M_{aX}(t) = M_X(at)$$

Second, setting $a = 1$ gives

$$M_{X+b}(t) = e^{bt}M_X(t)$$

3.7.3 Uniqueness and Inversion

From the standpoint of a transform, the moment-generating-function methodology can be represented as

$$X \xrightarrow{\mathcal{M}} M_X(t)$$

where $\mathcal{M}$ represents the transform. As pointed out previously, the paramount question concerns uniqueness: If $M_X(t) = M_Y(t)$, can it be asserted that X and Y possess the same probability distribution? Fortunately, the answer is in the affirmative. Of course, we do have to pay attention to the domains over which the moment-generating functions are defined. As the next theorem states, they need only agree on some neighborhood of the origin. The theorem is stated without proof.

THEOREM 3.16
(Uniqueness)

If $M_X(t) = M_Y(t)$ for all t in some open interval containing $t = 0$, then X and Y are identically distributed.

Numerous interrelated distributions play important roles in statistical analysis. Some of the most important will be discussed in Chapter 4. With the aid of Theorem 3.16, moment-generating functions will be employed to determine relationships between distributions. The next example illustrates the methodology.

EXAMPLE 3.38

Consider the density

$$f_X(x) = \begin{cases} \dfrac{1}{\sqrt{2\pi}} x^{-1/2} e^{-x/2}, & \text{if } x > 0 \\ 0, & \text{if } x < 0 \end{cases}$$

It will be shown in Section 4.6.5 that if X possesses the density $f_X(x)$, then

$$M_X(t) = (1 - 2t)^{-1/2}$$

for $t < \frac{1}{2}$.

Now consider the density

$$f_Z(z) = \frac{1}{\sqrt{2\pi}} e^{-z^2/2}$$

for all real z, known as the **standard normal distribution.** If we let $Y = Z^2$ and apply the result of Exercise 3.33, then $f_Y(y) = 0$ for $y < 0$ and, for $y > 0$,

$$f_Y(y) = \frac{f_Z(-\sqrt{y}) + f_Z(\sqrt{y})}{2\sqrt{y}}$$
$$= \frac{(2\pi)^{-1/2} e^{-y/2} + (2\pi)^{-1/2} e^{-y/2}}{2\sqrt{y}}$$
$$= (2\pi)^{-1/2} y^{-1/2} e^{-y/2}$$

Thus, $f_Y = f_X$, so that Z^2 and X are identically distributed.

We will now apply the method of moment-generating functions to show the same result, first computing the moment-generating function of $Y = Z^2$. Applying the definition, in conjunction with Theorem 3.10, gives

$$M_Y(t) = E[e^{tZ^2}]$$
$$= \int_{-\infty}^{\infty} e^{tz^2} f_Z(z)\, dz$$

$$= \frac{1}{\sqrt{2\pi}} \int_{-\infty}^{\infty} e^{tz^2} e^{-z^2/2} dz$$

$$= \frac{1}{\sqrt{2\pi}} \int_{-\infty}^{\infty} e^{-(z^2/2)(1-2t)} dz$$

which is finite only when $t < \frac{1}{2}$. The substitution

$$x = z(1 - 2t)^{1/2}$$

yields

$$dx = (1 - 2t)^{1/2} dz$$

Moreover, the integral is still from $-\infty$ to ∞. Consequently,

$$M_X(t) = (1 - 2t)^{-1/2} \frac{1}{\sqrt{2\pi}} \int_{-\infty}^{\infty} e^{-x^2/2} dx = (1 - 2t)^{-1/2}$$

where the final equality results from the fundamental integral identity stated prior to Example 3.7 in Section 3.2.2. Thus,

$$M_{Z^2}(t) = M_X(t)$$

By uniqueness (Theorem 3.16), X and Z^2 are identically distributed, the same conclusion that was reached by examining the densities directly.

Ultimately it will be seen that the conclusion of Example 3.38 is important in its own right; however, more important at this stage is the moment-generating-function methodology for the identification of various densities. This method will be employed extensively in the future.

At this juncture, one should stop and take notice of the various ways we have for specifying a probability distribution.

1. Give the definition of the random variable as it is defined on a sample space.
2. Specify the density.
3. Specify the probability distribution function.
4. Give the moment-generating function.

No distinction is made among the four possible specifications.

3.7.4 Finding Moments Using the Moment-Generating Function

As its name implies, the moment-generating function of a distribution can be employed to find the moments of the distribution.

THEOREM 3.17

For $k = 1, 2, \ldots$, the kth moment of the random variable X is given by

$$\mu'_k = E[X^k] = M_X^{(k)}(0)$$

where $M_X^{(k)}(0)$ denotes the kth derivative of $M_X(t)$ with respect to t, evaluated at $t = 0$.

Proof: We provide the proof for the continuous case. Let X possess the density $f(x)$. Then

$$\frac{d}{dt} M_X(t) = \frac{d}{dt} E[e^{tX}]$$

$$= \frac{d}{dt} \int_{-\infty}^{\infty} e^{tx} f(x)\, dx$$

$$= \int_{-\infty}^{\infty} \frac{\partial}{\partial t} [e^{tx} f(x)]\, dx$$

$$= \int_{-\infty}^{\infty} x e^{tx} f(x)\, dx$$

where we have assumed in the third equality that the derivative can be legitimately brought inside the integral (which it can be for all distributions discussed in this text). Letting $t = 0$ gives

$$M_X^{(1)}(0) = \int_{-\infty}^{\infty} x f(x)\, dx = E[X] = \mu_1'$$

Now take the derivative with respect to t of the first derivative to obtain

$$\frac{d^2}{dt^2} M_X(t) = \frac{d}{dt} \int_{-\infty}^{\infty} x e^{tx} f(x)\, dx$$

$$= \int_{-\infty}^{\infty} \frac{\partial}{\partial t} [x e^{tx} f(x)]\, dx$$

$$= \int_{-\infty}^{\infty} x^2 e^{tx} f(x)\, dx$$

Letting $t = 0$ gives

$$M_X^{(2)}(0) = \int_{-\infty}^{\infty} x^2 f(x)\, dx = E[X^2] = \mu_2'$$

Continuing recursively, for $k = 1, 2, \ldots$,

$$M_X^{(k)}(t) = \int_{-\infty}^{\infty} x^k e^{tx} f(x)\, dx$$

and letting $t = 0$ gives the desired result for all k. ∎

EXAMPLE 3.39   As discussed in Example 3.35, the moment-generating function for the exponential density is

$$M_X(t) = \frac{b}{b - t} = b(b - t)^{-1}$$

Differentiation with respect to t yields

$$M_X'(t) = b(b - t)^{-2}$$

A second differentiation yields

$$M_X''(t) = 2b(b - t)^{-3}$$

Proceeding recursively,

$$M_X^{(k)}(t) = k!b(b - t)^{-k-1}$$

for $k = 1, 2, \ldots$. Letting $t = 0$ yields

$$\mu_k' = E[X^k] = M_X^{(k)}(0) = \frac{k!}{b^k}$$

and thus all the moments of X have been found. In particular, for $k = 1$ and $k = 2$, the first and second moments are $1/b$ and $2/b^2$, respectively, as found previously.

EXERCISES

SECTION 3.1

3.1. Two fair dice are tossed and the random variable X gives the sum of the faces. Find the codomain of X and all probabilities of the form $P(X = x)$, where x is an element of the codomain. Find the following probabilities:
 a) $P(3 < X \le 6)$
 b) $P(X \le 7)$
 c) $P(9.5 < X)$
 d) $P(-4 < X \le 4)$

3.2. A fair coin is tossed 5 times and X counts the number of heads. Find the codomain of X and all probabilities of the form $P(X = x)$, where x is an element of the codomain. Find the following probabilities:
 a) $P(X < 5)$
 b) $P(X = 2 \text{ or } X = 4)$
 c) $P(1 < X \le 3)$

3.3. Five vacuum tubes are randomly selected with replacement from a box of 100, the box containing 4 defective and 96 nondefective tubes. If X counts the number of nondefective tubes selected, what is $P(X \le 4)$?

3.4. Suppose the selection is done without replacement in Exercise 3.3. Find $P(X = 2)$.

3.5. For the random variable of Example 3.4, find
 a) $P(X > 4)$

 b) $P(1 < X \le 5)$
 c) $P(X = 3)$

3.6. Prove

$$P(a \le X \le b) = P(a < X < b) + P(X = a) + P(X = b)$$

SECTION 3.2

3.7. Draw the probability mass functions for the random variables of Exercises 3.1 and 3.2.

3.8. X, which counts the number of passengers (including the driver) in cars passing through a certain intersection, possesses the density

x	1	2	3	4	5	6	7
$f(x)$	0.52	0.27	0.11	0.05	0.02	0.02	0.01

Find $P(2 < X \le 5)$.

3.9. A box contains 20 fuses, two of which are defective. Five fuses are uniformly randomly selected from the box without replacement. Find the probability mass function of the random variable X that counts the number of defective fuses drawn.

3.10. A fair coin is tossed until a head appears. Let X count the number of tosses. Find the probability mass function for X.

3.11. Suppose it has been empirically determined that the probability mass function for the random variable counting the number of I/O errors for a particular system in a particular time frame is given by

$$f(x) = \frac{e^{-1}}{x!}$$

Find the probability of observing
a) Exactly two errors
b) Two or more errors

3.12. X measures the lifetime (in years) of a drivebelt on a generator and X possesses the continuous density

$$f(x) = \begin{cases} 3e^{-3x}, & \text{if } x \geq 0 \\ 0, & \text{if } x < 0 \end{cases}$$

What is the probability the belt will survive at least 4 years?

3.13. Find the constant c so that

$$f(x) = \begin{cases} cxe^{-x/4}, & \text{if } x \geq 0 \\ 0, & \text{if } x < 0 \end{cases}$$

is a legitimate density. Then find $P(X > 2)$.

3.14. The pH of a certain reactant is measured by the random variable X whose density is given by

$$f(x) = \begin{cases} 25(x - 3.8), & \text{if } 3.8 \leq x \leq 4 \\ -25(x - 4.2), & \text{if } 4 < x \leq 4.2 \\ 0, & \text{otherwise} \end{cases}$$

Find the probability that the pH is between 3.90 and 4.05.

3.15. The error X on a class of instrument readings measured in milligrams possesses the density

$$f(x) = \begin{cases} \dfrac{3(1 - x^2)}{4}, & \text{if } -1 \leq x \leq 1 \\ 0, & \text{otherwise} \end{cases}$$

a) What is the probability that the error is negative?

b) What is the probability that the absolute error is less than $\frac{1}{2}$?

3.16. Can a constant c be found so the function $f(x) = c/|x|$, $-\infty < x < +\infty$, is a density? Explain your answer.

SECTION 3.3

In Exercises 3.17 through 3.22, find and graph the probability distribution function for the density of the cited exercise.

3.17. Exercise 3.2.
3.18. Exercise 3.8.
3.19. Exercise 3.11.
3.20. Exercise 3.13.
3.21. Exercise 3.14.
3.22. Exercise 3.15.

3.23. The function

$$F(x) = \begin{cases} 1 - \dfrac{1}{x^4}, & \text{if } x \geq 1 \\ 0, & \text{if } x < 1 \end{cases}$$

satisfies the conditions of Theorem 3.6. Find the corresponding density.

3.24. Find the density corresponding to the probability distribution function graphed in Figure 3.28.

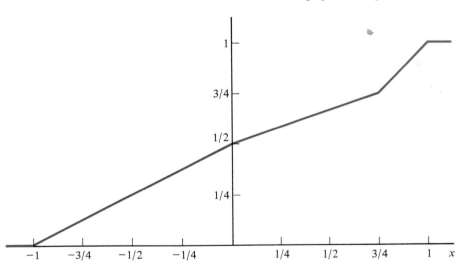

FIGURE 3.28
Probability distribution function for Exercise 3.24

3.25. Find the density corresponding to the probability distribution function graphed in Figure 3.29.

3.26. Find the probability distribution function for the **Laplace distribution,** which has two parameters and is defined by the density

$$f(x) = \frac{b}{2} e^{-b|x - a|}$$

where $b > 0$ and $-\infty < x < +\infty$.

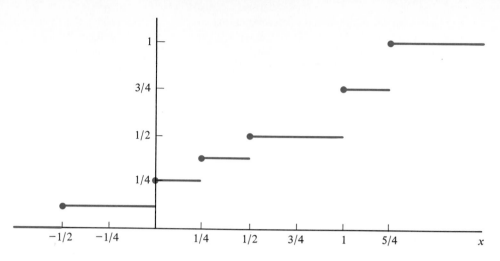

FIGURE 3.29
Probability distribution function for Exercise 3.25

$$-1/2 \qquad -1/4 \qquad \qquad 1/4 \qquad 1/2 \qquad 3/4 \qquad 1 \qquad 5/4 \qquad\qquad x$$

3.27. Referring to Exercise 3.26, suppose it has been found that the cruising speed (in knots) of a ship possesses a Laplace distribution with $a = 20$ knots and $b = \sqrt{2}$. What is the probability that a given observation will find the ship moving in excess of 22 knots?

3.28. Prove parts (iii) and (iv) of Theorem 3.5.

SECTION 3.4

3.29. Let X possess discrete density $f(x)$ and $Y = g(X) = e^X$. Express $f_Y(y)$ in terms of f. Find $f_Y(y)$ for the densities of Exercises 3.2 and 3.8.

3.30. Repeat Exercise 3.29, except let $Y = g(X) = X^3$.

3.31. Since $Y = g(X) = X^2$ is not generally one-to-one, Theorem 3.8 might not apply; however, if the codomain of X consists solely of nonnegative numbers, then the transformation is one-to-one. Recognizing this proviso, apply Theorem 3.8 to $g(X) = X^2$ and the density of Exercise 3.11.

3.32. Define the random variable X in the following manner. A fair coin is tossed twice, and X equals the number of heads, unless two tails occur, in which case $X = -1$. Find the densities of X and $Y = X^2$. Note that Theorem 3.8 does not apply in this instance.

3.33. Suppose X is a continuous random variable and $Y = X^2$. Show that

$$f_Y(y) = \begin{cases} \dfrac{f_X(-\sqrt{y}) + f_X(\sqrt{y})}{2\sqrt{y}}, & \text{if } y > 0 \\ 0, & \text{if } y \leq 0 \end{cases}$$

Apply the result to the uniform density over the interval $[-2, 3]$.

3.34. Assuming $X \geq 0$, find the density for Y in terms of the density for X when $Y = X^{1/2}$. Apply the result to the uniform density over the interval $[4, 9]$.

3.35. By iterating the results of Examples 3.14 and 3.15, find the density for $Y = a|X| + b$, where $a \neq 0$.

3.36. The sinusoidal function $y = a \sin x$ occurs often in the modeling of waveforms. Letting the angle X be

a random phenomenon, the model takes the form $Y = a \sin X$. Given that X falls between $-\pi/2$ and $\pi/2$, find the density for Y in terms of the density of X. Apply the result to the uniform density over the interval $[-\pi/2, \pi/2]$.

3.37. A binary quantization problem was addressed in Example 3.16. Now suppose the quantization involves two bits and is defined by

$$Y = \begin{cases} 3, & \text{if } X \geq 4 \\ 2, & \text{if } 2 \leq X < 4 \\ 1, & \text{if } 0 \leq X < 2 \\ 0, & \text{if } X < 0 \end{cases}$$

Express the density of Y in terms of the density for X. Apply the result to the uniform density over $[-2, 8]$.

3.38. Apply Theorem 3.9 to the transformation $Y = g(X) = \sinh X$. Apply the result to the exponential density.

3.39. Assuming $X > 0$, find the density for $Y = \log X$ in terms of the density for X. Apply the result to the exponential density.

SECTION 3.5

In Exercises 3.40 through 3.46, find the expected value of the random variable discussed in the cited exercise.

3.40. Exercise 3.2.

3.41. Exercise 3.8.

3.42. Exercise 3.11.

3.43. Exercise 3.13.

3.44. Exercise 3.14.

3.45. Exercise 3.15.

3.46. Exercise 3.23.

3.47. A coin is loaded so that the probability of a head is $2/3$. The coin is tossed five times and the random variable X is defined to be $+1$ or -1, depending on whether the number of heads exceeds the number of tails, or vice versa. Find $E[X]$.

3.48. X possesses an exponential density with parameter $b = 3$. Find
   a) $E[X^3]$
   b) $E[e^X]$
   c) $E[|X|]$

3.49. X possesses a uniform density over the interval $[-2\pi, 2\pi]$. Find
   a) $E[4X + 1]$
   b) $E[X^2]$
   c) $E[e^X]$
   d) $E[\sin X]$

3.50. For the density of Exercise 3.13, find $E[X^2 + 1]$.

3.51. Referring to the density of Exercise 3.15, find the mean absolute error, $E[|X|]$.

3.52. A measure of central tendency of a continuous distribution that is sometimes employed instead of the mean is the **median.** The median is that value $\tilde{\mu}$ for which

$$P(X \le \tilde{\mu}) = P(X \ge \tilde{\mu})$$

(see Figure 3.30). Find the medians of the following distributions:
   a) Exponential
   b) Uniform
   c) Distribution of Exercise 3.23
   d) Distribution of Exercise 3.24

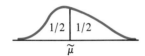

FIGURE 3.30 Illustration of the median

SECTION 3.6

In Exercises 3.53 through 3.60, find the second and third moments for the random variables in the cited exercises. Also find the variance and standard deviation.

3.53. Exercise 3.2.
3.54. Exercise 3.8.
3.55. Exercise 3.11.
3.56. Exercise 3.13.
3.57. Exercise 3.14.
3.58. Exercise 3.15.
3.59. Exercise 3.23.
3.60. Exercise 3.24.

3.61. In Exercise 2.3, a robot selects a block from a collection of blocks numbered 1 to 30. Let X be the number of the block selected and suppose the selection process is uniformly random. Show that $E[X] = 31/2$ and Var $[X] = 899/12$. Generalize the result to the case where there are n blocks by showing that $E[X] = (n + 1)/2$ and Var $[X] = (n^2 - 1)/12$.

3.62. Assuming X to be uniformly distributed on the interval $[0, 1]$, find

$$E[3X^4 + 5X^3 + 4X^2 - 8X^1 + 1]$$

3.63. A vision system must choose from among seven geometric shapes, four polygonal and three nonpolygonal. Four selections are made without replacement and without regard to order. Let X count the number of polygonal figures selected. Assume the system displays no "intelligence," so that when commanded to select four polygonal shapes, it selects them in a uniformly random manner. Find the following:
   a) A closed-form expression for the density of X.
   b) The first and second moments of X.
   c) Var $[X]$.

3.64. The **mean deviation** of a discrete random variable X is defined by

$$MD(X) = \sum_{f_X(x) > 0} |x - \mu_X| f_X(x)$$

Find the mean deviations of the random variables of Exercises 3.1 and 3.2.

3.65. For a continuous distribution, the mean deviation is defined by

$$MD(X) = \int_{-\infty}^{\infty} |x - \mu_X| f_X(x)\, dx$$

Find the mean deviation for the exponential and uniform distributions.

3.66. Show that for any random variable X possessing mean μ and variance σ^2,

$$E\left[\frac{X - \mu}{\sigma}\right] = 0$$

$$\text{Var}\left[\frac{X - \mu}{\sigma}\right] = 1$$

3.67. If X possesses nonzero mean μ and variance σ^2, then $V = \sigma/\mu$ is called the **coefficient of variation.** It provides a relative measure of dispersion and is therefore beneficial in comparing the dispersion of distributions defined in terms of disparate scales. Find the coefficient of variation for the exponential and uniform distributions.

3.68. Show

$$\mu_3 = \mu_3' - 3\mu\mu_2' + 2\mu^3$$
$$\mu_4 = \mu_4' - 4\mu\mu_3' + 6\mu^2\mu_2' - 3\mu^4$$

3.69. Referring to the three densities of Figure 3.31, it can be seen that the one in part (a) is symmetrical about its mean, the one in part (b) has a longer tail on the right, and the one in part (c) has a longer tail on the left. The distributions in parts (b) and (c) are said to be **skewed to the right** and **skewed to the left,** respectively. A commonly employed measure of skewness is the **coefficient of skewness,** which is defined by

$$\alpha_3 = \frac{\mu_3}{\sigma^3}$$

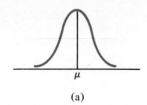

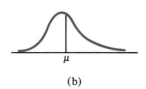

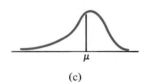

FIGURE 3.31
Illustration of skewness

(a)

(b)

(c)

FIGURE 3.32
Illustration of kurtosis

The coefficient is a dimensionless quantity that will be positive, negative, or zero, depending upon whether the distribution is skewed to the right, skewed to the left, or symmetrical, respectively. Find the coefficient of skewness for the following distributions:

a) Exponential distribution
b) Uniform distribution
c) Distribution of Exercise 3.3
d) Distribution of Exercise 3.15

3.70. Referring to the three densities of Figure 3.32, it can be seen that the one in part (a) is more peaked about its mean than is the one in part (b) and that the one in part (c) is less peaked about its mean than the one in part (b). The peakedness of a density is known as **kurtosis.** A commonly employed measure of kurtosis is the **coefficient of kurtosis:**

$$\alpha_4 = \frac{\mu_4}{\sigma^4}$$

The coefficient is dimensionless and gives a relative measure of kurtosis, the standard for comparison be-

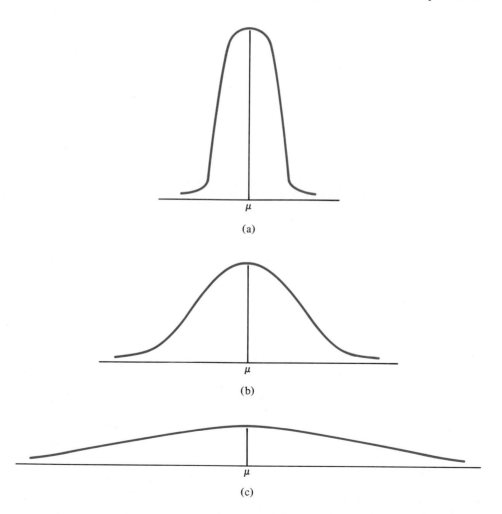

(a)

(b)

(c)

ing 3, the kurtosis measure for the standard normal curve to be discussed in Section 4.5. If α_4 is greater, less than, or equal to 3, the distribution is said to be **leptokurtic, platykurtic,** or **mesokurtic,** respectively. Find the coefficient of kurtosis for the uniform distribution over the interval [0, 1] and for the discrete random variable possessing the probability mass function

x	0	1	2	3	4	5	6
$f(x)$	$\frac{1}{16}$	$\frac{1}{16}$	$\frac{3}{16}$	$\frac{6}{16}$	$\frac{3}{16}$	$\frac{1}{16}$	$\frac{1}{16}$

3.71. For a continuous distribution, the value x_p such that

$$P(X \le x_p) = p$$

is called the $100p$th **percentile.** Whereas 50% of the probability mass lies to the left of the median, $100p\%$ of the mass lies to the left of x_p. For instance, 20% of the probability mass lies to the left of the 20th percentile. Find the 10th and 20th percentiles for the uniform distribution over the interval [0, 1] and for the exponential distribution with $b = 1$.

3.72. A measure of dispersion sometimes employed instead of the variance is the **semi-interquartile range,** defined by $(x_{0.75} - x_{0.25})/2$. If the difference of the percentiles is not divided by 2, then it is called the **interquartile range.** Find the semi-interquartile range for the uniform distribution and the distribution of Exercise 3.15.

3.73. Apply both forms of Chebyshev's inequality to the following distributions. In each case, compare the bound given by the inequality to the actual value.
a) Exponential
b) Uniform
c) Distribution of Exercise 3.2
d) Distribution of Exercise 3.14

3.74. Find the mean and variance of the Laplace distribution defined in Exercise 3.26.

3.75. The **Pareto distribution** possesses the positive parameters r and a and is defined by the density

$$f(x) = \begin{cases} \dfrac{ra^r}{x^{r+1}}, & \text{if } x \ge a \\ 0, & \text{otherwise} \end{cases}$$

Show that the Pareto distribution possesses a kth moment if and only if $k < r$. Assuming they exist, show that the mean and variance of the Pareto distribution are given by

$$E[X] = \frac{ra}{r-1}$$

and

$$\text{Var}[X] = \frac{ra^2}{(r-1)^2(r-2)}$$

SECTION 3.7

In Exercises 3.76 through 3.79, find the moment-generating function of the random variable in the cited exercise. Use the moment-generating function to find the mean and second moment.

3.76. Exercise 3.2.
3.77. Exercise 3.13.
3.78. Exercise 3.15.
3.79. Exercise 3.26.

3.80. Find the moment-generating function of $2X + 4$, when X is exponentially distributed and when it is uniformly distributed over the interval [0, 1].

3.81. Find the moment-generating function for the discrete density

$$f(x) = \binom{5}{x}(0.1)^x(0.9)^{5-x}$$

where $x = 0, 1, 2, 3, 4,$ or 5. Use the moment-generating function to find the mean and second moment.

3.82. Assuming the order of integration and summation can be interchanged, show that for a continuous random variable X,

$$M_X(t) = \sum_{k=0}^{\infty} E[X^k] \frac{t^k}{k!}$$

Assuming the order of summation can be interchanged, show the same result for a discrete random variable. *Hint:* In both instances, expand e^{tx} as a Maclaurin series.

IMPORTANT PROBABILITY DISTRIBUTIONS

The present chapter studies some important families of probability distributions. Each family is determined by a density in which one or more parameters must be specified in order to determine a particular member of the family. The distributions have been chosen because of their prominent role in applications. Consequently, one should pay attention not only to their mathematical properties, but also to the types of applications they model.

The first four sections discuss discrete distributions. From the perspectives of both application and theory, the binomial and the Poisson distributions are the most significant. The remaining sections concern continuous distributions, the most significant of these being the normal and gamma distributions.

4.1 BINOMIAL DISTRIBUTION

This section introduces the binomial distribution, a discrete distribution that is easy to use and has found wide application in many scientific areas. It is often applicable to experiments that employ repeated binary trials.

4.1.1 Bernoulli Trials and the Binomial Density

Many problems can be modeled by with-replacement selection of balls from an urn; indeed, many such instances have already appeared in the text. If an urn contains balls of only two colors, say black and white, and the probability of selecting a black ball on a given pick is p, while the probability of selecting a white ball is $q = 1 - p$, then each selection can be represented by the sample space $\{b, w\}$. If n balls are to be selected, then an appropriate sample space is $\{b, w\}^n$, which consists of all n-tuples of elements from $\{b, w\}$. Because of the with-replacement protocol, the selections are independent. Consequently, the multiplication property (Theorem 2.15) leads to a suitable probability measure on $\{b, w\}^n$ defined by

$$P((u_1, u_2, \ldots, u_n)) = p^x q^{n-x}$$

where x of the components in the n-tuple $(u_1, u_2, \ldots, u_n)$ are b and $n - x$ are w.

Suppose the random variable X counts the number of black balls selected. Then the codomain of X is

$$\Omega_X = \{0, 1, 2, \ldots, n\}$$

To determine the distribution of X, we need to find the probabilities

$$f(x) = P(X = x) = P(E_{\{x\}})$$

for $x = 0, 1, \ldots, n$, where $E_{\{x\}}$ is the event in $\{b, w\}^n$ consisting of all n-tuples having a b in x components. Since each of these n-tuples possesses probability $p^x q^{n-x}$, we need only count the number of n-tuples in $E_{\{x\}}$.

To determine the cardinality of $E_{\{x\}}$, imagine labeling each of the n components 1 through n and randomly selecting x of them to label b. Since the labeling of the x selected components does not depend upon the order in which they are selected for labeling, the theory of combinations applies and there are $C(n, x)$ ways to choose x components (out of n) to be labeled b. But these $C(n, x)$ selections precisely constitute the ways in which x of the components can contain a b while the remaining $n - x$ contain a w. Thus,

$$\text{card}(E_{\{x\}}) = C(n, x) = \binom{n}{x}$$

and the probability mass function for X is

$$f(x) = P(X = x) = P(E_{\{x\}}) = \text{card}(E_{\{x\}})p^x q^{n-x} = \binom{n}{x} p^x q^{n-x}$$

The preceding model was employed in the transistor-selection problem of Example 3.3. There, defective and nondefective transistors played the roles of the black and white balls, respectively, n was equal to 4, p was equal to 0.2, and the probability mass function was found by a tree-diagram analysis (Figure 3.2). Rather than go through the individual analysis as was done there, one could simply apply the general result just deduced to obtain the discrete density

$$f(x) = \binom{4}{x} (0.2)^x (0.8)^{4-x}$$

for the transistor-selection problem.

In general, we have the following definition, in which the notation $b(x; n, p)$ is introduced to represent the values of the **binomial density.**

DEFINITION 4.1   **Binomial Distribution.** A random variable X is said to possess a binomial distribution and to be **binomially distributed** if it has the discrete density

$$f(x) = b(x; n, p) = \binom{n}{x} p^x q^{n-x}$$

for $x = 0, 1, \ldots, n$, where $0 < p < 1$ and $q = 1 - p$.

Each member of the binomial family is determined by the specification of two parameters, n and p, as indicated by the notation $b(x; n, p)$. For instance,

$$b(2; 5, 0.3) = \binom{5}{2} (0.3)^2 (0.7)^3$$

and the density of Example 3.3 can be denoted by $b(x; 4, 0.2)$. Figure 4.1 illustrates the probability mass functions of several members of the binomial family.

The probability distribution function corresponding to the binomial density is given by

FIGURE 4.1   Various members of the binomial family

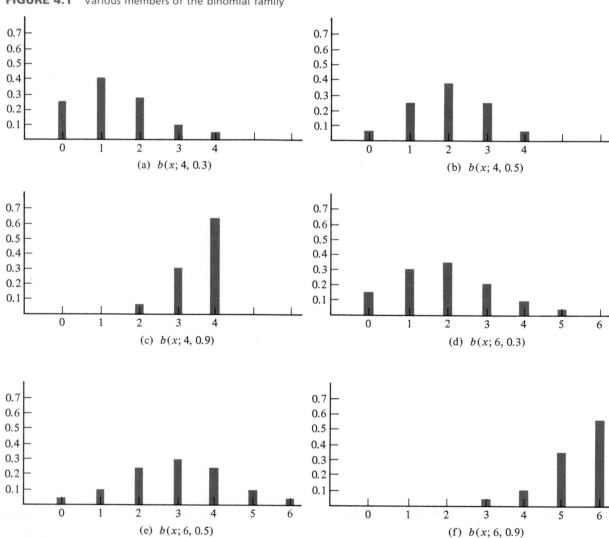

(a) $b(x; 4, 0.3)$

(b) $b(x; 4, 0.5)$

(c) $b(x; 4, 0.9)$

(d) $b(x; 6, 0.3)$

(e) $b(x; 6, 0.5)$

(f) $b(x; 6, 0.9)$

$$F_X(x) = P(X \le x) = B(x; n, p) = \sum_{k=0}^{x} b(k; n, p)$$

Table A.2 in the appendix gives values for $B(x; n, p)$.

For applications involving the binomial model, it is useful to generalize the urn-model analysis that led to the density. Leaving the particulars of the urn aside, a salient point is the repetitive nature of the **trials** that constitute the overall experiment. Each n-tuple results from n independent trials, each trial possessing a sample space of two outcomes. Abstracting from the specifics results in a definition that characterizes a class of experiments leading to the binomial distribution.

DEFINITION 4.2

Bernoulli Trials. Suppose an experiment consists of n trials, $n > 0$. The trials are called Bernoulli trials if three conditions are satisfied:

1. Each trial is defined by a sample space $\{S, F\}$ consisting of two outcomes, S to be called **success** and F to be called **failure.**
2. There exists a number p, $0 < p < 1$, such that for each trial, $P(S) = p$ and $P(F) = q$, where $q = 1 - p$.
3. The trials are independent.

The urn-model analysis that resulted in the binomial density applies directly to a sequence of n Bernoulli trials, the sample space being $\{S, F\}^n$. The key, besides two possible outcomes on each trial, is the independence of the trials. It is not sufficient that the probability of a success be the same on each trial. The next example makes this point.

EXAMPLE 4.1

Consider an urn with two black and two white balls and suppose two picks are made without replacement. The experiment consists of two trials, each trial having two possible outcomes. Thus, a suitable sample space is

$$\{b, w\}^2 = \{(b, b), (b, w), (w, b), (w, w)\}$$

Let the events B_1 and B_2 denote the selections of a black ball on the first and second picks, respectively. Clearly, $P(B_1) = 1/2$. Applying the multiplication property (and referring to Figure 4.2) yields

$$P(B_2) = \frac{1}{2} \times \frac{1}{3} + \frac{1}{2} \times \frac{2}{3} = \frac{1}{2}$$

so that the first two requirements for Bernoulli trials are satisfied. However, the trials are certainly not independent.

When using the Bernoulli-trial model in conjunction with the binomial distribution, it is customary to let the random variable count the number of successes. The choice of which experimental outcome to call a success and which to call a failure is purely arbitrary.

EXAMPLE 4.2

A multiprocessing system utilizes 12 processors and is configured so that it can satisfy all functions, with perhaps some loss of speed, so long as nine processors are functioning. If the probability of each processor functioning for the duration of the project is 0.8 and processor failures are independent, what is the probability that the system will remain up until completion

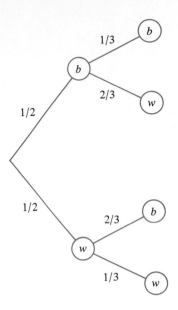

FIGURE 4.2
Tree diagram
corresponding to
Example 4.1

of the project? The functioning of each processor over the duration can be viewed as a trial with the probability of a success being $P(S) = 0.8$. If X counts the number of processors that last the duration, then the desired probability is

$$P(X \geq 9) = 1 - P(X \leq 8) = 1 - F_X(8)$$
$$= 1 - B(8; 12, 0.8) = 1 - 0.2054 = 0.7946$$

where the value of $B(8; 12, 0.8)$ has come from Table A.2.

EXAMPLE 4.3

Suppose 20% of all cars approaching a certain intersection make a left turn. If 20 cars are observed, what is the probability that between 5 and 11, inclusive, will turn left? Under the assumption that the turns are independent, let X count the number of cars turning left. The trials are Bernoulli and, applying Theorem 3.5, the desired probability is given by

$$P(5 \leq X \leq 11) = \sum_{x=5}^{11} b(x; 20, 0.2)$$
$$= B(11; 20, 0.2) - B(5; 20, 0.2) + b(5; 20, 0.2)$$
$$= B(11; 20, 0.2) - B(4; 20, 0.2)$$
$$= 0.9999 - 0.6296 = 0.3703$$

EXAMPLE 4.4

It has long been observed that small particles suspended in a fluid display erratic movements resulting from the impacts of the fluid molecules upon the particles. The motion of such particles is called **Brownian motion.** A rudimentary model for the description of Brownian motion is the **random walk.** In this example we describe the one-dimensional random walk, in which a particle, starting at the origin, moves a unit length to the right or to the left. The movements are taken independently and, for each movement, the probabilities of going right and left are p and $q = 1 - p$, respectively. Letting Y denote the number of units the particle has moved to the right after n movements, we desire the probability mass function for Y.

Before considering the general case, let us examine the particular case where $n = 6$. For $n = 6$, the codomain of Y is

$$\Omega_Y = \{-6, -4, -2, 0, 2, 4, 6\}$$

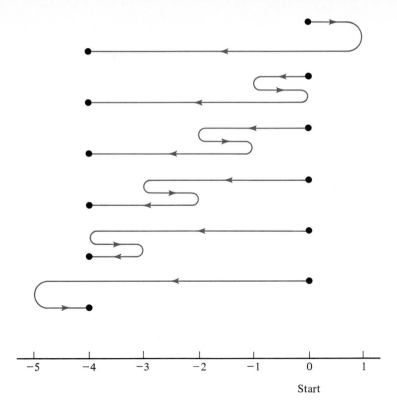

FIGURE 4.3
Illustration of random walk

-5   -4   -3   -2   -1   0   1

Start

For instance, $Y = -4$ if there are 5 movements left and 1 movement right. Letting S and F denote movements to the right and left, respectively, and letting X denote the binomial random variable with parameters $n = 6$ and p,

$$P(Y = -4) = P(X = 1) = b(1; 6, p) = \binom{6}{1} p q^5$$

(Figure 4.3 illustrates the 6 random walks that lead to $Y = -4$.) The probability that the particle is 2 units to the right after 6 movements is

$$P(Y = 2) = P(X = 4) = b(4; 6, p) = \binom{6}{4} p^4 q^2$$

More generally, for $n = 6$, the probability mass function for Y is

$$f_Y(y) = P(Y = y) = P\left(X = \frac{6 + y}{2}\right) = b\left(\frac{6 + y}{2}; 6, p\right)$$

For an arbitrary number of movements, say n, if Y counts the number of units the particle moves to the right and X is the binomial random variable for n Bernoulli trials, then

$$\Omega_Y = \{-n, -n + 2, \ldots, n - 2, n\}$$

and

$$f_Y(y) = P(Y = y)$$

$$= P\left(X = \frac{n + y}{2}\right)$$

$$= b\left(\frac{n + y}{2}; n, p\right)$$

$$= \binom{n}{\frac{n + y}{2}} p^{(n+y)/2} q^{(n-y)/2}$$

For instance, if $p = 0.6$, after 9 movements the probability of the particle being 3 units to the right is

$$f_Y(3) = b(6; 9, 0.6) = \binom{9}{6} (0.6)^6 (0.4)^3 = 0.2508$$

4.1.2 Properties of the Binomial Distribution

Our purpose here is to find the moment-generating function, mean, and variance of a binomial distribution.

THEOREM 4.1

If X is binomially distributed with discrete density $f(x) = b(x; n, p)$ and $q = 1 - p$, then the moment-generating function, mean, and variance of X are respectively given by

(i) $M_X(t) = (pe^t + q)^n$
(ii) $\mu = np$
(iii) $\sigma^2 = npq$

Proof: We first compute the moment-generating function and then apply Theorems 3.17 and 3.12 to find the mean and variance. First, by Theorem 3.10,

$$M_X(t) = E[e^{tX}]$$

$$= \sum_{x=0}^{n} e^{tx} \binom{n}{x} p^x q^{n-x}$$

$$= \sum_{x=0}^{n} \binom{n}{x} (pe^t)^x q^{n-x}$$

$$= (pe^t + q)^n$$

the last equality simply being the binomial expansion for $(a + b)^n$ applied to $a = pe^t$ and $b = q$.

The first moment, second moment, and variance are found by employing the first and second derivatives of the moment-generating function in conjunction with the fact that $p + q = 1$. Proceeding,

$$M_X'(t) = npe^t(pe^t + q)^{n-1}$$

$$M_X''(t) = np[e^t(pe^t + q)^{n-1} + e^t(n - 1)pe^t(pe^t + q)^{n-2}]$$

$$\mu = E[X] = M_X'(0) = np(p + q)^{n-1} = np$$

$$\mu_2' = E[X^2] = M_X''(0)$$
$$= np[(p + q)^{n-1} + (n - 1)p(p + q)^{n-2}]$$
$$= np[1 + (n - 1)p]$$
$$\sigma^2 = \mu_2' - \mu^2 = np + n(n - 1)p^2 - n^2p^2 = np(1 - p) = npq \qquad \blacksquare$$

Together with the defining density, the mean and variance of the binomial distribution are listed in Table 4.1.

EXAMPLE 4.5

A robot eye rotates at a constant rate and its field of vision is $45°$ (see Figure 4.4). Suppose 50 like objects, recognizable to the robot, are randomly and independently placed momentarily in the area surveyed by the eye. If X counts the number of objects seen by the robot, then X can be modeled as a binomial distribution with $p = 1/8$. The expected value of X is

$$E[X] = np = 50 \times 0.125 = 6.25$$

and the variance is

$$\sigma^2 = npq = 50 \times 0.125 \times 0.875 = 5.47$$

so that the standard deviation is $\sigma = 2.34$.

The probability that the number of observed objects will be within 2 standard deviations of the mean is

$$P(|X - \mu| \le 2\sigma) = P(6.25 - 4.68 \le X \le 6.25 + 4.68)$$
$$= P(1.57 \le X \le 10.93)$$
$$= \sum_{x=2}^{10} b(x; 50, 0.125)$$

TABLE 4.1 IMPORTANT DISCRETE DISTRIBUTIONS

Distribution	Density	Mean	Variance
Binomial	$\binom{n}{x} p^x q^{n-x}$	np	npq
Hypergeometric	$\dfrac{\binom{k}{x}\binom{N-k}{n-x}}{\binom{N}{n}}$	$\dfrac{nk}{N}$	$\dfrac{nk(N-k)(N-n)}{N^2(N-1)}$
Negative binomial	$\binom{x + k - 1}{k - 1} p^k q^x$	$\dfrac{kq}{p}$	$\dfrac{kq}{p^2}$
Geometric	pq^x	$\dfrac{q}{p}$	$\dfrac{q}{p^2}$
Poisson	$\dfrac{e^{-\lambda}\lambda^x}{x!}$	λ	λ

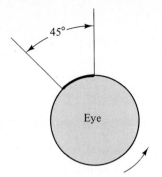

FIGURE 4.4
Robot eye

$$= 0.0315 + 0.0720 + 0.1209 + 0.1589 + 0.1702$$
$$+ 0.1528 + 0.1174 + 0.0782 + 0.0458$$
$$= 0.9477$$

Note how much stronger a result has been achieved using a known density than is achieved by means of Chebyshev's inequality, which provides a lower bound of 0.75 for the probability of being within 2 standard deviations of the mean (see Figure 3.26).

4.2 HYPERGEOMETRIC DISTRIBUTION

The present section considers a model related to unordered nonreplacement sampling. As will be illustrated in the examples, it is useful in examining the acceptability of a large batch of items based upon the testing of a small sample.

4.2.1 The Hypergeometric Density

Like the binomial distribution, the hypergeometric distribution arises from the consideration of an urn model involving the selection of balls. Consider an urn with N balls, k of which are black (b) and $N - k$ of which are white (w). Of the N balls, n are selected without replacement and without regard to order. The random variable X counts the number of black balls selected. Obviously, if x black balls are selected, then $n - x$ white balls are selected. Note, however, that there are two constraints: $x \leq k$ and $n - x \leq N - k$. In words, no matter what the number of balls selected, there cannot be more black balls selected than there are black balls in the urn or more white balls selected than there are white balls in the urn. Given that the constraints on x and $n - x$ are satisfied, we desire the probability $P(X = x)$.

Assuming the selection process is uniformly random, so that the hypothesis of equal probability applies, if S is the sample space of all possible selections of n balls and $E_{\{x\}}$ is the event consisting of those selections for which there are x black balls and $n - x$ white balls, then

$$P(X = x) = P(E_{\{x\}}) = \frac{\text{card } (E_{\{x\}})}{\text{card } (S)}$$

According to the theory of combinations (Section 2.3.4), the x black balls can be selected in $C(k, x)$ ways and the $n - x$ white balls can be selected in $C(N - k, n - x)$ ways. Applying the fundamental principle of counting (Section 2.3.3),

$$\text{card } (E_{\{x\}}) = C(k, x)C(N - k, n - x) = \binom{k}{x}\binom{N - k}{n - x}$$

Moreover, the theory of combinations shows there are $C(N, n)$ elements in the sample space. Thus,

$$P(X = x) = \frac{\binom{k}{x}\binom{N - k}{n - x}}{\binom{N}{n}}$$

As a matter of reference, note that, except for the sample space, this exact problem was analyzed in Example 2.15, where there was a collection of 26 "balls," 14 of whom were electrical engineers and 12 of whom were computer scientists, and where 5 "balls" were selected, 3 of whom were electrical engineers and 2 of whom were computer scientists. As was deduced there, the number of committees composed of 3 electrical engineers and 2 computer scientists is 24,024, and the probability of selecting such a committee by making uniformly random selections is

$$P(X = 3) = \frac{\binom{14}{3}\binom{12}{2}}{\binom{26}{5}} = \frac{24{,}024}{65{,}780} = 0.365$$

where X counts the number of electrical engineers selected.

The formula just derived for $P(X = x)$ is the **hypergeometric density.** As with the binomial distribution, the terminology **success** and **failure** is commonly employed when the hypergeometric model is employed, so that X counts the number of successes.

DEFINITION 4.3

Hypergeometric Distribution. A random variable X is said to possess a hypergeometric distribution if it has the density

$$f(x) = h(x; N, n, k) = \frac{\binom{k}{x}\binom{N - k}{n - x}}{\binom{N}{n}}$$

for $x = 0, 1, \ldots, n$, $x \leq k$, and $n - x \leq N - k$, where N, n, and k are positive integers such that $k \leq N$ and $n \leq N$.

According to Definition 4.3, particular members of the hypergeometric family are determined by the specification of three parameters, N, n, and k. The probability distribution function for a hypergeometrically distributed random variable X is

$$F_X(x) = P(X \leq x) = H(x; N, n, k) = \sum_{i=0}^{x} h(i; N, n, k)$$

EXAMPLE 4.6

For $N = 15$, $n = 5$, and $k = 12$, find the values of the hypergeometric distribution. According to the definition,

$$f(x) = \frac{\binom{12}{x}\binom{3}{5-x}}{\binom{15}{5}} = \frac{\binom{12}{x}\binom{3}{5-x}}{3003}$$

The constraint $n - x \leq N - k$ yields $5 - x \leq 3$ so that the codomain is

$$\Omega_X = \{2, 3, 4, 5\}$$

The probabilities are

$$f(2) = \frac{\binom{12}{2}\binom{3}{3}}{3003} = \frac{66}{3003} = 0.0220$$

$$f(3) = \frac{\binom{12}{3}\binom{3}{2}}{3003} = \frac{660}{3003} = 0.2198$$

$$f(4) = \frac{\binom{12}{4}\binom{3}{1}}{3003} = \frac{1485}{3003} = 0.4945$$

$$f(5) = \frac{\binom{12}{5}\binom{3}{0}}{3003} = \frac{792}{3003} = 0.2637$$

Figure 4.5 gives the graph of $f(x)$.

EXAMPLE 4.7

A shipment of 60 highly sensitive accelerometers is to be accepted or rejected based on the testing of 5 chosen randomly from the lot. The shipment will be rejected if more than 1 of the 5 fail. What is the probability the shipment will be accepted if it is known that 10% of the shipment does not meet the specifications? Let X denote the number of acceptable units

FIGURE 4.5
Hypergeometric density for Example 4.6

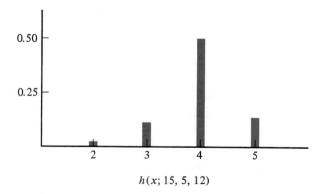

$h(x; 15, 5, 12)$

in the 5 selected for testing. Then X is a hypergeometric random variable with probability mass function

$$f(x) = h(x; 60, 5, 54) = \frac{\binom{54}{x}\binom{6}{5-x}}{\binom{60}{5}}$$

The probability that the shipment will be accepted is

$$f(4) + f(5) = 0.3474 + 0.5791 = 0.9265$$

EXAMPLE 4.8

A purchaser buys electrical relays in large lots. From each lot 15 units are selected, and the lot is accepted if there are fewer than 3 defective relays. Given the manufacturer knows the production process yields acceptable units at the rate of 90%, what is the probability that a lot will be rejected? At first glance, it appears the problem is not well posed. It seems our approach should be to let X count the number of nondefective units, so that the probability the lot will be rejected is $P(X \le 12)$. Assuming a nonreplacement methodology, X would then be hypergeometric with density

$$f(x) = h\left(x; N, 15, \frac{9N}{10}\right)$$

where N is the unknown lot size.

Although we cannot use the hypergeometric distribution, the problem can still be solved if a different model is employed. First of all, what is meant by the manufacturer's claim that 90% of the relays are nondefective? Does the claim really mean that there are 90% nondefective relays in each and every lot, or does it mean that, given an arbitrary relay, its probability of being nondefective is 0.9? If the latter interpretation is applied, then the selection of the 15 test units from the lot is an experiment consisting of 15 Bernoulli trials. Thus, the solution to the problem involves the binomial distribution and is given by

$$P(X \le 12) = F_X(12) = B(12; 15, 0.9) = 0.1814$$

On the other hand, suppose we do assume the 90% rating means there are actually 90% nondefective units in each lot. Then, under the condition that the lots are "large," the binomial distribution is still a satisfactory model, the reason being that the initial probabilities of 0.9 for nondefective and 0.1 for defective remain essentially unchanged from without-replacement pick to without-replacement pick and the selections can be modeled as being independent, even though they are not strictly so. Indeed, the point of modeling is not necessarily to select a mathematically exact model, but to choose one that works. For instance, in the present case, if there are 1000 items in a lot, then the binomial model yields the probability of selecting 10 nondefective relays to be

$$P(X = 10) = b(10; 15, 0.9) = \binom{15}{10}(0.9)^{10}(0.1)^5 = 0.01047$$

whereas the hypergeometric model yields

$$P(X = 10) = \frac{\binom{900}{10}\binom{100}{5}}{\binom{1000}{15}} = 0.01000$$

The difference, 0.00047, is negligible.

4.2.2 Properties of the Hypergeometric Distribution

Example 4.8 demonstrated that if n is small in comparison with N, then the binomial distribution can be used to approximate the hypergeometric distribution. This relationship can be seen in the mean and variance of the hypergeometric distribution. These are given in the next theorem, whose proof, which is an exercise in the manipulation of sums and factorials, is omitted.

THEOREM 4.2

If X possesses a hypergeometric distribution with discrete density $f(x) = h(x; N, n, k)$, then its mean and variance are respectively given by

(i) $\mu = \dfrac{nk}{N}$

(ii) $\sigma^2 = \dfrac{nk(N - k)(N - n)}{N^2(N - 1)}$

EXAMPLE 4.9

For the distribution of Example 4.6, which possesses the density $h(x; 15, 5, 12)$ and is graphed in Figure 4.5, the mean and variance are given by

$$\mu = \frac{5 \times 12}{15} = 4$$

$$\sigma^2 = \frac{5 \times 12 \times 3 \times 10}{15^2 \times 14} = 0.5714$$

Suppose we let $p = k/N$, so that the hypergeometric distribution takes the form

$$h(x; N, n, pN)$$

Then the mean is given by

$$\mu = \frac{nk}{N} = \frac{npN}{N} = np$$

which agrees with the mean of the binomial distribution. Indeed, if the binomial distribution is employed as an approximation to the hypergeometric distribution, then p is chosen to be the ratio k/N of the number of successes to the total number of objects from which the selections are to be made.

Regarding the variance, letting $p = k/N$ and $q = 1 - p$ yields

$$\sigma^2 = \frac{npN(N - pN)(N - n)}{N^2(N - 1)} = \frac{npq(N - n)}{N - 1}$$

Holding n fixed and letting $N \longrightarrow \infty$ yields npq, the variance of the binomial distribution. Consequently, the variance of the hypergeometric distribution is approximated by the variance of the binomial distribution if N is substantially greater than n.

EXAMPLE 4.10

A cosmetic must satisfy numerous requirements, including acceptable color, odor, and moisture content. If a manufacturer claims that 95% of the jars in a particular large batch are satisfactory and if 200 jars from the batch are tested, what is the expected number of satisfactory jars?

Letting X denote the number of satisfactory jars and assuming X to be binomial,

$$\mu = np = 200 \times 0.95 = 190$$

The variance of X is

$$\sigma^2 = npq = 200 \times 0.95 \times 0.05 = 9.5$$

Suppose the batch size were known to be 1500 jars, so that a hypergeometric model could be employed with $N = 1500$, $n = 200$, and $k = 0.95 \times 1500 = 1425$. Then, according to Theorem 4.2,

$$\mu = \frac{nk}{N} = \frac{200 \times 1425}{1500} = 190$$

which agrees with the binomial model, as it must, and

$$\sigma^2 = \frac{nk(N - k)(N - n)}{N^2(N - 1)} = \frac{200 \times 1425 \times 75 \times 1300}{2{,}250{,}000 \times 1499} = 8.2388$$

which is fairly close to the variance obtained using the binomial model.

4.3 NEGATIVE BINOMIAL DISTRIBUTION

In Example 3.6 we considered a random variable X that counted the number of jobs serviced from queue A prior to the servicing of the job at the head of queue B. The problem can be generalized so as to count the number of jobs serviced from queue A prior to the servicing of k jobs from queue B. The negative binomial distribution models this generalization.

4.3.1 The Negative Binomial Density

Consider a sequence of Bernoulli trials in which the probability of a success (S) on a given trial is p and the probability of a failure (F) on a given trial is $q = 1 - p$. The trials will be run until k successes have occurred and X will count the failures prior to the kth success. For any $x = 0, 1, 2, \ldots$, $E_{\{x\}}$ consists of all trial sequences of length $k + x$ for which there are x failures, there are k successes, and the $(k + x)$th trial is a success. In terms of product sets, $E_{\{x\}}$ consists of all $(k + x)$-tuples having $k - 1$ S's and x F's in the first $k + x - 1$ components and an S in the last component. For instance, if $k = 4$ and $x = 2$, then (F, S, F, S, S, S) and (S, S, S, F, F, S) are elements of $E_{\{2\}}$.

To find $P(X = x) = P(E_{\{x\}})$, let A denote the event that there are $k - 1$ successes on the first $x + k - 1$ trials and B the event that an S occurs on the $(x + k)$th trial. Then, by independence,

$$
\begin{aligned}
P(X = x) &= P(A \cap B) \\
&= P(A)P(B) \\
&= p \cdot b(k - 1; x + k - 1, p) \\
&= p \binom{x + k - 1}{k - 1} p^{k-1}(1 - p)^x \\
&= \binom{x + k - 1}{k - 1} p^k(1 - p)^x
\end{aligned}
$$

Negative Binomial Distribution. The random variable X is said to possess a negative binomial distribution if it has the discrete density

$$f(x) = b\hat{}(x; k, p) = \binom{x + k - 1}{k - 1} p^k q^x$$

for $x = 0, 1, 2, \ldots$, where k is a positive integer, $0 < p < 1$, $p + q = 1$, and $b\hat{}$ denotes the negative binomial density.

The "negative binomial" terminology derives from the fact that the probabilities $b\hat{}(x; k, p)$ correspond to the terms in the expansion $p^k(1 - q)^{-k}$.

For the special case, $k = 1$, the negative binomial density takes the form

$$f(x) = b\hat{}(x; 1, p) = pq^x$$

where $q = 1 - p$. The corresponding random variable is said to possess a **geometric distribution**. Referring to Example 3.6 (in the nonpriority case), this was the result obtained, with only the roles of p and q being reversed.

EXAMPLE 4.11

In an assembly process, the finished items are inspected by a vision sensor, the image data is processed, and a determination is made by computer as to whether or not a unit is satisfactory. If it is assumed that 2% of the units will be rejected, then what is the probability that the thirtieth unit observed will be the second rejected unit? If X denotes the number of satisfactory units observed prior to the observation of the second unsatisfactory unit, then X can be modeled as a negative binomial distribution with $k = 2$, $p = 0.02$, and $q = 0.98$. Since X counts failures in the negative binomial model, in the present context a failure is a satisfactory unit. The desired probability is

$$P(X = 28) = b\hat{}(28; 2, 0.02) = \binom{29}{1} (0.02)^2(0.98)^{28} = 0.0066$$

The probability that at least 3 satisfactory units are observed prior to the observation of 2 unsatisfactory units is

$$
\begin{aligned}
P(X \geq 3) &= 1 - P(X \leq 2) \\
&= 1 - P(X = 0) - P(X = 1) - P(X = 2) \\
&= 1 - b\hat{}(0; 2, 0.02) - b\hat{}(1; 2, 0.02) - b\hat{}(2; 2, 0.02) \\
&= 1 - \binom{1}{1} (0.02)^2 + \binom{2}{1} (0.02)^2(0.98) + \binom{3}{1} (0.02)^2(0.98)^2 \\
&= 1 - 0.0004 - 0.0008 - 0.0012 = 0.9976
\end{aligned}
$$

4.3.2 Properties of the Negative Binomial Distribution

The next theorem gives the moment-generating function, mean, and variance of the negative binomial distribution. The proof makes use of the negative binomial expansion.

THEOREM 4.3

If X possesses a negative binomial distribution, with density $f(x) = b\hat{}(x; k, p)$, then its moment-generating function, mean, and variance are given by

(i) $M_X(t) = p^k(1 - qe^t)^{-k}$

(ii) $\mu = \dfrac{kq}{p}$

(iii) $\sigma^2 = \dfrac{kq}{p^2}$

where $q = 1 - p$.

Proof: The moment-generating function is

$$M_X(t) = E[e^{tX}]$$

$$= \sum_{x=0}^{\infty} e^{tx} \binom{k + x - 1}{k - 1} p^k q^x$$

$$= p^k \sum_{x=0}^{\infty} \binom{k + x - 1}{k - 1} (qe^t)^x$$

Since the sum is the binomial expansion of $(1 - qe^t)^{-k}$,

$$M_X(t) = p^k(1 - qe^t)^{-k}$$

The mean and variance are deduced from the moment-generating function:

$$M_X'(t) = p^k qke^t(1 - qe^t)^{-k-1}$$

$$M_X''(t) = p^k qk[e^t(1 - qe^t)^{-k-1} + q(k + 1)e^{2t}(1 - qe^t)^{-k-2}]$$

$$E[X] = M_X'(0) = p^k qk(1 - q)^{-k-1} = qkp^{-1}$$

$$E[X^2] = M_X''(0) = p^k qk[(1 - q)^{-k-1} + q(k + 1)(1 - q)^{-k-2}]$$

$$= qk[p^{-1} + q(k + 1)p^{-2}]$$

$$\sigma^2 = E[X^2] - E[X]^2 = qk[p^{-1} + q(k + 1)p^{-2}] - q^2k^2p^{-2} = qkp^{-2} \quad \blacksquare$$

EXAMPLE 4.12

Referring to Example 4.11, where $p = 0.02$, $q = 0.98$, and $k = 2$, the expected number of satisfactory units prior to the observation of two unsatisfactory units is

$$E[X] = \frac{2 \times 0.98}{0.02} = 98$$

The variance is

$$\sigma^2 = \frac{2 \times 0.98}{0.0004} = 4900$$

so that the standard deviation is $\sigma = 70$.

For a geometric distribution, $k = 1$, and therefore Theorem 4.3 takes the form:

$$\boxed{\begin{aligned} &\text{(i)} \ \ M_X(t) = p(1 - qe^t)^{-1} \\[4pt] &\text{(ii)} \ \ \mu = \frac{q}{p} = \frac{1 - p}{p} \\[4pt] &\text{(iii)} \ \ \sigma^2 = \frac{q}{p^2} = \frac{1 - p}{p^2} \end{aligned}}$$

EXAMPLE 4.13

A binary switching device must be replaced upon a single failure. The assumption is made that its probability of failure on a given usage does not depend on its previous functioning. Moreover, its probability of properly functioning on any given usage is 0.995. If X denotes the number of times the switch functions properly prior to replacement, then X can be modeled as a geometric distribution with $p = 0.005$. The expected number of proper functionings is

$$E[X] = \frac{q}{p} = \frac{0.995}{0.005} = 199$$

and the variance is

$$\sigma^2 = \frac{0.995}{(0.005)^2} = 39{,}800$$

Note that in the parlance of the negative binomial distribution, a failure of the switch is called a "success" and a proper functioning of the switch is called a "failure."

4.4 POISSON DISTRIBUTION

The Poisson distribution is one of the most important probability distributions in engineering and computer science. Whereas the binomial, hypergeometric, and negative binomial distributions can be arrived at by means of selection protocols involving finite sample spaces, it is significantly more difficult to rigorously demonstrate a mathematical model yielding the Poisson distribution. Nevertheless, it is possible to characterize certain types of phenomena to which the Poisson distribution is applicable.

4.4.1 Poisson Experiments and the Poisson Density

Consider an experiment in which an observer is counting the number of happenings in a time interval. These might be the number of telephone calls received at a station, the number of radioactive particles counted by a geiger counter, the number of jobs arriving for service at a CPU, or the number of units being rejected at a quality control checking station. Suppose, moreover, that the following three conditions are satisfied:

1. If $I_1, I_2, \ldots, I_n$ are nonoverlapping intervals, then the numbers of happenings in the intervals are independent.
2. There exists a constant q such that the probability of exactly 1 happening occurring during an infinitesimal interval of length dt is approximately equal to $q \cdot dt$.
3. The probability that 2 or more happenings will occur during an infinitesimal interval is approximately 0.

If the three conditions are satisfied, then the experiment is called a **Poisson experiment.** For such an experiment, if X counts the number of happenings during any given interval, then it can be shown that X possesses a Poisson distribution.

Poisson Distribution. A discrete random variable X is said to possess a Poisson distribution if it has the discrete density

$$f(x) = p(x; \lambda) = \frac{e^{-\lambda}\lambda^x}{x!}$$

for $x = 0, 1, 2, \ldots$, where λ is a positive constant.

Figure 4.6 gives the probability mass functions for various choices of λ.

The three Poisson conditions require that the happenings are independent and occur at a constant rate. If the happenings occur in groups or if the rate varies, then the Poisson distribution is not appropriate. Although we will not pursue the matter, the infinitesimal constant-rate condition can be made mathematically rigorous by using the theory of differentials.

If the three Poisson conditions do hold and if X counts the number of happenings during some specific time interval of duration t, then X is Poisson distributed with $\lambda = qt$. Thus, for $x = 0, 1, 2, \ldots$,

$$P(X = x) = p(x; qt) = \frac{e^{-qt}(qt)^x}{x!}$$

Before proceeding, it is necessary to demonstrate that the Poisson density is a legitimate density, namely, that the sum of the probabilities is 1. Because the binomial, hypergeometric, and negative binomial densities were each derived directly from a probability space, such a check was not necessary. However, in the present case we are relying on Theorem 3.3 to guarantee that $p(x; \lambda)$ is a legitimate discrete density. The first two conditions of that theorem are clearly satisified. Moreover,

$$\sum_{x=0}^{\infty} p(x; \lambda) = \sum_{x=0}^{\infty} \frac{e^{-\lambda}\lambda^x}{x!} = e^{-\lambda} \sum_{x=0}^{\infty} \frac{\lambda^x}{x!}$$

which equals 1, since the sum is precisely the Taylor expansion for e^λ.

The probability distribution function for the Poisson distribution is

$$F_X(x) = P(X \leq x) = \Pi(x; \lambda) = \sum_{k=0}^{x} p(x; \lambda) = \sum_{k=0}^{x} \frac{e^{-\lambda}\lambda^k}{k!}$$

Values of $\Pi(x; \lambda)$ are tabulated in Table A.3.

In computing values of the Poisson density, the recursion formula

$$p(x + 1; \lambda) = \frac{\lambda p(x; \lambda)}{x + 1}$$

can be helpful. The validity of the recursion is shown by

$$p(x + 1; \lambda) = \frac{e^{-\lambda}\lambda^{x+1}}{(x + 1)!} = \frac{\lambda e^{-\lambda}\lambda^x}{(x + 1)x!} = \frac{\lambda}{x + 1} p(x; \lambda)$$

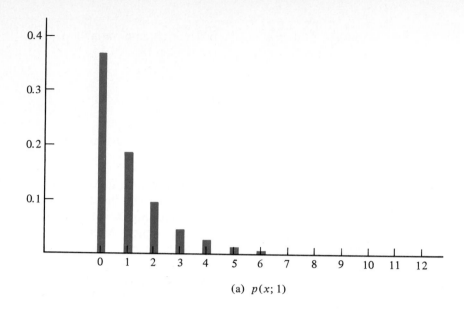

(a) $p(x; 1)$

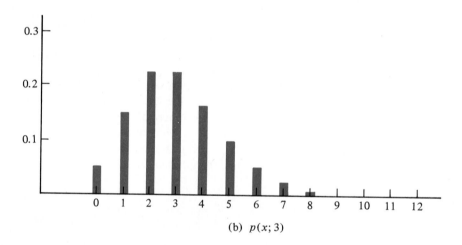

(b) $p(x; 3)$

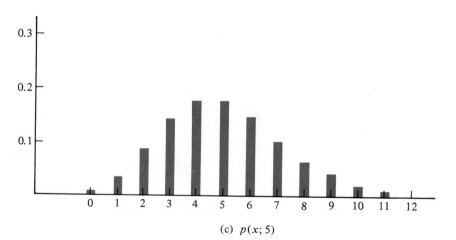

(c) $p(x; 5)$

FIGURE 4.6
Various members of the
Poisson family

4.4.2 Properties of the Poisson Distribution

The moment-generating function, mean, and variance for the Poisson distribution are computed in the next theorem. Note that the mean is equal to the variance.

If X possesses a Poisson distribution with parameter λ, then the moment-generating function, mean, and variance of X are respectively given by

(i) $M_X(t) = e^{\lambda(e^t - 1)}$

(ii) $\mu = \lambda$

(iii) $\sigma^2 = \lambda$

Proof: First compute the moment-generating function:

$$M_X(t) = E[e^{tX}]$$

$$= \sum_{x=0}^{\infty} \frac{e^{tx}e^{-\lambda}\lambda^x}{x!}$$

$$= e^{-\lambda} \sum_{x=0}^{\infty} \frac{(\lambda e^t)^x}{x!}$$

$$= e^{-\lambda}e^{\lambda e^t}$$

the last equality following from the fact that the series is the Taylor expansion for the exponential function. Simplification yields the moment-generating function in (i). Theorems 3.17 and 3.12 yield the mean and variance:

$$M_X'(t) = e^{\lambda(e^t - 1)}\lambda e^t$$

$$M_X''(t) = e^{\lambda(e^t - 1)}[\lambda e^t + \lambda^2 e^{2t}]$$

$$\mu = M_X'(0) = \lambda$$

$$\mu_2' = M_X''(0) = \lambda + \lambda^2$$

$$\sigma^2 = \mu_2' - \mu^2 = \lambda \qquad \blacksquare$$

The mean of a Poisson random variable is particularly revealing insofar as the mathematical Poisson model is concerned. If X counts the happenings in an interval of length t, then, as noted previously, $P(X = x) = p(x; qt)$. Consequently, by Theorem 4.4(i), $E[X] = qt$, which says that the expected number of happenings in an interval of length t is qt, or, that the **rate** of the happenings is q.

On average, a gram of radium emits 3.57×10^{10} alpha particles per second. If we assume that the emissions satisfy the three Poisson conditions and let X be the number of emissions observed in 1 nanosecond (10^{-9} second), then X possesses a Poisson distribution with

$$\lambda = (3.57 \times 10^{10}) \times 10^{-9} = 35.7 \text{ particles per nanosecond}$$

The probability that exactly 35 alpha particles will be emitted in a single nanosecond is

$$P(X = 35) = \frac{e^{-35.7}(35.7)^{35}}{35!} = 0.0668$$

The probability that inclusively between 27 and 29 alpha particles will be emitted in 1 nanosecond is

$$P(27 \leq X \leq 29) = \sum_{x=27}^{29} p(x; 35.7)$$
$$= p(27; 35.7) + p(28; 35.7) + p(29; 35.7)$$
$$= 0.0240 + 0.0306 + 0.0377 = 0.0923$$

EXAMPLE 4.15   In an interactive time-sharing environment it is found that, on average, a job arrives for CPU service every 6 seconds. If the arrivals satisfy the three Poisson conditions, then what is the probability that there will be less than or equal to 4 arrivals in a given minute? Let X count the number of arrivals in the minute. The arrival rate is $q = 10$ jobs per minute and $\lambda = q \times 1 = 10$. According to Table A.3,

$$P(X \leq 4) = \Pi(4; 10) = 0.0293$$

The probability that there will be inclusively between 8 and 12 jobs arriving is

$$P(8 \leq X \leq 12) = F_X(12) - F_X(7)$$
$$= \Pi(12; 10) - \Pi(7; 10)$$
$$= 0.7916 - 0.2202 = 0.5714$$

By Theorem 4.3(iii), the standard deviation is

$$\sigma = \sqrt{\lambda} = \sqrt{10} = 3.1623$$

EXAMPLE 4.16   In planning highway construction it is necessary to take into account the arrival distribution at certain key points. Suppose it has been determined that 90 vehicles per minute arrive at a proposed bridge crossing and the arrivals can be satisfactorily modeled by a Poisson distribution. Letting X count the number of arrivals in a fixed time interval of length $t = 6$ seconds, X possesses the density

$$f(x) = p(x; 90t) = p(x; 90 \times 0.1) = p(x; 9) = \frac{e^{-9}9^x}{x!}$$

X has mean and variance 9. The probability that strictly more than 12 vehicles arrive in a 6-second span is

$$P(X > 12) = 1 - P(X \leq 12) = 1 - \Pi(12; 9) = 1 - 0.8758 = 0.1242$$

4.4.3 The Poisson Approximation to the Binomial Distribution

Under certain circumstances, the Poisson distribution can be used as an approximation to the binomial distribution. If, in the binomial density $b(x; n, p)$, the mean np is held fixed, then, as is shown in the next theorem, the limit as n tends to ∞ of $b(x; n, p)$ is $p(x; np)$. Letting $np = \lambda$, the sequence of binomial densities approaching the Poisson density each have mean λ, which agrees with the mean of the limiting Poisson density. Note that, since $n \longrightarrow \infty$, $p \longrightarrow 0$. Moreover, $np = \lambda$ implies $p = \lambda/n$.

For $x = 0, 1, 2, \ldots$,

$$\lim_{n \to \infty} b\left(x; n, \frac{\lambda}{n}\right) = p(x; \lambda)$$

Proof: From the definition of the binomial distribution,

$$b\left(x; n, \frac{\lambda}{n}\right) = \binom{n}{x} \left(\frac{\lambda}{n}\right)^x \left(1 - \frac{\lambda}{n}\right)^{n-x}$$

$$= \frac{n(n-1) \cdots (n-x+1)}{x!} \left(\frac{\lambda}{n}\right)^x \left(1 - \frac{\lambda}{n}\right)^{n-x}$$

$$= \frac{n(n-1) \cdots (n-x+1) \lambda^x}{n^x \quad x!} \left(1 - \frac{\lambda}{n}\right)^n \left(1 - \frac{\lambda}{n}\right)^{-x}$$

the last equality resulting from a rearrangement of terms. Now,

$$\lim_{n \to \infty} \frac{n(n-1) \cdots (n-x+1)}{n^x} = \lim_{n \to \infty} 1\left(1 - \frac{1}{n}\right) \cdots \left(1 - \frac{x-1}{n}\right) = 1$$

$$\lim_{n \to \infty} \left(1 - \frac{\lambda}{n}\right)^{-x} = 1$$

and, using the limit form of the definition of the exponential,

$$\lim_{n \to \infty} \left(1 - \frac{\lambda}{n}\right)^n = e^{-\lambda}$$

Consequently, returning to the limit of $b(x; n, \lambda/n)$ and using the fact that a limit of a product is the product of the limits,

$$\lim_{n \to \infty} b\left(x; n, \frac{\lambda}{n}\right) = 1 \cdot \frac{\lambda^x}{x!} \cdot e^{-\lambda} \cdot 1$$

and the theorem is proven. ∎

For large n and small p, Theorem 4.5 is useful in approximating values of the binomial density with values of the Poisson density. To see why, replace λ/n in the statement of the theorem by p. The theorem then states that

$$\lim_{n \to \infty} b(x; n, p) = p(x; np)$$

as stated originally. To use the limit as an approximation, we require p to be small so that np is not too large, even though n is large. The reason is that the proof depends on three limits and convergence to the limits is too slow if $\lambda = np$ is large in comparison with n.

EXAMPLE 4.17

Communication channels do not always transmit the correct signal. Suppose that for a particular channel the *error rate* is 1 incorrect transmission per 100 messages. If 200 messages are sent in a given week and it is assumed that their transmissions are independent, what is the probability there will be at least 3 errors? In essence, there are 200 Bernoulli trials and the probability of a success, which is actually an erroneous transmission, is $p = 0.01$. Since $np = 2$ is small in comparison to $n = 200$, a Poisson approximation to the binomial probability is in order. Consequently, letting X count the number of erroneous transmissions in the 200 messages,

$$
\begin{aligned}
P(X \geq 3) &= 1 - P(X \leq 2) \\
&= 1 - B(2; 200, 0.01) \\
&\approx 1 - \Pi(2; 2) \\
&= 1 - 0.6767 = 0.3233
\end{aligned}
$$

Without using the Poisson approximation, the result is

$$
\begin{aligned}
P(X \geq 3) &= 1 - B(2; 200, 0.01) \\
&= 1 - b(0; 200, 0.01) - b(1; 200, 0.01) - b(2; 200, 0.01) \\
&= 1 - 0.1340 - 0.2707 - 0.2720 = 0.3233
\end{aligned}
$$

Hence, in this instance the approximation is excellent.

4.5 NORMAL DISTRIBUTION

We now turn our attention to continuous distributions, beginning with the normal distribution. Historically, the normal distribution has played a central role in the development of probability and statistics. The reasons for this preeminence are both practical and theoretical. The numerical measurements of many diverse phenomena can be modeled by the normal distribution—for instance, the quantitative analysis of errors (see Example 3.7). The central limit theorem, to be discussed in Section 7.2.3, shows that the normal distribution holds a special place among all probability distributions.

4.5.1 The Normal Density

The normal density leads to the familiar bell-shaped curve. As with other continuous probability distributions, the applicability of the normal distribution depends on the extent to which it fits the observational data. A normal density was employed in Example 3.7, since it fit the data rather well (see Figure 3.7).

There are two parameters, μ and σ, in the definition of the normal density, a particular member of the normal family being determined by a specification of these. For instance, in Example 3.7, $\mu = 0$ and $\sigma = 1/2$. The reason for using the Greek letters μ and σ in the definition of the normal density is that these parameters are the mean and standard deviation, respectively, of the distribution. Owing to the prominence of the normal family, it has long been customary to use the same symbols to denote its parameters as are generally employed to denote the mean and standard deviation of an arbitrary distribution.

DEFINITION 4.6

Normal Distribution. A continuous random variable X is said to possess a normal distribution if it has the continuous density

$$f(x) = \frac{1}{\sqrt{2\pi}\sigma} e^{-\frac{1}{2}\left(\frac{x-\mu}{\sigma}\right)^2}$$

for $-\infty < x < \infty$, where $-\infty < \mu < \infty$ and $\sigma > 0$. A normal distribution is also called a **Gaussian** distribution.

If X possesses a normal distribution with parameters μ and σ, then we will write $X \sim N(\mu, \sigma)$, the notation being meant to imply that X is a particular member of the normal family. We will also write $n(x; \mu, \sigma)$ to denote the normal density. Figure 4.7 illustrates various members of the normal family. In the special case where $\mu = 0$ and $\sigma = 1$, we write Z to denote the corresponding random variable, so that $Z \sim N(0, 1)$. Z is said to possess a **standard normal distribution**. The graph of the standard normal density, together with some important probabilities, is given in Figure 4.8. The standard normal density is given by

$$f_Z(z) = \frac{1}{\sqrt{2\pi}} e^{-z^2/2}$$

for $-\infty < z < \infty$.

Theoretically, our first task is to show that the normal density is a legitimate density. According to Theorem 3.4, we need to demonstrate that the total integral of the density is 1. The basic step in that direction is to show that the integral of the standard normal density is 1, which is precisely the result quoted in Section 3.2.2 prior to Example 3.7. To begin with, let

$$I = \frac{1}{\sqrt{2\pi}} \int_{-\infty}^{\infty} e^{-z^2/2} \, dz$$

Then

$$I^2 = \left(\frac{1}{\sqrt{2\pi}} \int_{-\infty}^{\infty} e^{-x^2/2} \, dx\right) \left(\frac{1}{\sqrt{2\pi}} \int_{-\infty}^{\infty} e^{-y^2/2} \, dy\right)$$

$$= \frac{1}{2\pi} \int_{-\infty}^{\infty} \int_{-\infty}^{\infty} e^{-(x^2+y^2)/2} \, dy \, dx$$

where the product of the single integrals has been written as a double integral. Now change to polar coordinates, letting $y = r \cos \theta$ and $x = r \sin \theta$. Then

$$r^2 = x^2 + y^2$$

and the rectangular area element $dA = dy \, dx$ becomes $r \, dr \, d\theta$. Since the integral is over the whole plane, r goes from 0 to ∞ and θ goes from 0 to 2π. Thus, the integral I^2 becomes

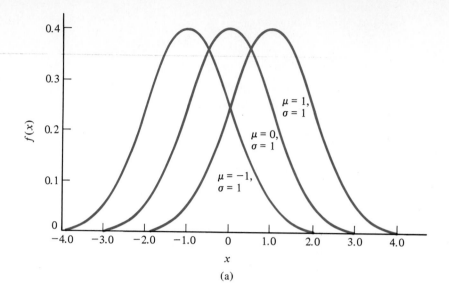

(a)

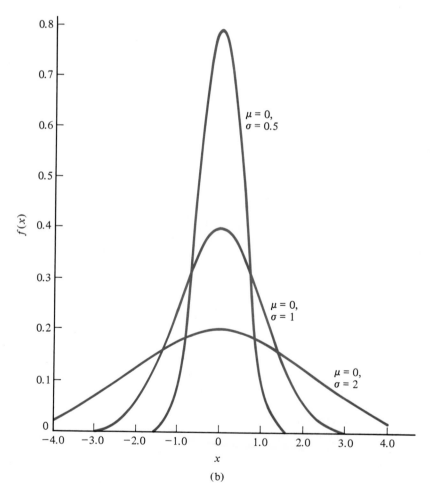

FIGURE 4.7
Various members of the normal family (Hahn/ Shapiro, *Statistical Models in Engineering*, © 1967 by John Wiley & Sons, Inc.)

(b)

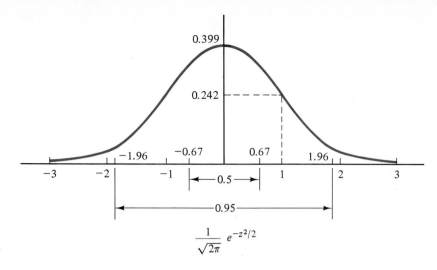

FIGURE 4.8
Standard normal density

$$\frac{1}{\sqrt{2\pi}} \, e^{-z^2/2}$$

$$
\begin{aligned}
I^2 &= \frac{1}{2\pi} \int_0^{2\pi} \int_0^{\infty} e^{-r^2/2} r \, dr \, d\theta \\
&= \frac{1}{2\pi} \int_0^{2\pi} d\theta \int_0^{\infty} e^{-r^2/2} r \, dr \\
&= \int_0^{\infty} e^{-r^2/2} r \, dr = 1
\end{aligned}
$$

Taking square roots yields $I = 1$.

Having established that $f_Z(z)$ is a density, it is straightforward to show that $n(x; \mu, \sigma)$ is a density for any choices of μ and σ. To do so, we use the substitution

$$z = \frac{x - \mu}{\sigma}$$

for which $dz = dx/\sigma$. Proceeding,

$$
\int_{-\infty}^{\infty} n(x; \mu, \sigma) \, dx = \frac{1}{\sqrt{2\pi}\,\sigma} \int_{-\infty}^{\infty} e^{-\frac{1}{2}\left(\frac{x-\mu}{\sigma}\right)^2} dx
$$

$$
= \frac{1}{\sqrt{2\pi}\,\sigma} \int_{-\infty}^{\infty} e^{-z^2/2}\, \sigma \, dz
$$

which is precisely the integral from $-\infty$ to ∞ of the standard normal density.

The probability distribution function for a normal random variable, $X \sim N(\mu, \sigma)$, is given by

$$
F_X(x) = \frac{1}{\sqrt{2\pi}\,\sigma} \int_{-\infty}^{x} e^{-\frac{1}{2}\left(\frac{y-\mu}{\sigma}\right)^2} dy
$$

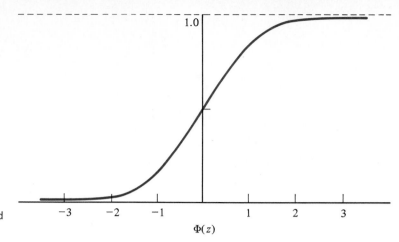

FIGURE 4.9
Probability distribution function for the standard normal density

$\Phi(z)$

Especially important in calculations is the probability distribution function of the standard normal variable, which we denote by $\Phi(z)$ and is defined by

Standard Normal Probability Distribution Function

$$\Phi(z) = F_Z(z) = \frac{1}{\sqrt{2\pi}} \int_{-\infty}^{z} e^{-y^2/2}\, dy$$

$\Phi(z)$ is graphed in Figure 4.9, and Table A.1 gives values for $\Phi(z)$. By Theorem 3.5, since Z is a continuous random variable,

$$P(a < Z < b) = \Phi(b) - \Phi(a)$$

for any a and b such that $a < b$ (see Figure 4.10).

EXAMPLE 4.18

If a random phenomenon can be modeled by the standard normal distribution, then what is the probability that an observation of the phenomenon will lie between -1.43 and 0.88? Using Table A.1, the desired probability is given by

$$P(-1.43 < Z < 0.88) = \Phi(0.88) - \Phi(-1.43) = 0.8106 - 0.0764 = 0.7342$$

(see Figure 4.11).

FIGURE 4.10
Illustration of the identity $P(a < Z < b) = \Phi(b) - \Phi(a)$

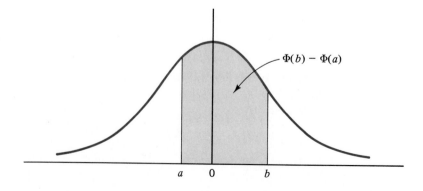

$\Phi(b) - \Phi(a)$

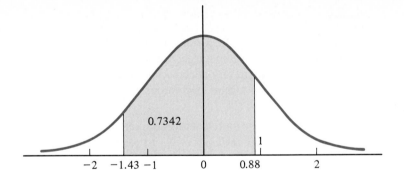

FIGURE 4.11
Illustration
corresponding to
Example 4.18

TABLE 4.2   IMPORTANT CONTINUOUS DISTRIBUTIONS

Distribution	Density	Mean	Variance
Normal	$\dfrac{1}{\sqrt{2\pi}\,\sigma}\,e^{-\frac{1}{2}\left(\frac{x-\mu}{\sigma}\right)^2}$	μ	σ^2
Uniform	$\dfrac{1}{b-a},\ a \le x \le b$	$\dfrac{a+b}{2}$	$\dfrac{(b-a)^2}{12}$
Gamma	$\dfrac{\beta^{-\alpha}}{\Gamma(\alpha)}\,x^{\alpha-1}e^{-x/\beta},\ x>0$	$\alpha\beta$	$\alpha\beta^2$
Erlang	$\dfrac{\beta^{-k}}{(k-1)!}\,x^{k-1}e^{-x/\beta},\ x>0$	$k\beta$	$k\beta^2$
Exponential	$be^{-bx},\ x>0$	$1/b$	$1/b^2$
Chi-square	$\dfrac{2^{-\nu/2}}{\Gamma(\nu/2)}\,x^{(\nu/2)-1}e^{-x/2},\ x>0$	ν	2ν
Beta	$\dfrac{1}{B(\alpha,\,\beta)}\,x^{\alpha-1}(1-x)^{\beta-1},\ 0<x<1$	$\dfrac{\alpha}{\alpha+\beta}$	$\dfrac{\alpha\beta}{(\alpha+\beta)^2(\alpha+\beta+1)}$
Weibull	$\alpha\beta^{-\alpha}x^{\alpha-1}e^{-(x/\beta)^\alpha},\ x>0$	$\beta\Gamma\left(\dfrac{\alpha+1}{\alpha}\right)$	$\beta^2\left[\Gamma\left(\dfrac{\alpha+2}{\alpha}\right)-\Gamma^2\left(\dfrac{\alpha+1}{\alpha}\right)\right]$
Rayleigh	$\eta^{-2}xe^{-(x/\eta)^2/2},\ x>0$	$\sqrt{\dfrac{\pi}{2}}\,\eta$	$\dfrac{4-\pi}{2}\,\eta^2$
Lognormal	$\dfrac{1}{\sqrt{2\pi}\,\zeta x}\,e^{-\frac{1}{2}\left(\frac{\log x-\zeta}{\kappa}\right)^2},\ x>0$	$e^{\zeta+\kappa^2/2}$	$e^{2\zeta+\kappa^2}(e^{\kappa^2}-1)$
Laplace	$\dfrac{b}{2}\,e^{-b|x-a|}$	a	$2/b^2$
Pareto	$\dfrac{ra^r}{x^{r+1}},\ x \ge a$	$\dfrac{ra}{r-1}\quad(r>1)$	$\dfrac{ra^2}{(r-1)^2(r-2)}\quad(r>2)$
Cauchy	$\dfrac{1}{\pi b\left[1+\left(\dfrac{x-a}{b}\right)^2\right]}$	Nonexistent	Nonexistent

4.5.2 Properties of the Normal Distribution

There are a number of important properties concerning the normal distribution. We begin by finding the moment-generating function and proving that μ and σ are truly the mean and standard deviation, respectively. Table 4.2 gives the densities, means, and variances of important continuous distributions.

THEOREM 4.6

If X is normally distributed with parameters μ and σ, then the moment-generating function, mean, and variance of X are respectively given by

(i) $M_X(t) = e^{\mu t + \sigma^2 t^2/2}$

(ii) $E[X] = \mu$

(iii) $\text{Var}\,[X] = \sigma^2$

Proof: By Theorem 3.10,

$$E[e^{tX}] = \frac{1}{\sqrt{2\pi}\,\sigma} \int_{-\infty}^{\infty} e^{tx} e^{-\frac{1}{2}\left(\frac{x-\mu}{\sigma}\right)^2} dx$$

$$= \frac{1}{\sqrt{2\pi}\,\sigma} \int_{-\infty}^{\infty} e^{-(\sigma^{-2}/2)(x^2 - 2\mu x + \mu^2 - 2\sigma^2 tx)} dx$$

Completing the square in the exponent yields

$$M_X(t) = \frac{1}{\sqrt{2\pi}\,\sigma} \int_{-\infty}^{\infty} e^{-(\sigma^{-2}/2)\{[x-(\mu + t\sigma^2)]^2 - 2t\mu\sigma^2 - t^2\sigma^4\}} dx$$

$$= e^{\mu t + t^2\sigma^2/2} \frac{1}{\sqrt{2\pi}\,\sigma} \int_{-\infty}^{\infty} e^{-\frac{1}{2}\left[\frac{x-(\mu + t\sigma^2)}{\sigma}\right]^2} dx$$

The substitution

$$y = \frac{x - (\mu + t\sigma^2)}{\sigma}$$

reduces the integral (together with the constant $1/\sqrt{2\pi}\sigma$) to the integral of the standard normal density, which is 1, thereby yielding the moment-generating function claimed in the statement of the theorem. As usual, the mean and variance are found by using the moment-generating function:

$$M_X'(t) = (\mu + \sigma^2 t)e^{\mu t + \sigma^2 t^2/2}$$

$$M_X''(t) = \sigma^2 e^{\mu t + \sigma^2 t^2/2} + (\mu + \sigma^2 t)^2 e^{\mu t + \sigma^2 t^2/2}$$

$$E[X] = M_X'(0) = \mu$$

$$E[X^2] = M_X''(0) = \sigma^2 + \mu^2$$

$$\text{Var}\,[X] = E[X^2] - E[X]^2 = \sigma^2 + \mu^2 - \mu^2 = \sigma^2 \qquad \blacksquare$$

The mean and variance are clearly reflected in the geometry of a **normal curve,** which is the name commonly employed for the graph of a normal density. As can be seen in Figure 4.7, a normal curve is symmetrically centered at μ and its peakedness is determined by σ: the greater σ, the less peaked the curve. Greater peakedness means that the probability mass of the corresponding random variable is more tightly packed about the mean. The maximum of the curve is attained at $x = \mu$, and thus the mean agrees with the **mode,** the latter being the x value of the maximum value of a density. Finally, an application of maximum-minimum theory from elementary calculus shows the curve to have two points of inflection, at $\mu - \sigma$ and $\mu + \sigma$.

A significant fact regarding normal random variables is that their probabilities can be evaluated by means of the standard normal variable. If X is normally distributed with mean μ and standard deviation σ, then, as demonstrated in the next theorem, the transformation $Z = (X - \mu)/\sigma$ yields a standard normal distribution.

THEOREM 4.7

If X is normally distributed with mean μ and variance σ^2, then the transformation

$$Z = \frac{X - \mu}{\sigma}$$

yields a standard normal variable. In particular, for any a and b, $a < b$ (including $-\infty$ and ∞),

$$P(a < X < b) = P\left(\frac{a - \mu}{\sigma} < Z < \frac{b - \mu}{\sigma}\right) = \Phi\left(\frac{b - \mu}{\sigma}\right) - \Phi\left(\frac{a - \mu}{\sigma}\right)$$

Proof: The transformation

$$Z = g(X) = \frac{X}{\sigma} - \frac{\mu}{\sigma}$$

is affine with $a = \sigma^{-1}$ and $b = -\mu/\sigma$. Thus, according to the results of Example 3.14,

$$
\begin{aligned}
f_Z(z) &= \frac{1}{|\sigma^{-1}|} f_X\left(\frac{z + \frac{\mu}{\sigma}}{\sigma^{-1}}\right) \\
&= \sigma f_X(\sigma z + \mu) \\
&= \frac{\sigma}{\sqrt{2\pi}\,\sigma} e^{-\frac{1}{2}\left(\frac{\sigma z + \mu - \mu}{\sigma}\right)^2} \\
&= \frac{1}{\sqrt{2\pi}} e^{-z^2/2}
\end{aligned}
$$

which is the standard normal density.

To establish the equality claimed in the statement of the theorem, we once again turn to Example 3.14. For the transformation under consideration, the result established in Example 3.14 becomes

$$F_Z(z) = F_X(\sigma z + \mu)$$

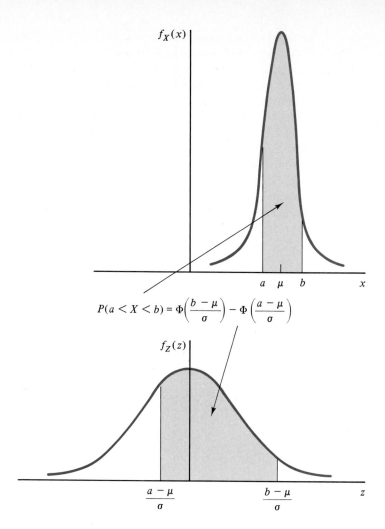

$$P(a < X < b) = \Phi\left(\frac{b - \mu}{\sigma}\right) - \Phi\left(\frac{a - \mu}{\sigma}\right)$$

FIGURE 4.12
Illustration of Theorem
4.7

Letting $x = \sigma z + \mu$ yields

$$F_X(x) = F_Z\left(\frac{x - \mu}{\sigma}\right)$$

Hence, Theorem 3.5 yields

$$P(a < X < b) = F_X(b) - F_X(a) = F_Z\left(\frac{b - \mu}{\sigma}\right) - F_Z\left(\frac{a - \mu}{\sigma}\right)$$

But $F_Z = \Phi$. ∎

  Figure 4.12 illustrates the relationship between the probabilities of a normal random variable $X \sim N(\mu, \sigma)$ and the standard normal variable Z.

EXAMPLE 4.19 The potential difference between two points is found to be a normally distributed random variable X with mean 12.00 volts and standard deviation 0.20 volts. We wish to determine the probability that an arbitrary measurement is (a) between 11.92 and 12.27 volts, (b) is greater than 12.45 volts, and (c) is less than 11.70 volts. In each case we appeal to Theorem 4.7 and Table A.1. For (a),

$$P(11.92 < X < 12.27) = \Phi\left(\frac{12.27 - 12.00}{0.20}\right) - \Phi\left(\frac{11.92 - 12.00}{0.20}\right)$$

$$= \Phi(1.35) - \Phi(-0.40)$$
$$= 0.9115 - 0.3446 = 0.5669$$

For (b),

$$P(X > 12.45) = 1 - P(X < 12.45)$$

$$= 1 - \Phi\left(\frac{12.45 - 12.00}{0.20}\right)$$

$$= 1 - \Phi(2.25)$$

$$= 1 - 0.9878 = 0.0122$$

For (c),

$$P(X < 11.70) = \Phi\left(\frac{11.70 - 12.00}{0.20}\right) = \Phi(-1.50) = 0.0668$$

EXAMPLE 4.20 Suppose the speeds of cars passing a particular point on a highway are measured and it is determined that the random variable X giving the speed of a passing car is normally distributed. What is the probability that an arbitrarily observed car will be moving at a speed within two standard deviations of the mean? Note that the problem is stated without specification of either the mean or the variance. Thus, the solution, once found, is generic relative to the normal family. Applying Theorem 4.7 with $X \sim N(\mu, \sigma)$ yields

$$P(|X - \mu| < 2\sigma) = P(\mu - 2\sigma < X < \mu + 2\sigma)$$

$$= \Phi\left(\frac{\mu + 2\sigma - \mu}{\sigma}\right) - \Phi\left(\frac{\mu - 2\sigma - \mu}{\sigma}\right)$$

$$= \Phi(2) - \Phi(-2)$$

$$= 0.9772 - 0.0228 = 0.9544$$

In terms of the relative-frequency interpretation of probability, 95.44% of all cars are traveling within two standard deviations of the mean. Compare this result with the lower bound of 0.75 given by Chebyshev's inequality (see Figure 3.26).

One should recognize from this example that the probabilities given in Figure 4.8 for the standard normal density apply to all normally distributed random variables, the only change being that, in the general case, we consider the values $\mu \pm k\sigma$ on the x axis rather than just $\pm k$.

Very often one is given the probability (expressed as) $1 - \alpha$ that a random variable X is within r units of its mean and wishes to determine r, given μ, σ, and α. If the variable is normally distributed, then the methodology is straightforward and r is obtained by solving the equation

$$P(|X - \mu| < r) = 1 - \alpha$$

which is illustrated in Figure 4.13 and is equivalent to the equation

$$P(|X - \mu| > r) = \alpha$$

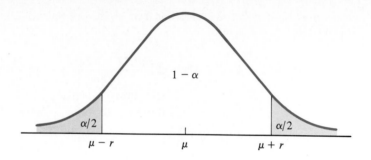

FIGURE 4.13
Illustration of
$P(|X - \mu| < r) = 1 - \alpha$

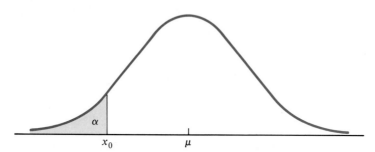

FIGURE 4.14
Illustration of
$P(X < x_0) = \alpha$

Analogously, one might wish to find the point x_0 such that $P(X < x_0) = \alpha$ (see Figure 4.14), or such that $P(X > x_0) = \alpha$ (see Figure 4.15). The next theorem, which is of practical significance, gives the solutions to each of these problems in terms of Φ^{-1}, the inverse of the probability distribution function of the standard normal variable. Its proof utilizes the useful relation

$$\Phi(-z) = 1 - \Phi(z)$$

which follows from the symmetry of the standard normal curve.

THEOREM 4.8

If X is normally distributed with mean μ and standard deviation σ, then

(i) $P(|X - \mu| > r) = \alpha$ possesses the solution $r = \sigma\Phi^{-1}(1 - \alpha/2)$.
(ii) $P(X < x_0) = \alpha$ possesses the solution $x_0 = \mu + \sigma\Phi^{-1}(\alpha)$.
(iii) $P(X > x_0) = \alpha$ possesses the solution $x_0 = \mu + \sigma\Phi^{-1}(1 - \alpha)$.

Proof: For (i), the equation to be solved is equivalent to

$$P(\mu - r < X < \mu + r) = 1 - \alpha$$

FIGURE 4.15
Illustration of
$P(X > x_0) = \alpha$

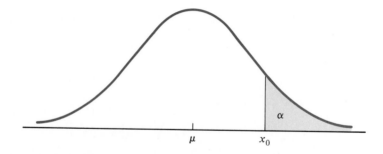

or, changing to the standard normal variable,

$$P\left(-\frac{r}{\sigma} < Z < \frac{r}{\sigma}\right) = 1 - \alpha$$

By Theorem 3.5, the following sequence of equations results:

$$\Phi\left(\frac{r}{\sigma}\right) - \Phi\left(-\frac{r}{\sigma}\right) = 1 - \alpha$$

$$\Phi\left(\frac{r}{\sigma}\right) - \left[1 - \Phi\left(\frac{r}{\sigma}\right)\right] = 1 - \alpha$$

$$2\Phi\left(\frac{r}{\sigma}\right) = 2 - \alpha$$

$$r = \sigma\Phi^{-1}\left(1 - \frac{\alpha}{2}\right)$$

As for (ii), changing to Z yields the equation

$$P\left(Z < \frac{x_0 - \mu}{\sigma}\right) = \alpha$$

which is simply the equation

$$\Phi\left(\frac{x_0 - \mu}{\sigma}\right) = \alpha$$

Inversion yields the result stated in the theorem. For part (iii), changing the given equation to an equation in Z yields

$$P\left(Z > \frac{x_0 - \mu}{\sigma}\right) = \alpha$$

or, equivalently,

$$1 - \Phi\left(\frac{x_0 - \mu}{\sigma}\right) = \alpha$$

Inversion gives the desired result. ∎

EXAMPLE 4.21

In the production of piston and connecting-rod assemblies for an internal combustion engine, the weights exhibit variability. Too great a variability results in loss of performance. Suppose a particular manufacturer rejects the 3% of the rods that are the lightest and the 3% that are the heaviest. If the mean weight is 4.72 lb, the standard deviation is 0.006 lb, and the distribution of the weights is Gaussian, then it is possible to determine the maximum and minimum acceptable weights. Letting X denote the random weight variable, we need to solve the equation

$$P(|X - 4.72| < r) = 0.94$$

for r. The solution will give the deviation from the mean that will exclude the 3% **tails** of the curve. By Theorem 4.8(i),

$$r = 0.006 \times \Phi^{-1}\left(1 - \frac{0.06}{2}\right)$$
$$= 0.006 \times \Phi^{-1}(0.97)$$
$$= 0.006 \times 1.88 = 0.0113$$

where Φ^{-1} is found by reading Table A.1 inversely. In terms of the original problem, the manufacturer accepts all assemblies weighing between $4.72 - 0.0113$ lb and $4.72 + 0.0113$ lb.

EXAMPLE 4.22

In a binary system, information is represented by electrical signals—for instance, voltages. One voltage represents the bit 0 and the other the bit 1. Let bits 0 and 1 be represented by 2 and 3 volts, respectively. Because of voltage fluctuation in a circuit, the input terminal of a digital circuit does not receive the intended voltage; instead, the signal it receives is a random phenomenon, the original signal being distorted by **channel noise.** Very often, the noise is modeled as a normally distributed random variable and is termed **Gaussian noise.** Suppose the noise is Gaussian with mean $\mu = 0$ and standard deviation $\sigma = 0.22$, and the input terminal recognizes the bit 0 if the received voltage is less than 2.6 and recognizes the bit 1 if the received voltage is greater than 2.6. Our problem regards the probability that the input terminal (receiver) will recognize the bit incorrectly.

Two possible errors can occur. First, the bit 0 can be received as bit 1. This occurs if the transmitted signal is 2 volts and the received signal is greater than 2.6 volts. For this to occur, the noise X must be greater than 0.6. If the intended bit is 0, then the probability of this type of error is

$$P(X > 0.6) = P\left(Z > \frac{0.60}{0.22}\right) = 1 - \Phi(2.73) = 1 - 0.9968 = 0.0032$$

On the other hand, the bit 1 can be received as bit 0. This occurs if the transmitted signal is 3 volts and the received signal is less than 2.6 volts. For this to occur, the noise X must be less than -0.4. Thus, the probability of this type of error is

$$P(X < -0.4) = P\left(Z < \frac{-0.40}{0.22}\right) = \Phi(-1.82) = 0.0344$$

4.5.3 The Normal Approximation to the Binomial Distribution

We will now consider a classical result of probability theory, the DeMoivre-Laplace limit theorem, which is concerned with the approximation of the binomial distribution by the normal distribution.

Figure 4.16 provides the graphs of the binomial probability mass function for $p = 0.3$ and $n = 5, 10, 15,$ and 20. As n grows, so does the symmetry of the mass functions. Moreover, the shape becomes increasingly like that of a normal curve. In fact, in Figure 4.16(d) the graph of the normal density

$$n(x; \ 6, \ 2.05) = \frac{1}{\sqrt{2\pi}(2.05)} e^{-\frac{1}{2}\left(\frac{x-6}{2.05}\right)^2}$$

has been superimposed, and the resemblance is striking. This normal density has been formed by taking as its mean and standard deviation the mean and standard deviation of the binomial density $b(x; \ 20, \ 0.3)$, these being

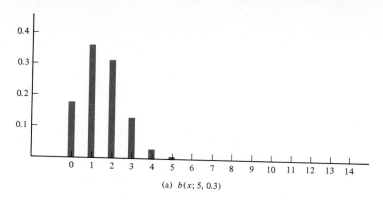

(a) $b(x; 5, 0.3)$

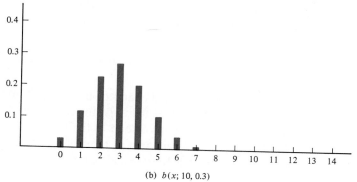

(b) $b(x; 10, 0.3)$

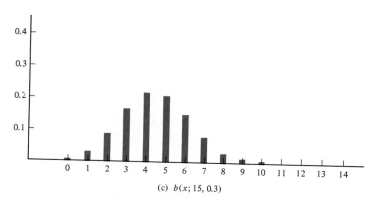

(c) $b(x; 15, 0.3)$

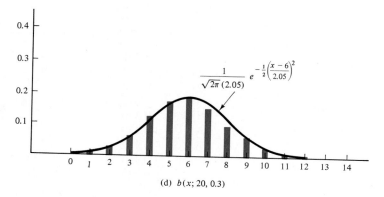

$$\frac{1}{\sqrt{2\pi}\,(2.05)}\,e^{-\frac{1}{2}\left(\frac{x-6}{2.05}\right)^2}$$

(d) $b(x; 20, 0.3)$

FIGURE 4.16
Binomial probability
mass functions for
$p = 0.3$ and $n = 5$,
10, 15, and 20

$$\mu = np = 20 \times 0.3 = 6$$

and

$$\sigma = \sqrt{npq} = \sqrt{20 \times 0.3 \times 0.7} = 2.05$$

The corresponding values of $b(x; 20, 0.3)$ and $n(x; 6, 2.05)$ are provided in Table 4.3. The point is that the normal density having its mean and variance in agreement with the corresponding moments for the binomial density provides a close approximation to the binomial density, so long as n is sufficiently large. This can be seen in terms of areas under the curve in Figure 4.17, where the curve of $n(x; 6, 2.05)$ has been superimposed upon the histogram of $b(x; 20, 0.3)$. As indicated in the figure, if X and Y denote the corresponding binomial and normal random variables, then

$$P(a \leq X \leq b) \approx P(a \leq Y \leq b)$$

where $P(a \leq X \leq b)$ is simply the sum of the areas of all the shaded rectangles and $P(a \leq Y \leq b)$ is the area under the superimposed normal curve. (The fact that the extreme left and right rectangles extend to the left of a and right of b, respectively, is not of importance in the statement of the DeMoivre-Laplace theorem; however, it does make a slight difference in practical settings, and thus we will discuss the matter subsequently.)

A rigorous mathematical statement of the normal approximation to the binomial distribution can be given in terms of probability distribution functions. Figure 4.18 shows the close similarity between the probability distribution functions of $b(x; 20, 0.3)$ and $n(x; 6, 2.05)$. If we again let X and Y denote the corresponding binomial and normal variables, and consider

$$X^* = \frac{X - 6}{2.05}$$

and

TABLE 4.3 COMPARISON OF BINOMIAL AND NORMAL DENSITIES

x	$b(x; 20, 0.3)$	$n(x; 6, 2.05)$
0	0.001	0.003
1	0.007	0.010
2	0.028	0.029
3	0.072	0.067
4	0.130	0.121
5	0.179	0.173
6	0.192	0.195
7	0.164	0.173
8	0.114	0.121
9	0.065	0.067
10	0.031	0.029
11	0.012	0.010
12	0.004	0.003

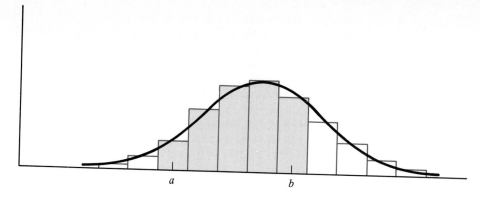

$$Y^* = \frac{Y - 6}{2.05} = Z$$

then the probability distribution function for the transformed binomial random variable X^* will be closely approximated by $\Phi(z)$.

More generally, if X_n denotes the binomially distributed random variable with parameters n and p, and if Y_n denotes the normally distributed random variable with mean $\mu = np$ and variance $\sigma^2 = npq$, then our contention is that Y_n can serve as an approximation to X_n. Making the same transformation as above,

$$X_n^* = \frac{X_n - np}{\sqrt{npq}}$$

$$Y_n^* = \frac{Y_n - np}{\sqrt{npq}} = Z$$

FIGURE 4.18
Probability distribution
functions for $b(x; 20, 0.3)$
and $n(x; 6, 2.05)$

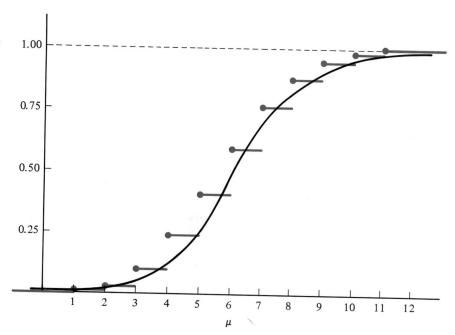

Our contention concerning the normal variable approximating the binomial variable can be interpreted in terms of the probability distribution function of X_n^* approaching Φ, the probability distribution function of $Y_n^* = Z$. It is precisely in these terms that the DeMoivre-Laplace theorem can be stated.

Before stating the theorem, we note that, for any random variable X, binomially distributed or otherwise, the transformed variable

$$X^* = \frac{X - \mu_X}{\sigma_X}$$

is called the **standardized** variable and possesses mean 0 and variance 1. If X happens to be normal, then $X^* = Z$.

THEOREM 4.9

(DeMoivre-Laplace). Suppose the random variable X_n is binomially distributed with parameters n and p. Then, as n tends to infinity, the probability distribution functions of the standardized random variables

$$X_n^* = \frac{X_n - np}{\sqrt{npq}}$$

approach the probability distribution function Φ of the standard normal variable.

In terms of the distribution functions themselves, if F_n denotes the distribution function of the standardized variable X_n^*, then Theorem 4.9 takes the form

$$\lim_{n\to\infty} F_n(z) = \Phi(z)$$

From an applications standpoint, a central concern is the approximation of the binomial probabilities by the standard normal probabilities derived from Table A.1. In this direction, the theorem tells us that, for large n,

$$P(X_n^* \leq z) = F_n(z) \approx \Phi(z)$$

or, letting

$$z = \frac{x - np}{\sqrt{npq}}$$

that

$$P\left(X_n^* \leq \frac{x - np}{\sqrt{npq}}\right) \approx \Phi\left(\frac{x - np}{\sqrt{npq}}\right)$$

Substituting the definition of X_n^* into the relation yields

$$P(X_n \leq x) \approx \Phi\left(\frac{x - np}{\sqrt{npq}}\right)$$

Applying Theorem 3.5, we obtain the desired result: for any integers a and b, $a < b$,

$$P(a \leq X_n \leq b) \approx \Phi\left(\frac{b - np}{\sqrt{npq}}\right) - \Phi\left(\frac{a - np}{\sqrt{npq}}\right)$$

for large n.

While the foregoing discussion might at first appear somewhat circuitous, it is necessary in order to give a rigorous statement of the approximation methodology. In fact, the last equation is precisely the form of the approximation indicated in Figure 4.17. To see this, just note that

$$\Phi\left(\frac{b - np}{\sqrt{npq}}\right) - \Phi\left(\frac{a - np}{\sqrt{npq}}\right) = P(a \leq Y_n \leq b)$$

where Y_n is the normally distributed random variable with mean np and variance npq.

As previously parenthetically noted, in Figure 4.17 the extreme left and right rectangles extend beyond the area properly under the normal curve. These extensions do not affect Theorem 4.9; however, they can have some impact when employing finite approximations. Thus, it is customary to employ a **continuity correction** by taking the area under the approximating curve between $a - 0.5$ and $b + 0.5$. The approximation then becomes

Corrected Normal Approximation to Binomial Probabilities

$$P(a \leq X_n \leq b) \approx \Phi\left(\frac{b - np + 0.5}{\sqrt{npq}}\right) - \Phi\left(\frac{a - np - 0.5}{\sqrt{npq}}\right)$$

EXAMPLE 4.23

For the binomial random variable X possessing density $b(x; 20, 0.3)$, use of Table A.2 shows

$$P(4 \leq X \leq 10) = B(10; 20, 0.3) - B(3; 20, 0.3) = 0.9829 - 0.1071 = 0.8758$$

The corrected normal approximation yields

$$P(4 \leq X \leq 10) \approx \Phi\left(\frac{10 - 6 + 0.5}{2.05}\right) - \Phi\left(\frac{4 - 6 - 0.5}{2.05}\right)$$
$$= \Phi(2.195) - \Phi(-1.220)$$
$$= 0.9859 - 0.1112 = 0.8747$$

EXAMPLE 4.24

To check a person's telepathic ability, a screen is placed between the investigator and the subject, and the subject is asked to choose the one among five cards to which the investigator is pointing. If the subject lacks telepathic ability, it is reasonable to assume that the probability of a success on any trial is 0.2. If there are 200 trials and the trials are assumed to be Bernoulli, then what is the probability that the subject will get more than 55 correct responses? The corrected normal approximation to the binomial distribution with $n = 200$, $p = 0.2$, $np = 40$, and $\sqrt{npq} = 5.657$ yields

$$P(X > 55) = 1 - P(X \leq 55)$$
$$\approx 1 - \Phi\left(\frac{55 - 40 + 0.5}{5.657}\right)$$
$$= 1 - \Phi(2.74) = 0.0031$$

If the subject does get more than 55 correct responses out of 200, then the investigator might have to reassess his hypothesis concerning the reasonableness of the success probability $p = 0.2$.

In general, the goodness of the normal approximation to the binomial distribution depends on the degree to which the normal curve fits the particular density of interest. For p near 1/2, the binomial density is fairly symmetrical and the fit is quite good for moderately large n. If the binomial density is skewed markedly because p is either close to 0 or to 1, then a much larger value of n is necessary for a decent approximation. A generally accepted rule of thumb is that the approximation is acceptable when

$$\min(np, nq) \geq 5$$

4.6 GAMMA DISTRIBUTION

The gamma distribution has found wide application in engineering. For instance, it has been successfully employed in the study of system reliability and it possesses an important relationship to the Poisson distribution. By varying the two parameters in the distribution, the density takes on a variety of shapes. Moreover, particular choices of the parameters result in the exponential, chi-square, and Erlang distributions.

4.6.1 The Gamma Function

Since the gamma density involves the gamma function, we begin with a brief discussion of this function, which is mainly recognized as an extension to real values of the factorial function.

DEFINITION 4.7

Gamma Function. For $x > 0$, $\Gamma(x)$ is defined by

$$\Gamma(x) = \int_0^\infty t^{x-1} e^{-t}\, dt$$

FIGURE 4.19
Gamma function

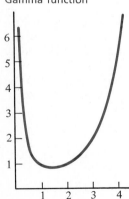

It is immediate from the definition that $\Gamma(1) = 1$. Moreover, although we will not prove it, $\Gamma(1/2) = \sqrt{\pi}$. Of greater consequence is that for any $x > 0$,

$$\Gamma(x + 1) = x\Gamma(x)$$

In particular, if x is an integer, then $\Gamma(x + 1) = x!$. The gamma function is graphed in Figure 4.19.

Closely related to the gamma function is the **beta function,** which is defined for all positive x and y by

$$B(x, y) = \int_0^1 t^{x-1}(1 - t)^{y-1}\, dt$$

It can be shown that

$$B(x, y) = \frac{\Gamma(x)\Gamma(y)}{\Gamma(x + y)}$$

4.6.2 The Gamma Density

The gamma density possesses two parameters, α and β. Choices of these parameters yield the various members of the gamma family.

DEFINITION 4.8

Gamma Distribution. A random variable X is said to possess a gamma distribution with parameters $\alpha > 0$ and $\beta > 0$ if it has the density

$$f(x) = \begin{cases} \dfrac{\beta^{-\alpha}}{\Gamma(\alpha)} x^{\alpha-1} e^{-x/\beta}, & \text{if } x > 0 \\ 0, & \text{if } x \le 0 \end{cases}$$

Various members of the gamma family are illustrated in Figure 4.20. Quite a variety of shapes can be described by different choices of α and β. For instance, if $\alpha < 1$, then the graph of the density possesses a negative derivative, has no points of inflection, and has both the x and y axes as asymptotes. If $\alpha > 1$, then the curve has a single maximum at $x = \beta(\alpha - 1)$ and has the x axis as an asymptote. Generally, the shape of the curve is determined by α and the scale is determined by β.

Since the gamma density is nonnegative, it follows that it is a legitimate density, since its total integral is 1. To see this, make the substitution $u = x/\beta$. Then $x = \beta u$, $dx = \beta\, du$, and

$$\int_0^\infty f(x)\, dx = \frac{\beta^{-\alpha}}{\Gamma(\alpha)} \int_0^\infty x^{\alpha-1} e^{-x/\beta}\, dx$$

$$= \frac{\beta^{-\alpha+1}}{\Gamma(\alpha)} \int_0^\infty (\beta u)^{\alpha-1} e^{-u}\, du$$

$$= \frac{1}{\Gamma(\alpha)} \int_0^\infty u^{\alpha-1} e^{-u}\, du = 1$$

since the last integral is $\Gamma(\alpha)$.

The probability distribution function for the gamma distribution is given by $F(x) = 0$ for $x \le 0$ and

$$F(x) = \frac{\beta^{-\alpha}}{\Gamma(\alpha)} \int_0^x t^{\alpha-1} e^{-t/\beta}\, dt$$

for $x > 0$. Letting $u = t/\beta$ yields $t = \beta u$ and $dt = \beta\, du$. Under this substitution, u goes from 0 to x/β as t goes from 0 to x and the probability distribution function takes the form

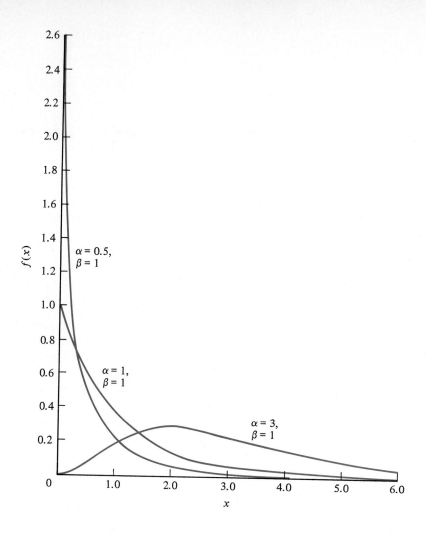

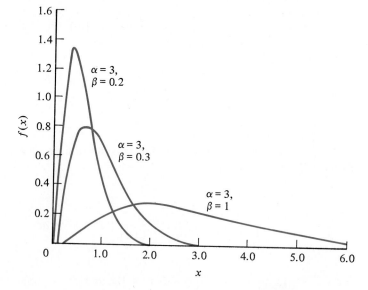

FIGURE 4.20
Various members of the gamma family (Hahn/ Shapiro, *Statistical Models in Engineering,* © 1967 by John Wiley & Sons, Inc.)

$$F(x) = \frac{\beta^{-\alpha}}{\Gamma(\alpha)} \int_0^{x/\beta} (\beta u)^{\alpha-1} e^{-u\beta} \, du$$

$$= \frac{1}{\Gamma(\alpha)} \int_0^{x/\beta} u^{\alpha-1} e^{-u} \, du$$

The function

$$\gamma(x; \alpha) = \frac{1}{\Gamma(\alpha)} \int_0^x t^{\alpha-1} e^{-t} \, dt$$

is known as the **incomplete gamma function.** Thus, the probability distribution function of the gamma distribution is given by

$$F(x) = \gamma\left(\frac{x}{\beta}; \alpha\right)$$

Table A.4 provides values for the incomplete gamma function.

EXAMPLE 4.25

A piece of sensitive laboratory equipment must be put through a rigorous recalibration after ten usages. It is used at the rate of twice a day, and the demands for its use are independent. The measurement of the amount of time between recalibrations is modeled as a gamma distribution with $\alpha = 10$ and $\beta = 1/2$. Since $\Gamma(10) = 9!$,

$$f_X(x) = \frac{2^{10}}{9!} x^9 e^{-2x}$$

for $x > 0$. The probability that a specific intercalibration time will be between 2.5 and 7 days is

$$P(2.5 < X < 7) = F_X(7) - F_X(2.5)$$
$$= \gamma(14; 10) - \gamma(5; 10)$$
$$= 0.891 - 0.032 = 0.859$$

THEOREM 4.10

If X possesses a gamma distribution with parameters α and β, then its moment-generating function, mean, variance, and kth-order moments are respectively given by

(i) $M_X(t) = (1 - \beta t)^{-\alpha}$, for $t < 1/\beta$

(ii) $\mu = \alpha\beta$

(iii) $\sigma^2 = \alpha\beta^2$

(iv) $\mu_k' = \dfrac{\beta^k \Gamma(\alpha + k)}{\Gamma(\alpha)}$

Proof: The moment-generating function is

$$M_X(t) = E[e^{tX}]$$

$$= \frac{\beta^{-\alpha}}{\Gamma(\alpha)} \int_0^\infty e^{tx} x^{\alpha-1} e^{-x/\beta} \, dx$$

$$= \frac{\beta^{-\alpha}}{\Gamma(\alpha)} \int_0^\infty x^{\alpha-1} e^{-(1/\beta - t)x} \, dx$$

which is finite so long as $t < 1/\beta$. The substitution

$$u = \left(\frac{1}{\beta} - t\right) x$$

yields

$$x = \left(\frac{1}{\beta} - t\right)^{-1} u$$

$$dx = \left(\frac{1}{\beta} - t\right)^{-1} du$$

and

$$M_X(t) = \frac{\beta^{-\alpha}}{\Gamma(\alpha)} \int_0^\infty \left[\left(\frac{1}{\beta} - t\right)^{-1} u \right]^{\alpha-1} e^{-u} \left(\frac{1}{\beta} - t\right)^{-1} du$$

$$= \frac{\beta^{-\alpha}}{\Gamma(\alpha)} \left(\frac{1}{\beta} - t\right)^{-\alpha} \int_0^\infty u^{\alpha-1} e^{-u} \, du$$

which, using the definition of $\Gamma(\alpha)$, reduces to the desired result.

To obtain the moments, we compute the successive derivatives of the moment-generating function:

$$M_X'(t) = \beta\alpha(1 - \beta t)^{-\alpha-1}$$

$$M_X''(t) = \beta^2(\alpha + 1)\alpha(1 - \beta t)^{-\alpha-2}$$

and, proceeding recursively,

$$M_X^{(k)}(t) = \beta^k(\alpha + k - 1) \cdots (\alpha + 1)\alpha(1 - \beta t)^{-\alpha-k}$$

Letting $t = 0$ gives

$$\mu_k' = M_X^{(k)}(0) = \beta^k(\alpha + k - 1) \cdots (\alpha + 1)\alpha$$

This expression can be simplified. Applying the factorial property $\Gamma(x + 1) = x\Gamma(x)$ recursively gives

$$\Gamma(\alpha + k) = (\alpha + k - 1)\Gamma(\alpha + k - 1)$$

$$= (\alpha + k - 1)(\alpha + k - 2)\Gamma(\alpha + k - 2)$$
$$\vdots$$
$$= (\alpha + k - 1)(\alpha + k - 2) \cdots (\alpha + 1)\alpha\Gamma(\alpha)$$

Dividing both sides of the equation by $\Gamma(\alpha)$ yields

$$(\alpha + k - 1)(\alpha + k - 2) \cdots (\alpha + 1)\alpha = \frac{\Gamma(\alpha + k)}{\Gamma(\alpha)}$$

and thus

$$\mu_k' = \frac{\beta^k \Gamma(\alpha + k)}{\Gamma(\alpha)}$$

Letting $k = 1$ gives the mean

$$\mu = \frac{\beta\Gamma(\alpha + 1)}{\Gamma(\alpha)} = \alpha\beta$$

Letting $k = 2$ gives the second moment

$$\mu_2' = \frac{\beta^2\Gamma(\alpha + 2)}{\Gamma(\alpha)} = \frac{\beta^2(\alpha + 1)\alpha\Gamma(\alpha)}{\Gamma(\alpha)} = \beta^2(\alpha + 1)\alpha$$

Finally, Theorem 3.12 yields

$$\sigma^2 = \mu_2' - \mu^2 = \beta^2(\alpha + 1)\alpha - \alpha^2\beta^2 = \alpha\beta^2 \qquad \blacksquare$$

EXAMPLE 4.26

Telephone calls arrive at a particular station at the constant rate of 2000 per hour and their arrivals are independent. A gamma distribution is employed with $\alpha = 3000$ and $\beta = 0.0005$ to describe the measurement X of elapsed time between a particular call and the 3000th subsequent call. According to Theorem 4.10, the mean elapsed time in hours is

$$\mu = 0.0005 \times 3000 = 1.5$$

the variance is

$$\sigma^2 = 3000 \times (0.0005)^2 = 0.00075$$

and the standard deviation is $\sigma = \sqrt{0.00075} = 0.0274$.

In applications, it is often the subfamilies of the gamma family that prove to be valuable. These are generated by restricting α and β. We discuss, in turn, the Erlang, exponential, and chi-square distributions.

4.6.3 Erlang Distribution

The **Erlang distribution,** which plays a key role in the theory of queuing, is simply the special case of the gamma distribution in which $\alpha = k$ is an integer. For a given value of k, it is common to refer to the distribution as a **k-Erlang** distribution. The density takes the form

$$f(x) = \frac{\beta^{-k}}{(k-1)!} x^{k-1} e^{-x/\beta}$$

for $x > 0$, and Theorem 4.10 holds with k in place of α.

The distributions of Examples 4.25 and 4.26 were both Erlang. In each, a k-Erlang distribution was employed to model the amount of time required for the occurrence of k events. It was stated that the rate of occurrence of the events was constant and that the occurrences of the events were independent. If we make the latter language more precise by assuming that the events constitute a Poisson experiment, then there is good reason to employ the k-Erlang model. In fact, if the rate of occurrence is q and the number of events occurring in any time interval of length t is a Poisson random variable with rate qt, which is precisely the situation when the events constitute a Poisson experiment, then the random variable Y that measures the time to the occurrence of the kth event possesses a k-Erlang distribution. Thus, the choice of models was dictated by the tacit assumption that the underlying distributions were Poisson. The next theorem formalizes these remarks. Since q is the rate of occurrence of the events, in observations per time unit, the number $1/q$ appearing in the theorem is the amount of time per occurrence.

THEOREM 4.11

Suppose events are occurring in such a way that the number of events occurring in any given time interval of length t is described by a Poisson distribution with mean qt. If Y denotes the random variable that, starting at time 0, gives the time at which the kth event occurs, then Y possesses a k-Erlang distribution with parameter $\beta = 1/q$.

Proof: Let $P_x(t)$ denote the probability of x events occurring in an interval of length t and $N(t)$ be the random variable that counts the number of events occurring in the interval $[0, t]$. Then the probability distribution function of Y is found as follows:

$$
\begin{aligned}
F_Y(t) &= P(Y \le t) \\
&= P[N(t) \ge k] \\
&= \sum_{x=k}^{\infty} P[N(t) = x] \\
&= \sum_{x=k}^{\infty} P_x(t) \\
&= \sum_{x=k}^{\infty} \frac{e^{-qt}(qt)^x}{x!}
\end{aligned}
$$

where the second equality follows from the fact that $Y \le t$ if and only if at least k events occur in the interval $[0, t]$. The density of Y is found by differentiating F_Y with respect to t:

$$
\begin{aligned}
f_Y(t) &= \frac{d}{dt} F_Y(t) \\
&= \sum_{x=k}^{\infty} \frac{1}{x!} [e^{-qt} x q (qt)^{x-1} - q e^{-qt}(qt)^x] \\
&= \sum_{x=k}^{\infty} \frac{q e^{-qt}(qt)^{x-1}}{(x-1)!} - \sum_{x=k}^{\infty} \frac{q e^{-qt}(qt)^x}{x!}
\end{aligned}
$$

Letting $j = x - 1$ in the first sum and $j = x$ in the second yields

$$f_Y(t) = \sum_{j=k-1}^{\infty} \frac{qe^{-qt}(qt)^j}{j!} - \sum_{j=k}^{\infty} \frac{qe^{-qt}(qt)^j}{j!}$$

The first series is the same as the second except there is an extra term, $j = k - 1$, in the first series. The difference of the sums is simply that extra term. Consequently,

$$f_Y(t) = \frac{qe^{-qt}(qt)^{k-1}}{(k-1)!}$$

which is a k-Erlang distribution with $\beta = 1/q$. ■

Examples 4.25 and 4.26 involved applications of Theorem 4.11 with $q = 2$ and $q = 2000$, respectively.

EXAMPLE 4.27

In a batch processing environment, the number of jobs arriving for service is 9 per hour. If the arrival process satisfies the requirements of a Poisson experiment, then Theorem 4.11 applies and the elapsed time X between a given arrival and the fifth subsequent arrival possesses a 5-Erlang distribution with $\beta = 1/9$. The probability that the elapsed time is less than 10 minutes is

$$P\left(X < \frac{1}{6}\right) = F_X\left(\frac{1}{6}\right) = \gamma\left(\frac{1}{6} \div \frac{1}{9}; 5\right) = \gamma(1.5; 5) = 0.0285$$

where we have employed linear interpolation in Table A.4. According to Theorem 4.10, the expected elapsed time in hours is

$$E[X] = 5 \times \frac{1}{9} = \frac{5}{9} = 0.5556$$

4.6.4 Exponential Distribution

The exponential distribution was introduced in Section 3.4.2 as Definition 3.8. However, the exponentially distributed random variables form a subclass of the gamma distributed variables. Letting $\alpha = 1$ and $b = 1/\beta$ in the gamma distribution yields the exponential distribution. In particular, an exponentially distributed random variable is 1-Erlang with $\beta = 1/b$. According to Theorem 4.10, if X is exponentially distributed with parameter b, then its moment-generating function, mean, and variance are given by $M_X(t) = b(b - t)^{-1}$, $\mu = b^{-1}$, and $\sigma^2 = b^{-2}$, respectively. These were derived independently in Examples 3.25, 3.20, and 3.31, respectively.

Since the exponential distribution is 1-Erlang, Theorem 4.11 applies with $k = 1$ and takes the form given in the next theorem.

THEOREM 4.12

Suppose events are occurring in such a way that the number of events occurring in any given time interval of length t is described by a Poisson distribution with mean qt. If Y denotes the random variable that measures the interevent time, then Y possesses an exponential distribution with parameter $b = q$.

EXAMPLE 4.28   Referring to Example 4.27, where the rate of arrival was 9 jobs per hour, the interarrival distribution is

$$f(t) = 9e^{-9t}$$

for $t \geq 0$. The mean interarrival time is $1/b = 1/9$, which states that a job arrives every 1/9 of an hour.

EXAMPLE 4.29   The exponential distribution is often employed to model the time to failure of a system. In such a case, the random variable X gives the length of time a system functions before it fails. For instance, if a system possesses a time-to-failure distribution that is exponential with mean $\mu = 6$ months, then, in years, X has the density $f(x) = 2e^{-2x}$ for $x \geq 0$ and has standard deviation 1/2 year. The probability the system functions without failure for at least 1 year is

$$P(X \geq 1) = 1 - F_X(1) = 1 - (1 - e^{-2}) = 0.1353$$

EXAMPLE 4.30   Service times can also be modeled by exponential distributions. For instance, consider an automatic control system that controls the ignition spark to a gasoline engine. Based upon sensor information, an algorithm in a microprocessor updates the timing. Suppose an updating only takes place if some threshold condition is exceeded by the changes in the sensor readings. Given a change in condition sufficient to result in an updating, suppose the average service time of the microprocessor is 0.02 seconds. Then a suitable service-time distribution might be exponential with mean 0.02, so that $b = 50$. As an illustration, the probability that the processor will require greater than 0.05 seconds is

$$P(X > 0.05) = \int_{0.05}^{\infty} 50e^{-50x} \, dx = e^{-(50)(0.05)} = 0.0821$$

From the perspective of modeling, one of the salient properties of an exponentially distributed random variable X is that it is memoryless. Intuitively, memorylessness can be understood by considering a system whose time to failure is exponentially distributed. The probability that such a system will function for longer than t hours is the same as the probability that the system will function for longer than $t + s$ hours, given that it has already functioned for at least s hours. In other words, our perception of the reliability of the system is not modified by the fact that we know it has been in service for some length of time. While it might at first appear that such a model is "unreasonable," observational data supports the suitability of such models in numerous circumstances. Moreover, there may be an engineering reason why a certain system should be viewed as memoryless.

In terms of conditional probabilities, a random variable X is **memoryless** if, for all nonnegative s and t,

$$P(X > t + s \mid X > s) = P(X > t)$$

It can be shown that a continuous random variable X is memoryless if and only if it is exponentially distributed.

4.6.5 Chi-Square Distribution

Letting $\alpha = \nu/2$ and $\beta = 2$ in the gamma density yields the chi-square density. This density plays an important role in statistical inference.

Chi-square Distribution. The random variable X is said to possess a chi-square distribution with ν **degrees of freedom** if it has the density

$$f(x) = \begin{cases} \dfrac{2^{-\nu/2}}{\Gamma(\nu/2)} x^{(\nu/2)-1} e^{-x/2}, & \text{if } x > 0 \\ 0, & \text{if } x \leq 0 \end{cases}$$

If X possesses a chi-square distribution with ν degrees of freedom, an expression that will be explained when the distribution is applied to statistical inference, we often write $X \sim \chi_\nu^2$. The next theorem follows immediately from Theorem 4.10 by letting $\alpha = \nu/2$ and $\beta = 2$ in that theorem.

If X possesses a chi-square distribution with ν degrees of freedom, then its moment-generating function, mean, and variance are respectively given by

(i) $M_X(t) = (1 - 2t)^{-\nu/2}$
(ii) $\mu = \nu$
(iii) $\sigma^2 = 2\nu$

In Example 3.38, the random variable X possessed a chi-square density with $\nu = 1$ and it was shown that X was identically distributed with the square of the standard normal distribution. This relationship between the chi-square distribution and the normal distribution will prove to be significant in later chapters.

4.7 BETA DISTRIBUTION

The beta distribution is a continuous distribution for which the values of the corresponding random variable are restricted to an interval of finite length. Its two parameters, α and β, both of which affect the shape of the distribution, can be varied to give rise to a wide variety of shapes, thus allowing manipulation of the distribution to fit various empirical data.

The definition of the beta density involves the beta function, which was introduced in Section 4.6.1. It is immediate from the definition of the beta function that the total integral of the beta density is 1 for any choice of α and β.

Beta Distribution. A random variable X is said to possess a beta distribution if it has the density

$$f(x) = \begin{cases} \dfrac{1}{B(\alpha, \beta)} x^{\alpha-1}(1 - x)^{\beta-1}, & \text{if } 0 < x < 1 \\ 0, & \text{otherwise} \end{cases}$$

where $\alpha > 0$ and $\beta > 0$.

As illustrated in Figure 4.21, the beta distribution takes on various shapes. If $\alpha < 1$ and $\beta < 1$, it is U-shaped; if $\alpha < 1$ and $\beta \geq 1$, it is reverse J-shaped; if

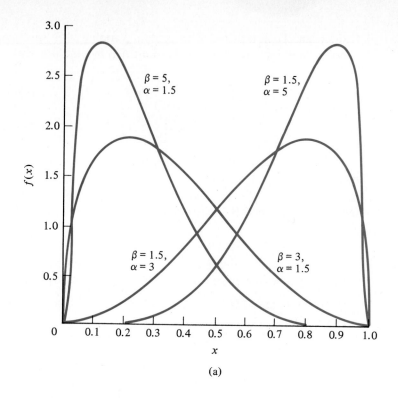

(a)

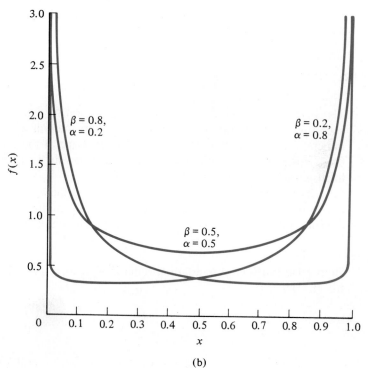

FIGURE 4.21
Various members of the beta family (Hahn/ Shapiro, Statistical Models in Engineering, © 1967 by John Wiley & Sons, Inc.)

(b)

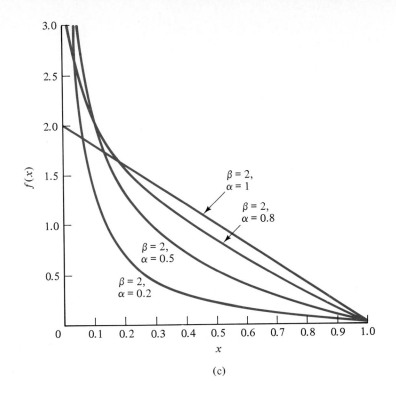

(c)

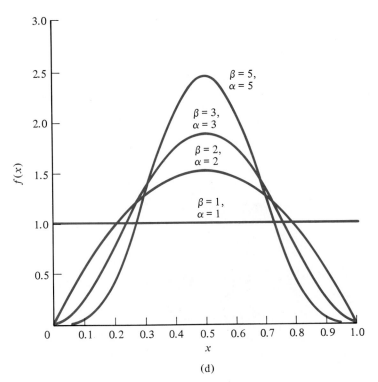

(d)

$\alpha \geq 1$ and $\beta < 1$, it is J-shaped; and if $\alpha > 1$ and $\beta > 1$, it possesses a single maximum at

$$x = \frac{\alpha - 1}{\alpha + \beta - 2}$$

If $\alpha = \beta$, then the graph of the density is symmetrical.

THEOREM 4.14

If X possesses a beta distribution, then its mean, variance, and higher-order moments are respectively given by

(i) $\mu = \dfrac{\alpha}{\alpha + \beta}$

(ii) $\sigma^2 = \dfrac{\alpha\beta}{(\alpha + \beta)^2(\alpha + \beta + 1)}$

(iii) $\mu'_k = \dfrac{\Gamma(\alpha + \beta)\Gamma(\alpha + k)}{\Gamma(\alpha)\Gamma(\alpha + \beta + k)}$

Proof: We begin by finding the kth moment directly from the definition. Applying the definition of $B(x, y)$ yields

$$E[X^k] = \frac{1}{B(\alpha, \beta)} \int_0^1 x^{\alpha + k - 1}(1 - x)^{\beta - 1}\, dx$$

$$= \frac{B(\alpha + k, \beta)}{B(\alpha, \beta)}$$

$$= \frac{\Gamma(\alpha + \beta)\Gamma(\alpha + k)\Gamma(\beta)}{\Gamma(\alpha)\Gamma(\beta)\Gamma(\alpha + k + \beta)}$$

which reduces to (iii). As a result,

$$E[X] = \frac{\Gamma(\alpha + \beta)\Gamma(\alpha + 1)}{\Gamma(\alpha)\Gamma(\alpha + \beta + 1)}$$

$$= \frac{\Gamma(\alpha + \beta)\alpha\Gamma(\alpha)}{\Gamma(\alpha)(\alpha + \beta)\Gamma(\alpha + \beta)}$$

which reduces to (i). The second moment is

$$E[X^2] = \frac{\Gamma(\alpha + \beta)\Gamma(\alpha + 2)}{\Gamma(\alpha)\Gamma(\alpha + \beta + 2)}$$

$$= \frac{\Gamma(\alpha + \beta)(\alpha + 1)\alpha\Gamma(\alpha)}{\Gamma(\alpha)(\alpha + \beta + 1)(\alpha + \beta)\Gamma(\alpha + \beta)}$$

$$= \frac{\alpha(\alpha + 1)}{(\alpha + \beta + 1)(\alpha + \beta)}$$

Finally,

$$\sigma^2 = \mu_2' - \mu^2 = \frac{\alpha(\alpha + 1)}{(\alpha + \beta + 1)(\alpha + \beta)} - \frac{\alpha^2}{(\alpha + \beta)^2}$$

which simplifies to (ii). ∎

Because of the great diversity of shapes exhibited by beta densities, the distribution has found wide application in modeling measurements that are restricted to a finite interval.

EXAMPLE 4.31

Suppose n independent random observations of a random phenomenon Y are made and the observations are listed in order from y_1 to y_n. In addition, let y_r and y_{n-s+1} be the rth smallest and sth greatest observations, respectively. If X is the random variable that denotes the proportion of the population described by Y that falls betwen y_r and y_{n-s+1}, then it can be shown that X possesses a beta distribution with $\alpha = n - r - s + 1$ and $\beta = r + s$.

As an illustration, consider an analytical chemist who plans to randomly select 10 units and weigh the active ingredients in each unit. What is the probability that at least 95% of the total units produced possess an active-ingredient weight between the lowest and highest weights found by the chemist? If X measures the proportion of the population between the two weights, then, since $n = 10$, $r = 1$, and $s = 1$, X possesses a beta distribution with $\alpha = 9$ and $\beta = 2$. Thus,

$$P(X > 0.95) = 1 - P(X \leq 0.95)$$

$$= 1 - \int_0^{0.95} f_X(x)\, dx$$

$$= 1 - \frac{\Gamma(11)}{\Gamma(9)\Gamma(2)} \int_0^{0.95} x^8(1 - x)\, dx$$

$$= 1 - \frac{10!}{8!1!} \left(\frac{x^9}{9} - \frac{x^{10}}{10} \right) \Bigg|_0^{0.95} = 0.091$$

The conclusion is that there is only a 0.091 probability that 95% of the production is within the limits set by the extremes of the sample. Had, instead, the sample size been 50 units, then the result would have been

$$P(X > 0.95) = 1 - \frac{50!}{48!} \int_0^{0.95} x^{48}(1 - x)\, dx = 0.7207$$

EXAMPLE 4.32

The beta distribution can be generalized so as to cover the interval (a, b). This **generalized beta distribution** possesses the density

$$f(x) = \frac{(x - a)^{\alpha - 1}(b - x)^{\beta - 1}}{(b - a)^{\alpha + \beta - 1} B(\alpha, \beta)}$$

for $a < x < b$, and $f(x) = 0$ elsewhere. If $a = 0$ and $b = 1$, this generalized beta density reduces to the standard beta density. The mean and variance for the generalized distribution are

$$\mu = a + \frac{(b - a)\alpha}{\alpha + \beta}$$

and

$$\sigma^2 = \frac{(b - a)^2 \alpha\beta}{(\alpha + \beta)^2(\alpha + \beta + 1)}$$

respectively.

An application of the generalized beta distribution is in the estimated time to complete a project phase. The methodology is known as PERT (Program Evaluation and Reporting Technique). The time X required for completion is modeled as a generalized beta distribution, with a and b being respectively the **optimistic** and **pessimistic** predicted times to completion, these times being engineering modeling assumptions.

As an illustration, suppose, based on experience (empirical considerations), the PERT model is employed with $a = 2$ weeks, $b = 7$ weeks, $\alpha = 3$, and $\beta = 2$. Then the expected time to completion is

$$E[X] = 2 + \frac{(7 - 2)3}{3 + 2} = 5 \text{ weeks}$$

The probability that the project phase will be completed in less than 6 weeks is

$$P(X < 6) = \frac{4!}{5^4 2! 1!} \int_2^6 (x - 2)^2(7 - x)\, dx = 0.8192$$

4.8 WEIBULL DISTRIBUTION

The Weibull distribution is commonly employed in time-to-failure models as an alternative to the exponential distribution. It possesses two parameters, α and β, and reduces to the exponential distribution when $\alpha = 1$ and $\beta = 1/b$. In Section 6.3.2 we will specifically consider its role in reliability.

4.8.1 The Weibull Density

DEFINITION 4.11

Weibull Distribution. A random variable X is said to possess a Weibull distribution if it has the density

$$f(x) = \begin{cases} \alpha\beta^{-\alpha}x^{\alpha-1}e^{-(x/\beta)^\alpha}, & \text{if } x > 0 \\ 0, & \text{if } x \le 0 \end{cases}$$

where $\alpha > 0$ and $\beta > 0$.

Figure 4.22 illustrates various members of the Weibull family.

The probability distribution function of a Weibull density can be expressed in the convenient closed-form expression

$$F(x) = 1 - e^{-(x/\beta)^\alpha}$$

for $x > 0$. In fact, a straightforward integration gives

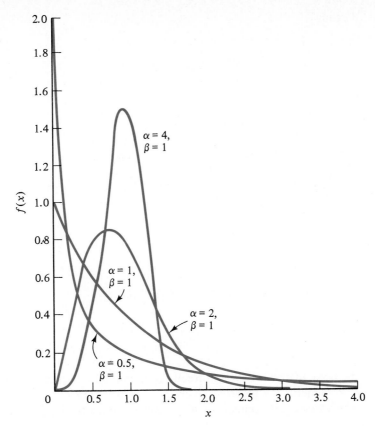

FIGURE 4.22
Various members of the Weibull family (Hahn/ Shapiro, Statistical Models in Engineering, © 1967 by John Wiley & Sons, Inc.)

$$F(x) = \alpha\beta^{-\alpha} \int_0^x t^{\alpha-1} e^{-(t/\beta)^\alpha} \, dt$$

$$= -e^{-(t/\beta)^\alpha} \Big|_0^x$$

$$= 1 - e^{-(x/\beta)^\alpha}$$

THEOREM 4.15

If X possesses a Weibull distribution, then its mean, variance, and higher-order moments are respectively given by

(i) $\mu = \beta\Gamma\left(\dfrac{\alpha+1}{\alpha}\right)$

(ii) $\sigma^2 = \beta^2\left[\Gamma\left(\dfrac{\alpha+2}{\alpha}\right) - \Gamma^2\left(\dfrac{\alpha+1}{\alpha}\right)\right]$

(iii) $\mu_k' = \beta^k\Gamma\left(\dfrac{\alpha+k}{\alpha}\right)$

Proof: First find the kth moment:

$$E[X^k] = \alpha\beta^{-\alpha} \int_0^\infty x^{\alpha+k-1} e^{-(x/\beta)^\alpha} \, dx$$

Letting

$$u = \left(\frac{x}{\beta}\right)^{\alpha}$$

gives

$$x = \beta u^{1/\alpha}$$

$$dx = (\beta/\alpha)u^{(1/\alpha)-1} \, du$$

and

$$E[X^k] = \alpha\beta^{-\alpha} \int_0^\infty (\beta u^{1/\alpha})^{\alpha+k-1} e^{-u} \left(\frac{\beta}{\alpha}\right) u^{(1/\alpha)-1} \, du$$

$$= \beta^k \int_0^\infty u^{k/\alpha} e^{-u} \, du$$

$$= \beta^k \, \Gamma\left(1 + \frac{k}{\alpha}\right)$$

the last equality following directly from the definition of the gamma function. Letting $k = 1$ yields the mean as stated in (i) and letting $k = 2$ gives the second moment. The variance, as stated in (iii), follows at once from Theorem 3.12. ∎

4.8.2 Rayleigh Distribution

A useful subfamily of the Weibull family is determined by letting $\alpha = 2$ and $\beta = \sqrt{2}\eta$. These changes yield the Rayleigh distribution.

DEFINITION 4.12

Rayleigh Distribution. A random variable X is said to possess a Rayleigh distribution if it has the density

$$f(x) = \begin{cases} \eta^{-2} x e^{-(x/\eta)^2/2}, & \text{if } x > 0 \\ 0, & \text{if } x \leq 0 \end{cases}$$

where $\eta > 0$.

Applying Theorem 4.15 to the Rayleigh distribution gives the next proposition.

THEOREM 4.16

If X possesses a Rayleigh distribution, then its mean, variance, and higher-order moments are respectively given by

(i) $\mu = \sqrt{\dfrac{\pi}{2}}\,\eta$

(ii) $\sigma^2 = \dfrac{4 - \pi}{2}\,\eta^2$

(iii) $\mu_k' = \dfrac{k}{2}(\sqrt{2}\eta)^k\Gamma\left(\dfrac{k}{2}\right)$

Proof: Letting $\alpha = 2$ and $\beta = \sqrt{2}\eta$ in Theorem 4.15 yields

$$\mu = \sqrt{2}\eta\Gamma\left(\dfrac{3}{2}\right)$$

which results in (i) since $\Gamma(3/2) = \sqrt{\pi}/2$. As for the variance,

$$\sigma^2 = 2\eta^2\left[\Gamma(2) - \Gamma\left(\dfrac{3}{2}\right)^2\right] = 2\eta^2\left(1 - \dfrac{\pi}{4}\right)$$

Lastly,

$$\mu_k' = (\sqrt{2}\eta)^k\Gamma\left(1 + \dfrac{k}{2}\right) = \dfrac{k}{2}(\sqrt{2}\eta)^k\Gamma\left(\dfrac{k}{2}\right)$$ ∎

Since a Rayleigh-distributed random variable X is a Weibull variable with $\alpha = 2$ and $\beta = \sqrt{2}\eta$, its probability distribution function derives from the distribution function of the Weibull distribution. Consequently,

$$F_X(x) = 1 - e^{-(x/\eta)^2/2}$$

EXAMPLE 4.33

Consider a guidance system in which there are two orthogonal accelerometers, one each along the x and y axes, to measure x and y accelerations, respectively. If the accelerometer errors are independent and the error of each accelerometer is normally distributed with mean 0 and standard deviation 0.5, then it can be shown that the radial error X possesses a Rayleigh distribution with $\eta = 0.5$ (see Example 5.17). According to Theorem 4.16, the expected error is

$$E[X] = 0.5 \times \sqrt{\dfrac{\pi}{2}} = 0.627$$

The probability that the error will be less than 0.85 is

$$P(X < 0.85) = F_X(0.85) = 1 - e^{-(0.85/0.5)^2/2} = 0.764$$

EXERCISES

SECTION 4.1

4.1. For a binomial distribution with $n = 15$ and $p = 0.2$, find
   a) $P(X < 5)$
   b) $P(X \geq 6)$
   c) $P(X = 7)$
   d) $P(2 < X \leq 5)$

4.2. A fair coin is tossed 20 times. Find the following:
   a) The probability that at least 8 heads appear.

   b) The probability that between (inclusively) 7 and 13 heads appear.
   c) The expected number of heads.
   d) $P(|X - \mu| \leq 2\sigma)$, where X counts the number of heads, μ is the mean, and σ is the standard deviation.

4.3. Repeat Exercise 4.2, except assume the probability of a head is 0.7.

4.4. An automatic stamping machine requires sheets to be in a particular orientation. If 4% of the plates are not placed in the proper orientation and it is assumed the placements are independent, what is the probability that less than 5 out of 40 plates will not be stamped correctly?

4.5. In a certain interactive programming environment, 12% of the Pascal programs arriving to be compiled will not do so due to syntactical errors. If the programmer population is sufficiently large, it can be assumed that failures to compile are independent and, relative to S (compiles) and F (does not compile), the process is Bernoulli. For an arbitrary collection of 25 submitted programs, what is the probability that at least 5 will fail to compile? If X counts the number of programs of the 25 that do compile, what are the mean and variance of X?

4.6. A quality control inspector is supposed to find unsafe welds in a fuselage by randomly selecting and visually inspecting welds. If the inspector misidentifies 3% of the welds, how many welds need to be inspected for the probability of at least one misidentification to exceed 0.95?

4.7. The probability of a transmitted bit reaching its destination unaltered is 0.999. Assuming transmission errors to be independent, if a message contains 10,000 bits, what is the expected number of bit errors? Find the standard deviation.

4.8. Referring to Exercise 4.7, find the probability that there is at least one bit error.

4.9. A production process is partitioned into two independent subprocesses. The probabilities of a defective component in the first and second subprocesses are 0.01 and 0.02, respectively. If 50 units are produced, what is the probability there will be less than 2 defective units?

4.10. Reconsider the Brownian motion problem of Example 4.4 under the condition that there is a *barrier* at the origin so that when the particle is at the origin, any movement left is abbreviated and the particle remains at the origin. Find the probability mass function for $n = 6$.

4.11. Use the moment-generating function to find the third moment of the binominal distribution. Use this moment to show that the coefficient of skewness is

$$\alpha_3 = \frac{1 - 2p}{(npq)^{1/2}}$$

(see Exercises 3.68 and 3.69).

4.12. Apply the result of Exercise 4.11 to the tossing of an unfair coin six times, the probability of a head being 0.7.

4.13. Use the moment-generating function to find the fourth moment of the binomial distribution. Use this moment to show that the coefficient of kurtosis is

$$\alpha_4 = 3 + \frac{1 - 6pq}{npq}$$

(see Exercises 3.68 and 3.70).

4.14. Apply the result of Exercise 4.13 to the tossing of an unfair coin six times, the probability of a head being 0.7.

4.15. X is binomially distributed with parameters n and p. Find the densities for $Y = 2X + 1$ and $Y = X^2$.

4.16. Find the mean of the binomial distribution directly from the definition without appealing to the moment-generating function.

4.17. Prove the recursion relation

$$b(x + 1; n, p) = \frac{(n - x)p}{(x + 1)q} b(x; n, p)$$

Find the probability mass function for the binomial distribution with $n = 6$ and $p = 0.3$ by using the recursion relation.

4.18. An inspector checks 30 throttle springs from a large shipment to see if they are within specified tolerance limits for tension. If more than 2 are found to be outside the designated limits, then the entire shipment is rejected; otherwise, it is accepted. Assuming 10% of the springs in the entire shipment are outside the prescribed range, what is the probability the shipment will be rejected? By supposing the shipment to be ''large,'' we are making the assumption that the sampling process can be treated as though it were with replacement, even though the springs are to be subjected to destructive testing and therefore cannot be placed back into the general population of springs. The question of largeness will be addressed in more detail in the next section.

SECTION 4.2

4.19. For $N = 12$, $n = 6$, and $k = 9$, find the values of the hypergeometric distribution.

4.20. A committee of 5 is to be randomly chosen from a group of 8 men and 6 women. Let X count the number of women selected. Write down the density for X and then find the probability that there will be less than 3 men on the committee. What is the expected number of women on the committee?

4.21. In a box of 20 minifloppies, it is known that 5 have not been formatted. If 6 disks are drawn at random, what is the expected number of formatted disks that will be drawn? What is the probability that no more than 3 formatted disks will be drawn?

4.22. A store purchases 50 batteries and a subset of 5 are chosen for testing. If 12 of the shipped batteries do not possess the specified cold cranking power, what is the expected number of test batteries that will not possess the specified cold cranking power? What is the probability that no more than 2 of the test batteries will not possess the desired power?

4.23. An urn contains 6 white and 4 black balls. Five are picked at random without replacement. Let X count the number of black balls. Find the mean and variance of X and then find $P(|X - \mu| \geq 2\sigma)$. Compare the actual value to the bound given by Chebyshev's inequality.

4.24. In a shipment of 5000 hydrometers, 4% are defective. If 250 are tested, what is the expected number of defective units in the sample? Find the variance.

4.25. Referring to Exercise 4.24, approximate the hypergeometric distribution by the binomial distribution and find both the mean and variance of the approximating binomial model.

4.26. Chemical fertilizer arrives in three different shipments, A, B, and C, each containing 40 batches. Four batches from each shipment are randomly selected and tested for nitrogen content. If two or more of the tested batches from any one shipment are deficient in nitrogen, then all three shipments are rejected. If shipments A, B, and C contain 3, 2, and 5 deficient batches, respectively, what is the probability that the entire shipment will be rejected?

SECTION 4.3

4.27. Balls are selected with replacement from an urn containing 3 white and 5 black balls. What is the probability that 7 white balls will be selected prior to the selection of the fourth black ball? What is the probability that the seventh black ball will be selected on the tenth pick?

4.28. In a large lot of polished steel shafts, 5% have surfaces that are not sufficiently smooth. Assuming the lot is sufficiently large that the sampling process can be treated as a sequence of Bernoulli trials, what is the probability that 10 satisfactory shafts will be selected prior to the selection of one that is not satisfactory? What is the probability that the third unsatisfactory shaft will be chosen on the thirtieth selection?

4.29. A copier gets jammed on approximately 0.7% of the documents fed into it. Assuming the jamming to be independent, what is the expected number of documents that will be copied prior to jamming? What is the variance of the random variable counting the number of documents copied prior to jamming?

4.30. Two long (infinite) lines of customers are waiting for service at a billing department. Although the customers believe each line to be independently serviced, in fact there is only one filing system working, so that behind the scenes only one customer is being serviced. Rather than service the customers alternately, the billing clerk has decided to flip a fair coin, a head meaning that a person from queue A will be served and a tail meaning that a person from queue B will be served.

For the customer fifth in line in queue B, answer the following questions:
a) What is the probability that 8 customers in queue A will be served prior to her being served?
b) What is the probability that at least 2 customers from queue A will be served first?
c) What is the expected number of customers from queue A who will be served prior to her?

4.31. Redo Exercise 4.30 assuming the coin is loaded so that the probability of a head is 2/3.

4.32. Use the moment-generating function to find the third moment for the negative binomial distribution. Use this moment to show that the coefficient of skewness is

$$\alpha_3 = \frac{2 - p}{[k(1 - p)]^{1/2}}$$

4.33. Apply the result of Exercise 4.32 to the geometric distribution. Further particularize the result by applying it to the geometric distribution with $p = 1/2$.

SECTION 4.4

4.34. For $\lambda = 5$, compute the first 9 values of the Poisson density.

4.35. Jobs arrive at a repair shop at the rate of 24 per (8-hour) day. Assuming the arrival process is independent and can be modeled as a Poisson experiment, what is the expected number of jobs that arrive in any given hour? What is the probability that more than 5 jobs will arrive in any given hour?

4.36. Data arrive at an input buffer at the rate of 4 pages per second, and the arrival process is modeled as Poisson. What is the expected number of page arrivals in a 3-second period? What is the probability of more than 15 pages arriving in a 3-second period?

4.37. Electrical power failures in a workplace have been modeled as a Poisson experiment with a rate of 1.5 per month.
a) What is the expected number of power failures in a year?
b) What is the probability of having more than 20 failures in a year?
c) What is the probability that the number of failures in a year will differ by more than a standard deviation from the expected number?

4.38. In a batch processing environment, jobs are completed at the rate of 1.4 per minute, and it is assumed that the completions satisfy the Poisson conditions. Let X count the number of jobs completed in a 10-minute span. Find the mean and variance of X. Find the probability that more than 20 jobs will be completed in a 10-minute span.

4.39. An automatic text reader commits errors at the rate of 3 letters per page. Assuming the number of errors can be modeled as a Poisson distribution with mean *qt*, where *t* is the number of pages, what is the expected number of errors on a 5-page report? What is the probability that the number of errors on the report will be between 12 and 20?

4.40. On a large construction site the number of workers who do not report to work on a given day is modeled as a Poisson distribution, the rate of absenteeism being 8 per day. Note that the model might be questioned, since there may be some degree of dependence from day to day regarding absentees; nontheless, it will be assumed that data support the model. Over a five-day period, what is the probability that the number of absentees will exceed 10 on each day? What is the probability that the number of absentees will be less than 5 on at least two of the days? (The daily numbers of absentees are independently distributed.)

4.41. The percentage of substandard rivets on a bridge is 1% and the allowable percentage is 0.6%. If 400 rivets are randomly selected for inspection, what is the probability of the conclusion being drawn that the rivet population is substandard? Use the Poisson approximation of the binomial distribution.

4.42. Use the moment-generating function of the Poisson distribution to find its third moment and then show that the coefficient of skewness is $\alpha_3 = \lambda^{-1/2}$.

4.43. Use the moment-generating function of the Poisson distribution to find its fourth moment and then show that the coefficient of kurtosis is $\alpha_4 = 3 + 1/\lambda$.

4.44. The concept of a Poisson experiment can be generalized from intervals to areas and volumes. Random points distributed in the plane are said to constitute a Poisson experiment with rate q if (1) for any collection of disjoint regions $R_1, R_2, \ldots, R_n$, the random variables $N(R_1), N(R_2), \ldots, N(R_n)$ that count the number of points in the regions are independent and (2) for any region R of finite area, the random variable $N(R)$ possesses a Poisson distribution with mean $qA(R)$, where $A(R)$ is the area of R. The number of automobile accidents occurring in a region over the period of a day is modeled as a Poisson experiment with rate $q = 0.02$ per square mile. What is the probability that more than 3 accidents will occur in a given 100-square-mile region?

4.45. Using integration by parts, show that the probability distribution function of a Poisson variable with parameter λ is given by the formula

$$\Pi(x; \lambda) = \frac{1}{x!} \int_{\lambda}^{+\infty} t^x e^{-t}\, dt$$

SECTION 4.5

4.46. *X* is normally distributed with mean 2 and variance 9. Find
a) $P(X < 2.45)$
b) $P(X < 0.14)$

c) $P(3.22 < X < 5.26)$
d) $P(-1.56 < X < 0.39)$
e) $P(X > -2.33)$
f) $P(X > 4.02)$
g) $P(|X - \mu| > 5.77)$

4.47. *X* is normally distributed with mean -3.2 and variance 8.5. Find
a) $P(X < -8.33)$
b) $P(X > 1.23)$
c) $P(X > -5.77)$
d) $P(X < 4.93)$
e) $P(-7.26 < X < -3.88)$
f) $P(-6.99 < X < -1.52)$
g) $P(4.22 < X < 4.69)$
h) $P(|X - \mu| < 2.48)$

4.48. For the random variable of Exercise 4.46, find $P(|X - \mu| > k\sigma)$ for $k = 1, 2$, and 3.

4.49. Referring to Exercise 4.48, compare the probabilities found there to the bounds given by Chebyshev's inequality.

4.50. The mean diameter of bolts being produced in a given production process is 0.207 inches and the variance is 0.0001. If the diameters are normally distributed, what is the probability that a bolt selected at random will possess a diameter between 0.19 and 0.21 inches?

4.51. Light bulbs have an advertised rating of 1750 lumens. If measurements are normally distributed with mean 1784 and standard deviation 38, what percentage of bulbs exceed the advertised rating?

4.52. The time to completion of a certain repair is normally distributed with mean 6.32 hours and standard deviation 1.55 hours. What is the probability that the repair will take less than 4 hours? What is the probability that the repair time will be within 3 hours of the mean?

4.53. The mean CPU time for a class of programs on a certain mainframe computer is 2.52 minutes and the standard deviation is 0.37 minutes. What is the probability that a randomly selected program will require between 2 and 4 minutes of CPU time? What is the probability that the CPU time will be within 1 minute of the mean?

4.54. The tolerance limits for a circuit breaker are 40 ± 0.5 amperes. This means that any circuit breaker that breaks at an amperage less than 39.5 breaks at too low a level and any that breaks at an amperage greater than 40.5 breaks at too high a level. If a shipment of circuit breakers possesses break points that are normally distributed with mean 39.3 and standard deviation 0.2, then what percentage of the shipment is defective (outside the specified tolerance limits)?

4.55. A battery is rated at 650 amperes cold cranking power. If a lot of batteries possesses normally distributed cold cranking powers with mean 674 and variance 400, what is the probability that a given battery will not satisfy the rating? What is the probability that a battery will possess cold cranking power between 625 and 675 amperes?

4.56. A temperature sending unit is supposed to signal a fan switch at 185 degrees Fahrenheit. Testing shows that the mean temperature at which the signal is sent is 187.4 degrees and that the distribution of signal temperatures is normally distributed with variance 5.6. If the tolerance limits on the sending unit are ±3 degrees, what is the probability that a unit will send a signal within the specified limits?

4.57. Nails are sold by the pound. The mean weight of a nominal 100 lb box is 100.6 lb; however, the variance of the weights is 2.5.
   a) What percentage of the boxes exceed the nominal weight?
   b) If the variance is kept the same, to what weight must the mean be increased so that 98% of the boxes exceed the nominal weight?
   c) If the mean is kept the same, to what must the variance be reduced so that 98% of the boxes exceed the nominal weight?

4.58. Tires are supposed to provide 40,000 miles of service before tread thickness falls below an unsafe limit. If tire service is normally distributed with standard deviation 5460 miles, then what must be the mean service so that 95% of all tires will exceed the 40,000-mile requirement?

4.59. Flight time of an air carrier is normally distributed with mean 2.34 hours and standard deviation 0.28 hours. If it is desired that the flight arrive at its destination on time 95% of the time, what flying time should be allowed?

4.60. The lifetime of a washing machine is a normally distributed random variable with mean 6.4 years and variance 2.3. Given that a machine has survived 5 years, what is the conditional probability that it will survive 8 years?

4.61. Referring to Example 4.21, not only is it important that all rod weights be within certain limits, but in a high-performance engine it is also important for the rods of the particular engine to possess uniform weight. If 4 rods are selected at random (and independently), what is the probability that the weights of all the rods will fall within 2 standard deviations of the mean?

4.62. A circuit is to employ two resistors, each rated at 12 ohms. Assuming the resistors from which the two have been (randomly) selected possess normally distributed resistances with mean 11.9 ohms and variance 0.04, what is the probability that both resistors possess resistances between 11.7 and 12.3 ohms? What is the probability that at least one resistor possesses a resistance greater than 12.4 ohms?

4.63. The probability of error for a single bit is 10^{-8}. If errors are independent and one billion bits are trans-

mitted, what is the probability of less than or equal to 15 errors?

4.64. For the sake of security, 5% of all messages passed over a network are bogus. A receiver with the appropriate key will know if the message is bogus. Of 400 messages sent, what is the expected number of bogus messages, assuming that they are randomly interspersed among the valid ones? What is the probability that the number of bogus messages will exceed 15 but be less than 25?

4.65. Find the semi-interquartile range for a normally distributed variable with mean μ and variance σ^2 (see Exercise 3.72).

4.66. By using the moment-generating function, find the third moment of the normal distribution. Show that the coefficient of skewness is 0.

4.67. By using the moment-generating function, find the fourth moment of the normal distribution. Show that the coefficient of kurtosis is 3.

4.68. From the definition of the density, use standard calculus maximum-minimum techniques to prove that the normal density possesses a maximum at μ and points of inflection at $\mu - \sigma$ and $\mu + \sigma$.

4.69. Suppose X possesses a Poisson distribution with parameter λ. Show that the limit as λ tends to infinity of the moment-generating function of the standardized Poisson variable

$$X* = \frac{X - \lambda}{\sqrt{\lambda}}$$

approaches the moment-generating function of the standard normal distribution.

SECTION 4.6

4.70. Referring to Example 4.27, find the probability that the interarrival time is greater than 15 minutes.

4.71. The time to failure (in years) of a radar system is modeled as a gamma distribution with $\beta = 1/2$ and $\alpha = 1.5$. What is the probability the system functions for at least 1 year prior to failure? What is the probability it will survive between 1 and 2 years?

4.72. Use an Erlang distribution to model the time to the arrival of the fifth job at the repair shop of Exercise 4.35. Find the probability that the fifth job does not arrive in the first 2 hours.

4.73. Referring to the power-failure model of Exercise 4.37, given a power failure, what is the probability that the next failure will occur within 2 months? What is the probability it will occur prior to 3 months but not in the next 2 months?

4.74. The time to failure of a computer network is modeled by an exponential distribution with mean $\mu = 0.72$ year. What is the probability that the network will not fail within a year?

4.75. The number of orders being placed at a parts distribution center is modeled as a Poisson random variable with a placement rate of 34 per hour. Model the time to the tenth order. Find the probability that the time to the tenth order will be less than fifteen minutes. What is the expected time to the tenth order?

4.76. Trucks arriving at a depot satisfy the conditions of a Poisson experiment with rate 6 per hour. Use the exponential distribution to model the interarrival time. If a truck arrives at 10:15 A.M., what is the probability that the next truck will not arrive until after 10:45 A.M.?

4.77. For the truck arrivals of Exercise 4.76, model the time between an arrival and the eighth following arrival. What is the expected time? What is the variance? What is the probability that the eighth arrival will occur within 2 hours?

4.78. In a certain environment, output time is modeled as an exponential distribution with mean $\mu = 16$ seconds. If a particular output process has been running for 20 seconds, what is the expected remaining output time?

4.79. Use the moment-generating function to find the third moment for the gamma distribution and then show that the coefficient of skewness is $\alpha_3 = 2\alpha^{-1/2}$.

4.80. Apply the result of Exercise 4.79 to the chi-square distribution.

4.81. Use the moment-generating function to find the fourth moment for the gamma distribution and then show that the coefficient of kurtosis is $\alpha_4 = 3(1 + 2/\alpha)$.

4.82. Apply the result of Exercise 4.81 to the chi-square distribution.

SECTION 4.7

4.83. The error of a pressure gage has been normalized to be between 0 and 1. This having been done, the error is modeled by a beta density with parameters $\alpha = 2$ and $\beta = 3$. What is the probability that the normalized error will be between 0.2 and 0.8? What is the expected normalized error?

4.84. The speeds of 30 randomly selected cars are observed as they pass a point on a highway. Find the probability that at least 95% of the cars passing the observation point are moving between the lowest and highest observed speeds. Find the probability that at least 98% are traveling between the extreme observed speeds.

4.85. The breaking strengths (in pounds) of 25 randomly selected cement briquettes are tested. Find the probability that at least 90% of all briquettes produced possess breaking strengths between the second lowest and the highest observed strengths.

4.86. The time to completion for the development of a new software package is modeled using PERT, with the optimistic time being 8 months, the pessimistic time being 15 months, $\alpha = 2$, and $\beta = 3$. Find the probability that the project will be completed within 1 year. Find the expected time to completion.

4.87. The time to completion of a project to develop a new infrared sensor is modeled using PERT, with the optimistic time being 2 years, the pessimistic time being 3.5 years, $\alpha = 4$, and $\beta = 2$. Find the probability that the project will not be completed for at least 3 years. Find the expected time to completion.

4.88. Derive the formulas given in Example 4.32 for the mean and variance of the generalized beta distribution.

4.89. Show that the coefficient of skewness for the beta distribution is

$$\alpha_3 = \frac{2(\beta - \alpha)(\alpha + \beta + 1)^{1/2}}{\alpha^{1/2}\beta^{1/2}(\alpha + \beta + 2)}$$

SECTION 4.8

4.90. The time to failure (in years) of a needle-bearing assembly possesses a Weibull distribution with $\alpha = 2$ and $\beta = 5$.
   a) What is the expected lifetime of an assembly?
   b) What is the variance?
   c) Find the probability that a bearing assembly will remain functional for more than 9 years.

4.91. The time (in years) to the first major resurfacing of a roadway with a certain surface possesses a Weibull distribution with $\alpha = 2$ and $\beta = 3.6$. Find the probability that the first major resurfacing will have to be done in less than 7 years but not in the first 3 years.

4.92. An automatically guided and self-propelled vehicle uses sensors that follow guide wires embedded in the plant floor. The radial distance by which a particular vehicle misses its proposed destination is given by a Rayleigh distribution with $\eta = 0.08$ meters
   a) What is the probability that the vehicle misses its targeted destination by more than 0.2 meters?
   b) What is the expected radial error?
   c) What is the standard deviation of the error?

4.93. The radial distance of a bore from the proposed center is a random variable possessing a Rayleigh distribution with $\eta = 0.6$ mm.
   a) What is the probability that the bore will be within 1 mm of its targeted center?
   b) What is the expected radial error?
   c) What is the standard deviation of the error?

5

||||||||||||||||||||||||||||||||||||||||||||||||||||||||||||||||||||||||||||||||||||||||||||||||||||||||||||||||||

MULTIVARIATE DISTRIBUTIONS

Rather than deal with a single random variable (measurement) in isolation, it is common to consider two or more random variables jointly. In such a circumstance, it is not merely the probability of this or that random variable taking on a particular value that is of interest, but rather the likelihood of a number of random variables taking on a number of values. More precisely, the variables possess a joint distribution. Of particular interest in random sampling is the notion of two random variables being independent of one another, or, more generally, a whole set of variables being independent of one another.

5.1 JOINTLY DISTRIBUTED RANDOM VARIABLES

Thus far, we have considered measurement processes involving a single random variable. Very often, however, the description of behavior requires a number of measurements. Typically, a system is understood in terms of input and output variables, where the latter are functions of the former. Should there be appreciable variability in the determination of the input variables, then they must be considered to be random. The resulting situation is depicted in Figure 5.1, where there are n input and m output random variables. As in the single-variable case discussed in Section 3.4, we need to determine the probability characteristics of the output variables in terms of the input variables. Indeed, it is precisely this determination that constitutes our understanding of the effects of the system.

What makes the **multivariate** case substantially different is that we cannot always consider each input variable in isolation from the others. If X and Y are two random variables, then, of necessity, probability statements about X and Y must take into account the **joint** characteristics of the variables. For instance, due to channel noise it is often prudent to view the reception of a signal as a random phenomenon. If we view a signal as a waveform in the time domain, then at each moment of time

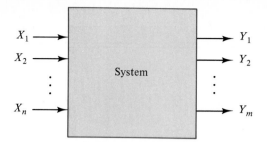

FIGURE 5.1
A system with input and output random variables

there is a distinct random variable. It is certainly reasonable to expect there to be a close relationship between variables close in time. Indeed, this relationship can be used to filter the received signal in order to obtain a more ''pure'' replication of the transmitted signal.

5.1.1 Discrete Multivariate Distributions

In a manner similar to the **univariate** (single-variable) case, jointly distributed discrete random variables can be described by a probability mass function; however, in the present circumstance the mass function will exhibit the manner in which the random variables are **jointly distributed.** To both illustrate and motivate the forthcoming definitions, we will study a simple case of two random variables resulting from an urn model.

Consider an urn in which there are six balls, 3 black, 2 white, and 1 red. The experiment is to uniformly randomly select 3 balls with replacement. Let X and Y respectively denote the numbers of black and white balls selected. Both X and Y are binomially distributed, with X possessing the density $f_X(x) = b(x; 3, 1/2)$ and Y possessing the density $f_Y(y) = b(y; 3, 1/3)$. Insofar as we consider X alone, $f_X(x)$ provides our full understanding of the variable. A similar statement applies to Y and its density $f_Y(y)$. Yet what of the relationship between the variables? For instance, what is the probability that $X = 2$ and $Y = 1$? Here, we are interested not merely in a probability statement involving a single variable; rather, the statement

$$P(X = 2, Y = 1) = p$$

expresses quantitative information concerning the variables jointly. As such, it represents a **joint probability.**

Referring to the tree diagram in Figure 5.2, exactly three paths, b–b–w, b–w–b, and w–b–b, result in the selection of 2 black balls and 1 white ball. Thus,

$$P(X = 2, Y = 1) = 3 \times \frac{1}{2} \times \frac{1}{2} \times \frac{1}{3} = \frac{1}{4}$$

In terms of the sample space, which is the product space $\{b, w, r\}^3$, the event resulting in $X = 2$ and $Y = 1$ is

$$\{(b, b, w), (b, w, b), (w, b, b)\}$$

each triple in the event having probability 1/12.

A full description of the joint behavior of X and Y must include the probabilities of all possible joint outcomes. These are given in Table 5.1. Note that there are

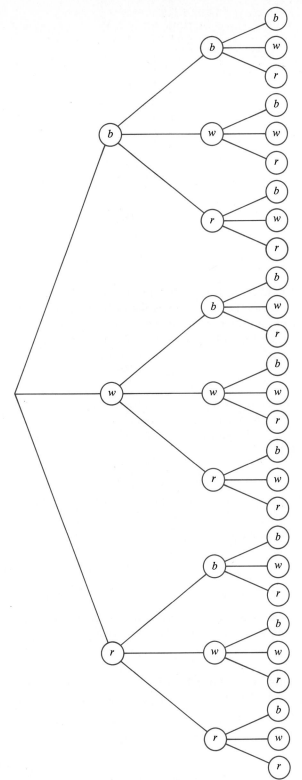

FIGURE 5.2
Tree diagram for the
selection of black, white,
and red balls

TABLE 5.1 PROBABILITY MASS FUNCTION FOR SELECTING BLACK AND WHITE BALLS WITH REPLACEMENT

$f(x, y)$		y			
		0	1	2	3
	0	$\frac{1}{216}$	$\frac{1}{36}$	$\frac{1}{18}$	$\frac{1}{27}$
x	1	$\frac{1}{24}$	$\frac{1}{6}$	$\frac{1}{6}$	
	2	$\frac{1}{8}$	$\frac{1}{4}$		
	3	$\frac{1}{8}$			

certain impossible outcomes, for instance $X = 2$ and $Y = 2$, and that these are not listed in the table. Table 5.1 gives the distribution of the joint probabilities of X and Y: it describes the manner in which the probability mass is jointly distributed. As a function, it is a function of two variables, x and y, and takes the form

$$f(x, y) = P(X = x, Y = y)$$

where the ordered pair (x, y) lies in the codomain $\Omega_{X,Y}$ of the pair (X, Y) of random variables, the codomain being the set of possible values taken on by the random-variable pair. In the case at hand, the elements of $\Omega_{X,Y}$ correspond to the 10 entries in Table 5.1:

$$\Omega_{X,Y} = \{(0, 0), (0, 1), (0, 2), (0, 3), (1, 0),$$
$$(1, 1), (1, 2), (2, 0), (2, 1), (3, 0)\}$$

Note that the function defined by Table 5.1 can be expressed in closed form as

$$f(x, y) = \frac{3!}{x!y!(3 - x - y)!} \left(\frac{1}{2}\right)^x \left(\frac{1}{3}\right)^y \left(\frac{1}{6}\right)^{3-x-y}$$

The derivation of this closed-form expression will be discussed in Example 5.3. For now, note the similarity to the binomial density.

DEFINITION 5.1

Discrete Joint (Bivariate) Density. If X and Y are two discrete random variables, then their joint density, also known as their **joint probability mass function,** is defined by

$$f_{X,Y}(x, y) = P(X = x, Y = y)$$

for every pair (x, y) in the codomain $\Omega_{X,Y}$.

The joint density is also known as the **joint distribution** of the random variables; however, this term is usually used more loosely to describe the random variables and

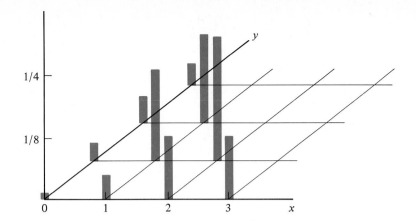

FIGURE 5.3
Density of Table 5.1

the density. As usual, if the variables X and Y are clear from the context, we will drop the subscript X, Y and simply write $f(x, y)$.

The graph of a discrete joint density is similar to that of a discrete density for a single random variable, except that the points (x, y) lie in the plane. The graph for the distribution of Table 5.1 is given in Figure 5.3.

In the univariate setting, the probabilities of interest are $P(X \in B)$, where B is a subset of the real line R. In the bivariate setting, these probabilities take the form $P[(X, Y) \in B]$, where B is a subset of the Euclidean plane R^2. Corresponding to Figure 3.1, which describes the meaning of $P(X \in B)$, is Figure 5.4. Relative to B in R^2,

$$P[(X, Y) \in B] = P(E_B)$$

Probabilities involving jointly distributed random variables are concerned with the position of the pair (X, Y) in the plane. Consequently, (X, Y) is often referred to as a **random vector.**

EXAMPLE 5.1

For the discrete distribution of Table 5.1 we wish to determine the probabilities of the following events: the random vector (X, Y) falls in the closed unit circle [see Figure 5.5(a)], the number of black balls exceeds 1 [see Figure 5.5(b)], and the random vector (X, Y) falls in the region A illustrated in Figure 5.5(c). For the first event,

$$\begin{aligned} P(X^2 + Y^2 \leq 1) &= P[\{(0, 0), (0, 1), (1, 0)\}] \\ &= P[(0, 0)] + P[(0, 1)] + P[(1, 0)] \\ &= \frac{1}{216} + \frac{1}{36} + \frac{1}{24} = \frac{2}{27} \end{aligned}$$

For the second event,

FIGURE 5.4
Illustration of E_B

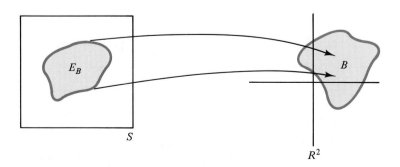

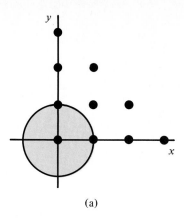

(a)

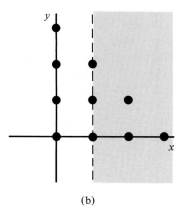

(b)

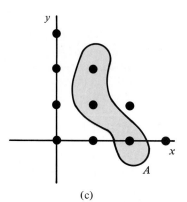

FIGURE 5.5
Events corresponding to
Example 5.1

(c)

$$P(X > 1) = P[\{(2, 0), (2, 1), (3, 0)\}] = \frac{1}{2}$$

Finally, reference to Figure 5.5(c) yields

$$P(X \in A) = P[\{(1, 1), (1, 2), (2, 0)\}] = \frac{11}{24}$$

EXAMPLE 5.2   A robot is equipped with an arm, gripper, and vision system. To check the vision system, 7 blocks, each having a geometric pattern supposedly recognizable to the robot, are placed on a table in a random fashion. Two of the blocks are identical isosceles triangles, two are identical squares, and three are identical circular disks. The robot is directed to select a triangle and a square. If the robot is to exhibit "intelligence," its selections should not be uniformly random (though they may indeed be random).

Letting X and Y denote the number of triangles and squares, respectively, we can construct their joint probability mass function as though the selections were uniformly random. The codomain of (X, Y) is

$$\Omega_{X,Y} = \{(0, 0), (0, 1), (0, 2), (1, 0), (1, 1), (2, 0)\}$$

Since the robot is given no instructions regarding the order of selection, the theory of combinations is employed to obtain the joint probabilities. Since there are 7 objects from which to choose and the robot is to choose 2, there are $C(7, 2)$ possible selections. These constitute the sample space corresponding to the selection process.

The probabilities are calculated as in the single-variable case. For instance, letting $E_{\{(0,1)\}}$ denote the event corresponding to the pair $(0, 1)$ in $\Omega_{X,Y}$,

$$f(0, 1) = P(X = 0, Y = 1) = P(E_{\{(0,1)\}}) = \frac{\text{card }(E_{\{(0,1)\}})}{\binom{7}{2}}$$

We must count the number of elements in the event. Proceeding, 0 triangles must be chosen from 2, 1 square must be chosen from 2, and 1 disk must be chosen from 3. Thus, by the fundamental principle of counting,

$$P(E_{\{(0,1)\}}) = \frac{\binom{2}{0}\binom{2}{1}\binom{3}{1}}{\binom{7}{2}} = \frac{6}{21}$$

Applying the same reasoning in general, the probability mass function for (X, Y) is given by

$$f(x, y) = \frac{\binom{2}{x}\binom{2}{y}\binom{3}{2 - x - y}}{\binom{7}{2}}$$

The density values are given in Table 5.2 and the density is illustrated in Figure 5.6.

Returning to the robot, uniformly random behavior would yield the probability 4/21 of selecting one triangle and one square. In the event the robot selects the proper blocks 75% of the time, it is a statistical question as to whether or not some affirmative conclusion regarding "intelligence" is warranted.

Another question of interest might be whether or not the robot can distinguish between polygonal and nonpolygonal figures. For such a question, we would be interested in the probability of uniformly randomly selecting only polygonal figures, which is given by

$$P(X + Y \geq 2) = P[\{(0, 2), (1, 1), (2, 0)\}] = \frac{6}{21}$$

Just as Definition 3.3 led to a collection of properties regarding discrete random variables and their densities, Definition 5.1 leads to a set of similar properties for discrete random vectors. Due to the similarity of development, we will forego a

TABLE 5.2 PROBABILITY MASS FUNCTION FOR SELECTION OF BLOCKS

$f(x, y)$		y 0	1	2
	0	$\dfrac{1}{7}$	$\dfrac{2}{7}$	$\dfrac{1}{21}$
x	1	$\dfrac{2}{7}$	$\dfrac{4}{21}$	
	2	$\dfrac{1}{21}$		

detailed discussion in the present context and content ourselves with a listing of the basic properties relevant to a discrete joint density:

1. $f(x, y) \geq 0$
2. There exists a denumerable (discrete) set of ordered pairs (x, y) such that $f(x, y) > 0$. These are the elements of the codomain $\Omega_{X,Y}$.
3. $\displaystyle\sum_{f(x,y)>0} f(x, y) = 1$
4. For any set B in the plane,

$$P((X, Y) \in B) = \sum_{\substack{(x,y)\in B \\ f(x,y)>0}} f(x, y)$$

5. If $f(x, y)$ is any function satisfying properties 1, 2, and 3, then there exists a pair of discrete random variables X and Y such that $f(x, y)$ is the probability mass function for X and Y.

Definition 5.1, which applies to bivariate distributions, generalizes at once to joint distributions involving n discrete random variables.

DEFINITION 5.2

Discrete Joint (Multivariate) Density. If $X_1, X_2, \ldots, X_n$ are n discrete random variables, then their joint density is defined by

$$f(x_1, x_2, \ldots, x_n) = P(X_1 = x_1, X_2 = x_2, \ldots, X_n = x_n)$$

FIGURE 5.6
Density for Example 5.2

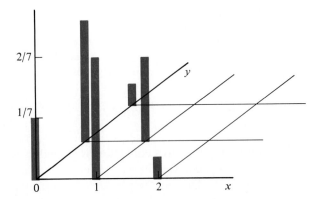

If it is desired to draw attention to the random variables to which a multivariate density applies, then the density can be written as

$$f_{X_1, X_2, \ldots, X_n}(x_1, x_2, \ldots, x_n)$$

EXAMPLE 5.3

As noted previously, the closed-form expression of the probability mass function for the distribution of Table 5.1 bears a resemblance to the binomial density. In fact, both that density and the binomial density are special cases of a general multivariate distribution known as the **multinomial distribution.** This distribution results from an experiment in which the following three conditions are satisfied:

1. The experiment consists of n independent trials.
2. For each trial, there are r possible outcomes, $w_1, w_2, \ldots, w_r$.
3. There exist numbers $p_1, p_2, \ldots, p_r$ such that, for $j = 1, 2, \ldots, r$, on each trial the probability of outcome w_j is equal to p_j.

(Note the similarity to Definition 3.2, in which a sequence of Bernoulli trials was defined.) For $j = 1, 2, \ldots, r$, the random variable X_j is the number of times outcome w_j occurs during the n trials. The sample space for the experiment is the set of n-tuples whose components are chosen from the collection $\{w_1, w_2, \ldots, w_r\}$ of trial outcomes. Assuming the hypothesis of equal probability, an n-tuple having x_j components with w_j, $j = 1, 2, \ldots, r$, has probability

$$p_1^{x_1} p_2^{x_2} \cdots p_r^{x_r}$$

of occurring, where $x_1 + x_2 + \cdots + x_r = n$. Like the analysis of the binomial density (Section 4.1.1), finding

$$P(X_1 = x_1, X_2 = x_2, \ldots, X_r = x_r)$$

requires counting all such n-tuples. In the binomial case, the construction of the appropriate n-tuples was equivalent to partitioning the n components into two classes, one for successes and the other for failures. In the present situation, there are r possible outcomes per trial. Thus, the n components must be partitioned among r cells, where the jth cell contains all components holding the trial outcome w_j. For $j = 1, 2, \ldots, r$, there are x_j components holding the outcome w_j. Consequently, we are partitioning n objects (components) so that x_j objects are in cell j. According to Theorem 2.6, this can be accomplished in

$$\frac{n!}{x_1! x_2! \cdots x_r!}$$

ways. Hence, the joint density of $X_1, X_2, \ldots, X_r$ is

$$f(x_1, x_2, \ldots, x_r) = \frac{n!}{x_1! x_2! \cdots x_r!} p_1^{x_1} p_2^{x_2} \cdots p_r^{x_r}$$

The corresponding distribution is called the multinomial distribution because the coefficients are the multinomial coefficients, which appear in algebra as coefficients in the multinomial expansion.

Although we have developed the multinomial distribution for r random variables when there are r possible outcomes on each trial, in fact, the rth variable is redundant, since

$$X_r = n - X_1 - X_2 - \cdots - X_{r-1}$$

Moreover,

$$p_r = 1 - p_1 - p_2 - \cdots - p_{r-1}$$

and

$$x_r = n - x_1 - x_2 - \cdots - x_{r-1}$$

Upon making these substitutions, the multinomial density is a function of $r - 1$ variables:

$$f_{X_1, X_2, \ldots, X_{r-1}}(x_1, x_2, \ldots, x_{r-1})$$

$$= \frac{n!}{x_1! x_2! \cdots (n - x_1 - x_2 - \cdots - x_{r-1})!} \cdot p_1^{x_1} p_2^{x_2} \cdots (1 - p_1 - p_2 - \cdots - p_{r-1})^{n - x_1 - x_2 - \cdots - x_{r-1}}$$

The distribution of Table 5.1 is multinomial with $n = 3$, $p_1 = 1/2$, and $p_2 = 1/3$. Given these parameters,

$$p_3 = 1 - p_1 - p_2 = \frac{1}{6}$$

is determined.

5.1.2 Discrete Marginal Distributions

Given the joint distribution of two discrete random variables X and Y, it is always possible to find the individual densities $f_X(x)$ and $f_Y(y)$. As an illustration, consider the joint distribution given in Table 5.1. There, $P(X = 0)$ is readily found, since the set of outcomes for which $X = 0$ is given by

$$E_{\{X=0\}} = \{(0, 0), (0, 1), (0, 2), (0, 3)\}$$

Consequently,

$$
\begin{aligned}
f_X(0) &= P(X = 0) \\
&= P[(0, 0)] + P[(0, 1)] + P[(0, 2)] + P[(0, 3)] \\
&= \frac{1}{216} + \frac{1}{36} + \frac{1}{18} + \frac{1}{27} = \frac{1}{8}
\end{aligned}
$$

which is simply the sum of the probabilities in the first row of Table 5.1. Using summation notation,

$$f_X(0) = \sum_{y=0}^{3} f(0, y)$$

Similarly, $f_X(1) = 3/8$, $f_X(2) = 3/8$, and $f_X(3) = 1/8$ can be found by summing the succeeding rows of the table. Analogously, the probability mass function for Y can be found by summing the columns of Table 5.1: $f_Y(0) = 8/27$, $f_Y(1) = 4/9$, $f_Y(2) = 2/9$, and $f_Y(3) = 1/27$. Recall that at the outset we recognized from the experiment that both X and Y were binomially distributed. Summing the rows and columns of Table 5.1 has yielded $f_X(x) = b(x; 3, 1/2)$ and $f_Y(y) = b(y; 3, 1/3)$, respectively.

Relative to the joint density $f_{X,Y}(x, y)$, the densities of X and Y, $f_X(x)$ and $f_Y(y)$, are called the **marginal densities.** From a practical perspective, the marginal distributions describe the individual behavior of the measurements X and Y, but they often lack, even when taken together, information regarding the joint behavior of the measurements. The latter point is crucial: although the marginal densities can be recovered from the joint density, in general, the joint density cannot be constructed from a knowledge of the marginal densities. The next theorem, whose proof is just a rigorous formulation of the summing of rows and columns in a probability table, summarizes the discussion regarding recovery of the marginal distributions.

THEOREM 5.1

If X and Y are discrete random variables possessing the joint density $f(x, y)$, then the marginal densities of X and Y are given by

$$f_X(x) = \sum_y f(x, y) \qquad \text{and} \qquad f_Y(y) = \sum_x f(x, y)$$

respectively, where the first sum is taken over all y such that $f(x, y) > 0$ and the second sum is taken over all x such that $f(x, y) > 0$.

Proof: We give the proof for $f_X(x)$. The proof rests upon the fact that, for fixed x, the union over all y of the events $E_{\{(x,y)\}}$, which give rise to the joint density values, is equal to $E_{\{X=x\}}$, the event giving rise to $f_X(x)$. Proceeding,

$$\sum_y f(x, y) = \sum_y P(X = x, Y = y)$$

$$= \sum_y P(E_{\{(x,y)\}})$$

$$= P\left(\bigcup_y E_{\{(x,y)\}}\right)$$

$$= P(E_{\{X=x\}})$$

$$= P(X = x)$$

$$= f_X(x)$$

where the third equality follows from additivity. ∎

EXAMPLE 5.4

For the joint distribution of Example 5.2, whose joint density is given in Table 5.2, the marginal densities are given by $f_X(0) = 10/21$, $f_X(1) = 10/21$, and $f_X(2) = 1/21$, each value having been obtained by summing the appropriate row in Table 5.2, and $f_Y(0) = 10/21$, $f_Y(1) = 10/21$, and $f_Y(2) = 1/21$, each value having been obtained by summing the appropriate column in Table 5.2.

EXAMPLE 5.5

It was noted prior to the statement of Theorem 5.1 that the joint density cannot, in general, be recovered from the marginal densities. To see this important point, consider the joint distribution defined by

$$f_{X,Y}(x, y) = b\left(x; 3, \frac{1}{2}\right) b\left(y; 3, \frac{1}{3}\right)$$

$$= \binom{3}{x} \binom{3}{y} \left(\frac{1}{2}\right)^x \left(\frac{1}{2}\right)^{3-x} \left(\frac{1}{3}\right)^y \left(\frac{2}{3}\right)^{3-y}$$

$$= \binom{3}{x} \binom{3}{y} \frac{2^{3-y}}{216}$$

TABLE 5.3 PROBABILITY MASS FUNCTION FOR EXAMPLE 5.5

$f(x, y)$		y 0	1	2	3	$f_X(x)$
	0	$\frac{1}{27}$	$\frac{1}{18}$	$\frac{1}{36}$	$\frac{1}{216}$	$\frac{1}{8}$
x	1	$\frac{1}{9}$	$\frac{1}{6}$	$\frac{1}{12}$	$\frac{1}{72}$	$\frac{3}{8}$
	2	$\frac{1}{9}$	$\frac{1}{6}$	$\frac{1}{12}$	$\frac{1}{72}$	$\frac{3}{8}$
	3	$\frac{1}{27}$	$\frac{1}{18}$	$\frac{1}{36}$	$\frac{1}{216}$	$\frac{1}{8}$
$f_Y(y)$		$\frac{8}{27}$	$\frac{4}{9}$	$\frac{2}{9}$	$\frac{1}{27}$	

The values of $f(x, y)$ are given in Table 5.3, along with the computation of the marginal probabilities for X and Y. Although the joint density is different from the one given in Table 5.1, the marginal densities are the same. For future reference, note that the present joint density has been formed by taking the product of the marginal densities resulting from Table 5.1.

In the case of more than 2 jointly distributed discrete random variables, Theorem 5.1 applies directly. Specifically, if $X_1, X_2, \ldots, X_n$ are jointly distributed with joint probability mass function $f(x_1, x_2, \ldots, x_n)$, then, for $k = 1, 2, \ldots, n$, the marginal density for X_k is given by

$$f_{X_k}(x_k) = \sum_{x_1} \sum_{x_2} \cdots \sum_{x_{k-1}} \sum_{x_{k+1}} \cdots \sum_{x_n} f(x_1, x_2, \ldots, x_n)$$

where the $(n - 1)$-fold summation notation means to hold x_k fixed and sum over all possible arguments involving the remaining variables. Although the expression is somewhat forbidding, in practice its implementation is straightforward. For instance, in the case of three random variables X, Y, and W, the probability table is three-dimensional and the marginal densities are found by summing over "slices" of the table. The next example illustrates the technique.

EXAMPLE 5.6

Suppose X, Y, and W possess the trivariate distribution illustrated in the three-dimensional probability table given in Figure 5.7. Each of the random variables can take on the values 0 and 1. The value $f_X(0)$ of the marginal density for X is determined by summing over the bottom horizontal slice, while $f_X(1)$ is found by summing over the top horizontal slice:

$$f_X(0) = \frac{1}{8} + \frac{1}{16} + \frac{1}{8} + \frac{1}{16} = \frac{3}{8}$$

$$f_X(1) = \frac{1}{4} + \frac{1}{4} + \frac{1}{16} + \frac{1}{16} = \frac{5}{8}$$

The marginal density for Y is determined by summing first over the left slice and then over the right slice:

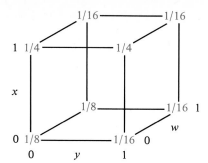

FIGURE 5.7
Density for Example 5.6

$$f_Y(0) = \frac{1}{8} + \frac{1}{4} + \frac{1}{8} + \frac{1}{16} = \frac{9}{16}$$

$$f_Y(1) = \frac{1}{16} + \frac{1}{4} + \frac{1}{16} + \frac{1}{16} = \frac{7}{16}$$

The marginal density for W is determined by summing first over the front slice and then over the rear slice:

$$f_W(0) = \frac{1}{16} + \frac{1}{8} + \frac{1}{4} + \frac{1}{4} = \frac{11}{16}$$

$$f_W(1) = \frac{1}{16} + \frac{1}{8} + \frac{1}{16} + \frac{1}{16} = \frac{5}{16}$$

5.1.3 Continuous Multivariate Distributions

In the multivariate setting, the definition of a continuous density is a direct extension of the univariate definition, the only change being that the single integral in Definition 3.4 becomes a multiple integral.

DEFINITION 5.3

Continuous Joint (Bivariate) Density. The random variables X and Y are said to be **jointly continuous** and to possess the joint density $f_{X,Y}(x, y)$ if, for any set B in the Euclidean plane,

$$P[(X, Y) \in B] = \iint_B f_{X,Y}(x,y) \, dy \, dx$$

Geometrically, probabilities for jointly continuous random variables appear as volumes under the surface defined by the density (see Figure 5.8). The total volume under the density surface must be 1.

In the univariate case it is common to find a legitimate density and then employ Theorem 3.4 to insure there exists a corresponding random variable. A similar technique is employed in the multivariate case. Specifically, if $f(x, y)$ is a nonnegative function of two variables such that

$$\int_{-\infty}^{\infty} \int_{-\infty}^{\infty} f(x, y) \, dy \, dx = 1$$

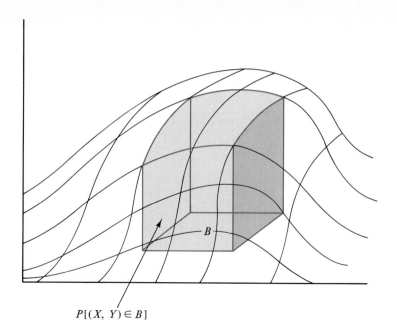

FIGURE 5.8
Illustration of a
continuous joint density

$$P[(X, Y) \in B]$$

then there exists a uniquely distributed continuous random vector (X, Y) possessing the joint density $f(x, y)$.

EXAMPLE 5.7

Suppose two happenings are to occur during a given time frame, which for convenience we will take as 1 hour. Let X and Y denote the random variables which specify the times at which the happenings occur. For instance, Y might be the arrival time of a particular module required in a production process, while X is the time at which a call is made for the module. Let us make the assumptions that both X and Y occur uniformly randomly within the hour and that they occur independently. Then X and Y are both uniformly distributed over the interval $[0, 1]$ and, as will be seen in Section 5.2, their joint density must be

$$f(x, y) = \begin{cases} 1, & \text{if } 0 \leq x, y \leq 1 \\ 0, & \text{otherwise} \end{cases}$$

so that the support of the density is the square $[0, 1] \times [0, 1]$. Given this density, numerous questions can be answered. For instance, the probability that both happenings occur in the last half hour is

$$P\left(X > \frac{1}{2}, Y > \frac{1}{2}\right) = \int_{1/2}^{\infty} \int_{1/2}^{\infty} f(x, y) \, dy \, dx$$

$$= \int_{1/2}^{1} \int_{1/2}^{1} dy \, dx = \frac{1}{4}$$

Figure 5.9 illustrates the region of the plane determined by the conditions $X > 1/2$ and $Y > 1/2$. Figure 5.10 illustrates the volume giving the desired probability.

If the random variables X and Y represent demand and arrival times, respectively, an important probability is given by

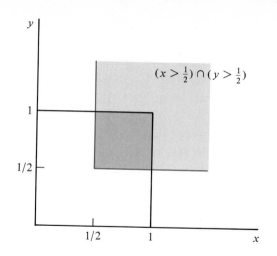

FIGURE 5.9
The event $(X > 1/2) \cap$
$(Y > 1/2)$ of Example 5.7

$$P(Y < X) = \int_{-\infty}^{\infty} \int_{-\infty}^{x} f(x, y)\, dy\, dx$$

$$= \int_{0}^{1} \int_{0}^{x} dy\, dx = \frac{1}{2}$$

Figure 5.11 illustrates the region of the plane determined by the condition $(Y < X)$, and Figure 5.12 illustrates the volume (probability) determined by the density.

In Example 5.7 the density was constant over a support region and zero elsewhere. In general, the random variables X and Y are said to be **jointly uniformly distributed** if they possess a joint density of the form

$$f(x, y) = \begin{cases} c, & \text{if } (x, y) \in A \\ 0, & \text{otherwise} \end{cases}$$

Since total probability must be 1, c must be the reciprocal of the area of A.

FIGURE 5.10
Probability of the event
$(X > 1/2) \cap (Y > 1/2)$
represented as a volume

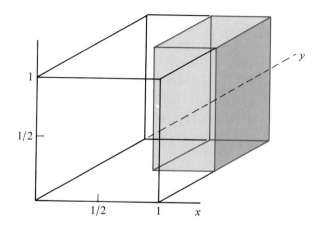

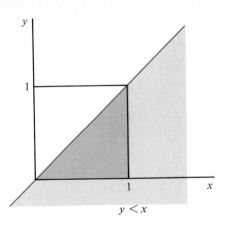

FIGURE 5.11
The event of Example 5.7
determined by the
condition $Y < X$

$y < x$

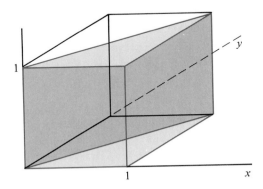

FIGURE 5.12
The probability $P(Y < X)$
represented as a volume

EXAMPLE 5.8

Let X and Y represent the times to failure, in years, of subsystems A and B, respectively. For instance X and Y might represent the times to failure of two distinct computers on board an aircraft. Suppose X and Y possess the joint density

$$f(x, y) = \begin{cases} 2e^{-(x+2y)}, & \text{if } x, y \geq 0 \\ 0, & \text{otherwise} \end{cases}$$

A straightforward double integration shows that

$$\int_0^\infty \int_0^\infty 2e^{-(x+2y)} \, dy \, dx = 1$$

The probability that both subsystems survive at least 1 year is

$$P(X \geq 1, Y \geq 1) = \int_1^\infty \int_1^\infty 2e^{-(x+2y)} \, dy \, dx$$

$$= \int_1^\infty e^{-x} \, dx \int_1^\infty 2e^{-2y} \, dy$$

$$= e^{-1}e^{-2} = 0.0498$$

The probability that subsystem A survives longer than subsystem B is

$$P(Y < X) = \int_0^\infty \int_0^x 2e^{-(x+2y)} \, dy \, dx$$

$$= \int_0^\infty e^{-x} \, dx \int_0^x 2e^{-2y} \, dy$$

$$= \int_0^\infty e^{-x}(1 - e^{-2x}) \, dx = \frac{2}{3}$$

The next definition extends the notion of joint continuity to n random variables. The generalization is straightforward.

DEFINITION 5.4

Continuous Joint (Multivariate) Density. The n random variables $X_1, X_2, \ldots, X_n$ are said to be jointly continuous and to possess the joint density $f(x_1, x_2, \ldots, x_n)$ if, for any set B in n-dimensional Euclidean space,

$$P[(X_1, X_2, \ldots, X_n) \in B] = \int \cdots \int_B f(x_1, x_2, \ldots, x_n) \, dx_n \cdots dx_1$$

where the integral is n-fold.

5.1.4 Continuous Marginal Distributions

As in the discrete case, it is possible to recover the densities of X and Y when given the joint distribution. Once again these are called **marginal densities.** Whereas in the discrete case the marginal density $f_X(x)$ is found by holding x fixed and summing over y, in the continuous case it is found by holding x fixed and integrating over y. An analogous comment applies to finding $f_Y(y)$.

THEOREM 5.2

If X and Y are jointly continuous with joint density $f(x, y)$, then the marginal densities of X and Y are given by

$$f_X(x) = \int_{-\infty}^\infty f(x, y) \, dy \qquad \text{and} \qquad f_Y(y) = \int_{-\infty}^\infty f(x, y) \, dx$$

Proof: We give the proof for $f_X(x)$. Let

$$h(x) = \int_{-\infty}^\infty f(x, y) \, dy$$

and let (a, b) be any interval on the real line. Then

$$\int_a^b h(x)\, dx = \int_a^b \int_{-\infty}^{\infty} f(x,\, y)\, dy\, dx$$
$$= P(a < X < b,\ -\infty < Y < \infty)$$
$$= P(a < X < b)$$

where the last two equalities follow from the definition of a joint density and the fact that it is certain that Y will be between $-\infty$ and ∞. Since the probabilities for X are given by integrating $h(x)$, $h(x)$ is a density for X. But that is precisely what we wished to show. ∎

EXAMPLE 5.9

For the jointly uniform density given in Example 5.7,

$$f_X(x) = \int_0^1 dy = 1$$

for $0 \le x \le 1$. If x is outside [0, 1], then $f(x,\, y) = 0$ and hence the integral of $f(x,\, y)$ with respect to y is 0. Thus, X is uniformly distributed over [0, 1], which is exactly what was stated in the example. A similar computation and remark hold for $f_Y(y)$ and Y.

EXAMPLE 5.10

As in Example 5.7, consider a demand-arrival model where the two random variables X and Y are jointly uniformly distributed; however, this time make the supposition that $Y \le X$, which means the demand must occur after the arrival. A suitable density is

$$f(x,\, y) = \begin{cases} 2, & \text{if } 0 \le x \le 1 \quad \text{and} \quad 0 \le y \le x \\ 0, & \text{otherwise} \end{cases}$$

the support region of which is illustrated in Figure 5.13. For $0 \le x \le 1$,

$$f_X(x) = \int_0^x 2\, dy = 2x$$

FIGURE 5.13
Support region for the joint distribution of Example 5.10

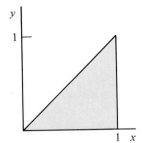

since, as is illustrated in Figure 5.14, for fixed x, $f(x,\, y)$ is only nonzero for y between 0 and x. Moreover, $f_X(x) = 0$ for x outside [0, 1].

As for Y, if $0 \le y \le 1$, then

$$f_Y(y) = \int_y^1 2\, dx = 2(1 - y)$$

since, as is illustrated in Figure 5.15, for fixed y, $f(x,\, y)$ is only nonzero for x between y and 1. Moreover, $f_Y(y) = 0$ for y outside [0, 1]. Although the joint distribution is uniform, the marginal distributions are not.

EXAMPLE 5.11

For the joint distribution of Example 5.8, the marginal density for X is

$$f_X(x) = \int_0^\infty 2e^{-(x+2y)}\, dy = e^{-x} \int_0^\infty 2e^{-2y}\, dy = e^{-x}$$

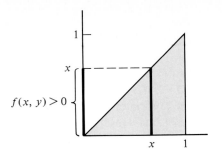

FIGURE 5.14
Illustration for the marginal density $f_X(x)$ of Example 5.10

for $x \geq 0$, and $f_X(x) = 0$ for $x < 0$. Similarly,

$$f_Y(y) = \begin{cases} 2e^{-2y}, & \text{if } y \geq 0 \\ 0, & \text{if } y < 0 \end{cases}$$

In terms of the time-to-failure model, taken individually, each subsystem possesses an exponential time-to-failure distribution.

Theorem 5.2 generalizes at once to multivariate continuous distributions. Specifically, if $X_1, X_2, \ldots, X_n$ are jointly continuous with joint density $f(x_1, x_2, \ldots, x_n)$, then, for any $k = 1, 2, \ldots, n$, the kth marginal density is given by

$$f_{X_k}(x_k) = \int_{-\infty}^{\infty} \cdots \int_{-\infty}^{\infty} f(x_1, x_2, \ldots, x_n) \, dx_n \cdots dx_{k+1} \, dx_{k-1} \cdots dx_1$$

the integral being $(n-1)$-fold over all variables excluding x_k.

In the case of a trivariate density $f_{X,Y,W}(x, y, w)$, the marginal density for X is given by

$$f_X(x) = \int_{-\infty}^{\infty} \int_{-\infty}^{\infty} f_{X,Y,W}(x, y, w) \, dw \, dy$$

In the case of more than two variables, besides univariate marginal densities there are **joint marginal densities.** These are obtained by integrating out some of the variables. For instance, in the case of four variables X, Y, U, and V, the joint marginal density for X and Y is given by

FIGURE 5.15
Illustration for the marginal density $f_Y(y)$ of Example 5.10

$$f_{X,Y}(x, y) = \int_{-\infty}^{\infty} \int_{-\infty}^{\infty} f_{X,Y,U,V}(x, y, u, v) \, dv \, du$$

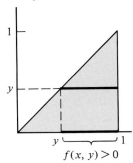

5.1.5 Joint Probability Distribution Functions

In considering the probability distribution functions of jointly distributed random variables, we will restrict our attention to the bivariate case. The same definition applies to both the discrete and continuous cases.

Joint Probability Distribution Function. If X and Y are jointly distributed, then their joint probability distribution function is defined for all x and y by

$$F_{X,Y}(x, y) = P(X \leq x, Y \leq y)$$

Figure 5.16 illustrates the portion of the plane that contributes to the probability defining $F(x, y)$, where, as usual, the subscripts are suppressed. In the discrete case,

$$F(x, y) = \sum_{r \leq x} \sum_{s \leq y} f(r, s)$$

and in the continuous case,

$$F(x, y) = \int_{-\infty}^{x} \int_{-\infty}^{y} f(r, s) \, ds \, dr$$

In the continuous case, taking partial derivatives with respect to x and y yields

$$\frac{\partial^2}{\partial x \, \partial y} F(x, y) = f(x, y)$$

at all points of continuity of $f(x, y)$. This is analogous to the situation in the univariate setting, except there a single ordinary derivative yielded the density.

If one takes into account the differences that result from having two variables, then many other univariate properties have bivariate analogues. For instance, in the univariate continuous case,

$$P(a < X < b) = F(b) - F(a)$$

In the bivariate continuous case,

$$P(a < X < b, c < Y < d) = F(b, d) - F(a, d) - F(b, c) + F(a, c)$$

The proof of the relationship follows directly from the application of additivity in Figure 5.17. Other similar relations hold, including those in the discrete case, where the values at the edges need to be taken into consideration. In sum, Theorem 3.5 can be generalized. There also exists an analogue of Theorem 3.6; however, the matter will not be pursued here.

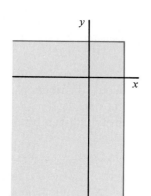

FIGURE 5.16

Portion of the plane contributing to $F_{X,Y}(x, y)$

FIGURE 5.17

Rectangle to illustrate the joint probability distribution function

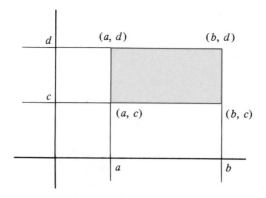

EXAMPLE 5.12 Let X and Y be the random variables discussed in Example 5.8. Then, for $x \geq 0$ and $y \geq 0$,

$$F(x, y) = \int_0^x \int_0^y 2e^{-(r + 2s)} \, ds \, dr$$

$$= \int_0^x e^{-r} \, dr \int_0^y 2e^{-2s} \, ds$$

$$= (1 - e^{-x})(1 - e^{-2y})$$

For points such that either x or y is negative, $F(x, y) = 0$.

5.2 INDEPENDENT RANDOM VARIABLES

The measurement processes represented by two or more random variables are sometimes independent, in the sense that knowledge concerning one does not alter our perception of the other. To give mathematical meaning to the notion of independence, it must be expressed in terms of the distributions of the random variables concerned. This can be accomplished by framing the problem in terms of the independence of events.

If X and Y are discrete random variables, then their mutual distribution is determined by their joint density $f(x, y)$, whereas their individual distributions are determined by the marginal densities $f_X(x)$ and $f_Y(y)$. A reasonable approach to independence is to say that X and Y are independent if, for all x and y in the respective codomains of X and Y, the events that determine $P(X = x)$ and $P(Y = y)$ are independent in the sense of Definition 2.7. In other words, we require $E_{\{X=x\}}$ and $E_{\{Y=y\}}$ to be independent, which means that

$$P(E_{\{X=x\}} \cap E_{\{Y=y\}}) = P(E_{\{X=x\}})P(E_{\{Y=y\}})$$
$$= P(X = x)P(Y = y)$$
$$= f_X(x)f_Y(y)$$

However,

$$P(E_{\{X=x\}} \cap E_{\{Y=y\}}) = P(E_{\{X=x, Y=y\}})$$
$$= P(X = x, Y = y)$$
$$= f_{X,Y}(x, y)$$

so that the preceding equation can be written as

$$f_{X,Y}(x, y) = f_X(x)f_Y(y)$$

In words, the joint density is equal to the product of the marginal densities.

The preceding considerations concerning the discrete case lead to the following definition, which applies whether or not the random variables are discrete.

DEFINITION 5.6

Independent Random Variables. The jointly distributed random variables X_1, X_2, . . . , X_n, possessing multivariate density $f(x_1, x_2, \ldots, x_n)$, are independent if

$$f(x_1, x_2, \ldots, x_n) = f_{X_1}(x_1)f_{X_2}(x_2) \cdots f_{X_n}(x_n)$$

In the case of two variables, X and Y, Definition 5.6 reduces to

$$f_{X,Y}(x, y) = f_X(x)f_Y(y)$$

If a collection of random variables is not independent, then the variables are said to be **dependent**.

The random variables X and Y of Example 5.5 are independent; in fact, the joint density was defined as the product of two binomial densities, $b(x; 3, 1/2)$ and $b(y; 3, 1/3)$, so that these are the marginal densities. Note that the joint distribution of Table 5.1 possesses the same marginal densities; however, there the random variables X and Y are not independent, the joint density not being the product of the marginal densities. The point is that the independence of X and Y is not a function of X and Y in isolation; rather, it is a function of their interaction. Indeed, only when they are independent is their joint density determined by their individual densities. Otherwise, joint behavior is not determined by marginal behavior.

In Example 5.7 it was noted that the assumption of independence dictated that the joint density of X and Y, each uniformly distributed over [0, 1], must be the uniform density over the square [0, 1] $\times$ [0, 1]. In fact, in the example,

$$\begin{aligned} f_{X,Y}(x, y) &= I_{[0,1] \times [0,1]}(x, y) \\ &= I_{[0,1]}(x)I_{[0,1]}(y) \\ &= f_X(x)f_Y(y) \end{aligned}$$

so that the variables are independent in the sense of Definition 5.6. The reasoning here is important in applications: mathematically, independence is proven by demonstrating the joint density to be a product of the marginal densities; however, in practice, often a modeling decision regarding independence is made, thereby determining the appropriate joint density.

The random variables of Example 5.11 are independent, since

$$f_{X,Y}(x, y) = 2e^{-(x+2y)} = e^{-x}2e^{-2y} = f_X(x)f_Y(y)$$

for $x, y \geq 0$, and both sides of the equation are zero otherwise.

Given a collection of independent random variables, every subcollection is also independent. We state this formally as a theorem.

THEOREM 5.3   If $X_1, X_2, \ldots, X_n$ are independent, then any subcollection containing more than a single random variable is also independent.

Proof: We give the proof for the case of three independent variables X, Y, and W. Extension to n variables is merely an exercise in manipulating notation. If X, Y, and W are independent, then

$$f_{X,Y}(x, y) = \int_{-\infty}^{\infty} f_{X,Y,W}(x, y, w) \, dw$$

$$= \int_{-\infty}^{\infty} f_X(x)f_Y(y)f_W(w) \, dw$$

$$= f_X(x)f_Y(y) \int_{-\infty}^{\infty} f_W(w) \, dw$$

$$= f_X(x)f_Y(y)$$

which means that X and Y are independent. ∎

EXAMPLE 5.13

Suppose n subsystems of a large system each possess an exponential time-to-failure distribution with parameter b. Under the assumption that the failures of the subsystems are independent, the joint density is

$$f(x_1, x_2, \ldots, x_n) = b^n e^{-b(x_1 + x_2 + \cdots + x_n)}$$

for $x_1, x_2, \ldots, x_n \geq 0$, and 0 otherwise. For $k = 1, 2, \ldots, n$,

$$f_{X_k}(x_k) = \begin{cases} b^{-bx_k}, & \text{if } x_k \geq 0 \\ 0, & \text{if } x_k < 0 \end{cases}$$

We began the section looking at the probabilities of a discrete density and recognizing how event independence leads naturally to a notion of random-variable independence. The next theorem brings us full circle.

THEOREM 5.4

If X and Y are independent random variables, then, for any sets A and B,

$$P(X \in A, Y \in B) = P(X \in A)P(Y \in B)$$

Proof: We give the proof for the continuous case. The discrete case is essentially the same. According to the definitions of joint density and independence,

$$P(X \in A, Y \in B) = \int_A \int_B f_{X,Y}(x, y) \, dy \, dx$$

$$= \int_A \int_B f_X(x)f_Y(y) \, dy \, dx$$

$$= \int_A f_X(x) \, dx \int_B f_Y(y) \, dy$$

$$= P(X \in A)P(Y \in B) \qquad ∎$$

Theorem 5.4 holds for more than two independent random variables.

EXAMPLE 5.14

For the time-to-failure model of Example 5.13, what is the probability that all n subsystems function for at least q years? Assuming the variables to be in years and applying Theorem 5.4 yields

$$P(X_1 > q, X_2 > q, \ldots, X_n > q) = \prod_{k=1}^{n} P(X_k > q)$$

$$= \prod_{k=1}^{n} [1 - F_{X_k}(q)]$$

$$= e^{-nbq}$$

5.3 FUNCTIONS OF SEVERAL RANDOM VARIABLES

As noted at the outset of Section 5.1 (and illustrated in Figure 5.1), a major concern is the study of functions of more than a single random variable. In particular, sums of random variables play a key role in statistics. If $X_1, X_2, \ldots, X_n$ are jointly distributed with joint density $f(x_1, x_2, \ldots, x_n)$ and

$$Y = g(X_1, X_2, \ldots, X_n)$$

where $g(x_1, x_2, \ldots, x_n)$ is a piecewise continuous function of n variables (as it will always be in this text), it is important to determine characteristics of the output variable Y in terms of characteristics of the input variables. Finding the density of Y gives a complete description; however, in many circumstances this is pragmatically impossible, so that we may have to content ourselves with certain moments of Y, such as its mean and variance.

For both practical and theoretical reasons, of most concern are **linear** functions

$$Y = a_1 X_1 + a_2 X_2 + \cdots + a_n X_n + b$$

where $a_1, a_2, \ldots, a_n$ are constants and b is either zero, the **homogeneous** case, or nonzero, the **nonhomogeneous** case. In statistics, a key role is played by the homogeneous linear case when the random variables $X_1, X_2, \ldots, X_n$ are assumed to be independent.

5.3.1 Distribution of a Function of Several Random Variables

If $Y = g(X_1, X_2, \ldots, X_n)$ is a function of n random variables for which the joint density $f(x_1, x_2, \ldots, x_n)$ of the variables is known, then a primary goal is to derive the density of Y. The present subsection attacks the problem directly, in the discrete case using the probability mass function and in the continuous case employing the probability distribution function.

EXAMPLE 5.15

Suppose the random variables X and Y possess the joint distribution given in Table 5.1 and they are input into a multiplier, so that the output is the function

$$U = g(X, Y) = XY$$

The codomain of the output random variable is $\Omega_U = \{0, 1, 2\}$. For any u in Ω_U,

$$f_U(u) = P(U = u) = P(E_{\{U = u\}})$$

Thus, the probabilities in Table 5.1 yield

$$f_U(0) = P[\{(0, 0), (0, 1), (0, 2), (0, 3), (1, 0), (2, 0), (3, 0)\}]$$

$$= \frac{1}{216} + \frac{1}{36} + \frac{1}{18} + \frac{1}{27} + \frac{1}{24} + \frac{1}{8} + \frac{1}{8} = \frac{5}{12}$$

$$f_U(1) = P[\{(1, 1)\}] = \frac{1}{6}$$

$$f_U(2) = P[\{(1, 2), (2, 1)\}] = \frac{1}{6} + \frac{1}{4} = \frac{5}{12}$$

Now suppose the random variables are input into an adder, so that the output is $V = X + Y$. Then

$$\Omega_V = \{0, 1, 2, 3\}$$

$$f_V(0) = P[(0, 0)] = \frac{1}{216}$$

$$f_V(1) = P\{(0, 1), (1, 0)\}] = \frac{1}{36} + \frac{1}{24} = \frac{15}{216}$$

$$f_V(2) = \frac{1}{18} + \frac{1}{6} + \frac{1}{8} = \frac{75}{216}$$

$$f_V(3) = \frac{1}{27} + \frac{1}{6} + \frac{1}{4} + \frac{1}{8} = \frac{125}{216}$$

EXAMPLE 5.16

Let X and Y be independent random variables, each being uniformly distributed over $[0, 1]$. We will find the density of the weighted average

$$W = \frac{1}{3}X + \frac{2}{3}Y$$

Since X and Y are independent, they are jointly uniformly distributed over the square $[0, 1] \times [0, 1]$. Moreover, because X and Y vary between 0 and 1, W also varies between 0 and 1. To find $f_W(w)$, first find the probability distribution function of W. Assuming $0 \leq w \leq 1$,

$$F_W(w) = P(W \leq w)$$

$$= P\left(\frac{X}{3} + \frac{2Y}{3} \leq w\right)$$

$$= \iint\limits_{x/3 + 2y/3 \leq w} f_{X,Y}(x, y)\, dy\, dx$$

The support of the joint density is the square $[0, 1] \times [0, 1]$ and the integral is taken over the region determined by

$$y \leq \frac{3w}{2} - \frac{x}{2}$$

Since the joint density is 1 on its support, the value of the integral is the area of the support region falling within the domain of integration. As depicted in Figure 5.18, there are three distinct ranges of w to consider:

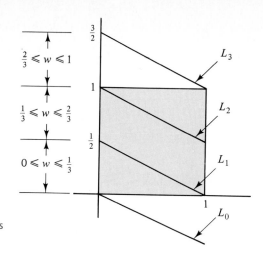

FIGURE 5.18
The three distinct ranges of w in Example 5.16

$$0 \le w \le \frac{1}{3}$$

$$\frac{1}{3} \le w \le \frac{2}{3}$$

$$\frac{2}{3} \le w \le 1$$

FIGURE 5.19
Computation of $F_W(w)$ in Example 5.16

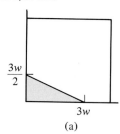

(a)

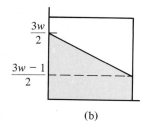

(b)

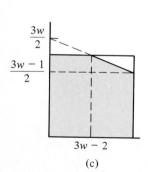

(c)

For w in the first range, the line $y = 3w/2 - x/2$ lies between the lines L_0 and L_1. Therefore, as shown in Figure 5.19(a), the required area is

$$F_W(w) = \frac{1}{2} \times \frac{3w}{2} \times 3w = \frac{9w^2}{4}$$

For w in the second range, the line $y = 3w/2 - x/2$ lies between the lines L_1 and L_2. Therefore, as shown in Figure 5.19(b), the required area is

$$F_W(w) = \frac{1}{4} + \frac{3w - 1}{2} = \frac{6w - 1}{4}$$

For w in the third range, the line $y = 3w/2 - x/2$ lies between the lines L_2 and L_3. Therefore, as shown in Figure 5.19(c), the required area is

$$F_W(w) = 1 - \frac{1}{2}\left[1 - \frac{3w - 1}{2}\right][1 - (3w - 2)] = 1 - \frac{9(1 - w)^2}{4}$$

Consequently,

$$F_W(w) = \begin{cases} 0, & \text{if } w \le 0 \\[2mm] \dfrac{9w^2}{4}, & \text{if } 0 \le w \le \dfrac{1}{3} \\[2mm] \dfrac{6w - 1}{4}, & \text{if } \dfrac{1}{3} \le w \le \dfrac{2}{3} \\[2mm] 1 - \dfrac{9(1 - w)^2}{4}, & \text{if } \dfrac{2}{3} \le w \le 1 \\[2mm] 1, & \text{if } w \ge 1 \end{cases}$$

Taking the derivative of $F_W(w)$ yields

$$f_W(w) = \begin{cases} \dfrac{9w}{2}, & \text{if } 0 < w < \dfrac{1}{3} \\[2ex] \dfrac{3}{2}, & \text{if } \dfrac{1}{3} < w < \dfrac{2}{3} \\[2ex] \dfrac{9(1-w)}{2}, & \text{if } \dfrac{2}{3} < w < 1 \\[2ex] 0, & \text{otherwise} \end{cases}$$

which is graphed in Figure 5.20.

EXAMPLE 5.17

In Example 4.33 it was remarked that, if X and Y are independent normal random variables with mean 0 and standard deviation σ (in that instance measuring errors), then the radius

$$W = (X^2 + Y^2)^{1/2}$$

possesses a Rayleigh distribution. We will now demonstrate that W is Rayleigh distributed with parameter $\eta = \sigma$ (see Definition 4.12). Since X and Y are independent, their joint density is the product of the marginal densities:

$$f(x, y) = \frac{1}{\sqrt{2\pi}\,\sigma} e^{-(x/\sigma)^2/2} \cdot \frac{1}{\sqrt{2\pi}\,\sigma} e^{-(y/\sigma)^2/2}$$

$$= \frac{1}{2\pi\sigma^2} e^{-(x^2+y^2)/2\sigma^2}$$

Consequently, for $w > 0$,

$$F_W(w) = P[(X^2 + Y^2)^{1/2} \le w]$$

$$= P(X^2 + Y^2 \le w^2)$$

$$= \iint\limits_{x^2+y^2\le w^2} \frac{1}{2\pi\sigma^2} e^{-(x^2+y^2)/2\sigma^2}\, dy\, dx$$

Since the integral is being taken over the entire disk of radius w, changing to polar coordinates yields

$$F_W(w) = \frac{1}{2\pi\sigma^2} \int_0^{2\pi} d\theta \int_0^w e^{-(r/\sigma)^2/2} r\, dr$$

$$= \sigma^{-2} \int_0^w e^{-(r/\sigma)^2/2} r\, dr$$

$$= 1 - e^{-(w/\sigma)^2/2}$$

Differentiation of $F_W(w)$ with respect to w gives

$$f_W(w) = w\sigma^{-2} e^{-(w/\sigma)^2/2}$$

which, as claimed, is a Rayleigh distribution with $\eta = \sigma$.

FIGURE 5.20
The density $f_W(w)$ for Example 5.16

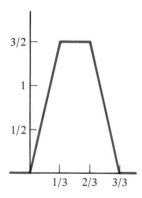

3/2

1

1/2

1/3   2/3   3/3

5.3.2 Distribution of a Sum of Independent Random Variables—The Moment-Generating-Function Method

The next theorem, which states that the moment-generating function of a sum of independent random variables is the product of the moment-generating functions, is the basis of the transform technique for finding distributions of sums of independent random variables.

THEOREM 5.5

If $X_1, X_2, \ldots, X_n$ are independent random variables and

$$Y = X_1 + X_2 + \cdots + X_n$$

then the moment-generating function of Y is given by

$$M_Y(t) = M_{X_1}(t)M_{X_2}(t) \cdots M_{X_n}(t)$$

Proof: As usual, we give the proof for the continuous case, the discrete case being derived by employing sums instead of integrals. Since the random variables are independent, their joint density is given by

$$f(x_1, x_2, \ldots, x_n) = f_{X_1}(x_1)f_{X_2}(x_2) \cdots f_{X_n}(x_n)$$

Thus,

$$
\begin{aligned}
M_Y(t) &= E[e^{tY}] \\
&= E[e^{t(X_1 + X_2 + \cdots + X_n)}] \\
&= \int_{-\infty}^{\infty} \cdots \int_{-\infty}^{\infty} e^{t(x_1 + x_2 + \cdots + x_n)} f(x_1, x_2, \ldots, x_n) \, dx_n \cdots dx_1 \\
&= \int_{-\infty}^{\infty} \cdots \int_{-\infty}^{\infty} e^{tx_1}f_{X_1}(x_1)e^{tx_2}f_{X_2}(x_2) \cdots e^{tx_n}f_{X_n}(x_n) \, dx_n \cdots dx_1 \\
&= \prod_{k=1}^{n} \int_{-\infty}^{\infty} e^{tx_k}f_{X_k}(x_k) \, dx_k \\
&= \prod_{k=1}^{n} M_{X_k}(t)
\end{aligned}
$$

∎

Note that independence is crucial in the proof of Theorem 5.5, since it allows the factoring of the multiple integral into a product of single integrals, each of which defines a moment-generating function.

Used in conjunction with Theorem 3.16, Theorem 5.5 can be employed to obtain some important practical results.

EXAMPLE 5.18

If X and Y are independent binomially distributed random variables with X having parameters n and p and Y having parameters m and p, then it is intuitively clear that the sum $X + Y$ is binomially distributed with parameters $n + m$ and p, the reason being that from a modeling perspective $X + Y$ counts the successes in $n + m$ Bernoulli trials. To obtain a mathematically rigorous proof of this result, we could compute the joint density

$$f_{X,Y}(x, y) = b(x; n, p)b(y; m, p)$$

of the random vector (X, Y), the assumption of independence assuring us that the joint density is the product of the marginal densities. We can also employ Theorem 5.5. According to Theorem 4.1, the moment-generating functions of X and Y are

$$M_X(t) = (pe^t + q)^n$$

and

$$M_Y(t) = (pe^t + q)^m$$

respectively. By Theorem 5.5,

$$\begin{aligned} M_{X+Y}(t) &= M_X(t)M_Y(t) \\ &= (pe^t + q)^n(pe^t + q)^m \\ &= (pe^t + q)^{n+m} \end{aligned}$$

which is the moment-generating function of a binomial random variable with parameters $m + n$ and p. Thus, by uniqueness (Theorem 3.16), $X + Y$ is such a variable.

THEOREM 5.6

If $X_1, X_2, \ldots, X_n$ are independently distributed Poisson random variables having the parameters $\lambda_1, \lambda_2, \ldots, \lambda_n$, respectively, then the sum

$$Y = X_1 + X_2 + \cdots + X_n$$

is Poisson distributed with parameter $\lambda_1 + \lambda_2 + \cdots + \lambda_n$.

Proof: First consider the case of two independent Poisson random variables X and Y with parameters λ and τ, respectively. According to Theorem 4.4, their moment-generating functions are

$$M_X(t) = e^{\lambda(e^t - 1)}$$

and

$$M_Y(t) = e^{\tau(e^t - 1)}$$

respectively. By Theorem 5.5,

$$\begin{aligned} M_{X+Y}(t) &= M_X(t)M_Y(t) \\ &= e^{\lambda(e^t - 1)}e^{\tau(e^t - 1)} \\ &= e^{(\lambda + \tau)(e^t - 1)} \end{aligned}$$

which is the moment-generating function of a Poisson variable with parameter $\lambda + \tau$. By uniqueness, $X + Y$ is such a variable. The result extends by induction to n variables. ∎

Whereas Theorem 5.6 showed a sum of independent Poisson random variables to be a Poisson variable, the next theorem gives a similar property of the gamma distribution.

Suppose $X_1, X_2, \ldots, X_n$ are independent and possess gamma distributions, with X_k having parameters α_k and β, for $k = 1, 2, \ldots, n$. Then the sum

$$Y = X_1 + X_2 + \cdots + X_n$$

is gamma distributed with parameters $\alpha_1 + \alpha_2 + \cdots + \alpha_n$ and β.

Proof: The moment-generating function of X_k is given by

$$M_{X_k}(t) = (1 - \beta t)^{-\alpha_k}$$

and thus, by Theorem 5.5,

$$M_Y(t) = (1 - \beta t)^{-\alpha_1}(1 - \beta t)^{-\alpha_2} \cdots (1 - \beta t)^{-\alpha_n}$$
$$= (1 - \beta t)^{-(\alpha_1 + \alpha_2 + \cdots + \alpha_n)}$$

which is the moment-generating function of a gamma variable with parameters $\alpha_1 + \alpha_2 + \cdots + \alpha_n$ and β. By uniqueness, Y is such a variable. ∎

An immediate corollary of Theorem 5.7 is that the sum of n independent identically distributed exponential random variables, each possessing parameter b, is n-Erlang with parameter $\beta = 1/b$.

Another immediate corollary of Theorem 5.7 applies to an independent sum of chi-square distributions. Here, $\alpha_k = \nu_k/2$ and $\beta = 2$. By Theorem 5.7, the sum is a gamma distribution with $\alpha = (\nu_1 + \nu_2 + \cdots + \nu_n)/2$ and $\beta = 2$, which is a chi-square distribution with $\nu_1 + \nu_2 + \cdots + \nu_n$ degrees of freedom.

EXAMPLE 5.19

Suppose $X_1, X_2, \ldots, X_n$ are identically distributed independent gamma variables having parameters α and β. If Y is the sum of the X_k, then, according to Theorem 5.7, Y is gamma distributed with parameters $n\alpha$ and β. In particular, if the X_k are chi-square variables, each possessing ν degrees of freedom, then the sum is a chi-square variable possessing $n\nu$ degrees of freedom.

Applied in conjunction with the result of Example 3.38, Example 5.19 leads to a proposition of some importance in statistics.

THEOREM 5.8

If $X_1, X_2, \ldots, X_n$ are independent normally distributed random variables, each possessing mean μ and standard deviation σ, then the sum

$$Y = \sum_{k=1}^{n} \left(\frac{X_k - \mu}{\sigma} \right)^2$$

possesses a chi-square distribution with $\nu = n$ degrees of freedom.

Proof: According to Theorem 4.7,

228 CHAP. 5: MULTIVARIATE DISTRIBUTIONS

$$Z_k = \frac{X_k - \mu}{\sigma}$$

is a standard normal variable for $k = 1, 2, \ldots, n$. As demonstrated in Example 3.38, the square of a standard normal variable is chi-square distributed with $\nu = 1$. Thus, by Example 5.19,

$$Z_1^2 + Z_2^2 + \cdots + Z_n^2$$

possesses a chi-square distribution with $\nu = n$ degrees of freedom. ■

In many statistical applications it is important to know the distribution of a **linear combination**

$$a_1 X_1 + a_2 X_2 + \cdots + a_n X_n$$

of independent normally distributed random variables $X_1, X_2, \ldots, X_n$, the a_k, $k = 1, 2, \ldots, n$, being constants. The next theorem, which results from another application of Theorem 5.5, answers the question completely.

THEOREM 5.9

If $X_1, X_2, \ldots, X_n$ form a collection of independent normally distributed random variables such that, for $k = 1, 2, \ldots, n$, X_k has mean μ_k and standard deviation σ_k, then, for any constants $a_1, a_2, \ldots, a_n$, the linear combination

$$Y = a_1 X_1 + a_2 X_2 + \cdots + a_n X_n$$

is normally distributed with mean

$$\mu_Y = a_1 \mu_1 + a_2 \mu_2 + \cdots + a_n \mu_n$$

and variance

$$\sigma_Y^2 = a_1^2 \sigma_1^2 + a_2^2 \sigma_2^2 + \cdots + a_n^2 \sigma_n^2$$

Proof: According to Theorem 4.6, for each k, X_k has moment-generating function

$$M_{X_k}(t) = e^{\mu_k t + \sigma_k^2 t^2 / 2}$$

Applying Theorem 3.15 yields

$$M_{a_k X_k}(t) = M_{X_k}(a_k t) = e^{a_k \mu_k t + a_k^2 \sigma_k^2 t^2 / 2}$$

Thus, by Theorem 5.5,

$$M_Y(t) = \prod_{k=1}^{n} M_{a_k X_k}(t)$$

$$= \prod_{k=1}^{n} e^{a_k \mu_k t + a_k^2 \sigma_k^2 t^2 / 2}$$

$$= e^{\sum_{k=1}^{n} a_k \mu_k t + a_k^2 \sigma_k^2 t^2/2}$$

$$= e^{\left(\sum_{k=1}^{n} a_k \mu_k\right)t + \left(\sum_{k=1}^{n} a_k^2 \sigma_k^2\right)t^2/2}$$

which is the moment-generating function of a normal random variable with the claimed mean and variance. ∎

EXAMPLE 5.20

Measurements of resistance in a particular lot of 5-ohm resistors are normally distributed with mean 5.01 ohms and standard deviation 0.02 ohm. In a lot of 8-ohm resistors, measurements yield a mean of 8.02 ohms and a standard deviation of 0.03 ohm. Given a 5-ohm and an 8-ohm resistor in series, the resistance of the two resistors is given by

$$Y = U + V$$

where $U \sim N(5.01, 0.02)$ and $V \sim N(8.02, 0.03)$. Thus, by Theorem 5.9, Y is normally distributed with mean

$$\mu_Y = 5.01 + 8.02 = 13.03$$

ohms and variance

$$\sigma_Y^2 = 0.0004 + 0.0009 = 0.0013$$

5.4 JOINT DISTRIBUTIONS OF OUTPUT RANDOM VARIABLES

As illustrated in Figure 5.1, there may be several output random variables resulting from the input of several random variables into a system. Thus far we have considered the distribution of a single output random variable. At this point we wish to consider the joint distribution of the output random variables.

5.4.1 Discrete Multivariate Output Distributions

Consider the situation in which there are two input random variables, X and Y, and two output random variables, U and V. If the variables are discrete, then the joint density for U and V is

$$f_{U,V}(u, v) = P(U = u, V = v)$$

As outputs of the system, both U and V are functions of X and Y. Thus, there exist functions g and h such that

$$U = g(X, Y)$$
$$V = h(X, Y)$$

Consequently,

$$f_{U,V}(u, v) = P[g(X, Y) = u, h(X, Y) = v]$$

Assuming the joint density for X and Y is known, this equation can be employed to find the bivariate distribution of U and V.

EXAMPLE 5.21

In Example 5.15 we found the densities of two functions of the random variables whose distribution is given in Table 5.1:

$$U = g(X, Y) = XY$$

$$V = h(X, Y) = X + Y$$

Now consider the joint density of the random vector (U, V). The codomain of (U, V) is

$$\Omega_{U,V} = \{(0, 0), (0, 1), (0, 2), (0, 3), (1, 2), (2, 3)\}$$

The joint density $f_{U,V}(u, v)$ is computed in the following manner:

$$f_{U,V}(0, 0) = P(XY = 0, X + Y = 0) = P[\{(0, 0)\}] = \frac{1}{216}$$

$$f_{U,V}(0, 1) = P(XY = 0, X + Y = 1) = P[\{(0, 1), (1, 0)\}] = \frac{15}{216}$$

$$f_{U,V}(0, 2) = P(XY = 0, X + Y = 2) = P[\{(0, 2), (2, 0)\}] = \frac{39}{216}$$

$$f_{U,V}(0, 3) = P(XY = 0, X + Y = 3) = P[\{(0, 3), (3, 0)\}] = \frac{35}{216}$$

$$f_{U,V}(1, 2) = P(XY = 1, X + Y = 2) = P[\{(1, 1)\}] = \frac{36}{216}$$

$$f_{U,V}(2, 3) = P(XY = 2, X + Y = 3) = P[\{(1, 2), (2, 1)\}] = \frac{90}{216}$$

Note that U and V are not independent. For instance,

$$f_{U,V}(0, 0) = \frac{1}{216} \neq \frac{5}{12} \times \frac{1}{216} = f_U(0)f_V(0)$$

$f_U(0)$ and $f_V(0)$ having been found in Example 5.15.

The transformation $(X, Y) \longrightarrow (U, V)$ results from two functions, each possessing two independent variables and a single dependent variable:

$$u = g(x, y)$$

$$v = h(x, y)$$

If the mapping is one-to-one, then the preceding pair of equations can be uniquely solved for x and y to yield the pair of equations:

$$x = r(u, v)$$

$$y = s(u, v)$$

The mapping $(x, y) \longrightarrow (u, v)$ is called a **vector mapping** and its **inverse mapping** is $(u, v) \longrightarrow (x, y)$. Under the condition of one-to-oneness, there exists a bivariate analogue of Theorem 3.8 that provides a direct formula for the joint discrete density of U and V in terms of the joint density of X and Y.

THEOREM 5.10

Suppose X and Y are discrete random variables with joint density $f(x, y)$, $U = g(X, Y)$, $V = h(X, Y)$, and the vector mapping

$$\binom{u}{v} = \binom{g(x, y)}{h(x, y)}$$

possesses the inverse vector mapping

$$\binom{x}{y} = \binom{r(u, v)}{s(u, v)}$$

Then the joint density for U and V is given by

$$f_{U,V}(u, v) = f_{X,Y}[r(u, v), s(u, v)]$$

Proof: For fixed u and v, there exists a unique pair of points x and y such that

$$
\begin{aligned}
f_{U,V}(u, v) &= P(U = u, V = v) \\
&= P[U = g(x, y), V = h(x, y)] \\
&= P(A)
\end{aligned}
$$

where A is the set of outcomes w in the sample space such that $U(w) = u$ and $V(w) = v$ (see Figure 5.21). Since the random vector (U, V) is the composition of the vector mapping (g, h) with the random vector (X, Y), $U(w) = u$ and $V(w) = v$ if and only if

FIGURE 5.21
Diagram corresponding to the proof of Theorem 5.10

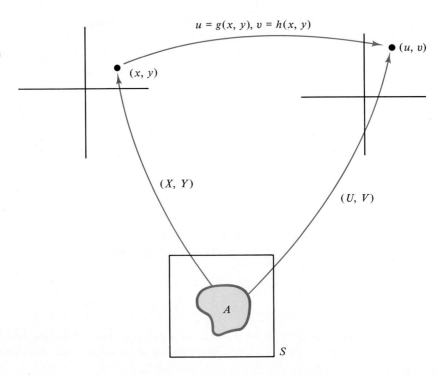

$$X(w) = x = r(u, v)$$

and

$$Y(w) = y = s(u, v)$$

so that

$$P(A) = P[X = r(u, v), Y = s(u, v)] = f_{X,Y}[r(u, v), s(u, v)] \qquad \blacksquare$$

EXAMPLE 5.22 Suppose X and Y are independent binomially distributed random variables, X having parameters n and p and Y having parameters m and ρ. Then their joint density is the product of the marginal densities:

$$f(x, y) = f_X(x)f_Y(y) = \binom{n}{x}\binom{m}{y} p^x \rho^y (1 - p)^{n-x}(1 - \rho)^{m-y}$$

Suppose the output random vector is defined by

$$U = g(X, Y) = X + Y$$
$$V = h(X, Y) = X - Y$$

The equation pair

$$u = x + y$$
$$v = x - y$$

possesses the unique solution

$$x = r(u, v) = \frac{u + v}{2}$$

$$y = s(u, v) = \frac{u - v}{2}$$

Therefore, according to Theorem 5.10,

$$f_{U,V}(u, v) = \binom{n}{\dfrac{u+v}{2}} \binom{m}{\dfrac{u-v}{2}} p^{(u+v)/2} \rho^{(u-v)/2} (1 - p)^{n - [(u+v)/2]}(1 - \rho)^{m - [(u-v)/2]}$$

Note that there are constraints on the variables u and v, these being determined by the constraints on x and y. Table 5.4 gives the input and output codomains for the case $n = m = 2$.

5.4.2 Continuous Multivariate Output Distributions

For continuous jointly distributed random variables, the problem is somewhat more difficult. Although we could employ the joint probability distribution function of the output random vector (U, V), such an approach is mathematically cumbersome and will not be pursued. Fortunately, for the case of one-to-one vector mappings there

TABLE 5.4 INPUT AND OUTPUT CODOMAINS

Input		Output	
x	y	u	v
0	0	0	0
0	1	1	-1
0	2	2	-2
1	0	1	1
1	1	2	0
1	2	3	-1
2	0	2	2
2	1	3	1
2	2	4	0

is a continuous analogue of Theorem 5.10. Although the proof is beyond the level of the present text, we can state the theorem and give examples.

THEOREM 5.11

Suppose the random variables X and Y are jointly continuous with joint density $f(x, y)$ and suppose the vector mapping $(x, y) \longrightarrow (u, v)$ is defined by

$$u = g(x, y)$$
$$v = h(x, y)$$

where both $g(x, y)$ and $h(x, y)$ possess continuous partial derivatives and where the mapping is one-to-one on the set $A_{X,Y}$ of all values (x, y) in the codomain of the random vector (X, Y) for which $f(x, y) > 0$. If $A_{U,V}$ denotes the set of all points (u, v) corresponding to points in $A_{X,Y}$ and if the equation pair

$$x = r(u, v)$$
$$y = s(u, v)$$

represents the inverse mapping $(u, v) \longrightarrow (x, y)$, then the joint density of the random vector

$$\begin{pmatrix} U \\ V \end{pmatrix} = \begin{pmatrix} g(X, Y) \\ h(X, Y) \end{pmatrix}$$

is given by

$$f_{U,V}(u, v) = \begin{cases} f_{X,Y}[r(u, v), s(u, v)]|J(u, v)|, & \text{if } (u, v) \in A_{U,V} \\ 0, & \text{otherwise} \end{cases}$$

where

$$J(u, v) = \det \begin{bmatrix} \dfrac{\partial x}{\partial u} & \dfrac{\partial x}{\partial v} \\ \dfrac{\partial y}{\partial u} & \dfrac{\partial y}{\partial v} \end{bmatrix}$$

is the **Jacobian** of the mapping.

EXAMPLE 5.23

Let X and Y be uniformly jointly distributed on the square $[0, 1] \times [0, 1]$ and consider the random-vector mapping

$$U = X + Y$$
$$V = X - Y$$

The equation pair

$$u = x + y$$
$$v = x - y$$

possesses the unique solution pair

$$x = \frac{u + v}{2}$$

$$y = \frac{u - v}{2}$$

Using the notation of Theorem 5.11,

$$A_{X,Y} = [0, 1] \times [0, 1]$$

and $A_{U,V}$ is found by solving the variable-constraint pair

$$0 \leq \frac{u + v}{2} \leq 1$$

$$0 \leq \frac{u - v}{2} \leq 1$$

Both $A_{X,Y}$ and $A_{U,V}$ are depicted in Figure 5.22. Since

$$\frac{\partial x}{\partial u} = \frac{\partial x}{\partial v} = \frac{\partial y}{\partial u} = \frac{1}{2}$$

and

$$\frac{\partial y}{\partial v} = -\frac{1}{2}$$

the Jacobian is

$$J(u, v) = \det \begin{bmatrix} \dfrac{1}{2} & \dfrac{1}{2} \\ \dfrac{1}{2} & -\dfrac{1}{2} \end{bmatrix} = -\frac{1}{2}$$

so that the absolute value of the Jacobian is the constant 1/2. Thus, according to Theorem 5.11,

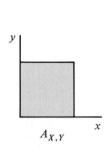

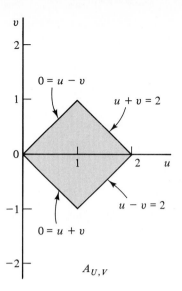

FIGURE 5.22
Support regions $A_{X,Y}$ and $A_{U,V}$ for Example 5.23

$$f_{U,V}(u, v) = 1 \times \frac{1}{2} = \frac{1}{2}$$

in the region depicted in Figure 5.22 and $f_{U,V}(u, v) = 0$ for all other (u, v).

EXAMPLE 5.24

Let X and Y possess the joint density

$$f_{X,Y}(x, y) = \begin{cases} 2e^{-(x+2y)}, & \text{if } x, y > 0 \\ 0, & \text{otherwise} \end{cases}$$

and consider the random-vector mapping defined by the pair

$$U = \frac{X}{Y}$$

$$V = X + Y$$

Both $A_{X,Y}$ and $A_{U,V}$ are the first quadrant of the plane, not including the axes. Moreover, the vector mapping

$$u = \frac{x}{y}$$

$$v = x + y$$

possesses the unique solution pair

$$x = \frac{uv}{1 + u}$$

$$y = \frac{v}{1 + u}$$

The required partial derivatives are

$$\frac{\partial x}{\partial u} = \frac{v}{(1 + u)^2}$$

$$\frac{\partial x}{\partial v} = \frac{u}{1 + u}$$

$$\frac{\partial y}{\partial u} = -\frac{v}{(1 + u)^2}$$

$$\frac{\partial y}{\partial v} = \frac{1}{1 + u}$$

and the Jacobian determinant is

$$J(u, v) = \frac{\partial x}{\partial u}\frac{\partial y}{\partial v} - \frac{\partial x}{\partial v}\frac{\partial y}{\partial u} = \frac{v}{(1 + u)^2}$$

Since the Jacobian is always positive, Theorem 5.11 yields

$$f_{U,V}(u, v) = \frac{2v}{(1 + u)^2} e^{-(2 + u)v/(1 + u)}$$

for $u > 0$ and $v > 0$, and $f_{U,V}(u, v) = 0$ otherwise.

Thus far, we have considered the case of two input and two output random variables. In fact, the generalizations of both Theorems 5.10 and 5.11 to n input and n output random variables are straightforward, although both cases do involve increased computational burdens. In the case of three input variables X, Y, and Z and three output variables U, V, and W, the vector mapping is defined by three functions of three variables:

$$u = g(x, y, z)$$

$$v = h(x, y, z)$$

$$w = k(x, y, z)$$

If the mapping is one-to-one, then there are three functions of u, v, and w determining the inversion:

$$x = r(u, v, w)$$

$$y = s(u, v, w)$$

$$z = t(u, v, w)$$

In the continuous case, the Jacobian is the three-by-three determinant

$$J(u, v, w) = \det \begin{bmatrix} \dfrac{\partial x}{\partial u} & \dfrac{\partial x}{\partial v} & \dfrac{\partial x}{\partial w} \\[2mm] \dfrac{\partial y}{\partial u} & \dfrac{\partial y}{\partial v} & \dfrac{\partial y}{\partial w} \\[2mm] \dfrac{\partial z}{\partial u} & \dfrac{\partial z}{\partial v} & \dfrac{\partial z}{\partial w} \end{bmatrix}$$

and the joint output density of U, V, and W is given by

$$f_{U,V,W}(u, v, w) = f_{X,Y,Z}[r(u, v, w), s(u, v, w), t(u, v, w)]|J(u, v, w)|$$

If the joint density of U and V in terms of the density of the input variables X, Y, and Z is desired, it can be found by first finding $f_{U,V,W}(u, v, w)$ and then integrating out w to obtain the desired joint marginal density.

5.5 EXPECTED VALUE OF A FUNCTION OF SEVERAL RANDOM VARIABLES

Like the case of a single random variable, if the joint density of the input variables is known, then the expectation of a function of several random variables can be found without first finding the density of the output. This fundamental result is the analogue of Theorem 3.10 and we state it without proof.

THEOREM 5.12

Suppose $X_1, X_2, \ldots, X_n$ are jointly distributed discrete random variables possessing joint probability mass function $f(x_1, x_2, \ldots, x_n)$ and

$$y = g(x_1, x_2, \ldots, x_n)$$

is any real-valued function. Then

$$E[g(X_1, X_2, \ldots, X_n)] = \sum_{(x_1, x_2, \ldots, x_n)} g(x_1, x_2, \ldots, x_n)f(x_1, x_2, \ldots, x_n)$$

where the sum is taken over all $(x_1, x_2, \ldots, x_n)$ such that $f(x_1, x_2, \ldots, x_n) > 0$. If, on the other hand, the random variables $X_1, X_2, \ldots, X_n$ are continuous and possess joint density $f(x_1, x_2, \ldots, x_n)$, then, so long as g is piecewise continuous,

$$E[g(X_1, X_2, \ldots, X_n)]$$

$$= \int_{-\infty}^{\infty} \cdots \int_{-\infty}^{\infty} g(x_1, x_2, \ldots, x_n)f(x_1, x_2, \ldots, x_n) \, dx_n \cdots dx_1$$

where the integral is n-fold.

For the case where X and Y are continuous and possess the bivariate density $f(x, y)$ and where $w = g(x, y)$ is a function of two variables, Theorem 5.12 takes the form

$$E[g(X, Y)] = \int_{-\infty}^{\infty} \int_{-\infty}^{\infty} g(x, y)f(x, y) \, dy \, dx$$

Theorem 5.12 has an immediate corollary that is among the most important propositions in the theory of probability. Put simply, the expected value of a sum is the sum of the expected values. Note that independence of the variables is not required.

For any finite collection of random variables $X_1, X_2, \ldots, X_n$,

$$E[X_1 + X_2 + \cdots + X_n] = E[X_1] + E[X_2] + \cdots + E[X_n]$$

Proof: We give the proof in the continuous case for two random variables. The discrete case is proven similarly except for the use of sums instead of integrals. Moreover, once the proposition is shown for two random variables, the general case follows by induction. For two continuous random variables X and Y,

$$E[X + Y] = \int_{-\infty}^{\infty} \int_{-\infty}^{\infty} (x + y)f(x, y)\, dy\, dx$$

$$= \int_{-\infty}^{\infty} \int_{-\infty}^{\infty} xf(x, y)\, dy\, dx + \int_{-\infty}^{\infty} \int_{-\infty}^{\infty} yf(x, y)\, dy\, dx$$

$$= \int_{-\infty}^{\infty} xf_X(x)\, dx + \int_{-\infty}^{\infty} yf_Y(y)\, dy$$

$$= E[X] + E[Y] \qquad\blacksquare$$

EXAMPLE 5.25

The mean of a Poisson random variable with parameter λ is λ. If $X_1, X_2, \ldots, X_n$ are Poisson distributed with parameters $\lambda_1, \lambda_2, \ldots, \lambda_n$, then, according to Theorem 5.13,

$$E[X_1 + X_2 + \cdots + X_n] = \lambda_1 + \lambda_2 + \cdots + \lambda_n$$

This conclusion holds whether or not the random variables are independently distributed. Although the same conclusion can be drawn from Theorem 5.6, in which the actual distribution of the sum is found, Theorem 5.6 requires independence, whereas Theorem 5.13 does not. In effect, the conclusion of Theorem 5.6 is stronger; however, it requires a special condition, one which is often not demonstrable.

As noted subsequent to Theorem 3.11 in Section 3.6.1, for any random variable X and constant a, $E[aX] = aE[X]$. As a result, Theorem 5.13 shows that the expected-value operator E is a **linear operator:**

$$E[a_1X_1 + a_2X_2 + \cdots + a_nX_n] = a_1E[X_1] + a_2E[X_2] + \cdots + a_nE[X_n]$$

for any random variables $X_1, X_2, \ldots, X_n$ and for any constants $a_1, a_2, \ldots, a_n$.

5.6 COVARIANCE

Moments can be generalized to jointly distributed random variables, a particularly important instance being the covariance. The latter plays a key role in the analysis of linear systems.

5.6.1   Moments of a Bivariate Distribution

Although moments can be generalized to a collection of n random variables, our immediate interest lies with bivariate moments, and thus the definition is given for the case of two random variables.

DEFINITION 5.8

Product Moments. Given two random variables X and Y and two nonnegative integers p and q, the $(p + q)$-**order product moment** is

$$\mu'_{pq} = E[X^p Y^q]$$

The $(p + q)$-**order central moment** is

$$\mu_{pq} = E[(X - \mu_X)^p (Y - \mu_Y)^q]$$

In the continuous case, application of Theorem 5.12 yields

$$\mu'_{pq} = \int_{-\infty}^{\infty} \int_{-\infty}^{\infty} x^p y^q f_{X,Y}(x, y) \, dy \, dx$$

and

$$\mu_{pq} = \int_{-\infty}^{\infty} \int_{-\infty}^{\infty} (x - \mu_X)^p (y - \mu_Y)^q f_{X,Y}(x, y) \, dy \, dx$$

Similar expressions apply in the discrete case, except there are double summations instead of double integrals. As in the case of ordinary moments, it is tacitly assumed in the definition of the $(p + q)$-order product moment that the integral is absolutely convergent:

$$\int_{-\infty}^{\infty} \int_{-\infty}^{\infty} |x|^p |y|^q f_{X,Y}(x, y) \, dy \, dx < \infty$$

It is also assumed in the definition of the central moment that the integral is absolutely convergent. For discrete random variables, the corresponding summations are assumed to be absolutely convergent.

If $q = 0$, then the product moment reduces to the pth moment for X. Correspondingly, if $p = 0$, then the product moment reduces to the qth moment for Y.

Our main interest is with **second-order moments.** These are the second moments of each variable individually and the product moments μ'_{11} and μ_{11}, especially the latter.

DEFINITION 5.9

Covariance. The covariance of two jointly distributed random variables X and Y is their second-order central moment:

$$\text{Cov}\,[X, Y] = E[(X - \mu_X)(Y - \mu_Y)]$$

The covariance is also denoted by σ_{XY}^2, this notation being appropriate since, if $Y = X$, then Cov $[X, Y]$ reduces to the variance of X. The covariance can be expressed in terms of the second-order product moment and the product of the expected values.

THEOREM 5.14

Assuming it exists, the covariance is given by

$$\text{Cov }[X, Y] = E[XY] - \mu_X\mu_Y$$

Proof: By the linearity of the expected-value operator,

$$\text{Cov }[X, Y] = E[(X - \mu_X)(Y - \mu_Y)]$$
$$= E[XY - \mu_Y X - \mu_X Y + \mu_X\mu_Y]$$
$$= E[XY] - \mu_Y E[X] - \mu_X E[Y] + \mu_X\mu_Y$$

Since $\mu_X = E[X]$ and $\mu_Y = E[Y]$, the result follows. ■

If $X = Y$, then Theorem 5.14 reduces to Theorem 3.12.

THEOREM 5.15

If X and Y are independent random variables, then

$$E[XY] = E[X]E[Y]$$

In particular, Cov $[X, Y] = 0$.

Proof: We prove the theorem for continuous random variables. If X and Y possess the joint density $f(x, y)$, then

$$E[XY] = \int_{-\infty}^{\infty} \int_{-\infty}^{\infty} xyf(x, y) \, dy \, dx$$

$$= \int_{-\infty}^{\infty} \int_{-\infty}^{\infty} xyf_X(x)f_Y(y) \, dy \, dx$$

$$= \int_{-\infty}^{\infty} xf_X(x) \, dx \int_{-\infty}^{\infty} yf_Y(y) \, dy$$

$$= E[X]E[Y]$$

A direct application of Theorem 5.14 shows that Cov $[X, Y] = 0$. ■

5.6.2 Correlation

As will be seen shortly, the covariance provides a measure of the linear relationship between random variables; however, the deviations $X - \mu_X$ and $Y - \mu_Y$, from which the covariance is derived, are dependent upon the units in which X and Y are measured. The correlation coefficient provides a normalized measure.

Correlation Coefficient. The correlation coefficient of the random variables X and Y is defined by

$$\rho_{XY} = \frac{\text{Cov}\,[X,\,Y]}{\sigma_X \sigma_Y}$$

If the correlation coefficient, which is also denoted by Corr $[X,\,Y]$, is 0, then the random variables X and Y are said to be **uncorrelated.** From the definitions, it is immediate that X and Y are uncorrelated if and only if Cov $[X,\,Y] = 0$. By Theorem 5.15, if X and Y are independent, then they are uncorrelated. As the next example shows, the converse does not hold.

EXAMPLE 5.26

Let X and Y be discrete random variables possessing the joint density $f(x,\,y)$ defined by

$$f(1,\,0) = f(0,\,1) = f(-1,\,0) = f(0,\,-1) = \frac{1}{4}$$

Then

$$f_X(x) = \begin{cases} 1/4, & \text{if } x = -1 \\ 1/2, & \text{if } x = 0 \\ 1/4, & \text{if } x = 1 \\ 0, & \text{otherwise} \end{cases}$$

and $f_Y(y)$ is similarly defined. In addition,

$$E[X] = E[Y] = E[XY] = 0$$

so that Cov $[X,\,Y] = 0$ and the variables are uncorrelated. However,

$$f_X(0)f_Y(0) = \frac{1}{4} \neq 0 = f(0,\,0)$$

so that X and Y are not independent.

The next example illustrates the manner in which the correlation coefficient measures the linear relationship between the variables. Very often, practical interpretation rests on the type of graphical analysis employed in the example.

EXAMPLE 5.27

Suppose X and Y are discrete random variables whose joint outcomes are equally likely and whose codomain consists of the points illustrated in Figure 5.23. Each dot in the figure represents a mass point of equal probability. Since the codomain has 10 points, the probability mass function is $f(x,\,y) = 0.1$ for each $(x,\,y)$ in the codomain. Moreover, the equal likelihood of the joint probabilities means that the marginal density for X is simply

$$f_X(x) = \sum_y f(x,\,y) = \frac{\text{card }\{y\colon (x,\,y) \in \Omega_{X,Y}\}}{10}$$

For instance, since there are 3 points in the codomain for which $x = 6$, $f_X(6) = 0.3$. Consequently, if we label the points of the codomain $(x_j,\,y_j)$, for $j = 1, 2, \ldots, 10$, then the kth moment of X is given by

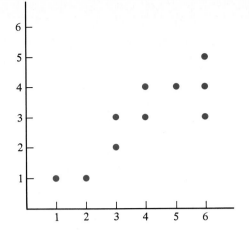

FIGURE 5.23
Codomain for the bivariate distribution of Example 5.27

$$E[X^k] = \sum_x x^k f_X(x) = \frac{1}{10} \sum_{j=1}^{10} x_j^k$$

Analogous expressions apply to the product moments and to the moments for Y.

To compute the correlation coefficient, $E[X]$, $E[Y]$, $E[XY]$, Var $[X]$, Var $[Y]$, Cov $[X, Y]$, σ_X, and σ_Y are needed. Due to the simple form of the moments, it is convenient to list the relevant quantities, including the codomain points, in a column format and to sum the columns. This information, which is given in Table 5.5, yields

$$E[X] = \frac{1}{10} \sum_{j=1}^{10} x_j = 4.0$$

$$E[Y] = \frac{1}{10} \sum_{j=1}^{10} y_j = 3.0$$

$$E[X^2] = \frac{1}{10} \sum_{j=1}^{10} x_j^2 = 18.8$$

$$E[Y^2] = \frac{1}{10} \sum_{j=1}^{10} y_j^2 = 10.6$$

$$E[XY] = \frac{1}{10} \sum_{j=1}^{10} x_j y_j = 13.8$$

$$\text{Var } [X] = E[X^2] - E[X]^2 = 18.8 - 16.0 = 2.8$$
$$\text{Var } [Y] = E[Y^2] - E[Y]^2 = 10.6 , - 9.0 = 1.6$$
$$\text{Cov } [X, Y] = E[XY] - E[X]E[Y] = 13.8 - 12.0 = 1.8$$
$$\rho_{XY} = \frac{\text{Cov } [X, Y]}{\sigma_X \sigma_Y} = \frac{1.8}{1.6733 \times 1.2649} = 0.85$$

Note in Figure 5.23 that the distribution of the points tends to lie along a straight line with positive slope.

Like Figure 5.23, Figures 5.24, 5.25, and 5.26 also represent equally likely discrete distributions having 10 possible outcomes. Their correlation coefficients are $\rho = 1$, $\rho = 0.20$, and $\rho = -0.95$, respectively. The four figures and their respective correlation coefficients indicate the manner in which the correlation coefficient describes the degree to which the probability mass of a joint distribution is concentrated

TABLE 5.5 INFORMATION FOR CORRELATION COEFFICIENT

j	x_j	y_j	$x_j y_j$	x_j^2	y_j^2
1	1	1	1	1	1
2	2	1	2	4	1
3	3	2	6	9	4
4	3	3	9	9	9
5	4	3	12	16	9
6	4	4	16	16	16
7	5	4	20	25	16
8	6	3	18	36	9
9	6	4	24	36	16
10	6	5	30	36	25
Sum	40	30	138	188	106

along a straight line. For values of ρ near $+1$, there is a strong positive linear relationship between the random variables, with values of ρ greater than 0.8 usually being taken to mean substantial linearity. Values of ρ near 0 indicate no linear relationship between the variables. Values near -1 give evidence of a strong negative linear relationship.

To obtain a rigorous quantification of the preceding remarks, suppose $Y = aX + b$, so that Y is a nonhomogeneous linear function of X. Then

$$
\begin{aligned}
\text{Cov}\,[X, Y] &= E[XY] - E[X]E[Y] \\
&= E[X(aX + b)] - E[X]E[aX + b] \\
&= aE[X^2] + bE[X] - (aE[X]^2 + bE[X]) \\
&= a(E[X^2] - E[X]^2) \\
&= a\,\text{Var}\,[X]
\end{aligned}
$$

Moreover, according to Theorem 3.13,

$$
\text{Var}\,[Y] = \text{Var}\,[aX + b] = a^2\,\text{Var}\,[X]
$$

Applying the definition of the correlation coefficient gives

FIGURE 5.24
Discrete bivariate distribution possessing equally likely outcomes with $\rho = 1$

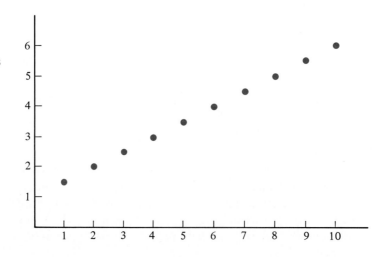

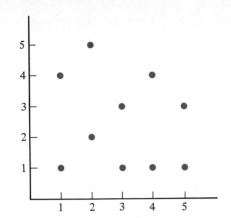

FIGURE 5.25
Discrete bivariate distribution possessing equally likely outcomes with $\rho = 0.20$

$$\rho_{XY} = \frac{a}{|a|} = \begin{cases} 1, & \text{if } a > 0 \\ -1, & \text{if } a < 0 \end{cases}$$

Thus, a full linear relationship between the variables yields a correlation coefficient of ± 1, where the sign indicates whether or not the slope of the line determined by the relationship is positive or negative. The next theorem, provided without further proof, gives the converse. (For those who have studied advanced calculus, the theorem is simply a statement of the Cauchy-Schwarz inequality in terms of the correlation coefficient.)

THEOREM 5.16

For any random variables X and Y,

$$-1 \le \rho_{XY} \le 1$$

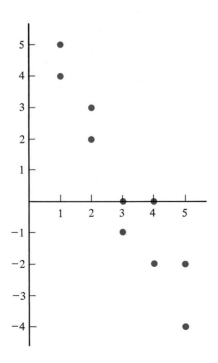

FIGURE 5.26
Discrete bivariate distribution possessing equally likely outcomes with $\rho = -0.95$

and $|\rho_{XY}| = 1$ if and only if there exist constants a and b, $a \neq 0$, such that $Y = aX + b$.

EXAMPLE 5.28

Prior to the application of a recognition algorithm, a speech recognizer must often filter the signals it receives in order to suppress unwanted ambient noise or to enhance salient characteristics of the signal. Suppose two filters are tested and counts of recognition errors are taken on sets of various test patterns, each set consisting of ten patterns. Let X and Y count the errors made when employing the two respective filters, and suppose it is speculated that their joint discrete distribution is given by Table 5.6 (in which the marginal densities have been calculated). For instance, on an arbitrary set of ten speech patterns,

$$P(X = 1, Y = 2) = 0.05$$

According to Table 5.6, $E[X] = 0.79$, $E[Y] = 0.81$, $E[X^2] = 1.37$, $E[Y^2] = 1.69$, $E[XY] = 0.62$,

$$\text{Var}\,[X] = 1.37 - 0.62 = 0.75$$

$$\text{Var}\,[Y] = 1.69 - 0.66 = 1.03$$

$$\text{Cov}\,[X, Y] = 0.62 - (0.79)(0.81) = -0.02$$

$$\rho_{XY} = \frac{\text{Cov}\,[X, Y]}{\sqrt{\text{Var}\,[X]\,\text{Var}\,[Y]}} = -0.02$$

Thus, there is essentially no linear correlation between the error counts. The lack of linear correlation is evident in the graph of the probability mass function, which is given in Figure 5.27.

EXAMPLE 5.29

Suppose X and Y measure the traffic densities on two communication channels and each is rated between 0 and 1, 0 meaning no traffic and 1 meaning full capacity. Moreover, suppose X and Y are uniformly distributed over the region A depicted in Figure 5.28. Since the area of the region is 1/3, the joint density is

$$f(x, y) = \begin{cases} 3, & \text{if } (x, y) \in A \\ 0, & \text{otherwise} \end{cases}$$

The marginal density for X is

TABLE 5.6 PROBABILITY MASS FUNCTION FOR RECOGNITION ERRORS

$f(x, y)$		0	1	y 2	3	4	$f_X(x)$
x	0	0.24	0.13	0.04	0.03	0.01	0.45
	1	0.16	0.10	0.05	0.04	0.01	0.36
	2	0.08	0.05	0.01			0.14
	3	0.02	0.02	0.01			0.05
$f_Y(y)$		0.50	0.30	0.11	0.07	0.02	

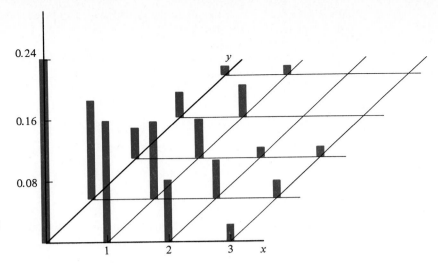

FIGURE 5.27
Probability mass function for the distribution of Table 5.6

$$f_X(x) = 3 \int_{x^2}^{x^{1/2}} dy = 3(x^{1/2} - x^2)$$

if $0 \le x \le 1$, and $f_X(x) = 0$ otherwise. The marginal density for Y is

$$f_Y(y) = 3 \int_{y^2}^{y^{1/2}} dx = 3(y^{1/2} - y^2)$$

if $0 \le y \le 1$, and $f_Y(y) = 0$ otherwise. Furthermore,

$$E[X] = \int_0^1 3x(x^{1/2} - x^2)\,dx = \frac{9}{20}$$

$$E[X^2] = \int_0^1 3x^2(x^{1/2} - x^2)\,dx = \frac{9}{35}$$

$$\text{Var}\,[X] = \frac{9}{35} - \left(\frac{9}{20}\right)^2 = 0.0546$$

$$E[XY] = 3 \int_0^1 \int_{x^2}^{x^{1/2}} xy\,dy\,dx = \frac{1}{4}$$

FIGURE 5.28
Support region for the first distribution of Example 5.29

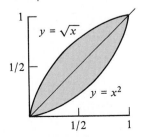

and, since X and Y possess the same density (except for the name of the variable), Y has the same moments as X. Therefore,

$$\text{Cov}\,[X, Y] = \frac{1}{4} - \left(\frac{9}{20}\right)^2 = 0.0475$$

$$\rho_{XY} = \frac{0.0475}{0.0546} = 0.87$$

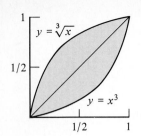

FIGURE 5.29
Support region for the second distribution of Example 5.29

From Figure 5.28, it is apparent that the variables are distributed about the line $y = x$ (although not symmetrically). The correlation coefficient $\rho_{XY} = 0.87$ indicates a rather strong linear relationship between X and Y.

Now consider the same problem, but instead assume that the joint distribution is uniform over the support region depicted in Figure 5.29. Then

$$E[X] = E[Y] = 0.4571$$

$$\text{Var } [X] = \text{Var } [Y] = 0.0578$$

$$E[XY] = 0.25$$

$$\text{Cov } [X, Y] = 0.0410$$

$$\rho_{XY} = 0.71$$

The next proposition concerns the manner in which the variance operates on a linear combination of random variables.

THEOREM 5.17

If $X_1, X_2, \ldots, X_n$ are random variables and

$$Y = a_1X_1 + a_2X_2 + \cdots + a_nX_n$$

then

$$\text{Var } [Y] = \sum_{j,k=1}^{n} a_ja_k \text{ Cov } [X_j, X_k]$$

In particular, if the random variables $X_1, X_2, \ldots, X_n$ are uncorrelated, then

$$\text{Var } [Y] = a_1^2 \text{ Var } [X_1] + a_2^2 \text{ Var } [X_2] + \cdots + a_n^2 \text{ Var } [X_n]$$

Proof: Applying, in order, Theorem 3.12, the linearity of E, some algebra, the linearity of E, the term-by-term subtraction of two finite sums, and Theorem 5.14, we obtain

$$\text{Var } [Y] = E[Y^2] - E[Y]^2$$

$$= E[(a_1X_1 + a_2X_2 + \cdots + a_nX_n)^2]$$

$$\quad - (a_1E[X_1] + a_2E[X_2] + \cdots + a_nE[X_n])^2$$

$$= E\left[\sum_{j,k=1}^{n} a_ja_kX_jX_k\right] - \sum_{j,k=1}^{n} a_ja_kE[X_j]E[X_k]$$

$$= \sum_{j,k=1}^{n} a_ja_kE[X_jX_k] - \sum_{j,k=1}^{n} a_ja_kE[X_j]E[X_k]$$

$$= \sum_{j,k=1}^{n} a_ja_k(E[X_jX_k] - E[X_j]E[X_k])$$

$$= \sum_{j,k=1}^{n} a_ja_k \text{ Cov } [X_j, X_k]$$

In particular, if the variables are uncorrelated, then

$$\text{Cov}[X_j, X_k] = 0$$

unless $j = k$, in which case

$$\text{Cov}[X_k, X_k] = \text{Var}[X_k] \qquad \blacksquare$$

5.7 CONDITIONAL DISTRIBUTIONS

A basic problem in the design of engineering control systems is the extraction of information about one random variable when given information about the behavior of another. This leads to the problem of **conditioning.** Individually, probabilistic knowledge about the variable Y in the random vector (X, Y) is contained within its density, which can be derived from the joint density of X and Y. A natural question arises: What can be said about Y, given we know the outcome of X?

5.7.1 Conditional Densities

If X and Y are discrete and possess the joint density $f(x, y)$, then conditional probabilities regarding Y take the form of conditional probabilities concerning the relevant events. According to Definition 2.6, the probability that $Y = y$ given that $X = x$ is

$$P(Y = y \mid X = x) = \frac{P(Y = y, X = x)}{P(X = x)} = \frac{f(x, y)}{f_X(x)}$$

Intuitively, the observation $X = x$ is recorded and we are left with probabilistic knowledge concerning the random variable Y. Since Y is discrete, only the conditional probabilities of its possible outcomes need be specified.

In the case of jointly distributed continuous random variables, a problem arises, since $P(X = x) = 0$ for any point x. Nevertheless, so long as $f_X(x) \neq 0$, $f(x, y)/f_X(x)$ is still defined. This expression is taken as the general definition of the conditional density.

DEFINITION 5.11

Conditional Density. If X and Y possess joint density $f(x, y)$, then, for all x such that $f_X(x) > 0$, the conditional density of Y given $X = x$ is defined by

$$f(y \mid x) = \frac{f(x, y)}{f_X(x)}$$

For given x, $f(y \mid x)$ is a function of y. That $f(y \mid x)$ is a legitimate density follows from Theorem 3.4, since

$$\int_{-\infty}^{\infty} f(y \mid x)\, dy = \frac{1}{f_X(x)} \int_{-\infty}^{\infty} f(x, y)\, dy = 1$$

the latter integral being equal to $f_X(x)$. The random variable associated with the conditional density is called the **conditional random variable of Y given x** and is denoted by $Y|x$.

Definition 5.11 is symmetric in the sense that interchanging the variables gives the conditional density $f(x \mid y)$ of X given $Y = y$. The corresponding conditional random variable is $X|y$.

EXAMPLE 5.30

In Example 5.28 X and Y counted speech-recognition errors in the case of two filters. The joint-density information given in Table 5.6 yields the conditional density $f(y \mid 0)$:

$$f(0 \mid 0) = \frac{f(0, 0)}{f_X(0)} = \frac{0.24}{0.45} = 0.5333$$

$$f(1 \mid 0) = \frac{f(0, 1)}{f_X(0)} = \frac{0.13}{0.45} = 0.2889$$

$$f(2 \mid 0) = \frac{f(0, 2)}{f_X(0)} = \frac{0.04}{0.45} = 0.0889$$

$$f(3 \mid 0) = \frac{f(0, 3)}{f_X(0)} = \frac{0.03}{0.45} = 0.0667$$

$$f(4 \mid 0) = \frac{f(0, 4)}{f_X(0)} = \frac{0.01}{0.45} = 0.0222$$

The interpretation is clear: if $X = 0$, then there is high probability that Y will be near 0. For instance,

$$P(Y|0 \le 1) = f(0 \mid 0) + f(1 \mid 0) = 0.8222$$

Now consider the conditional density $f(x \mid 0)$:

$$f(0 \mid 0) = \frac{f(0, 0)}{f_Y(0)} = \frac{0.24}{0.50} = 0.48$$

$$f(1 \mid 0) = \frac{f(1, 0)}{f_Y(0)} = \frac{0.16}{0.50} = 0.32$$

$$f(2 \mid 0) = \frac{f(2, 0)}{f_Y(0)} = \frac{0.08}{0.50} = 0.16$$

$$f(3 \mid 0) = \frac{f(3, 0)}{f_Y(0)} = \frac{0.02}{0.50} = 0.04$$

It is clear from the conditional density that the probability mass of the conditional random variable $X|0$ is concentrated near $x = 0$. For instance,

$$P(X|0 \le 1) = f(0 \mid 0) + f(1 \mid 0) = 0.80$$

The interpretation is that if 0 errors are recorded using the filter measured by Y, then it is likely there will be at most 1 error using the filter measured by X.

As a third illustration using the same joint distribution, consider the conditional density $f(y \mid 3)$, which gives the probabilities that $Y = y$ given that $X = 3$:

$$f(0 \mid 3) = \frac{f(3, 0)}{f_X(3)} = \frac{0.02}{0.05} = 0.4$$

$$f(1 \mid 3) = \frac{f(3, 1)}{f_X(3)} = \frac{0.02}{0.05} = 0.4$$

$$f(2 \mid 3) = \frac{f(3, 2)}{f_X(3)} = \frac{0.01}{0.05} = 0.2$$

If 3 errors are recorded when employing the filter measured by X—an unlikely occurrence—then there is a high probability that there will at most 1 error recorded when using the filter measured by Y.

EXAMPLE 5.31

For the traffic density problem of Example 5.29, in which the random variables X and Y are uniformly distributed over the region in Figure 5.28, the conditional density for Y given $X = x$ is defined for $0 < x < 1$, and is given by

$$f(y \mid x) = \frac{f(x, y)}{f_X(x)} = \frac{3}{3(x^{1/2} - x^2)} = (x^{1/2} - x^2)^{-1}$$

if $x^2 \le y \le x^{1/2}$ and $f(y \mid x) = 0$ otherwise. Consequently, the conditional random variable $Y|x$ is uniformly distributed over the interval $[x^2, x^{1/2}]$, which makes intuitive sense, since if $X = x$, then (X, Y) must fall on the vertical line at x in Figure 5.30. For instance, given $X = 1/2$, Y is uniformly distributed between 0.25 and 0.71.

EXAMPLE 5.32

Suppose X and Y possess the joint density

$$f(x, y) = \begin{cases} \dfrac{1 + xy}{3}, & \text{if } 0 < x < 2 \text{ and } 0 < y < 1 \\ 0, & \text{otherwise} \end{cases}$$

The marginal density for X is

$$f_X(x) = \int_0^1 \frac{1 + xy}{3} \, dy = \frac{2 + x}{6}$$

for $0 < x < 2$ and $f_X(x) = 0$ elsewhere. The marginal density for Y is

$$f_Y(y) = \int_0^2 \frac{1 + xy}{3} \, dx = \frac{2 + 2y}{3}$$

for $0 < y < 1$ and $f_Y(y) = 0$ otherwise. The conditional density for $Y|x$ is defined for $0 < x < 2$ and is given by

$$f(y \mid x) = \begin{cases} \dfrac{1 + xy}{1 + x/2}, & \text{if } 0 < y < 1 \\ 0, & \text{otherwise} \end{cases}$$

FIGURE 5.30
Illustration corresponding to $Y|x$ in Example 5.31

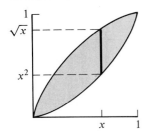

The conditional density for $X|y$ is defined for $0 < y < 1$ and is given by

$$f(x \mid y) = \begin{cases} \dfrac{1 + xy}{2(1 + y)}, & \text{if } 0 < x < 2 \\ 0, & \text{otherwise} \end{cases}$$

In general terms, the conditional density contains information about the manner in which the outcome of one random variable is conditioned by the occurrence of another. If X and Y are independent, then

$$f(y \mid x) = \frac{f_{X,Y}(x, y)}{f_X(x)} = \frac{f_X(x)f_Y(y)}{f_X(x)} = f_Y(y)$$

so that the conditional density for Y is equal to the marginal density for Y.

5.7.2 Conditional Expectation

For fixed x, the conditional random variable $Y|x$ possesses an expected value $E[Y|x]$, called the **conditional expectation of Y given x.** It is also known as the **conditional mean of Y given x** and is denoted by $\mu_{Y|x}$. In the continuous case, $E[Y|x]$ is computed by employing the conditional density $f(y \mid x)$:

$$E[Y|x] = \int_{-\infty}^{\infty} yf(y \mid x)\, dy$$

Analogously, the conditional expectation of X given y is

$$E[X|y] = \int_{-\infty}^{\infty} xf(x \mid y)\, dx$$

Similar expressions apply in the discrete case with a summation replacing the integral.

EXAMPLE 5.33

Referring to the conditional random variable $Y|3$ whose density was found in Example 5.30,

$$\begin{aligned} E[Y|3] &= 0 \times f(0 \mid 3) + 1 \times f(1 \mid 3) + 2 \times f(2 \mid 3) \\ &= 0 \times 0.4 + 1 \times 0.4 + 2 \times 0.2 = 0.8 \end{aligned}$$

In words, if 3 errors result from using the filter measured by X, then the expected number of errors using the filter measured by Y is 0.8.

EXAMPLE 5.34

In Example 5.10 we considered the uniformly jointly distributed random variables X and Y, the support region being illustrated in Figure 5.13. The marginal densities were computed to be

$$f_X(x) = \begin{cases} 2x, & \text{if } 0 \le x \le 1 \\ 0, & \text{otherwise} \end{cases}$$

and

$$f_Y(y) = \begin{cases} 2(1 - y), & \text{if } 0 \le y \le 1 \\ 0, & \text{otherwise} \end{cases}$$

The modeling assumptions were that the joint density $f(x, y)$ described the demand-arrival pair (X, Y) and that $Y \le X$. The conditional densities are required to find the conditional expectations $E[Y|x]$, for some fixed x, and $E[X|y]$, for some fixed y. The conditional density for $Y|x$ is defined for $0 < x \le 1$ and is given by

$$f(y \mid x) = \frac{f(x, y)}{f_X(x)} = \frac{2}{2x} = \frac{1}{x}$$

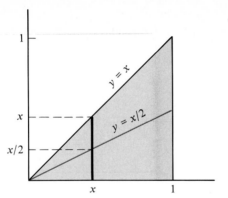

FIGURE 5.31
Curve of regression for
Example 5.34

for $0 \leq y \leq x$ and $f(y \mid x) = 0$ otherwise. The conditional density for $X|y$ is defined for $0 \leq y < 1$ and is given by

$$f(x \mid y) = \frac{f(x, y)}{f_Y(y)} = \frac{2}{2(1 - y)} = \frac{1}{1 - y}$$

for $y \leq x \leq 1$ and $f(x \mid y) = 0$ otherwise. Thus, $Y|x$ and $X|y$ are uniformly distributed over the intervals $[0, x]$ and $[y, 1]$, respectively (see Figures 5.14 and 5.15). Consequently,

$$E[Y|x] = \frac{x}{2}$$

$$E[X|y] = \frac{1 + y}{2}$$

In terms of the demand-arrival model, if it is known that demand occurs at time x, then the expected arrival time is $x/2$. In the other direction, if it is known that arrival occurs at time y, then the expected demand time is $(1 + y)/2$.

Consider Figure 5.31, in which the support region of the joint uniform density of Example 5.34 is illustrated. If x is determined, then the random vector (X, Y) must fall on the vertical line indicated in the figure. Since (X, Y) is uniformly distributed, it is intuitively clear that $Y|x$ will be uniformly distributed, which is exactly the result deduced in Example 5.34. Moreover, the center of mass of the distribution, $E[Y|x] = x/2$, agrees with intuition, since the midpoint of the vertical line is $(x, x/2)$. If all points x in the interval $[0, 1]$ are considered, then the locus of points $(x, E[Y|x])$ is the straight line $y = x/2$ illustrated in the figure. No matter what the outcome of X, the conditional expectation of $Y|x$ lies on this line.

DEFINITION 5.12

Curve of Regression. The locus of points $(x, E[Y|x])$ is called the curve of regression of Y on X. In addition, the conditional expectation $E[Y|x]$ is also called the **regression of Y on x.**

As noted prior to the definition, the curve of regression locates the conditional means. It is the graph of $E[Y|x]$ as a function of x. In Example 5.34 the curve of regression is a straight line; in the next, it will be parabolic.

EXAMPLE 5.35

Let X and Y be uniformly distributed over the support region illustrated in Figure 5.32. Then $f(x, y) = 3$ on the depicted region and 0 elsewhere. The marginal density for X is

$$f_X(x) = \int_0^{x^2} 3 \, dy = 3x^2$$

for $0 \leq x \leq 1$ and $f_X(x) = 0$ for x outside $[0, 1]$. The marginal density for Y is

$$f_Y(y) = \int_{y^{1/2}}^{1} 3 \, dx = 3(1 - y^{1/2})$$

for $0 \leq y \leq 1$ and $f_Y(y) = 0$ for y outside $[0, 1]$. The conditional densities $f(y \mid x)$, defined for $0 < x \leq 1$, and $f(x \mid y)$, defined for $0 \leq y < 1$, are given by

$$f(y \mid x) = \begin{cases} x^{-2}, & \text{if } 0 \leq y \leq x^2 \\ 0, & \text{otherwise} \end{cases}$$

and

$$f(x \mid y) = \begin{cases} (1 - y^{1/2})^{-1}, & \text{if } y^{1/2} \leq x \leq 1 \\ 0, & \text{otherwise} \end{cases}$$

respectively. Thus, both $Y|x$ and $X|y$ are uniformly distributed. In addition,

$$E[Y|x] = \frac{x^2}{2}$$

$$E[X|y] = \frac{1 + y^{1/2}}{2}$$

The curve of regression of Y on X is illustrated in Figure 5.32.

Although Definition 5.12 is given in terms of the regression curve of Y on X, the definition is clearly symmetric, so that the curve of regression of X on Y is given by the locus of points $(y, E[X|y])$.

Since, for fixed x, $Y|x$ is a random variable, it (possibly) possesses other moments besides the mean. Of particular interest is the **conditional variance,**

$$\text{Var}\,[Y|x] = E[(Y|x - \mu_{Y|x})^2]$$

denoted also by $\sigma^2_{Y|x}$. Var $[Y|x]$ measures the spread of the conditional density $f(y \mid x)$. According to Theorem 3.12, it is given by

$$\text{Var}\,[Y|x] = E[(Y|x)^2] - \mu^2_{Y|x}$$

where $E[(Y|x)^2]$ is the second moment of $Y|x$.

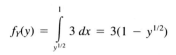

FIGURE 5.32
Curve of regression for
Example 5.35

EXAMPLE 5.36

For the joint distribution of Example 5.35, it was found that $Y|x$ is uniformly distributed over the interval $[0, x^2]$. Thus, its variance is

$$\text{Var}\,[Y|x] = \frac{x^4}{12}$$

It was also found that $X|y$ is uniformly distributed over the interval $[y^{1/2}, 1]$. Consequently,

$$\text{Var}[X|y] = \frac{(1 - y^{1/2})^2}{12}$$

5.8 BIVARIATE NORMAL DISTRIBUTION

The bivariate normal distribution is a generalization of the univariate normal distribution to two variables. It plays an important role in regression analysis. In particular, its curve of regression serves as the archetype for an entire class of regression models.

DEFINITION 5.13

Bivariate Normal Density. The bivariate normal density is defined by

$f(x, y)$

$$= \frac{\exp\left\{-\frac{1}{2(1 - \rho^2)}\left[\left(\frac{x - \mu_X}{\sigma_X}\right)^2 - 2\rho\left(\frac{x - \mu_X}{\sigma_X}\right)\left(\frac{y - \mu_Y}{\sigma_Y}\right) + \left(\frac{y - \mu_Y}{\sigma_Y}\right)^2\right]\right\}}{2\pi\sigma_X\sigma_Y(1 - \rho^2)^{1/2}}$$

where μ_X, μ_Y, σ_X, σ_Y, and ρ are parameters with the constraints $\sigma_X > 0$, $\sigma_Y > 0$, and $-1 < \rho < 1$. If the random variables X and Y possess the joint density $f(x, y)$, then they are said to possess a **bivariate normal (Gaussian) distribution.**

There are 5 parameters determining the bivariate normal family of densities. Since the notation for these parameters is standard, we need not mention them specifically when referring to the bivariate normal distribution. The reason for the notation employed in designating these parameters will become clear in the next theorem, which, among other things, states that the marginal distributions of the bivariate normal distribution are themselves normal.

THEOREM 5.18

If X and Y are jointly normally distributed with parameters μ_X, μ_Y, σ_X, σ_Y, and ρ, then the marginal distributions for X and Y are normal, with $X \sim N(\mu_X, \sigma_X)$ and $Y \sim N(\mu_Y, \sigma_Y)$. Moreover, the correlation coefficient of X and Y is ρ.

Proof: We prove that X is normally distributed with parameters μ_X and σ_X. Let $u = (x - \mu_X)/\sigma_X$ to simplify the expression for the bivariate normal density. In addition, make the substitution $v = (y - \mu_Y)/\sigma_Y$ in the integral defining $f_X(x)$. Since $dv = dy/\sigma_Y$, the integral becomes

$$f_X(x) = \frac{1}{2\pi\sigma_X\sqrt{1 - \rho^2}} \int_{-\infty}^{\infty} e^{-\{1/[2(1 - \rho^2)]\}(u^2 - 2\rho uv + v^2)} \, dv$$

Next, complete the square in the quadratic in the exponent:

$$u^2 - 2\rho uv + v^2 = v^2 - 2\rho uv + \rho^2 u^2 - \rho^2 u^2 + u^2$$
$$= (v - \rho u)^2 + (1 - \rho^2)u^2$$

Inserting the completed square into the integral yields

$$f_X(x) = \frac{e^{-u^2/2}}{2\pi\sigma_X\sqrt{1-\rho^2}} \int_{-\infty}^{\infty} e^{-(v-\rho u)^2/[2(1-\rho^2)]} \, dv$$

Next, let

$$z = (1-\rho^2)^{-1/2} (v - \rho u)$$

Then

$$dz = (1-\rho^2)^{-1/2} \, dv$$

$$dv = (1-\rho^2)^{1/2} \, dz$$

and

$$f_X(x) = \frac{e^{-u^2/2}}{2\pi\sigma_X} \int_{-\infty}^{\infty} e^{-z^2/2} \, dz$$

$$= \frac{1}{\sqrt{2\pi}\sigma_X} e^{-(1/2)[(x-\mu_X)/\sigma_X]^2} \frac{1}{\sqrt{2\pi}} \int_{-\infty}^{\infty} e^{-z^2/2} \, dz$$

$$= \frac{1}{\sqrt{2\pi}\sigma_X} e^{-(1/2)[(x-\mu_X)/\sigma_X]^2}$$

Thus, $f_X(x)$ is a normal density with parameters μ_X and σ_X. An analogous argument applies to $f_Y(y)$. Finally, although we will not go through the details, an integration shows that

$$E[XY] = \rho\sigma_X\sigma_Y + \mu_X\mu_Y$$

so that, according to Theorem 5.14,

$$\text{Cov}\,[X, Y] = \rho\sigma_X\sigma_Y$$

and Corr $[X, Y] = \rho$. ∎

From the form of the bivariate normal density, it is apparent that X and Y are independent if and only if $\rho = 0$, in which case the bivariate normal density is simply a product of univariate normal densities. Consequently, jointly normally distributed random variables are uncorrelated if and only if they are independent.

Of particular importance in applications are the conditional density $f(y \mid x)$ of $Y|x$ and the conditional mean $\mu_{Y|x}$.

THEOREM 5.19

If X and Y possess a bivariate normal distribution, then, for any fixed x, the conditional random variable $Y|x$ is normally distributed with conditional mean

$$\mu_{Y|x} = \mu_Y + \rho \frac{\sigma_Y}{\sigma_X} (x - \mu_X)$$

and conditional variance

$$\sigma_{Y|x}^2 = \sigma_Y^2 (1 - \rho^2)$$

Proof: Algebraic simplification of

$$f(y \mid x) = \frac{f(x, y)}{f_X(x)}$$

yields

$$f(y \mid x) = \frac{1}{\sqrt{2\pi}\sigma_Y\sqrt{1 - \rho^2}} \exp\left\{-\frac{1}{2}\left[\frac{y - \left(\mu_Y + \rho\frac{\sigma_Y}{\sigma_X}(x - \mu_X)\right)}{\sigma_Y\sqrt{1 - \rho^2}}\right]^2\right\}$$

which is a univariate normal density with mean and variance agreeing with those in the statement of the theorem. ∎

Because of symmetry in the bivariate normal density, $X|y$ is normal with conditional mean and variance being found by simply interchanging the roles of x and y in Theorem 5.19.

EXAMPLE 5.37

It has been determined that, subsequent to a given initial time of operation, the internal temperature Y of an automatically adjusted variable-speed drive and the number of revolutions X of the output shaft are jointly normally distributed with the following parameters:

$$\mu_X = 4500 \text{ rpm}$$

$$\sigma_X = 950 \text{ rpm}$$

$$\mu_Y = 95°$$

$$\sigma_Y = 2°$$

$$\rho = 0.8$$

According to Theorem 5.19, if the observed number of revolutions is $X = 6000$, then the conditional distribution $Y|6000$ is normally distributed with mean

$$\mu_{Y|6000} = 95 + (0.8)\left(\frac{2}{950}\right)(6000 - 4500) = 97.53$$

and variance

$$\sigma_{Y|6000}^2 = 2^2[1 - (0.8)^2] = 1.44$$

Now suppose the internal temperature must remain below 100° for safe operation. According to Theorem 4.7, given an observation of $X = 6000$ rpm, the probability that the machine is in a safe state is

$$P(Y|6000 < 100) = \Phi\left(\frac{100 - 97.53}{1.2}\right) = 0.9803$$

Figure 5.33 illustrates the surface of a bivariate normal density, this surface being known as a **bivariate normal surface.** According to Theorem 5.19, the curve

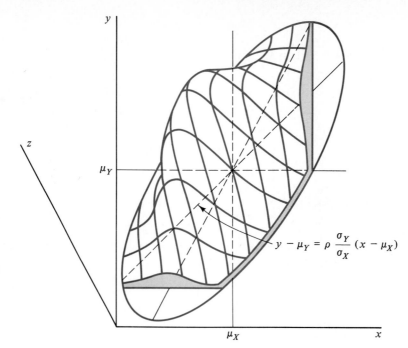

FIGURE 5.33.
Bivariate normal surface with the regression line (Hoel, *Introduction to Mathematical Statistics*, 4/e, © 1984 by John Wiley & Sons, Inc.)

$$y - \mu_Y = \rho \frac{\sigma_Y}{\sigma_X}(x - \mu_X)$$

of regression is a straight line and this line is depicted in the figure. In addition, since for fixed x the conditional density is normal, a plane through x that is parallel to the yz plane will cut the surface in such a manner as to form a curve

$$z = f(x, y)$$

that is a normal-like curve. It is not the graph of a normal density, since

$$z = f_X(x)f(y \mid x)$$

and $f(y \mid x)$ is a normal density. Specifically, the conditional density $f(y \mid x)$ possesses a graph that results from slicing the normal surface with a plane parallel to the yz plane through x and then normalizing by dividing by $f_X(x)$.

According to the definition of the regression curve, $\mu_{Y|x}$ lies on the curve; indeed, insofar as the curve is concerned, $\mu_{Y|x}$ is a function of x, called the **regression function.** Moreover, Theorem 5.19 shows the standard deviations of the conditional variables $Y|x$ to be identical, each being given by $\sigma_Y\sqrt{1 - \rho^2}$.

EXERCISES

SECTION 5.1

5.1. An urn contains 5 white, 3 black, and 2 red balls. Four balls are selected in a uniformly random manner and without replacement. The random variables X and Y count the number of black and white balls selected, respectively.

a) Express the joint distribution of X and Y in a table.

b) Obtain a general expression for the joint probability mass function.

c) Find $P(X < 3)$.

d) Find $P(X + Y < 4)$.

e) Find $P(-1 \le X - Y < 3)$.

f) Find $P(X < Y)$.

g) Find $P(XY < 3)$.

5.2. Repeat Exercise 5.1 under the condition that the selections are done with replacement.

5.3. An urn contains 6 black, 4 white, 2 red, and 8 green

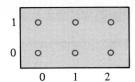

FIGURE 5.34
Sensor array for Exercise 5.5

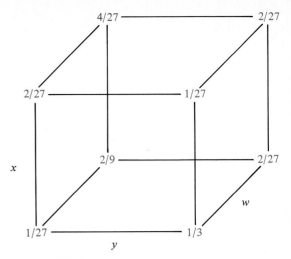

FIGURE 5.35
Trivariate distribution for Exercise 5.13

balls. Five balls are selected in a uniformly random manner and without replacement. The random variables X, Y, U, and V count the number of black, white, red, and green balls selected, respectively.
a) Express the joint distribution of X, Y, U, and V in closed form.
b) Find $P(X = 2)$.
c) Find $P(Y > 2)$.
d) Find $P(X + Y + U \le 3)$.
e) Find $P(X + Y \le V)$.

5.4. Repeat Exercise 5.3 under the condition that the selections are done with replacement.

5.5. Consider the 2×3 sensor array depicted in Figure 5.34. Each sensor is either activated or not activated, depending on whether the sensed light intensity exceeds a predetermined threshold. Assuming that each of the possible 64 sensor configurations is equally likely, U counts the number of activated sensors. Moreover, V equals 0 or 1, depending on whether the sensor at the origin is nonactivated or activated. Find the joint probability distribution for U and V, and express it in a table. Find $P(U \le V + 1)$.

5.6. In a two-phase production process, defects are attributed to either the first or second phase. Table 5.7 gives the probabilities $P(X = x, Y = y)$, where x and y are respectively the numbers of first- and second-phase defects observed on a random unit. Find
a) $P(X < Y)$
b) $P(X = Y)$
c) $P(X^2 \le Y)$

In Exercises 5.7 through 5.12, find the marginal densities for the random variables X *and* Y (U *and* V *in Exercise 5.11*) *in the cited exercises.*

5.7. Exercise 5.1
5.8. Exercise 5.2
5.9. Exercise 5.3
5.10. Exercise 5.4
5.11. Exercise 5.5
5.12. Exercise 5.6

5.13. Referring to the trivariate distribution given in Figure 5.35, find the marginal densities for X, Y, and W.

5.14. X and Y possess a joint uniform density over the region depicted in Figure 5.36. Find
a) $P(X < 1)$
b) $P(Y - X < -1)$
c) $P(1/2 < Y < 3/2)$
d) $P(Y < X^2)$

5.15. X and Y possess a joint uniform density over the region depicted in Figure 5.37. Find
a) $P(X < 1/2)$
b) $P(Y > 1/2)$
c) $P(Y < X)$
d) $P(Y > X^2)$

5.16. X and Y are jointly uniformly distributed over the unit disk centered at the origin (see Figure 5.38). Find

TABLE 5.7 JOINT DISTRIBUTION FOR EXERCISE 5.6

$f(x, y)$		0	1	2	3	4
				y		
	0	0.14	0.13	0.10	0.04	0.03
	1	0.14	0.12	0.08	0.03	0.03
x	2	0.04	0.02	0.02	0.01	0
	3	0.03	0.02	0.01	0.01	0

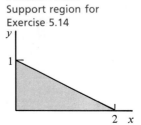

FIGURE 5.36
Support region for Exercise 5.14

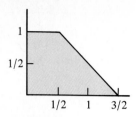

FIGURE 5.37
Support region for
Exercise 5.15

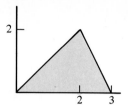

FIGURE 5.39
Support region for
Exercise 5.18

a) $P(X < Y)$
b) $P(X^2 + Y^2 < 1/4)$
c) $P(\max\{X, Y\} < 1/2)$

5.17. X, Y, and W are jointly uniformly distributed over the unit sphere centered at the origin. Find
a) $P(X < Y)$
b) $P(X^2 + Y^2 + W^2 < 1/8)$
c) $P(\max\{|X|, |Y|, |W|\} < 1/2)$

5.18. X and Y are jointly uniformly distributed over the region depicted in Figure 5.39. Find
a) $P(X < 1/2)$
b) $P(Y < X^2)$
c) $P(X < Y^2)$

5.19. Find the constant c so that

$$f(x, y) = \begin{cases} ce^{-(2x+4y)}, & \text{if } x, y \geq 0 \\ 0, & \text{otherwise} \end{cases}$$

is a bivariate density. For the random variable possessing density $f(x, y)$, find
a) $P(X > 2)$
b) $P(X > Y)$
c) $P(X + Y < 1)$

5.20. Find the constant c so that

$$f(x, y) = \begin{cases} c(x^2 + xy), & \text{if } 0 \leq x \leq 2 \text{ and } 0 \leq y \leq 3 \\ 0, & \text{otherwise} \end{cases}$$

is a bivariate density. For the random variable possessing density $f(x, y)$, find $P(X < Y)$.

FIGURE 5.38
Support region for
Exercise 5.16

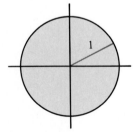

5.21. Find the constant c so that

$$f(x, y) = \begin{cases} cxe^{-(x+y)}, & \text{if } x, y \geq 0 \\ 0, & \text{otherwise} \end{cases}$$

is a bivariate density.

In Exercises 5.22 through 5.28, find the marginal densities for the random variables X *and* Y *in the cited exercises.*

5.22. Exercise 5.14
5.23. Exercise 5.15
5.24. Exercise 5.16
5.25. Exercise 5.18
5.26. Exercise 5.19
5.27. Exercise 5.20
5.28. Exercise 5.21

5.29. Find the bivariate marginal densities $f_{X,Y}(x, y)$, and $f_{X,W}(x, w)$, and $f_{Y,W}(y, w)$ for the trivariate density of Exercise 5.13.

5.30. Find the constant c so that

$$f(x, y, w) = \begin{cases} ce^{-(2x+y+3w)}, & \text{if } 0 \leq x, y, w \\ 0, & \text{otherwise} \end{cases}$$

is a trivariate density.

5.31. For the random variables X, Y, and W corresponding to the trivariate density of Exercise 5.30, find the bivariate marginal densities $f_{X,Y}(x, y)$, $f_{X,W}(x, w)$, and $f_{Y,W}(y, w)$ and the univariate marginal densities $f_X(x)$, $f_Y(y)$, and $f_W(w)$.

5.32. Find the six marginal densities, univariate and bivariate, for the random vector (X, Y, W) possessing a joint uniform distribution over the unit cube $[0, 1]^3$.

In Exercises 5.33 through 5.35, find the probability distribution functions for the joint distributions of the cited exercises.

5.33. Exercise 5.14
5.34. Exercise 5.19
5.35. Exercise 5.20

SECTION 5.2

In Exercises 5.36 through 5.42, determine whether or not the random variables of the cited exercise are independent. Prove your claim.

5.43. X, which gives the heights of males in a certain county, is normally distributed with mean 69.3 inches and variance 6.2. Y, which gives the miles per gallon attained by automobiles in the same county, is normally distributed with mean 24.4 mpg and variance 21.6. Assuming X and Y are independent, give their joint density.

5.44. X is binomially distributed with $n = 5$ and $p = 1/3$. Y is binomially distributed with $n = 5$ and $p = 1/4$. Assuming X and Y to be independent, find their joint distribution.

5.45. X is binomially distributed with $n = 4$ and $p = 0.3$. Y is binomially distributed with $n = 1$ and $p = 0.3$. Assuming X and Y to be independent, find their joint distribution.

5.46. X and Y are independent Poisson random variables with parameters $\lambda_X = 2$ and $\lambda_Y = 5$, respectively. Find their joint density.

SECTION 5.3

5.47. For the random variables of Exercise 5.1, find the densities for the following functions of X and Y:
a) $X + Y$
b) $X - Y$
c) XY

5.48. Repeat Exercise 5.47 for the random variables of Exercise 5.2.

5.49. Find the density for the total number of production defects in Exercise 5.6.

5.50. Find the density for $X/2 + Y/2$ for the random variables of Exercise 5.14.

5.51. Show that if X and Y are independent geometric random variables with parameter p, then $X + Y$ possesses a negative binomial distribution with parameters p and $k = 2$.

5.52. Customers arrive in three independent streams at three service centers. Each arrival process satisfies the three conditions for a Poisson experiment and the rates of arrival are 10, 12, and 6 per hour, respectively. What is the distribution of the sum of the arrivals over a one-hour period? How many customers can be expected to arrive in total during a 20-minute period?

5.53. Machines break down at five different production locations at the rates of 1.2, 3.2, 0.7, 3.5, and 2.4 per day. The breakdown processes at the various sites are independent, and each process is assumed to satisfy the conditions for a Poisson experiment. What is the probability that the total number of breakdowns on a given day will exceed 12?

5.54. The number of airplane crashes for a certain class of plane is a Poisson random variable with rate 4 per year. The number of crashes of a second class of plane is also a Poisson random variable, but with rate 7 per year. If it is assumed that the numbers of crashes for the two types of planes are independent, then, following a crash of a plane in either class, what is the expected length of time until the fifth following crash of a plane in either class? What is the probability that the fifth crash will not occur within the next 4 months?

5.55. Two ignition systems are employed in a spacecraft as a redundant feature to help insure ignition. If both systems possess exponentially distributed survival times, the first with mean 3 years and the second with mean 4 years, what is the probability that at least one will survive 2 years?

5.56. For the machines of Exercise 5.53, following a breakdown at any site, what is the probability that the next breakdown at any site will occur within the next 2 hours? What is the expected length of time until the tenth succeeding breakdown?

5.57. The decibel level of a faulty exhaust system is found to possess a normal distribution with mean 90.4 decibels and variance 5.8. If averages of two measurements are taken, what is the distribution of the averages? What has happened to the variance when averaging two measurements instead of just taking a single measurement?

5.58. The diameters of shafts produced for a certain application are normally distributed with mean 1.500 inches and standard deviation 0.02. Collars for the shafts have inside diameters that are normally distributed with mean 1.552 inches and standard deviation 0.01. Given a shaft and collar selected at random, what is the probability that the shaft will not fit the collar? Use the fact that the difference of two normally distributed random variables is normally distributed.

5.59. A transmission shift assembly is operated by means of a series of rods. These rods form a linkage which allows the action of the gearshift to properly maneuver the gears into position. Assume there are four rods that form the assembly, each rod being selected from a different stock. If the rods possess the respective normal distributions $N(12, 0.08)$, $N(4, 0.02)$, $N(14, 0.09)$, and $N(6, 0.02)$, find the probability that the sum of the lengths will lie between 35.7 and 36.3 inches. Apply Theorem 5.9.

5.60. Find the density for $X + Y$, where X and Y are independent exponential random variables with parameters b_1 and b_2, respectively.

SECTION 5.4

5.61. For the random variables of Exercise 5.1, find the joint density of the random variables

$$U = X + Y$$
$$V = XY$$

Simply list the values of the joint density.

5.62. Repeat Exercise 5.61, except employ the random variables

$$U = X - Y$$
$$V = 2X$$

5.63. For the random variables whose joint distribution is given in Table 5.1, find the joint density of the random variables

$$U = 2X + Y$$
$$V = X - Y$$

5.64. Let X and Y be independent Poisson random variables with parameters λ_1 and λ_2, respectively. Their joint density is then the product of their individual densities. Find the joint density of the random variables

$$U = X + Y$$
$$V = X - Y$$

5.65. Repeat Exercise 5.64, except let

$$U = X + 2Y$$
$$V = 3X$$

5.66. Let X and Y be independent binomial distributions possessing parameters n and p and m and p, respectively. Find the joint density for the random variables

$$U = X + 2Y$$
$$V = X + Y$$

5.67. X and Y possess the joint continuous density

$$f(x, y) = \begin{cases} 2 - x - y, & \text{if } 0 < x, y < 1 \\ 0, & \text{otherwise} \end{cases}$$

Find the joint density for the random variables

$$U = X + Y$$
$$V = X - Y$$

5.68. Let X and Y be uniformly distributed over the square region $(0, 1)^2$. Find the joint density for the random variables

$$U = X$$
$$V = X - Y$$

5.69. Let X and Y possess independent standard normal distributions. Find the joint density for the random variables

$$U = X + Y$$
$$V = X$$

5.70. After finding the appropriate value of c, let X and Y possess the joint density

$$f(x, y) = \begin{cases} cxy, & \text{if } 0 < x, y < 1, \text{ and } x + y < 1 \\ 0, & \text{otherwise} \end{cases}$$

Find the joint density for the random variables

$$U = X^2$$
$$V = XY$$

SECTION 5.5

5.71. X and Y possess a jointly uniform distribution over the rectangular region $[2, 4] \times [0, 6]$. Find $E[X^2(Y - 1)]$.

5.72. For the jointly distributed random variables of Exercise 5.67, find $E[e^X - Y]$.

5.73. X and Y are independent and binomially distributed with $n = 5$ and $p = 0.2$. Find $E[X^2Y^2]$.

5.74. For the jointly distributed random variables of Exercise 5.68, find $E[X^4Y^2]$.

5.75. Suppose X, Y, and W possess binomial, Poisson, and geometric distributions, respectively, with parameters $\lambda = 3$, $n = 7$ and $p_Y = 2/3$, and $p_W = 4/5$, respectively. Find the expected value of $2X + 4Y - W/2$.

5.76. Three programs must be run sequentially. Each comes from a collection of programs possessing an Erlangian run time; however, owing to the manner in which the programs are selected, their run times are not considered independent. Provide an expression for the expected total run time.

SECTION 5.6

5.77. Find the covariance and correlation coefficient for the jointly distributed random variables whose distribution is defined in Table 5.1.

5.78. The random vector (X, Y) possesses eight equally likely outcomes, these being indicated in Figure 5.40. Find the correlation coefficient of the random vector.

5.79. Repeat Exercise 5.78 for the equally likely outcomes of Figure 5.41.

5.80. A construction engineer decides to check the correlation between the number of tardy arrivals among members of the crew and the number of "serious" flaws in workmanship discovered per day. The probabilities that result from the tabulation of relative frequencies appear in Table 5.8, where X is the num-

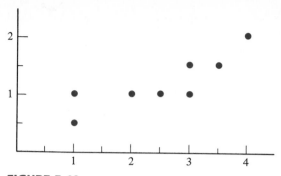

FIGURE 5.40
Outcomes for Exercise 5.78

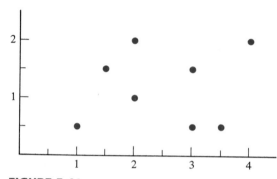

FIGURE 5.41
Outcomes for Exercise 5.79

ber of tardy employees and Y is the number of serious flaws. Find the correlation coefficient between X and Y.

5.81. Find the covariance for the jointly uniformly distributed random variables X and Y over the unit square $[0, 1]^2$. Find the correlation coefficient.

5.82. Find the covariance and correlation coefficient for the jointly uniformly distributed random variables over the region depicted in Figure 5.42.

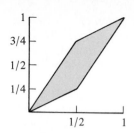

FIGURE 5.42
Support region for
Exercise 5.82

SECTION 5.7

In Exercises 5.83 through 5.88, find the conditional density $f(y \mid x)$ *for the distribution of the cited exercise.*

5.83. Exercise 5.1
5.84. Exercise 5.2
5.85. Exercise 5.14
5.86. Exercise 5.15
5.87. Exercise 5.19
5.88. Exercise 5.67

In Exercises 5.89 through 5.94, find the conditional expectation $E[Y|x]$ *corresponding to the cited exercise.*

5.89. Exercise 5.1
5.90. Exercise 5.2
5.91. Exercise 5.14
5.92. Exercise 5.15
5.93. Exercise 5.19
5.94. Exercise 5.67

5.95. Find $E[X|y]$ for the distribution of Exercise 5.14.
5.96. Find the conditional variance Var $[Y|x]$ for the distribution of Exercise 5.14.

SECTION 5.8

5.97. Give the expressions for the conditional means $E[Y|x]$ and $E[X|y]$ of a jointly normally distributed random vector (X, Y) with parameters $\mu_X = 2$, $\mu_Y = 4.2$, $\sigma_X = 1.2$, $\sigma_Y = 2.6$, and $\rho = 0.4$. Also give the densities for the marginal and conditional distributions.

TABLE 5.8 DISTRIBUTION FOR EXERCISE 5.80

	$f(x, y)$	0	1	2	*y* 3	4	5	6
	0	0.05	0.03	0.01	0.01			
	1	0.02	0.10	0.25	0.16	0.06	0.02	
x	2	0.01	0.02	0.02	0.05	0.08	0.03	0.01
	3			0.01	0.02	0.01	0.02	0.01

5.98. It is believed that the duration of heat treatment, X, and the depth of hardening, Y, for a specific lot of gears are jointly normally distributed with $\mu_X = 18$ seconds, $\mu_Y = 0.31$ inch, $\sigma_X = 4.8$, $\sigma_Y = 0.08$, and $\rho = 0.87$.

a) Give the regression function $E[Y|x]$.
b) Find the conditional distribution given the duration of treatment is observed to be 15 seconds.
c) Given $X = 15$, what is the probability that the depth of hardening will be between 0.23 and 0.39?
d) Given the observation 0.30 for the depth of hardening, what is the expected duration?

5.99. It is hypothesized that the shear strength of welds, Y, is jointly normally distributed with the weld diameters, X. The model has the parameters $\mu_Y = 2240$ psi, $\mu_X = 0.201$ inch, $\sigma_Y = 342$, $\sigma_X = 0.046$, and $\rho = 0.75$.

a) Give the regression function $E[Y|x]$.
b) If the diameter 0.184 is observed, then what is the conditional distribution of the shear strength?
c) Given $X = 0.184$, what is the probability that the shear strength will exceed 1500 psi?

6

TOPICS IN APPLIED PROBABILITY

Several topics that are fundamental to contemporary applications of probability in the physical and computing sciences are covered in this chapter. Each section is independent of the others (although it would be beneficial to read Section 6.1.1 prior to reading Section 6.2), and the entire chapter can be omitted without loss of continuity. Topics include Markov chains, single-server queues, system reliability, entropy, and the computer simulation of system behavior. Markov chains play an important role in modeling systems that transition between discrete states in a nondeterministic fashion. Queuing theory studies congestion resulting from service systems in which service and arrival processes are random. The analysis of reliability plays a key role in the development of survivable systems. The study of entropy involves a quantification of uncertainty. Finally, behavior of large systems is often simulated when analytic description poses mathematically intractable problems.

6.1 MARKOV CHAINS

Many nondeterministic physical processes vary randomly over time. For instance, the total number of cars having arrived at an intersection is a random function of the duration (time) of observation. The position of a particle whose behavior exhibits Brownian motion is a random function of time. A signal that cannot be described deterministically is a random function of time. What is common to these processes is that they are treated as **random processes** in time: their behavior is affected by so many influences that they cannot be adequately modeled by deterministic variables. Even if it is assumed that a signal to be transmitted is deterministic, the transmission causes the signal to appear as a random phenomenon. Over time, the signal is a **random time function.** The present section introduces the notion of a random process and considers in detail a specific class of processes known as Markov processes.

6.1.1 Stochastic Processes

To gain some insight into the character of a random function, consider counting the number of defective units being produced by an assembly line. Assuming each unit is inspected and a unit arrives at the end of the line every minute, the process is defined over discrete time and is parameterized by the nonnegative integers. Specifically, for $k = 0, 1, 2, \ldots$, there is a random variable $X(k)$ that counts the number of defective units having been produced through the first k minutes, where $X(0) = 0$ is the nonrandom initial condition. On any given production run, there will be a function determined by the values of the random variables $X(1), X(2), X(3), \ldots$, say $X(1) = x_1, X(2) = x_2, X(3) = x_3, \ldots$. But this observed function does not describe the process, since the value at each k is a random phenomenon. Another run will likely yield a different set of observed values. In fact, there are many possible deterministic functions that might result from observation, and it is the collection of these possible functions, together with their probabilistic description, that constitutes a random function.

While the preceding illustration concerns a stochastic process defined on the discrete-time values $0, 1, 2, \ldots$, a random process might also be defined over continuous time.

DEFINITION 6.1

Stochastic Process (Random Function). A stochastic process is a collection of random variables (assumed to be defined on a common sample space). If there are denumerably many variables, then the process is denoted by $\{X(k): k = 0, 1, 2, \ldots\}$, or simply by $\{X(k)\}$, and is called a **discrete-time process.** If the random variables are indexed over continuous time, $t \geq 0$, then the process is denoted by $\{X(t): t \geq 0\}$, or simply by $\{X(t)\}$, and is called a **continuous-time process.** In addition, any particular set of observations of the process is called a **realization** of the process.

When dealing with stochastic processes, it is important to keep in mind that the situation is multivariate. For each instant of time t, $X(t)$ is a random variable, but for any collection of time points, say $t_1, t_2, \ldots, t_n$, the random variables $X(t_1)$, $X(t_2), \ldots, X(t_n)$ possess a multivariate distribution.

EXAMPLE 6.1

Consider the defective-count process discussed previously and suppose the probability of a defective unit is 0.02, this probability remaining constant throughout the duration of the process. If the process continues indefinitely, then it consists of an infinite number of realizations. For the sake of simplicity, suppose a total of $K = 4$ observations are made. Then the process (random function) consists of the 16 realizations (deterministic functions) listed in Table 6.1, where the initial condition $X(0) = 0$ is omitted. The realization in row 10 of the table is graphed in Figure 6.1, the graph points having been connected by a polygonal arc to illustrate the flow of the process.

Since the process possesses a finite number of realizations, a complete description of the process will include the probability of each particular function's occurring. In the present example, each realization results from a set of four observations. In effect, the four observations constitute a sequence of four Bernoulli trials. Letting 1 denote a defective unit and 0 a nondefective unit, each realization results from a string of zeros and ones, there being 16 such strings. For instance, the realization graphed in Figure 6.1 results from the outcome (1, 0, 1, 0), which, when written as a string, takes the form 1010 and represents the binary encoding of the integer 10. Each row of Table 6.1 results from an outcome of the Bernoulli trials that gives the binary encoding of the row number. The salient point is that the probability of each realization's occurring is found from the Bernoulli-trial outcomes from which it derives. Thus, realization 10 (in row 10) possesses the probability

TABLE 6.1 REALIZATIONS OF THE DEFECTIVE-COUNT PROCESS

Realization	x_1	x_2	x_3	x_4
0	0	0	0	0
1	0	0	0	1
2	0	0	1	1
3	0	0	1	2
4	0	1	1	1
5	0	1	1	2
6	0	1	2	2
7	0	1	2	3
8	1	1	1	1
9	1	1	1	2
10	1	1	2	2
11	1	1	2	3
12	1	2	2	2
13	1	2	2	3
14	1	2	3	3
15	1	2	3	4

$$(0.02)^2(0.98)^2 = 0.0004$$

of occurring, while realization 0 possesses the probability $(0.98)^4 = 0.9224$ of occurring.

Should the process not be restricted to a finite number of observations, then each realization would be an infinite string (a deterministic function defined on the set of nonnegative integers). The first 10 points in the graph of one such realization are depicted in Figure 6.2.

As a random variable, $X(k)$ counts the number of defective units in the first k observations. Since the trials are assumed to be Bernoulli, $X(k)$ is a binomial random variable with $n = k$ and $p = 0.02$.

The set of possible values of a stochastic process is known as the **state space** of the process. We will consider only denumerable (discrete) state spaces. A process possessing a discrete state space is called a **chain.**

In Example 6.1, if the number of observations is K, then the state space is finite (and thus denumerable) and is given by

$$S = \{0, 1, 2, \ldots, K\}$$

If the number of observations is infinite, then the state space is denumerably infinite and consists of the set of nonnegative integers. In either case, the process is a chain.

FIGURE 6.1
A realization of the process of Example 6.1, $K = 4$

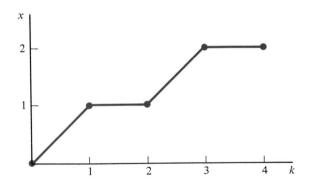

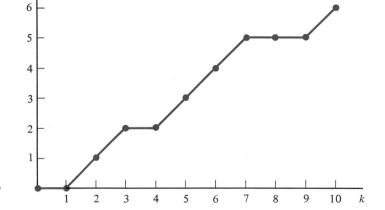

FIGURE 6.2
A realization of the
process of Example 6.1,
$K = 10$

Consider a stochastic process with discrete state space

$$S = \{0, 1, 2, \ldots\}$$

(which may actually be finite). A key question arising in the analysis of random processes concerns the probability that the process is in state n at time t. This probability, denoted by $P_n(t)$, is given in terms of $X(t)$; specifically,

$$P_n(t) = P[X(t) = n]$$

EXAMPLE 6.2

Suppose aircraft arrive at an airport at the constant rate of 2 arrivals per minute and the arrivals are independent. Specifically, suppose the arrivals satisfy the conditions of a Poisson experiment (Section 4.4.1). Let $\{X(t): t \geq 0\}$ be the continuous-time process that counts the number of arrivals in the time interval $[0, t]$. Then the state space is the set of nonnegative integers and, as stated in Section 4.4.1, $X(t)$ possesses a Poisson distribution with parameter $2t$, its probability mass function being given, for $n = 0, 1, 2, \ldots$, by

$$P_n(t) = P[X(t) = n] = f_{X(t)}(n) = \frac{e^{-2t}(2t)^n}{n!}$$

The process $\{X(t)\}$ is called a **Poisson process.** Two realizations of the process are given in Figure 6.3.

6.1.2 The Markov Property

If a system is observed over time, then its state at any given moment depends upon its history. For conditioning, all observations must be taken into account. There are, however, fundamentally important processes that are conditioned solely by the most recent observation.

Consider a stochastic process $\{X(t)\}$ possessing discrete state space $S = \{0, 1, 2, \ldots\}$. For any two points in time, say t and t', and for any two states, say m and n, define the **transition probability**

$$p_{m,n}^{t,t'} = P[X(t') = n \mid X(t) = m]$$

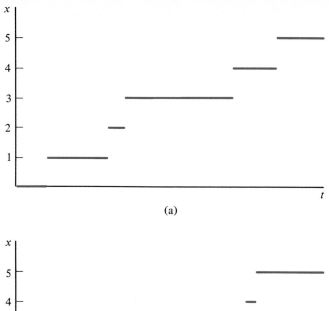

FIGURE 6.3
Two realizations of the
Poisson process

which is the conditional probability that the process is in state n at time t', given it is in state m at time t. In the case of a discrete-time process $\{X(i)\}$, the transition probabilities take the form

$$p_{m,n}^{i,j} = P[X(j) = n \,|\, X(i) = m]$$

which is the conditional probability of the process being in state n at time j, given it is in state m at time i.

In general, given a sequence of time points

$$t_1 < t_2 < \cdots < t_{k-1} < t_k$$

the conditional probability that the process is in state m_k at time t_k, given it has been in states m_1 through m_{k-1} at times t_1 through t_{k-1}, respectively, depends upon the values of the process at times t_1, t_2, . . . , t_{k-1}. However, in some cases, the probability only depends upon the most recent value of t prior to t_k. In such a case, the analysis of the process is greatly simplified.

DEFINITION 6.2

Markov Chain. A stochastic process is said to be a Markov chain if it possesses a discrete state space and, if given a set of time points

$$t_1 < t_2 < \cdots < t_{k-1} < t_k$$

then it satisfies the **Markov property**

$$P[X(t_k) = m_k \mid X(t_{k-1}) = m_{k-1}, \ldots, X(t_2) = m_2, X(t_1) = m_1]$$
$$= P[X(t_k) = m_k \mid X(t_{k-1}) = m_{k-1}]$$
$$= p_{m_{k-1}, m_k}^{t_{k-1}, t_k}$$

EXAMPLE 6.3

In Example 6.1 we considered a discrete-time process that counts the number of defective units. The process is clearly Markovian: the probability that there are m_k defective units at time i_k given there are m_1 through m_{k-1} defective units at times i_1 through i_{k-1}, respectively, depends only on the most recent defective-error count.

The transition probabilities can be computed in a straightforward manner. To go from state m at time i to state n at time j, we must observe $n - m$ defective units in $j - i$ observations. Recalling that $p = 0.02$, this is simply the binomial probability

$$b(n - m; j - i, 0.02)$$

so that

$$p_{m,n}^{i,j} = \binom{j - i}{n - m} (0.02)^{n-m}(0.98)^{j-i-(n-m)}$$

Of particular import is that the transition probabilities are not dependent upon the actual times or states, but only on the difference in time $j - i$ and the incremental number of states $n - m$.

EXAMPLE 6.4

Consider a Poisson process with parameter qt (see Example 6.2). According to the conditions of a Poisson experiment, the happenings in nonoverlapping intervals are independent. As a result, for $t < t' < t''$, the **increments** $X(t'') - X(t')$ and $X(t') - X(t)$ are independent. The Markovian nature of the process is evident from the model, and a rigorous demonstration will not be given. According to the model, the number of happenings in any time interval $[t, t']$ possesses a Poisson distribution with parameter $q(t' - t)$. The transition probabilities are precisely the probabilities of the increments:

$$p_{m,n}^{t,t'} = P[X(t') - X(t) = n - m] = \frac{e^{-q(t'-t)}(qt' - qt)^{n-m}}{(n - m)!}$$

As in Example 6.3, the transition probabilities depend only on the differences $n - m$ and $t' - t$.

6.1.3 Transition Probability Matrix

We now consider discrete-time Markov chains possessing a finite state space, which for notational convenience we take to be $S = \{1, 2, \ldots, N\}$. Besides the transition probabilities $p_{m,n}^{i,j}$, there are **state probabilities**

$$p_n^j = P[X(j) = n]$$

and **initial probabilities**

$$p_n = p_n^0 = P[X(0) = n]$$

The next theorem gives an important relation among the transition probabilities.

(Chapman-Kolmogorov equations). Given a discrete-time Markov chain, for any times $i < r < j$ and states m and n,

$$p_{m,n}^{i,j} = \sum_{k=1}^{N} p_{m,k}^{i,r} p_{k,n}^{r,j}$$

the sum being taken over all states.

Proof: Only an intuitive argument will be presented. For any state k, the transition $X(i) = m$ to $X(j) = n$ can be accomplished by a concatenation of transitions, $X(i) = m$ to $X(r) = k$ followed by $X(r) = k$ to $X(j) = n$. Under the Markovian assumption, the transitions are independent. Moreover, the transition $X(i) = m$ to $X(j) = n$ is the mutually exclusive union of all (over k) such transitions. Therefore, the probability of the transition $X(i) = m$ to $X(j) = n$ can be written as a sum of products over all k, which is precisely what is stated in the theorem. ∎

Transition Probability Matrix. Consider a discrete-time Markov chain with finite state space $S = \{1, 2, \ldots, N\}$. For times i and j, its transition probability matrix is given by

$$\mathbf{P}(i, j) = \begin{pmatrix} p_{1,1}^{i,j} & p_{1,2}^{i,j} & \cdots & p_{1,N}^{i,j} \\ p_{2,1}^{i,j} & p_{2,2}^{i,j} & \cdots & p_{2,N}^{i,j} \\ \vdots & \vdots & & \vdots \\ p_{N,1}^{i,j} & p_{N,2}^{i,j} & \cdots & p_{N,N}^{i,j} \end{pmatrix}$$

From the definition of the transition probabilities, it is immediate that the sum of the probabilities in any row of $\mathbf{P}(i, j)$ must be 1, since the events from which the row probabilities are derived consist of all possible transitions from a particular state in the time frame i to j. Moreover, according to the definition of matrix multiplication, the Chapman-Kolmogorov equations reduce to

$$\mathbf{P}(i, j) = \mathbf{P}(i, r)\mathbf{P}(r, j)$$

Proceeding recursively,

$$\begin{aligned} \mathbf{P}(i, j) &= \mathbf{P}(i, j - 1)\mathbf{P}(j - 1, j) \\ &= \mathbf{P}(i, j - 2)\mathbf{P}(j - 2, j - 1)\mathbf{P}(j - 1, j) \\ &\vdots \\ &= \mathbf{P}(i, i + 1)\mathbf{P}(i + 1, i + 2) \cdots \mathbf{P}(j - 1, j) \end{aligned}$$

In other words, $\mathbf{P}(i, j)$ can be expressed as a matrix product of **one-step transition probability matrices.** Now let

$$\mathbf{p}(j) = \begin{pmatrix} p_1^j \\ p_2^j \\ \vdots \\ p_N^j \end{pmatrix}$$

be the jth **state vector,** whose components are the j-state probabilities. The next proposition shows that the jth state vector can be expressed in terms of the **initial vector** $\mathbf{p}(0)$, whose components are the initial probabilities p_n, $n = 1, 2, \ldots, N$, in conjunction with the first j one-step transition probability matrices.

THEOREM 6.2

For any discrete-time Markov chain with finite state space,

$$\mathbf{p}(j) = [\mathbf{P}(0, 1)\mathbf{P}(1, 2) \cdots \mathbf{P}(j - 1, j)]^T \mathbf{p}(0)$$

where T represents the matrix transpose.

Proof: We first show the claim

$$\mathbf{p}(j) = \mathbf{P}(0, j)^T \mathbf{p}(0)$$

The kth component of the vector obtained from the product $\mathbf{P}(0, j)^T \mathbf{p}(0)$ is given by the dot product of the kth column of $\mathbf{P}(0, j)$ with $\mathbf{p}(0)$, which is

$$
\begin{pmatrix} p_{1,k}^{0,j} \\ p_{2,k}^{0,j} \\ \vdots \\ p_{N,k}^{0,j} \end{pmatrix} \bullet \begin{pmatrix} p_1 \\ p_2 \\ \vdots \\ p_N \end{pmatrix} = \sum_{n=1}^{N} p_{n,k}^{0,j} p_n
$$

$$
= \sum_{n=1}^{N} P[X(j) = k \mid X(0) = n]P[X(0) = n]
$$

$$
= \sum_{n=1}^{N} P[X(j) = k, X(0) = n]
$$

$$
= P[X(j) = k]
$$

$$
= p_k^j
$$

where the next to last equality follows from Theorem 2.12, since running over all n in the state space creates a partition of the ways in which $X(j)$ can equal k. But p_k^j is the kth component of the state vector $\mathbf{p}(j)$ and thus the claim follows. As for the theorem, it follows at once, since

$$\mathbf{P}(0, j) = \mathbf{P}(0, 1)\mathbf{P}(1, 2) \cdots \mathbf{P}(j - 1, j) \qquad \blacksquare$$

A process is said to be **stationary** (or **homogeneous**) if the transition probabilities depend only on the time difference (number of steps) $j - i$, as was the case in Example 6.3. Letting $r = j - i$ be the number of steps in a stationary process, the transition probabilities take the form

$$p_{m,n}^r = p_{m,n}^{j-i} = p_{m,n}^{i,j}$$

and these are called the *r*-step probabilities. Of particular use are the **one-step probabilities**

$$p_{m,n} = P[X(i + 1) = n \mid X(i) = m]$$

The Chapman-Kolmogorov equations (Theorem 6.1) can be rewritten in the stationary case as

$$p_{m,n}^{r+s} = \sum_{k=1}^{N} p_{m,k}^r p_{k,n}^s$$

In words, the probability of transitioning from state m to state n in $r + s$ steps is the sum over all states k of the products of the probabilities of transitioning from state m to k in r steps and from k to n in s steps.

In the stationary form, the Chapman-Kolmogorov equations have a useful graphical interpretation in terms of the one-step probabilities. Figure 6.4 gives an illustration of a three-state stationary Markov chain in which each circle represents a state, the arrows between states represent possible transitions, and the labels on the arrows denote the transition probabilities. Although all possible transition arrows are drawn, those possessing transition probability zero can be omitted. The arrows having no "tail states" represent the initial distribution. The diagram itself is called a **Markov diagram.** It can obviously be extended to any finite number of states.

Referring to Figure 6.4, we can see the meaning of the stationary form of the Chapman-Kolmogorov equations in terms of "arrow chasing." For instance, to transition from state 1 to state 3 in two steps, the process can take any one of three paths: 1–1–3, 1–2–3, or 1–3–3. The strings represent mutually exclusive events, and therefore

$$p_{1,3}^2 = P(1\text{–}1\text{–}3) + P(1\text{–}2\text{–}3) + P(1\text{–}3\text{–}3)$$

$$= p_{1,1}p_{1,3} + p_{1,2}p_{2,3} + p_{1,3}p_{3,3}$$

which is precisely what is stated by the Chapman-Kolmogorov equations for $m = 1$, $n = 3$, $r = 1$, and $s = 1$.

The next theorem gives a representation of the state probabilities

$$p_n^j = P[X(j) = n]$$

of a stationary Markov chain in terms of the one-step and initial probabilities. The theorem is stated in terms of the **one-step transition probability matrix**

$$\mathbf{P} = \mathbf{P}(0,\ 1) = \begin{pmatrix} p_{1,1} & p_{1,2} & \cdots & p_{1,N} \\ p_{2,1} & p_{2,2} & \cdots & p_{2,N} \\ \vdots & \vdots & & \vdots \\ p_{N,1} & p_{N,2} & \cdots & p_{N,N} \end{pmatrix}$$

FIGURE 6.4

Three-state Markov diagram

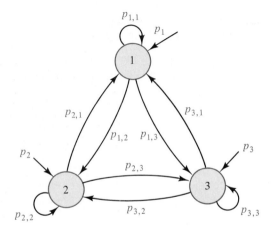

Owing to stationarity,

$$\mathbf{P} = \mathbf{P}(i, i + 1)$$

for all i.

If $\{X(i)\}$ is a stationary discrete-time Markov chain with state space $S = \{1, 2, \ldots, N\}$, then

$$\mathbf{p}(j) = [\mathbf{P}^j]^T \mathbf{p}(0)$$

Proof: According to Theorem 6.2,

$$\mathbf{p}(j) = [\mathbf{P}(0, 1)\mathbf{P}(1, 2) \cdots \mathbf{P}(j - 1, j)]^T \mathbf{p}(0)$$

In the stationary case,

$$\mathbf{P}(0, 1)\mathbf{P}(1, 2) \cdots \mathbf{P}(j - 1, j) = \mathbf{P}^j \qquad \blacksquare$$

EXAMPLE 6.5

In Example 4.4 we considered a one-dimensional random walk and found the distribution of the random variable giving the number of units the particle is to the right of the origin after a given number of steps. Let $\{X(i)\}$ denote the process that gives the positive distance from the origin after i steps. Then, according to the results of Example 4.4, the state probabilities are given by

$$p_n^i = P[X(i) = n] = \begin{pmatrix} i \\ \dfrac{i + n}{2} \end{pmatrix} p^{(i+n)/2} \, q^{(i-n)/2}$$

where p and $q = 1 - p$ are the probabilities of steps to the right and left, respectively. As stated, the process is a Markov chain, but its state space is infinite.

Consider the case where the random walk has **barriers**, so that the particle can move only so far left and so far right. Assume there are barriers at 0 and 3, so that the first movement must be to the right. Consequently, the state space is

$$S = \{0, 1, 2, 3\}$$

The Markov diagram in Figure 6.5 illustrates the states, possible transitions, and one-step transition probabilities. Since the only condition pertaining to the transition probabilities is the present state of the particle, the process is Markovian. In addition, the process is stationary, the r-step probabilities depending only on the initial and final states of the transition and the number of steps, not on the actual times at which the steps are taken. The one-step transition probability matrix is

FIGURE 6.5
Markov diagram for random walk with barriers

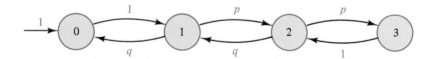

$$\mathbf{P} = \begin{pmatrix} p_{0,0} & p_{0,1} & p_{0,2} & p_{0,3} \\ p_{1,0} & p_{1,1} & p_{1,2} & p_{1,3} \\ p_{2,0} & p_{2,1} & p_{2,2} & p_{2,3} \\ p_{3,0} & p_{3,1} & p_{3,2} & p_{3,3} \end{pmatrix} = \begin{pmatrix} 0 & 1 & 0 & 0 \\ q & 0 & p & 0 \\ 0 & q & 0 & p \\ 0 & 0 & 1 & 0 \end{pmatrix}$$

Assuming the particle starts at the origin, the initial vector is

$$\mathbf{p}(0) = \begin{pmatrix} p_0 \\ p_1 \\ p_2 \\ p_3 \end{pmatrix} = \begin{pmatrix} 1 \\ 0 \\ 0 \\ 0 \end{pmatrix}$$

According to Theorem 6.3, the state vector for time $j = 2$ is given by

$$\mathbf{p}(2) = [\mathbf{P}^2]^T \mathbf{p}(0) = \begin{pmatrix} q & 0 & q^2 & 0 \\ 0 & q(1+p) & 0 & q \\ p & 0 & p(1+q) & 0 \\ 0 & p^2 & 0 & p \end{pmatrix} \begin{pmatrix} 1 \\ 0 \\ 0 \\ 0 \end{pmatrix} = \begin{pmatrix} q \\ 0 \\ p \\ 0 \end{pmatrix}$$

Thus, the state probabilities for $j = 2$ are

$$P[X(2) = 0] = q$$

$$P[X(2) = 2] = p$$

$$P[X(2) = 1] = P[X(2) = 3] = 0$$

Chasing the arrows in Figure 6.5 shows these to be correct (as they must be).

6.1.4 Two-State Markov Chains

Consider a two-state stationary Markov chain with state space $\{0, 1\}$ and one-step transition matrix

$$\mathbf{P} = \begin{pmatrix} p_{0,0} & p_{0,1} \\ p_{1,0} & p_{1,1} \end{pmatrix} = \begin{pmatrix} a & 1-a \\ 1-d & d \end{pmatrix}$$

where $a = p_{0,0}$, $d = p_{1,1}$,

$$p_{0,0} + p_{0,1} = 1$$

and

$$p_{1,0} + p_{1,1} = 1$$

Figure 6.6 gives the appropriate Markov diagram. Squaring $\mathbf{P}$ yields

$$\mathbf{P}^2 = \begin{pmatrix} a^2 + (1-a)(1-d) & a(1-a) + d(1-a) \\ a(1-d) + d(1-d) & (1-a)(1-d) + d^2 \end{pmatrix}$$

More generally, letting

$$u = 2 - a - d$$

it can be shown by mathematical induction that, if

FIGURE 6.6
Two-state Markov diagram

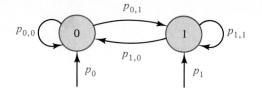

$$|1 - u| < 1$$

then

$$\mathbf{P}^j = u^{-1} \begin{pmatrix} 1 - d & 1 - a \\ 1 - d & 1 - a \end{pmatrix} + u^{-1}(1 - u)^j \begin{pmatrix} 1 - a & a - 1 \\ d - 1 & 1 - d \end{pmatrix}$$

Thus,

$$\lim_{j \to \infty} \mathbf{P}^j = u^{-1} \begin{pmatrix} 1 - d & 1 - a \\ 1 - d & 1 - a \end{pmatrix}$$

where we have used the fact that

$$\lim_{j \to \infty} (1 - u)^j = 0$$

In the stationary case, $\mathbf{P}(0, j) = \mathbf{P}^j$ and hence the matrix limit can be rewritten in terms of the individual components of $\mathbf{P}(0, j)$:

Steady-State Distribution

$$p^* = \lim_{j \to \infty} p^j_{0,0} = \lim_{j \to \infty} p^j_{1,0} = \frac{1 - p_{1,1}}{2 - p_{0,0} - p_{1,1}}$$

$$1 - p^* = \lim_{j \to \infty} p^j_{0,1} = \lim_{j \to \infty} p^j_{1,1} = \frac{1 - p_{0,0}}{2 - p_{0,0} - p_{1,1}}$$

The pair $(p^*, 1 - p^*)$ is called the **steady-state distribution** of the process. From a practical perspective, the existence of the steady-state probabilities means that, for large j, we can treat the transition probabilities as constants, independent of j. For instance, for large j, the probability of transitioning from state 1 to state 0 in j steps is approximately p^*, no matter the precise value of j. Moreover, for large j,

$$p^j_{0,0} \approx p^* \approx p^j_{1,0}$$

so that, after j steps, the probability of the process being in state 0 is essentially independent of the state of the process prior to commencing the j steps. Similarly, for large j,

$$p^j_{0,1} \approx 1 - p^* \approx p^j_{1,1}$$

so that, after j steps, the probability of the process being in state 1 is also essentially independent of the starting state.

EXAMPLE 6.6

On numerous occasions we have employed the binomial distribution to model the selection of defective and nondefective units of output by a production process. This choice utilizes the modeling assumption that the trials are independent. However, due to the nature of the physical process, it might well be that the occurrence of a defective unit is significant in that it is more likely to be followed by another defective unit than is a nondefective unit to be followed by a defective unit. Going further, it may be that the longer the current string of defective units, the ever more likely that the next unit produced will be defective. For the moment, let us make the simplifying assumption that the only unit of historical relevance is the most recent, which means the process is a two-state Markov chain. Let 0 and 1 denote a defective and a nondefective unit, respectively, and let the transition probabilities be given by $p_{0,0} = 0.20$, $p_{0,1} = 0.80$, $p_{1,0} = 0.01$, and $p_{1,1} = 0.99$. Then the one-step transition probability matrix is

$$\mathbf{P} = \begin{pmatrix} 0.20 & 0.80 \\ 0.01 & 0.99 \end{pmatrix}$$

and the steady-state probabilities are

$$p^* = \frac{1 - 0.99}{2 - 0.20 - 0.99} = 0.0123$$

$$1 - p^* = 0.9877$$

Thus, for large j and for any i,

$$P[X(i + j) = 0 \,|\, X(i) = 0] \approx P[X(i + j) = 0 \,|\, X(i) = 1]$$

$$\approx p^* = 0.0123$$

$$P[X(i + j) = 1 \,|\, X(i) = 0] \approx P[X(i + j) = 1 \,|\, X(i) = 1]$$

$$\approx 1 - p^* = 0.9877$$

6.2 A SINGLE-SERVER QUEUING MODEL (M/M/1)

Whenever there are jobs to be serviced, variability in job arrivals and in service times creates varying degrees of congestion, the result being a waiting line, or **queue.** A number of waiting-line models have been introduced thus far in the examples. The applications are far-reaching. In production, queues result from the arrival of partially completed products at the various assembly stages, the service at each stage being the specific subassembly required at the stage. In transportation, queues result at bottlenecks because of limited throughput. In communications, transmissions must await open lines and service at switching stations. In computer science, multiprogramming results in jobs awaiting space in memory and CPU service.

The present section treats the **first-in, first-out** queuing model, also known as **FIFO.** The salient feature of the model is that jobs are serviced in the order of their arrival. The queue will also be assumed to be of unlimited length, so that no jobs are turned away and all arriving jobs get on line. Results will be presented without supporting theoretical details, the majority of which are beyond the level of the present text.

6.2.1 Service and Arrival Processes

Starting at some time subsequent to time 0, jobs begin arriving asynchronously for service, the incoming jobs forming what is known as the **arrival stream.** A fixed **arrival rate** of q (jobs per minute, jobs per second, or whatever) is assumed, so that the mean time between arrivals is $1/q$. The service time will also be variable, whether the service is by a CPU, a ticket agent, or a machinist taking jobs as they arrive. A fixed **service rate** u will be assumed, so that the mean service time is $1/u$. We will also assume the interarrival and service times are independent and that the arrival and service rates are independent of the length of the queue.

Our main interest is the number of jobs in the system, the number being served (0 or 1) plus the number in the queue. At any time $t \geq 0$, the random variable $N(t)$ gives the number of jobs in the system. $\{N(t): t \geq 0\}$ is a continuous-time stochastic process with state space

$$S = \{0, 1, 2, \ldots\}$$

For any $t \geq 0$,

$$P_n(t) = P[N(t) = n]$$

denotes the probability that the process is in state n at time t, which means, for $n > 0$, that 1 job is in service and $n - 1$ jobs are in the queue.

Although a number of service distributions have been applied in practice, our attention will be restricted to the commonly employed exponential service distribution. Specifically, it will be assumed that the service-time random variable Y possesses an exponential distribution with parameter u. Then the mean service time is $1/u$ and, as specified previously, u is the service rate. Since Y possesses the exponential density, it is memoryless, which means that

$$P(Y > t + s \mid Y > s) = P(Y > t)$$

for all nonnegative s and t. Equivalently,

$$P(Y \leq t + s \mid Y > s) = P(Y \leq t)$$

which says that the probability of a job's completing service in the next increment of time is not dependent upon how long the job has been in service.

Our assumption regarding the arrival process, which we denote by $\{V(t)\}$, is that it is Poisson. As a result, the number of arrivals in any given interval of time depends only on the length of the interval and the arrival rate q (see Section 4.4.1 and Example 6.2). Moreover, if $A_n(t)$ denotes the probability of there being n arrivals in an interval of length t, then

$$A_n(t) = P[V(t) = n] = \frac{(qt)^n e^{-qt}}{n!}$$

Of particular importance is the **interarrival-time** distribution. According to Theorem 4.12, the arrival process being Poisson implies that the random variable measuring the interarrival time possesses an exponential distribution with parameter q. Thus, if an arrival is observed, then the time to the next arrival is exponentially distributed with mean $1/q$.

Figure 6.7 gives a graphic illustration of the queuing model under consideration. It has a single server, a queue of unlimited length, a FIFO discipline, and arrival and service rates of q and u, respectively. In the terminology of queuing theory, a queue with a Poisson arrival process and an exponential service distribution is called an **M/M/s queue,** where s is the number of servers. Thus, in the present section we

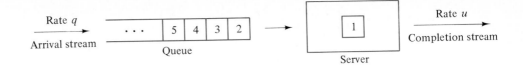

FIGURE 6.7
Single-server queue
model

are considering **M/M/1** queues. Such queues are well studied and have served as models in various physical settings.

6.2.2 Steady-State Probabilities

The process $\{N(t)\}$ gives the number of jobs in the system at time t. We assume that the arrival rate is less than the service rate, namely, that $q/u < 1$, for if not, the queue would grow without bound. The ratio q/u is called the **utilization factor** of the system and will be denoted by r. It can be shown that if $r < 1$, then there exists a **steady state** for the system. Specifically, for all n, there exists the limit

$$\lim_{t \to \infty} P_n(t) = P_n$$

which means that for large t the probability of the process being in state n can be assumed to be independent of t. Since the state of the process is the number in the system, the steady-state condition means that the probability of there being n jobs in the system is essentially independent of t, so long as t is sufficiently large. While there will be fluctuations of the state probabilities for small t, these will be transient and (essentially) vanish for large t. Ultimately, our concern is with the operation of the system in the steady state, which is the period when the system has been in operation long enough to insure that $P_n(t)$ can be treated as constant relative to t and can be replaced by P_n, the steady-state limit. Thus, our concern is with the **steady-state probabilities**

$$P_0, P_1, P_2, \ldots$$

(whose sum must be 1). Note that the process $\{N(t)\}$ can be replaced with the random variable N when in the steady state. In the case of an unlimited M/M/1 queue, it can be shown that, for $n \geq 0$,

Steady-State Probabilities

$$\boxed{P_n = (1 - r)r^n}$$

EXAMPLE 6.7

During a certain period, cars approach a toll bridge in a Poisson arrival stream, the arrivals being independent and at the rate of 3 per minute. There is a single collector at the bridge, and the service time is exponentially distributed with mean 15 seconds. Assuming the waiting line to be of unlimited length, the M/M/1 model applies with $q = 3$, $u = 4$, and $r = 3/4$. Consequently, the steady-state probabilities are given by

$$P_n = \left(\frac{1}{4}\right)\left(\frac{3}{4}\right)^n = \frac{3^n}{4^{n+1}}$$

For instance, with N denoting the steady-state random variable giving the number of jobs in the system, the probability of there being 0 jobs in the system is

$$P(N = 0) = P_0 = 0.25$$

the probability of there being no queue is

$$P(N \le 1) = P_0 + P_1 = 0.25 + 0.19 = 0.44$$

and the probability of there being at least 2 jobs in the queue is

$$P(N \ge 3) = 1 - P(N \le 2) = 1 - (P_0 + P_1 + P_2) = 0.42$$

6.2.3 System Characteristics

Given the steady-state state probabilities, various measures of importance that hold in the steady state can be readily obtained. For instance, the expected number of jobs in the system is

$$E[N] = \sum_{n=0}^{\infty} nP_n = \sum_{n=1}^{\infty} n(1 - r)r^n = \frac{r}{1 - r}$$

where we have summed the series to obtain the result. As the utilization factor approaches 1, the expected number of jobs in the system approaches infinity.

Numerous other system characteristics can be developed. The variance of the number of jobs in the system is

$$\mathrm{Var}\,[N] = E[N^2] - E[N]^2 = \frac{r}{(1 - r)^2}$$

As with $E[N]$, $\mathrm{Var}\,[N]$ approaches infinity as the utilization factor approaches 1. If Q denotes the number of jobs in the queue, then it can be shown that

$$E[Q] = \frac{r^2}{1 - r}$$

which also is very large when r is close to 1. Note that

$$E[Q] = rE[N]$$

From the perspective of an arriving job, if it arrives at time t, it spends some **waiting time** $W(t)$ in the queue prior to being given service. Assuming the system to be in the steady state, the waiting time W is not a function of t and the expected waiting time can be shown to be

$$E[W] = \frac{E[Q]}{q} = \frac{E[N]}{u} = \frac{r}{u(1 - r)}$$

Note that the probability of having to wait is given by

$$P(N > 0) = 1 - P(N = 0) = 1 - P_0 = 1 - (1 - r) = r$$

Next consider the expected time a job spends in the system, again assuming the system is in the steady state when the job arrives. The time T in the system is the sum of the waiting time in the queue plus the time in service. Hence, by the linearity of the expected-value operator,

$$E[T] = E[W] + \frac{1}{u} = \frac{1}{u(1-r)}$$

Note that

$$E[N] = qE[T]$$

a property known as **Little's law.**

EXAMPLE 6.8

Consider a single-CPU multiprogramming environment with a FIFO discipline, unlimited queue length, Poisson arrivals, and exponential service. Although FIFO is usually impractical for CPU service, a FIFO analysis can often provide a reasonable indication of the degree of congestion one might expect, even if the discipline is modified to achieve certain objectives. Indeed, some of the FIFO characteristics apply without change to other disciplines.

Assuming an arrival rate of $q = 15$ jobs per minute and a service rate of $u = 30$ jobs per minute, the utilization factor is $r = 0.5$. The expected number of jobs in the system, when in the steady state, is

$$E[N] = \frac{0.5}{1 - 0.5} = 1$$

and the variance is

$$\text{Var}[N] = \frac{0.5}{(1 - 0.5)^2} = 2$$

The expected number of jobs in the queue is

$$E[Q] = rE[N] = 0.5 \times 1 = 0.5$$

The expected waiting time in the queue is

$$E[W] = \frac{E[Q]}{q} = \frac{0.5}{15} = 0.03333$$

minutes (2 seconds). The expected time in the system is

$$E[T] = \frac{E[N]}{q} = \frac{1}{15} = 0.06667$$

minutes (4 seconds).

Suppose jobs reaching the CPU begin to increase in length so that u decreases to 20 jobs per minute, the average time of service now being 3 seconds instead of 2 seconds. Under the new loading, $r = 0.75$, the expected number of jobs in the system is

$$E[N] = \frac{0.75}{1 - 0.75} = 3$$

the expected number of jobs in the queue is

$$E[Q] = rE[N] = 0.75 \times 3 = 2.25$$

and the expected waiting time is

$$E[W] = \frac{E[N]}{u} = \frac{3}{20} = 0.15$$

minutes (9 seconds), which under certain circumstances might be considered a significant increase of the 2-second mean waiting time when u was 30 jobs per minute. Another change of interest is the increase in the variance of N. For $u = 30$, Var $[N] = 2$; with $u = 20$,

$$\text{Var } [N] = \frac{0.75}{(1 - 0.75)^2} = 12$$

so that the length of the queue is starting to show increasing variability, which means that time in the system is beginning to be less predictable. From the perspective of the user, service is beginning to degrade.

In addition to the increased running time of the programs, suppose more programs begin to arrive, so that the arrival rate increases slightly to $q = 18$ jobs per minute. If u remains at 20, then $r = 0.9$; the expected number of jobs in the system is

$$E[N] = \frac{0.9}{1 - 0.9} = 9$$

the expected number of jobs in the queue is

$$E[Q] = rE[N] = 0.9 \times 9 = 8.1$$

and the expected time in the system is

$$E[T] = \frac{E[N]}{q} = \frac{9}{18} = 0.5$$

minutes (30 seconds). In addition, the variance of the number of the jobs in the system increases to

$$\text{Var } [N] = \frac{0.9}{(1 - 0.9)^2} = 90$$

Not only has the expected number of jobs in the system now increased to 9, but the standard deviation of N is now 9.49. To see the effect of this increased variability, consider a job that arrives when there are

$$\mu_N + 2\sigma_N \approx 28$$

jobs in the system. Since each of the 28 jobs, including the one in service, has an expected service time of 3 seconds, the expected time in the system for such a job is 87 seconds. Of course, the high variability means that some jobs have short stays in the system, but such high variability is certainly not desirable.

6.3 RELIABILITY

Large-scale systems are comprised of many and sundry components. The reliability of such systems in the face of various conditions and heavy loading is a prime concern of system designers. Time-to-failure distributions are employed to model both component and system survivability, the latter usually being expressed in terms of subsystem reliabilities. For example, the reliability of a computer system might be modeled in terms of constituent subsystems such as CPUs, memory modules, I/O devices, software, and network communication. The present section concerns system reliability and the concomitant problem of facilitating system fault tolerance through the introduction of redundant components.

6.3.1 Reliability and Hazard Functions

The basic building block of a reliability model is the **time-to-failure distribution** of a system (or subsystem or component), time to failure being measured by a nonnegative random variable T. Given T, a system's **reliability** is characterized by its reliability function.

DEFINITION 6.4

Reliability Function. The reliability function corresponding to a continuous time-to-failure random variable T possessing density $f(t)$ is given by

$$R(t) = P(T > t) = 1 - F(t) = \int_t^\infty f(x)\, dx$$

$F(t)$ being the probability distribution function of T.

In accordance with Theorem 3.6, a reliability function $R(t)$ is monotonically decreasing, $R(0) = 1$, and

$$\lim_{t \to \infty} R(t) = 0$$

When comparing the reliability of two systems S_1 and S_2 having reliability functions $R_1(t)$ and $R_2(t)$, it would be easy to judge system S_1 more reliable than system S_2 if $R_1(t)$ were to be greater than or equal to $R_2(t)$ for all t. In practice, such an approach does not generally work because, as depicted in Figure 6.8, there may not be strict ordering between the reliability functions. To overcome this problem, the **mean time to failure (MTTF)** is often employed as a measure of reliability. For the time-to-failure random variable T with density $f(t)$ and reliability function $R(t)$, the MTTF is simply the expected value of T. The MTTF can be expressed in terms of either the time-to-failure density $f(t)$ or the reliability function:

$$E[T] = \int_0^\infty tf(t)\, dt = \int_0^\infty R(t)\, dt$$

the latter integral being obtained from the former by an application of integration by parts. When employing the MTTF, the system with the greater MTTF is judged to be more reliable.

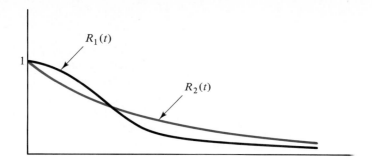

FIGURE 6.8
Unordered reliability
functions

EXAMPLE 6.9

Suppose systems S_1 and S_2 possess exponential time-to-failure distributions with parameters q_1 and q_2, respectively. Then

$$R_1(t) = e^{-q_1 t}$$

$$R_2(t) = e^{-q_2 t}$$

Moreover, the MTTFs are given by

$$E[T_1] = \frac{1}{q_1}$$

$$E[T_2] = \frac{1}{q_2}$$

If $q_1 > q_2$, then, as judged by the MTTF, S_2 is more reliable than S_1, and conversely.

The probability that a system will fail in the time interval t to $t + \Delta t$ under the condition that it has survived to time t is given by the conditional probability

$$P(t < T < t + \Delta t \mid T > t) = \frac{P(t < T < t + \Delta t, T > t)}{P(T > t)}$$

$$= \frac{P(t < T < t + \Delta t)}{P(T > t)}$$

$$= \frac{F(t + \Delta t) - F(t)}{R(t)}$$

Dividing by Δt yields the average rate of failure in the interval $[t, t + \Delta t]$, given the system has survived to time t. Letting $\Delta t \to 0$ yields the **instantaneous failure rate**

$$h(t) = \lim_{\Delta t \to \infty} \frac{P(t < T < t + \Delta t \mid T > t)}{\Delta t}$$

$$= \lim_{\Delta t \to \infty} \frac{F(t + \Delta t) - F(t)}{\Delta t \, R(t)}$$

Since the derivative of the probability distribution function is the density, taking the limit yields

$$h(t) = \frac{F'(t)}{R(t)} = \frac{f(t)}{R(t)}$$

where it should be recognized that $F'(t) = f(t)$ at points of continuity of $f(t)$.

DEFINITION 6.5

Hazard Function. For a time-to-failure random variable T possessing density $f(t)$ and reliability function $R(t)$,

$$h(t) = \frac{f(t)}{R(t)}$$

is called the hazard function.

As demonstrated prior to Definition 6.5, the hazard function gives the instantaneous failure rate.

A typical hazard function is illustrated in Figure 6.9. The rate of failure tends to be higher at the outset of operation because, in addition to random failures, there tend to be early failures resulting from defectively produced components. Following this wear-in stage, failures tend to occur at a rather constant rate, the second stage covering the duration of the serviceable life of the system. A third stage occurs when failures begin to occur due to wear-out. It is sometimes possible to flatten out the curve at the beginning by utilizing system components that have been ''burnt in'' prior to being brought on line. Such an approach is particularly feasible when employing electronic components.

The next theorem demonstrates that the reliability function can be expressed in terms of the hazard function.

THEOREM 6.4

If $f(t)$, $h(t)$, and $R(t)$ are the density, hazard function, and reliability function, respectively, of a time-to-failure random variable T, then

$$R(t) = e^{-\int_0^t h(x)dx}$$

and

FIGURE 6.9
Typical hazard function

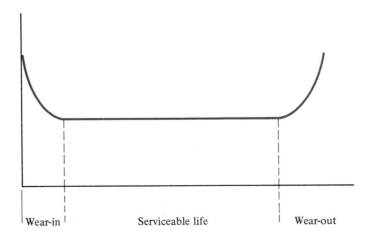

Wear-in   Serviceable life   Wear-out

$$f(t) = h(t)e^{-\int_0^t h(x)dx}$$

Proof: Since $f(t) = F'(t)$ and $R(t) = 1 - F(t)$,

$$h(t) = \frac{f(t)}{R(t)} = -\frac{R'(t)}{R(t)}$$

Integrating both sides from 0 to t while recalling that $R(0) = 1$ yields

$$\int_0^t h(x)\, dx = -\int_0^t \frac{R'(x)}{R(x)}\, dx = -\log[R(x)]\bigg|_0^t = -\log[R(t)]$$

Applying the exponential gives the desired expression of $R(t)$ in terms of $h(t)$. The identity $f(t) = h(t)R(t)$ yields the second expression stated in the theorem. ∎

6.3.2 Exponential and Weibull Time-To-Failure Models

Consider a situation in which the hazard function is constant, say $h(t) = q$. Intuitively, the assumption is that strictly wear-in failures have been eliminated and that the system does not reach a wear-out stage. As is often the case, the model reflects a pragmatic idealization of the process; nonetheless, in the case of electronic components, where early-age failures are customarily removed by burn-in, and wear-out does not occur until t is very large, the constant failure-rate assumption is often warranted. This is especially so in the computer industry, where obsolescence is very likely to occur before wear-out. In any event, according to Theorem 6.4, the time-to-failure distribution is exponential with

$$f(t) = qe^{-qt}$$

and

$$R(t) = e^{-qt}$$

As seen in Example 6.9, the MTTF is $E[T] = 1/q$. The density, reliability function, and hazard function for the constant failure-rate case are depicted in Figure 6.10. In all three cases, the functions are zero for negative t.

Since the exponential distribution is memoryless, in the exponential time-to-failure model the MTTF is independent of the time the system has already been in operation. In other words, if we begin to observe the system at any time after it has begun operation, so long as it is still in operation, its expected survival time under our scrutiny is q.

To obtain another perspective on the exponential time-to-failure model, consider a procedure in which each time a component fails it is instantly replaced by an identical component. Moreover, suppose the failures are observed to obey the conditions of a Poisson experiment with rate q. Then, as discussed in Section 4.4.1, the random variable that counts the number of failures in a time interval of length t possesses a Poisson distribution with parameter qt, where q is the rate of failure occurrences. By Theorem 4.12, the interfailure-time distribution is exponential with parameter q. Since

FIGURE 6.10

Density, reliability function, and hazard function under constant failure rate q

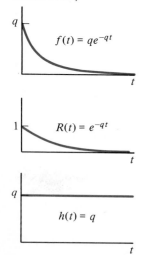

each occurrence of a failure results from the first (and only) failure of a component, the resulting interfailure-time distribution agrees with the individual time-to-failure distributions, and thus the assumption of Poisson failures with rate q has resulted in a component exponential time-to-failure model with failure rate q. Although we will not do so, a converse can be proven, namely, that an exponential interfailure-time distribution implies that the number of failures in the component-replacement model must possess a Poisson distribution. In sum, the exponential failure model is often appropriate because it is often reasonable to assume that the failure rate of replacements remains constant over time and that replacement failures are independent, so that the conditions of a Poisson experiment apply.

A more general assumption than a constant failure rate is one given by a constant times a power of t. In such a case, the rate might either be constant, increase smoothly with time, or decrease smoothly with time. The next theorem shows that such a hazard function arises if and only if the time-to-failure distribution is Weibull.

<hr/>

THEOREM 6.5

The hazard function is

$$h(t) = \alpha \beta^{-\alpha} t^{\alpha - 1}$$

for $t > 0$, if and only if the time-to-failure distribution is Weibull with parameters α and β. In such a case, the reliability function is

$$R(t) = e^{-(t/\beta)^{\alpha}}$$

<hr/>

Proof: First suppose the time-to-failure distribution is Weibull, so that

$$f(t) = \alpha \beta^{-\alpha} t^{\alpha - 1} e^{-(t/\beta)^{\alpha}}$$

for $t > 0$. As demonstrated in Section 4.8.1, the probability distribution function is then

$$F(t) = 1 - e^{-(t/\beta)^{\alpha}}$$

for $t > 0$. Thus, as claimed in the statement of the theorem,

$$R(t) = 1 - F(t) = e^{-(t/\beta)^{\alpha}}$$

Straightforward algebra shows

$$h(t) = \frac{f(t)}{R(t)} = \alpha \beta^{-\alpha} t^{\alpha - 1}$$

Conversely, suppose $h(t)$ is as just given. We wish to show that $f(t)$ is the appropriate Weibull density. Expressing $R(t)$ in terms of $h(t)$, as derived in Theorem 6.4, yields

$$R(t) = e^{-\int_0^t h(x)dx}$$
$$= e^{-\alpha \beta^{-\alpha} \int_0^t x^{\alpha - 1} dx}$$
$$= e^{-\beta^{-\alpha} t^{\alpha}}$$

Therefore $f(t) = h(t)R(t)$ is the desired Weibull density. ∎

Letting $\alpha = 1$ and $\beta = 1/q$ in the Weibull density gives the exponential density with mean $1/q$, and the hazard function in Theorem 6.5 reduces to the constant $h(t) = q$, thus agreeing with the previous discussion.

In the case of a Weibull time-to-failure distribution, Theorem 4.15 gives the MTTF to be

$$E[T] = \beta \Gamma\left(\frac{\alpha + 1}{\alpha}\right)$$

EXAMPLE 6.10

Mechanical components such as shafts, bearings, and seals begin wearing out from the outset of operation, although the rate of deterioration may not be a simple matter to deduce. For instance, a brass bush on a shaft wears more rapidly as time in operation passes, one reason being that, as the bush wears, the fit becomes less snug and "play" develops, thereby increasing the rate of deterioration. Since a mechanical unit, such as a transmission, or an electrome-chanical unit, such as a generator, contains many parts that wear continuously over time, it might be expected that their hazard functions increase over time.

Suppose a particular unit possesses the linear hazard function

$$h(t) = \frac{t}{5000}$$

where t refers to the number of hours in operation. According to Theorem 6.5, the time-to-failure density is Weibull with $\alpha = 2$ and $\beta = 100$. The reliability function is

$$R(t) = e^{-t^2/10,000}$$

6.3.3 Series Systems

Thus far, we have discussed the reliability of a single system or subsystem. We now consider the reliability of a system in terms of its subsystem (or component) relia-bilities. The analysis rests on the manner in which the survivability of the overall system depends upon the survivability of the individual subsystems.

As discussed in Section 2.6.3, one manner of component arrangement is in a **series** configuration, as depicted in Figure 2.23. The overall system will function if and only if all its components continue to function: it cannot tolerate any component failures. Referring to Figure 2.23, suppose that, for $k = 1, 2, \ldots, m$, the time-to-failure random variable for component A_k is T_k and the corresponding reliability function is $R_k(t)$. Moreover, let T and $R(t)$ be the time-to-failure variable and reliability function, respectively, for the overall system. Then, assuming the component times to failure to be independent,

$$
\begin{aligned}
R(t) &= P(T > t) \\
&= P(T_1 > t, T_2 > t, \ldots, T_m > t) \\
&= P(T_1 > t)P(T_2 > t) \cdots P(T_m > t) \\
&= R_1(t)R_2(t) \cdots R_m(t)
\end{aligned}
$$

so that the reliability of the system is the product of the component reliabilities. Using product notation,

Reliability Function for Series System

$$R(t) = \prod_{k=1}^{m} R_k(t)$$

Since $0 \leq R_k \leq 1$ for each k, it follows that $R(t) \leq R_k(t)$ for all k, so that in a series configuration, system reliability is worse than that of its individual subsystems. Figure 6.11 provides a graphic of a serial system in terms of the reliability functions.

As a special case, if each of the components possesses the same reliability function, say $R_c(t)$, then the system reliability function is given by

$$R(t) = R_c(t)^m$$

EXAMPLE 6.11

Suppose each of the components in a series configuration possesses a constant failure rate, the rate for component A_k being q_k, $k = 1, 2, \ldots, m$. Then the reliability of the overall series system is given by

$$R(t) = \prod_{k=1}^{m} e^{-q_k t} = e^{-\left(\sum_{k=1}^{m} q_k\right)t}$$

Letting

$$q_s = \sum_{k=1}^{m} q_k$$

denote the **system failure rate,** we obtain

$$R(t) = e^{-q_s t}$$

The MTTF is

$$E[T] = \int_0^\infty R(t)\, dt = \frac{1}{q_s}$$

As an illustration, suppose an *RLC* circuit contains 1 inductor, 2 capacitors, and 4 resistors, all of which are required for proper functioning. Furthermore, suppose that in a particular operating environment, each component possesses a constant failure rate: the inductor has failure rate $q_1 = 2.1 \times 10^{-5}$ failures per hour, each capacitor has failure rate $q_2 = 3.5 \times 10^{-7}$ failures per hour, and each resistor has failure rate $q_3 = 5.0 \times 10^{-6}$ failures per hour. Assuming the failures of the components to be independent, we wish to find the circuit reliability function and the circuit MTTF. Since the failure rates are constant, the exponential model applies with

$$q_s = 2.1 \times 10^{-5} + 2 \times 3.5 \times 10^{-7} + 4 \times 5.0 \times 10^{-6} = 4.17 \times 10^{-5}$$

Thus,

$$R(t) = e^{-(0.0000417)t}$$

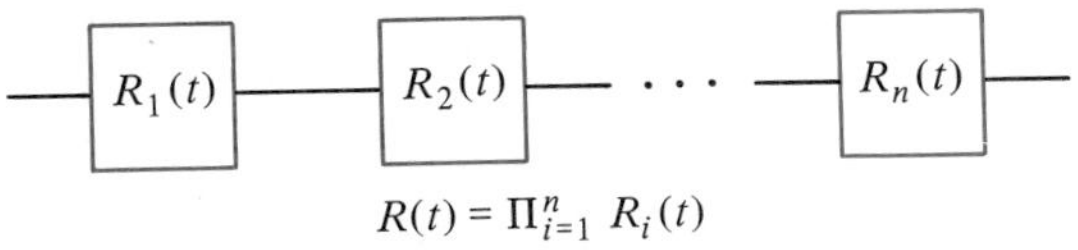

FIGURE 6.11
Serial configuration (Dougherty/Giardina, *Mathematical Methods for Artificial Intelligence and Autonomous Systems*, © 1988, p. 248. Reprinted by permission of Prentice Hall, Inc., Englewood Cliffs, New Jersey.)

and the MTTF is

$$R(t) = \Pi_{i=1}^{n}\, R_i(t)$$

$$E[T] = \frac{1}{0.0000417} = 2.3981 \times 10^4$$

EXAMPLE 6.12

If all m components in a series system possess the same constant failure rate, say q, then it is immediate from the results of Example 6.11 that the system reliability function is

$$R(t) = e^{-mqt}$$

As an illustration, suppose a 12-bit register is formed by concatenating three 4-bit integrated-circuit registers, each having a constant failure rate of 2×10^{-6} per hour. Assuming IC failures to be independent, the reliability function of the 12-bit register is

$$R(t) = e^{-(0.000006)t}$$

and the MTTF is

$$E[T] = \frac{1}{0.000006} = 1.6667 \times 10^5$$

6.3.4 Parallel Systems and Fault Tolerance

Very often, an engineer wishes to design a system that will continue to function despite the failure of one or more of its subsystems. In the event a system is autonomous, survivability in the face of subsystem faults is especially important. The achievement of such **fault tolerance** depends upon the judicious utilization of replicated components: when a particular component fails, a duplicate assumes its function. The system tolerates faults because redundancy has been designed into it. Of course, there must be in place the necessary mechanisms for the detection and isolation of the failed components.

Operationally, redundancy is achieved through the use of **parallel** component configurations. Such configurations were briefly discussed in Section 2.6.3, and the general parallel survivability model is depicted in Figure 2.24. Referring to the figure, assume that, for $k = 1, 2, \ldots, m$, subsystem A_k possesses the time-to-failure random variable T_k and reliability function $R_k(t)$. Assume the overall system possesses time-to-failure random variable T and reliability function $R(t)$. Moreover, assume the system can tolerate $m - 1$ faults, so that it will continue to function (possibly in a less efficient manner) so long as a single subsystem is functioning. If the time-to-failure distributions are independent, then the overall system reliability function is given by

$$\begin{aligned}
R(t) &= P(T > t) \\
&= 1 - P(T \leq t) \\
&= 1 - P(T_1 \leq t, T_2 \leq t, \ldots, T_m \leq t) \\
&= 1 - P(T_1 \leq t)P(T_2 \leq t) \cdots P(T_m \leq t) \\
&= 1 - [1 - R_1(t)][1 - R_2(t)] \cdots [1 - R_m(t)]
\end{aligned}$$

where the third equality follows from the fact that the system fails prior to time t if and only if all subsystems fail prior to time t. In product notation form,

Reliability Function for Parallel System

$$R(t) = 1 - \prod_{k=1}^{m} [1 - R_k(t)]$$

FIGURE 6.12
Parallel configuration
(Dougherty/Giardina,
*Mathematical Methods
for Artificial Intelligence
and Autonomous
Systems*, © 1988, p. 248.
Reprinted by permission
of Prentice Hall, Inc.,
Englewood Cliffs, New
Jersey.)

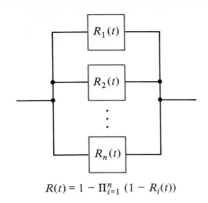

$$R(t) = 1 - \Pi_{i=1}^{n} (1 - R_i(t))$$

For any particular k,

$$1 - R(t) = [1 - R_k(t)] \prod_{j \neq k} [1 - R_j(t)]$$

Since, for any j, $0 \leq 1 - R_j(t) \leq 1$,

$$1 - R(t) \leq 1 - R_k(t)$$

Thus, $R(t) \geq R_k(t)$. Consequently, the reliability of a parallel system is greater than or equal to each of its component reliabilities, and parallelism increases system survivability. In particular, the MTTF of a parallel system is greater than the MTTF of any of its constituent subsystems. Figure 6.12 gives an illustration of a parallel system in terms of reliability functions.

In the special case where each component possesses the same reliability function $R_c(t)$, the reliability function of a parallel system is given by

$$R(t) = 1 - [1 - R_c(t)]^m$$

EXAMPLE 6.13

Consider a parallel system in which the survivabilities of the subsystems are independent and, for $k = 1, 2, \ldots, m$, subsystem A_k possesses the constant failure rate q_k. The system reliability function is then

$$R(t) = 1 - \prod_{k=1}^{m} (1 - e^{-q_k t})$$

In particular, if each subsystem possesses failure rate q, then

$$R(t) = 1 - (1 - e^{-qt})^m$$

For instance, suppose three processors are networked in such a manner that whenever one or more of them are down, the remaining ones carry the full burden. Assuming processor failures to be independent and the common failure rate to be 0.2 failure per year, the reliability function of the network, insofar as the processors are concerned, is given by

$$R(t) = 1 - (1 - e^{-t/5})^3$$

For each individual processor, the MTTF is $1/q = 5$ years. For the system as a whole, the MTTF is

$$E[T] = \int_0^\infty [1 - (1 - e^{-t/5})^3]\, dt$$

$$= \int_0^\infty (3e^{-t/5} - 3e^{-2t/5} + e^{-3t/5})\, dt = \frac{55}{6}$$

Regarding Example 6.13, in general it can be shown that if there exist m independent components in parallel, with each component possessing constant failure rate q, then the MTTF of the system is given by

$$E[T] = \frac{1}{q}\left(1 + \frac{1}{2} + \cdots + \frac{1}{m}\right)$$

Note that for large m, the addition of redundant components yields ever-decreasing advantage.

Most systems in which redundancy exists are not strictly parallel; the system might be mixed, in that a parallel system might possess series subsystems or a series system might possess parallel subsystems. Indeed, there can be subsystems within subsystems within subsystems, and so on. The next example compares the reliability of two such systems.

EXAMPLE 6.14

Based upon a larger MTTF, we will compare the reliabilities of computer systems A and B of Figure 6.13. Component failures are assumed to be independent, and each component is assumed to possess the reliability function

$$R_c(t) = \begin{cases} 1 - \dfrac{t}{3}, & \text{if } 0 \le t \le 3 \\ 0, & \text{if } t > 3 \end{cases}$$

FIGURE 6.13
Computers for Example 6.14 (Dougherty/ Giardina, *Mathematical Methods for Artificial Intelligence and Autonomous Systems*, © 1988, p. 250. Reprinted by permission of Prentice Hall, Inc., Englewood Cliffs, New Jersey.)

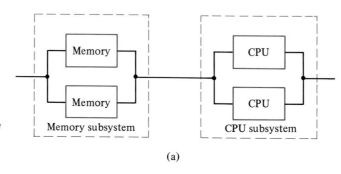

(a)

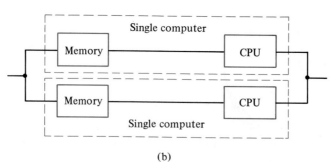

(b)

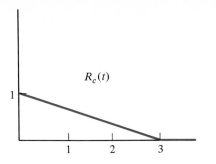 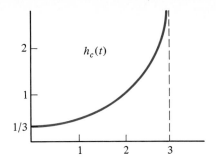

FIGURE 6.14
Component reliability and hazard functions for Example 6.14

For each component, the hazard function is defined for $t < 3$ and is given by

$$h_c(t) = -\frac{R'_c(t)}{R_c(t)} = \frac{1}{3-t}$$

The hazard function is not defined for $t \geq 3$, which makes sense, since the probability that a component will survive more than 3 years is zero. The component reliability and hazard functions are illustrated in Figure 6.14.

Based upon the diagrams in Figure 6.13, system B is comprised of two computers in parallel, where redundancy exists at a high level. On the contrary, system A incorporates redundancy at a low level by having two memories in parallel and two CPUs in parallel. It will be seen that the latter approach leads to a higher MTTF. Of course, system A is more complicated, in that each CPU has access to either memory. Proceeding, the reliability function of the two memories in parallel is

$$R_1(t) = 1 - \left[1 - \left(1 - \frac{t}{3}\right)\right]^2 = 1 - \frac{t^2}{9}$$

for $0 \leq t \leq 3$, and zero otherwise. The same reliability function applies to the two computers in parallel. Since these subsystems are in series, these reliabilities are multiplied together to give the reliability of the overall system A. Thus,

$$R_A(t) = \begin{cases} \left(1 - \dfrac{t^2}{9}\right)^2, & \text{if } 0 \leq t \leq 3 \\ 0, & \text{if } t > 3 \end{cases}$$

The corresponding MTTF is

$$E[T_A] = \int_0^3 \left(1 - \frac{t^2}{9}\right)^2 dt = \frac{8}{5}$$

As for system B, each computer consists of a memory and a CPU in series. Hence, each possesses reliability

$$R_2(t) = \begin{cases} \left(1 - \dfrac{t}{3}\right)^2, & \text{if } 0 \leq t \leq 3 \\ 0, & \text{if } t > 3 \end{cases}$$

Since the overall system consists of two computers in parallel, the reliability of system B is

$$R_B(t) = 1 - [1 - R_2(t)]^2 = 2R_2(t) - R_2(t)^2$$

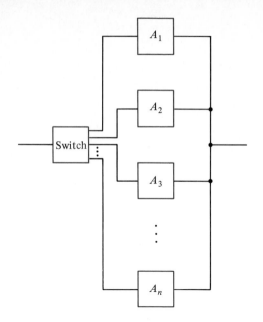

FIGURE 6.15
Standby parallel
configuration

The MTTF of system B is therefore

$$E[T_B] = \int_0^3 \left(1 - \frac{4t^2}{9} + \frac{4t^3}{27} - \frac{t^4}{81} \right) dt = \frac{7}{5}$$

Consequently, based upon the MTTF, system A is more reliable.

Thus far, we have assumed a parallel model in which all subsystems commence operation simultaneously at time $t = 0$. Redundancy is **active** in that all subsystems are active at the outset. Another possibility is **standby** redundancy, where a system requires the operation of only a single subsystem at any given time and when the active subsystem fails, another is switched on. Figure 6.15 gives an illustration of a standby parallel configuration. Component A_1 begins operation at time $t = 0$ and remains in operation until it fails, at which point the switch brings component A_2 on line. A_2 operates until it fails and then component A_3 takes over. The procedure continues until all m components have failed, at which point the system fails.

Assuming the m components to be identical and to operate independently, the interfailure-time distribution of the system is the same as the individual time-to-failure distributions. In particular, the time-to-failure random variable T of the system is given by the sum of the m component time-to-failure variables T_k, $k = 1$, $2, \ldots , m$:

$$T = T_1 + T_2 + \cdots + T_m$$

If each component possesses the constant failure rate q, then each possesses an exponential time-to-failure distribution with parameter q. It is an immediate corollary of Theorem 5.7 that the sum T possesses an m-Erlang distribution with parameter $\beta = 1/q$. Thus, the MTTF of the entire system is

$$E[T] = \frac{m}{q}$$

and the variance of the system time to failure is

$$\text{Var}\,[T] = \frac{m}{q^2}$$

Since, for $t > 0$, the probability distribution function for the m-Erlang distribution is given by the incomplete gamma function $\gamma(t/\beta; m)$ (see Section 4.6.2), the reliability function of the standby redundant system is

$$R(t) = 1 - F(t) = 1 - \gamma(qt; m)$$

EXAMPLE 6.15

An autonomous vehicle contains four identical power cells, each possessing a constant failure rate $q = 0.02$ failure per hour of operation, where failure is determined by the unit's not being able to maintain a voltage in excess of some predetermined threshold value. (Although the hazard function for such cells is likely to increase with time, for the sake of simplicity we have assumed it to be constant.) Assuming the cells are arranged in a standby parallel configuration, the MTTF of the power-cell system is

$$E[T] = \frac{4}{0.02} = 200$$

hours. The variance of time to failure is

$$\text{Var}\,[T] = \frac{4}{0.0004} = 10,000$$

so that the standard deviation of time to failure is $\sigma_T = 100$ hours. The reliability function for the system is

$$R(t) = 1 - \gamma\left(\frac{t}{50}; 4\right)$$

For instance, the probability that the power system will survive at least 250 hours is

$$R(250) = 1 - \gamma(5; 4) = 1 - 0.735 = 0.265$$

where we have employed Table A.4 to evaluate $\gamma(5; 4)$.

6.4 ENTROPY

While it is true that we lack certainty in predicting the outcome of a random variable, the degree of uncertainty is not the same in all cases; rather, it depends upon the random variable in question. For instance, consider a random variable X that can take on two values, say 0 and 1, with probabilities $P(X = 1) = p$ and $P(X = 0) = q$. If $p = 0.99$ and $q = 0.01$, the observer feels less uncertain than if the probabilities were $p = 0.6$ and $q = 0.4$. Intuitively, the observer's uncertainty is maximized when the probabilities are equal ($p = q = 0.5$) and minimized when one of the probabilities is zero ($p = 0$ and $q = 1$, or $p = 1$ and $q = 0$). Of course, in the latter case there appears to be certainty.

More generally, suppose a random variable X possessing probability mass function $f(x)$ can take on the n values $x_1, x_2, \ldots, x_n$, with $f(x_i) = p_i$, for $i = 1, 2,$

. . . , n. Intuitively, at least, it appears the uncertainty of observation is increased when the probabilities p_i are more alike than when one or two of them carry most of the probability mass. More specifically, it appears reasonable that a measure of uncertainty should satisfy the following criteria:

1. Uncertainty is nonnegative and equal to zero if and only if there exists i such that $p_i = 1$.
2. Uncertainty is maximum when the outcomes of X are equally likely.
3. If the random variables X and Y possess n and m equally likely outcomes, respectively, with $n < m$, then the uncertainty of X is less than the uncertainty of Y.
4. Uncertainty is a continuous function of $p_1, p_2, \ldots, p_n$; that is, a slight change in the probabilities results in only a slight change in uncertainty.

Definition 6.6 provides a formalization of uncertainty which satisfies the four preceding criteria.

DEFINITION 6.6

Entropy. Let X be a discrete random variable possessing codomain $\{x_1, x_2, \ldots, x_n\}$ and probability mass function

$$f(x_i) = P(X = x_i) = p_i$$

for $i = 1, 2, \ldots, n$. Then the entropy of X, also called the **uncertainty** of X, is

$$H[X] = -\sum_{i=1}^{n} p_i \log_2 p_i = \sum_{i=1}^{n} p_i \log_2 (p_i^{-1})$$

where the convention is adopted that $p_i \log_2 p_i = 0$ if $p_i = 0$. $H[X]$ is measured in **bits.**

The definition of entropy does not utilize all information regarding the random variable X. Specifically, it only involves the probability point masses, not the manner in which they are distributed on the x axis. Consequently, any two random variables possessing the same outcome probabilities are indistinguishable from the perspective of entropy. The actual codomains play no role. As a function, it would be appropriate to write entropy as

$$H = H(p_1, p_2, \ldots, p_{n-1})$$

because the sum of the probabilities is 1 and therefore p_n is not an independent variable but is instead a function of $p_1, p_2, \ldots, p_{n-1}$.

As defined, $H[X]$ satisfies the four criteria listed prior to the definition. Since $0 \le p_i \le 1$ for all i, it is immediate from the definition that $H[X] \ge 0$ and $H[X] = 0$ if and only if there exists i such that $p_i = 1$. Because the logarithm is a continuous function, H is certainly a continuous function of $p_1, p_2, \ldots, p_n$. In the event the outcomes of X are equally likely—that is, if

$$p_1 = p_2 = \cdots = p_n = \frac{1}{n}$$

—then

$$H[X] = \sum_{i=1}^{n} \frac{1}{n} \log_2 n = \log_2 n$$

so that, in the case of equally likely outcomes, entropy is an increasing function of n. To complete the verification of the four criteria, we need only demonstrate that $H[X]$ is maximized in the case of equally likely outcomes. We provide a demonstration in the case of two outcomes, where the two probabilities are p and $1 - p$. Writing the entropy as a function of p and recognizing that a base 2 logarithm can be expressed in terms of a natural logarithm, we have

$$H(p) = -\frac{1}{\log 2} [p \log p + (1 - p) \log (1 - p)]$$

where it is assumed that $0 < p < 1$, for otherwise the entropy is 0. Taking the derivative yields

$$H'(p) = -\frac{1}{\log 2} [\log p - \log (1 - p)]$$

Setting the derivative equal to 0 and solving for p yields $p = 1/2$. Since $H''(p)$ is negative, a maximum is attained at $p = 1/2$.

EXAMPLE 6.16

Let the random variable X possess a discrete density defined by $f(0) = 1/8$, $f(1) = 1/4$, $f(2) = 1/8$, and $f(3) = 1/2$. Then

$$H[X] = \frac{1}{8} \log_2 8 + \frac{1}{4} \log_2 4 + \frac{1}{8} \log_2 8 + \frac{1}{2} \log_2 2 = \frac{7}{4}$$

If a random variable Y has four equally likely outomes, no matter what they are, then the entropy is maximized for the case of four outcomes and

$$H[Y] = \log_2 4 = 2$$

EXAMPLE 6.17

To encode the integers 1, 2, . . . , n in binary form, each integer is associated with a string of zeros and ones. When measuring the efficiency of a particular encoding, it is assumed that the probability of selecting each integer is known. To measure the transmission efficiency of the code, let B denote the random variable that counts the number of bits transmitted for an arbitrary integer and find $E[B]$—the smaller $E[B]$, the more efficient the code. The code will be assumed to be a *prefix code*, which means that no code corresponding to a specific integer may be obtained from the code of a distinct integer by simply adjoining zeros and ones. For instance, if 001 is the code of an integer, then 00110 cannot be the code of a different integer. If $p_1, p_2, \ldots, p_n$ are the probabilities of occurrence of the n integers, then, in effect, there is a random variable X with codomain $\{1, 2, \ldots, n\}$ and probability mass function $f(k) = p_k$ for $k = 1, 2, \ldots, n$. Given an outcome of X, B counts the number of bits in the encoding of the outcome.

Consider encoding 0, 1, 2, and 3 relative to the random variable X of Example 6.16. One acceptable code is given by

$$
\begin{array}{ccc}
0 & \longleftrightarrow & 1 \\
1 & \longleftrightarrow & 01 \\
2 & \longleftrightarrow & 001 \\
3 & \longleftrightarrow & 000
\end{array}
$$

For this code, B has the possible outcomes 1, 2, and 3. Moreover,

$$P(B = 1) = P(X = 0) = f(0) = \frac{1}{8}$$

$$P(B = 2) = P(X = 1) = f(1) = \frac{1}{4}$$

$$P(B = 3) = P(X = 2) + P(X = 3) = f(2) + f(3) = \frac{1}{8} + \frac{1}{2} = \frac{5}{8}$$

Thus,

$$E[B] = 1 \times \frac{1}{8} + 2 \times \frac{1}{4} + 3 \times \frac{5}{8} = \frac{5}{2}$$

A better code would take into account the various probabilities of X and assign longer bit codes to the outcomes possessing less likely occurrences. For instance, the code

$$0 \longleftrightarrow 001$$
$$1 \longleftrightarrow 01$$
$$2 \longleftrightarrow 000$$
$$3 \longleftrightarrow 1$$

yields $E[B] = 7/4$.

It can be shown generally that if the codomain of X is $\{1, 2, \ldots, n\}$ and B is the random variable that counts the number of bits in a prefix encoding of the codomain, then

$$E[B] \geq H[X]$$

Moreover, there exists at least one prefix code such that

$$E[B] \leq H[X] + 1$$

Thus, in the present example, the second code is optimal, since, for it, $E[B] = H[X]$.

6.5 MONTE CARLO SIMULATION

A system's characteristics are determined by the characteristics of its components. This theme has been especially evident in the study of queues, parallel systems, and reliability. From a strictly mathematical perspective, one can hope to apply the analytic methods of Chapter 5 to express system behavior in terms of component behavior, namely, to express the distribution of a function of random variables in terms of the distributions of the input variables. Practically, however, direct analytic methods have severe limitations. The performance of large systems often depends on many inter-dependent components, and an attempt to represent system behavior explicitly in terms of subsystem behavior can lead to intractable mathematical difficulties. One way of handling such performance analyses is to employ a technique known as **Monte Carlo simulation.**

Rather than explicitly represent a system distribution by an analytic approach, one might decide to observe the system in action, take measurements, and employ estimation techniques to arrive at estimates of important characteristics. For instance, rather than produce a mathematical model of a computer network and use a probabilistic analysis to arrive at the expected time to failure, we might actually observe the system in action, collect failure data, and find an estimate of the time to failure. Similarly, rather than employ analytic means to determine the expected time an arriving job must wait for CPU service, we might observe an actual CPU and use numerical data to provide an estimate of waiting time. The impracticality of such an empirical approach is evident: it can be both expensive and time consuming. The problem is especially acute in large, complex systems, but it is precisely with such systems that a strict mathematical approach is likely to be intractable.

One way around the dilemma is to build a computer model of the system and to simulate its behavior, the approach being to use knowledge of component distributions to provide **synthetic** system data. Instead of observing an actual system, we simulate component behavior by computer generation of component data and utilize these synthetic data in conjunction with the relationship between the components and overall system performance to produce synthetic system data. Whereas the gathering of actual performance data may be time consuming and expensive, the production of synthetic data is accomplished by computer, thereby making it possible to quickly and cheaply gather a large collection of observations. Performance is simulated, and statistics resulting from the simulation provide information regarding system performance. Naturally, the worth of the entire procedure depends on the degree to which the component distributions represent actual component behavior and the degree to which the relationship between the components and overall performance has been effectively modeled. Another problem, and one to which we now turn our attention, is the manner in which the computer is to simulate observations governed by a particular random variable.

The key to the computer generation of **random values,** which, by definition, are outcomes of a random variable, is the simulation of the uniform distribution U defined on $(0, 1)$. Taking the view that an outcome u of U is a nonterminating decimal, the generation of values of U involves the uniformly random selection of digits between 0 and 9, inclusively, and the concatenation of a string of such digits. The result will be a finite decimal expansion, our view being that such finite expansions constitute the outcomes of U. Of course, since the strings will be finite, say of length r, the process does not actually generate all possible outcomes of U; nevertheless, as in all computer-generated processes, we satisfy ourselves with a given degree of approximation, in this case the number of digits r. Thus, a typical generated value of U takes the form $u = 0.d_1 d_2 \ldots d_r$, where, for $i = 1, 2, \ldots, r$, d_i is an outcome of a random variable possessing equally likely outcomes between 0 and 9.

The digits $d_1, d_2, \ldots, d_r$ can be generated by selecting, with replacement, balls numbered 0 through 9 from an urn, or, more practically, employing electronic means to implement the uniformly random selection process. In practice, the digits are often generated by a nonrandom process whose outcomes, called **pseudorandom values,** simulate actual randomness. Another way to proceed is to employ a table of random values that has been previously generated by a random process. Table A.12 is such a table. To use it, simply place a decimal point in front of each of the five-digit integers to obtain a random-value table with $r = 5$.

Besides generating random values for the uniform distribution U, it is also useful to generate random values for the standard normal distribution Z. Although we will not go into details concerning the manner of generation, Table A.13 contains randomly generated values of Z.

Given a method to generate random values corresponding to U, we can employ these values to generate values of other distributions. While there are many ways of doing so, the following theorem proves to be useful in a number of circumstances.

Suppose F is a continuous probability distribution function which is strictly increasing and U is the uniform random variable over $(0, 1)$. Then

$$X = F^{-1}(U)$$

is a random variable possessing the probability distribution function F.

Proof: Since F is increasing, $F^{-1}(U) \leq x$ if and only if $U \leq F(x)$. Thus,

$$F_X(x) = P(X \leq x) = P[F^{-1}(U) \leq x] = P[U \leq F(x)] = F(x)$$

the last equality following since U is uniformly distributed over $(0, 1)$ and $0 \leq F(x) \leq 1$. ■

Figure 6.16 provides a graphic illustration of Theorem 6.6. To apply it to the generation of values of X, simply plot the generated value of U on the vertical axis and use the graph of F to find the corresponding value of X on the horizontal axis. One can also use the explicit representation $x = F^{-1}(u)$ to obtain the random X values if the representation can be achieved in closed form. Otherwise, graphical or approximation techniques are required to arrive at a simulation of X.

EXAMPLE 6.18

To simulate an exponential random variable X with parameter b, consider its probability distribution function

$$u = F(x) = 1 - e^{-bx}$$

Solving for x in terms of u gives

$$x = -b^{-1} \log (1 - u)$$

Therefore, according to Theorem 6.6,

$$X = -b^{-1} \log (1 - U)$$

possesses an exponential distribution with mean $1/b$. To generate a sequence of values corresponding to the exponential distribution, generate a sequence corresponding to U and apply the expression of X in terms of U. Since $1 - U$ is uniformly distributed, the representation of X can be simplified to

$$X = -b^{-1} \log U$$

FIGURE 6.16
Graphic illustration of Theorem 6.6

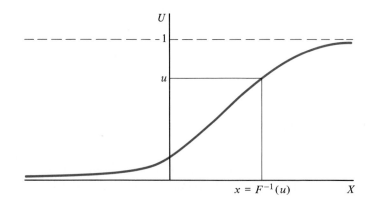

EXAMPLE 6.19

To simulate a uniform random variable X possessing a uniform distribution over the interval (a, b), note that the probability distribution function of X is given by

$$F(x) = \begin{cases} 1, & \text{if } x \geq b \\ \dfrac{x - a}{b - a}, & \text{if } a \leq x \leq b \\ 0, & \text{otherwise} \end{cases}$$

On $[a, b]$, F is strictly increasing. Thus, Theorem 6.6 applies with

$$u = \frac{x - a}{b - a}$$

on $[a, b]$. Solving gives

$$x = (b - a)u + a$$

so that X possesses the representation

$$X = (b - a)U + a$$

EXAMPLE 6.20

The sum of n independent identically distributed exponential random variables possessing common parameter $1/\beta$ is an n-Erlang distribution with parameter β. If $U_1, U_2, \ldots, U_n$ are independent uniform random variables on $(0, 1)$, then, according to Example 6.18,

$$X = -\sum_{k=1}^{n} \beta \log U_k$$

provides a representation of an n-Erlang distribution in terms of uniform variables.

Although an analogue of Theorem 6.6 can be given for discrete random variables, it is often simpler to develop representations directly from an understanding of the random variable. The next example illustrates this approach.

EXAMPLE 6.21

Suppose X possesses a binomial distribution with parameters n and p. Then X is identically distributed with a sum of n independent binomial variables $X_1, X_2, \ldots, X_n$, each possessing parameters 1 and p (see Example 5.18). For each k, let

$$X_k = \begin{cases} 1, & \text{if } U_k < p \\ 0, & \text{otherwise} \end{cases}$$

where $U_1, U_2, \ldots, U_n$ comprise an independent collection of random variables, each uniformly distributed over $(0, 1)$. Then

$$X = \sum_{k=1}^{n} X_k$$

provides a representation of X that can be simulated using generated random values of U.

EXAMPLE 6.22

To simulate a Poisson random variable with parameter q, recall that, according to Theorem 4.12, the interevent time of a Poisson process with mean qt possesses an exponential distribution with parameter q. For $t = 1$, the Poisson variable X counts the number of events in the time interval [0, 1]. A value x of X can be generated by summing interevent times $T_1, T_2, \ldots$ and letting x be the value for which

$$\sum_{i=1}^{x} T_i \leq 1$$

and

$$\sum_{i=1}^{x+1} T_i > 1$$

By Example 6.18, each interevent time is of the form $-q^{-1} \log U_i$, where $U_1, U_2, \ldots$ are independent random variables possessing uniform distributions over $(0, 1)$.

Having presented some illustrations of random-variable expressions that are conducive to computer-generated random values, we next give an example demonstrating the manner in which Monte Carlo simulation can be applied to practical problems.

TABLE 6.2 SYNTHETIC SAMPLE

x	y	$x + y$
2.293	1.724	4.017
0.980	2.682	3.662
2.474	1.623	4.097
0.010	5.334	5.344
2.055	1.864	3.919
0.415	3.414	3.829
1.169	2.506	3.675
0.159	4.048	4.207
0.453	3.347	3.800
0.304	3.637	3.941
0.015	5.176	5.191
2.137	1.815	3.952
0.181	3.972	4.153
0.120	4.210	4.330
0.004	5.647	5.651
0.423	3.398	3.821
0.222	3.846	4.068
0.296	3.654	3.950
0.358	3.522	3.880
2.313	1.712	4.025
0.089	4.372	4.461
0.219	3.854	4.073
0.819	2.852	3.671
2.075	1.852	3.927
0.452	3.348	3.800
0.491	3.284	3.775
1.866	1.984	3.850
0.856	2.810	3.666
1.447	2.278	3.725
0.056	4.603	4.659

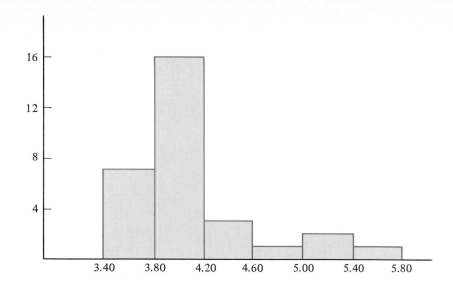

FIGURE 6.17
Histogram for Example 6.23

EXAMPLE 6.23

Consider a situation in which a product must proceed through two independent subassembly stages, the first being described by a time-to-completion random variable X possessing an exponential distribution with parameter $b = 1$ and the second being described by a time-to-completion random variable Y possessing a normal distribution with mean 3 and variance 1, all times being specified in hours. Overall system performance is measured by full-assembly time to completion $X + Y$. If an investigator wishes to study $X + Y$, one possibility would be to find its distribution. Another would be to use Monte Carlo simulation to obtain a synthetically derived empirical distribution. A precise statistical analysis of the relationship between the empirical distribution and the theoretical distribution of $X + Y$ would require the methods of Chapter 7; nonetheless, given that a quantification of the relationship is possible, at this point we can at least recognize that the synthetic empirical distribution can be employed to make inferences concerning the theoretical distribution.

Random values corresponding to X are obtained from Table A.12 by means of the relation $X = -\log U$ derived in Example 6.18. Random values corresponding to Y are obtained from Table A.13 by means of the relation $Y = Z + 3$, where Z is the standard normal random variable. Table 6.2 gives the random values x, y, and $x + y$ obtained by using the first columns of Tables A.12 and A.13, the sample size being 30. A histogram for the values $x + y$ is provided in Figure 6.17. The linearity of the expected-value operator yields $E[X + Y] = 1 + 3 = 4$. The raw data of Table 6.2 possess the empirical mean $\overline{x + y} = 4.104$. In a sense to be rigorously defined in Chapter 7, 4.104 can serve as an estimate of the true mean $E[X + Y]$. Of course, one might argue that the true mean is already known. Yes, but there are many other distributional properties of $X + Y$ that may be inferred (by the theory of statistical inference) from the empirical distribution that has been computer generated.

In Example 6.23 the system variable was merely a sum of two independent random variables, and one might clearly argue that simulation was unnecessary. However, suppose the system consisted of a few hundred components and these were not all independent. Under such circumstances, simulation can be a very useful tool. Indeed, because a computer can generate synthetic samples much larger than 30, with a number such as 1000 being obtained in a very short time, quite accurate results can be derived.

EXERCISES

SECTION 6.1

6.1. A unit coming off the production line can have 0, 1, or 2 major defects. For a randomly selected unit, the probabilities of 0, 1, and 2 defects are 0.52, 0.40, and 0.08, respectively. Consider the stochastic process $\{X(k)\}$ that counts the total number of production defects up through the completion of the kth unit. Construct a table of all possible realizations up through $k = 3$ and include realization probabilities in the table.

6.2. A fair coin is flipped repeatedly. If, on a flip, a head appears, the player receives one dollar; if a tail appears, the player loses one dollar. The stochastic process $\{X(k)\}$ keeps track of the player's total winnings (or losses) up to and including the kth flip. Construct a table of all possible realizations up through $k = 4$ and include the realization probabilities in the table. Find the probability that winnings exceed two dollars at the completion of the fourth flip. (See Example 4.4.)

6.3. Referring to Example 6.2, suppose there exists a second independent airport at which the arrival rate is 3 per minute. Let $\{X(t)\}$ be the process that counts the total number of arrivals (at the two airports) up to time t. Give the expression for $P_n(t)$.

6.4. Starting from the bottom left and proceeding counterclockwise, the corners of a square are labeled 1 through 4. A ball is selected from an urn containing three balls labeled number 1, two balls labeled number 2, four balls labeled number 3, and one ball labeled number 4. A ball is selected at random and placed back into the urn. A token is placed on the corner of the square bearing the same number as the selected ball. The process of picking balls with replacement is then continued. Each time an even number is picked, the token is moved one corner in the counterclockwise direction; each time an odd number is picked, the token is moved one corner in the clockwise direction.
   a) Draw the Markov diagram.
   b) Give the initial vector and the one-step transition matrix.
   c) Find the state vector for $k = 3$.
   d) What is the probability that the token will be in corner number 1 following the selection of the fourth ball (including the initial selection)?

6.5. Consider the problem of a rat being placed in the maze of Figure 6.18. A die is tossed and the rat is initially placed into the room bearing the same number as on the die. Thereafter, at successive time intervals, the rat may remain in the room in which it is located or may exit at one of the doors, the probability of staying in the room being 1/2 and the remaining prob-

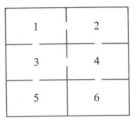

FIGURE 6.18
Maze for Exercise 6.5

ability being equally distributed among the doors to the room.
   a) Find the one-step transition matrix.
   b) Find the state vector at the conclusion of the second time interval.
   c) What is the probability of the rat's being in room 2 at the completion of the second time interval?

6.6. The scenario of Exercise 6.5 is repeated; however, the rat is placed initially into room number 1 (there being no die toss). Find the state vector at the conclusion of the third time interval.

6.7. Assume initially that a voter will vote Republican or Democrat with equal probability. Having voted Republican, in the next election the probability of voting Republican is 0.82. Having voted Democrat, in the next election the probability of voting Republican is 0.15. For the purposes of the model, assume that a voter votes either Republican or Democrat and that 0 denotes Republican and 1 denotes Democrat.
   a) Find the one-step transition matrix.
   b) Find the state vector upon completion of the fourth vote.
   c) Find the steady-state distribution of the process.

6.8. Reconsider the barrier problem of Example 6.5. This time assume that if the particle is in the 0 position, then it can either stay there with probability q or it can go right with probability p, and if it is in the 3 position, then it can either stay there with probability p or it can go left with probability q. Find the one-step transition matrix and the state vector $\mathbf{p}(2)$.

6.9. A letter is to be chosen randomly from among the letter set $\{b, c, u, r, p\}$ and then replaced. The process is repeated with the intent being to spell the word *up*. Model the recognition scheme using three states, realizing that once the word is successfully spelled, the process stays in the "fully spelled" state.
   a) Find the one-step transition matrix.
   b) Find the state vector $\mathbf{p}(3)$.

6.10. A simple model for describing the state of a program in execution involves defining the states *running*, *ready*, and *blocked*. A program is running if it has the CPU, is ready if it could use the CPU if the CPU

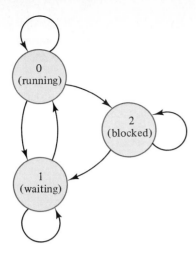

FIGURE 6.19
State diagram for
Exercise 6.10

were available, and is blocked if it is awaiting some external event such as the completion of I/O. Referring to the state diagram of Figure 6.19, it can be seen that there are seven possible transitions. Using the labels of the diagram, give the one-step transition matrix. Note that $\{X(k)\}$ gives the state at the completion of the kth clock pulse.

6.11. A lathe is rated 0 if it is operative at the end of the day shift and 1 otherwise. If it is operative at the end of the day shift, then the probability is 0.95 that it will be operative at the end of the next day shift; if it is inoperative at the end of the day shift, then the probability is 0.35 that it will be operative at the end of the next day shift. Assume the initial condition of the lathe is that it is operative.
a) Give the one-step transition matrix.
b) Find the state vector $\mathbf{p}(5)$.
c) Find the steady-state distribution.
6.12. In Exercise 6.11, the process is considered to be stationary because the age of the lathe has not been taken into account. Modify the model so that the probability of remaining in the operative state is $(0.95)k^{-1/8}$ and the probability of remaining in the inoperative state is still 0.65. The process is no longer stationary.
a) Find the first four one-step transition matrices.
b) Assuming the process to be initially in the operative state, find the probability that it is in the inoperative state after the fourth day shift.
6.13. Urn A contains 2 black and 2 white balls and urn B contains 2 black and 1 white ball. The first two selections are made from urn A. Thereafter, if balls of the same color have been picked on each of the two preceding picks, then the next ball is selected from urn B; otherwise, it is selected from urn A. Assume

all selections are made under the hypothesis of equal probability and they are made with replacement. Define the discrete-time stochastic process $\{X(k)\}$ by $X(k) = 0$ if the kth ball selected is black and $X(k) = 1$ if the kth ball selected is white. Explain why the process is not Markovian. Nevertheless, find $P[X(4) = 0]$.

SECTION 6.2

In Exercises 6.14 through 6.17, find the following:
a) The steady-state probabilities for $n = 0, 1, 2, 3, 4$.
b) The probability there are at least four jobs in the system.
c) The probability the queue is empty.
d) The probability there are at least four jobs in the queue.
e) The expected number of jobs in the system.
f) The variance of the number of jobs in the system.
g) The mean number of jobs in the queue.
h) The mean waiting time in the queue.
i) The mean time in the system.
In all cases the arrival stream is Poisson, the service time is exponentially distributed, no arriving jobs turn away, and the queue is of unlimited length.

6.14. A telephone operator receives inquiries at the rate of 24 per hour, and the mean service time is 2 minutes.
6.15. Data packets arrive for input into a file system at the rate of 12 per minute and the mean input time is 3.6 seconds.
6.16. Packages arrive at a conveyor belt at the rate of 30 per hour and the robot loading the packages on the belt operates with mean loading time 1.8 minutes.
6.17. Football fans arrive at a ticket window at the rate of 3 per minute and the mean ticketing time at the single window is 15 seconds.

6.18. Referring to Exercise 6.17, what would happen if the mean ticketing time were 21 seconds?
6.19. Referring to Exercise 6.17, describe the operation of the system if the arrival process and service time were deterministic, with arrivals being exactly 20 seconds apart and every service taking exactly 15 seconds.

SECTION 6.3

6.20. Find the reliability function corresponding to a time-to-failure random variable that possesses a uniform distribution over the interval $[0, 4]$.
6.21. Find the reliability function corresponding to a time-to-failure random variable possessing the density

$$f(t) = \begin{cases} 3(t + 1)^{-4}, & \text{if } t \geq 0 \\ 0, & \text{otherwise} \end{cases}$$

6.22. Systems S_1 and S_2 possess the reliability functions

$$R_1(t) = \begin{cases} -\dfrac{t}{2} + 1, & \text{if } 0 \leq t \leq 2 \\ 0, & \text{otherwise} \end{cases}$$

and

$$R_2(t) = \begin{cases} -2t + 1, & \text{if } 0 \le t \le \dfrac{1}{2} \\ 0, & \text{otherwise} \end{cases}$$

respectively. Based upon the MTTF, which system is more reliable?

6.23. Systems S_1 and S_2 possess the reliability functions

$$R_1(t) = \begin{cases} -\dfrac{t}{8} + 1, & \text{if } 0 \le t \le 8 \\ 0, & \text{otherwise} \end{cases}$$

and

$$R_2(t) = (t + 1)^{-2}$$

Compare the reliabilities of the systems based upon maximum MTTF.

6.24. Find the hazard function for the time-to-failure random variable whose density is given in Exercise 6.20.

6.25. Find the hazard function for the time-to-failure random variable whose density is given in Exercise 6.21.

6.26. Find the hazard functions for the random variables whose reliability functions are given in Exercise 6.22.

6.27. Find the hazard functions for the random variables whose reliability functions are given in Exercise 6.23.

6.28. Find the reliability function and density corresponding to the hazard function $h(t) = 3t^2$.

6.29. To make a prediction regarding the weather, a computer requires statistical data from three sources. Assuming the three sources possess the respective hazard functions $h_1(t) = 4t^6$, $h_2(t) = 0.8$, and $h_3(t) = (0.6)t$, where t is measured in months, find the reliability function. Assume source failures are independent.

6.30. In a sequential assembly procedure, there are four stations through which the product must pass from inception to completion. The four subassembly stations possess respective failure rates (in failures per week) $q_1 = 1.2$, $q_2 = 0.5$, $q_3 = 0.6$, and $q_4 = 0.9$.

Find the reliability function and the MTTF for the system under the assumption of independent failures.

6.31. A pipeline processor possesses five serial segments, each of which must process data in turn to produce the final output of the processor. If each segment possesses a constant failure rate of 0.23 failure per year, segment failures being independent, find the reliability function and MTTF for the system.

6.32. A reconnaissance system employs four satellites, each possessing the constant failure rate of 0.1 failure per year. Assuming the system functions only when all four satellites are in operation, what is its MTTF? What is the probability that the system will fail within one year? Assume satellite failures are independent.

6.33. Suppose the system of Exercise 6.32 can function with only a single satellite. Find its MTTF. What is the probability that the system will fail within one year?

6.34. Suppose the system of Exercise 6.32 can function so long as at least two satellites are operable. Find the MTTF.

6.35. For the weather-prediction problem of Exercise 6.29, suppose the prediction can be made (albeit a bit less accurately) so long as at least one source remains operative. Find the reliability function for the system.

6.36. Company A can buy a particular part from any one of two suppliers, and both suppliers can supply the needed number to maintain production. Suppose the suppliers' production processes possess reliability functions $R_1(t)$ and $R_2(t)$ of Exercise 6.22, t being in years. Moreover, suppose that once a supplier loses production, its production does not resume. What is the expected time until company A is unable to secure the needed part? Assume failures for the suppliers are independent.

6.37. The computer depicted in Figure 6.20 possesses a single CPU and three parallel memory modules. Assuming processor and memory failures to be independent, the processor failure rate to be 0.1 per year,

FIGURE 6.20
Computer for Exercise 6.37

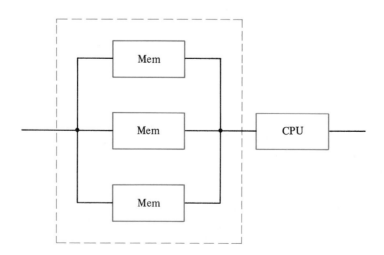

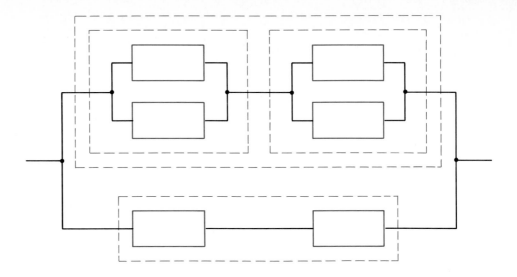

FIGURE 6.21
System for
Exercise 6.39

and each memory module to possess a failure rate of 0.06 per year, find the MTTF of the system.

6.38. To improve the fault tolerance of the system of Figure 6.20, it is reconfigured so as to include three parallel processors in the CPU subsystem. What is the new MTTF?

6.39. Each component in the system of Figure 6.21 possesses the reliability function

$$R(t) = \begin{cases} 1 - t, & \text{if } 0 \le t \le 1 \\ 0, & \text{otherwise} \end{cases}$$

Find the MTTF of the system.

6.40. Each component in the system of Figure 6.22 possesses the reliability function

$$R(t) = \begin{cases} 1 - t^2, & \text{if } 0 \le t \le 1 \\ 0, & \text{otherwise} \end{cases}$$

Find the MTTF of the system.

6.41. A construction company is making repairs on a bridge and is employing a single welder; however, to increase the reliability of the welding operation, the welder begins the job with three arc-welding units, each possessing a failure rate of 0.32 per month. Assume unit failures are independent and the welder utilizes the units in a standby mode. Find the MTTF, the standard deviation of time to failure, and the probability the welder will not be without a welding unit for the duration of the job (six months). Assume that once a unit breaks down, it is not brought back to the project after repair.

6.42. A battery-operated radio requires a single battery whose failure rate is 0.5 per month. If there are 5 batteries, 1 to be used initially and 4 to be employed in a standby mode, what is the MTTF of the radio? Find the standard deviation of the time to failure and the probability that the radio will be operative (at least insofar as batteries are concerned) for at least 2 months. Assume battery-failure times are independent.

SECTION 6.4

6.43. Find the entropy for a binomially distributed random variable with $n = 5$ and $p = 3/4$.

6.44. Find the entropy for a binomially distributed random variable with $n = 5$ and $p = 2/3$.

6.45. X, Y, and W are discrete random variables respectively possessing 4, 5, and 6 equally likely outcomes. Find the entropy of each.

6.46. Find the entropy for the random variable of Example 3.19.

6.47. Find the entropy for the random variable of Exercise 3.8.

6.48. Consider the encoding

$$
\begin{aligned}
0 &\longleftrightarrow 00000 \\
1 &\longleftrightarrow 1 \\
2 &\longleftrightarrow 01 \\
3 &\longleftrightarrow 001 \\
4 &\longleftrightarrow 00001 \\
5 &\longleftrightarrow 0001
\end{aligned}
$$

Under the condition that the relative frequencies of the symbols 0, 1, 2, 3, 4, and 5 are governed by the random variable of Exercise 6.43, find $E[B]$, where B is the random variable that counts the number of bits transmitted for a single symbol.

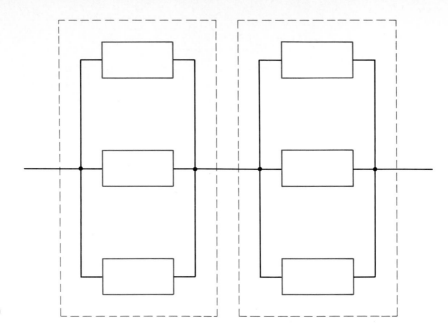

FIGURE 6.22
System for Exercise 6.40

6.49. Referring to Exercise 6.48, find the prefix encoding that minimizes $E[B]$.

6.50. Repeat Exercise 6.48, except let $p = 1/4$ in the governing binomial distribution.

6.51. Referring to Exercise 6.50, find the prefix encoding that minimizes $E[B]$.

6.52. The discrete random variable X possesses a density defined by $f(1) = f(2) = f(6) = 0.1$, $f(3) = f(4) = 0.2$, and $f(5) = 0.3$. Find the binary prefix encoding of the symbols 1 through 6 that minimizes $E[B]$.

SECTION 6.5

For the exercises of the current section, employ Tables A.12 and A.13 beginning with the first column of each (so that answers will be uniform). All empirical means are to be computed directly from the sample data.

6.53. An assembly process consists of three independent subassembly stages. The mean times to completion for the three stages are exponentially distributed with means 0.5, 1.0, and 2 hours, respectively. Find a synthetic sample of size $n = 20$ for total assembly time, find the mean of the synthetic sample, and compare the empirical mean with the theoretical mean.

6.54. A rectangle possesses dimensions X by Y, where $X \sim N(30, 4)$ and $Y \sim N(20, 1)$. Find a synthetic sample of size 25 for the area $A = XY$ of the rectangle. Find the empirical mean.

6.55. Repeat Exercise 6.54 under the assumption that X possesses a uniform distribution on [2.9, 3.1] and Y possesses a uniform distribution on [1.9, 2.1].

6.56. The number of vehicles arriving at intersection A in a given hour possesses a Poisson distribution with mean 2.1, whereas the number arriving at intersection B in the same hour possesses a Poisson distribution with mean 1.5. Simulate a single value for the total number of vehicles arriving at the two intersections during the hour.

6.57. Apply the model of Example 6.23 under the assumption that X possesses a 3-Erlang distribution with parameter $\beta = 8$. Again employ a sample size of 30, this time finding the synthetic sample and the empirical mean. Compare the latter with the theoretical mean.

6.58. Simulate a single value of a binomial distribution with parameters $p = 1/4$ and $n = 24$.

7

ESTIMATION

The statistical method involves the use of data to make inferences concerning a population, or some collection of populations. Given that a random phenomenon is modeled by a family of distributions, it is necessary to estimate the values of the distribution parameters that apply to the particular phenomenon. The estimation procedure requires an estimation rule that can be applied to data that have been collected according to some acceptable procedure. The present chapter discusses the notion of an estimator and introduces key estimators for the mean and variance. The goodness of estimators is also investigated. In particular, the class of minimum-variance unbiased estimators is studied. Two important methods for arriving at parametric estimation rules are considered. Discussions of interval and Bayesian estimation follow. Section 7.8.3 introduces the *t* distribution, one of the most important sampling distributions in statistics.

7.1 ESTIMATORS

A random variable serves as a model to describe a measurement process, and its distribution is viewed as a law governing the behavior of the phenomenon under study. Thus far, we have paid attention to the description of such laws and the practical consequences derivable from them. Little attention has been paid to the construction and testing of probability models, both of which depend upon experimental data. The process of utilizing observations of random phenomena for the purpose of determining the underlying probability laws governing the relevant random variables is known as **statistical inference.**

7.1.1 Parametric Estimation

Several inference problems can be readily articulated. Employing observations of a random phenomenon, we might wish to find a density that suitably fits (describes) the observational data. Often, based upon scientific principles, expert opinion, heuristics, or hunches, the phenomenon is assumed to be described by a distribution belonging to some well-studied family of distributions, such as the normal, gamma, or Poisson. An assumption of this nature reduces the problem from determining the form of the distribution to estimating certain parameters that characterize the distribution. For instance, if the random variable in question is assumed to be a member of the univariate normal family $N(\mu, \sigma)$, then two parameters, μ and σ, need to be estimated. Once these are estimated, a unique member of the normal family is determined. In the case of the Poisson distribution, one need only estimate λ, the members of the Poisson family being specified by a single parameter. The process of utilizing observational data to determine parameters used in characterizing families of distributions is called **parametric estimation,** where the term ''parameter'' includes descriptive parameters, such as the mean and variance, as well as those constants specified in the defining density. For example, if a random variable is believed to belong to the gamma family, then one might wish to estimate one or both of the defining parameters α and β, or to perhaps estimate the mean, which is itself expressed in terms of α and β.

When speaking of the estimation of distribution parameters, the exact meaning of the term ''estimation'' is a tricky matter. In everyday parlance, to estimate a quantity means to put forth a best guess as to the actual value of the quantity. There is the tacit assumption that the true value is known to someone, or is at least knowable. In statistics, the problem is much more subtle. Unless one is dealing with a finite set and can collect data on all members of the set, there is no way to determine an actual parameter. For example, when describing an arrival process by a Poisson random variable, we do not mean to imply that the arrival rate can somehow be determined by making a sufficient number of observations. Rather, the Poisson variable serves as a model that provides a convenient and comprehensible descriptive framework for the process. There is no intention to claim that there is some ultimately determinable rate. In such a circumstance, the rate ''estimate'' obtained by the application of some inferential statistical procedure can better be described as a contingent working hypothesis, one which will be employed so long as there is not sufficient reason to believe it is not in conformity with the observational data. Ultimately, the meaning of the estimates derived from statistical inference procedures lies within the theory and methodology of the procedures themselves. For this reason we will be careful to distinguish between the numerical results of estimation and the procedures by which those results are obtained.

Parametric inferences can be divided into three classes. Estimation of the mean of a Poisson random variable involves an attempt to settle on a suitable value to utilize as the mean. This type of inference is known as **point estimation.** Another approach might be to find an interval believed, to some degree of confidence, to contain the parameter of interest. This type of inference is known as **interval estimation.** In a third scenario, it might be necessary to choose between two alternative hypotheses regarding a parameter. For instance, one hypothesis might be that the mean is equal to some conjectured value μ_0, whereas the competing hypothesis is that it is not equal to μ_0. This latter inference problem comes under the domain of **hypothesis testing.** All three types of problems will be treated extensively in the remainder of the text.

7.1.2 Random Samples

A random variable X is a mathematical entity used to model a measurement process relating to some physical phenomenon. From an intuitive perspective, one can imagine there being some collection of entities to be observed and each observation results in a measurement yielding an outcome in the codomain of X. For instance, if 10,000 cases of fuses are being produced, each box containing five 20-amp fuses, and we wish to count the number of defective fuses in a box, then there are 10,000 possible observations, and X possesses the codomain

$$\Omega_X = \{0, 1, 2, 3, 4, 5\}$$

Such an illustration is atypical, because in many modeling situations there is no well-defined collection of observations to which the probability distribution refers. Nevertheless, even in such circumstances it is sometimes convenient to speak as if there were such a collection, the reason being that certain intuitions arising in finite models sometimes (but certainly not always) apply universally. Consequently, we often speak of the **population** of observations and refer to the **population random variable,** keeping in mind that the so-called population may be a figment of our imaginations.

As discussed in Section 7.1.1, our goal is to form estimates based upon collections of observations of random phenomena. If the estimates are to be indicative of the entire class of phenomena under consideration, then the observation procedure must not overly weight any particular subclass of the phenomena. In statistical parlance, we wish to consider **samples** drawn from the population and, in some manner, to utilize the resulting measurement data to form estimates of **population parameters** (more precisely, the unknown parameters in the density of the population random variable). Should the observations not be indicative of the population as a whole, then it is likely that the estimates will be inaccurate. Thus, each observation should, in a probabilistic sense, be identical with the underlying random variable and, in addition, not be influenced by other observations.

To obtain a mathematical description of the preceding considerations, let X be the population random variable and suppose n observations are to be made. Since each observation is itself random, the n observations correspond to n random variables, $X_1, X_2, \ldots, X_n$. The requirement that each observation reflect the underlying measurement process means that each X_k must be identically distributed with X, and the requirement that each observation not be influenced by the others means that $X_1, X_2, \ldots, X_n$ must be independent. We are led to the following definition.

DEFINITION 7.1

Random Sample. A random sample of size n corresponding to the random variable X is a collection of n independent random variables $X_1, X_2, \ldots, X_n$ such that, for $k = 1, 2, \ldots, n$, X_k is identically distributed with X.

If $X_1, X_2, \ldots, X_n$ comprise a random sample for X, and X possesses density $f(x)$, then the conditions of independence and identical distribution with X mean that the joint density of $X_1, X_2, \ldots, X_n$ is given by

$$f(x_1, x_2, \ldots, x_n) = f(x_1)f(x_2) \cdots f(x_n)$$

A sampling procedure whose probability model is a random sample is called **random sampling,** and any data set obtained by random sampling is also called a **random sample.** The data themselves are called **sample values.**

It should be borne in mind that Definition 7.1 is a mathematical definition. Whether or not a sampling procedure satisfies the twin conditions is another question, one which itself can be judged with reference to data. In practice, heuristic considerations are often employed to justify the assumption that a sampling procedure is random. Under such circumstances, should conclusions be based upon theory that is itself dependent upon random sampling when, in fact, the sampling procedure is not random, then the worth of the conclusions must be held in question. Practically speaking, we are once again confronted by the modeling dilemma: the precise degree to which the model represents the physical process must, of necessity, remain an open question.

7.1.3 Estimation Rules and Statistics

The key to point estimation is the formulation of rules by which a set of sample values of a random variable might be used to estimate the values of unknown parameters. Functions must be designed whose input variables are the outcomes of some experiment and whose output variables can be employed as "reasonable" estimates for the unknown parameters. If the sample values are assumed to result from random sampling, then the function depends on n independent, identically distributed random variables.

To formalize the principle of point estimation, suppose X possesses the density $f(x; \theta)$, where the inclusion of θ as a second argument in the density is meant to imply that θ is unknown. Moreover, let $X_1, X_2, \ldots, X_n$ constitute a random sample arising from X. What is needed is a function of the n random variables,

$$\hat{\theta} = \hat{\theta}(X_1, X_2, \ldots, X_n)$$

so that for a particular set of sample values, say $x_1, x_2, \ldots, x_n$, evaluating $\hat{\theta}$ at these values produces a point estimate of θ, namely, $\hat{\theta}(x_1, x_2, \ldots, x_n)$. As a function of random variables, $\hat{\theta}(X_1, X_2, \ldots, X_n)$ is itself a random variable and is called an **estimator** of θ. The term **estimate** is reserved for a specific value of the random variable $\hat{\theta}$, and we employ the notation

$$\overset{*}{\theta} = \hat{\theta}(x_1, x_2, \ldots, x_n)$$

to denote an estimate. It is important to keep in mind that an estimator is a random variable and an estimate is one of the elements in the codomain of this variable. What is shared by both an estimator and a resulting estimate is the function rule defining $\hat{\theta}$, this rule being known as the **estimation rule.**

Whether or not an estimator serves its purpose is a question that must ultimately be addressed. Put simply, does it produce estimates that can be assumed to be close to the unknown parameter? As the definition stands, that question is left open. Indeed, from a strictly mathematical perspective, an estimator is merely a function of the random sample. How to go about finding a good estimator, as well as giving meaning to the notion of goodness, is an important topic that must be addressed.

At this point, no restriction is placed on the type of functions that can be employed as estimators; however, we do insist that an estimation rule not depend on any unknown parameters. In terms of the next definition, an estimator must be a **statistic.**

DEFINITION 7.2

Statistic. A function of the random variables comprising a random sample is called a statistic if it does not depend on any unknown parameters.

$\hat{\theta} = \hat{\theta}(X_1, X_2, \ldots, X_n)$ is a statistic if the estimation rule defining $\hat{\theta}$ is free of unknown parameters. This does not mean that the distribution of $\hat{\theta}$ is free of unknown parameters, only the rule that produces $\hat{\theta}$.

EXAMPLE 7.1

Based on experience, it is assumed that the service time X at a highway toll possesses an exponential distribution but the parameter b is unknown. Thus, the density of interest is

$$f(x; b) = \begin{cases} be^{-bx}, & \text{if } x \geq 0 \\ 0, & \text{if } x < 0 \end{cases}$$

To estimate b, a random sample of n service times will be taken. The problem is to find an appropriate estimator. One way to proceed is to recognize that $E[X] = 1/b$, so that $b = 1/\mu$, where μ denotes the mean of X. If a decent estimator of μ can be found, then its reciprocal can be employed as an estimator of b. Proceeding heuristically, μ is the center of mass of the distribution, and for a given set of sample values the empirical mean $\bar{x}$ gives the center of mass. If we now make an inductive leap and suppose the normalized relative frequency distribution of the data set to be approximately equal to the theoretical distribution of the random variable, then it appears plausible that $\bar{x}$ will provide a decent estimate of μ. Since, for a particular set of sample values $x_1, x_2, \ldots, x_n$,

$$\bar{x} = \frac{1}{n}(x_1 + x_2 + \cdots + x_n)$$

we are led to consider the mean estimator

$$\hat{\mu} = \frac{1}{n}(X_1 + X_2 + \cdots + X_n)$$

so that

$$\hat{b} = \frac{1}{\hat{\mu}} = \frac{n}{X_1 + X_2 + \cdots + X_n}$$

Mathematically, each estimator results from an estimation rule, $\hat{\mu}$ from

$$\hat{\mu}(\xi_1, \xi_2, \ldots, \xi_n) = \frac{1}{n}(\xi_1 + \xi_2 + \cdots + \xi_n)$$

and $\hat{b}$ from

$$\hat{b}(\xi_1, \xi_2, \ldots, \xi_n) = \frac{n}{\xi_1 + \xi_2 + \cdots + \xi_n}$$

At this point a judgment concerning the worth of these estimators is not possible, since no criterion of goodness has yet been formulated.

As an illustration of the use of these estimators, suppose a random sample of size 6 is taken and the following times (in seconds) are recorded:

$$10 \quad 28 \quad 13 \quad 12 \quad 15 \quad 12$$

Then the estimate of μ (for this particular data set) is

$$\overset{*}{\mu} = \bar{x} = 15$$

and the corresponding estimate of b is

$$\overset{*}{b} = \frac{1}{\bar{x}} = \frac{1}{15}$$

Had different empirical values been obtained when the sample was taken, then different estimates of μ and b would have resulted.

Suppose it is desired to estimate two parameters associated with a random variable X, say θ_1 and θ_2. Then the density is expressed as $f(x; \theta_1, \theta_2)$ and a pair of estimators is required:

$$\hat{\theta}_1 = \hat{\theta}_1(X_1, X_2, \ldots, X_n)$$

$$\hat{\theta}_2 = \hat{\theta}_2(X_1, X_2, \ldots, X_n)$$

More generally, an estimation problem can involve m parameters $\theta_1, \theta_2, \ldots, \theta_m$. In such a case, the density is written as

$$f(x; \theta_1, \theta_2, \ldots, \theta_m)$$

and m estimators

$$\hat{\theta}_k = \hat{\theta}_k(X_1, X_2, \ldots, X_n)$$

$k = 1, 2, \ldots, m$, are required. Often, the m unknown parameters are written as a vector $\boldsymbol{\theta}$, the estimator random variables $\hat{\theta}_1, \hat{\theta}_2, \ldots, \hat{\theta}_m$ are written as a random-vector estimator $\hat{\boldsymbol{\theta}}$, and the density is written $f(x; \boldsymbol{\theta})$.

EXAMPLE 7.2

Suppose it is known that the resistance across a resistor is normally distributed but the mean and variance are unknown. Then the density takes the form

$$f(x; \boldsymbol{\theta}) = f(x; \mu, \sigma^2) = \frac{1}{\sqrt{2\pi}\, \sigma} e^{-\frac{1}{2}\left(\frac{x-\mu}{\sigma}\right)^2}$$

and the estimation problem is to devise two estimation rules, one to form an estimator of μ and the other to form an estimator of σ^2. Based upon the center-of-mass reasoning of Example 7.1, let

$$\hat{\mu} = \frac{1}{n}(X_1 + X_2 + \cdots + X_n)$$

To arrive at an estimator of the variance, we reason (not so plausibly in this case) as follows: since σ^2 gives a measure of dispersion for the theoretical distribution while the empirical variance

$$s^2 = \frac{1}{n-1} \sum_{k=1}^{n} (x_k - \bar{x})^2$$

gives a measure of dispersion of a particular set of sample values $x_1, x_2, \ldots, x_n$, we might try the estimator

$$\hat{\sigma^2} = \hat{\sigma^2}(X_1, X_2, \ldots, X_n) = \frac{1}{n-1} \sum_{k=1}^{n} (X_k - \hat{\mu})^2$$

Whereas in the case of the mean, a geometrically credible center-of-mass argument was employed, to this point no correspondence between the theoretical dispersion of a probability distribution and the empirical dispersion, as measured by s^2, has been demonstrated. In fact, the goodness of the estimator $\widehat{\sigma^2}$ will be established in Theorem 7.7. In any event, the foregoing reasoning results in the random-vector estimator

$$\hat{\boldsymbol{\theta}} = \begin{pmatrix} \hat{\mu} \\ \widehat{\sigma^2} \end{pmatrix}$$

Suppose five measurements are taken, these being (in ohms)

$$8.04 \quad 8.02 \quad 8.07 \quad 7.99 \quad 8.03$$

Based upon the estimators $\hat{\mu}$ and $\widehat{\sigma^2}$, the resulting estimates are

$$\overset{*}{\mu} = \bar{x} = 8.03$$

ohms and

$$\overset{*}{\sigma^2} = \frac{1}{4}\left[(0.01)^2 + (0.01)^2 + (0.04)^2 + (0.04)^2 + 0^2\right] = 0.00085$$

In vector form,

$$\overset{*}{\boldsymbol{\theta}} = \begin{pmatrix} 8.03 \\ 0.00085 \end{pmatrix}$$

Estimators are statistics (functions of random variables comprising a random sample) employed for the purpose of obtaining a working value for an unknown parameter. Consider an arbitrary statistic

$$Y = h(X_1, X_2, \ldots, X_n)$$

defined on the random sample $X_1, X_2, \ldots, X_n$. If Y were to be employed to find a parameter associated with some probability distribution, then it would be called an estimator and written in different notation; nonetheless, mathematically Y is a function of several random variables. Thus, the statistic Y possesses its own distribution.

DEFINITION 7.3

Sampling Distribution. The distribution of a statistic is called a sampling distribution.

EXAMPLE 7.3

In Example 7.1 we considered the statistic

$$\hat{\mu} = \frac{1}{n}(X_1 + X_2 + \cdots + X_n)$$

where, for $k = 1, 2, \ldots, n$, X_k possesses an exponential density with parameter b. If U denotes the sum of the sample variables, then Theorem 5.7 shows that U possesses a gamma density with $\alpha = n$ and $\beta = 1/b$. Thus,

$$f_U(x) = \frac{b^n}{(n-1)!} x^{n-1} e^{-bx}$$

for $x > 0$ (see Definition 4.8). According to the result of Example 3.14, the density for $\hat{\mu}$ is given by

$$f_{\hat{\mu}}(y) = nf_U(ny) = \frac{(nb)^n}{(n-1)!} y^{n-1} e^{-nby}$$

for $y > 0$, which is a gamma density with $\alpha = n$ and $\beta = 1/nb$. Thus, $\hat{\mu}$ possesses a gamma sampling distribution. We might certainly expect judgments concerning the goodness of $\hat{\mu}$ as an estimator to involve properties of the gamma distribution. For instance,

$$E[\hat{\mu}] = \alpha\beta = \frac{1}{b}$$

so that the expected value of the estimator is the mean of the population density, the precise parameter for which it is to serve as an estimator.

Note in Example 7.3 that the sampling distribution depends on the sample size n. In particular, the variance of the estimator is given by

$$\text{Var}\,[\hat{\mu}] = \alpha\beta^2 = n\left(\frac{1}{nb}\right)^2 = \frac{1}{nb^2}$$

7.2 SAMPLE MEAN

The most fundamental parameter of a distribution is its mean. In the present section we discuss in detail a mean estimator of profound importance from the perspectives of both theory and application.

7.2.1 Point Estimation of the Mean

The mean estimator considered in Examples 7.1, 7.2, and 7.3, which is simply the unweighted average of the sample variables, is known as the sample mean.

DEFINITION 7.4

Sample Mean. Given a random sample $X_1, X_2, \ldots, X_n$ arising from a population random variable X, the statistic

$$\bar{X} = \frac{1}{n}(X_1 + X_2 + \cdots + X_n)$$

is called the sample mean and its distribution is called the **sampling distribution of the mean.**

More generally, consider a weighted average of the sample variables, say

$$Y = a_1X_1 + a_2X_2 + \cdots + a_nX_n$$

where the coefficients a_k are nonnegative and sum to 1. Such a linear combination of the random variables $X_1, X_2, \ldots, X_n$ is called a **convex combination.** Suppose the population random variable possesses mean μ and variance σ^2. Due to the linearity of the expected-value operator, the mean of Y is

$$\mu_Y = a_1\mu + a_2\mu + \cdots + a_n\mu = (a_1 + a_2 + \cdots + a_n)\mu = \mu$$

In particular, $\mu_{\bar{X}} = \mu$. Moreover, due to Theorem 5.17, the variance of Y is

$$\sigma_Y^2 = a_1^2\sigma^2 + a_2^2\sigma^2 + \cdots + a_n^2\sigma^2 = (a_1^2 + a_2^2 + \cdots + a_n^2)\sigma^2 \leq \sigma^2$$

An obvious question arises: How can the coefficients in a convex combination be chosen so as to minimize the variance? To answer the question, first suppose the convex combination possesses only two summands. Then Y can be written as

$$Y = aX_1 + (1 - a)X_2$$

so that

$$\sigma_Y^2 = a^2\sigma^2 + (1 - a)^2\sigma^2$$

To apply maximum-minimum theory from elementary calculus, differentiate the variance of Y with respect to a to obtain

$$\frac{d}{da}\sigma_Y^2 = 2a\sigma^2 - 2(1 - a)\sigma^2$$

Setting the derivative equal to 0 and solving for a yields $a = 1/2$. Since the second derivative is always positive, $a = 1/2$ is the desired minimum. In sum, the variance of Y is minimal when both coefficients are equal to 1/2. A similar argument applies to the case of n summands. Thus, for a random sample of size n, the convex combination with minimum variance is the sample mean. Letting $a_k = 1/n$, the previously found expressions for the mean and variance of Y yield the following theorem.

<table>
<tr>
<td>THEOREM 7.1</td>
<td>Suppose $X_1, X_2, \ldots, X_n$ comprise a random sample arising from the random variable X. Then the sample mean possesses the minimum variance among all convex combinations of the sample variables. In addition, the mean and variance of the sample mean are

$$\mu_{\bar{X}} = \mu_X$$

$$\sigma_{\bar{X}}^2 = \frac{\sigma_X^2}{n}$$
</td>
</tr>
</table>

The expected value of the sample mean, as well as of any convex combination of random variables comprising a random sample, is equal to the original mean. Such a property is certainly desirable if $\bar{X}$ is to be employed as an estimator of the mean, since it suggests that the values of the estimator will center about the parameter it is to estimate.

Unbiased Estimator. The statistic $\hat{\theta}$ is an unbiased estimator of θ if $E[\hat{\theta}] = \theta$; otherwise, $\hat{\theta}$ is said to be a **biased** estimator, in which case

$$E[\hat{\theta}] - \theta$$

is called the **bias** of $\hat{\theta}$.

Implicit in the definition of unbiasedness is that $E[\hat{\theta}] = \theta$ no matter what the value of θ. Generally (although not always), if forced to choose between an unbiased and a biased estimator, our choice would be the unbiased one.

From the perspective of estimation, the significance of the mean and variance formulae of Theorem 7.1 is quite transparent. If an estimate of the mean is desired, we might make a single observation of the random variable X and utilize the single datum resulting from the observation as an estimate of the mean μ. Such an approach yields the mean estimator $\hat{\mu} = X$. Since $E[\hat{\mu}] = E[X]$, $\hat{\mu}$ is unbiased. Now consider the estimator $\hat{\mu} = \bar{X}$, where $\bar{X}$ is the sample mean arising from a random sample of size n. Taking n observations and averaging again leads to an estimator $\hat{\mu}$ whose expected value is μ; however, this time, since the variance of the sample-mean estimator is the original variance divided by n, the probability $P(|\hat{\mu} - \mu| \leq r)$ of $\hat{\mu}$ lying within a prescribed distance r from the true mean is increased. The **precision** of the estimator is enhanced.

For the situation in which the underlying random variable X is normal, not only does Theorem 7.1 apply, but so too does Theorem 5.9. Consequently, $\bar{X}$ is normal. As is illustrated in Figure 7.1 for sample size n, the sampling distribution of the mean possesses greater peakedness. Precision increases as n increases. For the case of normal random variables, the next theorem provides an easy-to-apply quantification of this precision.

If X is normally distributed with mean μ and variance σ^2 and if the sample mean $\bar{X}$ arises from a random sample of size n, then, for any $r > 0$,

$$P(|\bar{X} - \mu| < r) = \Phi\left(\frac{\sqrt{n}r}{\sigma}\right) - \Phi\left(\frac{-\sqrt{n}r}{\sigma}\right)$$

where $\Phi(z)$ is the probability distribution function of the standard normal variable.

Proof: Applying Theorems 7.1 and 4.7 yields

$$P(|\bar{X} - \mu| < r) = P(-r < \bar{X} - \mu < r)$$

$$= P\left(-\frac{r}{\sigma_{\bar{X}}} < \frac{\bar{X} - \mu_{\bar{X}}}{\sigma_{\bar{X}}} < \frac{r}{\sigma_{\bar{X}}}\right)$$

$$= P\left(-\frac{r}{\sigma/\sqrt{n}} < Z < \frac{r}{\sigma/\sqrt{n}}\right)$$

$$= \Phi\left(\frac{r}{\sigma/\sqrt{n}}\right) - \Phi\left(\frac{-r}{\sigma/\sqrt{n}}\right)$$

$$= \Phi\left(\frac{\sqrt{n}r}{\sigma}\right) - \Phi\left(\frac{-\sqrt{n}r}{\sigma}\right)$$

$\blacksquare$

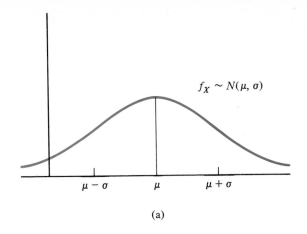

$$f_X \sim N(\mu, \sigma)$$

$\mu - \sigma \qquad \mu \qquad \mu + \sigma$

(a)

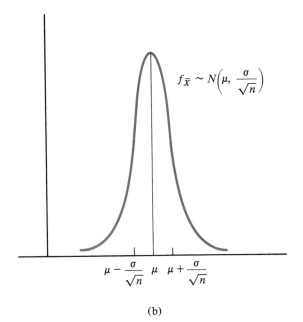

$$f_{\bar{X}} \sim N\left(\mu, \frac{\sigma}{\sqrt{n}}\right)$$

$\mu - \dfrac{\sigma}{\sqrt{n}} \quad \mu \quad \mu + \dfrac{\sigma}{\sqrt{n}}$

(b)

FIGURE 7.1
Comparison of the normal distribution and the corresponding distribution of the sample mean

Applying Thorem 7.2 and employing

$$A(n, r, \sigma) = P(|\bar{X} - \mu| < r)$$

as a measure of precision, certain conclusions can be drawn regarding the use of the sample-mean estimator in the case of normal random variables.

1. The larger the sample, the more precise the estimator: for fixed r and σ, if $n_1 < n_2$, then

$$A(n_1, r, \sigma) < A(n_2, r, \sigma)$$

2. For fixed r and σ, as n tends to infinity, the probability of $\bar{X}$ being within r of μ tends to 1:

$$\lim_{n \to \infty} A(n, r, \sigma) = 1$$

3. For fixed r and n, precision is a function of σ and is greater for smaller values of σ: if $\sigma_1 < \sigma_2$, then

$$A(n, r, \sigma_2) < A(n, r, \sigma_1)$$

Regarding the last property, if in the proof of Theorem 7.2 we do not replace the variance of $\bar{X}$ by its expression in terms of the variance of X, then the statement of the theorem takes the form

$$P(|\bar{X} - \mu| < r) = \Phi\left(\frac{r}{\sigma_{\bar{X}}}\right) - \Phi\left(\frac{-r}{\sigma_{\bar{X}}}\right)$$

and the dependence of the precision upon the variance of $\bar{X}$ is transparent. The standard deviation of the sample mean, $\sigma_{\bar{X}}$, is called the **standard error of the mean.**

EXAMPLE 7.4

One manner in which to differentiate analytically between different materials is to test for specific heat, Q, which is the number of calories required to raise the temperature of 1 gram of the material by 1 degree centigrade. Suppose nine 1-gram samples of a material are to be tested, and it is known from experience that the test measurements are normally distributed with the variance on any given test being 0.000049. From a modeling standpoint, the assumption is that variability results from the testing procedure and that this variability has remained constant over time. If the sampling process is random, then $X_1, X_2, \ldots, X_9$, comprise a random sample, where, for $k = 1, 2, \ldots, 9$, $X_k \sim N(\mu, 0.007)$. According to Theorem 7.2,

$$P(|\bar{X} - \mu| < r) = \Phi\left(\frac{3r}{0.007}\right) - \Phi\left(\frac{-3r}{0.007}\right)$$

If $r = 0.001$, then

$$P(|\bar{X} - \mu| < 0.001) = \Phi(0.43) - \Phi(-0.43) = 0.6664 - 0.3336 = 0.3328$$

If $r = 0.005$, then

$$P(|\bar{X} - \mu| < 0.005) = \Phi(2.14) - \Phi(-2.14) = 0.9838 - 0.0162 = 0.9676$$

The result of Example 7.4 for $r = 0.001$ can be heuristically interpreted in the following manner. If the experimental procedure is implemented on all possible nine-element samples from the population of all possible measurements, then approximately 33% of the resulting empirical means will fall within 0.001 of the actual mean, which is the specific heat of the material under consideration. Analogously, the second probability is interpreted to mean that approximately 97% of the resulting empirical means will fall within 0.005 of the actual mean.

Whether or not a specific empirical mean is or is not within the prescribed distance r of the mean can never be known. However, one point is certain: for a particular empirical mean $\bar{x}$ computed from a specific set of nine sample values, $\bar{x}$ is either definitely within or definitely not within the prescribed distance from the mean. For instance, suppose the observed sample values are

0.093   0.090   0.091   0.088   0.087   0.091   0.092   0.088   0.090

For these sample values, $\bar{x} = 0.090$. Whether or not $\bar{x}$ is within 0.005 of the actual mean is unknown. Should the material of interest be copper, whose specific heat is

$Q = 0.093$, then the resulting empirical mean is indeed within the desired limits. Of course, unless the experimenter already knows the material is copper, he or she cannot be certain of the actual (deterministic) accuracy of the obtained empirical mean.

Rather than proceed by computing the probability that $|\bar{X} - \mu| < r$ for a given sample size, it is wise to determine the necessary sample size according to the desired degree of precision, the assumption being that we are given both r and the desired probability, say $1 - \alpha$, that $\bar{X}$ is within r of μ. The next theorem gives the necessary sample size when X is normally distributed. The result will involve the point

$$z_{\alpha/2} = \Phi^{-1}\left(1 - \frac{\alpha}{2}\right)$$

Referring to Figure 7.2, it can be seen that $z_{\alpha/2}$ is the point on the z axis for which

$$P(Z > z_{\alpha/2}) = \frac{\alpha}{2}$$

From the symmetry of the standard normal curve, $-z_{\alpha/2} = \Phi^{-1}(\alpha/2)$ and

$$P(Z < -z_{\alpha/2}) = \frac{\alpha}{2}$$

Note also that

$$P(|Z| < z_{\alpha/2}) = 1 - \alpha$$

THEOREM 7.3.

Suppose X is normally distributed with mean μ and variance σ^2. If $\bar{X}$ is the sample mean resulting from a random sample of size n arising from X,

$$n \geq \left(\frac{\sigma}{r} z_{\alpha/2}\right)^2$$

then

$$P(|\bar{X} - \mu| < r) \geq 1 - \alpha$$

Proof: To find n, we solve the equation

$$P(-r < \bar{X} - \mu < r) \geq 1 - \alpha$$

FIGURE 7.2
Illustration of the point $z_{\alpha/2}$

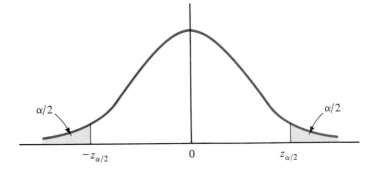

which, according to Theorem 7.2, reduces to

$$\Phi\left(\frac{\sqrt{n}r}{\sigma}\right) - \Phi\left(\frac{-\sqrt{n}r}{\sigma}\right) \geq 1 - \alpha$$

By the symmetry of the standard normal curve,

$$1 - 2\Phi\left(\frac{-\sqrt{n}r}{\sigma}\right) \geq 1 - \alpha$$

Solving for n yields

$$n \geq \left[\frac{\sigma}{r}\Phi^{-1}\left(\frac{\alpha}{2}\right)\right]^2$$

which is equivalent to the result stated in the theorem. ∎

EXAMPLE 7.5

Referring to Example 7.4, find the sample size required to ensure that

$$P(|\bar{X} - \mu| < 0.001) \geq 0.95$$

Since $\sigma/r = 7$ and (by Table A.1)

$$z_{\alpha/2} = z_{0.025} = 1.96$$

n must be chosen so that

$$n \geq [7(1.96)]^2 = 188.24$$

Thus, $n = 189$ will ensure the desired degree of precision.

7.2.2 Law of Large Numbers

Following Theorem 7.2, three conclusions were stated that could be drawn from the theorem. Of particular interest is the second, which states that

$$\lim_{n \to \infty} P(|\bar{X} - \mu| < r) = 1$$

where it must be kept in mind that $\bar{X}$ depends on the sample size n. From the standpoint of populations, the distribution of $\bar{X}$ represents the population of all possible averages

$$\frac{x_1 + x_2 + \cdots + x_n}{n}$$

where $x_1, x_2, \ldots, x_n$ represent possible sample values. In other words, the sample mean is a random variable representing the arithmetic (empirical) mean of repeated observations of the population random variable X. Thus, the preceding limit states that, as n increases, the arithmetic mean of the observations stabilizes near (within r of) the expected value of the population random vaiable. Since r is an arbitrary positive value, the mean of the observations will, with probability near 1, stabilize within any predetermined small neighborhood of the mean of X as n tends to infinity. Although Theorem 7.2 has been proven for the case in which X is normal, this stabilizing property of the sample mean can be shown to hold whenever X possesses a mean and variance. The property is called the **law of large numbers.**

THEOREM 7.4

(Law of Large Numbers). If X possesses mean μ and variance σ^2 and if $\bar{X}$ denotes the sample mean resulting from a random sample of size n arising from X, then, for any $r > 0$,

$$\lim_{n \to \infty} P(|\bar{X} - \mu| < r) = 1$$

Proof: According to Chebyshev's inequality (Theorem 3.14),

$$P(|\bar{X} - \mu_{\bar{X}}| < r) \geq 1 - \frac{\sigma_{\bar{X}}^2}{r^2}$$

Applying Theorem 7.1 yields

$$P(|\bar{X} - \mu| < r) \geq 1 - \frac{\sigma^2}{nr^2}$$

The theorem follows since

$$\lim_{n \to \infty} \frac{\sigma^2}{nr^2} = 0$$

∎

In the literature, Theorem 7.4 is known as the **weak law of large numbers.** There is a stronger form of the law, which is beyond the scope of the present text.

Theorem 7.4 has a particularly interesting interpretation if X represents the binomial random variable associated with a single trial in which the probability of a success is p. If $X_1, X_2, \ldots, X_n$ constitute a random sample arising from X, then $\bar{X}$ is the ratio of the total number of successes on n trials to the total number of trials and, according to Theorem 7.4, this ratio tends to stabilize around the mean of X, which is p. Thus, the probability of a success on a given trial is approximated (in a certain sense) by the ratio of successes to trials. Of course, the actual ratio of successes to observations recorded on any given sequence of trials might differ markedly from p. It is in the sense of the probabilistic measure of precision $P(|\bar{X} - \mu| < r)$ that the law of large numbers must be interpreted.

The limiting property exhibited by the sample mean in Theorem 7.4 is exhibited by other estimators. This property is formalized in the following definition, in which one should keep in mind that an estimator is indexed by the sample size.

DEFINITION 7.6

Consistent Estimator. An estimator $\hat{\theta}$ of θ is said to be consistent if, for any $r > 0$,

$$\lim_{n \to \infty} P(|\hat{\theta} - \theta| < r) = 1$$

Theorem 7.4 states that the sample mean is a consistent estimator of the mean. An examination of the proof of Theorem 7.4 shows it to depend only on the unbiasedness of the sample-mean estimator and the fact that Var $[\bar{X}]$ tends to 0 as n approaches infinity. Thus, we have the following sufficient (but not necessary) conditions for consistency.

THEOREM 7.5

If $\hat{\theta}$ is an unbiased estimator of θ for which

$$\lim_{n \to \infty} \text{Var } [\hat{\theta}] = 0$$

then $\hat{\theta}$ is a consistent estimator of θ.

7.2.3 Central Limit Theorem

In the event the random variable of interest is not normally distributed, Theorem 7.2 does not apply. Nevertheless, it is a remarkable fact that, even though the population random variable X is not normally distributed, the standardized sample mean will still possess an approximately standard normal distribution so long as n is sufficiently large. Consequently, since the proof of Theorem 7.2 depends only on the normality of $(\bar{X} - \mu_{\bar{X}})/\sigma_{\bar{X}}$ and Theorem 7.1, which applies whether or not X is normal, the methodology of Theorem 7.2 can be applied (in an approximate sense) for large n. The next theorem, known as the **central limit theorem** (or **Ljapunov's theorem**) provides the theoretical underpinnings for this so-called **large-sample** approach. The theorem is stated without proof.

THEOREM 7.6

(Central Limit Theorem). If $X_1, X_2, \ldots, X_n$ comprise a random sample arising from the random variable X possessing mean μ and variance σ^2, then, as n tends to infinity, the probability distribution functions of the standardized sample means

$$\bar{X}* = \frac{\bar{X} - \mu_{\bar{X}}}{\sigma_{\bar{X}}} = \frac{\bar{X} - \mu}{\sigma/\sqrt{n}}$$

approach the probability distribution function Φ of the standard normal variable.

In essence, Theorem 7.6 gives the approximation $\bar{X}* \approx Z$ for large n, where Z is the standard normal variable. In terms of the sample mean itself,

$$\bar{X} \approx \sigma_{\bar{X}} Z + \mu_{\bar{X}} = \frac{\sigma}{\sqrt{n}} Z + \mu$$

so that $\bar{X}$ is approximately normally distributed with mean μ and standard deviation $\sigma/\sqrt{n}$.

The probabilistic interpretation of Theorem 7.6 is essentially the same as the interpretation of Theorem 4.9 that was extensively discussed in Section 4.5.3. Specifically, for large n,

$$P\left(a < \frac{\bar{X} - \mu_{\bar{X}}}{\sigma_{\bar{X}}} < b\right) \approx P(a < Z < b)$$

In particular, the third equality in the proof of Theorem 7.2 becomes an approximation, and the conclusion of the theorem becomes

$$P(|\bar{X} - \mu| < r) \approx \Phi\left(\frac{\sqrt{n}r}{\sigma}\right) - \Phi\left(\frac{-\sqrt{n}r}{\sigma}\right)$$

As with Theorem 4.9, practical usage of Theorem 7.6 requires some rule of thumb regarding the necessary size of the random sample. In general, the closer the underlying random variable is to normality, the faster $\bar{X}^*$ approaches Z; however, if $n \geq 30$, one can usually assume the approximation to be quite good.

EXAMPLE 7.6

An industrial engineer wishes to determine the mean performance time of a group of workers performing a certain task and wishes to know the probability that the sample mean will be within 2 minutes of the actual mean when 50 observations are taken. It has been determined that 25 is a reasonable estimate of the variance. Although the population random variable is believed to be nonnormal, the sample size is greater than 30, and therefore Theorem 7.6 justifies the approximate use of Theorem 7.2. Thus,

$$P(|\bar{X} - \mu| < 2) \approx \Phi\left(\frac{2\sqrt{50}}{5}\right) - \Phi\left(\frac{-2\sqrt{50}}{5}\right) = 0.9977 - 0.0023 = 0.9954$$

It should be noted that Theorem 4.9 is actually a special case of Theorem 7.6. To see this, suppose X is binomially distributed with density $b(x; 1, p)$ and suppose $X_1, X_2, \ldots, X_n$ are independent and identically distributed with X. Then, according to the results of Example 5.18,

$$Y = X_1 + X_2 + \cdots + X_n$$

is binomially distributed with density $b(x; n, p)$. By Theorem 4.1, $\mu_Y = np$ and $\sigma_Y = \sqrt{npq}$. Since $Y = n\bar{X}$, Theorem 7.6 states that

$$\frac{\bar{X} - \mu_{\bar{X}}}{\sigma_{\bar{X}}} = \frac{\dfrac{Y}{n} - \dfrac{\mu_Y}{n}}{\dfrac{\sigma_Y}{n}} = \frac{Y - np}{\sqrt{npq}}$$

approaches the standard normal variable as $n \longrightarrow \infty$, which is precisely the conclusion of Theorem 4.9.

The central limit theorem is often used to explain the wide applicability of the normal variable in modeling physical phenomena. To see why, we need to give a more general form of the theorem. Suppose $X_1, X_2, \ldots$ form an infinite sequence of independent random variables with respective means $\mu_1, \mu_2, \ldots$ and respective standard deviations $\sigma_1, \sigma_2, \ldots$. For $n = 1, 2, \ldots$, let

$$Y_n = X_1 + X_2 + \cdots + X_n$$

By the linearity of the expected-value operator,

$$\mu_{Y_n} = \mu_1 + \mu_2 + \cdots + \mu_n$$

and, by Theorem 5.17,

$$\sigma_{Y_n}^2 = \sigma_1^2 + \sigma_2^2 + \cdots + \sigma_n^2$$

Now, let

$$Y_n^* = \frac{Y_n - \mu_{Y_n}}{\sigma_{Y_n}}$$

be the standardized version of Y_n. It can be shown under rather general conditions that the sequence of variables Y_n satisfies the conclusion of the central limit theorem, namely, that as n tends to infinity, the standardized variables Y_n^* approach the standard normal variable Z. In particular, it can be shown that if the values of X_n lie in a fixed bounded interval for all n and if

$$\lim_{n \to \infty} \sigma_{Y_n}^2 = \infty$$

then Y_n^* approaches Z as n approaches infinity. Consequently, for sufficiently large n, $Y_n^* \approx Z$ and

$$Y_n \approx \sigma_{Y_n} Z + \mu_{Y_n}$$

so that Y_n^* is approximately standard normally distributed and Y_n is approximately normally distributed.

A phenomena-based interpretation of the latter form of the central limit theorem is most enlightening insofar as the widespread occurrence of normally distributed random variables is concerned. Specifically, if a macrophenomenon is additively comprised of a large number of independent infinitesimal random microphenomena, then the macrophenomenon will possess an approximately normal distribution. Since many macrophenomena are so comprised, in that they apparently result from an infinite variety of causes, the central limit theorem makes the frequent occurrence of the normal variable seem reasonable. Of course, we must be careful to recognize that the central limit theorem is a theorem of mathematics, and therefore its relationship to material behavior cannot be thoroughly explained solely within the realms of either mathematics or science. Nevertheless, there is certainly heuristic justification for utilizing the normal variable when modeling a macrophenomenon known to result from a very large number of additive microphenomena.

7.3 SAMPLE VARIANCE

The most commonly employed estimator of the variance is known as the **sample variance.** Just as the distribution of the sample mean is determined by the population of all empirical means, the population of all empirical variances determines the distribution of the sample variance.

7.3.1 Estimation Using the Sample Variance

Based upon the supposition that the sample variance arises from the population of empirical variances, Definition 1.5 leads to the following definition of the sample variance.

DEFINITION 7.7

Sample Variance. If the random variables $X_1, X_2, \ldots, X_n$ constitute a random sample of size n arising from the population random variable X, then the statistic

$$S^2 = \frac{1}{n-1} \sum_{k=1}^{n} (X_k - \bar{X})^2$$

is called the sample variance of the random sample. Moreover, $S = \sqrt{S^2}$ is called the **sample standard deviation.**

It was the sample variance that was employed as the estimator of the variance in Example 7.2.

Essentially, S^2 is the average of the squared deviations of the sample variables from the sample mean; however, instead of being divided by n, the sum is divided by $n - 1$. Theorem 7.8 will make clear the reason for the $n - 1$ divisor: it results in S^2 being an unbiased estimator of the variance of the population random variable. Before proceeding to Theorem 7.8, we give a useful identity.

THEOREM 7.7

If $X_1, X_2, \ldots, X_n$ comprise a random sample of size n arising from the population random variable X with mean μ, then

$$\sum_{k=1}^{n} (X_k - \bar{X})^2 = \sum_{k=1}^{n} (X_k - \mu)^2 - n(\bar{X} - \mu)^2$$

Proof: A bit of algebraic manipulation yields

$$\sum_{k=1}^{n} (X_k - \bar{X})^2 = \sum_{k=1}^{n} [(X_k - \mu) - (\bar{X} - \mu)]^2$$

$$= \sum_{k=1}^{n} (X_k - \mu)^2 - 2(\bar{X} - \mu) \sum_{k=1}^{n} (X_k - \mu) + n(\bar{X} - \mu)^2$$

$$= \sum_{k=1}^{n} (X_k - \mu)^2 - 2(\bar{X} - \mu)n\bar{X} + 2n\mu(\bar{X} - \mu) + n(\bar{X} - \mu)^2$$

$$= \sum_{k=1}^{n} (X_k - \mu)^2 - n(\bar{X} - \mu)^2 \qquad \blacksquare$$

THEOREM 7.8

So long as X possesses a mean and variance, the sample variance S^2 is an unbiased estimator of the variance.

Proof: Let the underlying random variable X possess mean μ and variance σ^2 and let $X_1, X_2, \ldots, X_n$ be a random sample. Then

$$E[S^2] = E\left[\frac{1}{n-1} \sum_{k=1}^{n} (X_k - \bar{X})^2 \right]$$

$$= \frac{1}{n-1} \left(\sum_{k=1}^{n} E[(X_k - \mu)^2] - nE[(\bar{X} - \mu)^2] \right)$$

$$= \frac{1}{n-1} \left(\sum_{k=1}^{n} \text{Var}[X_k] - n\,\text{Var}[\bar{X}] \right)$$

$$= \frac{1}{n-1} (n\sigma^2 - \sigma^2) = \sigma^2$$

where we have employed, in order, the definition of S^2, Theorem 7.7 in conjunction with the linearity of the expected value, the assumption that X_k is identically distributed with X together with the fact that $\mu_{\bar{X}} = \mu$, the identity $\text{Var}[X] = n\,\text{Var}[\bar{X}]$, and straightforward factoring. ∎

Had S^2 been defined by dividing by n instead of $n-1$, the result would have been a biased estimator of the variance. Specifically,

$$E\left[\frac{1}{n}\sum_{k=1}^{n}(X_k - \bar{X})^2\right] = E\left[\frac{n-1}{n}S^2\right] = \frac{n-1}{n}E[S^2] = \frac{n-1}{n}\sigma^2$$

For large n, the bias is not significant; indeed, as n tends to infinity, the bias vanishes:

$$\lim_{n\to\infty} E\left[\frac{n-1}{n}S^2\right] = \lim_{n\to\infty}\frac{n-1}{n}\sigma^2 = \sigma^2$$

DEFINITION 7.8

Asymptotically Unbiased. An estimator $\hat{\theta}$ of θ is said to be asymptotically unbiased if

$$\lim_{n\to\infty} E[\hat{\theta}] = \theta$$

According to Definition 7.8, which says that the bias tends to zero as the sample size increases, the estimator of the variance obtained by dividing by n instead of $n-1$ is asymptotically unbiased.

In the mean-estimation methodology of Section 7.2, the variance of the distribution was assumed to be known. Because of the central limit theorem, so long as the sample is large, there need be no assumption of normality to (approximately) apply Theorem 7.2; however, the use of Theorem 7.2 to measure the precision of the estimator does depend on a knowledge of the variance of the population random variable, a condition that is often not satisfied in practice. Fortunately, if n is large, S^2 provides a decent estimator of the variance. Consequently, for large samples, neither normality nor knowledge of the variance need be assumed to apply, in an approximate sense, the methodology surrounding Theorem 7.2.

EXAMPLE 7.7

To find the mean speed of cars passing a particular point on a bridge, a highway engineer clocks the speeds of 100 randomly selected cars over a period of one week. The variance is unknown before the experiment, and it is noted that the frequency histogram does not appear normal; nevertheless, the large size of the sample warrants the use of both the central limit theorem and the empirical variance s^2 resulting from the sample. Thus, applying Theorem 7.2 yields

$$P(|\bar{X} - \mu| < r) \approx \Phi\left(\frac{\sqrt{n}r}{s}\right) - \Phi\left(\frac{-\sqrt{n}r}{s}\right)$$

where s is the empirical standard deviation. Assuming $r = 1$ and the 100 measurements possess standard deviation $s = 5.3$,

$$P(|\bar{X} - \mu| < 1) \approx \Phi(1.89) - \Phi(-1.89) = 0.9706 - 0.0294 = 0.9412$$

Now, suppose it is desired that

$$P(|\bar{X} - \mu| < 0.4) \geq 0.95$$

To determine n approximately using the approach of Example 7.5, it is necessary to estimate σ, but the estimate $s = 5.3$ is already in hand. Thus,

$$n \geq \left(\frac{5.3}{0.4} z_{0.025}\right)^2 = (13.25 \times 1.96)^2 = 674.44$$

A sample size of $n = 675$ will do.

It is important to note the difference between Example 7.5 and Example 7.7. In the former, σ^2 was known, whereas in the latter, a particular estimate s^2 of σ^2 was employed. Should s^2 be fairly close to the true variance, then the derived value of n will be approximately correct; if, on the other hand, despite the use of a large sample of 100, s^2 is somewhat distant from the true variance, then the result will be misleading. Unfortunately, there is no way to determine whether or not any particular estimate is satisfactory.

7.3.2 Variance of the Sample Variance

When the population random variable X is normally distributed, the variance of the sample variance can be found by employing the distribution of the random variable

$$\frac{n-1}{\sigma^2} S^2 = \frac{1}{\sigma^2} \sum_{k=1}^{n} (X_k - \bar{X})^2$$

where $X_1, X_2, \ldots, X_n$ comprise a random sample. This particular sampling distribution plays important roles in a number of statistical problems.

THEOREM 7.9

Let $\bar{X}$ and S^2 be the sample mean and sample variance resulting from a random sample of size n arising from a normal random variable X possessing mean μ and variance σ^2. Then

(i) $\bar{X}$ and S^2 are independent.
(ii) The random variable

$$\frac{n-1}{\sigma^2} S^2$$

possesses a chi-square distribution with $n - 1$ degrees of freedom.

Proof: The proof that $\bar{X}$ and S^2 are independent is beyond the level of the current text and will not be demonstrated. As for (ii), rearranging the terms in the identity of Theorem 7.7 and dividing by σ^2 yields

$$\sum_{k=1}^{n} \left(\frac{X_k - \mu}{\sigma}\right)^2 = \frac{n-1}{\sigma^2} S^2 + \left(\frac{\bar{X} - \mu}{\sigma/\sqrt{n}}\right)^2$$

According to Theorem 5.8, the sum on the left-hand side of the preceding equality possesses a chi-square distribution with n degrees of freedom. As demonstrated in

Example 3.38, the square of a standard normal variable possesses a chi-square distribution with 1 degree of freedom. Since $\bar{X}$ and S^2 are independent, the preceding equation takes the form

$$W = U + V$$

where U and V are independent and where W and V are chi-square distributed with n and 1 degrees of freedom, respectively. According to Theorem 5.7, applied to the chi-square case (see the comments following the proof of the theorem), the sum of independent chi-square variables is a chi-square variable and the degrees of freedom of the summands are added to obtain the degrees of freedom of the resulting chi-square variable. Thus, it certainly seems plausible that, in the sum $W = U + V$, U must be a chi-square variable with $n - 1$ degrees of freedom. A rigorous proof of this fact is outside the limits of the present text, and thus we will content ourselves with this heuristic argument. ∎

The variance of the sample variance arising from a normally distributed random variable can be derived from Theorem 7.9.

THEOREM 7.10

If X is normally distributed with mean μ and variance σ^2 and if S^2 is the sample variance corresponding to a random sample of size n derived from X, then

$$\text{Var}\,[S^2] = \frac{2\sigma^4}{n - 1}$$

Proof: Letting W denote the random variable of Theorem 7.9(ii),

$$\text{Var}\,[S^2] = \text{Var}\left[\frac{\sigma^2}{n - 1}\,W\right] = \frac{\sigma^4}{(n - 1)^2}\,\text{Var}\,[W]$$

which reduces to the result claimed in the statement of the theorem since, according to Theorems 4.13 and 7.9,

$$\text{Var}\,[W] = 2(n - 1) \qquad \blacksquare$$

In the case of a normal population random variable, Theorem 7.10 shows that

$$\lim_{n \to \infty} \text{Var}\,[S^2] = 0$$

Since S^2 is an unbiased estimator, Theorem 7.5 shows S^2 to be a consistent estimator.

The application of Theorem 7.9 requires the evaluation of probabilities involving the chi-square distribution. Table A.5 provides these probabilities. The probabilities in the table correspond to Figure 7.3, where

$$P(W > \chi^2_{\alpha,\nu}) = \alpha$$

and W is a chi-square variable with ν degrees of freedom. For instance, if W possesses $\nu = 10$ degrees of freedom, then

$$P(W > 3.940) = 0.95$$

and

$$P(W > 18.307) = 0.05$$

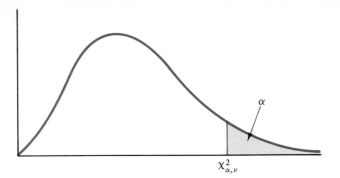

FIGURE 7.3
Illustration of the point
$\chi^2_{\alpha,\nu}$

$\chi^2_{\alpha,\nu}$

EXAMPLE 7.8

Consider a normally distributed random variable with variance σ^2. We wish to investigate the precision of the sample variance as an estimator of the variance when the sample size is 30. One possible measure of precision is the probability that the sample variance is within 20% of the true variance:

$$P[(0.8)\sigma^2 < S^2 < (1.2)\sigma^2]$$

Multiplying the terms of the inequality by $(n - 1)/\sigma^2$ yields

$$P[(n - 1)(0.8) < W < (n - 1)(1.2)]$$

where W is the chi-square statistic of Theorem 7.9(ii). Since $n = 30$, W possesses 29 degrees of freedom. Therefore (from Table A.5) the desired probability is

$$P(23.2 < W < 34.8) = P(W > 23.2) - P(W > 34.8)$$

$$= 0.77 - 0.21 = 0.56$$

The probability that S^2 is within 20% of the true variance is only 0.56 for a sample size of 30.

Although S^2 is a consistent estimator of σ^2, it is clear from Example 7.8 that there is a fair likelihood that a sample size of 30 will not produce a satisfactory estimate of the variance. On the other hand, a sample size of 30 is usually sufficient for the employment of the central limit theorem, and therefore such a sample size is often considered sufficient for large-sample estimation of the mean. When such estimation involves the use of the sample variance in place of σ^2, it is wise to employ a sample size larger than 30 in order to improve the precision of the variance estimation.

7.4 ORDER STATISTICS

There are numerous applications where the sample values resulting from a random sample are ordered from least to greatest, with a particular value in the ordering being of interest. For instance, one might be interested in the largest of the values, the smallest, or the one in the middle (the median if the sample size is odd).

DEFINITION 7.9

Order Statistics. If $X_1, X_2, \ldots, X_n$ comprise a random sample arising from the population random variable X, then the n order statistics corresponding to the random sample are the n random variables

$$Y_1 \leq Y_2 \leq \cdots \leq Y_n$$

which result from ordering the values of the sample variables from lowest to highest.

It is important to note that each order statistic can be expressed as a function of the sample variables. For instance,

$$Y_1 = \min \{X_1, X_2, \ldots, X_n\}$$

and

$$Y_n = \max \{X_1, X_2, \ldots, X_n\}$$

A number of statistics are defined in terms of the order statistics of a random sample. Using the notation of Definition 7.9, the **sample median** is

$$\tilde{X} = \begin{cases} Y_{(n+1)/2}, & \text{if } n \text{ is odd} \\ \dfrac{1}{2}(Y_{n/2} + Y_{(n/2)+1}), & \text{if } n \text{ is even} \end{cases}$$

From the perspective of populations, $\tilde{X}$ is the distribution of the population of all empirical medians resulting from the possible sample values of the random sample. It is sometimes employed as an estimator of the mean.

Another statistic defined in terms of order statistics is the **range,** which is simply the highest order statistic minus the smallest order statistic, $Y_n - Y_1$. Although it suffers because it uses only extreme values, the range is sometimes employed as a measure of dispersion in situations where there is a desire to avoid the time-consuming calculations involved in obtaining the sample standard deviation (see Sections 8.10.1 and 8.10.2).

An estimator of the mean that employs only the extreme order statistics is the **midrange,**

$$\frac{Y_1 + Y_n}{2}$$

Example 7.12 discusses a situation in which the midrange is a better estimator of the mean than the sample mean.

THEOREM 7.11

Suppose $X_1, X_2, \ldots, X_n$ is a random sample arising from the continuous random variable X and

$$Y_1 \leq Y_2 \leq \cdots \leq Y_n$$

are the n order statistics resulting from the sample. Then, for $k = 1, 2, \ldots, n$,

$$f_{Y_k}(y) = \frac{n!}{(n-k)!(k-1)!} F_X(y)^{k-1}[1 - F_X(y)]^{n-k}f_X(y)$$

In particular,

$$f_{Y_1}(y) = n[1 - F_X(y)]^{n-1}f_X(y)$$

$$f_{Y_n}(y) = nF_X(y)^{n-1}f_X(y)$$

and, if n is odd, the density for the sample median is given by

$$f_{\tilde{X}}(y) = \frac{n!}{\left[\left(\dfrac{n-1}{2}\right)!\right]^2} F_X(y)^{(n-1)/2}[1 - F_X(y)]^{(n-1)/2} f_X(y)$$

Proof: We prove only the cases for 1 and n. Since the value of Y_1 is the minimum of the sample values, its probability distribution function is given by

$$F_{Y_1}(y) = P(Y_1 \le y)$$

$$= P(\min \{X_1, X_2, \ldots, X_n\} \le y)$$

$$= P(X_1 \le y \quad \text{or} \quad X_2 \le y \quad \text{or} \quad \ldots \quad \text{or} \quad X_n \le y)$$

$$= 1 - P(X_1 > y, X_2 > y, \ldots, X_n > y)$$

$$= 1 - P(X > y)^n$$

$$= 1 - [1 - F_X(y)]^n$$

where we have used, in order, the definition of the probability distribution function, the definition of Y_1, the fact that the minimum is less than or equal to y if and only if at least one of the sample variables is less than or equal to y, Theorem 2.7, the independence of the sample variables in conjunction with the fact that X_k is identically distributed with X, and the definition of F_X. Differentiation with respect to y gives the density as claimed in the statement of the theorem.

As for Y_n, analogous reasoning shows that

$$F_{Y_n}(y) = P(Y_n \le y)$$

$$= P(\max \{X_1, X_2, \ldots, X_n\} \le y)$$

$$= P(X_1 \le y, X_2 \le y, \ldots, X_n \le y)$$

$$= P(X \le y)^n$$

$$= F_X(y)^n$$

Taking the derivative of the probability distribution function with respect to y yields the density of Y_n. ∎

EXAMPLE 7.9

To test the time to failure of a vacuum pump, a manufacturer randomly selects 5 pumps and runs them continuously until failure. Suppose the assumption prior to testing is that the time to failure is exponentially distributed with the mean time to failure being a third of a year. Moreover, suppose the population from which the pumps are chosen is sufficiently large to warrant the assumption that the sampling procedure is random in the sense of Definition 7.1, even though sampling is done without replacement. The result of the testing will be 5 times to failure. If placed in order from smallest to largest, these times to failure will represent the values of the 5 order statistics Y_1 through Y_5. Since X possesses an exponential distribution with parameter $b = 3$, Theorem 7.11 shows the density of the sample median $\tilde{X} = Y_3$ to be

$$f_{\tilde{X}}(y) = \frac{5!}{2!2!} 3(1 - e^{-3y})^2 e^{-9y}$$

for $y \geq 0$ and zero elsewhere. Thus, the probability distribution function of $\tilde{X}$ is

$$
\begin{aligned}
F_{\tilde{X}}(y) &= 90 \int_0^y (e^{-9t} - 2e^{-12t} + e^{-15t})\, dt \\
&= e^{-9y}(-10 + 15e^{-3y} - 6e^{-6y}) + 1
\end{aligned}
$$

for $y \geq 0$. For instance, the probability that at least 3 pumps fail within 2 months (1/6 of a year) is

$$
F_{\tilde{X}}\left(\frac{1}{6}\right) = e^{-1.5}(-10 + 15e^{-0.5} - 6e^{-1}) + 1 = 0.31
$$

The density for the minimum order statistic is

$$
f_{Y_1}(y) = 15e^{-15y}
$$

for $y \geq 0$, and therefore the probability that no pump fails during the first month is

$$
P\left(Y_1 > \frac{1}{12}\right) = \int_{1/12}^{\infty} f_{Y_1}(y)\, dy = 0.29
$$

7.5 MINIMUM-VARIANCE UNBIASED ESTIMATORS

Given an estimator $\hat{\theta}$ of a parameter θ, we need a measure of the **goodness** of the estimator. Since $\hat{\theta}$ is a random variable, it is impossible to measure a numerical error $|\hat{\theta} - \theta|$. Thus, the expected value of the square of the error is often employed as a measure of goodness.

DEFINITION 7.10   **Mean-Square Error of a Point Estimator.** If the statistic $\hat{\theta}$ is employed as a point estimator of the parameter θ, then the mean-square error associated with $\hat{\theta}$ is

$$
E[(\hat{\theta} - \theta)^2]
$$

Since the expected value is a linear operator,

$$
\begin{aligned}
E[(\hat{\theta} - \theta)^2] &= E[\hat{\theta}^2] - 2E[\hat{\theta}\theta] + E[\theta^2] \\
&= E[\hat{\theta}^2] - E[\hat{\theta}]^2 + E[\hat{\theta}]^2 - 2\theta E[\hat{\theta}] + \theta^2 \\
&= \text{Var}\,[\hat{\theta}] + (E[\hat{\theta}] - \theta)^2
\end{aligned}
$$

so that the mean-square error is equal to the variance of $\hat{\theta}$ plus the square of the bias resulting from utilizing $\hat{\theta}$ as an estimator of θ. If $\hat{\theta}$ is unbiased, then the mean-square error reduces to the variance of $\hat{\theta}$.

Given two estimators of θ, say $\hat{\theta}_1$ and $\hat{\theta}_2$, $\hat{\theta}_1$ is called a **better estimator,** or **more efficient estimator,** than $\hat{\theta}_2$ if

$$
E[(\hat{\theta}_1 - \theta)^2] < E[(\hat{\theta}_2 - \theta)^2]
$$

If attention is restricted to unbiased estimators, then $\hat{\theta}_1$ is better than $\hat{\theta}_2$ if and only if

$$\text{Var}[\hat{\theta}_1] < \text{Var}[\hat{\theta}_2]$$

A natural question arises: Among all unbiased estimators of a parameter θ, is it possible to find a "best" estimator? The next definition provides a rigorous definition of the most commonly employed criterion of bestness.

DEFINITION 7.11

Minimum-Variance Unbiased Estimator (MVUE). An unbiased estimator $\hat{\theta}$ is said to be a minimum-variance unbiased estimator of θ if, for any unbiased estimator $\hat{\hat{\theta}}$ of θ,

$$\text{Var}[\hat{\theta}] \leq \text{Var}[\hat{\hat{\theta}}]$$

$\hat{\theta}$ is also called a **best unbiased estimator** of θ.

Even if a minimum-variance unbiased estimator cannot be found, it might be possible to find a best unbiased estimator among all estimators in some restricted class of unbiased estimators of θ. For instance, if $\{\hat{\theta}_1, \hat{\theta}_2, \ldots\}$ is a collection of unbiased estimators of θ and, for $k = 1, 2, \ldots$,

$$\text{Var}[\hat{\theta}_1] \leq \text{Var}[\hat{\theta}_k]$$

then $\hat{\theta}_1$ is a best unbiased estimator of θ relative to the class under consideration.

EXAMPLE 7.10

Given a random sample $X_1, X_2, \ldots, X_n$ arising from a random variable X with mean μ, Theorem 7.1 shows that, among convex combinations of the X_k, the sample mean possesses minimum variance. Thus, relative to the class of convex combinations, $\bar{X}$ is the best unbiased estimator of the mean. We can actually go a bit further by considering the class of all **unbiased linear estimators** of the mean, these being of the form

$$\hat{\mu} = a_1 X_1 + a_2 X_2 + \cdots + a_n X_n$$

where $E[\hat{\mu}] = \mu$. By the linearity of E,

$$E[\hat{\mu}] = (a_1 + a_2 + \cdots + a_n)\mu$$

so that unbiasedness implies that

$$a_1 + a_2 + \cdots + a_n = 1$$

Thus, the only extra generality over the class of convex-combination estimators is that the a_k need not be between 0 and 1. Consequently, the differential calculus argument that led to Theorem 7.1 applies directly in the case of unbiased linear estimators, and therefore the sample mean is a **best linear unbiased estimator (BLUE)** of the mean. For future reference, note that no assumption has been made regarding the distribution of X.

Finding an MVUE is generally difficult; however, checking to see whether or not a particular unbiased estimator possesses minimum variance can often be accomplished with the aid of the next theorem, which gives a lower bound on the variances of all possible unbiased estimators of a given parameter. Given an estimator whose variance is equal to the lower bound provided in the statement of the theorem, that estimator must be an MVUE. Before stating the theorem, we point out that it holds under rather general conditions on the density of the population random variable X; however, a precise formulation of the conditions is beyond the level of the present text, and so these conditions are omitted here.

THEOREM 7.12

(Cramer-Rao Inequality). If $\hat{\theta}$ is an unbiased estimator of the parameter θ in the density $f(x; \theta)$ of the random variable X, then

$$\text{Var}\,[\hat{\theta}] \geq \frac{1}{nE\left[\left(\dfrac{\partial}{\partial\theta}\log f(X;\theta)\right)^2\right]}$$

where n is the size of the random sample.

EXAMPLE 7.11

Using Theorem 7.12, we will show that the sample mean is an MVUE for the mean of a normally distributed random variable. Let X be normally distributed with mean μ and variance σ^2. Since

$$f(x; \mu) = \frac{1}{\sqrt{2\pi}\,\sigma}\,e^{-\frac{1}{2}\left(\frac{x-\mu}{\sigma}\right)^2}$$

it is immediate that

$$\log f(x; \mu) = -\log(\sigma\sqrt{2\pi}) - \frac{1}{2}\left(\frac{x-\mu}{\sigma}\right)^2$$

Taking the partial derivative with respect to the parameter μ yields

$$\frac{\partial}{\partial\mu}\log f(x; \mu) = \frac{x-\mu}{\sigma^2}$$

Thus,

$$E\left[\left(\frac{\partial}{\partial\mu}\log f(X; \mu)\right)^2\right] = \sigma^{-4}E[(X-\mu)^2] = \sigma^{-2}$$

Consequently, the lower bound in the Cramer-Rao inequality is σ^2/n. Since all unbiased estimators of the mean must possess variances at least as large as σ^2/n, and since, according to Theorem 7.1, the sample mean possesses precisely this variance, the sample mean must be a minimum-variance unbiased estimator of the mean of a normal distribution.

There is a distinct difference between the result of Example 7.10 and that of Example 7.11. In Example 7.10, it was demonstrated that the sample mean is the BLUE of the mean for any random variable possessing a mean and variance; in Example 7.11, it was shown that the sample mean is an MVUE of the mean for a particular distribution. Thus, in the latter example the BLUE is also an MVUE. In general, since an MVUE possesses minimum variance over all unbiased estimators and a BLUE possesses minimum variance only over the class of all linear unbiased estimators, if $\hat{\theta}_1$ and $\hat{\theta}_2$ are an MVUE and a BLUE, respectively, then

$$\text{Var}\,[\hat{\theta}_1] \leq \text{Var}\,[\hat{\theta}_2]$$

It is certainly possible to have strict inequality. In particular, as illustrated in the next example, the sample mean is not always an MVUE.

EXAMPLE 7.12

Suppose the errors in a measurement process are uniformly distributed between $-1/2$ and $1/2$ grams. The problem of estimating the mass of a particular specimen is one of taking a random sample of weighings, defining a statistic on the sample, and then using the statistic as an estimator of the mean μ in the uniform density

$$f(x;\,\mu) = \begin{cases} 1, & \text{if } \mu - \dfrac{1}{2} \leq x \leq \mu + \dfrac{1}{2} \\ 0, & \text{otherwise} \end{cases}$$

For a sample size of 3, one possible unbiased estimator of μ is the sample mean

$$\bar{X} = \frac{1}{3}(X_1 + X_2 + X_3)$$

the variance of which is

$$\text{Var}\,[\bar{X}] = \frac{\text{Var}\,[X]}{3} = \frac{1}{36}$$

since, as seen in Example 3.32, Var $[X] = 1/12$.

 Letting

$$Y_1 \leq Y_2 \leq Y_3$$

denote the order statistics resulting from the random sample, a different estimator is the midrange

$$\hat{\mu} = \frac{1}{2}(Y_1 + Y_3)$$

For instance, if three weighings are taken and the resulting values are $x_1 = 177.24$, $x_2 = 177.15$, and $x_3 = 177.87$ grams, then the corresponding ordered listing of the outcomes is given by

$$y_1 = 177.15 < y_2 = 177.24 < y_3 = 177.87$$

and the corresponding midrange estimate is

$$\overset{*}{\mu} = \frac{1}{2}(177.15 + 177.87) = 177.51$$

Although we will not go through the details, it can be shown that $\hat{\mu}$ is an unbiased estimator of μ and that

$$\text{Var}\,[\hat{\mu}] = \frac{1}{40} < \frac{1}{36} = \text{Var}\,[\bar{X}]$$

Thus, $\hat{\mu}$ is a better unbiased estimator than the sample mean.

In reference to Example 7.12, if measurement error is normally distributed and an unbiased estimator is desired, then, as demonstrated in Example 7.11, the sample mean is the best choice; however, if the error is uniformly distributed, then the midrange is a better choice than the sample mean.

The next example discusses a situation in which one might choose not to employ an MVUE even though one exists that takes a particularly simple mathematical form.

EXAMPLE 7.13

Suppose the time-to-failure distribution of some population of devices is exponentially distributed but the mean is unknown. One obvious way to estimate the mean is to take a random sample $X_1, X_2, \ldots, X_n$ and utilize the sample mean, which is an MVUE (see Exercise 7.32). A salient problem with this approach is that it requires the investigator to wait until all n devices have failed. If one or two devices continue to function satisfactorily for a long period of time, there will likely be excessive costs in both labor expenses and delay. The problem is particularly acute if operational expenses are high, as they often are in the testing of complex systems. One way to cut these costs is to **truncate** the test procedure after the observation of m failures, where m is a fixed number of failures chosen beforehand. The technique is known as **censored sampling.** Under censored sampling, a mean estimator is defined by

$$\hat{\mu} = \frac{T}{m}$$

where T is the total accumulated lifetime of the devices up to the termination of the experiment, which occurs at the observation of the mth failure. Thus,

$$T = Y_1 + Y_2 + \cdots + Y_m + (n - m)Y_m$$

where $Y_1, Y_2, \ldots, Y_m$ are the first m order statistics. For instance, if 10 devices are tested and it has been determined beforehand to terminate the observations after the fifth failure, and if the observed failure times are 276, 288, 302, 368, and 400 hours, then the resulting censored estimate of the mean is

$$\overset{*}{\mu} = \frac{276 + 288 + 302 + 368 + 400 + (5)(400)}{10} = 363.4$$

hours. We will now show that $\hat{\mu}$ is an unbiased estimator of μ. Since all n devices survive until time Y_1, $n - 1$ devices survive until time $Y_2, \ldots,$ and $n - m + 1$ devices survive until time Y_m, T can be rewritten as

$$T = nY_1 + (n - 1)(Y_2 - Y_1) + (n - 2)(Y_3 - Y_2) + \cdots + (n - m + 1)(Y_m - Y_{m-1})$$

Since Y_1 is the minimum order statistic, Theorem 7.11 shows its density to be

$$f_{Y_1}(y) = n[1 - F_X(y)]^{n-1} f_X(y) = n e^{-(n-1)by} b e^{-by} = n b e^{-nby}$$

Thus, Y_1 is exponentially distributed with parameter nb, and $E[Y_1] = 1/nb$. Now, $Y_2 - Y_1$ is the minimum of the remaining times to failure. Due to the memoryless property of the exponential distribution, the remaining $n - 1$ times to failure each possess the original exponential distribution. Thus, we can view $Y_2 - Y_1$ as the minimum order statistic relative to the $n - 1$ remaining times to failure and so

$$E[Y_2 - Y_1] = \frac{1}{(n - 1)b}$$

Similar reasoning applies to $Y_k - Y_{k-1}$ for $k = 3, 4, \ldots, m$, and therefore, for these k,

$$E[Y_k - Y_{k-1}] = \frac{1}{(n - k + 1)b}$$

By the linearity of E,

$$E[T] = nE[Y_1] + (n - 1)E[Y_2 - Y_1] + \cdots + (n - m + 1)E[Y_m - Y_{m-1}] = \frac{m}{b}$$

there being m terms in the sum. Consequently,

$$E[\hat{\mu}] = \frac{E[T]}{m} = \frac{1}{b} = \mu$$

and $\hat{\mu}$ is unbiased.

Although the censored-sampling estimator $\hat{\mu}$ provides a cheaper-to-use unbiased alternative to the sample mean, it estimates the mean with less efficiency than the sample mean. Specifically, whereas Var $[\bar{X}] = 1/nb^2$ in the case of an exponential distribution, it can be shown that the variance of the censored-sampling estimator is Var $[\hat{\mu}] = 1/mb^2$, so that $m < n$ implies

$$\text{Var } [\bar{X}] < \text{Var } [\hat{\mu}]$$

The present section began by considering the mean-square error as a measure of estimator goodness. In the case of unbiased estimators, this error reduces to the variance of the estimator, and therefore two unbiased estimators are compared according to their variances. Now, suppose we wish to compare an unbiased estimator and a biased estimator. One cannot jump to the conclusion that the unbiased estimator is better in the mean-square sense simply because its expected value is equal to the parameter to be estimated. Specifically, if $\hat{\theta}_1$ is an unbiased estimator of θ and $\hat{\theta}_2$ is a biased estimator of θ with bias b, then

$$E[(\hat{\theta}_1 - \theta)^2] = \text{Var } [\hat{\theta}_1]$$

and, as seen at the outset of the section,

$$E[(\hat{\theta}_2 - \theta)^2] = \text{Var } [\hat{\theta}_2] + b^2$$

Thus, the biased estimator $\hat{\theta}_2$ is better than the unbiased estimator $\hat{\theta}_1$ in the mean-square sense if

$$\text{Var } [\hat{\theta}_2] < \text{Var } [\hat{\theta}_1] - b^2$$

7.6 MAXIMUM-LIKELIHOOD ESTIMATION

Given a distribution $f(x; \theta)$, with θ to be estimated, an obvious question is how to go about deciding upon an estimator $\hat{\theta}$ of θ such that $\hat{\theta}$ possesses desirable properties. The notion of an estimator is a rather general concept; indeed, given a random sample, any statistic can be employed as an estimator of θ, whether or not the statistic provides good results. What is needed are procedures leading to statistics that are appropriate to the distribution and to the parameter to be estimated.

7.6.1 Maximum-Likelihood Estimators

One of the most commonly employed estimation procedures is the method of maximum likelihood. The genesis of the method is quite intuitive, and the resulting estimators usually possess good qualities. Moreover, in practical situations the estimators can often be found by employing conventional optimization techniques, such as setting partial derivatives equal to zero.

DEFINITION 7.12

Likelihood Function. Suppose X is a random variable with density $f(x; \theta)$, θ to be estimated, and $X_1, X_2, \ldots, X_n$ is a random sample of size n. The joint density of the random variables comprising the sample is called the likelihood function of the sample and (due to the independence of the sample variables and the fact that each is identically distributed with X) is given by

$$L(x_1, x_2, \ldots, x_n; \theta) = f(x_1; \theta)f(x_2; \theta) \cdots f(x_n; \theta)$$

For the moment, suppose X is discrete. Then

$$L(x_1, x_2, \ldots, x_n; \theta) = \prod_{k=1}^{n} P(X = x_k; \theta)$$

where the notation $P(X = x_k; \theta)$ is used to denote the probability that $X = x_k$ relative to the unknown parameter θ. Suppose n observations are taken which yield a particular collection of sample values $x_1, x_2, \ldots, x_n$. We ask ourselves: Given these specific n observations, what value of θ would be the most reasonable choice? If there exists a value $\overset{*}{\theta}$ such that

$$L(x_1, x_2, \ldots, x_n; \overset{*}{\theta}) \geq L(x_1, x_2, \ldots, x_n; \theta)$$

for all possible choices of θ, then, based upon the meaning of the likelihood function in the discrete case, $\overset{*}{\theta}$ maximizes the probability

$$P(X_1 = x_1, X_2 = x_2, \ldots, X_n = x_n; \theta)$$

which is precisely the probability of obtaining the observations that were, in fact, obtained. Choosing the value of θ that makes it most likely that the data would be as obtained is certainly a reasonable approach.

Maximum-Likelihood Estimator. If a value $\overset{*}{\theta}$ can be found such that $\overset{*}{\theta}$ maximizes the likelihood function for a given set of sample values $x_1, x_2, \ldots, x_n$, then $\overset{*}{\theta}$ is called the **maximum-likelihood estimate** for the given set of sample values. Since the likelihood function is a function of θ and $x_1, x_2, \ldots, x_n$, the maximum-likelihood estimate will be a function of (will depend upon) the sample values. If the same functional relationship between the estimate and $x_1, x_2, \ldots, x_n$ holds for all possible choices of the x_k, then that functional relationship can be taken as an estimation rule, and the result will be an estimator $\hat{\theta}$ of θ, this estimator being known as the maximum-likelihood estimator or the **maximum-likelihood filter.**

To ease the notational burden, we will usually denote the likelihood function by $L(\theta)$, the sample values $x_1, x_2, \ldots, x_n$ being understood from the context. In addition, extensive use will be made of the fact that $L(\theta)$ is maximized as a function of θ if and only if $\log L(\theta)$ is maximized. Lastly, the examples which follow make repeated use of the identity

$$n\bar{x} = \sum_{k=1}^{n} x_k$$

$\bar{x}$ being the mean of $x_1, x_2, \ldots, x_n$.

EXAMPLE 7.14

Consider a normal distribution with known variance σ^2 and unknown mean μ. The likelihood function is

$$L(\mu) = \frac{1}{(2\pi\sigma^2)^{n/2}} \prod_{k=1}^{n} e^{-\frac{1}{2}\left(\frac{x_k - \mu}{\sigma}\right)^2}$$

$$= \frac{1}{(2\pi\sigma^2)^{n/2}} \exp\left[-\frac{1}{2}\sum_{k=1}^{n}\left(\frac{x_k - \mu}{\sigma}\right)^2\right]$$

Therefore,

$$\log L(\mu) = -\frac{n}{2}\log(2\pi\sigma^2) - \frac{1}{2}\sum_{k=1}^{n}\left(\frac{x_k - \mu}{\sigma}\right)^2$$

Taking the derivative with respect to μ yields

$$\frac{d}{d\mu}\log L(\mu) = \sigma^{-2}\sum_{k=1}^{n}(x_k - \mu) = \sigma^{-2}(n\bar{x} - n\mu)$$

Setting the derivative equal to zero gives the estimate

$$\overset{*}{\mu} = \bar{x}$$

Since the estimation procedure yields the same function of $x_1, x_2, \ldots, x_n$ no matter what the sample values, the procedure leads to the maximum-likelihood estimator

$$\hat{\mu} = \bar{X}$$

EXAMPLE 7.15

This time assume X possesses a normal distribution but that μ is known and σ^2 is unknown. The logarithm of the likelihood function is the same as in Example 7.14; however, in the present case we differentiate with respect to σ^2 to obtain

$$\frac{d}{d\sigma^2} \log L(\sigma^2) = -\frac{n}{2\sigma^2} + \frac{1}{2\sigma^4} \sum_{k=1}^{n} (x_k - \mu)^2$$

Setting the derivative equal to zero, solving for σ^2, and recognizing the independence of the procedure relative to the particular choice of the x_k leads to the maximum-likelihood estimator

$$\widehat{\sigma^2} = \frac{1}{n} \sum_{k=1}^{n} (X_k - \mu)^2$$

It is instructive to compare this estimator with the sample variance S^2. If μ is known, we certainly do not wish to employ the estimator $\bar{X}$ instead of μ when estimating the variance. Thus, the appearance of μ instead of $\bar{X}$ in the estimator $\widehat{\sigma^2}$ seems reasonable. Note also the divisor n instead of the divisor $n - 1$. Recalling Theorem 7.8, the division by $n - 1$ in the sample variance results from our desire to have S^2 be an unbiased estimator. An immediate question arises: Is the maximum-likelihood estimator for the variance of a normal random variable with known mean a biased estimator? It is not! In fact,

$$E[\widehat{\sigma^2}] = \frac{1}{n} \sum_{k=1}^{n} E[(X_k - \mu)^2] = \frac{1}{n} \sum_{k=1}^{n} \text{Var}[X_k] = \text{Var}[X]$$

the last equality following from the assumption that X_k is identically distributed with X for all k.

EXAMPLE 7.16

Consider a uniform distribution over the interval $[0, \beta]$, where β is unknown. The corresponding density is

$$f(x; \beta) = \frac{1}{\beta} I_{[0,\beta]}(x)$$

and the likelihood function is

$$L(\beta) = \frac{1}{\beta^n} \prod_{k=1}^{n} I_{[0,\beta]}(x_k)$$

$L(\beta)$ possesses the value β^{-n} if $x_k \leq \beta$ for all k and is zero otherwise. Here, the optimization methods of elementary calculus do not apply. Instead, we proceed in the following manner. For any set of sample values $x_1, x_2, \ldots, x_n$, should β be chosen so that one of the sample values, say x_i, is greater than β, then

$$I_{[0,\beta]}(x_i) = 0$$

and $L(\beta) = 0$, which certainly does not maximize $L(\beta)$. Thus, the maximum-likelihood estimate of β must be larger than all of the sample values. On the other hand, $1/\beta_*^n$ is maximized by selecting β as small as possible. Putting the two restraints together shows β to be the largest of the sample values. Since this choice is independent of the specific sample values, we obtain the maximum-likelihood estimator

$$\hat{\beta} = \max \{X_1, X_2, \ldots, X_n\}$$

so that $\hat{\beta}$ is the maximum order statistic.

Since $f_X(x) = \beta^{-1}I_{[0,\beta]}(x)$ and

$$F_X(x) = \begin{cases} 1, & \text{if } x > \beta \\ x/\beta, & \text{if } 0 \le x \le \beta \\ 0, & \text{if } x < 0 \end{cases}$$

Theorem 7.11 yields

$$f_{\hat{\beta}}(y) = \frac{ny^{n-1}}{\beta^n} I_{[0,\beta]}(y)$$

Hence,

$$E[\hat{\beta}] = \frac{n}{\beta^n} \int_0^\beta y^n \, dy = \frac{n}{n+1}\beta$$

Although in this instance the maximum-likelihood estimator is biased, it is asymptotically unbiased, and an unbiased estimator is given by

$$\frac{n+1}{n}\hat{\beta}$$

7.6.2 Estimation of More Than a Single Parameter

In general, there may be a number of parameters requiring estimation, not just one. In such a case, the density of the underlying distribution can be written as

$$f(x; \theta_1, \theta_2, \ldots, \theta_m)$$

and the likelihood function corresponding to a random sample $X_1, X_2, \ldots, X_n$ takes the form

$$L(x_1, x_2, \ldots, x_n; \theta_1, \theta_2, \ldots, \theta_m) = \prod_{k=1}^{n} f(x_k; \theta_1, \theta_2, \ldots, \theta_m)$$

For a particular set of sample values, say $x_1, x_2, \ldots, x_n$, a **maximum-likelihood estimate** of the parameter vector

$$\boldsymbol{\theta} = \begin{pmatrix} \theta_1 \\ \theta_2 \\ \vdots \\ \theta_m \end{pmatrix}$$

is a vector

$$\overset{*}{\boldsymbol{\theta}} = \begin{pmatrix} \overset{*}{\theta}_1 \\ \overset{*}{\theta}_2 \\ \vdots \\ \overset{*}{\theta}_m \end{pmatrix}$$

such that

$$L(x_1, x_2, \ldots, x_n; \overset{*}{\boldsymbol{\theta}}) \geq L(x_1, x_2, \ldots, x_n; \boldsymbol{\theta})$$

for all possible choices of $\boldsymbol{\theta}$. If the (vector) estimation rule defining $\overset{*}{\boldsymbol{\theta}}$ in terms of the sample values applies for all possible choices of the sample values, then the estimation rule defines the **maximum-likelihood estimator**

$$\hat{\boldsymbol{\theta}} = \begin{pmatrix} \hat{\theta}_1 \\ \hat{\theta}_2 \\ \vdots \\ \hat{\theta}_m \end{pmatrix}$$

In terms of the sample variables $X_1, X_2, \ldots, X_n$, the maximum-likelihood estimator is of the form

$$\hat{\boldsymbol{\theta}} = \hat{\boldsymbol{\theta}}(X_1, X_2, \ldots, X_n)$$

EXAMPLE 7.17

Example 7.14 discussed the maximum-likelihood estimator of the mean of a normal distribution under the assumption that the variance is known, and Example 7.15 discussed the maximum-likelihood estimator of the variance of a normal distribution under the assumption that the mean is known. Very often, neither the mean nor the variance is known. In such an instance, one needs to consider the maximum-likelihood vector-estimator

$$\hat{\boldsymbol{\theta}} = \begin{pmatrix} \hat{\mu} \\ \widehat{\sigma^2} \end{pmatrix}$$

The likelihood function is the same as in Examples 7.14 and 7.15, except now it is written as

$$L(\mu, \sigma^2)$$

Furthermore, we can utilize the derivative with respect to μ found in Example 7.14 and the derivative with respect to σ^2 found in Example 7.15, except now they are partial derivatives. Thus,

$$\frac{\partial}{\partial \mu} \log L(\mu, \sigma^2) = \frac{n(\bar{x} - \mu)}{\sigma^2}$$

$$\frac{\partial}{\partial \sigma^2} \log L(\mu, \sigma^2) = -\frac{n}{2\sigma^2} + \frac{1}{2\sigma^4} \sum_{k=1}^{n} (x_k - \mu)^2$$

Setting both partial derivatives equal to zero and solving simultaneously yields, for the particular sample values $x_1, x_2, \ldots, x_n$, the estimates

$$\overset{*}{\mu} = \bar{x}$$

$$\overset{*}{\sigma}{}^2 = \frac{1}{n} \sum_{k=1}^{n} (x_k - \bar{x})^2$$

Since the same procedure applies to any set of sample values, we obtain the maximum-likelihood estimators

$$\hat{\mu} = \bar{X}$$

$$\widehat{\sigma^2} = \frac{1}{n} \sum_{k=1}^{n} (X_k - \bar{X})^2 = \frac{n-1}{n} S^2$$

where S^2 is the sample variance. In vector form,

$$\hat{\theta} = \begin{pmatrix} \bar{X} \\ \dfrac{n-1}{n} S^2 \end{pmatrix}$$

Whereas the maximum-likelihood estimator of the variance is unbiased when the mean is known (Example 7.15), it is biased in the present case, where the mean is not known. Note, however, that it is asymptotically unbiased.

A useful property of maximum-likelihood estimators is the **invariance property**: If $\hat{\theta}$ is a maximum-likelihood estimator of the parameter θ and g is a one-to-one function, then the maximum-likelihood estimator of the parameter $\tau = g(\theta)$ is $\hat{\tau} = g(\hat{\theta})$. For instance, Exercise 7.42 states that the sample mean $\bar{X}$ is the maximum-likelihood estimator of the mean μ of an exponentially distributed random variable X. If b is the parameter of the exponential family, then $b = 1/\mu$. Thus, the maximum-likelihood estimator for b is $\hat{b} = 1/\bar{X}$. Functionally, $b = g(\mu)$ and $\hat{b} = g(\hat{\mu})$. A second example concerns estimation of the standard deviation. In Example 7.17 we saw that the maximum-likelihood estimator for the variance σ^2 of a normally distributed random variable X whose mean and variance are unknown is $\hat{\sigma^2} = (n-1)S^2/n$. By the invariance property, the maximum-likelihood estimator of the standard deviation is $\hat{\sigma} = \sqrt{(n-1)/n}\, S$.

The method of maximum-likelihood yields estimators possessing desirable properties, especially in the case of large samples. Although a detailed discussion is beyond the level of the present text, we can note that the maximum-likelihood method often (but certainly not always) yields minimum-variance unbiased estimators.

7.7 METHOD OF MOMENTS

An alternative estimation procedure to the method of maximum likelihood is the **method of moments,** which often requires less calculation, albeit sometimes at the cost of efficiency. The method makes use of the sample moments of a random sample.

DEFINITION 7.14

Sample Moments. Suppose $X_1, X_2, \ldots, X_n$ comprise a random sample arising from the population random variable X. The rth sample moment of the random sample is the random variable

$$M'_r = \frac{1}{n} \sum_{k=1}^{n} X_k^r$$

For $r = 1$, the sample moment M'_1 is the sample mean. For $r = 2$, the sample moment M'_2 can be expressed in terms of the sample variance and the sample mean. Indeed, since

$$\frac{1}{n} \sum_{k=1}^{n} (X_k - \bar{X})^2 = \frac{1}{n} \sum_{k=1}^{n} X_k^2 - \frac{2}{n} \sum_{k=1}^{n} X_k \bar{X} + \frac{1}{n} \sum_{k=1}^{n} \bar{X}^2$$

straightforward algebra yields

$$M_2' = \frac{n-1}{n} S^2 + \bar{X}^2$$

For $r = 1, 2, \ldots$, M_r' is the mean of the variables $X_1^r, X_2^r, \ldots, X_n^r$, which themselves constitute a random sample arising from the random variable X^r. Thus, Theorem 7.1 yields

$$E[M_r'] = E[X^r]$$

the rth moment about the origin of X, and

$$\text{Var}[M_r'] = \frac{\text{Var}[X^r]}{n}$$

Consequently, M_r' provides an unbiased estimator of $E[X^r]$, and the variance of the estimator tends to zero as n tends to infinity. According to Theorem 7.5, M_r' is a consistent estimator of $E[X^r]$.

Given any collection of observed values resulting from the random sample, say $x_1, x_2, \ldots, x_n$, the **empirical rth moment** corresponding to the sample is

$$m_r' = \frac{1}{n}(x_1^r + x_2^r + \cdots + x_n^r)$$

which is just the empirical mean of the rth powers of the observations. For any r, m_r' is the estimate of the rth moment of X resulting from the rth sample-moment estimator.

Suppose that a number of parameters of X, say $\theta_1, \theta_2, \ldots, \theta_p$, need to be estimated. Then $E[X^r]$ is a function of the unknown parameters—that is, there exists a function h_r such that

$$E[X^r] = h_r(\theta_1, \theta_2, \ldots, \theta_p)$$

Given the observations resulting from a random sample, method-of-moment estimation of the parameters $\theta_1, \theta_2, \ldots, \theta_p$ occurs by setting

$$E[X^r] = m_r'$$

which is reasonable since m_r' is employed to estimate $E[X^r]$. This leads to the system of equations

$$h_1(\theta_1, \theta_2, \ldots, \theta_p) = m_1'$$
$$h_2(\theta_1, \theta_2, \ldots, \theta_p) = m_2'$$
$$\vdots$$
$$h_N(\theta_1, \theta_2, \ldots, \theta_p) = m_N'$$

where N, the number of moments employed, is chosen so that a unique solution for $(\theta_1, \theta_2, \ldots, \theta_p)$ can be found. When a solution for $(\theta_1, \theta_2, \ldots, \theta_p)$ exists, this solution is denoted (as usual) by

$$\overset{*}{\mathbf{\theta}} = \begin{pmatrix} \overset{*}{\theta}_1 \\ \overset{*}{\theta}_2 \\ \vdots \\ \overset{*}{\theta}_p \end{pmatrix}$$

and $\overset{*}{\theta}_1, \overset{*}{\theta}_2, \ldots, \overset{*}{\theta}_p$ are the **method-of-moment estimates** corresponding to the set of sample values.

If, more generally, we can solve the system of equations

$$h_1(\theta_1, \theta_2, \ldots, \theta_p) = M_1'$$

$$h_2(\theta_1, \theta_2, \ldots, \theta_p) = M_2'$$

$$\vdots$$

$$h_N(\theta_1, \theta_2, \ldots, \theta_p) = M_N'$$

then the solutions $\hat{\theta}_1, \hat{\theta}_2, \ldots, \hat{\theta}_p$ are the **method-of-moment estimators** for the unknown parameters.

EXAMPLE 7.18

Estimating the parameter b in an exponential density requires only a single equation. Since

$$h_1(b) = E[X] = \frac{1}{b}$$

and $M_1' = \bar{X}$, solving the equation

$$h_1(b) = M_1'$$

yields the method-of-moment estimator $\hat{b} = 1/\bar{X}$, a result equivalent to the one obtained in Exercise 7.42 by the method of maximum likelihood.

EXAMPLE 7.19

The gamma distribution possesses two parameters, α and β. According to Theorem 4.10,

$$E[X] = \alpha\beta$$

$$E[X^2] = \frac{\beta^2\Gamma(\alpha + 2)}{\Gamma(\alpha)} = (\alpha + 1)\alpha\beta^2$$

The resulting system of equations is

$$\alpha\beta = M_1' = \bar{X}$$

$$(\alpha + 1)\alpha\beta^2 = M_2' = \frac{n-1}{n}S^2 + \bar{X}^2$$

Solving for α and β yields the estimators

$$\hat{\alpha} = \frac{\bar{X}^2}{\frac{n-1}{n}S^2}$$

$$\hat{\beta} = \frac{\frac{n-1}{n}S^2}{\bar{X}}$$

7.8 INTERVAL ESTIMATION

Rather than find a point estimate $\overset{*}{\theta}$ of an unknown parameter θ, it is sometimes more beneficial to locate an interval $(\overset{*}{\theta}_1, \overset{*}{\theta}_2)$ which is believed to contain θ. For instance, in the production of pipe fittings it might be more important to specify an interval believed to contain the mean of the inside diameters, rather than have a point estimate of the mean inside diameter. In other words, we may be more interested in the mean's being within some acceptable bounds than in having some estimate of its actual value. Although no estimation procedure can provide an interval that will certainly contain the desired parameter, it is possible to develop interval-estimation procedures for which one can specify the probability that an interval resulting from the procedure will contain the desired parameter.

7.8.1 Confidence Intervals

An **interval estimate** of the parameter θ is an interval of the form $(\overset{*}{\theta}_1, \overset{*}{\theta}_2)$, which we "hope" contains θ. The upper and lower limits, $\overset{*}{\theta}_1$ and $\overset{*}{\theta}_2$, respectively, must be found by some estimation procedure involving a random sample. Thus, they result from two statistics, $\hat{\theta}_1$ and $\hat{\theta}_2$. In practice, we consider some estimator of θ, say $\hat{\theta}$, and utilize the sampling distribution of $\hat{\theta}$ to determine the statistics $\hat{\theta}_1$ and $\hat{\theta}_2$. Whether or not a particular interval estimate $(\overset{*}{\theta}_1, \overset{*}{\theta}_2)$ actually contains the parameter θ is not knowable; however, what can be determined is the probability that the **interval estimator** $(\hat{\theta}_1, \hat{\theta}_2)$ contains θ.

DEFINITION 7.15

Confidence Interval. If $(\hat{\theta}_1, \hat{\theta}_2)$ is an interval estimator for which

$$P(\hat{\theta}_1 < \theta < \hat{\theta}_2) = 1 - \alpha$$

and $\overset{*}{\theta}_1$ and $\overset{*}{\theta}_2$ are estimates resulting from a particular set of sample values, then the interval $(\overset{*}{\theta}_1, \overset{*}{\theta}_2)$ is called a **$(1 - \alpha)100\%$ confidence interval,** $1 - \alpha$ is called the **confidence coefficient,** and the endpoints $\overset{*}{\theta}_1$ and $\overset{*}{\theta}_2$ are called the lower and upper **confidence limits,** respectively.

It is important to recognize in Definition 7.15 that the probability $1 - \alpha$ corresponds to the probability that θ lies between the statistics $\hat{\theta}_1$ and $\hat{\theta}_2$, and that once estimates are computed from a set of sample values, the resulting confidence interval either contains, or does not contain, the parameter θ. Our degree of confidence regarding the inclusion of θ in the confidence interval obtained from a specific sample is measured probabilistically in terms of the endpoint statistics. Using our intuition and employing a relative-frequency interpretation, $(\overset{*}{\theta}_1, \overset{*}{\theta}_2)$ is a $(1 - \alpha)100\%$

confidence interval if, among all possible confidence intervals resulting from the interval estimator $(\hat{\theta}_1, \hat{\theta}_2)$, $(1 - \alpha)100\%$ of them contain the parameter θ. From the perspective of precision, for a given confidence coefficient we would naturally prefer a shorter interval over a longer one; however, to attain a short interval possessing high confidence may require a prohibitively large sample.

7.8.2 A Confidence Interval for the Mean of a Normal Distribution with Known Variance

Consider finding an interval estimate of the mean μ of a normally distributed random variable X possessing a known variance σ^2. Referring to Figure 7.2,

$$P(-z_{\alpha/2} < Z < z_{\alpha/2}) = 1 - \alpha$$

where Z is the standard normal random variable. For a random sample of size n, the standardized version of the sample mean is identically distributed with Z. Thus,

$$P\left(-z_{\alpha/2} < \frac{\bar{X} - \mu}{\sigma/\sqrt{n}} < z_{\alpha/2}\right) = 1 - \alpha$$

Rearranging the terms of the inequality yields

$$P\left(\bar{X} - z_{\alpha/2} \cdot \frac{\sigma}{\sqrt{n}} < \mu < \bar{X} + z_{\alpha/2} \cdot \frac{\sigma}{\sqrt{n}}\right) = 1 - \alpha$$

which is an estimator equation of the type demanded by Definition 7.15, the lower and upper confidence-interval statistics being

$$\hat{\mu}_1 = \bar{X} - z_{\alpha/2} \cdot \frac{\sigma}{\sqrt{n}}$$

$$\hat{\mu}_2 = \bar{X} + z_{\alpha/2} \cdot \frac{\sigma}{\sqrt{n}}$$

An actual $(1 - \alpha)100\%$ confidence interval resulting from a set of sample values takes the form

$$(\overset{*}{\mu}_1, \overset{*}{\mu}_2) = \left(\bar{x} - z_{\alpha/2} \cdot \frac{\sigma}{\sqrt{n}}, \bar{x} + z_{\alpha/2} \cdot \frac{\sigma}{\sqrt{n}}\right)$$

where $\bar{x}$ is the empirical mean of the sample data. The next theorem summarizes the foregoing considerations.

THEOREM 7.13

If X is a normally distributed random variable with unknown mean μ and known variance σ^2, and if $\bar{x}$ is an empirical mean resulting from a random sample of size n, then

$$\left(\bar{x} - z_{\alpha/2} \cdot \frac{\sigma}{\sqrt{n}}, \bar{x} + z_{\alpha/2} \cdot \frac{\sigma}{\sqrt{n}}\right)$$

is a $(1 - \alpha)100\%$ confidence interval for μ.

EXAMPLE 7.20

A manufacturer measures the inside diameter (in inches) of 20 pipe fittings of a certain type that have been obtained by a random sampling procedure, perhaps by sampling with replacement or without replacement, in the latter case it being assumed that the overall population of fittings is sufficiently large that we can ignore the effect of nonreplacement. Suppose the variability is known and the data result from measurement and production processes that lead to normally distributed measurements. Then Theorem 7.13 is applicable. Specifically, the population random variable X is normally distributed with unknown mean μ (the desired average diameter) and known variance, say $\sigma^2 = 0.000144$. If the empirical mean of the 20 measurements is $\bar{x} = 1.988$ inches, then, since $z_{0.025} = 1.96$, the 95% confidence interval described in Theorem 7.13 is

$$\left(1.988 - \frac{1.96 \times 0.012}{4.47}, \, 1.988 + \frac{1.96 \times 0.012}{4.47} \right) = (1.983, 1.993)$$

Should another set of sample values be found, it is likely that a different 95% confidence interval will result. Indeed, if ten sample data sets are taken, the situation depicted in Figure 7.4 might arise. According to the figure, nine of the resulting confidence intervals contain μ and one does not. In fact, it is possible that one could take ten empirical samples with none of the resulting 95% confidence intervals containing the mean. Of course, this is unlikely, since the relative-frequency interpretation of 95% confidence intervals is that 95% of them will contain the mean. Not only is such an occurrence unlikely, it would be unknown to the experimenter, since he or she can never know whether or not a specific confidence interval contains the mean.

Instead of a sample of 20, suppose a sample of 30 is drawn which has empirical mean $\bar{x} = 1.991$. Then the resulting 95% confidence interval is $(1.987, 1.995)$. The larger sample size has resulted in a tighter interval estimate.

Confidence intervals resulting from the methodology of Theorem 7.13 are centered at the empirical means of the samples. Moreover, if

$$r = z_{\alpha/2} \cdot \frac{\sigma}{\sqrt{n}}$$

then the confidence interval of Theorem 7.13 takes the form

$$(\bar{x} - r, \bar{x} + r)$$

FIGURE 7.4

A collection of confidence intervals

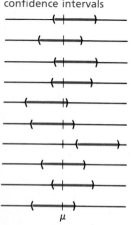

μ

Should one desire a confidence interval of a predetermined length $2r$ centered at the empirical mean, then the relation of r in terms of n can be solved to obtain

$$n = \left(\frac{\sigma}{r} z_{\alpha/2} \right)^2$$

where the equality means n is chosen to be the smallest integer greater than or equal to the expression on the right-hand side of the equation. This is precisely the expression given in Theorem 7.3, which should be expected, since

$$P(\bar{X} - r < \mu < \bar{X} + r) = P(|\bar{X} - \mu| < r)$$

As indicated in Figure 7.5, the error of estimation, $|\bar{x} - \mu|$, is less than r if and only if the confidence interval resulting from the estimate $\bar{x}$ contains μ.

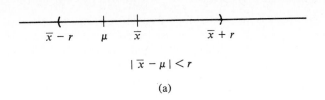

$$|\bar{x} - \mu| < r$$

(a)

FIGURE 7.5
The equivalence
between the error of
estimation being less
than r and the
confidence interval
resulting from $\bar{x}$
containing μ

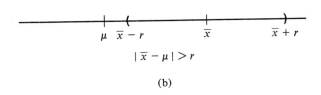

$$|\bar{x} - \mu| > r$$

(b)

EXAMPLE 7.21

A bulk packager of rivets desires a 95% confidence interval of unit length for the mean weight of what are believed to be 100-lb bags. From experience, the distribution of the weights is assumed to be normal and the variance is accepted to be $\sigma^2 = 0.81$. Since $r = 0.5$ and $z_{\alpha/2} = 1.96$,

$$n = \left(\frac{0.9 \times 1.96}{0.5}\right)^2 = 13$$

(in the sense that 13 is the smallest integer greater than 12.45). Note that because of the rounding up to get n, the actual value of r for any resulting confidence interval will be

$$r = 1.96 \times \frac{0.9}{\sqrt{13}} = 0.49$$

which is a bit tighter than demanded by the packager.

Suppose 13 bags are selected by a random sampling procedure and the empirical mean of their weights is $\bar{x} = 99.62$ lb. The resulting 95% confidence interval is

$$(99.62 - 0.49, 99.62 + 0.49) = (99.13, 100.11)$$

As stated, Theorem 7.13 applies to normal population variables; nevertheless, just as the central limit theorem (Theorem 7.6) justifies large-sample application of Theorem 7.2 to nonnormal random variables, the same theorem justifies use of the confidence-interval procedure established in Theorem 7.13 to obtain approximate results in a large-sample setting. In addition, if the variance is unknown and the sample is large, then the sample variance can be employed to estimate the true variance. Thus, for $n \geq 30$ (better yet, larger), Theorem 7.13 can be applied (in the approximate sense) to any random variable, the $(1 - \alpha)100\%$ confidence interval taking the form

$$\left(\bar{x} - z_{\alpha/2} \cdot \frac{s}{\sqrt{n}}, \bar{x} + z_{\alpha/2} \cdot \frac{s}{\sqrt{n}}\right)$$

where s is the empirical standard deviation of the sample values.

EXAMPLE 7.22

Once again consider the packaging problem of Example 7.21; however, this time assume that neither the form nor the variance of the population distribution is known. The expression of Theorem 7.3 yields $n = 13$; however, we cannot proceed, since $n = 13$ is not sufficient for either application of the central limit theorem or estimation of σ by s. Nonetheless, a sample size of 40 will exceed the number needed for a confidence-interval radius of no more than 0.5, will allow the use of the central limit theorem, and will very likely provide a decent estimate of the standard deviation. Proceeding, suppose 40 observations are made and the resulting empirical mean and standard deviation are found to be $\bar{x} = 99.71$ and $s = 0.88$, respectively. The methodology of Theorem 7.13 yields the approximate 95% confidence interval

$$\left(99.71 - 1.96 \times \frac{0.88}{\sqrt{40}}, 99.71 + 1.96 \times \frac{0.88}{\sqrt{40}}\right) = (99.44, 99.98)$$

Regarding methodology, the technique employed to construct the confidence interval of Theorem 7.13 is used often. In essence, it consists of finding some random variable U involving the desired parameter θ whose distribution is well studied and is not dependent upon unknown parameters. Examining the density of U, two values, u_1 and u_2, are chosen such that

$$P(u_1 < U < u_2) = 1 - \alpha$$

Hopefully, the inequality can be algebraically rearranged to obtain an equivalent expression

$$P(\hat{\theta}_1 < \theta < \hat{\theta}_2) = 1 - \alpha$$

where the statistics $\hat{\theta}_1$ and $\hat{\theta}_2$ are functions of U that do not depend upon θ. Given a set of sample values, $\hat{\theta}_1$ and $\hat{\theta}_2$ lead to estimates that constitute the lower and upper limits of an appropriate confidence interval, these limits being $\overset{*}{\theta}_1$ and $\overset{*}{\theta}_2$, respectively. In the case of Theorem 7.13, the methodology was employed with U being the standardized version of the sample mean. In the small-sample case, a different random variable will be employed. It should be noted that the method does not lead to a unique confidence interval, since the interval depends upon the choice of u_1 and u_2. In addition, its success depends upon a satisfactory rearrangement of the probability inequality.

7.8.3 A Confidence Interval for the Mean of a Normal Distribution with Unknown Variance— The t Distribution

Even when a population random variable possesses a normal distribution, application of Theorem 7.13 requires knowledge of the variance, which very often is unknown. If the sample is sufficiently large, then the variance can be estimated by the sample variance; however, in many cases, owing to cost, time, or considerations beyond the investigator's control, such a large-sample approach is not feasible. Even when it is, S may yield a poor estimate of σ. What is needed is an estimation technique that does not depend upon knowledge of the variance, a so-called **small-sample** approach.

The key to Theorem 7.13 is the fact that, for $X \sim N(\mu, \sigma)$,

$$\frac{\bar{X} - \mu}{\sigma/\sqrt{n}}$$

possesses a standard normal distribution. Extension of the methodology to the situation when σ is unknown involves the distribution of the statistic

$$\frac{\bar{X} - \mu}{S/\sqrt{n}}$$

First we introduce the t distribution, which, in the literature, is often called the **student's t distribution.**

DEFINITION 7.16

t Distribution. Suppose Z is the standard normal variable, W possesses a chi-square distribution with ν degrees of freedom, and Z and W are independent. Then the random variable

$$T = \frac{Z}{\sqrt{W/\nu}}$$

is said to possess a t distribution with ν **degrees of freedom.**

THEOREM 7.14

If T possesses a t distribution with ν degrees of freedom, then its density is given by

$$f(t) = \frac{\Gamma\left(\dfrac{\nu + 1}{2}\right)}{\sqrt{\pi\nu}\,\Gamma\left(\dfrac{\nu}{2}\right)}\left(1 + \frac{t^2}{\nu}\right)^{-(\nu + 1)/2}$$

for $-\infty < t < \infty$.

Proof: The proof involves an interesting application of Theorem 5.11. Let Z and W be defined as in Definition 7.16. Their independence means that their joint density is the product of their individual densities, namely,

$$f_{Z,W}(z, w) = \frac{1}{\sqrt{2\pi}} e^{-z^2/2} \frac{1}{2^{\nu/2}\,\Gamma\left(\dfrac{\nu}{2}\right)} w^{(\nu/2)-1} e^{-w/2}$$

for $w > 0$ and $-\infty < z < \infty$, and $f_{Z,W}(z, w) = 0$ elsewhere. Introduce a second output random variable $U = W$. According to Definition 7.16, the random variable T is defined in terms of the random variables Z and W by means of the function

$$t = g(z, w) = \frac{z}{\sqrt{w/\nu}}$$

U is defined in terms of Z and W by means of the function

$$u = h(z, w) = w$$

The vector mapping $(z, w) \longrightarrow (t, u)$ possesses the inverse mapping $(t, u) \longrightarrow (z, w)$ defined by the equation pair

$$z = r(t, u) = t\sqrt{\frac{u}{v}}$$

$$w = s(t, u) = u$$

Thus, the Jacobian of the mapping is given by

$$J(t, u) = \det \begin{bmatrix} \sqrt{\dfrac{u}{v}} & \dfrac{t}{2\sqrt{uv}} \\ 0 & 1 \end{bmatrix} = \sqrt{\frac{u}{v}}$$

The density $f_{Z,W}(z, w)$ is nonzero on the upper half of the zw plane and, under the mapping $(z, w) \longrightarrow (t, u)$, the upper half of the zw plane is mapped into the upper half of the tu plane. Thus, according to Theorem 5.11, the joint density for T and U is given by

$$f_{T,U}(t, u) = f_{Z,W}[r(t, u), s(t, u)]|J(t, u)|$$

$$= f_{Z,W}\left(t\sqrt{\frac{u}{v}}, u\right)\sqrt{\frac{u}{v}}$$

$$= \frac{\sqrt{\dfrac{u}{v}}}{\sqrt{2\pi}\,2^{v/2}\,\Gamma\left(\dfrac{v}{2}\right)}\, u^{(v/2)-1}\, e^{-(u/2)[1 + (t^2/v)]}$$

for $-\infty < t < \infty$ and $u > 0$ (the upper half of the tu plane), and $f_{T,U}(t, u) = 0$ otherwise. To obtain the density for T, integrate the joint density over all u:

$$f_T(t) = \int_{-\infty}^{\infty} f_{T,U}(t, u)\, du$$

$$= \frac{1}{\sqrt{2\pi v}\,2^{v/2}\,\Gamma\left(\dfrac{v}{2}\right)} \int_0^{\infty} u^{(v-1)/2}\, e^{-(u/2)[1 + (t^2/v)]}\, du$$

Letting

$$y = \frac{u\left(1 + \dfrac{t^2}{v}\right)}{2}$$

yields

$$u = 2\left(1 + \frac{t^2}{v}\right)^{-1} y$$

$$du = 2\left(1 + \frac{t^2}{v}\right)^{-1} dy$$

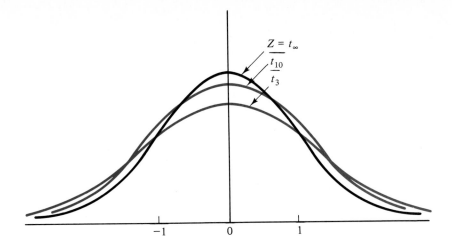

FIGURE 7.6
Z density and two *t* densities

and

$$f_T(t) = \frac{1}{\sqrt{2\pi\nu}2^{\nu/2}\,\Gamma\!\left(\dfrac{\nu}{2}\right)} \int_0^\infty \left(\frac{2y}{1+\dfrac{t^2}{\nu}}\right)^{(\nu-1)/2} \frac{2e^{-y}}{1+\dfrac{t^2}{\nu}}\,dy$$

$$= \frac{1}{\sqrt{\pi\nu}\,\Gamma\!\left(\dfrac{\nu}{2}\right)}\left(1+\frac{t^2}{\nu}\right)^{-(\nu+1)/2} \int_0^\infty y^{[(\nu+1)/2]-1}e^{-y}\,dy$$

According to Definition 4.7, the preceding integral is $\Gamma[(\nu+1)/2]$ and the theorem follows. ∎

If T possesses a t distribution, then its density is very much like the density of the standard normal variable Z, the likeness becoming greater for increasing values of ν. This likeness is evident in Figure 7.6, where t densities are superimposed with the standard normal density. In the limit, the t distribution approaches the standard normal distribution, and it is common to call Z the t distribution with $\nu = \infty$ degrees of freedom. To evaluate probabilities involving the t distribution, we employ Table A.6, which provides points $t_{\alpha,\nu}$ on the t axis for which

$$P(T > t_{\alpha,\nu}) = \alpha$$

ν being the number of degrees of freedom (see Figure 7.7). For instance, $t_{0.05,7} = 1.895$ is the point on the t axis for which

$$P(T > 1.895) = 0.05$$

T possessing a t distribution with 7 degrees of freedom. Since the t distribution is symmetric about the origin, Table A.6 need not include negative t values. Specifically,

$$t_{1-\alpha,\nu} = -t_{\alpha,\nu}$$

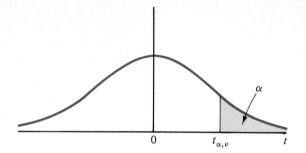

FIGURE 7.7

Illustration of $t_{\alpha,\nu}$

For example, if T possesses a t distribution with $\nu = 12$ degrees of freedom, then

$$t_{0.975,12} = -t_{0.025,12} = -2.179$$

and

$$P(T > -2.179) = 1 - 0.025 = 0.975$$

Owing to the symmetry of its density, if T possesses a t distribution with ν degrees of freedom, then $E[T] = 0$. Moreover, for $\nu > 2$,

$$\text{Var}\,[T] = \frac{\nu}{\nu - 2} > 1$$

so that the dispersion of T is greater than the dispersion of Z, although $\text{Var}\,[T]$ does approach $\text{Var}\,[Z] = 1$ as $\nu \longrightarrow \infty$.

It is the next theorem which makes the t distribution important in the construction of small-sample confidence intervals for the mean.

THEOREM 7.15

If X is normally distributed with mean μ, then the statistic

$$\frac{\bar{X} - \mu}{S/\sqrt{n}}$$

possesses a t distribution with $n - 1$ degrees of freedom, n being the size of the random sample leading to $\bar{X}$ and S.

Proof: Straightforward algebra yields

$$\frac{\bar{X} - \mu}{S/\sqrt{n}} = \frac{\dfrac{\bar{X} - \mu}{\sigma/\sqrt{n}}}{\sqrt{\dfrac{(n-1)S^2}{\sigma^2(n-1)}}} = \frac{\dfrac{\bar{X} - \mu}{\sigma/\sqrt{n}}}{\sqrt{\dfrac{W}{n-1}}}$$

where

$$W = \frac{(n-1)S^2}{\sigma^2}$$

According to Theorem 7.9, $\bar{X}$ and S^2 are independent, so that the standardized version of $\bar{X}$ and W are independent. In addition, the same theorem states that W is a chi-square variable with $n - 1$ degrees of freedom. Thus, the present theorem follows from Definition 7.16. ∎

The degrees-of-freedom terminology in the t distribution corresponds to the form of the t distribution in Theorem 7.15. The sample standard deviation S is expressed in terms of the sum of n deviations from $\bar{X}$, namely, $X_1 - \bar{X}$, $X_2 - \bar{X}$, . . . , $X_n - \bar{X}$; however, since

$$\sum_{k=1}^{n} (X_k - \bar{X}) = 0$$

only $n - 1$ of the deviations are **freely determined,** because, given any $n - 1$ of the deviations, the remaining one is also known.

Using the t distribution, we can employ the methodology articulated in Section 7.8.2 to create confidence intervals for the mean when the underlying distribution is normal and the variance is unknown. Whereas the point $z_{\alpha/2}$ is employed when using the standard normal distribution, the point $t_{\alpha/2,n-1}$ is employed when using the t distribution.

THEOREM 7.16

If $\bar{x}$ and s are the empirical mean and standard deviation resulting from a set of sample values obtained from a random sample of size n in which the population random variable is normally distributed with unknown mean μ and unknown variance σ^2, then

$$\left(\bar{x} - t_{\alpha/2,n-1} \cdot \frac{s}{\sqrt{n}}, \bar{x} + t_{\alpha/2,n-1} \cdot \frac{s}{\sqrt{n}} \right)$$

is a $(1 - \alpha)100\%$ confidence interval for the mean.

Proof: Letting T denote the t-distributed random variable with $n - 1$ degrees of freedom,

$$P(-t_{\alpha/2,n-1} < T < t_{\alpha/2,n-1}) = 1 - \alpha$$

(see Figure 7.8). Therefore, by Theorem 7.15,

$$P\left(-t_{\alpha/2,n-1} < \frac{\bar{X} - \mu}{S/\sqrt{n}} < t_{\alpha/2,n-1} \right) = 1 - \alpha$$

FIGURE 7.8
t values for a $(1 - \alpha)$ 100% confidence interval for the mean

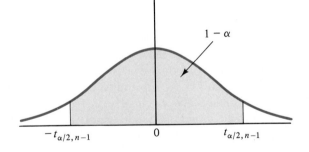

Algebraic rearrangement yields

$$P\left(\bar{X} - t_{\alpha/2,n-1} \cdot \frac{S}{\sqrt{n}} < \mu < \bar{X} + t_{\alpha/2,n-1} \cdot \frac{S}{\sqrt{n}}\right) = 1 - \alpha$$

The theorem follows at once from Definition 7.15. ∎

In Theorem 7.16 the points $-t_{\alpha/2,n-1}$ and $t_{\alpha/2,n-1}$ have been chosen so that the resulting confidence interval is symmetric about the empirical mean of the sample values. As statistics, the lower and upper interval endpoints are

$$\hat{\mu}_1 = \bar{X} - t_{\alpha/2,n-1} \cdot \frac{S}{\sqrt{n}}$$

$$\hat{\mu}_2 = \bar{X} + t_{\alpha/2,n-1} \cdot \frac{S}{\sqrt{n}}$$

Each is a function of two statistics, $\bar{X}$ and S. μ actually lies in the resulting confidence interval if and only if

$$|\bar{x} - \mu| < t_{\alpha/2,n-1} \cdot \frac{s}{\sqrt{n}}$$

EXAMPLE 7.23

The standard procedure for testing the compressive strength of hardened concrete is by loading cylinders in compression perpendicular to the axis of the cylinder, the standard test cylinder being 6 inches in diameter, 12 inches high, and 28 days of age. A manufacturer, who has reason to believe that compressive strength is normally distributed, desires a 90% confidence interval based on six sample values, these being (in pounds per square inch)

$$4052 \quad 4120 \quad 4095 \quad 4188 \quad 3978 \quad 4090$$

Theorem 7.16 applies with $\bar{x} = 4087.2$, $s = 69.96$, $\sqrt{n} = 2.45$, and $t_{0.05,5} = 2.015$ (see Table A.6). The resulting 90% confidence-interval endpoints are

$$4087.2 \pm (2.015)\left(\frac{69.96}{2.45}\right)$$

Thus, the desired confidence interval is (4029.7, 4144.7).

In comparing the standard normal and t distribution confidence-interval methodologies for the mean, it is important to recognize the limitations of each. If the underlying distribution is known to be normal and the variance is known, then either method can be employed. For given n and α, the Z method yields confidence intervals of fixed length $2z_{\alpha/2}\sigma/\sqrt{n}$. On the other hand, confidence-interval length in the t method is the random variable $2t_{\alpha/2,n-1}S/\sqrt{n}$.

If the variance is unknown (or the presumed value of the variance is in doubt), then one might choose the t method even though the large-sample standard normal approach can be used with s being substituted as an estimate for σ, the reason being that s is only an estimate. The catch here is the assumption that the population distribution is known to be normal. If it is not, and the sample is large, then the central limit theorem justifies the use of the standard normal approach; if it is not, then, whether or not the sample is large, the *t* method is inappropriate, since it requires normality in all cases. Given these provisions, one might ask what happens in the event he or she is constrained to a small sample and the assumption of normality is questionable. Fortunately, the *t* method is **robust,** which in statistical parlance means that the method holds approximately for modest deviations of *X* from normality. Thus, if there is reason to believe that the distribution of *X* is basically bell-shaped, then, with caution, one can apply the *t* method for the construction of approximate confidence intervals. If it is believed that *X* differs markedly from normality, then the small-sample method should not be utilized.

7.8.4 An Approximate Confidence Interval for a Proportion

In a binomial experiment consisting of a single trial, the probability *p* (of a success) represents a proportion. For instance, in an urn model, if there are 3 black and 6 white balls, with *X* being the number of selected black balls (0 or 1), then $p = 3/9 = 1/3$ is the proportion of black balls in the urn. According to Exercise 7.39, the maximum-likelihood estimator of *p* is $\hat{p} = \bar{X}$. In the present circumstance, $\bar{X}$ is simply the number of successes in *n* Bernoulli trials divided by *n*. According to Theorems 7.1 and 4.1,

$$E[\hat{p}] = E[\bar{X}] = E[X] = p$$

$$\text{Var}\,[\hat{p}] = \text{Var}\,[\bar{X}] = \frac{\text{Var}\,[X]}{n} = \frac{pq}{n}$$

where $q = 1 - p$ and $\hat{q} = 1 - \hat{p}$. For a given set of sample values, the estimator $\hat{p}$ leads to the estimate $\overset{*}{p} = \bar{x}$.

If *n* is sufficiently large, then the central limit theorem assures us that $\hat{p}$ is approximately normal. As a rule of thumb, *n* should be large enough that $n\overset{*}{p} \geq 5$ and $n\overset{*}{q} \geq 5$. For such *n*, we can emulate the methodology used to arrive at Theorem 7.13 to obtain an approximate $(1 - \alpha)100\%$ confidence interval. Proceeding, choose $z_{\alpha/2}$ so that

$$P\left(-z_{\alpha/2} < \frac{\hat{p} - p}{\sqrt{pq/n}} < z_{\alpha/2}\right) = 1 - \alpha$$

Simplifying the probability inequality yields

$$P\left(\hat{p} - z_{\alpha/2}\sqrt{\frac{pq}{n}} < p < \hat{p} + z_{\alpha/2}\sqrt{\frac{pq}{n}}\right) = 1 - \alpha$$

For large *n*, $\hat{p}$ and $\hat{q}$ are good estimators of *p* and *q*. Thus, the empirical values $\overset{*}{p}$ and $\overset{*}{q}$ obtained from the sample can be substituted for *p* and *q*, this further approximation being analogous to the use of *s* in place of σ in the large-sample methodology for the mean. The following approximate confidence interval is thereby obtained:

Approximate Confidence Interval for a Proportion

$$\left(\overset{*}{p} - z_{\alpha/2}\sqrt{\frac{\overset{**}{pq}}{n}}, \overset{*}{p} + z_{\alpha/2}\sqrt{\frac{\overset{**}{pq}}{n}}\right)$$

EXAMPLE 7.24

From a relative-frequency perspective, estimating the probability p of selecting a defective unit is tantamount to estimating the proportion of defective units in the population. As discussed previously, sampling a large lot without replacement can be modeled as random sampling, the justification being that the total population of units is so vast as compared to the sample size that the selections are, to a practical degree, independent.

Suppose an air-conditioner manufacturer considers a nominally rated 5000-BTU air conditioner to be defective if its test rating is less than 4800 BTU. To determine the proportion of defective units being produced, a random sample of 100 is selected and the units tested. Assuming all but 6 pass the test,

$$\overset{*}{p} = \frac{6}{100} = 0.06$$

Since $n\overset{*}{p} = 6 \geq 5$ and $n\overset{*}{q} = 94 \geq 5$, the preceding approximate confidence-interval method can be applied. In particular, the upper and lower limits of an approximate 95% confidence interval are given by

$$0.06 \pm (1.96)\sqrt{\frac{(0.06)(0.94)}{100}}$$

so that the resulting 95% confidence interval is $(0.01, 0.11)$.

7.8.5 A Confidence Interval for the Variance of a Normal Distribution

A confidence interval for the variance σ^2 of a normal random variable can be obtained from the statistic

$$W = \frac{(n-1)S^2}{\sigma^2}$$

which, according to Theorem 7.9, possesses a chi-square distribution with $n - 1$ degrees of freedom. In deriving the subsequent theorem, we will reference Figure 7.9, which illustrates a chi-square density with $n - 1$ degrees of freedom, and in which the points $\chi^2_{1-\alpha/2,n-1}$ and $\chi^2_{\alpha/2,n-1}$ are chosen so that

$$P(W < \chi^2_{1-\alpha/2,n-1}) = \frac{\alpha}{2}$$

$$P(W > \chi^2_{\alpha/2,n-1}) = \frac{\alpha}{2}$$

FIGURE 7.9
χ^2 values for a $(1 - \alpha)$ 100% confidence interval for the variance

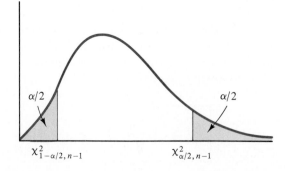

Suppose a set of sample values obtained from a random sample of size n arising from a normal random variable possessing variance σ^2 has empirical variance s^2. Then

$$\left(\frac{(n-1)s^2}{\chi^2_{\alpha/2,n-1}}, \frac{(n-1)s^2}{\chi^2_{1-\alpha/2,n-1}} \right)$$

is a $(1-\alpha)100\%$ confidence interval for the variance σ^2. A $(1-\alpha)100\%$ confidence interval for the standard deviation is obtained by taking the square roots of the lower and upper limits of the confidence interval for the variance.

Proof: Referring to Figure 7.9 and using the fact that the aforementioned random variable W possesses a chi-square distribution with $n-1$ degrees of freedom,

$$P\left(\chi^2_{1-\alpha/2,n-1} < \frac{(n-1)S^2}{\sigma^2} < \chi^2_{\alpha/2,n-1} \right) = 1 - \alpha$$

Rearranging the probability inequality yields

$$P\left(\frac{(n-1)S^2}{\chi^2_{\alpha/2,n-1}} < \sigma^2 < \frac{(n-1)S^2}{\chi^2_{1-\alpha/2,n-1}} \right) = 1 - \alpha$$

which, according to Definition 7.15, is precisely what is meant by the statement of the theorem. ∎

EXAMPLE 7.25

Readings on a temperature gage derive from a sending unit immersed in the medium whose temperature the gage reflects. A company wishes to estimate the variability of a new low-cost sending unit at a critical temperature. Using Theorem 7.17, we will construct a 95% confidence interval of the variance based upon a sample of 30 units that yielded the empirical variance $s^2 = 0.85$. Since (Table A.5) $\chi^2_{0.025,29} = 45.722$ and $\chi^2_{0.975,29} = 16.047$, the desired 95% confidence interval is

$$\left(\frac{29 \times 0.85}{45.722}, \frac{29 \times 0.85}{16.047} \right) = (0.54, 1.54)$$

The corresponding 95% confidence interval for the standard deviation is $(0.73, 1.24)$. Had $s^2 = 0.85$ been obtained from a sample size of 50, then, based on the chi-square values $\chi^2_{0.025,49} = 70.222$ and $\chi^2_{0.975,49} = 31.555$, the 95% confidence interval for σ^2 given in Theorem 7.17 would have been $(0.59, 1.32)$.

Two points should be noted regarding Theorem 7.17. First, the confidence interval is not symmetric about the estimate s^2. Second, the population random variable must be normally distributed (or closely so). Should one have reason to doubt normality, then the confidence interval of Theorem 7.17 should not be employed.

7.9 BAYESIAN ESTIMATION

In discussing estimation, we have thus far taken the approach that the parameter to be estimated is unknown and that all knowledge concerning the parameter must be extracted from observational data. Such an approach is consistent with the relative-frequency interpretation of probability, namely, that the underlying probability distribution is in some degree mirrored in the normalized relative-frequency distribution. The scientist remains outside the experiment, and decisions are made through a suitable modeling of the observational data. Decision methods that adhere to this principle are called **objective** or **classical** methods.

A different approach is to consider the experimenter's prior understanding of the phenomenon being studied. Here, the state of nature is not something he or she necessarily views for the first time in a state of ignorance. Rather, the investigator possesses prior understanding regarding the phenomenon to be observed. Although the value of the parameter to be estimated is not known, neither is it fully unknown. From experience, the investigator might be aware of some distributional behavior concerning the parameter. If so, then it is appropriate to model the state of nature, not as an unknown constant to be estimated, but instead as a random variable whose specific value appropriate to the sample at hand is to be estimated. This approach is said to be **subjective,** in that it takes into account the investigator's beliefs concerning the parameter (state of nature).

7.9.1 Prior Distributions and Bayes Risk

If a parameter to be estimated is modeled as a random variable, then it is written in upper-case format Θ, and a sample value of the parameter is denoted by θ. As a random variable, Θ possesses a density $\pi(\theta)$, called the **prior density** or **prior distribution,** which quantifies the experimenter's degree of belief concerning the distribution of Θ prior to the observation of sample values. If the experimenter's belief concerning the likely value of the parameter is of a high degree, then the prior density will possess a small variance; conversely, if his or her belief is rather weak, then the prior density will possess a large variance.

As an illustration, suppose one wishes to estimate the mean service time required for a group of technicians to perform a certain task. The classical methodology would be to take a random sample of service times and estimate the mean by applying some estimation rule to the sample values. For instance, if the service-time distribution is exponential, then, according to Exercise 7.42, the maximum-likelihood estimator is the sample mean and $\bar{x}$ is employed as the estimate. Suppose, however, that the mean service time has been observed over a long period and has been seen to fluctuate in a Gaussian manner. Although the engineering staff cannot deterministically weight the factors that cause this fluctuation, they ascribe it to variations in worker performance. To take into account such experiential insight, it might be wise to model the mean of the exponential service time as a random variable which reflects the changing state of nature. Since the (random) mean is believed to be normally distributed, the appropriate prior density is

$$\pi(\mu) = \frac{1}{\sqrt{2\pi}\,\sigma_0} e^{-\frac{1}{2}\left(\frac{\mu - \mu_0}{\sigma_0}\right)^2}$$

where μ_0 and σ_0 are the mean and standard deviation, respectively, of the (random) mean M (upper-case mu) of the service-time distribution. Given a random sample, the resulting estimate of the mean would then incorporate the observational data in conjunction with the a priori understanding of the mean's behavior.

To understand the manner in which Θ is to be estimated, it is necessary to look more closely at the mean-square error of a point estimator (Definition 7.10). Given a random sample $X_1, X_2, \ldots, X_n$, the mean-square error associated with the estimator

$\hat{\theta}$ of θ is a function of θ, the true value of the parameter. In the present context, this error is called the **risk function** (of θ) and is written

$$\mathcal{R}(\theta, \hat{\theta}) = E[(\theta - \hat{\theta})^2]$$

Since $\hat{\theta} = \hat{\theta}(X_1, X_2, \ldots, X_n)$, in the continuous case $\mathcal{R}(\theta, \hat{\theta})$ is given by the n-fold integral

$$\mathcal{R}(\theta, \hat{\theta}) = \int_{-\infty}^{+\infty} \cdots \int_{-\infty}^{+\infty} [\theta - \hat{\theta}(x_1, x_2, \ldots, x_n)]^2 f(x_1, x_2, \ldots, x_n; \theta) \, dx_1 \, dx_2 \cdots dx_n$$

where $f(x_1, x_2, \ldots, x_n; \theta)$ is the joint density of the sample random variables. Since the sample variables are independent, the density is the product of the marginal densities. Hence,

$$\mathcal{R}(\theta, \hat{\theta}) = \int_{-\infty}^{+\infty} \cdots \int_{-\infty}^{+\infty} [\theta - \hat{\theta}(x_1, x_2, \ldots, x_n)]^2 \left[\prod_{k=1}^{n} f(x_k; \theta) \right] dx_1 \, dx_2 \cdots dx_n$$

In the event that the population random variable is discrete, the integrals in the preceding expressions are replaced by summations.

According to the definitions of Section 7.5, $\hat{\theta}_1$ is a better estimator of θ than $\hat{\theta}_2$ if

$$E[(\theta - \hat{\theta}_1)^2] < E[(\theta - \hat{\theta}_2)^2]$$

Such an inequality is problematic, since it may hold for some values of θ and not for others. The problem was mitigated in Section 7.5 by restricting attention to unbiased estimators, in which case the inequality reduces to

$$\text{Var}\,[\hat{\theta}_1] < \text{Var}\,[\hat{\theta}_2]$$

The result was the MVUE of Definition 7.11.

In the subjective approach, the parameter Θ is random and the risk function takes the form

$$\mathcal{R}(\Theta, \hat{\theta}) = E_{\hat{\theta}}[(\Theta - \hat{\theta})^2]$$

where the subscript $\hat{\theta}$ denotes that the expected value is taken with respect to $\hat{\theta}$. To compare the quality of two estimators of Θ, it appears reasonable to compare their risk functions. However, since Θ is a random variable, the n-fold integral $E_{\hat{\theta}}[(\Theta - \hat{\theta})^2]$ is also a random variable. Therefore the expected values of the risk functions are compared.

DEFINITION 7.17

Bayes Risk. Suppose Θ is a random variable possessing prior density $\pi(\theta)$. Then the Bayes risk (also called the **mean risk**) of the estimator $\hat{\theta}$ is given by

$$B(\hat{\theta}) = E[\mathcal{R}(\Theta, \hat{\theta})]$$

where the expected value is taken with respect to the distribution of Θ.

If Θ is a continuous random variable, then the Bayes risk takes the form

$$B(\hat{\theta}) = \int_{-\infty}^{+\infty} \mathcal{R}(\theta, \hat{\theta})\pi(\theta)\, d\theta$$

If Θ is discrete, then

$$B(\hat{\theta}) = \sum \mathcal{R}(\theta, \hat{\theta})\pi(\theta)$$

where the sum is taken over all θ in the codomain of Θ.

Given a prior distribution, $B(\hat{\theta})$ is a real number expressing the expected risk. Estimators can be compared in terms of their Bayes risks: $\hat{\theta}_1$ is judged **better** than $\hat{\theta}_2$ if

$$B(\hat{\theta}_1) < B(\hat{\theta}_2)$$

To find a **best** estimator among a class of estimators, select one possessing minimal Bayes risk.

DEFINITION 7.18

Bayes Estimator. For a given prior distribution $\pi(\theta)$, the estimator $\hat{\theta}_0$ is called the Bayes estimator in the class C of estimators if

$$B(\hat{\theta}_0) = \min_{\hat{\theta} \in C} B(\hat{\theta})$$

7.9.2 The Posterior Distribution

In general, parametric estimation concerns the estimation of an unknown parameter in a probability distribution. In the classical approach to estimation, if X is the random variable of interest, $f(x)$ its density, and θ the unknown parameter, then the density is written as $f(x; \theta)$ in order to indicate the dependency of the distribution on the parameter to be estimated. In the Bayesian setting, the parameter Θ is a random variable, so that the density of X is conditioned on the value of Θ. Thus, it is appropriate to consider the density as a conditional density and write $f(x \mid \theta)$.

Given a random sample $X_1, X_2, \ldots, X_n$, the sample random variables can be considered to be jointly distributed with the parameter random variable Θ. Thus, there exists a joint probability density

$$f_{X_1, X_2, \ldots, X_n, \Theta}(x_1, x_2, \ldots, x_n, \theta)$$

To simplify notation we will suppress the random-variable subscript and simply write $f(x_1, x_2, \ldots, x_n, \theta)$. Since our ultimate goal is to form an estimator that gives a value of Θ dependent on observations of the sample random variables, it should be anticipated that the conditional density of Θ given $X_1, X_2, \ldots, X_n$ will play a role. This density, denoted by

$$f(\theta \mid x_1, x_2, \ldots, x_n)$$

is called the **posterior density.** The next theorem expresses the posterior density in terms of the prior density and the density of the population random variable.

If $X_1, X_2, \ldots, X_n$ is a random sample arising from the random variable X possessing density $f(x \mid \theta)$ and Θ possesses prior density $\pi(\theta)$, then the posterior density is

$$f(\theta \mid x_1, x_2, \ldots, x_n) = \frac{\pi(\theta) \prod\limits_{k=1}^{n} f(x_k \mid \theta)}{\int\limits_{-\infty}^{+\infty} \left[\pi(\theta) \prod\limits_{k=1}^{n} f(x_k \mid \theta) \right] d\theta}$$

Proof: Due to the independence of the sample variables and the fact that they are identically distributed with X, their conditional density with respect to Θ is

$$f(x_1, x_2, \ldots, x_n \mid \theta) = \prod_{k=1}^{n} f(x_k \mid \theta)$$

Thus, according to the definition of a conditional density, the joint density of the sample variables and the random parameter is

$$f(x_1, x_2, \ldots, x_n, \theta) = \pi(\theta) f(x_1, x_2, \ldots, x_n \mid \theta)$$

$$= \pi(\theta) \prod_{k=1}^{n} f(x_k \mid \theta)$$

The joint density $f(x_1, x_2, \ldots, x_n)$ of the sample variables can be obtained as a marginal density of $f(x_1, x_2, \ldots, x_n, \theta)$ by integrating over θ from $-\infty$ to $+\infty$:

$$f(x_1, x_2, \ldots, x_n) = \int_{-\infty}^{+\infty} f(x_1, x_2, \ldots, x_n, \theta) \, d\theta$$

$$= \int_{-\infty}^{+\infty} \left[\pi(\theta) \prod_{k=1}^{n} f(x_k \mid \theta) \right] d\theta$$

Consequently, the conditional density of Θ given $X_1, X_2, \ldots, X_n$ takes the form

$$f(\theta \mid x_1, x_2, \ldots, x_n) = \frac{f(x_1, x_2, \ldots, x_n, \theta)}{f(x_1, x_2, \ldots, x_n)}$$

$$= \frac{\pi(\theta) \prod\limits_{k=1}^{n} f(x_k \mid \theta)}{\int\limits_{-\infty}^{+\infty} \left[\pi(\theta) \prod\limits_{k=1}^{n} f(x_k \mid \theta) \right] d\theta}$$

∎

EXAMPLE 7.26

Thus far, we have taken the view that the proportion of defective units resulting from a manufacturing process is a constant. Now, however, consider the situation in which the proportion of defective units fluctuates due to uncertainties in the production process, such as worker efficiency, machine wear, and lack of component uniformity caused by the need to utilize various suppliers. Suppose numerous defective proportions have been recorded and they tend to possess a beta distribution with parameters α and β. It might then be wise to consider the proportion of defective units as a random variable P possessing prior density

$$\pi(p) = \frac{p^{\alpha-1}(1-p)^{\beta-1}}{B(\alpha, \beta)}$$

for $0 \leq p \leq 1$ and $\pi(p) = 0$ otherwise. If X denotes the number of defective units selected on a single observation, then the density of X takes the form

$$f(x \mid p) = p^x(1-p)^{1-x}$$

where x is 0 or 1. If $X_1, X_2, \ldots, X_n$ is a random sample of n observations, then

$$\pi(p) \prod_{k=1}^{n} f(x_k \mid p) = \frac{p^{n\bar{x}+\alpha-1}(1-p)^{n-n\bar{x}+\beta-1}}{B(\alpha, \beta)}$$

where $n\bar{x} = x_1 + x_2 + \cdots + x_n$ is the number of defectives observed. Therefore,

$$\int_{-\infty}^{+\infty} \pi(p) \prod_{k=1}^{n} f(x_k \mid p) \, dp = \frac{1}{B(\alpha, \beta)} \int_{0}^{1} p^{n\bar{x}+\alpha-1}(1-p)^{n-n\bar{x}+\beta-1} \, dp$$

$$= \frac{B(n\bar{x} + \alpha, n - n\bar{x} + \beta)}{B(\alpha, \beta)}$$

the last equality following because the integral is a beta function with arguments $n\bar{x} + \alpha$ and $n - n\bar{x} + \beta$ (see Section 4.6.1). According to Theorem 7.18, the posterior density for P given $X_1, X_2, \ldots, X_n$ is

$$f(p \mid x_1, x_2, \ldots, x_n) = \frac{p^{n\bar{x}+\alpha-1}(1-p)^{n-n\bar{x}+\beta-1}}{B(n\bar{x} + \alpha, n - n\bar{x} + \beta)}$$

for $0 \leq p \leq 1$, which is a beta density with parameters $n\bar{x} + \alpha$ and $n - n\bar{x} + \beta$.

It is important to keep in mind that the posterior distribution expresses our probabilistic understanding regarding the random parameter Θ subsequent to observation of the random sample, whereas the prior distribution expresses that understanding before observation. Intuitively, the prior-posterior distribution pair represents an updating of our understanding regarding the distribution of the parameter (state of nature). While this approach might at first appear odd, it should be recognized that it is remarkably reflective of intelligence. An intelligent being does not act solely upon the most recent information; rather, the new data is incorporated into existing understanding to yield an updated understanding regarding the phenomenon being observed. The full implication of this updating in regard to estimation will soon become apparent.

7.9.3 Bayes Estimators

According to Definition 7.18, a Bayes estimator is determined by minimizing the Bayes risk relative to some prior distribution $\pi(\theta)$. The Bayes risk of an estimator $\hat{\theta}$ is given by

$$B(\hat{\theta}) = E[\mathscr{R}(\Theta, \hat{\theta})]$$

$$= \int_{-\infty}^{+\infty} \mathscr{R}(\theta, \hat{\theta})\pi(\theta)\, d\theta$$

$$= \int_{-\infty}^{+\infty} E_{\hat{\theta}}[(\theta - \hat{\theta}(X_1, X_2, \ldots, X_n))^2]\pi(\theta)\, d\theta$$

$$= \int_{-\infty}^{+\infty}\left[\int_{-\infty}^{+\infty}\cdots\int_{-\infty}^{+\infty} [\theta - \hat{\theta}(x_1, x_2, \ldots, x_n)]^2 \right.$$
$$\left. f(x_1, x_2, \ldots, x_n \mid \theta)\, dx_1\, dx_2 \cdots dx_n\right]\pi(\theta)\, d\theta$$

$$= \int_{-\infty}^{+\infty}\cdots\int_{-\infty}^{+\infty} [\theta - \hat{\theta}(x_1, x_2, \ldots, x_n)]^2$$
$$f(x_1, x_2, \ldots, x_n, \theta)\, dx_1\, dx_2 \cdots dx_n\, d\theta$$

where we have employed, in order, the definitions of Bayes risk, of expected value, of the risk function, of expected value, and of conditional density. Note that the final integral is $(n + 1)$-fold. Relative to $f(x_1, x_2, \ldots, x_n, \theta)$, the last integral is the expected value

$$E[(\Theta - \hat{\theta}(X_1, X_2, \ldots, X_n))^2]$$

Thus, the Bayes estimator is that function of the sample variables that minimizes this expectation.

Although a rigorous derivation of the Bayes estimator is outside the scope of the present text, we can provide a heuristic analysis of the minimization. As discussed at the conclusion of Section 3.6.3, the mean of a random variable X provides the best estimate of the outcome of X in the sense that it minimizes the mean-square error $E[(X - c)^2]$ over all possible choices of c. Analogously, if X and Y are jointly distributed, then, given an observation $X = x$, the best (mean-square) estimate of $Y|x$ is the conditional expectation $E[Y|x]$. In the case of Bayesian estimation, the variable Θ is estimated based upon observations $x_1, x_2, \ldots, x_n$. The concept of conditional expectation can be extended to more than a single variable and, analogous to that case, the best estimate of the variable Θ given $x_1, x_2, \ldots, x_n$ is the conditional expectation

$$\overset{*}{\theta} = E[\Theta \mid x_1, x_2, \ldots, x_n] = \int_{-\infty}^{+\infty} \theta f(\theta \mid x_1, x_2, \ldots, x_n)\, d\theta$$

Assuming that $\overset{*}{\theta}$ is expressed by the same functional relation for all possible choices of the sample values, an estimator $\hat{\theta}$ is defined. It can be shown that this estimator minimizes the Bayes risk

$$B(\hat{\theta}) = E[(\Theta - \hat{\theta}(X_1, X_2, \ldots, X_n))^2]$$

The next theorem formally states this fundamental proposition.

THEOREM 7.19

Suppose X is a random variable possessing density $f(x \mid \theta)$, where θ is considered to be a value of the random parameter Θ possessing prior density $\pi(\theta)$. If $X_1, X_2, \ldots, X_n$ is a random sample, then the Bayes estimator relative to the prior distribution $\pi(\theta)$ is given by the conditional expectation

$$\hat{\theta} = E[\Theta \mid X_1, X_2, \ldots, X_n]$$

In particular, for a set of sample values $x_1, x_2, \ldots, x_n$, the Bayes estimate is given in terms of the posterior density by

$$\overset{*}{\theta} = E[\Theta \mid x_1, x_2, \ldots, x_n] = \int_{-\infty}^{+\infty} \theta f(\theta \mid x_1, x_2, \ldots, x_n)\, d\theta$$

EXAMPLE 7.27

In Example 7.26 we found the posterior density in the case where the random parameter P corresponds to the proportion of defective units in some manufacturing process. According to Theorem 7.19, for any collection of sample values $x_1, x_2, \ldots, x_n$, the Bayes estimate of P relative to the beta prior distribution is

$$\overset{*}{p} = E[P \mid x_1, x_2, \ldots, x_n] = \int_{-\infty}^{+\infty} p f(p \mid x_1, x_2, \ldots, x_n)\, dp$$

which is simply the mean (with respect to p) of the posterior beta distribution found in Example 7.26. By Theorem 4.14

$$\overset{*}{p} = \frac{n\bar{x} + \alpha}{n + \alpha + \beta}$$

$n\bar{x}$ being the number of defective units in the sample. Thus, the Bayes estimator of the proportion is the statistic

$$\hat{p} = \frac{n\bar{X} + \alpha}{n + \alpha + \beta}$$

Note that the Bayes estimator depends on both the sample values and the prior distribution. Moreover, for large n the difference between the maximum-likelihood estimator $\bar{X}$ and the Bayes estimator is small. Indeed, the Bayes estimator can be rewritten as

$$\hat{p} = \frac{\bar{X} + \dfrac{\alpha}{n}}{1 + \dfrac{\alpha + \beta}{n}}$$

which approaches $\bar{X}$ as $n \longrightarrow \infty$.

The closeness of the Bayes estimator and the maximum-likelihood estimator that occurred for large n in the preceding example is typical of Bayes estimators. In fact, it can be shown in general that the difference between the two estimators is small when compared to $1/\sqrt{n}$. If n is not large and the sample values are compatible with the prior distribution, then the Bayes and maximum-likelihood estimators will still be close; however, if n is small and the sample values differ markedly from those that would be expected given the prior distribution, then the two estimators can differ considerably.

EXAMPLE 7.28

Suppose in the estimation of a proportion that the experimenter lacks any credible prior information regarding the state of nature. If a Bayesian approach is to be taken, then it would seem reasonable to employ a uniform prior distribution with the random proportion P being between 0 and 1. Thus

$$\pi(p) = \begin{cases} 1, & \text{if } 0 \le p \le 1 \\ 0, & \text{otherwise} \end{cases}$$

As in Example 7.26, the density governing a single observation is

$$f(x \mid p) = p^x(1 - p)^{1-x}$$

Therefore, according to Theorem 7.18, the posterior density is

$$f(p \mid x_1, x_2, \ldots, x_n) = \frac{p^{n\bar{x}}(1 - p)^{n - n\bar{x}}}{\displaystyle\int_0^1 p^{n\bar{x}}(1 - p)^{n - n\bar{x}} \, dp}$$

$$= \frac{p^{n\bar{x}}(1 - p)^{n - n\bar{x}}}{B(n\bar{x} + 1, n - n\bar{x} + 1)}$$

which is a beta density with parameters $n\bar{x} + 1$ and $n - n\bar{x} + 1$. By Theorem 7.19, the Bayes estimate corresponding to the sample values $x_1, x_2, \ldots, x_n$ is the mean of the posterior beta density, namely,

$$\overset{*}{p} = \frac{n\bar{x} + 1}{n + 2} = \frac{\bar{x} + \dfrac{1}{n}}{1 + \dfrac{2}{n}}$$

The Bayes estimator is

$$\hat{p} = \frac{\bar{X} + \dfrac{1}{n}}{1 + \dfrac{2}{n}}$$

EXERCISES

SECTION 7.1

7.1. Suppose X possesses an exponential distribution with mean $E[X] = 3$. Find the joint density of a random sample X_1, X_2, X_3, X_4.

7.2. Repeat Exercise 7.1 with $E[X] = 1/b$ and a random sample of size n.

7.3. Find the joint density for a random sample of size n arising from a Poisson distribution with mean λ.

7.4. Find the joint density for a random sample of size n arising from a gamma distribution with parameters α and β.

7.5. Find the joint density for a random sample of size n arising from a random variable possessing a uniform distribution over the interval $[a, b]$.

7.6. The following BTU measurements are taken relative to some chemical process:

$$1244 \quad 1198 \quad 1212 \quad 1235 \quad 1245$$
$$1190 \quad 1202 \quad 1220 \quad 1233 \quad 1208$$

It is assumed that the measurements are sample values corresponding to a random sample of size 10 arising from the random variable X. Employ the estimators discussed in Example 7.2 to obtain estimates of the mean and variance of X.

7.7. Apply the mean estimator $\hat{\mu} = (X_{10} + X_1)/2$ to the experiment of Exercise 7.6.

7.8. Apply the mean estimator $\hat{\mu} = (X_3 + X_4 + X_5 + X_6 + X_7 + X_8)/6$ to the experiment of Exercise 7.6.

7.9. Suppose X possesses a normal distribution with mean μ and variance σ^2. Find the sampling distribution and expected value of the mean estimator $\hat{\mu} = (X_1 + X_2 + \cdots + X_n)/n$.

7.10. Repeat Exercise 7.9 except let X possess a binomial distribution with parameters p and m.

7.11. For the estimator of Example 7.8, show that $E[\hat{\mu}] = \mu$.

SECTION 7.2

7.12. Find the mean and variance of the estimators of Exercises 7.7 and 7.8.

7.13. Given a random sample of size n, find the mean and variance of the estimator $\hat{\mu} = (X_n + X_1)/2$.

7.14. A random sample of 20 measurements is to be taken of the pressure in a pressurized chamber. It is assumed by the investigator that variation in the measurements can be ascribed to instrument error and that the measurements are normally distributed with variance $\sigma^2 = 0.0004$. What is the probability that the sample mean will be within $r = 0.01$ atmosphere of the true mean?

7.15. Redo Exercise 7.14 with $r = 0.008$.

7.16. Redo Exercise 7.14 with $n = 30$.

7.17. Redo Exercise 7.14 with $r = 0.008$ and $n = 30$.

7.18. To check the viscosity of a synthetic oil, a petroleum engineer plans to take a random sample (at a fixed temperature) of viscosity measurements from various production runs. Assuming the measurement process to be normally distributed with known variance $\sigma^2 = 0.0054$, what must the sample size n to insure that the probability of the sample mean's being within $r = 0.01$ of the mean is at least $p = 0.95$?

7.19. Redo Exercise 7.18 with $p = 0.98$.

7.20. Redo Exercise 7.18 with $p = 0.90$.

7.21. Redo Exercise 7.18 with $r = 0.004$.

7.22. Redo Exercise 7.18 with $r = 0.05$.

7.23. Consider the mean estimator

$$\hat{\mu} = \frac{1}{n-4} \sum_{k=3}^{n-2} X_k$$

where it is assumed that $X_1, X_2, \ldots, X_n$ comprise a random sample of size $n \geq 5$. Show that $\hat{\mu}$ is a consistent estimator of μ.

7.24. In a case-hardening process for machined shafts, an engineer wishes to estimate the mean depth of hardening (in inches). To do so, he or she makes the assumption that the variance is $\sigma^2 = 0.0005$; however, no assumption is made concerning the distribution of the depths. If 40 observations are to be taken, what is the probability that the sample mean will lie within 0.004 inch of the mean?

7.25. Referring to Exercise 7.24, what must the sample size be to make the probability of the sample mean's lying within 0.004 of the mean to be 0.99?

SECTION 7.3

7.26. The time to failure of a commercially available flashlight battery is 2.78 hours when employed in a particular flashlight, failure meaning that the luminous intensity drops below a certain threshold value. It is known that the distribution of the intensity is not normal. Given that 100 test measurements are made with a resulting standard deviation $s = 0.26$, what is the probability that the sample mean will be within 0.05 of the true mean?

7.27. Referring to Exercise 7.26, how many measurements need to be taken so that $P(|\bar{X} - \mu| < 0.05) \geq 0.98$? Assume $s = 0.26$ is a good approximation to σ.

7.28. To find the mean water temperature of a large lake on a given day, an experimenter decides to take 40 measurements at randomly selected locations. Assuming normality, what is the probability that the sample variance will be within 10% of the true variance?

SECTION 7.4

7.29. The revolutions per minute (rpm) of three randomly selected electric motors are measured. It is assumed that the rpm measurements are uniformly distributed over the interval [1540, 1560]. Find the densities and probability distribution functions of the three order statistics.

7.30. Referring to Exercise 7.29, what is the probability that at least two test pumps possess rpm ratings less than 1550 rpm?

7.31. A random variable X possesses the probability distribution function

$$F_X(x) = \begin{cases} 0, & \text{if } x < 0 \\ \dfrac{x}{2}, & \text{if } 0 \leq x \leq 1 \\ \dfrac{x}{4} + \dfrac{1}{4}, & \text{if } 1 \leq x \leq 3 \\ 1, & \text{if } x > 3 \end{cases}$$

For a random sample of four, find the densities of the least and maximum order statistics.

SECTION 7.5

7.32. Show that the sample mean is an MVUE for the mean of an exponential random variable.

7.33. Show that the sample mean is an MVUE for the mean of a Poisson random variable.

7.34. Suppose X is a binomial random variable with parameters 1 and p. Show that the sample mean is an MVUE of p.

7.35. Suppose $\hat{\mu}_1$ and $\hat{\mu}_2$ are sample means corresponding to independent random samples of sizes n_1 and n_2, respectively, arising from a normally distributed random variable possessing mean μ and variance σ^2. Show that, for any constant a between 0 and 1, $\hat{\mu} = a\hat{\mu}_1 + (1 - a)\hat{\mu}_2$ is an unbiased estimator. Show that this estimator possesses minimum variance when $a = n_1/(n_1 + n_2)$.

SECTION 7.6

7.36. Show that the sample mean is the maximum-likelihood estimator of the mean of a Poisson random variable.

7.37. For a random sample of size n arising from a gamma distribution with α known and β unknown, show that

$$\hat{\beta} = \frac{1}{n\alpha} \sum_{k=1}^{n} X_k$$

is the maximum-likelihood estimator of β.

7.38. Show that

$$\widehat{\eta^2} = \frac{1}{2n} \sum_{k=1}^{n} X_k^2$$

is the maximum-likelihood estimator of the parameter η^2 in the Rayleigh distribution.

7.39. Show that $\hat{p} = \bar{X}$ is the maximum-likelihood estimator for the parameter p in the binomial density $b(x; 1, p)$.

7.40. Find the maximum-likelihood estimator of the parameter of a geometric distribution.

7.41. For the density

$$f(x; a) = \begin{cases} e^{-(x-a)} & \text{if } x \geq a \\ 0, & \text{otherwise} \end{cases}$$

show that the maximum-likelihood estimator for a is the minimum sample value.

7.42. Show that $\bar{X}$ is the maximum-likelihood estimator of the mean of an exponential distribution.

SECTION 7.7

7.43. Use the method of moments to determine an estimator of the mean of a Poisson random variable.

7.44. Use the method of moments to find an estimator of the parameter of a geometric distribution.

7.45. Use the method of moments to find an estimator of the parameter p of a binary random variable.

7.46. Use the method of moments to find an estimator of the parameter α of a beta distribution, given $\beta = 1$.

7.47. Use the method of moments to find an estimator of the parameter a in the density of Exercise 7.41.

SECTION 7.8

7.48. The following measurements (in psi) are taken of the tensile strength of the alloy titanium 150-A:

151,320	149,832	150,458
150,020	149,263	149,908
149,395	150,956	150,003

Assuming the measurements to have been randomly selected from a normal population with standard deviation $\sigma = 442$, find both 90% and 95% confidence intervals for the mean tensile strength.

7.49. Fractures in metals occur over time, not instantaneously. The following rates of crack advance (in feet per second) are measured in a particular steel under a given set of test conditions:

3230	3106	3328	3065	3240	3132	3189
3092	3188	3250	3204	3268	3288	3300

Given that the measurements have been randomly selected from a normal population with standard deviation $\sigma = 53$, find both 99% and 85% confidence intervals for the mean rate of crack advance.

7.50. Referring to Exercise 7.48, what sample size is necessary to yield a 95% confidence interval of length 200?

7.51. Referring to Exercise 7.49, what sample size is necessary to yield a 98% confidence interval of length 10?

7.52. Accessing disk storage is extremely slow in comparison to the accessing of internal computer memory, because disk access requires mechanical movements. For a particular system, 100 measurements are taken of the time to access a record from a disk, the empirical mean and standard deviation being $\bar{x} = 0.032$ seconds and $s = 0.005$, respectively. Find a 95%

confidence interval corresponding to the data and explain why no distributional assumptions are necessary.

7.53. To find an interval estimate of the number of miles a certain brand of tire will last before the tread depth falls below a minimal safety threshold, a manufacturer tests 50 tires under various operating conditions. The recorded data yield the estimates $\bar{x} = 32{,}460$ miles and $s = 3106$. Find an 80% confidence interval for the mean number of miles.

7.54. Find $t_{0.025,10}$ and $t_{0.01,12}$.

7.55. Fatigue failure in metal results from the repeated application of small loads, which eventually produce a crack. To test a metal for its ability to withstand fatigue, repeated stress is applied and reversed for a large number of cycles until fracture, the number of cycles being a function of the applied stress. During a fatigue test, the following data (in 10^5 cycles) are randomly obtained:

$$4.39 \quad 4.75 \quad 4.20 \quad 4.66$$
$$4.72 \quad 4.28 \quad 4.40 \quad 4.48$$

Assuming the cycles-to-failure population is normally distributed with unknown mean and variance, find both 95% and 99% confidence intervals for the mean using the given data.

7.56. Using the data of Exercise 7.48, find a 95% confidence interval under the assumption that the variance is unknown.

7.57. Using the data of Exercise 7.49, find a 99% confidence interval under the assumption that the variance is unknown.

7.58. The recrystallization temperature is the temperature at which the first new grain appears during the annealing process of cold-worked metal. It is affected by prior deformation, annealing time, and soluble impurities. Consequently, test conditions must be specified. An engineer wishes to estimate the recrystallization temperature of aluminum under a given set of conditions. It is assumed that variation arises from measurement error and the temperature distribution is normally distributed. The following temperatures are recorded (in degrees centigrade):

$$148.73 \quad 149.43 \quad 149.98 \quad 150.40$$
$$151.01 \quad 149.10 \quad 150.00$$

Use the data to find both 95% and 99% confidence intervals for the mean.

7.59. A miniature autonomous aircraft is programmed to fly to a certain location, use its sensors to record terrain information, and return. Out of 140 flights, the aircraft completes its mission successfully 124 times. Find both 90% and 95% confidence intervals for the probability that a randomly observed mission will be successful.

7.60. Thirty electricians are randomly selected from the

membership of a certain union and each is given a test to determine whether his or her knowledge is up to an acceptable level. Assuming that ten fail the test, find a 90% confidence interval for the proportion of union members who are sufficiently knowledgeable of their trade.

7.61. To find an interval estimate for the proportion of defective diodes in a large lot, an investigator randomly selects 100 diodes and finds that two of them are defective. Can he or she apply the approximation technique discussed in the text?

7.62. Referring to Exercise 7.48 and assuming the variance to be unknown, use the data provided to find a 95% confidence interval for the variance.

7.63. Referring to Exercise 7.49 and assuming the variance to be unknown, use the data provided to find a 99% confidence interval for the variance.

SECTION 7.9

7.64. A manufacturer of lamps believes that the proportion of defective lamps satisfies a beta distribution with $\alpha = 1$ and $\beta = 12$.
   a) Find the mean and variance of the prior density.
   b) What is the posterior density corresponding to a random sample of 20 units?
   c) What are the mean and variance of the posterior distribution?

7.65. The manager of an appliance service center believes that the proportion P of dissatisfied customers possesses a beta distribution with $\alpha = 2$ and $\beta = 10$.
   a) Find the mean and variance of the prior distribution.
   b) What is the posterior density corresponding to a random sample of 10 customers?
   c) What are the mean and variance of the posterior distribution?

7.66. If jobs arriving for service at a station satisfy the conditions of a Poisson experiment and possess arrival rate q per hour, then the number X of jobs arriving in an hour possesses a Poisson distribution with parameter q. Now, suppose the arrival rate is not constant, but the operating engineer believes it to be a gamma-distributed random variable Q possessing a prior density with parameters α and β. Show that the posterior density relative to a single observation (of jobs arriving in a given hour) is gamma distributed.

7.67. Referring to Exercise 7.66 (and recalling that the exponential distribution is a member of the gamma family), suppose Q possesses an exponential prior distribution with mean 2. Given a single observation of the process over an hour interval, find the posterior density, together with its mean and standard deviation.

7.68. Suppose X possesses an exponential distribution with random parameter B that is uniformly distributed over the interval [0, 1]. Find the posterior distribution for a random sample of size n.

7.69. Suppose X possesses an exponential distribution with random parameter B that itself possesses an exponential distribution with parameter c. Find the posterior distribution for a random sample of size n.

7.70. Based on a single observation of the process, find the Bayes estimator of the random parameter Q of Exercise 7.66.

7.71. Find the Bayes estimator of the parameter B in Exercise 7.68 based on a random sample of size n.

7.72. Find the Bayes estimator of the parameter B in Exercise 7.69 based on a random sample of size n.

HYPOTHESIS TESTS CONCERNING
A SINGLE PARAMETER

A type of inference different from parameter estimation relates to the choice between two hypotheses concerning a population. The present chapter concerns hypothesis tests dealing with a single parameter of a population distribution. Since the outcome of any test will depend on the observation of random phenomena, an important part of the design of a test is the analysis of possible errors. The first three sections of the chapter provide a general discussion of hypothesis testing, with the third concentrating on the important subject of operating characteristic curves. The next four sections describe tests for the mean of a normal random variable, for a proportion, and for the variance of a normal random variable. The last three sections discuss the design of tests that are best (relative to a given criterion), distribution-free tests, and the construction of various types of control charts for product quality.

8.1 TESTING A HYPOTHESIS

A common problem in science is the empirical verification or rejection of a hypothesis (or hypothetical assertion) concerning a population. Like the theory of estimation, hypothesis testing falls within the domain of inferential statistics. In both cases a decision is made. In estimation theory the decision is to utilize a parameter estimate $\overset{*}{\theta}$ in place of a population parameter θ; in hypothesis testing the decision is to choose between two competing hypotheses regarding the population random variable.

8.1.1 Tests Involving Two Simple Hypotheses

A **statistical hypothesis** is a conjecture regarding a random variable or the probability distribution of a random variable. Testing a statistical hypothesis involves two things: (1) determination of a test statistic and (2) utilization of a sample value of the test statistic to choose between a given hypothesis, called the **null hypothesis** (denoted by H_0) and a competing hypothesis, called the **alternative hypothesis** (denoted by H_1).

Implicit in the formulation of a testing procedure is the recognition that a statistic is a random variable defined as a function of several random variables $X_1, X_2, \ldots, X_n$ comprising a random sample, say

$$V = V(X_1, X_2, \ldots, X_n)$$

The testing procedure is to observe V and, based upon the observation, decide which hypothesis, H_0 or H_1, is to be adopted. In terms of the sampling distribution of the test statistic V, two regions are defined: the **acceptance region,** comprised of those values of V resulting in the adoption of H_0, and the **critical (rejection) region,** comprised of those values of V resulting in the adoption of H_1. Note that the terminology relates to *whether the null hypothesis H_0 is* **accepted** *or* **rejected.**

Since there are competing hypotheses regarding the population random variable X, the distribution of the test statistic V is not fully known; indeed, it is the observation of V that is to lead to a decision regarding which hypothesis is to be adopted. For instance, if the density of X contains an unknown parameter θ, which is known to possess one of two values, θ_0 or θ_1, then the density of X is of the form $f(x; \theta)$. Our choice of which value of θ to accept is determined by an observation of the test statistic. If $\theta = \theta_0$ is the null hypothesis, then we write

$$H_0: \theta = \theta_0$$

$$H_1: \theta = \theta_1$$

to represent the hypothesis-testing problem. The possible outcomes of the chosen test statistic are divided into two mutually exclusive classes: those in the acceptance region A and those in the critical (rejection) region C. Upon observation, if V falls in A, then H_0 is accepted; if V falls in C, then H_0 is rejected.

To better appreciate the foregoing concepts, we will consider in depth the design of a particular hypothesis test.

EXAMPLE 8.1

Having purchased a shipment of synthetic rubber seals for disk-brake calipers, a truck manufacturer suspects that substandard seals have been substituted for the ones ordered. From experience the manufacturer knows that, when subjected to ultrarigorous testing conditions, only 10% of the ordered seals will fail but 30% of the substandard seals will fail. To detect whether or not there has been a bogus shipment, the manufacturer plans to test 20 seals and make a determination based upon the number of brake failures due to seal rupture.

From the perspective of hypothesis testing, there is a population described by a binomial random variable X, with p, the probability of a failure, being unknown. Assuming the shipment to be large in number, the sampling procedure can be treated as random even though the testing will be done without replacement. If the test statistic V counts the number of seals that fail the testing procedure, then

$$V = X_1 + X_2 + \cdots + X_{20}$$

where $X_1, X_2, \ldots, X_{20}$ comprise a random sample, and the sampling distribution of V is binomial with parameters $n = 20$ and p. The two competing hypotheses can be described by

$$H_0: p = 0.3$$

$$H_1: p = 0.1$$

The null hypothesis represents the conjecture that there has been a bogus shipment.

Having chosen a test statistic, the next problem is to determine a critical region. The greater the number of failures, the more reason there is to accept the null hypothesis (that there has been a bogus shipment). The fewer the failures, the more reason there is to reject the null hypothesis and accept the alternative hypothesis (that the seals shipped are the genuine ones ordered). A reasonable choice for the critical region might be

$$C = \{0, 1, 2, 3, 4\}$$

where up to 4/20 or 0.2 of the test seals fail. The corresponding acceptance region would be

$$A = \{5, 6, 7, 8, 9, 10, 11, 12, 13, 14, 15, 16, 17, 18, 19, 20\}$$

If 4 or fewer seals fail the test, then the manufacturer rejects the null hypothesis (H_0: $p = 0.3$) and accepts the alternative hypothesis (H_1: $p = 0.1$). On the other hand, if 5 or more seals fail, then the manufacturer accepts the null hypothesis (that there has been a bogus shipment).

Now suppose 20 seals are subjected to the ultrarigorous testing procedure and only 3 fail. Then, according to the hypothesis test just designed, the manufacturer rejects the null hypothesis and concludes that the shipment does not consist of substandard seals. Is the manufacturer *certain* of this conclusion? No. The conclusion is an *inference* based upon the outcome of an experiment. A different outcome might yield a different inference.

In Example 8.1 it has been assumed that the density of the population random variable X is binomial with $n = 1$ and p unknown, so that the density takes the form

$$f_X(x; p) = b(x; 1, p)$$

In addition, each hypothesis, the null and the alternative, specifies a particular value of the unknown parameter p.

More generally, a statistical hypothesis might concern a number of unknown parameters relating to the distribution of the population random variable. Specifically, if $\theta_1, \theta_2, \ldots, \theta_m$ are unknown parameters, then the density of X can be written as

$$f_X(x; \theta_1, \theta_2, \ldots, \theta_m)$$

and a hypothesis might involve conjectures regarding the m unknown parameters. A hypothesis is a **simple hypothesis** if it specifies the value of each parameter in the density; otherwise, it is said to be a **composite hypothesis.** In the brake-seal problem, there were two simple hypotheses relating to a single unknown parameter. Now, consider a normally distributed population random variable with unknown mean and standard deviation. Its density takes the form $f_X(x; \mu, \sigma)$. A *simple null hypothesis* would be

$$H_0: \mu = 2, \quad \sigma = 3$$

which concerns the specification of two unknown parameters. A *composite null hypothesis* would be

$$H_0: \mu = 2, \quad \sigma < 3$$

For now we shall focus on simple hypotheses. Composite hypotheses regarding a single parameter will be addressed in Section 8.2.

8.1.2 Two Types of Error

Two types of error can result from the procedure used in Example 8.1. First, it might be that the null hypothesis H_0 is true, but the procedure (together with the actual observation) leads us to conclude H_0 is false. In other words, the actual **state of nature** is given by $\theta = \theta_0$, but the observed value of the test statistic falls in the critical region C, so we wrongly conclude that $\theta = \theta_1$. Second, it might be that H_0 is false, but the hypothesis test leads us to conclude it is true. In other words, the state of nature is given by $\theta = \theta_1$, but the observed value of the test statistic falls in the acceptance region A, so we wrongly conclude that $\theta = \theta_0$.

In effect, then, there are four possibilities, of which two lead to correct decisions and two lead to incorrect decisions. These four possibilities are characterized in Table 8.1, and the following definitions apply.

DEFINITION 8.1

Type I Error. An error of type I occurs if the null hypothesis H_0 reflects the true state of nature but the test statistic falls in the critical region, thereby leading to a rejection of H_0.

DEFINITION 8.2

Type II Error. An error of type II occurs if the null hypothesis H_0 does not reflect the true state of nature but the test statistic falls in the acceptance region, thereby leading to an acceptance of H_0.

Crucial to the construction of a suitable hypothesis test is the choice of a critical region that results in tolerable type I and type II errors. Since we can never know with certainty whether or not an error has been committed, we must concern ourselves with *error probabilities*. The probability of making a type I error, given the null hypothesis reflects the true state of nature, is called the **size of type I error** (also called the size of the critical region). The size of type I error is customarily denoted by α and is defined by

$$\alpha = P(V \in C; H_0)$$

where V is the test statistic, C is the critical region, and the notation "$P(E; H)$" is used to denote the probability of the event E given the condition H. The size of type II error is the probability of making a type II error, given the alternative hypothesis reflects the true state of nature. It is customarily denoted by β and is defined by

TABLE 8.1 TYPES OF ERROR

		True State of Nature	
		H_0	H_1
Decision	Reject H_0	Type I error	Correct decision
	Accept H_0	Correct decision	Type II error

$$\beta = P(V \in A; H_1)$$

where A is the acceptance region.

EXAMPLE 8.2

Consider the test statistic, critical region, and acceptance region of Example 8.1. To find the type I error, let $p = 0.3$, so that the test statistic V is binomially distributed with $n = 20$ and $p = 0.3$. Table A.2 shows the size of type I error to be

$$\begin{aligned} \alpha &= P(V \in C; H_0) \\ &= P(V \leq 4; p = 0.3) \\ &= B(4; 20, 0.3) = 0.2375 \end{aligned}$$

The table shows the size of type II error to be

$$\begin{aligned} \beta &= P(V \in A; H_1) \\ &= P(V \geq 5; p = 0.1) \\ &= 1 - P(V \leq 4; p = 0.1) \\ &= 1 - B(4; 20, 0.1) \\ &= 1 - 0.9568 = 0.0432 \end{aligned}$$

It is substantially more likely that H_0 will be rejected incorrectly than it will be accepted incorrectly.

From the standpoint of brake safety, there is a serious problem here. If H_0 is rejected (and H_1 accepted) when, in fact, H_0 represents the true state of nature, then substandard brake seals will be employed in the manufacturer's trucks. Specifically, given there has been a bogus shipment of seals, the probability is approximately 0.24 that this fact will not be detected. Since substandard seals present a serious safety hazard, the manufacturer should recognize that the size of type I error must be reduced.

Referring to Figure 8.1, which illustrates the probability mass function of the test statistic V under the assumption H_0: $p = 0.3$, we see that α, the probability of type I error given $p = 0.3$, is simply the probability mass to the left of 4.5. Decreasing the size of the critical region will reduce α. For instance, suppose the critical region is chosen as

$$C = \{0, 1, 2, 3\}$$

Then

$$\begin{aligned} \alpha &= P(V \in C; p = 0.3) \\ &= P(V \leq 3; p = 0.3) \\ &= B(3; 20, 0.3) = 0.1071 \end{aligned}$$

Because the sample value 4 is now included in the acceptance region, the corresponding size of type II error is

$$\begin{aligned} \beta &= P(V \in A; p = 0.1) \\ &= P(V \geq 4; p = 0.1) \\ &= 1 - B(3; 20, 0.1) = 0.1330 \end{aligned}$$

The size of type I error has been reduced, but at the cost of increased type II error. It is now much less likely that a substandard shipment will be accepted but much more likely that a shipment of satisfactory quality will be rejected. Of course, if safety is the prime concern, then the present situation is superior.

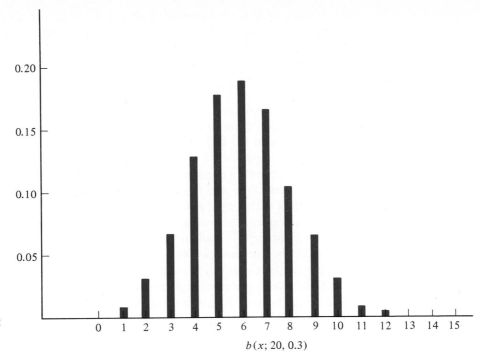

FIGURE 8.1

Distribution of the test statistic V of Example 8.2 under the null hypothesis H_0: $p = 0.3$

$b(x; 20, 0.3)$

If the manufacturer is still not satisfied, the size of type I error can be further reduced. For example, if the critical region is reduced to $C = \{0, 1, 2\}$, then $\alpha = 0.0355$ and $\beta = 0.3231$. Once again, α is reduced at the cost of increasing β.

The dilemma faced in Example 8.2 is customary in hypothesis testing. If we maintain the same sample size and attempt to reduce the size of type I error by reducing the critical region, we thereby increase the acceptance region. The result is usually an increase in the size of type II error. One way out of the dilemma is to increase the sample size. Unfortunately, however, in many circumstances the sample size is fixed by external constraints, such as cost.

If, in Example 8.2, the manufacturer's main concern is to *gain confidence* that the seals meet satisfactory standards and are therefore safe, then he or she most likely desires a small type I error. The manufacturer is more willing to risk the rejection of a good shipment than to risk the installation of substandard seals.

In general, the sizes of type I and type II errors are measures of risk, α measuring the risk involved in wrongly deciding to reject the null hypothesis and β measuring the risk involved in wrongly deciding to reject the alternative hypothesis. From a relative-frequency perspective, if α is very small, say $\alpha = 0.05$, only 5% of the time will the test statistic fall in the critical region if H_0 is true. Consequently, if the statistic does fall in the critical region, we are led to question the believability of H_0. Moreover, in rejecting H_0, there is little risk of making an error. On the other hand, should α be relatively large, say $\alpha = 0.2$, then, given that H_0 reflects the true state of nature, the wrong decision will be made 20% of the time, thereby entailing a much greater risk. Analogous remarks apply to the size of type II error.

In setting up an experiment, it is common to choose the null hypothesis so that its acceptance represents a weak decision, whereas its rejection represents a strong decision. Specifically, the size of type I error is chosen so as to reduce to a tolerable risk the chance of incorrectly accepting the alternative hypothesis. Typically, values of α are chosen to be less than or equal to 0.10, with $\alpha = 0.05$ and $\alpha = 0.01$ being the most commonly employed sizes. If the sample value of the test statistic does fall in the critical region, one is quite confident that it falls there due to the falseness of the null hypothesis rather than due to chance. From a scientific perspective, the intent is to provide compelling experimental evidence that a particular conjecture regarding the population is valid. In rejecting the null hypothesis with low risk, one is at the same time accepting the alternative hypothesis with low risk.

Based upon these considerations, it is understandable why α is called the **level of significance** of the test—the smaller α, the greater the significance of accepting the alternative hypothesis. Logically (and this point is crucial to the suitable application of a hypothesis test), a decision to reject the null hypothesis with a small-sized critical region is to conclude that the alternative hypothesis is worthy of adoption. On the other hand, a decision to accept the null hypothesis is merely to take the position that there is inadequate evidence for accepting the alternative hypothesis. Of course, one must be prudent in applying this logical principle in real-world situations, where there might be burdensome costs in maintaining the null hypothesis. While such a principle can be employed in situations where to make no strong decision means no action need take place, in a practical situation such luxury is often lacking. For instance, in Example 8.2, if the size of the critical region is chosen to be as close to 0.05 as possible, then $\alpha = 0.0355$ and $\beta = 0.3231$. While there is little risk here regarding safety, there is certainly a high risk regarding business considerations. If the supplier of the seals has been honest, then there is a 32% chance the manufacturer will erroneously conclude the seals are substandard. The consequences of such an erroneous decision include loss of time, a souring of business relations, further testing by an independent laboratory, and possible lawsuits. Because of the consequences of a type II error, it is often necessary to increase the sample size in order to reduce β. If both α and β are small, then both rejection and acceptance of the null hypothesis are strong decisions.

EXAMPLE 8.3

Suppose the truck manufacturer in Examples 8.1 and 8.2 finds the risk involved with $\beta = 0.3231$ intolerable but does not want to increase the size of type I error to $\alpha = 0.1071$. If the laboratory testing of the seals is not overly expensive or time consuming, the manufacturer has an option to increase the sample size—say, to 50. Then the test statistic, which is binomially distributed, can be treated as an approximately normal random variable and Theorem 4.9 can be applied. Under the assumption that the null hypothesis is valid, the test statistic V has an approximately normal distribution with mean

$$\mu = np = 50 \times 0.3 = 15$$

and standard deviation

$$\sigma = \sqrt{npq} = \sqrt{10.5} = 3.24$$

Referring to Figure 8.2 and recognizing that H_0 is to be rejected for small values of the test statistic, we find a critical region of size $\alpha = 0.05$ by selecting on the v axis the appropriate sample value v_0 that separates the critical and acceptance regions. The value v_0 is found by solving the probability equation

$$P(V < v; p = 0.3) = 0.05$$

where $p = 0.3$ is the null hypothesis. Changing to the standard normal variable Z yields

$$P\left(Z < \frac{v - 15}{3.24}\right) = 0.05$$

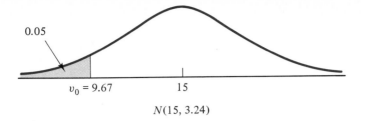

FIGURE 8.2
Critical region of size
$\alpha = 0.05$ in Example 8.3

$v_0 = 9.67$ 15

$N(15, 3.24)$

By Table A.1,

$$\frac{v - 15}{3.24} = -1.645$$

so that the desired value is $v_0 = 9.6702$. Thus, the null hypothesis will be rejected at a 0.05 level of significance if the critical region is to the left of 9.67. Correspondingly, the size of type II error is

$$\beta = P(V > 9.67; p = 0.1)$$

Applying Theorem 4.9 with $\mu = np = 5$ and $\sigma = \sqrt{npq} = \sqrt{4.5} = 2.12$ leads to

$$\beta = P\left(Z > \frac{9.67 - 5}{2.12}\right) = 1 - \Phi(2.20) = 0.0139$$

which is an extremely small value of β.

Since the outcomes of the experiment are discrete, the actual critical region consists of nonnegative integers less than or equal to 9 and the acceptance region consists of integers between 10 and 50, inclusive. Thus, the actual values of α and β are slightly different than those shown. Since we are using an approximation to begin with and since the resulting type I and type II errors are quite small, such slight differences are of minor consequence.

Whenever a value v_0 can be found that separates the critical and acceptance regions, v_0 is called a **critical value.** In some cases there might be more than a single critical value.

Having demonstrated in Example 8.3 that a larger sample size greatly reduces the risk resulting from type II error, let us return to a sample size of 20 to attack the problem from a different angle. Suppose the manufacturer is at first not interested in demonstrating safety, but rather in demonstrating that the shipment consists of substandard seals. Instead of making the null hypothesis represent a substandard shipment and the alternative hypothesis represent a shipment of satisfactory quality, we reverse the roles and consider the pair of hypotheses

$$H_0: p = 0.1$$

$$H_1: p = 0.3$$

Now, for a low value of α, the rejection of the null hypothesis leads to a strong conclusion that the shipment consists of substandard seals, and the critical region will be chosen so as to include higher numbers of observed failures. Whether or not acceptance of H_0 is a strong conclusion depends upon β. The key point is that the formulation of the hypotheses reflects the logical intent of the investigator.

8.2 COMPOSITE HYPOTHESES

Although at times it is appropriate to test one simple hypothesis against another, more often than not, one is trying to detect a meaningful change in a given parameter. For instance, a materials engineer might be interested in demonstrating the worth of a new steel-tempering process by detecting an increase in hardness, a petroleum engineer might wish to demonstrate that improved performance results from a new gasoline additive by detecting increased acceleration in test cars, or an agricultural specialist might wish to demonstrate the benefit of applying a new fertilizer by detecting increased crop yields. Each of these scientists is trying to empirically establish an overall increase in the value of some parameter without necessarily providing a specific alternative value of the parameter.

One might also be interested in demonstrating there has been a decrease in the value of a parameter. For example, an industrial engineer might wish to show the import of a change in a production process by providing evidence that production time has decreased, a computer programmer might wish to verify the benefit of a newly encoded algorithm over one presently in use by showing a decrease in run time, or a pharmaceutical chemist might wish to demonstrate the superior worth of a new drug by showing its use results in shorter recovery times.

In all of the preceding cases, new parameters are not being specified; instead, the intent is to detect change in a given direction. In each case, the null and alternative hypotheses must be formulated to accurately reflect the condition the investigator wishes to reject and the alternative he or she hopes to establish.

In a third scenario, the experimenter might merely wish to detect change in a parameter, the direction of the change being unimportant. For instance, since the bore of a cylinder must be centered according to a fixed set of coordinates, an industrial engineer must make certain that the mean horizontal and vertical distances of the center of the bore do not vary too widely from the prespecified coordinates. If X measures the horizontal error, rejection of the null hypothesis $\mu_X = 0$ results in the detection of horizontal error.

When the purpose of a hypothesis test is to detect a change in a population parameter, the hypothesis testing methodology is called **detection.** A fundamental question concerns the degree to which detection or nondetection is meaningful. Ultimately, meaningfulness is a heuristic notion determined by the engineer and then used by the statistician in the construction of a worthwhile test.

8.2.1 One-Tailed Tests

An appropriate test to demonstrate that a parameter θ has undergone a meaningful positive change from a given value θ_0 is given by

$$H_0: \theta = \theta_0$$

$$H_1: \theta > \theta_0$$

where H_1 is a composite hypothesis. From the perspective of detection, the null hypothesis $\theta = \theta_0$ might represent a working (contingent) hypothesis that has been employed for some time, whereas the alternative hypothesis represents a positive change in the parameter. From the perspective of scientific proof, a decision to accept the alternative hypothesis might represent verification that a new process is superior to the old.

Given the null and alternative hypotheses in the forms just specified, the stat-

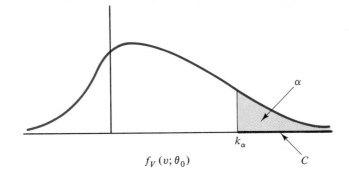

FIGURE 8.3
Critical region of size α for an upper one-tailed test

$f_V(v; \theta_0)$

k_α

C

istician must formulate a test statistic whose outcome can be employed to make a decision as to which hypothesis to accept. As in the case of competing simple hypotheses, the codomain of the test statistic must be decomposed into an acceptance region and a critical (rejection) region. Once again there are two types of possible error, and Table 8.1 applies. As before, α will denote the size of type I error and β the size of type II error. There is a difference, however, in the meaning of β, since the probability of the test statistic's falling in the acceptance region, given the truth of H_1, ultimately depends upon the value of θ, which is unspecified. Mathematically, the size of type II error will be a function of θ, $\beta = \beta(\theta)$. It is therefore necessary to pay careful attention to the empirical significance of possible type II errors when considering a particular critical region. Section 8.3 will be dedicated to this problem.

In testing the alternative hypothesis $\theta > \theta_0$ against the null hypothesis $\theta = \theta_0$, it is reasonable to select a test statistic V and a critical value k_α (α being the size of type I error), so that V falling to the left of k_α indicates the null hypothesis should be accepted and V falling to the right of k_α indicates the null hypothesis should be rejected. Specifically, the critical region is given by

$$C = \{v : v \geq k_\alpha\}$$

and the acceptance region is given by

$$A = \{v : v < k_\alpha\}$$

Assuming that H_0 reflects the true state of nature, the relationship between the critical region and the density $f_V(v; \theta_0)$ of the test statistic V is illustrated in Figure 8.3. According to the figure, α is determined by the probability mass under the density lying to the right of k_α—namely, the probability mass under the upper tail of the distribution. Consequently, the resulting hypothesis test is called an **upper one-tailed test.** Rejection occurs if $V \geq k_\alpha$; acceptance occurs if $V < k_\alpha$.

EXAMPLE 8.4

Up until a certain point, called the *proportional limit*, the strain of a ductile metal is proportional to the stress (Hooke's law). For a very small range past the proportional limit, the metal maintains its elasticity; however, it very soon reaches its *yield point*, the point at which the deformation is no longer reversible. Yield points are generally measured in kips per square inch (ksi), where 1 kip equals 1000 pounds.

Suppose a structural engineer has been experimenting with a new alloy that may increase the yield point in a certain high-strength structural steel. The engineer confronts the problem of demonstrating the superior yield point of the new alloy as opposed to the old. From experience it is believed that measurements of yield points in the class of alloys under consideration are normally distributed with known standard deviation $\sigma = 0.75$ ksi. The engineer plans to test nine rods composed of the new alloy, hoping to demonstrate that the previous mean yield point of 46 ksi can be rejected in favor of a higher yield point. The appropriate hypothesis test is given by

$$H_0: \mu = 46$$

$$H_1: \mu > 46$$

Since the sample mean is an unbiased estimator of the mean, it seems plausible that the alternative hypothesis $H_1: \mu > 46$ will be indicated by high values of $\bar{X}$ and not by low values of $\bar{X}$. In Section 8.8 we will be concerned with the construction of "best tests"; however, for the present we will content ourselves with a heuristic argument in favor of the sample-mean test statistic, assuming, of course, that the 9 test rods are selected in accordance with a random sampling procedure.

Given the engineer desires a critical region of size $\alpha = 0.05$, the appropriate critical value of k_α needs to be found. Since the critical region lies to the right of the critical value, k_α is found by solving the equation

$$P(\bar{X} \geq k_\alpha; H_0) = 0.05$$

(see Figure 8.4). Under the null hypothesis, $\mu = 46$. Therefore, changing to the standard normal variable yields the equivalent equation

$$P\left(Z > \frac{k_\alpha - 46}{0.75/\sqrt{9}}\right) = 0.05$$

which is itself equivalent to

$$\Phi\left(\frac{k_\alpha - 46}{0.25}\right) = 0.95$$

According to Table A.1,

$$\frac{k_\alpha - 46}{0.25} = 1.645$$

so that $k_\alpha = 46.41$ and $[46.41, \infty)$ is the desired critical region of size $\alpha = 0.05$. To illustrate the meaning of the critical value, if the yield points of 9 randomly selected rods are measured and the resulting empirical mean is $\bar{x} = 46.89$, then the alternative hypothesis is accepted and it is concluded that the yield point has been increased. If, on the other hand, an empirical mean of $\bar{x} = 46.22$ is obtained, then the null hypothesis is not rejected. Even

FIGURE 8.4
Critical region of size $\alpha = 0.05$ for Example 8.4

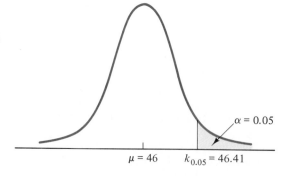

$\mu = 46$ $k_{0.05} = 46.41$

$\alpha = 0.05$

though, in the latter case, the sample value of the test statistic is greater than 46, it is not sufficiently greater to warrant acceptance of the alternative hypothesis at the 0.05 level of significance.

Let us now turn our attention to the size of type II error, which in this case is a function of the true mean. To appreciate the import of type II error, one must consider the scientific meaning of the test. In the present case, the experimenter will conclude that the new yield point is greater than the old value of 46 ksi if the sample value $\bar{x}$ of the test statistic $\bar{X}$ is greater than or equal to 46.41 ksi. Logically, type II error results from a wrong decision to accept the null hypothesis when the true value of the mean is actually greater than 46 ksi. Practically, such an error may or may not be of consequence. For instance, if an increase of 0.2 ksi is not important from a structural-engineering standpoint, a mistaken acceptance of the null hypothesis when the true mean is 46.2 ksi is of little consequence. On the other hand, if an increase of 0.6 ksi is structurally important, then a mistaken acceptance of the null hypothesis when the true mean is 46.6 ksi is of consequence. Indeed, a mistaken acceptance of the null hypothesis in the latter case results in the continued use of the inferior alloy and an erroneous conclusion that the experimental work of the investigative team has gone for naught.

Recognizing the size of type II error to be a function $\beta = \beta(\mu)$, we examine the two type II errors just discussed. First,

$$
\begin{aligned}
\beta(46.2) &= P(\bar{X} \in A; \mu = 46.2) \\
&= P(\bar{X} < 46.41; \mu = 46.2) \\
&= P\left(Z < \frac{46.41 - 46.2}{0.75/\sqrt{9}}\right) \\
&= \Phi(0.84) = 0.7995
\end{aligned}
$$

Second,

$$
\begin{aligned}
\beta(46.6) &= P(\bar{X} \in A; \mu = 46.6) \\
&= P(\bar{X} < 46.41; \mu = 46.6) \\
&= P\left(Z < \frac{46.41 - 46.6}{0.25}\right) \\
&= \Phi(-0.76) = 0.2236
\end{aligned}
$$

Both $\beta(46.2)$ and $\beta(46.6)$ are illustrated in Figure 8.5.

While the size of type II error when $\mu = 46.2$ is very high, its significance depends upon engineering (or management) judgment regarding the importance of an increase of 0.2 ksi. If the detection of a 0.2-ksi increase is not considered consequential, then the high value of β is not significant.

Interpretation of $\beta(46.6) = 0.2236$ is more complicated. Just as in the case of competing simple hypotheses, α and β represent risk. The structural engineer might very well have fixed $\alpha = 0.05$ to satisfy acceptable industry standards for reaching an affirmative conclusion regarding the alternative hypothesis. The appearance of a type II error of size 0.2236 may or may not be a risk the engineer is willing to take. If an improvement of 0.6 ksi is significant from an engineering perspective, then $\beta(46.6) = 0.2236$ is likely to be considered too high; however, if 46.6 ksi represents only a moderate benefit over 46.0 ksi, then $\beta(46.6) = 0.2236$ might be a risk the engineer is willing to bear.

If it is felt that probability 0.2236 is too high a risk, then one way to proceed is to increase the sample size, thereby reducing the variance of the sample mean. For example, if $n = 25$, then $\sigma_{\bar{x}} = 0.15$, so that the critical value for $\alpha = 0.05$ is found by solving

$$
P\left(Z > \frac{k_\alpha - 46}{0.15}\right) = 0.05
$$

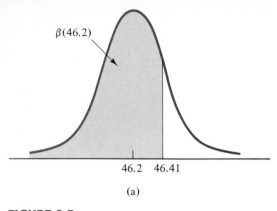
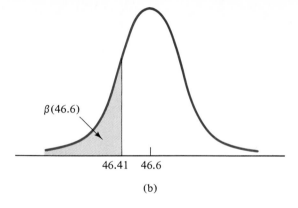

FIGURE 8.5
Two type II errors for Example 8.4

With the help of Table A.1, we find the critical value $k_{0.05} = 46.25$, so that

$$\beta(46.6) = P(\bar{X} < 46.25; \mu = 46.6) = P\left(Z < \frac{46.25 - 46.6}{0.15}\right) = 0.0099$$

which is a very small type II error.

As formulated in Example 8.4, the competing hypotheses do not exhaust the possibilities for μ. How is it known that μ is not less than 46 ksi? Certainly, it is more realistic to consider the hypothesis test

$$H_0: \mu \leq 46$$

$$H_1: \mu > 46$$

The intent is to demonstrate that the new mean yield point is greater than 46 and the logical negation of $\mu > 46$ is $\mu \leq 46$. However, we must proceed with care. The size α of type I error is the probability of the test statistic falling in the critical region given H_0. If H_0 is composite, then $\alpha = \alpha(\mu)$ is a function of μ, where $\mu \leq 46$. Referring to Figure 8.6, if $\mu' < 46$ and k_α is a given critical value, then

$$\alpha(\mu') = P(\bar{X} > k_\alpha; \mu = \mu')$$
$$< P(\bar{X} > k_\alpha; \mu = 46)$$
$$= \alpha(46)$$

In words, the size of type I error is maximized when $\mu = 46$. Consequently, we can reformulate the null hypothesis to be composite ($\mu \leq 46$) and utilize $\alpha = 46$ when finding the critical value. For the resulting critical value k_α, given H_0, the probability of type I error will be less than or equal to α no matter what the actual value of μ. Intuitively, any large sample value of the test statistic that appears highly unlikely under the assumption that the mean yield point is 46 ksi will appear even more unlikely under the assumption that the mean yield point is less than 46 ksi. It is precisely the unlikeliness of the test statistic falling in the critical region that leads us to reject the null hypothesis.

More generally, to show a parameter θ to be greater than a specific value θ_0, we can employ the hypothesis test

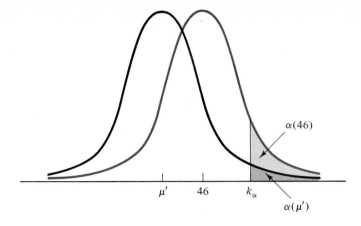

FIGURE 8.6
Illustration of maximum
type I error

$$H_0: \theta \leq \theta_0$$
$$H_1: \theta > \theta_0$$

The determination of a critical value is based upon the boundary value $\theta = \theta_0$, and $\beta = \beta(\theta)$ will be determined relative to that critical value.

Whereas the one-tailed methodology was employed in Example 8.4 to construct a test to detect an increased yield point, it is very possible that a structural engineer might desire to show more. For instance, he or she might wish to demonstrate that the new mean yield point is not merely greater than 46 ksi but that, in fact, it is greater than 47 ksi, which would indicate an improvement of at least 1 ksi. An appropriate test for this stronger conjecture is given by

$$H_0: \mu \leq 47$$
$$H_1: \mu > 47$$

Once again the sample mean is employed as the test statistic.

Lower one-tailed tests are constructed in a manner analogous to upper one-tailed tests, the desire being to demonstrate $\theta < \theta_0$. The test takes the form

$$H_0: \theta \geq \theta_0$$
$$H_1: \theta < \theta_0$$

and analogous remarks apply with regard to α and β.

EXAMPLE 8.5

In an interactive multiprogramming environment, *response time* is defined as the time between a user's pressing the ENTER key and the commencement of the response, whether that response be in hard copy or on a screen. Suppose that to decrease the response time in a particular system the database has been reconfigured. Prior to the reconfiguration, the mean response time was 5.42 seconds, and it is claimed that the new mean response time is less than 3 seconds. To statistically demonstrate the validity of the claim, 50 response times are to be randomly checked over a period of one week and the lower one-tailed test

$$H_0: \mu \geq 3$$
$$H_1: \mu < 3$$

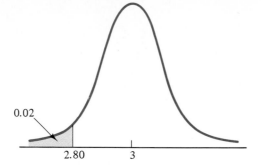

FIGURE 8.7
Critical region of size
$\alpha = 0.02$ for Example 8.5

0.02

2.80   3

is to be employed. If X denotes the random variable measuring response times, then the sample mean $\bar{X}$ will be the test statistic. Although X is not presumed to be normally distributed, the large sample size justifies application of the central limit theorem, and therefore $\bar{X}$ can be treated as normally distributed. It will be assumed that the standard deviation of X is known to be $\sigma = 0.68$. For a type I error of size $\alpha = 0.02$, the critical value k_α is found by solving the equation

$$P(\bar{X} < k_{0.02}; \mu = 3) = 0.02$$

(see Figure 8.7). Changing to the standard normal variable yields

$$P\left(Z < \frac{k_{0.02} - 3}{0.68/\sqrt{50}}\right) = 0.02$$

which, according to Table A.1, reduces to

$$\frac{k_{0.02} - 3}{0.096} = -2.054$$

Solving for the critical value gives $k_{0.02} = 2.80$, so that any value of the sample mean less than or equal to 2.80 results in acceptance of the alternative (desired) hypothesis.

Some values of β corresponding to the critical value 2.80 are

$$\beta(2.9) = P(\bar{X} > 2.8; \mu = 2.9) = P\left(Z > \frac{2.8 - 2.9}{0.096}\right) = 1 - \Phi(-1.04) = 0.8508$$

$$\beta(2.8) = P(\bar{X} > 2.8; \mu = 2.8) = P(Z > 0) = 0.5$$

$$\beta(2.7) = P(\bar{X} > 2.8; \mu = 2.7) = 1 - \Phi(1.04) = 0.1492$$

If one feels that these values of β are too high, there are two possible remedies: (1) increase the sample size or (2) increase the size of type I error. The latter choice means tolerating a greater risk of erroneously accepting the claim regarding the new response time.

8.2.2 Two-Tailed Tests

Rather than to show an increase or decrease in a particular parameter, sometimes it is simply necessary to show that the parameter is not equal to some stated value, where the intent is to detect a meaningful difference. As an illustration, consider detecting the deviation of a cylinder bore from a prescribed position, where specifications require the horizontal distance of the bore center from the origin of the coordinate system to be 4.3 cm. An appropriate hypothesis test to determine whether there has been significant deterioration of the alignment process in the horizontal direction is given by

$$H_0: \mu = 4.3$$

$$H_1: \mu \neq 4.3$$

For such a test, the acceptance region consists of an interval about $\mu = 4.3$ and the critical region consists of the half-axes to either side of the acceptance interval. The test is said to be **two-tailed** because the probability mass of the critical region consists of the sum of the areas under both the left and right tails of the test-statistic distribution. Specifically, for level of significance α, there will be two critical values, $\mu - r_{\alpha/2}$ and $\mu + r_{\alpha/2}$. The acceptance and critical regions will be given by

$$A = (\mu - r_{\alpha/2}, \mu + r_{\alpha/2})$$

and

$$C = (-\infty, \mu - r_{\alpha/2}] \cup [\mu + r_{\alpha/2}, \infty)$$

respectively.

EXAMPLE 8.6 In designing a service system, it is beneficial to have some working hypothesis regarding the number of units to be serviced in a given time frame. Too low an estimate will result in inadequate service, and too high an estimate will result in a waste of resources. Although a detailed analysis of service problems requires the theory of queuing (Chapter 6), hypothesis testing can be applied to determine the worth of contingent hypotheses concerning the mean number of units served over a period of time.

Suppose a company's allocation of computer resources has been based upon a mean number of 110 interactive users during any given hour and it is desired to check if there has been relevant deviation from that number. A suitable hypothesis test is

$$H_0: \mu = 110$$

$$H_1: \mu \neq 110$$

Letting X denote the random variable measuring the number of users in an hour, we will employ the sample mean as the test statistic. Although X is a discrete random variable, if past observations of X have shown its density to be closely approximated by a normal curve, it is appropriate to treat the distribution of X as being normally distributed. In addition, it will be assumed that the standard deviation of X is known to be $\sigma = 8.4$. If a sample size of 25 observation hours is employed, then the acceptance region is determined by the equation

$$P(\mu - r_{\alpha/2} < \bar{X} < \mu + r_{\alpha/2}; \mu = 110) = 1 - \alpha$$

which, by the symmetry of the normal distribution, reduces to

$$P(\bar{X} < 110 - r_{\alpha/2}) = \frac{\alpha}{2}$$

Letting $\alpha = 0.05$, changing to the standard normal variable, and putting the result in terms of $\Phi(z)$ yields

$$\Phi\left(\frac{-r_{0.025}}{8.4/5}\right) = 0.025$$

FIGURE 8.8
Two-tailed critical region
of size α = 0.05 for
Example 8.6

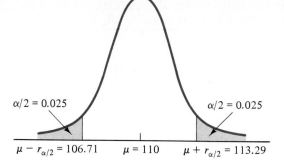

$\alpha/2 = 0.025$ $\alpha/2 = 0.025$

$\mu - r_{\alpha/2} = 106.71$ $\mu = 110$ $\mu + r_{\alpha/2} = 113.29$

From Table A.1,

$$\frac{-r_{0.025}}{1.68} = -1.96$$

so that $r_{0.025} = 3.29$. Thus, as illustrated in Figure 8.8, the acceptance region is the interval

$$A = (106.71, 113.29)$$

and the critical region is

$$C = (-\infty, 106.71] \cup [113.29, \infty)$$

Should the test statistic $\bar{X}$ fall in the critical region, then the null hypothesis that the mean number of interactive users is 110 is rejected and the conclusion is drawn that the system is either over- or underutilized. On the other hand, if $\bar{X}$ falls in the acceptance region, then the conclusion is drawn that there is no compelling reason to alter the working hypothesis.

Insofar as the size of type II error is concerned, some representative values are given by

$$\beta(112) = P(\bar{X} \in A; \mu = 112)$$
$$= P(106.71 < \bar{X} < 113.29; \mu = 112)$$
$$= P\left(\frac{106.71 - 112}{1.68} < Z < \frac{113.29 - 112}{1.68}\right)$$
$$= \Phi(0.77) - \Phi(-3.15)$$
$$= 0.7794 - 0.0008 = 0.7786$$

$$\beta(114) = P(106.71 < \bar{X} < 113.29; \mu = 114)$$
$$= P\left(\frac{106.71 - 114}{1.68} < Z < \frac{113.29 - 114}{1.68}\right)$$
$$= \Phi(-0.42) - \Phi(-4.34) = 0.3372$$

$$\beta(116) = P(106.71 < \bar{X} < 113.29; \mu = 116)$$
$$= \Phi(-1.61) - \Phi(-5.53) = 0.0537$$

The computations of these type II errors are illustrated in Figure 8.9. By the symmetry of the standard normal density, for a value less than the hypothesized mean of 110, say $110 - d$,

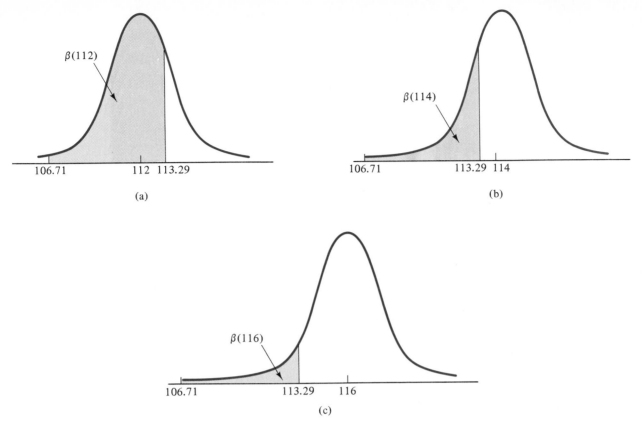

FIGURE 8.9
Three type II errors for Example 8.6

$$\beta(110 - d) = P(106.71 < \bar{X} < 113.29; \mu = 110 - d)$$

$$= P\left(\frac{d - 3.29}{1.68} < Z < \frac{d + 3.29}{1.68}\right)$$

$$= P\left(\frac{-d - 3.29}{1.68} < Z < \frac{-d + 3.29}{1.68}\right)$$

$$= P(106.71 < \bar{X} < 113.29; \mu = 110 + d)$$

$$= \beta(110 + d)$$

(see Figure 8.10). In other words, the size of type II error depends only on the absolute deviation of the true mean from the hypothesized mean. Thus, using the values of β already computed,

$$\beta(108) = \beta(112) = 0.7786$$

$$\beta(106) = \beta(114) = 0.3372$$

$$\beta(104) = \beta(116) = 0.0537$$

The β values are interpreted in the following manner. If the true mean hourly number of users differs by 2 from the hypothesized value, then there is a high probability of type II error with $\alpha = 0.05$ and $n = 25$, which means it is likely that the deviation will not be detected. Since a deviation of 2 users is of little practical significance, so too is the high size of type II error.

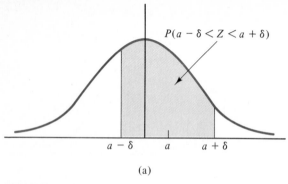

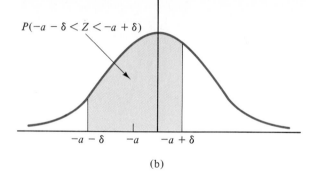

FIGURE 8.10
The identity $P(a - \delta < Z < a + \delta) = P(-a - \delta < Z < -a + \delta)$

Regarding a deviation of 6 users, the type II error is quite low. Thus, even if such a deviation is considered important (an unlikely supposition), it is highly probable that it will be detected.

From the examples thus far examined, it should be clear that a worthwhile interpretation of either a one- or two-tailed hypothesis test depends on both α and β. While it is usually true that α takes priority, because the goal is often detection of a deviation from the null hypothesis, it is also true that β measures risk. More to the point, β indicates the degree to which the test is not likely to detect precisely the thing it is designed to detect. If lack of detection is not important in some range of the parameter, then high β values are not significant in that range; if, on the other hand, lack of detection is costly in some range of the parameter, then low values of β must be demanded in that range. Of course, the relative importance of detection is a matter to be determined by the engineering specialist, not by the statistician. In the next section we will make a detailed study of the relationship between α, β, and the sample size n. A convenient methodology will be provided by which an investigator can quickly grasp this relationship. Subsequent sections will introduce standard hypothesis-testing methodologies.

8.3 OPERATING CHARACTERISTIC CURVES

A careful analysis of type II error is essential to the selection and interpretation of an appropriate hypothesis test. Even if α is fixed by an industry standard or simply by custom, an analysis of the risk involved in making a wrong decision to accept the null hypotheses can lead to a suitable sample size. Even if the sample size is limited by economic or time constraints, a careful analysis of β can enhance understanding of the practical implications of the test result. The mechanism employed to analyze β is the operating characteristic curve.

DEFINITION 8.3

Operating Characteristic (OC) Curve. For a hypothesis test involving the null hypothesis

$$H_0:\ \theta = \theta_0$$

the operating characteristic curve is the graph of the function of θ defined by the probability of accepting the null hypothesis given θ represents the true state of nature.

Mathematically, if V is the test statistic being used to test H_0: $\theta = \theta_0$ against an alternative hypothesis and A is the acceptance region for the test, then the OC curve is defined in terms of the function

$$H(\theta) = P(V \in A; \theta)$$

For a two-tailed test having alternative hypothesis

$$H_1: \theta \neq \theta_0,$$

acceptance region A, and level of significance α, the OC curve is the graph of the function $H(\theta)$ defined, for $\theta = \theta_0$, by

$$\begin{aligned} H(\theta_0) &= P(V \in A; \theta = \theta_0) \\ &= P(V \in A; H_0) \\ &= 1 - P(V \notin A; H_0) \\ &= 1 - \alpha \end{aligned}$$

and, for $\theta \neq \theta_0$, by

$$\boxed{H(\theta) = P(V \in A; \theta \neq \theta_0) = \beta(\theta)}$$

Thus, in the two-tailed case, the OC curve takes the value $1 - \alpha$ at the hypothesized value of the parameter and the size of the type II error at other values of the parameter.

Two points should be recognized. First, the size of β is not independent of α. Consequently, a prior determination of α will affect the OC curve, which itself will reflect the relationship between α and β. Second, the size of β depends upon the sample size n, so that the OC curve will reflect the risk of mistakenly rejecting the alternative hypothesis for a given sample size. Indeed, the interplay between the OC curve and n provides a basis for choosing a sample size conducive to the desired level of risk.

EXAMPLE 8.7

For the two-tailed hypothesis test of Example 8.6, where the sample mean is the test statistic, $\alpha = 0.05$, and $n = 25$, the function defining the OC curve is given by

$$H(110) = 1 - 0.05 = 0.95$$

and, for $\theta \neq 110$,

$$\begin{aligned} H(\mu) &= \beta(\mu) \\ &= P(106.71 < \bar{X} < 113.29; \mu) \\ &= P\left(\frac{106.71 - \mu}{1.68} < Z < \frac{113.29 - \mu}{1.68}\right) \\ &= \Phi\left(\frac{1139.29 - \mu}{1.68}\right) - \Phi\left(\frac{106.71 - \mu}{1.68}\right) \end{aligned}$$

The OC curve is illustrated in Figure 8.11. Notice how the symmetry of the test statistic in conjunction with the symmetry of the acceptance region has resulted in a symmetric OC curve.

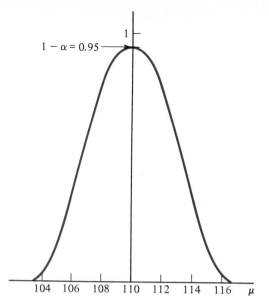

FIGURE 8.11
OC curve for Examples
8.6 and 8.7 with $n = 25$

A change in α results in a new OC curve; however, since α is often fixed by outside circumstances, the more useful relationship is usually between the OC curve and the sample size n. For instance, if we again turn our attention to Example 8.6 but do not fix the sample size, then the equation defining $r_{0.025}$ implicitly contains the variable n. Reducing the defining probability equation to an expression involving the probability distribution function of the standard normal variable yields the relationship

$$\Phi\left(\frac{-r_{0.025}}{(8.4)/\sqrt{n}}\right) = 0.025$$

Employing Table A.1 and solving for $r_{0.025}$ yields

$$r_{0.025} = \frac{(1.96)(8.4)}{\sqrt{n}} = \frac{16.46}{\sqrt{n}}$$

The resulting acceptance region is

$$A_n = \left(110 - \frac{16.46}{\sqrt{n}}, 110 + \frac{16.46}{\sqrt{n}}\right)$$

As n tends to infinity, the length of A_n tends to zero. If H_n denotes the function defining the OC curve for sample size n, then, for $\mu \neq 110$,

$$H_n(\mu) = P(\bar{X} \in A_n; \mu)$$
$$= P\left(110 - \frac{16.46}{\sqrt{n}} < \bar{X} < 110 + \frac{16.46}{\sqrt{n}}; \mu\right)$$
$$= P\left(\frac{110 - \mu}{8.4/\sqrt{n}} - \frac{16.46}{8.4} < Z < \frac{110 - \mu}{8.4/\sqrt{n}} + \frac{16.46}{8.4}\right)$$
$$= P\left(\frac{(110 - \mu)\sqrt{n}}{8.4} - 1.96 < Z < \frac{(110 - \mu)\sqrt{n}}{8.4} + 1.96\right)$$

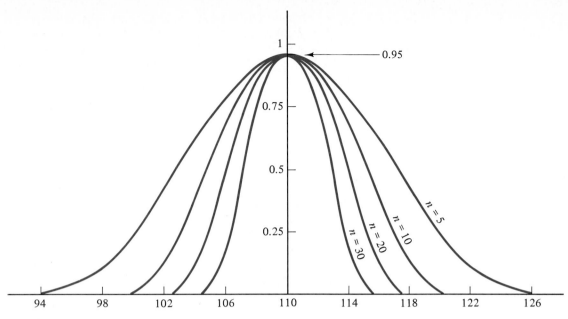

FIGURE 8.12
OC-curve family for Examples 8.6 and 8.7

which tends to zero as n approaches infinity. Thus, for fixed $\mu \neq 110$,

$$\lim_{n \to \infty} H_n(\mu) = 0$$

The salient point concerning this limit is that as n grows, the OC curves fall off ever more rapidly. This behavior of the OC-curve family is depicted in Figure 8.12. Since for each n the corresponding OC curve gives the size of type II error for $\mu \neq 110$, the falling off of the OC curves means that it is theoretically possible to select a curve from an OC-curve family for which the corresponding sample size n eliminates as much risk as desired in regard to the erroneous acceptance of the null hypothesis. In selecting a suitable sample size it is important to keep in mind that the goal is not simply to reject the null hypothesis; rather, it is to construct a test that will, with high probability, result in the rejection of the null hypothesis when the data reveal a meaningful deviation, meaningfulness being a practical and scientific category.

As an illustration of the OC-curve methodology, suppose in Example 8.6 that a deviation of 8 users per hour is considered sufficient to warrant attention. It is then necessary to reduce the size of type II error to a tolerable level when the true state of nature is $\mu = 102$ or $\mu = 118$. If it is decided that a level of $\beta(102) = \beta(118) = 0.10$ is tolerable, then an appropriate sample size can be found by selecting the OC curve in Figure 8.12 that passes through the point $(118, 0.10)$. A quick inspection shows the desired curve to be the one corresponding to sample size $n = 10$.

Operating characteristic curves also result from one-tailed tests. We will consider the upper one-tailed test

$$H_0: \theta \leq \theta_0$$

$$H_1: \theta > \theta_0$$

the analysis of lower one-tailed tests being analogous. As in the two-tailed case, if $H(\theta)$ is the function defining the OC curve, then $H(\theta_0) = 1 - \alpha$. For $\theta > \theta_0$, $H(\theta)$ gives the size of type II error. However, in the present case, while $H(\theta)$ is defined for $\theta < \theta_0$, it does not give the size of type II error for such θ; instead, for $\theta < \theta_0$, H_0 is presumed to be true according to the design of the test, and so

$$
\begin{aligned}
H(\theta) &= P(V \in A; \theta) \\
&= 1 - P(V \notin A; \theta) \\
&= 1 - \alpha(\theta)
\end{aligned}
$$

If this equation appears somewhat strange, it is because the assumption was made in Section 8.2.1 that a null hypothesis of the form $\theta \leq \theta_0$ would be treated as though it were $\theta = \theta_0$ insofar as the size of type I error is concerned. In all cases with which we are concerned, the "boundary value" θ_0 results in the largest type I error among all θ satisfying the null hypothesis. Nevertheless, the OC curve is still defined for $\theta < \theta_0$. In essence, we are adopting the convention that the level of significance of a test is the maximum among all sizes of type I error.

EXAMPLE 8.8

Example 8.4 concerned a hypothesis test involving the yield point of a ductile steel alloy. Although the null hypothesis was stated as H_0: $\mu = 46$, it could just as well have been stated more realistically as H_0: $\mu \leq 46$. Using the data of Example 8.4, the function defining the OC curve is

$$
\begin{aligned}
H(\mu) &= P(\bar{X} \in A; \mu) \\
&= P(\bar{X} < 46.41; \mu) \\
&= P\left(Z < \frac{46.41 - \mu}{0.25}\right)
\end{aligned}
$$

FIGURE 8.13
OC curve for Examples 8.4 and 8.8

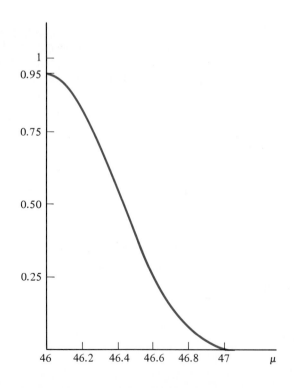

The OC curve is graphed in Figure 8.13. Note that

$$H(46) = P(Z < 1.64) = 0.95 = 1 - 0.05$$

and

$$\lim_{\mu \to \infty} H(\mu) = 0$$

As in the case of two-tailed tests, one-tailed OC-curve families can be employed to determine sample sizes based upon tolerable levels of risk at parameter values included under the alternative hypothesis. Rather than give an example of the technique at this point, we will wait until the next section, when a normalization of the method will be presented.

8.4 TESTING THE MEAN OF A NORMAL DISTRIBUTION WITH KNOWN VARIANCE

A number of hypothesis-testing problems have been considered in which the sample mean was chosen to be the test statistic and, owing to either a normal population random variable or the central limit theorem, the sample mean was assumed to be normally distributed. In all cases the acceptance and critical regions were selected according to the principle that a parameter should be rejected if the sample value of the test statistic falls in a region in which, given the null hypothesis, it is unlikely to fall. Moreover, in each example utilizing the sample mean, the critical values depended upon a reduction to the standard normal variable. In the present section we will standardize the approach by utilizing the standard normal variable as the test statistic. The result will be standardized critical values, standardized OC curves, and a more universally recognizable methodology.

8.4.1 Two-Tailed Tests Involving the Standard Normal Variable

Consider the two-tailed test of the mean

$$H_0: \mu = \mu_0$$

$$H_1: \mu \neq \mu_0$$

where the population random variable X is normal and possesses a known variance σ^2 and where the sample size is presumed to be n. Based upon the intuitive reasoning employed previously, the sample mean is chosen as the test statistic and the acceptance region is taken to be the interval

$$A = (\mu_0 - r_{\alpha/2}, \mu_0 + r_{\alpha/2})$$

$r_{\alpha/2}$ depending upon α. Specifically, $r_{\alpha/2}$ is determined by the probability equation

$$P(\bar{X} \in A; H_0) = 1 - \alpha$$

which, under the null hypothesis $\mu = \mu_0$, is analytically equivalent to the equation

$$P(\mu_0 - r_{\alpha/2} < \bar{X} < \mu_0 + r_{\alpha/2}) = 1 - \alpha$$

Standardizing $\bar{X}$ gives

$$P\left(\frac{-r_{\alpha/2}}{\sigma/\sqrt{n}} < \frac{\bar{X} - \mu_0}{\sigma/\sqrt{n}} < \frac{r_{\alpha/2}}{\sigma/\sqrt{n}}\right) = 1 - \alpha$$

Letting

$$z_{\alpha/2} = \frac{r_{\alpha/2}}{\sigma/\sqrt{n}}$$

results in the equivalent equation

$$P\left(-z_{\alpha/2} < \frac{\bar{X} - \mu_0}{\sigma/\sqrt{n}} < z_{\alpha/2}\right) = 1 - \alpha$$

As a result, rather than taking $(\mu_0 - r_{\alpha/2}, \mu_0 + r_{\alpha/2})$ as the acceptance region, with the sample mean as the test statistic, we can standardize the procedure by utilizing the standardized sample mean as the test statistic. The resulting two-tailed Z test is expressed in terms of the standard normal distribution, with the acceptance interval having the standardized critical values $-z_{\alpha/2}$ and $z_{\alpha/2}$ as endpoints. The standardized procedure is illustrated in Figure 8.14.

A test statistic having been decided upon, it is necessary to develop the corresponding expression for $\beta(\mu)$, $\mu \neq \mu_0$. This expression will lead at once to the OC-curve family, since, in the two-tailed case, $H(\mu) = \beta(u)$ for $\mu \neq \mu_0$ and $H(\mu_0) = 1 - \alpha$. For $\mu \neq \mu_0$,

$$
\begin{aligned}
\beta(\mu) &= P\left(-z_{\alpha/2} < \frac{\bar{X} - \mu_0}{\sigma/\sqrt{n}} < z_{\alpha/2}; \mu_X = \mu\right) \\
&= P\left(\mu_0 - z_{\alpha/2} \cdot \frac{\sigma}{\sqrt{n}} < \bar{X} < \mu_0 + z_{\alpha/2} \cdot \frac{\sigma}{\sqrt{n}}; \mu_X = \mu\right) \\
&= P\left(-z_{\alpha/2} + \frac{\mu_0 - \mu}{\sigma/\sqrt{n}} < Z < z_{\alpha/2} + \frac{\mu_0 - \mu}{\sigma/\sqrt{n}}\right) \\
&= \Phi\left(z_{\alpha/2} + \frac{\mu_0 - \mu}{\sigma/\sqrt{n}}\right) - \Phi\left(-z_{\alpha/2} + \frac{\mu_0 - \mu}{\sigma/\sqrt{n}}\right)
\end{aligned}
$$

where the third equality has been obtained by standardizing the sample mean under the assumptions $\mu_X = \mu$ and $\sigma_X = \sigma$. Treating n as a variable yields a family of OC curves.

FIGURE 8.14
Two-tailed Z test for the mean of a normal distribution

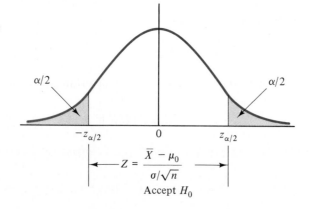

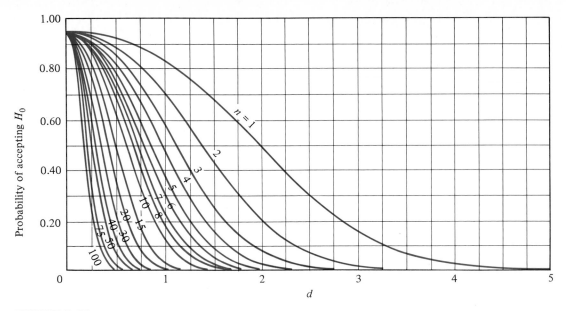

FIGURE 8.15
OC curves for the two-tailed Z test at level $\alpha = 0.05$ (Reprinted by permission of the Institute of Mathematical Statistics.)

Fortunately, for a given value of α, it is necessary to consider only a single family of OC curves, not a different family for every hypothesized mean and possible value of the standard deviation. Specifically, we need consider only the OC curves defined by the normalized deviation

$$d = \frac{|\mu_0 - \mu|}{\sigma}$$

where, due to symmetry, the OC curves need be defined only for nonnegative values of $(\mu_0 - \mu)/\sigma$. Figures 8.15 and 8.16 give the OC-curve families for levels of significance 0.05 and 0.01, respectively, in each case the probability of accepting the null hypothesis being plotted against d. Once a desired level of significance is selected, to apply either OC-curve family one need only choose the alternative value of the mean, say μ_1, whose detection is thought to be important, decide upon a tolerable level of risk $\beta(\mu_1)$ regarding the mistaken acceptance of the null hypothesis, and then choose the OC curve closest to the point $(d, \beta(\mu_1))$ when passing through the vertical line determined by d. The parameter n associated with the chosen curve gives the desired sample size for levels α and $\beta(\mu_1)$. Once the sample size has been determined, other important values of β can be found from the appropriate OC curve.

Rather than find n by the OC-curve methodology, an alternate approach in the present Z-test setting is to use the approximation formula

$$n \approx \frac{(z_{\alpha/2} + z_\beta)^2 \sigma^2}{(\mu_0 - \mu_1)^2}$$

where z_β is the z point for which

$$P(Z > z_\beta) = \beta(\mu_1)$$

Two points regarding this approximation should be recognized. First, it does not totally negate the value of the OC-curve family for the given α, since the OC curves

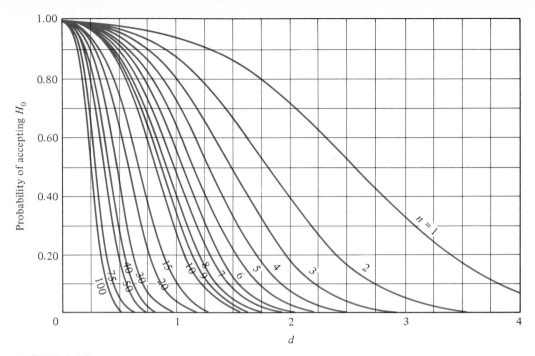

FIGURE 8.16
OC curves for the two-tailed Z test at level $\alpha = 0.01$ (Reprinted by permission of the Institute of Mathematical Statistics.)

still can be used to facilitate a suitable interpretation of the result. Second, the approximation is only reasonably good when

$$P\left(Z < -z_{\alpha/2} - \frac{|\mu_0 - \mu_1|}{\sigma/\sqrt{n}}\right)$$

is small in comparison to $\beta(\mu_1)$.

Table 8.2 contains the most pertinent information regarding the Z test of the mean of a normal population random variable. Included is information to be subsequently developed for one-tailed tests. Note that in all cases the null hypothesis is H_0: $\mu = \mu_0$; nevertheless, under our assumptions governing composite hypotheses, the same information applies to H_0: $\mu \leq \mu_0$ and to H_0: $\mu \geq \mu_0$ in the upper and lower one-tailed cases, respectively.

EXAMPLE 8.9

Consider a chemical production process in which it is necessary to maintain a proper pH level. Suppose a pH level of 8.8 is desirable and a chemist plans to take several measurements to detect whether there has been significant deviation from that level. Because a conclusion that the pH level varies too greatly from 8.8 can cause a production interruption, the chemist wishes to avoid a false detection. It is decided that a suitable significance level is $\alpha = 0.05$. In the other direction, it is recognized that a deviation of $0.4\ p$H can portend eventual problems, the conclusion being that given a deviation of 0.4, there should only be 0.10 probability of not rejecting the null hypothesis. From experience it is known that measurements tend to be normally distributed with standard deviation $\sigma = 0.3$. The appropriate hypothesis test is

$$H_0: \mu = 8.8$$
$$H_1: \mu \neq 8.8$$

TABLE 8.2 IMPORTANT INFORMATION RELATING TO THE Z TEST OF THE MEAN

Null Hypothesis: H_0: $\mu = \mu_0$

Test Statistic: $Z = \dfrac{\bar{X} - \mu_0}{\sigma/\sqrt{n}}$

Alternative Hypothesis (Two-Tailed): H_1: $\mu \neq \mu_0$

    Critical Region: $C = \{z: z \leq -z_{\alpha/2} \text{ or } z \geq z_{\alpha/2}\}$

    Values of β: $\beta(\mu) = \Phi\left(z_{\alpha/2} + \dfrac{\mu_0 - \mu}{\sigma/\sqrt{n}}\right) - \Phi\left(-z_{\alpha/2} + \dfrac{\mu_0 - \mu}{\sigma/\sqrt{n}}\right)$

    Expression of n in terms of α and $\beta(\mu_1)$: $n \approx \dfrac{(z_{\alpha/2} + z_\beta)^2 \sigma^2}{(\mu_0 - \mu_1)^2}$

Alternative Hypothesis (Upper One-Tailed): H_1: $\mu > \mu_0$

    Critical Region: $C = \{z: z \geq z_\alpha\}$

    Values of β: $\beta(\mu) = \Phi\left(z_\alpha + \dfrac{\mu_0 - \mu}{\sigma/\sqrt{n}}\right)$

    Expression of n in terms of α and $\beta(\mu_1)$: $n = \dfrac{(z_\alpha + z_\beta)^2 \sigma^2}{(\mu_0 - \mu_1)^2}$

Alternative Hypothesis (Lower One-Tailed): H_1: $\mu < \mu_0$

    Critical Region: $C = \{z: z \leq -z_\alpha\}$

    Values of β: $\beta(\mu) = 1 - \Phi\left(-z_\alpha + \dfrac{\mu_0 - \mu}{\sigma/\sqrt{n}}\right)$

    Expression of n in terms of α and $\beta(\mu_1)$: Same as upper one-tailed test.

the test statistic being

$$Z = \frac{\bar{X} - 8.8}{0.3/\sqrt{n}}$$

The size of type II error at the alternative mean $\mu_1 = 9.2$ is

$$\beta = \beta(9.2) = 0.10$$

Using

$$d = \frac{|8.8 - 9.2|}{0.3} = 1.33$$

a suitable sample size can be found by referring to Figure 8.15. Selecting the coordinate $(1.33, 0.10)$ in the figure yields the curve determined by $n = 6$. Thus, the test statistic becomes

$$Z = \frac{\bar{X} - 8.8}{0.1225}$$

Since $z_{\alpha/2} = 1.96$ and the test is two-tailed, the critical region is

$$C = \{z : z \leq -1.96 \text{ or } z \geq 1.96\}$$

Now suppose 6 measurements are taken and the empirical mean of the observations is $\bar{x} = 9.1$. Then the sample value of the test statistic is

$$z = \frac{9.1 - 8.8}{0.1225} = 2.45$$

which is greater than 1.96. Thus, the null hypothesis is rejected, and it is assumed that the pH level has deviated too widely from the specified level.

Had the empirical mean been $\bar{x} = 8.5$, then the sample value of the test statistic would have been $z = -2.45$ and there would have been a rejection of the null hypothesis based upon the left side of the critical region instead of the right side. Finally, if the approximation formula for n is employed instead of the OC-curve family, then the resulting approximate value of n is

$$n \approx \frac{(1.96 + 1.28)^2 (0.3)^2}{(0.4)^2} = 5.90$$

which rounds off to $n = 6$.

8.4.2 One-Tailed Tests Involving the Standard Normal Variable

We now turn our attention to one-tailed tests, continuing to assume a normal population random variable X possessing a known variance σ^2. We will make a detailed analysis of the upper one-tailed test and simply defer to Table 8.2 for the relationships relevant to the lower one-tailed test. Thus, consider the hypothesis test

$$H_0 : \ \mu \leq \mu_0$$

$$H_1 : \ \mu > \mu_0$$

Once again the sample mean is an intuitively reasonable test statistic; however, in the present setting the acceptance region is

$$A = (-\infty, k_\alpha)$$

where k_α depends upon the choice of α and is determined by the equation

$$P(\bar{X} \geq k_\alpha; \mu_X = \mu_0) = \alpha$$

Standardizing $\bar{X}$ yields

$$P\left(\frac{\bar{X} - \mu_0}{\sigma/\sqrt{n}} \geq \frac{k_\alpha - \mu_0}{\sigma/\sqrt{n}} \right) = \alpha$$

and letting

$$z_\alpha = \frac{k_\alpha - \mu_0}{\sigma/\sqrt{n}}$$

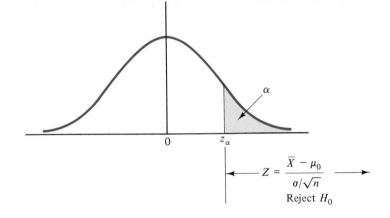

FIGURE 8.17
Upper one-tailed Z test for the mean of a normal distribution

results in the equation

$$P\left(\frac{\bar{X} - \mu_0}{\sigma/\sqrt{n}} \geq z_\alpha\right) = \alpha$$

Thus, instead of using the test statistic $\bar{X}$, we use the standardized version of $\bar{X}$, and the critical region becomes the half-infinite interval $[z_\alpha, \infty)$ on the z axis, the critical value z_α being determined by the upper tail of the standard normal distribution. The standardized procedure is illustrated in Figure 8.17.

As for type II error, for $\mu > \mu_0$,

$$\beta(\mu) = P\left(\frac{\bar{X} - \mu_0}{\sigma/\sqrt{n}} < z_\alpha; \mu_X = \mu\right)$$

$$= P\left(\bar{X} < \mu_0 + z_\alpha \cdot \frac{\sigma}{\sqrt{n}}; \mu_X = \mu\right)$$

$$= P\left(Z < z_\alpha + \frac{\mu_0 - \mu}{\sigma/\sqrt{n}}\right)$$

$$= \Phi\left(z_\alpha + \frac{\mu_0 - \mu}{\sigma/\sqrt{n}}\right)$$

where the third equality has been obtained by standardizing $\bar{X}$ under the assumptions $\mu_X = \mu$ and $\sigma_X = \sigma$. By varying n and using the fact that the OC curve is defined by $\beta(\mu)$ when the null hypothesis is false, the preceding equality can be employed to find the appropriate family of OC curves for a given α. As in the two-tailed case, the same family can be employed for any standard deviation σ by using

$$d = \frac{\mu - \mu_0}{\sigma}$$

Since the test is one-tailed and $\mu > \mu_0$ when the alternative hypothesis is true, d is always nonnegative. Figures 8.18 and 8.19 give the one-tailed Z-test OC-curve families for levels of significance 0.05 and 0.01, respectively. In each case the probability of accepting the null hypothesis is plotted against d. Given α, once $\beta(\mu_1)$ has been

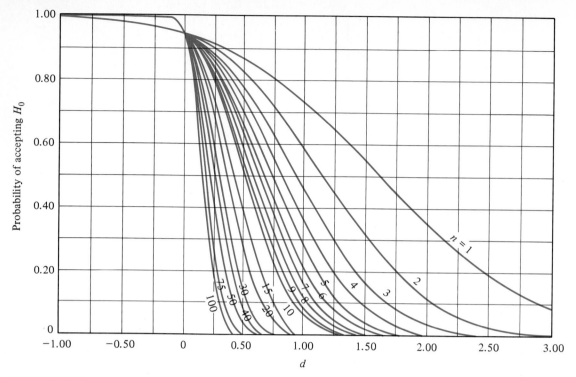

FIGURE 8.18
OC curves for the one-tailed Z test at level $\alpha = 0.05$ (Reprinted by permission of the Institute of Mathematical Statistics.)

FIGURE 8.19
OC curves for the one-tailed Z test at level $\alpha = 0.01$ (Reprinted by permission of the Institute of Mathematical Statistics.)

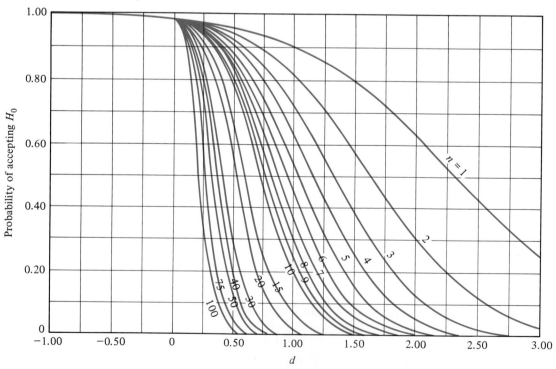

selected for some value μ_1 of the mean considered to be significantly greater than μ_0, the corresponding sample size n can be found by selecting the OC curve passing closest to the point $(d, \beta(\mu_1))$.

Whereas in the two-tailed case there was only an approximate analytic solution for n in terms of α and $\beta(\mu_1)$, in the present case n can be found directly from the preceding equation giving $\beta(\mu)$ in terms of Φ. Letting $\beta = \beta(\mu_1)$, the equality

$$\beta = \Phi\left(z_\alpha + \frac{\mu_0 - \mu_1}{\sigma/\sqrt{n}}\right)$$

is equivalent to writing

$$-z_\beta = z_\alpha + \frac{\mu_0 - \mu_1}{\sigma/\sqrt{n}}$$

Solving for n yields

$$n = \frac{(z_\alpha + z_\beta)^2 \sigma^2}{(\mu_0 - \mu_1)^2}$$

(see Table 8.2).

The development of the lower one-tailed methodology is completely analogous to the upper one-tailed case. Hence, we will forego the details and content ourselves with a summary. Again the test statistic is the standardized version of the sample mean; however, this time the alternative hypothesis is H_1: $\mu < \mu_0$, and the critical region consists of all z such that $z \leq -z_\alpha$. Again the OC curves of Figures 8.18 and 8.19 are applicable, but with

$$d = \frac{\mu_0 - \mu}{\sigma}$$

Finally, the analytic expression of n in terms of α and $\beta(\mu_1)$ is the same as in the upper one-tailed case.

EXAMPLE 8.10

In any combinational logic device there is *propagation delay*, which is the time difference between the change of the inputs to the device and the precipitated change in the output. For instance, should there be a change in either of the inputs to an OR gate, there will be a predetermined change in the output; however, that change, which depends upon the movement of electrons, will only be accomplished once the effect of the changed input states has propagated through the gate. Although, in fact, low-to-high delays can differ from high-to-low delays, we will ignore such differences in the present example.

Now, suppose a manufacturer claims that, owing to the use of a new semiconductor material, the logic gates produced by his or her company have decreased propagation delays. Specifically, among the combinational circuits required for a particular application, the claim is made that the mean propagation delay of the circuit with the slowest response time is less than 22 ns, the standard deviation being 0.4 ns. To check the claim, a number of randomly selected gates are to be tested. A statistical verification of the claim consists of rejecting the null hypothesis that the mean propagation delay is not less than 22 ns. A suitable hypothesis test is

$$H_0: \mu \geq 22$$

$$H_1: \mu < 22$$

Assuming the measurement of propagation delay (in the present case) to be normally distributed, an appropriate test statistic is the standardized sample mean. Since the slowest response time among the utilized combinational circuits contributes to a lower bound on cycle time, it is important that the alternative hypothesis not be mistakenly accepted. Consequently, it is decided that $\alpha = 0.01$ is a prudent level of significance. In addition, since it would be advantageous to use the new logic family if the claim is valid, a low value of β at the alternative mean $\mu_1 = 21.7$ is desired. It is decided that selection of the appropriate OC curve will depend upon the risk probability $\beta(21.7) = 0.05$. Using

$$d = \frac{22 - 21.7}{0.4} = 0.75$$

and locating the point $(0.75, 0.05)$ in Figure 8.19 yields the sample size $n = 28$. In the present case there exists an analytic expression for n, so the sample size can be directly determined by

$$n = \frac{(2.33 + 1.65)^2(0.4)^2}{(22.0 - 21.7)^2} = 28.16$$

which agrees with our OC-curve interpolation (unless one insists on rounding up so that β will be slightly lower than the prescribed value rather than slightly higher). Using $n = 28$ gives the sample statistic

$$Z = \frac{\bar{X} - 22}{0.4/\sqrt{28}} = \frac{\bar{X} - 22}{0.0756}$$

The desired critical region is

$$C = \{z: z \leq -2.33\}$$

Suppose, now, that a random sample of 28 combinational circuits is drawn and the resulting empirical mean is $\bar{x} = 21.85$. Then the corresponding sample value of the test statistic is

$$z = \frac{21.85 - 22.00}{0.0756} = -1.98$$

Since the sample value is not in the critical region, the null hypothesis is not rejected and the manufacturer's claim is not supported by the data, at least not to the degree demanded.

An examination of both the one- and two-tailed Z tests for the mean shows that the twin assumptions of normality and known variance are employed to guarantee the normality of the sample mean and to find the standardization of the sample mean, respectively. In the case of a large sample, the central limit theorem assures the approximate normality of the sample mean, and the sample variance estimator can be used to estimate the variance. Thus, the foregoing methodology can be applied to large samples in order to obtain approximate results. Consequently, the Z tests outlined in Table 8.2 are often referred to as **large-sample** methods.

EXAMPLE 8.11

Rather than applying a tensile test (as done in Example 8.4) when testing the yield point of a ductile metal, it is often more convenient to perform a hardness test, even though the results might not be as scientifically informative. *Hardness* is defined as the resistance of a metal to penetration. Most hardness tests are performed by pressing an indenter into the surface of the metal and then measuring the resulting impression. A large impression indicates a soft metal, whereas a small impression indicates a hard metal. The Brinell hardness test is performed by using a given load to force a hardened steel ball having a diameter of 10 mm into a metal. The diameter of the impression is then measured by a laser scanner. The *Brinell hardness* is the load divided by the area of the impression.

To demonstrate that a new treatment increases the hardness in a type of steel plate, a producer wishes to show the new Brinell hardness to be greater than the old, which happens to have been 240. The appropriate hypothesis test is

$$H_0: \ \mu \leq 240$$

$$H_1: \ \mu > 240$$

To minimize the risk of mistakenly concluding that the new treatment increases hardness, it is decided that the size of type I error should be $\alpha = 0.01$. Moreover, since an increase in the hardness of less than 3 points is not considered meaningful, the alternative mean value of 243 is chosen for the point at which to specify β, and it is demanded that $\beta(243) = 0.15$.

Suppose experience has shown that a change in hardness for the type of steel in question results in a change in the standard deviation of the measurement process. To employ the Z test using the standardized version of the sample mean, a sample size of at least 30 will be required, so that the sample standard deviation will act as a reasonably good estimator of the true standard deviation. In addition, since no claim has been made regarding the normality of the measurement process, $n = 30$ will allow the use of the central limit theorem, thereby ensuring that the standardized sample mean will possess an approximately standard normal distribution. Now, not only must n be sufficiently large to allow the use of the sample variance and the central limit theorem, it must also be sufficiently large to give the desired value of β at $\mu_1 = 243$. To employ the OC curves in Figure 8.19 (for level of significance $\alpha = 0.01$), it is necessary to know the standard deviation σ, which, however, is unknown. To circumvent the problem we will use a rough guess as to the true value of σ, the guess being based upon past experience. Keep in mind that an overly large estimate of σ will result in the sample size being unnecessarily large, whereas an undervalued estimate will result in too small a sample size for the desired β. (These statements are verified by simply recognizing that σ is in the denominator of d and the magnitude of d is inversely related to the required sample size.) Even though the standard deviation is believed to be in the vicinity of 3 points on the Brinell scale, to be conservative we will use $\sigma = 4$ in the computation of d. Thus,

$$d = \frac{243 - 240}{4} = 0.75$$

Locating the point (0.75, 0.15) in Figure 8.19 locates the OC curve determined by $n = 20$. Consequently, since the sample must consist of at least 30 observations to qualify as a "large" sample, the necessary sample size is $n = \max(30, 20) = 30$. Finally, since $z_{0.01} = 2.33$, the critical region is

$$C = \{z: z \geq 2.33\}$$

Suppose 30 observations yield the empirical mean $\bar{x} = 241.6$ and the empirical standard deviation $s = 3.5$. Then the sample value of the test statistic is

$$z \approx \frac{\bar{x} - 240}{s/\sqrt{n}} = \frac{241.6 - 240.0}{3.5/\sqrt{30}} = 2.50$$

where the approximation results from the use of s in place of σ. Since $2.50 \geq 2.33$, the alternative hypothesis is accepted, and it is concluded that the new treatment results in a Brinell hardness in excess of 240.

In Section 8.5 the t distribution will be used to perform hypothesis tests in the event the standard deviation is unknown; nevertheless, this will only partly alleviate the problem, since the t distribution results from a normally distributed population random variable. Moreover, use of the appropriate family of OC curves (for the t test) will still require a rough guess as to the value of σ when computing d. Of course, should the sample size be limited by physical or economic constraints, then the large-sample method fails, because S cannot be employed as an estimator of σ.

8.4.3 Comments on Interpretation and Methodology

Even when the population random variable is normal and the variance is known, if we are constrained to a very small sample, it will require a value of $\bar{X}$ that greatly differs from the null hypothesis in order to accept the alternative hypothesis. To see this in the upper one-tailed case, note that

$$\frac{\bar{X} - \mu_0}{\sigma/\sqrt{n}} \geq z_\alpha$$

if and only if

$$\bar{X} \geq \mu_0 + \frac{\sigma z_\alpha}{\sqrt{n}}$$

Consequently, should a small sample size result in acceptance of the null hypothesis even though the sample value of $\bar{X}$ appreciably differs from μ_0, it might be best to wait until a retest can be performed with a larger sample before making irreversible decisions based upon the outcome of the test.

From another perspective, the previous inequality shows that extremely large values of n will result in rejection of the null hypothesis even if $\bar{X}$ is only slightly larger then μ_0. While the latter decision might be mathematically sound, it may result in rejections when the true mean is only inconsequentially larger than μ_0. This is fine if rejection is desired in marginal cases; however, in some circumstances such a test might not serve the needs of the experimenter. One possible approach is to employ a hypothesis test of the form

$$H_0: \mu \leq \mu_0 + \delta$$

$$H_1: \mu > \mu_0 + \delta$$

where δ is positive and is selected in accordance with the intent of the test. Basically, the point is this: if an investigator possesses good insight into the engineering ramifications of the decisions that might result from a hypothesis test, then there is greater likelihood that he or she will design a test appropriate to his or her requirements.

The rejection or acceptance of the null hypothesis depends upon both the observational data and the structure of the test. For the upper one-tailed Z test of the mean, the critical value z_α determines whether or not a particular test-statistic sample value z results in rejection of H_0. If $z \geq z_\alpha$, then H_0 is rejected and z is said to be **significant.** Because the critical region is dependent upon α, significance is relative to the choice of α. If $\alpha = 0.01$, then $z_\alpha = 2.33$. Consequently, a sample value of 2.11 results in acceptance of H_0, which from the perspective of detection means nondetection. Had less stringent significance been required, say $\alpha = 0.05$, then the

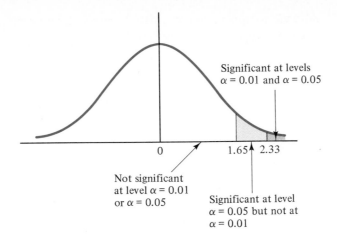

FIGURE 8.20
Comparison of
significance levels

critical value would have been $z_\alpha = 1.65$ and the sample value 2.11 would have been significant, thereby leading to detection.

More generally, suppose z is a sample value of the test statistic Z. If $z \geq 2.33$, then it is significant at both the levels of significance 0.01 and 0.05. If $1.65 \leq z < 2.33$, then it is significant at level 0.05 but not at level 0.01. Finally, if $z < 1.65$, then it is significant at neither level 0.05 nor level 0.01. These significance ranges are illustrated in Figure 8.20.

Returning to the test value 2.11, it can be seen from Table A.1 that 2.11 is significant at level $\alpha = 0.0174$ but not significant at any level $\alpha < 0.0174$. On the contrary, it is significant at any level $\alpha \geq 0.0174$. In sum, 0.0174 represents the lowest level of significance at which the test-statistic value 2.11 results in a rejection of the null hypothesis.

DEFINITION 8.4

P **Value.** The *P* value corresponding to an observed value of a test statistic is the lowest level of significance for which the test-statistic value results in rejection of the null hypothesis.

Most computer printouts for statistical packages include the *P* value corresponding to a given set of observations. An advantage of the *P* value is that it gives the investigator some insight into the degree of significance of a test result. For instance, for the upper one-tailed test just considered, the *P* value of the observed test-statistic value 2.11 is 0.0174. Thus, rejection is quite strong with respect to significance level $\alpha = 0.05$. In addition, given a *P* value, someone not directly involved in the experiment can discern at which levels the test-statistic value would be significant. Specifically, the test-statistic value is significant at significance levels higher than the *P* value and is not significant at levels lower than the *P* value.

8.4.4 General Outline for the Design and Implementation of a Hypothesis Test

With the tools presently developed, it is possible to provide a general outline for a systematic approach to the design and implementation of a hypothesis test involving a null hypothesis stating a specific value of a given parameter:

1. Identify the competing hypotheses, keeping in mind that common practice dictates that detection occurs when the null hypothesis is rejected. In the case of a two-tailed test, the hypothesis test will take the form

$$H_0: \theta = \theta_0$$
$$H_1: \theta \neq \theta_0$$

In the case of an upper one-tailed test,

$$H_0: \theta = \theta_0 \quad (\text{or } \theta \leq \theta_0)$$
$$H_1: \theta > \theta_0$$

In the case of a lower one-tailed test,

$$H_0: \theta = \theta_0 \quad (\text{or } \theta \geq \theta_0)$$
$$H_1: \theta < \theta_0$$

2. Determine the level of significance α. The determination of α reflects the degree of risk one is willing to take that there will be a mistaken rejection of the null hypothesis, or, in other words, a false detection.
3. Choose a suitable test statistic.
4. Decide on a value of the parameter, say θ_1, for which an erroneous acceptance of the null hypothesis poses serious consequences. For that value of the parameter, decide on a tolerable risk level $\beta(\theta_1)$. Using the family of OC curves corresponding to the chosen test statistic and the selected level of significance, determine the sample size n by selecting the appropriate OC curve. If n is fixed beforehand, then simply record it.
5. Determine the critical region.
6. Take a random sample of n observations and compute the corresponding sample value of the test statistic.
7. Decide whether or not to reject the null hypothesis: if the sample value of the test statistic lies in the critical region, then reject the null hypothesis; otherwise, accept it.
8. (*Optional*) State the P value corresponding to the sample value of the test statistic.

In the remainder of the text we will employ the numbering scheme 1 through 8 as a reporting device in the solution of numerous hypothesis-testing problems. For instance, a compact manner in which to present the results of Example 8.10 is given by the following format:

1. *Hypothesis test.* $H_0: \mu \geq 22$
     $H_1: \mu < 22$

2. *Level of significance.* $\alpha = 0.01$

3. *Test statistic.* $Z = \dfrac{\bar{X} - 22}{0.4/\sqrt{n}}$

4. *Sample size determination.* $\beta(21.7) = 0.05$

$$d = \frac{22 - 21.7}{0.4} = 0.75$$
$$n = 28$$

5. *Critical region.* $z \leq -2.33$

6. *Sample value.* $z = \dfrac{\bar{x} - 22}{0.4/\sqrt{28}} = \dfrac{21.85 - 22.00}{0.0756} = -1.98$

7. *Decision.* Accept H_0 because $z > -2.33$.
8. *P value.* 0.0239

Had the sample data been provided rather than the value of the sample mean, then step 6 would have included the computation of $\bar{x}$.

8.5 TESTING THE MEAN OF A NORMAL DISTRIBUTION WITH UNKNOWN VARIANCE

Two common occurrences often mitigate the usefulness of the Z test for testing the mean of a normal distribution. First, in many situations the variance is not known with a satisfactory degree of accuracy. This may result from a general lack of information regarding the population at hand or from uncertainty regarding the variational effects of a change whose effect on the mean is being tested. Second, given uncertainty concerning the variance, for practical reasons it might be impossible to make a large number of independent observations, thereby ruling out estimation of the variance by the sample variance. Such a restriction is certainly the case in settings where observations of the random variable are costly in either time or money. As was done in Section 7.8.3 in the case of interval estimation, when the variance is unknown and cannot be reliably estimated, we turn to the t distribution.

8.5.1 Two-Tailed Mean Tests Involving the t Distribution

As in the previous section, to perform the two-sided hypothesis test

$$H_0:\ \mu = \mu_0$$

$$H_1:\ \mu \neq \mu_0$$

it is natural to first look to the sample-mean test statistic. In the present setting, however, the variance of the population variable X is unknown. Hence, the reasoning of Section 8.4.1 cannot be replicated. Nevertheless, we can proceed in a somewhat similar manner by finding a test statistic whose probability mass is symmetrically centered at μ_0 when the true mean is μ_0. Such a statistic is

$$T = \frac{\bar{X} - \mu_0}{S/\sqrt{n}}$$

which, by Theorem 7.15, possesses a t distribution with $n-1$ degrees of freedom. The use of the t distribution should not be surprising in light of the fact that it was employed to provide confidence intervals for the means of normally distributed random variables with unknown variances.

The acceptance region when employing the t distribution in the two-tailed case is the interval

$$A = (-t_{\alpha/2, n-1},\ t_{\alpha/2, n-1})$$

where the critical value $t_{\alpha/2, n-1}$ depends upon the chosen level of significance α and the sample size n. Specifically, for $\mu = \mu_0$,

$$P\left(-t_{\alpha/2, n-1} < \frac{\bar{X} - \mu_0}{S/\sqrt{n}} < t_{\alpha/2, n-1} \right) = 1 - \alpha$$

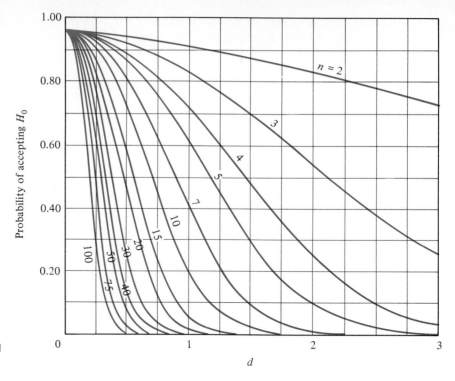

FIGURE 8.21

OC curves for the two-tailed t test at level $\alpha = 0.05$ (Reprinted by permission of the Institute of Mathematical Statistics.)

As with the Z test, it is convenient to employ OC curves to determine the value of β at various alternative choices of μ. For the levels $\alpha = 0.05$ and $\alpha = 0.01$, these families of curves are depicted in Figures 8.21 and 8.22, respectively. As with the OC curves corresponding to the two-tailed Z test, the OC curves for the two-tailed t test have been standardized according to the difference

$$ d = \frac{|\mu_0 - \mu|}{\sigma} $$

Since σ is unknown, use of the OC curves depends upon a crude estimate of σ. The anomalies here are evident. Given the sample size n and an alternative mean μ, one must make a determination of $\beta(\mu)$ based upon a rough estimate of σ, an estimate that in all likelihood will not come from the sample variance, since a small-sample limitation is being presumed. Analysis of Figures 8.21 and 8.22 shows that an overly large (conservative) estimate of σ yields an overly large estimate of β. Conversely, too small an estimate of σ yields too small an estimate of β.

In the other direction, suppose a risk level $\beta(\mu_1)$ is specified at the alternative mean μ_1. If the OC-curve family is to be employed to determine the proper sample size, then the same dilemma results; however, here the predicament is unimportant if we have assumed ourselves to be in a situation where n is either fixed in advance or is at least constrained to be small. It is important, however, if n is not constrained and we are employing the t test to avoid using an estimate of σ in evaluating the test statistic.

For the Z test it was possible to express $\beta(\mu)$ in terms of α, μ_0, σ, and n by means of the probability distribution function of Z. In the present setting, this approach is not feasible, since σ is unknown, and thus we are left with evaluating

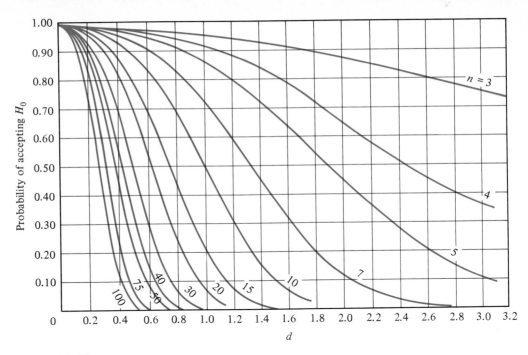

FIGURE 8.22
OC curves for the two-tailed t test at level $\alpha = 0.01$ (Reprinted by permission of the Institute of Mathematical Statistics.)

$$\beta(\mu) = P\left(-t_{\alpha/2,n-1} < \frac{\bar{X} - \mu_0}{S/\sqrt{n}} < t_{\alpha/2,n-1} \right)$$

under the condition that $\mu \neq \mu_0$. Consequently, the OC curves of Figures 8.21 and 8.22 have been computed numerically.

EXAMPLE 8.12

An insulation's capacity to block the flow of heat is customarily measured by its R value. To test a manufacturer's claim that a new insulating material possesses an R value of 3.8 per inch of thickness, 5 test measurements will be taken, each on a specimen from a different production batch. The manufacturer's claim will be rejected if the measurements indicate it to be either too high or too low. Although it will be assumed that the measurement process is normally distributed, the newness of the product makes it likely that a good estimate of the variance is lacking. Moreover, the small sample size does not permit use of the sample-variance estimator. We will present the hypothesis-test analysis in the format introduced in Section 8.4.4.

1. *Hypothesis test.* H_0: $\mu = 3.8$
   H_1: $\mu \neq 3.8$
2. *Level of significance.* Assuming there is no reason to demand an especially stringent level of significance, let $\alpha = 0.05$.
3. *Test statistic.* Normality allows use of the t test with $5 - 1 = 4$ degrees of freedom. Thus, the test statistic is

$$T = \frac{\bar{X} - 3.8}{S/\sqrt{n}}$$

4. *Sample size.* $n = 5$
5. *Critical region.* From Table A.6,

$$t_{\alpha/2, n-1} = t_{0.025, 4} = 2.776$$

The resulting critical region is

$$C = \{t: t \leq -2.776 \text{ or } t \geq 2.776\}$$

6. *Sample value.* Suppose the 5 measurements are

$$3.9, \ 3.8, \ 4.0, \ 4.1, \ 4.2$$

Then $\bar{x} = 4.0$ and $s = 0.158$. Consequently, the sample value of the test statistic is

$$t = \frac{\bar{x} - 3.8}{s/\sqrt{5}} = \frac{4.0 - 3.8}{0.158/\sqrt{5}} = 2.830$$

7. *Decision.* Reject H_0, since $t > 2.776$.

From the perspective of the buyer, the manufacturer's claim has been rejected because the product has performed better than claimed! One might preferably have framed a one-tailed test to decide whether the product performs at least as well as claimed. We will examine this matter shortly.

Let us now backtrack and consider type II error for the test as designed. If a deviation of 0.4 from the claimed value is considered consequential, then we need to refer to the OC curve in Figure 8.21 corresponding to $n = 5$ and locate the β value associated with

$$d = \frac{|3.8 - 4.2|}{\sigma}$$

To be conservative, let $\sigma = 0.25$, where from experience it is believed very unlikely that σ will be greater than 0.25. With $d = 1.6$ and $n = 5$, Figure 8.21 yields $\beta \approx 0.27$, so that when performing the test there is a reasonable possibility of erroneously accepting the claim of the manufacturer.

8.5.2 One-Tailed Mean Tests Involving the t Distribution

When X is a normally distributed random variable with unknown variance σ^2, the appropriate test statistic for the upper one-tailed hypothesis test

$$H_0: \ \mu \leq \mu_0$$

$$H_1: \ \mu > \mu_0$$

is the same t distribution as employed in the two-tailed case, the difference being the use of the one-tailed critical region

$$C = \{t: t \geq t_{\alpha, n-1}\}$$

The corresponding OC-curve families for $\alpha = 0.05$ and $\alpha = 0.01$ are given in Figures 8.23 and 8.24, respectively, where

$$d = \frac{\mu - \mu_0}{\sigma}$$

The comments of Section 8.5.1 regarding the estimation of σ apply directly in the present case.

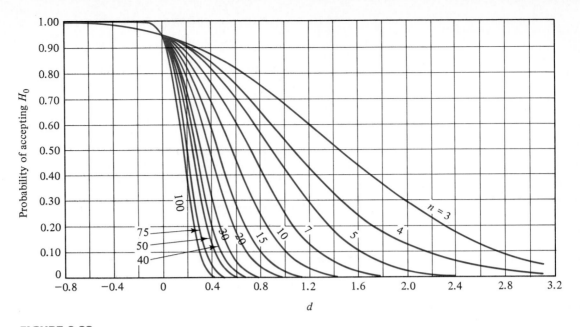

FIGURE 8.23
OC curves for the one-tailed t test at level $\alpha = 0.05$ (Reprinted by permission of the Institute of Mathematical Statistics.)

Analogous remarks apply to the lower one-tailed t test for the mean, and we will forego a detailed explanation, noting only that the OC-curve families are the same as in the upper one-tailed case except that

$$d = \frac{\mu_0 - \mu}{\sigma}$$

FIGURE 8.24
OC curves for the one-tailed t test at level $\alpha = 0.01$ (Reprinted by permission of the Institute of Mathematical Statistics.)

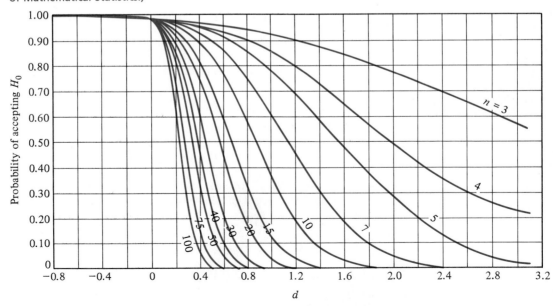

TABLE 8.3 IMPORTANT INFORMATION RELATING TO THE t TEST OF THE MEAN

Null Hypothesis: H_0: $\mu = \mu_0$

Test Statistic: $T = \dfrac{\bar{X} - \mu_0}{S/\sqrt{n}}$

Alternative Hypothesis (Two-Tailed): H_1: $\mu \neq \mu_0$

Critical Region: $C = \{t: t \leq -t_{\alpha/2, n-1} \text{ or } t \geq t_{\alpha/2, n-1}\}$

Alternative Hypothesis (Upper One-Tailed): H_1: $\mu > \mu_0$

Critical Region: $C = \{t: t \geq t_{\alpha, n-1}\}$

Alternative Hypothesis (Lower One-Tailed): H_1: $\mu < \mu_0$

Critical Region: $C = \{t: t \leq -t_{\alpha, n-1}\}$

The salient points regarding all three t tests of the mean are summarized in Table 8.3.

EXAMPLE 8.13

The environmental agency of a certain city claims the amount of hydrocarbons in the atmosphere has been reduced to a level of 1.8 ppm (parts per million). To verify the claim, a sample of 12 observations is to be taken at randomly selected times and locations over a period of one month. From past testing it is known that the standard deviation tends to vary from test to test; however, it is also known that measurements tend to be normally distributed. Thus, a t test is appropriate. We again employ the numbered format.

1. *Hypothesis test*: H_0: $\mu \geq 1.8$

   $\quad\quad H_1$: $\mu < 1.8$

2. *Level of significance.* To avoid any controversy regarding verification, the stringent level of significance $\alpha = 0.01$ will be employed.

3. *Test statistic.* $T = \dfrac{\bar{X} - 1.8}{S/\sqrt{n}}$

4. *Sample size.* $n = 12$

5. *Critical region.* Since the test is one-tailed, the critical value is

$$t_{\alpha, n-1} = t_{0.01, 11} = 2.718$$

so that the critical region is

$$C = \{t: t \leq -2.718\}$$

6. *Sample value.* Suppose the 12 readings yield the empirical mean $\bar{x} = 1.68$ and standard deviation $s = 0.17$. Then the test statistic takes the sample value

$$t = \frac{\bar{x} - 1.80}{s/\sqrt{n}} = \frac{1.68 - 1.80}{0.17/\sqrt{12}} = -2.445$$

7. *Decision.* Accept H_0 since $t > -2.718$.

8. *P value.* Using linear interpolation in Table A.5 for $\nu = 11$ results in the approximate P value

$$0.01 + \frac{2.718 - 2.445}{2.718 - 2.201}(0.025 - 0.01) = 0.018$$

The low P value reflects the fact that a t value of -2.445 is close to being significant at level $\alpha = 0.01$.

To see the interplay between sample size and significance level, suppose a mean value of 1.70 ppm is considered a meaningful reduction in hydrocarbon level and $\sigma = 0.2$ is considered to be a conservative estimate of σ. Then

$$d = \frac{1.8 - 1.7}{0.2} = 0.5$$

in Figure 8.24 and the OC curve for $n = 12$ gives $\beta(1.7) \approx 0.80$. To bring $\beta(1.7)$ down to 0.40 demands a sample size of approximately 30. Had the less stringent level of significance $\alpha = 0.05$ been employed, then for $n = 12$, $\beta(1.7) \approx 0.50$, which is still rather high (see Figure 8.23). Nevertheless, for $\alpha = 0.05$,

$$t_{\alpha, n-1} = t_{0.05, 11} = 1.796$$

and the same sample data result in acceptance of the claim. Indeed, this conclusion can be drawn at once from the P value, since $0.05 > 0.018$.

The t test is somewhat robust, in that mild deviations of the population random variable from normality do not have an overly adverse impact upon the method. Nevertheless, if it is suspected that the population variable deviates substantially from normality, then the test should not be employed. Instead, one should take a large sample and apply the large-sample Z test, thereby allowing the central limit theorem to guarantee approximate normality of the sample mean. If, on the other hand, the normality of the population variable is well established but the variance is unknown, then the t test is a better choice than applying the Z test with the sample variance estimating the variance, since even for fairly large n, such as $n = 50$, there is still substantial probability mass of S^2 away from its mean σ^2. As will be seen in Section 8.8 (Example 8.20), for a normal random variable with known variance, the Z test is "best" (in a sense to be defined).

8.6 TESTING A PROPORTION

Examples 8.1, 8.2, and 8.3 involved the testing of two competing simple hypotheses concerning the probability of a success for a binomial random variable. In the present section the methodology is extended to composite alternative hypotheses. The small-sample case is treated first and the large-sample case, where it is convenient to employ an approximation to the standard normal distribution, is treated next.

8.6.1 Small-Sample Testing of a Proportion

If a population is split into two mutually exclusive classes S (success) and F (failure), then the categorization is described by a binomial random variable X with binomial density $b(x; 1, p)$, where $p = P(S)$ and $q = 1 - p = P(F)$ are the probabilities of randomly selecting elements from S and F, respectively. As has been done in the interval estimation of a proportion, as well as in Example 8.1, if the size n of a sample is small in comparison with the total population, then the selection process can be modeled as random even if there is no replacement. Thus, the desired test statistic

$$V = X_1 + X_2 + \cdots + X_n$$

where $X_1, X_2, \ldots, X_n$ constitute the resulting random sample, is binomially distributed with mean $\mu = np$ and variance $\sigma^2 = npq$. V is simply the number of successes in the sample.

If the sample size is sufficiently large, then Theorem 4.9 is applicable; if not, then the analysis must proceed with a direct application of the binomial distribution.

For the small-sample case, consider the two-tailed hypothesis test

$$H_0: p = p_0$$
$$H_1: p \neq p_0$$

If the null hypothesis is true, then $E[V] = np_0$. Consequently, we choose an interval (a, b) containing np_0 as the acceptance region, where a and b are integers. If α' is the desired level of significance, then a and b are chosen so that, given $p = p_0$,

$$P(a < V < b) \approx 1 - \alpha'$$

the approximation resulting from the fact that V is discrete. The critical points a and b should be chosen so that the probability mass $1 - \alpha'$ is approximately equally distributed between the subintervals $(a, np_0]$ and $[np_0, b)$. If this is done, then the critical region will be given by

$$C = (-\infty, a] \cup [b, +\infty)$$

and

$$P(V \leq a) \approx P(V \geq b) \approx \frac{\alpha'}{2}$$

(see Figure 8.25). Since V is binomially distributed, for small n the relevant probabilities can be found from Table A.2. Specifically,

$$P(V \leq a) = B(a; n, p_0)$$
$$P(V \geq b) = 1 - B(b - 1; n, p_0)$$
$$P(a < V < b) = P(V \leq b - 1) - P(V \leq a)$$
$$= B(b - 1; n, p_0) - B(a; n, p_0)$$

Because V is discrete, the critical region has been chosen so that type I error is approximately equal to the desired level α'. The actual size of type I error is

$$\alpha = P(V \leq a) + P(V \geq b)$$

The size of type II error for $p \neq p_0$ is given by

$$\beta(p) = P(a < V < b; p) = B(b - 1; n, p) - B(a; n, p)$$

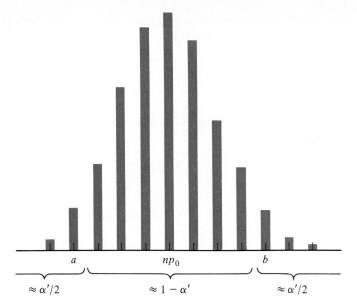

FIGURE 8.25
Small-sample testing of a proportion

This relation can be used in conjunction with Table A.2 to determine the sample size necessary to approximately produce a particular value of β at an alternative proportion p_1.

One-tailed tests of a proportion are performed with the same test statistic as are two-tailed tests, the distinction being in the critical regions. Because of the similarity between the upper and lower one-tailed tests, we will discuss in detail only the former. For the upper one-tailed test

$$H_0: \ p \leq p_0$$
$$H_1: \ p > p_0$$

with desired level of significance α', an integral critical value c is chosen so that, for $p = p_0$,

$$\alpha = P(V \geq c) \approx \alpha'$$

The critical region is the interval $[c, +\infty)$ and the acceptance region is $(-\infty, c)$. In terms of the probability distribution function of V, exact type I error is given by

$$\alpha = 1 - B(c - 1; n, p_0)$$

For $p > p_0$, the size of type II error is

$$\beta(p) = P(V < c; p) = B(c - 1; n, p)$$

If the error $\beta(p_1)$ is considered a serious level of risk at the alternative proportion p_1, then the preceding relation can be employed in conjunction with Table A.2 to determine an appropriate sample size.

EXAMPLE 8.14

In reference to Example 8.1, the supplier of the brake-caliper seals purports the seals to be of the agreed quality. This claim is analytically equivalent to a claim that only 10% of all seals will fail the ultrarigorous testing procedure. To demonstrate the falseness of the claim, an upper one-tailed hypothesis test

$$H_0: p \leq 0.1$$
$$H_1: p > 0.1$$

is appropriate. Assuming 20 seals are subjected to the test, the acceptance and critical regions take the forms

$$A = \{0, 1, \ldots, c - 1\}$$

and

$$C = \{c, c + 1, \ldots, 20\}$$

respectively. If the desired significance level is $\alpha' = 0.05$, then c must be chosen so that

$$0.05 \approx 1 - B(c - 1; 20, 0.1)$$

or

$$0.95 \approx B(c - 1; 20, 0.1)$$

According to Table A.2, $B(4; 20. 0.1) = 0.9568$. No other value in the 0.1 column is as close to 0.95. Hence, $c = 5$, and the desired critical region is

$$C = \{5, 6, \ldots, 20\}.$$

The actual size of type I error is

$$\alpha = 1 - 0.9568 = 0.0432$$

Now, suppose it is recognized that a 20% failure rate in the testing procedure indicates a serious problem. Then $\beta(0.2)$ is an important value, since it gives the probability of accepting the shipment as legitimate even though 20% of the seals fail the testing procedure. For the chosen critical region,

$$\beta(0.2) = P(V < 5; p = 0.2) = B(4; 20, 0.2) = 0.6296$$

which is quite high. Lowering $\beta(0.2)$ requires a larger sample. Most likely, the test should be redesigned using the large-sample approach to be introduced next.

8.6.2 Large-Sample Testing of a Proportion

Now consider a hypothesis test for p when the sample size is sufficiently large to employ Theorem 4.9. If $p = p_0$ and $q_0 = 1 - p_0$, then

$$\frac{V - np_0}{\sqrt{np_0 q_0}}$$

possesses an approximately standard normal distribution, and the Z tests of Section 8.4 can be applied (with the results being approximate). This is precisely the approach taken in Example 8.3 for competing simple hypotheses. Recall that the approximation corresponding to Theorem 4.9 should be used only if $np_0 \geq 5$ and $nq_0 \geq 5$.

As an illustration, consider the lower one-tailed test

$$H_0\text{: } p \geq p_0$$

$$H_1\text{: } p < p_0$$

Then, according to Table 8.2, for a given α the appropriate critical region is

$$C = \{z\text{: } z \leq -z_\alpha\}$$

For a particular set of sample values $x_1, x_2, \ldots, x_n$, each of which is 0 (failure) or 1 (success), the test sample value is

$$z = \frac{v - np_0}{\sqrt{np_0q_0}} = \frac{x_1 + x_2 + \cdots + x_n - np_0}{\sqrt{np_0q_0}}$$

For any $p \neq p_0$, $\beta(p)$ can be found directly. For instance, in the lower one-tailed case,

$$\beta(p) = P\left(\frac{V - np_0}{\sqrt{np_0q_0}} > -z_\alpha; p\right)$$

$$= P(V > np_0 - z_\alpha\sqrt{np_0q_0}; p)$$

$$= P\left(\frac{V - np}{\sqrt{npq}} > \frac{n(p_0 - p) - z_\alpha\sqrt{np_0q_0}}{\sqrt{npq}}\right)$$

$$\approx 1 - \Phi\left(\frac{n(p_0 - p) - z_\alpha\sqrt{np_0q_0}}{\sqrt{npq}}\right)$$

where the approximation follows from the fact that $(V - np)/\sqrt{npq} \approx Z$ for large n.

The third equality in the preceding computation can be solved to obtain n in terms of $\beta = \beta(p_1)$, where p_1 is an alternative proportion. Letting $\beta = P(Z > z_\beta)$, the third equality can be interpreted as

$$z_\beta \approx \frac{n(p_0 - p_1) - z_\alpha\sqrt{np_0q_0}}{\sqrt{np_1q_1}}$$

where again there is an approximation because $(V - np)/\sqrt{npq} \approx Z$. Solving for n yields

$$n \approx \left[\frac{z_\beta\sqrt{p_1q_1} + z_\alpha\sqrt{p_0q_0}}{p_0 - p_1}\right]^2$$

Table 8.4 contains the pertinent information regarding large-sample testing of a proportion, including the expressions for n in terms of α and β. Note that the solution for n in the two-tailed case can be a poor approximation even if np_1 and nq_1 are very large, the reason being that this expression suffers from the same lack of symmetry that affected the two-tailed solution for n in Table 8.2.

TABLE 8.4 IMPORTANT INFORMATION RELATED TO THE LARGE-SAMPLE TESTING OF A PROPORTION

Null Hypothesis: H_0: $p = p_0$

Test Statistic: $\dfrac{V - np_0}{\sqrt{np_0q_0}}$ (V is the number of successes.)

Alternative Hypothesis (Two-Tailed): H_1: $p \neq p_0$

    Critical Region: $C = \{z: z \leq -z_{\alpha/2} \text{ or } z \geq z_{\alpha/2}\}$

    Values of β: $\beta(p) = \Phi\left(\dfrac{n(p_0 - p) + z_{\alpha/2}\sqrt{np_0q_0}}{\sqrt{npq}}\right)$

                      $- \Phi\left(\dfrac{n(p_0 - p) - z_{\alpha/2}\sqrt{np_0q_0}}{\sqrt{npq}}\right)$

    Expression of n in terms of α and $\beta = \beta(p_1)$: $n \approx \left(\dfrac{z_\beta\sqrt{p_1q_1} + z_{\alpha/2}\sqrt{p_0q_0}}{p_0 - p_1}\right)^2$

Alternative Hypothesis (Upper One-Tailed): H_1: $p > p_0$

    Critical Region: $C = \{z: z \geq z_\alpha\}$

    Values of β: $\beta(p) = \Phi\left(\dfrac{n(p_0 - p) + z_\alpha\sqrt{np_0q_0}}{\sqrt{npq}}\right)$

    Expression of n in terms of α and $\beta = \beta(p_1)$: $n \approx \left(\dfrac{z_\beta\sqrt{p_1q_1} + z_\alpha\sqrt{p_0q_0}}{p_0 - p_1}\right)^2$

Alternative Hypothesis (Lower One-Tailed): H_1: $p < p_0$

    Critical Region: $C = \{z: z \leq -z_\alpha\}$

    Values of β: $\beta(p) = 1 - \Phi\left(\dfrac{n(p_0 - p) - z_\alpha\sqrt{np_0q_0}}{\sqrt{npq}}\right)$

    Expression of n in terms of α and $\beta = \beta(p_1)$: Same as for upper one-tailed test.

EXAMPLE 8.15

In Example 8.14 it was seen that a sample size of 20 is insufficient to lower β to a reasonable level at the alternative proportion $p_1 = 0.2$. If we now presume n to be variable and apply the large-sample technique with $\beta(0.2) = 0.1$, then

$$z_\alpha = z_{0.05} = 1.65$$

$$z_\beta = z_{0.10} = 1.28$$

and, according to Table 8.4,

$$n \approx \left[\frac{(1.28)\sqrt{(0.2)(0.8)} + (1.65)\sqrt{(0.1)(0.9)}}{0.1 - 0.2}\right]^2 = 101.40$$

Letting $n = 102$, both np_0 and nq_0 are greater than 5, so that the large-sample methodology applies. The test statistic is

$$\frac{V - np_0}{\sqrt{np_0 q_0}} = \frac{V - 10.2}{3.03}$$

and the critical region is the interval $[1.65, +\infty)$.

If the test is run and there are 17 failures, then the sample value is

$$\frac{v - 10.2}{3.03} = \frac{17.0 - 10.2}{3.03} = 2.24$$

which lies in the critical region. Note the increased sensitivity of the test. In the small-sample approach of Example 8.14, with $n = 20$, if a sample yields 20% failures ($v = 4$), then the null hypothesis is not rejected. In the present case, with $n = 102$, 17% failures ($v = 17$) causes a rejection of the null hypothesis.

8.7 TESTING THE VARIANCE OF A NORMAL DISTRIBUTION

The present section discusses variance testing of a normally distributed random variable X with unknown mean μ and unknown variance σ^2. The appropriate two-tailed hypothesis test is

$$H_0: \sigma^2 = \sigma_0^2$$

$$H_1: \sigma^2 \neq \sigma_0^2$$

According to Theorem 7.9, if S^2 is the sample variance corresponding to a random sample of size n arising from X and the null hypothesis reflects the true state of nature, then the statistic

$$W = \frac{(n - 1)S^2}{\sigma_0^2}$$

possesses a chi-square distribution with $n - 1$ degrees of freedom. Since S^2 is an unbiased estimator of σ^2, if $\sigma^2 = \sigma_0^2$, then (intuitively) the test statistic should be near $n - 1$. With $n - 1$ being the mean of the chi-square variable W (see Theorem 4.13), it seems reasonable to take the two tails of the chi-square distribution with $n - 1$ degrees of freedom as the critical region. Thus, for level of significance α, the acceptance region is

$$A = (\chi^2_{1 - \alpha/2, n - 1}, \chi^2_{\alpha/2, n - 1})$$

where, as is depicted in Figure 7.9, the left and right critical values are chosen so that $P(W \in A) = 1 - \alpha$. Both critical values can be obtained from Table A.5.

A key difference in testing the variance as opposed to testing the mean is the nature of the OC curves. In particular, the distribution of the test statistic is not symmetric, so that $\beta(\sigma^2)$ is not symmetric about σ_0^2. The lack of symmetry can be observed in Figures 8.26 and 8.27, which provide the OC-curve families for levels of significance $\alpha = 0.05$ and $\alpha = 0.01$, respectively. Normalization of OC-curve families in the case of the variance is accomplished by using

$$\lambda = \frac{\sigma}{\sigma_0}$$

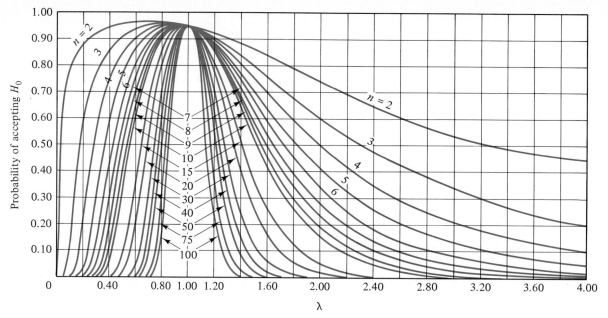

FIGURE 8.26
OC curves for the two-tailed χ^2 test at level
$\alpha = 0.05$ (Reprinted by permission of the Institute of Mathematical Statistics.)

Finally, note that the acceptance region for the null hypothesis is consistent with the interval estimation methodology employed for the variance in Section 7.8.5.

The upper and lower one-tailed tests for the variance of a normally distributed random variable take the forms

FIGURE 8.27
OC curves for the two-tailed χ^2 test at level
$\alpha = 0.01$ (Reprinted by permission of the Institute of Mathematical Statistics.)

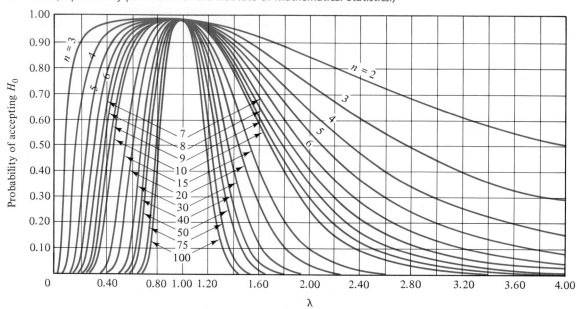

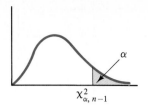

FIGURE 8.28
Critical region for the
upper one-tailed χ^2 test
of the variance

$$H_0: \sigma^2 \leqslant \sigma_0^2$$

$$H_1: \sigma^2 > \sigma_0^2$$

$$H_0: \sigma^2 \geqslant \sigma_0^2$$

$$H_1: \sigma^2 < \sigma_0^2$$

respectively. In both cases the test statistic is the same chi-square variable W utilized in the two-tailed test. For level of significance α, the acceptance and critical regions in the upper one-tailed case are

$$A = [0, \chi_{\alpha,n-1}^2)$$

$$C = [\chi_{\alpha,n-1}^2, +\infty)$$

As depicted in Figure 8.28, given the null hypothesis to be true, $P(W \in C) = \alpha$. Figures 8.29 and 8.30 give the OC-curve families for levels $\alpha = 0.05$ and $\alpha = 0.01$, respectively. As in the two-tailed setting, OC-curve normalization is given by $\lambda = \sigma/\sigma_0$.

For the lower one-tailed case, the acceptance and critical regions are

$$A = (\chi_{1-\alpha,n-1}^2, +\infty)$$

$$C = [0, \chi_{1-\alpha,n-1}^2]$$

FIGURE 8.29
OC curves for the upper one-tailed χ^2 test at level $\alpha = 0.05$ (Reprinted by permission of the Institute of Mathematical Statistics.)

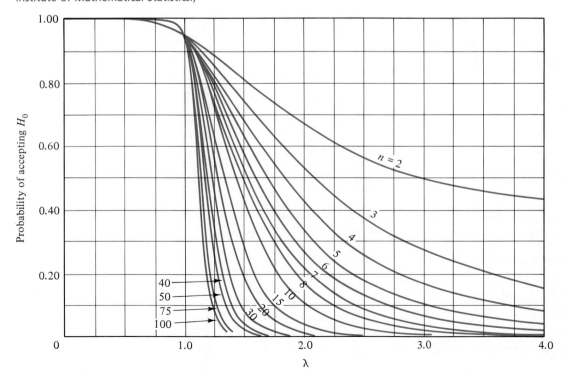

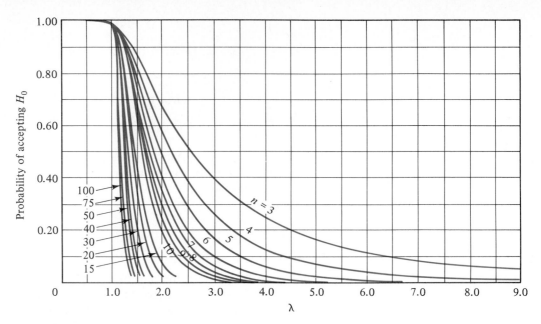

FIGURE 8.30
OC curves for the upper one-tailed χ^2 test at level $\alpha = 0.01$ (Reprinted by permission of the Institute of Mathematical Statistics.)

As illustrated in Figure 8.31, if the null hypothesis is true, then $P(W \in C) = \alpha$. Due to lack of symmetry, the same OC curves cannot be used in both the lower and upper one-tailed cases. The lower one-tailed OC-curve families for $\alpha = 0.05$ and $\alpha = 0.01$ are given in Figures 8.32 and 8.33, respectively. Once again normalization is given by $\lambda = \sigma/\sigma_0$.

Table 8.5 provides a summary of the important information concerning chi-square testing of the variance of a normal distribution.

EXAMPLE 8.16

In milling, which is one of the most versatile of all machine processes, surfaces are machined by the action of a rotating cutter. A common usage of milling occurs in the manufacturing of cylinder heads, where the surface of the head is "cut" an amount in accordance with the desired compression ratio of the engine. In such instances, a high degree of precision is important, especially in the case of high-performance engines. Even if a machine's mean cut matches specifications, the variance must remain satisfactorily small.

After a period of use, it is desired to check the variation occurring in a particular milling machine. To do so, test cuts are to be made, each set at 0.050 inch. It is assumed that the distribution of the cut sizes is normal. The machine must receive maintainance if the variance exceeds 0.000009. The problem will be approached in the standard hypothesis-testing format.

FIGURE 8.31

Critical region for the lower one-tailed χ^2 test of the variance

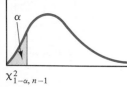

$\chi^2_{1-\alpha, n-1}$

1. *Hypothesis test.* Significant values are those indicating too great a variance. Thus, the test is upper one-tailed and is defined by

$$H_0: \sigma^2 \leq 0.000009$$

$$H_1: \sigma^2 > 0.000009$$

2. *Level of significance.* Because of the costs, both in service and down-time, the size of type I error must be kept very small. Thus, α is set to 0.01.

3. *Test statistic.* $W = \dfrac{(n-1)S^2}{0.000009}$

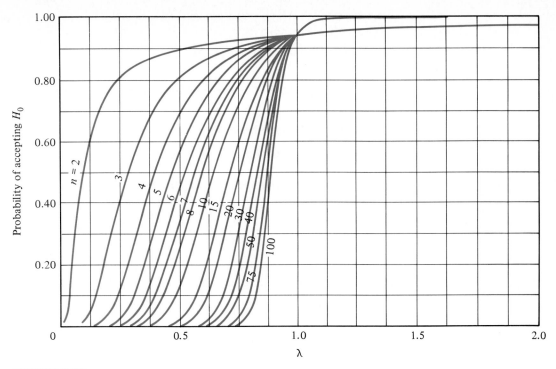

FIGURE 8.32
OC curves for the lower one-tailed χ^2 test at level $\alpha = 0.05$ (Reprinted by permission of the Institute of Mathematical Statistics.)

FIGURE 8.33
OC curves for the lower one-tailed χ^2 test at level $\alpha = 0.01$ (Reprinted by permission of the Institute of Mathematical Statistics.)

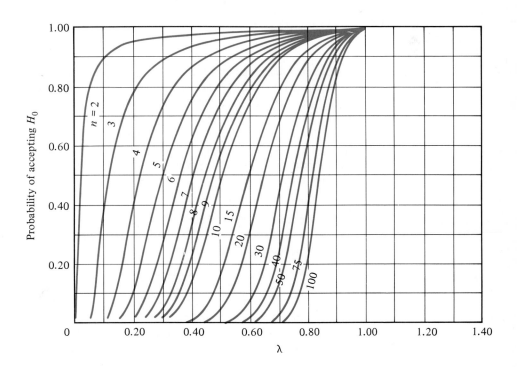

TABLE 8.5   IMPORTANT INFORMATION RELATED TO TESTING THE VARIANCE

Null Hypothesis: H_0: $\sigma^2 = \sigma_0^2$

Test Statistic: $W = \dfrac{(n-1)S^2}{\sigma_0^2}$ (Chi-square with $n-1$ degrees of freedom)

Alternative Hypothesis (Two-Tailed): H_1: $\sigma^2 \neq \sigma_0^2$

    Critical Region: $C = \{w: w \leq \chi_{1-\alpha/2,n-1}^2 \text{ or } w \geq \chi_{\alpha/2,n-1}^2\}$

Alternative Hypothesis (Upper One-Tailed): H_1: $\sigma^2 > \sigma_0^2$

    Critical Region: $C = \{w: w \geq \chi_{\alpha,n-1}^2\}$

Alternative Hypothesis (Lower One-Tailed): H_1: $\sigma^2 < \sigma_0^2$

    Critical Region: $C = \{w: w \leq \chi_{1-\alpha,n-1}^2\}$

4. *Sample size.* Deviations of 0.005 are considered serious, so that $\beta(0.000025)$ must be kept to 0.10. With

$$\lambda = \frac{0.005}{0.003} = 1.67$$

it can be seen from Figure 8.30 that a sample size of approximately 30 will yield $\beta(0.000025) = 0.10$.

5. *Critical region.* According to Table A.5, for $n = 30$ and $\alpha = 0.01$, the critical value is

$$\chi_{\alpha,n-1}^2 = \chi_{0.01,29}^2 = 49.558$$

and thus the critical region is

$$C = [49.558, +\infty)$$

6. *Sample value.* Suppose 30 randomly sampled observations are recorded, the empirical variance being $s^2 = 0.000014$. Then the sample value of the test statistic is

$$w = \frac{29 \times s^2}{0.000009} = 45.11$$

7. *Decision.* Since $w < 49.558$, the null hypothesis is not rejected and the machine is kept in operation.

EXAMPLE 8.17

Suppose the variational problem of Example 8.16 is reversed. The machine is kept in operation only if there is strong empirical evidence indicating that the variance is less than 0.000009. Then the appropriate test is lower one-tailed.

1. *Hypothesis test.* H_0: $\sigma^2 \geq 0.000009$

    H_1: $\sigma^2 < 0.000009$

2. *Level of significance.* In the present setting, a type I error occurs if, based upon the test, the machine is kept in service even though its precision is somewhat substandard. Depending upon the application, such an incorrect decision may or may not be of major consequence. Thus,

without additional information to guide us, the standard level of significance $\alpha = 0.05$ will be employed.

3. *Test statistic.* $W = \dfrac{(n-1)S^2}{0.000009}$

4. *Sample size.* A type II error means taking the milling machine out of service, so that type II error should be minimized. Suppose it is specified that $\beta(0.000004)$ should be kept to 0.10. Then

$$\lambda = \frac{0.002}{0.003} = 0.67$$

and a look at Figure 8.32 shows that the sample size must be in the neighborhood of 30.

5. *Critical region.* With $n = 30$, Table A.5 gives the critical value

$$\chi^2_{1-\alpha, n-1} = \chi^2_{0.95, 29} = 17.708$$

and the critical region is

$$C = [0, 17.708]$$

6. *Sample value.* If 30 observations are made with the resulting empirical variance being $s^2 = 0.000006$, then the sample value of the test statistic is

$$w = \frac{29 \times s^2}{0.000009} = 19.333$$

7. *Decision.* Since $w > 17.708$, the null hypothesis is accepted and the machine is taken out of production.

8.8 BEST TESTS

Thus far we have constructed hypothesis tests by intuitively choosing a test statistic and critical region. While it is true that critical values have been determined in accordance with levels of significance, the forms of the acceptance regions have depended upon the heuristic notion that the test statistic should fall near the parameter value specified by the null hypothesis when the null hypothesis reflects the true state of nature. The question naturally arises as to whether one test is better than another. As usual, to answer this question it must be framed in the context of a criterion of goodness.

Consider the hypothesis test

$$H_0: \theta = \theta_0$$

$$H_1: \theta = \theta_1$$

consisting of two simple hypotheses. Let α and n be fixed. Suppose V and U are two test statistics with critical regions C_V and C_U, respectively, and acceptance regions A_V and A_U, respectively. Given the truth of the null hypothesis,

$$P(V \in C_V; \theta = \theta_0) = P(U \in C_U; \theta = \theta_0) = \alpha$$

Conversely, given the truth of the alternative hypothesis, there is a size of type II error for each test statistic, namely,

$$\beta_V = P(V \in A_V; \theta = \theta_1)$$

and

$$\beta_U = P(U \in A_U; \theta = \theta_1)$$

The test involving V is said to be **better** than the test involving U if $\beta_V < \beta_U$. Such a criterion is certainly reasonable, since it simply states that, for a given level of significance, the V test possesses less likelihood of a wrongful acceptance of H_0 given H_1 than does the U test. More generally, we have the following definition.

DEFINITION 8.5

Best Test. For a hypothesis test comprised of two simple hypotheses, a test is said to be a best test if, among all tests with level of significance α, it minimizes the size of type II error.

Full appreciation of Definition 8.5 depends on a deeper look at the manner in which the observations constituting a random sample lead to a decision. Given the test statistic V, critical region C, and acceptance region A, to say that V falls in C is to say that the empirical observations $x_1, x_2, \ldots, x_n$ corresponding to the random sample $X_1, X_2, \ldots, X_n$ fall in some region C_n in n-dimensional space. Another way to put the matter is to say that V falls in C if and only if the random vector

$$\mathbf{X} = \begin{pmatrix} X_1 \\ X_2 \\ \vdots \\ X_n \end{pmatrix}$$

falls in C_n. Relative to the data, it is C_n which constitutes the critical region. The test statistic is only a means, albeit a practically useful one, by which we view the critical region. Consequently, Definition 8.5 can be stated in terms of critical regions, where these critical regions are subsets of n-dimensional space. The appropriate terminology is then a **best critical region,** and there need be no mention of the supporting test statistic. In terms of the definition, a best critical region is one which minimizes the size of type II error over all critical regions of a given size α.

EXAMPLE 8.18

According to the methodology of Section 8.4.2, the critical region for an upper one-tailed test of the mean of a normal random variable with known standard deviation σ is

$$C = \{z \colon z \geq z_\alpha\}$$

the level of significance being α. For a given set of observations $x_1, x_2, \ldots, x_n$, the test statistic, which in this case is the standardized sample mean, falls in C if and only if

$$\bar{x} \geq \mu_0 + z_\alpha \cdot \frac{\sigma}{\sqrt{n}}$$

But this inequality defines a region in n-dimensional space, namely,

$$C_n = \{(x_1, x_2, \ldots, x_n): x_1 + x_2 + \cdots + x_n \geq n\mu_0 + z_\alpha \sigma \sqrt{n}\}$$

For a random sample of size $n = 2$, the inequality becomes

$$x_1 + x_2 \geq 2\mu_0 + \sqrt{2}\, z_\alpha \sigma$$

and the resulting critical region C_2 is the region in the Cartesian plane illustrated in Figure 8.34.

Henceforth, we will no longer use the subscript notation to differentiate between a critical region defined directly in terms of the n-dimensional sample points $(x_1, x_2, \ldots, x_n)$ and the logically equivalent region defined in terms of the sample value of the corresponding test statistic.

The following theorem, which is stated without proof, provides a means of constructing best tests when the competing hypotheses are both simple. It makes use of the likelihood function

$$L(x_1, x_2, \ldots, x_n; \theta) = \prod_{i=1}^{n} f(x_i; \theta)$$

corresponding to the random sample $X_1, X_2, \ldots, X_n$, where θ is the parameter of interest and $f(x; \theta)$ is the density of the population random variable X. Attention should be paid to the fact that the theorem is stated in terms of a critical region C in n-dimensional space.

THEOREM 8.1

(Neyman-Pearson Lemma). Suppose there exists a critical region C of size α corresponding to the hypothesis test

$$H_0: \theta = \theta_0$$

$$H_1: \theta = \theta_1$$

and a constant k such that

$$\frac{L(x_1, x_2, \ldots, x_n; \theta_1)}{L(x_1, x_2, \ldots, x_n; \theta_0)} \geq k$$

for all points $(x_1, x_2, \ldots, x_n)$ in C and

FIGURE 8.34
Best critical region corresponding to Example 8.18

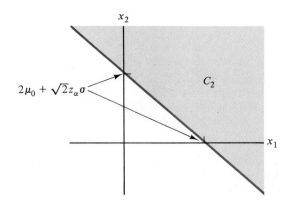

$$\frac{L(x_1, x_2, \ldots, x_n; \theta_1)}{L(x_1, x_2, \ldots, x_n; \theta_0)} \leq k$$

for all points $(x_1, x_2, \ldots, x_n)$ not in C. Then C is a best critical region of size α.

A little reflection shows the Neyman-Pearson lemma to be fully in line with the kind of intuitive reasoning heretofore employed to construct critical regions. Indeed, if we assume $f(x; \theta)$ to be a discrete density, then for any sample point $(x_1, x_2, \ldots, x_n)$, the ratio appearing in the statement of the theorem is the probability of obtaining the n observations $x_1, x_2, \ldots, x_n$, given $\theta = \theta_1$, divided by the probability of obtaining the n observations $x_1, x_2, \ldots, x_n$, given $\theta = \theta_0$. Observation vectors which make this ratio high indicate the truth of the alternative hypothesis; observation vectors which make this ratio small indicate the truth of the null hypothesis (at least intuitively).

EXAMPLE 8.19

Suppose X possesses an exponential distribution with unknown parameter b, but it is known that b is either 2 or 6. A decision as to which value reflects the true state of nature will be made according to the data resulting from a random sample of size n, the hypothesis test being

$$H_0: b = b_0 = 2$$

$$H_1: b = b_1 = 6$$

We will employ the Neyman-Pearson lemma to construct a best critical region of size α. The likelihood function is given by

$$L(b) = \prod_{i=1}^{n} be^{bx_i} = b^n e^{-b\sum_{i=1}^{n} x_i}$$

Thus, the ratio of the likelihood functions corresponding to H_1 and H_0 is

$$r = \frac{b_1^n e^{-b_1 n\bar{x}}}{b_0^n e^{-b_0 n\bar{x}}} = 3^n e^{-4n\bar{x}}$$

where $\bar{x}$ denotes the mean of $x_1, x_2, \ldots, x_n$. To apply the Neyman-Pearson lemma, it is necessary to find a region C and a constant k such that $r \geq k$ whenever $(x_1, x_2, \ldots, x_n)$ is an element of C and $r \leq k$ whenever $(x_1, x_2, \ldots, x_n)$ is not an element of C. Setting $r \geq k$ and solving for $\bar{x}$ yields the logically equivalent inequality

$$\bar{x} \leq \frac{\log (3^n/k)}{4n}$$

Let C be the set of all points $(x_1, x_2, \ldots, x_n)$ satisfying this inequality and select k such that

$$P\left(\bar{X} \leq \frac{\log (3^n/k)}{4n}; H_0\right) = \alpha$$

Then C will be a critical region of size α satisfying the conditions of the Neyman-Pearson lemma. Thus, the inequality involving $\bar{x}$ determines a best critical region of size α defined in terms of the sample mean $\bar{X}$.

Now, suppose $n = 5$ and $\alpha = 0.2$. Writing the probability inequality involving $\bar{X}$ and α in terms of the probability distribution function of $\bar{X}$ yields

$$F_{\bar{X}}\left(\frac{\log(3^n/k)}{4n}\right) = 0.2$$

As demonstrated in Example 7.3, $\bar{X}$ possesses a gamma distribution with parameters n and $1/nb$. In Section 4.6.2 it was shown that the probability distribution function of a gamma distribution is expressed in terms of the incomplete gamma function. In particular,

$$F_{\bar{X}}(t) = \gamma\left(\frac{t}{\beta}; n\right)$$

Given H_0, $\beta = 1/nb = 1/10$, and it can be seen from Table A.4 (by interpolation) that $\gamma(10t; 5) = 0.2$ when $10t = 3.08$. Therefore, in terms of the test statistic $\bar{X}$, the critical region is determined by the critical value 0.308. In other words, the null hypothesis is rejected at the 0.2 significance level if $\bar{x} \leq 0.308$.

To interpret the critical region of Example 8.19 directly in terms of the sample values of the observations, note that k is the solution to the equation

$$\frac{\log \dfrac{3^5}{k}}{20} = 0.308$$

Hence, $k = 0.513$, and the null hypothesis is rejected if the observation vector $(x_1, x_2, \ldots, x_5)$ falls in the region of 5-dimensional space determined by the inequality

$$\frac{L(x_1, x_2, x_3, x_4, x_5; 6)}{L(x_1, x_2, x_3, x_4, x_5; 2)} \geq 0.513$$

However, there is no logical difference between specifying the critical region in this manner and specifying it in terms of the test statistic discovered in the example. Indeed, as the computations performed in the example demonstrate, the inequality $\bar{x} \leq 0.308$ determines the same critical region in 5-dimensional space. The difference, of course, is that the expression in terms of the test statistic $\bar{X}$ appears more natural than the expression in terms of the ratio of the likelihood functions.

The latter point can be emphasized by examining the form of the critical region as expressed in terms of the sample mean. The null hypothesis is rejected if the sample mean falls below 0.308. Since the mean of the exponential distribution is $1/b$, H_0 is equivalent to $\mu = 0.5$ and H_1 is equivalent to $\mu = 0.17$. Had the problem been posed in terms of hypotheses concerning the mean, our first impression would have been to employ a lower one-tailed test with the sample mean as the test statistic. As just shown, this is exactly the course resulting from the application of Theorem 8.1.

A fundamental point to recognize is that it was not necessary to solve for k in Example 8.19. Once the form of the best critical region has been determined relative to a test statistic, the role of k is to make the critical region a predetermined size α. However, this task can be accomplished, as it was in Example 8.19, by an analysis of the critical region as given in terms of the test statistic.

EXAMPLE 8.20   Consider the construction of a best test with respect to the hypotheses

$$H_0: \mu = \mu_0$$

$$H_1: \mu = \mu_1 > \mu_0$$

where the population random variable X is normally distributed with unknown mean μ and known variance σ^2. The problem is to find a best critical region of size α by employing Theorem 8.1. Referring to Example 7.14 for the likelihood function, we find the ratio of the likelihood functions corresponding to μ_1 and μ_0 to be

$$r = \frac{\exp\left[-\dfrac{1}{2\sigma^2} \sum_{i=1}^{n} (x_i - \mu_1)^2\right]}{\exp\left[-\dfrac{1}{2\sigma^2} \sum_{i=1}^{n} (x_i - \mu_0)^2\right]}$$

$$= \exp\left\{\frac{1}{2\sigma^2} [n(\mu_0^2 - \mu_1^2) + 2(\mu_1 - \mu_0)n\bar{x}]\right\}$$

Setting $r \geq k$ and taking logarithms yields

$$2(\mu_1 - \mu_0)n\bar{x} \geq 2\sigma^2 \log k + n(\mu_1^2 - \mu_0^2)$$

Since $\mu_1 - \mu_0 > 0$,

$$\bar{x} \geq \left(\frac{\sigma^2}{\mu_1 - \mu_0}\right) \frac{\log k}{n} + \frac{\mu_1 + \mu_0}{2}$$

Let k^* denote the right-hand side of the inequality. Since $r \geq k$ if and only if $\bar{x} \geq k^*$, an appropriate critical region is the set of all observation vectors $(x_1, x_2, \ldots, x_n)$ for which $\bar{x} \geq k^*$. In accordance with the remarks made prior to the present example, we need not look for k, but rather utilize the equality

$$P(\bar{X} \geq k^*; H_0) = \alpha$$

to determine a best critical region in terms of the test statistic $\bar{X}$. This best test is upper one-tailed, the critical value k^* depending on α. Having found k^*, one could solve for k; however, there is no purpose in doing so once the form of the critical region has been determined relative to the test statistic $\bar{X}$.

Although the Neyman-Pearson lemma is stated in terms of simple hypotheses, it often leads to results concerning a simple hypothesis and a composite alternative hypothesis. For instance, in Example 8.20 the critical region does not depend upon the actual value of μ_1, requiring only that μ_1 is greater than μ_0. Consider the upper one-tailed hypothesis test

$$H_0: \mu = \mu_0$$

$$H_1: \mu > \mu_0$$

According to Example 8.20, for any alternative value μ_1 the size of type II error, $\beta(\mu_1)$, is minimized if $\bar{X}$ is employed as the test statistic. Thus, if V is any other test statistic, the size of type II error for V is greater than or equal to the size of type II error for $\bar{X}$ at all possible alternative means. Equivalently, over the domain $\mu > \mu_0$, the OC curve of $\bar{X}$ lies below the OC curve of V.

DEFINITION 8.6

Uniformly Most Powerful Test. Consider testing the simple null hypothesis H_0 against the composite alternative hypothesis H_1. For a fixed size α, a uniformly most powerful test is one whose OC curve lies below the OC curves of all other possible tests.

According to Definition 8.6 and the remarks prior to it, the upper one-tailed test for the mean of a normal distribution that was derived in Example 8.20 is uniformly most powerful. Although stated in a different form, the Z test given in Table 8.2 is equivalent to the sample-mean test of Example 8.20.

The terminology "powerful" in Definition 8.6 derives from the **power function** of a test concerning the parameter θ, which is defined as

$$P(\theta) = 1 - \beta(\theta)$$

Minimizing the size of type II error is equivalent to maximizing the power function. Thus, if $\beta(\theta)$ is minimized uniformly, $P(\theta)$ is maximized uniformly.

8.9 DISTRIBUTION-FREE TESTS CONCERNING CENTRALITY

8.9.1 The Nonparametric Approach

Hypothesis tests involving the Z and t distributions have required the underlying population variables to be normally distributed, or, in the case of Z tests, have depended on the central limit theorem for approximations in large-sample settings. Although various tests show varying degrees of robustness, in that slight departures from normality can be tolerated, these departures have the effect of increasing the size of type II error. Specifically, while Z and t tests might be best tests relative to type II error when the populations are strictly normal, the size of type II error can make the tests unreliable if normality is violated to an appreciable degree. This is especially the case when small samples are involved.

The present section concerns tests that do not assume specific distributions for the underlying population variables. Thus, they are often called **distribution-free tests.** Because the inferences drawn tend to be about the location of a distribution rather than about the value of the mean or variance, the tests are commonly called **nonparametric tests;** however, "nonparametric" may be a misnomer because distribution-free tests can involve inferences concerning parameters. Besides the lack of restriction to normality, or to other strict distributional assumptions, another common advantage of nonparametric tests is their ease of computation. Often they are fast and efficient to implement.

There are drawbacks, however, to nonparametric procedures. In general, if a parametric test, such as a t test, is employed when the distributional assumptions applying to the parametric test are satisfied by the population variable, then the parametric test will yield a lower type II error. Intuitively, type II error will be greater for the corresponding distribution-free test because, in not making a specific assumption regarding the underlying population distribution, a nonparametric test does not utilize all the information in the distribution. Of course, should there be excessive deviation from the underlying distributional assumptions, then it is very possible that the nonparametric test will be superior. Thus, in the absence of knowledge concerning the population distribution, a nonparametric test, if one is available, is sometimes more suitable. A second problem with nonparametric tests is that, very often, the computation of β is problematic. The decision as to whether to use a nonparametric test when a corresponding parametric test is available often depends on one's belief in the degree to which the population variable conforms to normality: if it is believed to be normal, then use the parametric test; if it is believed to substantially deviate from normality, then use the distribution-free alternative.

8.9.2 Sign Test

The sign test is a distribution-free alternative to the t test for the mean. It requires no assumption of normality and applies to all continuous distributions. In testing the null hypothesis H_0: $\mu = \mu_0$ for a random variable X, we are essentially testing to see whether the density of X is centered at μ_0. The upper one-tailed and lower one-tailed alternatives state that the density is centered to the right and to the left of μ_0, respectively. The two-tailed alternative simply states that the density is not centered at μ_0. Rather than utilize the mean as a measure of central tendency, we can instead employ the **median.** Assuming X to possess a continuous distribution, its median is a point $\tilde{\mu}$ such that

$$P(X \le \tilde{\mu}) = P(X \ge \tilde{\mu}) = \frac{1}{2}$$

The median splits the distribution in two, half of the probability mass lying on either side. The sign test involves testing the null hypothesis

$$H_0: \tilde{\mu} = \tilde{\mu}_0$$

against any one of three alternatives.

Let $X_1, X_2, \ldots, X_n$ be a random sample arising from the continuous random variable X. Under H_0, one might expect the sample values to distribute themselves somewhat equally on either side of $\tilde{\mu}_0$. On the other hand, if the true median is greater than $\tilde{\mu}_0$, it might be expected that most of the sample values will fall to the right of $\tilde{\mu}_0$. The sign test makes use of these straightforward observations by employing the test statistic V that counts the number of sample values for which $X_i - \tilde{\mu}_0 > 0$. V possesses a binomial distribution and, under H_0, the probability of a success is given by

$$p = P(X > \tilde{\mu}_0) = \frac{1}{2}$$

the assumption being made that no sample variable takes on the value $\tilde{\mu}_0$. (Although the probability of a continuous random variable taking on a specific value is zero, in practice, measurement restrictions might result in a value equal to $\tilde{\mu}_0$; if so, the value is deleted from the sample and the sample size is correspondingly reduced.) To test H_0 against any of the three alternatives

$$H_1: \tilde{\mu} > \tilde{\mu}_0$$

$$H_1: \tilde{\mu} < \tilde{\mu}_0$$

or

$$H_1: \quad \tilde{\mu} \neq \tilde{\mu}_0$$

V is employed as a test statistic for an upper one-tailed, lower one-tailed, or two-tailed test, respectively, of the binomial null hypothesis

$$H_0: \quad p = \frac{1}{2}$$

The test is implemented by the methods discussed in Section 8.6. Note that the large-sample approach of Section 8.6.2 applies for $n \geq 10$, since, for such n, $np_0 = nq_0 = n/2 \geq 5$.

If X possesses a symmetric density, as in the case of a normal random variable, then $\mu = \tilde{\mu}$ and the sign test concerns the mean. An exception occurs in the event that X is symmetric but does not possess a mean, an illustration of this anomaly being the Cauchy distribution (see Example 3.22). The sign test is applicable to the Cauchy distribution with the null hypothesis involving the median.

EXAMPLE 8.21

A chemist wishes to demonstrate that the melting point of a new compound exceeds 184.35 degrees centigrade. Although he or she would customarily assume melting-point measurements to be normally distributed, there is some doubt because of the manner in which the measurements have been taken. Thus, it is decided that a sign test might be appropriate, at least for a preliminary determination. Twelve measurements are taken, the data being recorded in Table 8.6. Included in the table is a column giving the values of $X - \tilde{\mu}_0$. V is simply the number of positive values in that column. We employ the hypothesis-test format.

1. *Hypothesis test.* H_0: $\tilde{\mu} \leq 184.35$
   $\qquad\qquad\quad H_1$: $\tilde{\mu} > 184.35$
2. *Level of significance.* Approximately 0.05
3. *Test statistic.* Sign-test statistic V
4. *Sample size.* $n = 12$
5. *Critical region.* The size of the critical region is given by $1 - B(c - 1; 12, 0.5)$, where c is the critical value. Using Table A.2, we see that

$$1 - B(8; 12, 0.5) = 0.073$$

TABLE 8.6 MELTING-POINT DATA

x_i	$x_i - \tilde{\mu}_0$
184.32	-0.03
184.41	$+0.06$
184.69	$+0.34$
184.92	$+0.57$
184.20	-0.15
185.06	$+0.71$
184.24	-0.11
184.85	$+0.50$
184.30	-0.05
184.69	$+0.34$
184.82	$+0.47$
185.01	$+0.66$

gives the closest approximation to the desired level 0.05. Thus the critical value is $c = 9$ and the actual significance level is $\alpha = 0.073$, the null hypothesis being rejected at that level if $V \geq 9$.

6. *Sample value.* $v = 8$
7. *Decision.* Accept H_0 because $v < 9$.

Although we chose not to do so, the large-sample approach could have been utilized in Example 8.21 since the sample size exceeded 10.

Finding the size of type II error when employing the sign test is problematic, because one must posit not only a specific competing median but also the form of the competing density. The latter point arises because the test is distribution-free and, except for the condition $\tilde{\mu} = \tilde{\mu}_0$, α is determined independently of distributional knowledge.

Since the normal distribution is symmetric, given normality, the sign test applies to the null hypothesis H_0: $\mu = \mu_0$. Should X possess a normal distribution, the type II error corresponding to the t test of the mean is less than that of the sign test, since, under the condition of normality, the t test is a best test. However, should X be markedly nonnormal, then the sign test can be employed in place of the t test. Keep in mind that if X merely varies slightly from normality, then the t test remains superior relative to type II error. Should it happen that X is nonnormal but possesses a symmetric distribution, then the signed-rank test of Section 8.9.3 is a better alternative to the t test than the sign test.

8.9.3 Signed-Rank Test

The domain of application of the sign test includes all continuous distributions. In the case of continuous distributions possessing symmetric densities, a better test for the null hypothesis H_0: $\tilde{\mu} = \tilde{\mu}_0$ is the **signed-rank** (or **Wilcoxon signed-rank**) test. This test utilizes the statistic S, which we now proceed to define. Suppose X_1, X_2, . . . , X_n comprise a random sample corresponding to a symmetric continuous population distribution with median $\tilde{\mu}$. S is found by ranking the values of $|X_i - \tilde{\mu}_0|$ in increasing order and then taking the sum of the ranks corresponding to positive values of $X_i - \tilde{\mu}_0$.

If the null hypothesis is true, one would expect there to be a somewhat uniform mixing of both positive and negative values of $x_i - \tilde{\mu}_0$ in the ranking. Since the sum of the first n integers is $n(n + 1)/2$, under the null hypothesis we might expect S to be around $n(n + 1)/4$. Correspondingly, for the upper and lower one-tailed alternatives, S would likely be nearer to $n(n + 1)/2$ and 0, respectively. For the two-tailed alternative, it would be expected that S would likely be nearer to $n(n + 1)/2$ or 0, rather than be near to $n(n + 1)/4$. These heuristic comments associated with the central tendency of S are formalized in the next theorem.

THEOREM 8.2

Under H_0: $\tilde{\mu} = \tilde{\mu}_0$,

$$E[S] = \frac{n(n + 1)}{4}$$

$$\text{Var} [S] = \frac{n(n + 1)(2n + 1)}{24}$$

Proof: Define the random variable U_k to be k if the kth rank corresponds to a positive difference and to be 0 otherwise. Under H_0, the median of X is $\tilde{\mu}_0$ and therefore

$$P(U_k = k) = \frac{1}{2}$$

for $k = 1, 2, \ldots, n$. Looking at the matter from a different perspective, for each k, U_k is equal to k times a binomial random variable with $n = 1$ and $p = 1/2$. Call this binomial variable Y_k. Since

$$S = U_1 + U_2 + \cdots + U_n$$

the linearity of the expected-value operator yields

$$E[S] = \sum_{k=1}^{n} E[U_k] = \sum_{k=1}^{n} kE[Y_k] = \sum_{k=1}^{n} \frac{k}{2} = \frac{n(n+1)}{4}$$

Since Var $[U_k] = k^2/4$ and $U_1, U_2, \ldots, U_n$ comprise an independent collection,

$$\text{Var } [S] = \sum_{k=1}^{n} \text{Var } [U_k] = \sum_{k=1}^{n} \frac{k^2}{4} = \frac{n(n+1)(2n+1)}{24} \qquad \blacksquare$$

Because S is discrete, for a desired level of significance, we must be satisfied with finding a critical value that yields approximately the desired level. For the upper one-tailed alternative H_1: $\tilde{\mu} > \tilde{\mu}_0$ at significance level α (and for sample size n), the critical value $s_{\alpha,n}$ is defined by

$$P(S \geq s_{\alpha,n}; H_0) = \alpha$$

the appropriate value of $s_{\alpha,n}$ being determined from Table A.10. H_0 is rejected if $S \geq s_{\alpha,n}$. Owing to the symmetry of the ranking scheme, H_0 is rejected in favor of the lower one-tailed alternative H_1: $\tilde{\mu} < \tilde{\mu}_0$ if

$$S \leq \frac{n(n+1)}{2} - s_{\alpha,n}$$

For the two-tailed alternative H_1: $\tilde{\mu} \neq \tilde{\mu}_0$, H_0 is rejected if $S \geq s_{\alpha/2,n}$ or

$$S \leq \frac{n(n+1)}{2} - s_{\alpha/2,n}$$

Under our assumptions, the population distribution is continuous and therefore the probability of any sample variable being $\tilde{\mu}_0$ is 0, as is the probability of tie rankings. However, due to measurement restrictions, both phenomena can occur in practice. Thus, two provisos must be added to the test procedure. First, if any sample value equals $\tilde{\mu}_0$, it must be discarded from the sample and n reduced accordingly. Second, in case of tie ranks, the assigned rank is obtained by averaging. Specifically, if m sample values are tied for the kth rank, then each is assigned the rank

$$\frac{k + (k+1) + \cdots + (k+m-1)}{m}$$

For instance, if two values tie for rank 6, each is assigned rank $(6 + 7)/2 = 6.5$.

EXAMPLE 8.22
In the melting-point test of Example 8.21, the data of Table 8.6 did not lead to a significant value of the sign-test statistic at the 0.05 level. Nevertheless, if we examine the data in the table, it certainly appears that it is strongly distributed to the right of $\bar{\mu}_0 = 184.35$. The weakness of the sign test is apparent: the differential between plus and minus signs is not sufficient to lead to rejection. The magnitudes of the positive differences greatly exceed the magnitudes of the negative differences, but the sign test does not take these magnitudes into account. The signed-rank test helps rectify this problem. Of course, it requires the assumption of population symmetry. We apply the signed-rank test to the melting-point data.

1. *Hypothesis test.* H_0: $\bar{\mu} \leq 184.35$
   $\qquad\qquad\qquad$ H_1: $\bar{\mu} > 184.35$
2. *Level of significance.* Approximately 0.05
3. *Test statistic.* Signed-rank statistic S
4. *Sample size.* $n = 12$
5. *Critical region.* From Table A.10, it can be seen that the closest level to 0.05 is 0.055, and therefore the critical region is defined by $S \geq s_{0.055,12} = 60$.
6. *Sample value.* Referring to Table 8.6 and ranking the absolute deviations from $\bar{\mu}_0$ in increasing order produces the following ranking scheme:

Rank:	1	2	3	4	5	6.5
Deviation:	-0.03	-0.05	$+0.06$	-0.11	-0.15	$+0.34$

Rank:	6.5	8	9	10	11	12
Deviation:	$+0.34$	$+0.47$	$+0.50$	$+0.57$	$+0.66$	$+0.71$

Summing the ranks corresponding to positive deviations yields the sample value $s = 66$.

7. *Decision.* Reject H_0 since $s \geq 60$.

When the sample size n is at least 20, under the null hypothesis the distribution of S is approximately normal. According to Theorem 8.2, the approximation yields the test statistic

$$Z = \frac{S - \dfrac{n(n+1)}{4}}{\sqrt{\dfrac{n(n+1)(2n+1)}{24}}}$$

The customary Z-test critical regions apply. In the event there are tie values, the variance of S needs to be adjusted; however, the adjustment is usually slight and will not be discussed.

In deciding whether to apply a t test or the signed-rank test, we must consider the degree to which the underlying population differs from normality. If it is either normal or close to normal, then the t test prevails with lower type II error; if it is markedly nonnormal, then the signed-rank test prevails. As is often the case, the decision in less extreme situations is unclear. Like the sign test, the computation of β for the signed-rank test depends upon the distribution and is therefore problematic. Nonetheless, it can be said that for large samples the performance of the signed-rank test is not that much worse than the t test even when the population is nearly normal.

8.10 QUALITY CONTROL CHARTS

The maintenance of quality is a fundamental part of any manufacturing process. Every process exhibits natural variability, even if the process is operating within acceptable constraints. Variation occurs for a number of reasons, including differences in raw materials, inconsistent worker performance, and machine wear. It can be revealed by testing various characteristics of the product, either on-line or off-line. Each test measures some aspect of the product, or partially completed product, with respect to a given standard of quality. If a marked loss of quality is detected, then the cause of the decline must be determined and corrective action taken.

A common technique for making decisions with regard to product quality involves the use of quality control charts. Since the use of these charts entails the establishment of regions of acceptability and the making of inferences based upon the observation of statistics, the quality-control-chart concept is related to the hypothesis-testing methodology discussed earlier in the chapter. However, the procedure is dynamic in that repeated samples are taken and the decision to investigate the production process is made whenever a sample value is unacceptably large or unacceptably small.

8.10.1 $\bar{X}$ Charts

Over time, observed process fluctuation relative to some quality measurement can be viewed in two lights. First, it may be that the variability results from unidentified causes such as those mentioned previously. Such variability is the product of any number of random fluctuations and often results in measurements that are normally distributed about some central value. For instance, if the quality control technician is measuring the diameter of a shaft, then the measurements often possess a normal distribution about an expected shaft diameter. So long as the measurements stay within some limits, the technician might be satisfied that the process is functioning acceptably well. On the other hand, the process might not be producing an acceptable product, and such a situation could very well be detected by the measurements' falling outside some pair of limits. In the first case, the process is said to be in **statistical control;** in the latter case, the process is out of control and it is assumed that there exist **assignable causes** producing the excess fluctuation.

The decision as to whether a process is in or out of control is often based on a **control chart.** Such a chart consists of a central value about which the test measurements fluctuate and two limits, the **upper control limit (UCL)** and the **lower control limit (LCL).** The two limits define a region in which quality measurements will fall if the process is in control. Samples are taken over time, and the sample means are plotted on the control chart. A typical control chart is illustrated in Figure 8.35, where the means (dots) of the samples are connected by a polygonal arc to illustrate the flow of the measurements. A control chart whose limits determine a range of acceptability for repeated sample means is called an $\bar{X}$ **chart.**

If it is assumed that when the process is in control the quality measurement X is normally distributed with known mean μ and known standard deviation σ, then a straightforward way of setting up a control chart is to let the center line be determined by μ and the upper and lower control limits be

$$UCL = \mu + \frac{3\sigma}{\sqrt{n}}$$

$$LCL = \mu - \frac{3\sigma}{\sqrt{n}}$$

n being the common size of the samples. If each sample is randomly selected, then the sample mean of each satisfies the relation

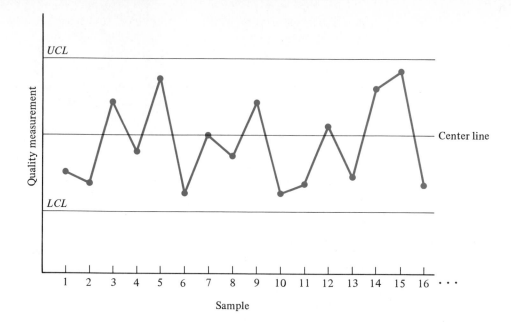

FIGURE 8.35
General form of a
quality control chart

$$P\left(\mu - \frac{3\sigma}{\sqrt{n}} \leq \bar{X} \leq \mu + \frac{3\sigma}{\sqrt{n}}\right) = 0.9974$$

(see Table A.1). Thus, the likelihood of any sample mean falling outside the interval [*LCL*, *UCL*] is negligible. If a sample mean should fall outside the control limits, then it is highly likely that the process has fallen out of control due to some assignable cause. When the process is believed to be out of statistical control, a decision must be rendered on how to best locate the cause and take the necessary corrective action.

The purpose of a control chart is to detect fluctuation due to assignable causes. As a result, it is important to select samples in a manner that enhances the likelihood of detection. Consequently, observations are usually grouped into samples in a de-signed manner. Samples are formed so that within-sample variation is likely due to random fluctuations and so that between-sample variation is likely due to assignable causes. Such groupings of observations are called **rational subgroups.** For instance, if a product is being produced in three assembly lines, then observational groupings can be by line, the result being groupings by source. Another possible method by which to form rational subgroups is to employ order of production. In any case, if the process is in control, then each rational subgroup represents a random sample from the underlying population, and it should be expected that each sample mean falls within the control limits. On the other hand, should the process be out of control, then the sample mean of some rational subgroup is somewhat more likely to fall outside the acceptable range because it represents a different population (with a different mean). Due to the rational-subgroup methodology, the cause should be more easily identified than had each sample been randomly drawn from among all units produced by the process.

EXAMPLE 8.23

Paint cans are filled automatically on four different production lines, each can possessing a nominal rating of 1 gallon. The process has been in effect for quite some time and is presumed to possess mean $\mu = 1.010$ gallons and standard deviation $\sigma = 0.015$ gallon. The data of Table 8.7 give the empirical means computed from 24 samples, each containing five tested cans. Rational subgroups have been constituted according to the production line: the samples have been taken in order from lines A, B, C, and D, the cycle of selection beginning again once a sample from D has been selected. The upper and lower control limits are

$$UCL = 1.010 + \frac{3 \times 0.015}{\sqrt{5}} = 1.030$$

$$LCL = 1.010 - \frac{3 \times 0.015}{\sqrt{5}} = 0.990$$

It can be seen from the control chart of Figure 8.36 that the process is out of control. Owing to the use of rational subgroups, plant engineers can look to line C for the problem; indeed, should it be necessary to halt line C, lines A, B, and D can continue to operate.

Note that in Table 8.7 we have employed the upper-case notation "$\bar{X}$" rather than "$\bar{x}$" to denote an empirical mean. Since it is commonplace in quality control to let upper-case letters denote the value of a random variable as well as the random variable itself, we will follow that custom in the present section.

As a practical matter, the mean and standard deviation of a quality measurement are likely to be unknown unless a process has been in effect for quite some time and it has appeared stable over the duration. Consequently, it is often necessary to estimate μ and σ in order to set up the control chart. Typically, each sample will consist of a small (common) number of observations, at least 4 and usually 5 or 6. To estimate the mean, it is common to employ at least 20 (typically 25 or more) preliminary samples. To proceed, suppose there are k preliminary samples, n is the common sample size, and $\bar{X}_i$ denotes the sample mean of the ith sample (rational subgroup). Then, under the assumption that the process is in control, the best unbiased estimator of the mean is

$$\hat{\mu} = \bar{\bar{X}} = \frac{1}{k} \sum_{i=1}^{k} \bar{X}_i$$

the average of the nk observations. Thus, the center line on the $\bar{X}$ control chart is determined by $\bar{\bar{X}}$.

Rather than employ the sample standard deviations to estimate σ, it is more common to employ the sample ranges, each sample range being the largest observation

TABLE 8.7 MEANS OF PAINT-CAN SAMPLES

Sample	$\bar{X}$	Sample	$\bar{X}$
1	1.024	13	1.020
2	1.002	14	1.010
3	1.012	15	1.001
4	0.995	16	1.028
5	1.028	17	1.005
6	0.991	18	1.030
7	0.996	19	0.992
8	1.018	20	1.020
9	1.000	21	1.015
10	1.004	22	0.996
11	1.000	23	0.980
12	1.025	24	1.022

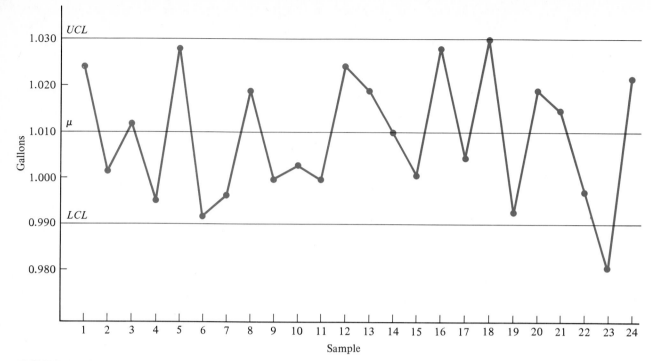

FIGURE 8.36
$\bar{X}$ chart for paint-can samples

minus the smallest observation in the sample. When the sample sizes are small, as is typically the case, the sample range is only a slightly less efficient estimator of the standard deviation of a normal population than is the sample standard deviation. Moreover, it is computationally less intensive. If we let

$$\bar{R} = \frac{1}{k} \sum_{i=1}^{k} R_i$$

denote the average of the sample ranges, then

$$\hat{\sigma} = \frac{\bar{R}}{d_2}$$

is an unbiased estimator of σ, the factor d_2 being given in Table A.14. Using the range approach, the upper and lower control limits of an $\bar{X}$ control chart are

Control Limits for $\bar{X}$ Chart

$$UCL = \bar{\bar{X}} + \frac{3\bar{R}}{d_2\sqrt{n}} = \bar{\bar{X}} + A_2\bar{R}$$

$$LCL = \bar{\bar{X}} - \frac{3\bar{R}}{d_2\sqrt{n}} = \bar{\bar{X}} - A_2\bar{R}$$

respectively, where, for convenience, the values of A_2 are tabulated in Table A.14. When setting up an $\bar{X}$ control chart based upon the computed control limits, it is common practice to treat the computed limits as trial limits. Specifically, the control chart is set up based upon the estimates $\bar{\bar{X}} \pm A_2\bar{R}$, and the sample means are plotted on the chart. If all points fall within the lines determined by the estimates, then the control chart is utilized as constructed for future observations. However, if a point falls outside the range determined by the trial control limits, then, as a practical matter, one should examine the rational subgroup leading to the extreme mean to determine if there is an assignable cause for the deviation. If a cause is determined, then the problem should be rectified and new control limits determined by discarding the extreme sample from the data. Should more than a single subgroup lead to excessive deviation, each should be investigated, the cause of variation determined, and new control limits estimated based upon the remaining samples. In any event, once the new control limits have been established, the sample means should once again be plotted to see if all fall within the new control limits. If they do, then the newly calculated control limits are used to set up the control chart; if they do not, then further investigation of the remaining preliminary samples is required and further recalculation of the control limits is necessary.

If no assignable cause can be found for an extreme deviation, then an expedient way to proceed is to simply discard the data from that sample anyway and to recompute the control-limit estimates based upon the remaining data. Obviously, such an approach is less satisfactory than one which results from a discovery of an assignable cause; nevertheless, in discarding the extreme sample we are simply recognizing the possibility that the extreme mean resulted from an assignable cause and are taking the precaution of not utilizing the corresponding observations in the determination of control limits that are assumed to result from a process that is under control.

EXAMPLE 8.24

A *coordinate measuring machine* (*CMM*) consists of a contact probe and a mechanism for positioning the probe in space in such a way that the probe can examine the features of a part. Using a CMM, it is possible to determine the angle between two edges in a part. Consider a new part, two of whose edges are designed to meet at a right angle. To set up an $\bar{X}$ chart, 22 preliminary samples of six parts are taken, the resulting mean angle measurements (in degrees) and ranges being given in Table 8.8. Based upon those data,

$$\sum_{i=1}^{22} \bar{X}_i = 1980.94$$

$$\sum_{i=1}^{22} R_i = 10.68$$

$$\bar{\bar{X}} = \frac{1980.94}{22} = 90.04$$

$$\bar{R} = \frac{10.68}{22} = 0.49$$

$$UCL = \bar{\bar{X}} + A_2\bar{R} = 90.04 + (0.483)(0.49) = 90.28$$

$$LCL = \bar{\bar{X}} - A_2\bar{R} = 90.04 - (0.483)(0.49) = 89.80$$

The resulting control chart is illustrated in Figure 8.37. The points corresponding to samples 5 and 13 lie outside the control limits. Suppose the two samples are investigated and assignable causes are discovered. Dropping the two samples from the preliminary samples yields the new estimates

$$\bar{\bar{X}} = \frac{1800.20}{20} = 90.01$$

$$\bar{R} = \frac{9.57}{20} = 0.48$$

The corresponding control limits are

$$UCL = 90.01 + (0.483)(0.48) = 90.24$$

$$LCL = 90.01 - (0.483)(0.48) = 89.78$$

All remaining points fall within the range determined by *LCL* and *UCL*. Hence, it is assumed that these are the appropriate control limits when the process is under control.

Example 8.24 stated that assignable causes were located which related to the extreme means in two of the samples. Based upon the convention cited prior to the example, even if assignable causes are not located, the samples should be discarded and the control limits recalculated.

8.10.2 R Charts

When a process is working acceptably well, not only should the sample means of quality measurements not fall too far from the center, but neither should the measurement variances fall too far from the expected variance. Consequently, quality can be monitored by forming a control chart relating to variability. Although one can monitor the standard deviations of samples, it is more common to monitor the sample ranges. The resulting chart is known as an **R chart.**

In general, the expected value of R is $E[R] = d_2\sigma$ and the standard deviation of R is $\sigma_R = d_3\sigma$, the values of d_2 and d_3 being given in Table A.14. If σ is known, then the center line of the R chart is determined by $d_2\sigma$ and the upper and lower control limits are

$$UCL_R = d_2\sigma + 3d_3\sigma = (d_2 + 3d_3)\sigma = D_2\sigma$$

$$LCL_R = d_2\sigma - 3d_3\sigma = (d_2 - 3d_3)\sigma = D_1\sigma$$

where $D_1 = d_2 - 3d_3$ and $D_2 = d_2 + 3d_3$ can be obtained from Table A.14.

TABLE 8.8 MEANS AND RANGES OF ANGLES BETWEEN TWO EDGES

Sample	$\bar{X}$	R	Sample	$\bar{X}$	R
1	90.18	0.66	12	90.03	0.32
2	89.85	0.42	13	90.42	0.67
3	89.88	0.50	14	89.88	0.24
4	90.00	0.53	15	89.82	0.44
5	90.32	0.44	16	90.16	0.38
6	90.10	0.40	17	90.11	0.52
7	90.20	0.37	18	89.84	1.00
8	90.00	0.40	19	90.10	0.44
9	89.85	0.48	20	89.82	0.41
10	90.22	0.63	21	90.18	0.38
11	90.02	0.50	22	89.96	0.55

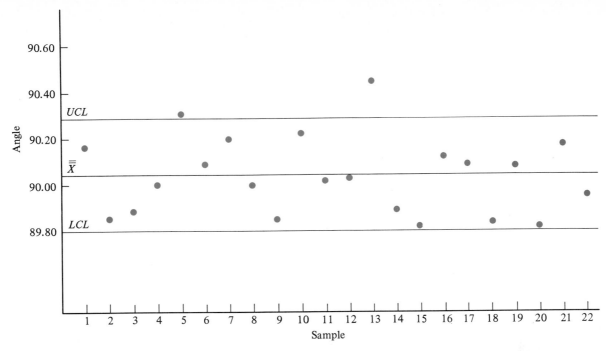

FIGURE 8.37
Trial $\bar{X}$ chart for angle samples

Unless a process has been running in a stable fashion for a long while, it cannot be presumed that σ is known. If σ is unknown, then it is necessary to employ $\bar{R}$ as an estimate of the mean of R, so that the center line of the R control chart is determined by $\bar{R}$. Since $\bar{R}/d_2$ is an unbiased estimator of σ, the upper and lower control limits take the forms

Control Limits for R Chart

$$UCL_R = \bar{R} + \frac{3d_3\bar{R}}{d_2} = D_4\bar{R}$$

$$LCL_R = \bar{R} - \frac{3d_3\bar{R}}{d_2} = D_3\bar{R}$$

where $D_3 = 1 - 3d_3/d_2$ and $D_4 = 1 + 3d_3/d_2$ can be obtained from Table A.14.

When setting up an R chart, a similar approach to the one taken when setting up an $\bar{X}$ chart should be employed. Once again, preliminary samples whose sample ranges fall outside the control limits should be examined in an effort to determine assignable causes. Since it is likely that an investigator would be setting up both the $\bar{X}$ and R control charts from the same preliminary samples, he or she should find both pairs of trial control limits, plot both the sample means and sample ranges on the respective charts, discard samples that appear extreme on either chart, and then reformulate the estimates based upon the remaining samples. If all remaining points fall between the new limits, then it can be assumed (perhaps unjustly) that, given the process is in control, the new estimates yield valid control limits. Should some remaining sample values (means or ranges) fall outside the new limits, these, too, must be discarded and new limits computed. The procedure can be repeated until all sample values fall within the prescribed control limits.

EXAMPLE 8.25

Suppose we wish to employ the data of Table 8.8 to set up both an $\bar{X}$ chart and an R chart of the coordinate measuring machine of Example 8.24. From the data of Table 8.8, the trial control limits for the R chart are

$$UCL_R = D_4\bar{R} = (2.004)(0.49) = 0.98$$

$$LCL_R = D_3\bar{R} = (0)(0.49) = 0$$

The resulting control chart is illustrated in Figure 8.38. Looking at the $\bar{X}$ chart of Figure 8.37 and the R chart of Figure 8.38, we see there are three extreme samples, these being samples 5, 13, and 18, the latter possessing too great a sample range. Upon discarding these three samples, there are 19 preliminary samples remaining, and we obtain $\bar{X} = 90.02$, $\bar{R} = 0.45$,

$$UCL_{\bar{x}} = 90.02 + (0.483)(0.45) = 90.24$$

$$LCL_{\bar{x}} = 90.02 - (0.483)(0.45) = 89.80$$

$$UCL_R = (2.004)(0.45) = 0.90$$

and $LCL_R = 0$. All sample values of $\bar{X}$ fall between $LCL_{\bar{x}}$ and $UCL_{\bar{x}}$, and all sample values of R fall between LCL_R and UCL_R. Thus, it is assumed that both pairs of control limits are appropriate when the process is in control.

8.10.3 p Charts

Thus far, product quality has been examined by taking actual measurements relating to specific product characteristics. Another way to proceed is to characterize a product by attributes. For instance, a unit can be rated as defective or nondefective relative

FIGURE 8.38

Trial R chart for angle samples

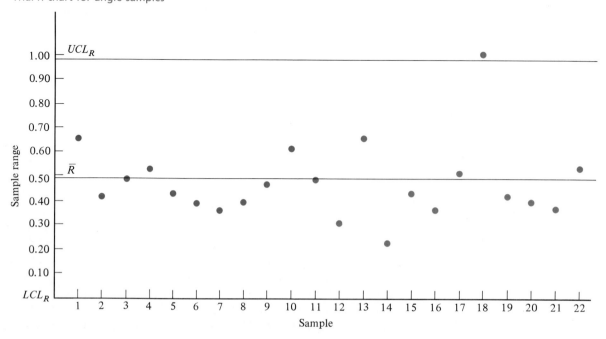

to some design standard. Actual measurements utilize continuous measuring devices such as micrometers, scales, and calipers. While it is true that the resulting data are more informative than a simple defective–nondefective classification, the latter may involve less cost. As a practical illustration, the diameter of a part may be checked by a simple *go/no-go* gage: if the diameter is less than a prespecified value, then it will pass through the gage opening; otherwise, it will not. In the present subsection, we will examine control charts for the proportion of defective units. These are called **p charts.** Sample sizes for p charts need to be quite a bit larger than for $\bar{X}$ charts and R charts.

If it can be presumed that the proportion p of defective units is known when the process is under control, then the expected number of defectives for a sample of size n is np and the variance is $np(1 - p)$. If D is the number of defectives in a sample, then D/n gives the proportion of defectives in the sample. Since $E[D/n] = p$ and Var $[D/n] = p(1 - p)/n$, the center line of the p chart is determined by p and the upper and lower control limits are

$$UCL_p = p + 3\left[\frac{p(1 - p)}{n}\right]^{1/2}$$

$$LCL_p = p - 3\left[\frac{p(1 - p)}{n}\right]^{1/2}$$

The interpretation of these limits is different than in the case of an $\bar{X}$ chart. Although the control limits represent a range of six standard deviations, one cannot assert, as in the case of an $\bar{X}$ chart, that 99.74% of all sample proportions will fall between LCL_p and UCL_p. The actual percentage depends upon the value of p. Of course, should the sample sizes be sufficiently large so that $np \geq 5$, then the central limit theorem asserts that the distribution is approximately normal and a similar interpretation applies.

In the event that p is not known, it can be estimated from each preliminary sample by $\hat{p}_i = D_i/n$, where D_i is the number of defectives in sample i. The estimator of p based upon all k samples is the average of the k sample estimators,

$$\bar{p} = \frac{1}{k}\sum_{i=1}^{k}\hat{p}_i = \frac{1}{kn}\sum_{i=1}^{k}D_i$$

The corresponding estimator of Var $[D/n]$ for a sample of size n is $\bar{p}(1 - \bar{p})/n$. Thus, the upper and lower control limits for the p chart are

Control Limits for p Chart

$$UCL_p = \bar{p} + 3\left[\frac{\bar{p}(1 - \bar{p})}{n}\right]^{1/2}$$

$$LCL_p = \bar{p} - 3\left[\frac{\bar{p}(1 - \bar{p})}{n}\right]^{1/2}$$

As with $\bar{X}$ and R charts, the control limits are treated as trial limits when employing the estimator $\bar{p}$; however, since p represents the proportion of defective units, the interpretation is slightly different. Specifically, if preliminary samples indicate that p is too high, then, from a practical perspective, the process must be examined even if the preliminary samples reveal the process to be in control. It is likely of little value to have a process in control if the proportion of defective units is unacceptably high. One possible approach would be a hypothesis test concerning p. If p appears to be acceptable, then the usual plot of the preliminary points is used to reveal whether or not the process is in control. As in the previous cases, if the data suggest that the process is out of control, one must discard samples yielding extreme values of D/n and search for assignable causes. One might argue that no reexamination is necessary for sample proportions falling below LCL_p, since such values indicate the process to be behaving significantly better than expected. Practically, such a passive approach would miss the opportunity of searching for assignable causes that may result in an improvement of the process. Moreover, values below LCL_p may indicate an unsatisfactory estimate of p. Finally, if p happens to be very small, then the lower control limit might be negative. In such a case, it is customary to let $LCL_p = 0$.

EXAMPLE 8.26

Table 8.9 gives the numbers of blemished (defective) tires in 24 preliminary samples, each comprised of 100 tires. Suppose it is desired to set up a p chart. Based upon the data,

$$\sum_{i=1}^{24} D_i = 97$$

$$\bar{p} = \frac{97}{24 \times 100} = 0.040$$

$$UCL_p = 0.040 + 3\left[\frac{(0.04)(1 - 0.04)}{100}\right]^{1/2} = 0.099$$

$$LCL_p = 0$$

Note that $LCL_p = 0$ since the computed limit is negative. The resulting p chart is illustrated in Figure 8.39. Only the proportion corresponding to sample 15 lies outside the control limits (above UCL_p). Thus, it is discarded, and the remaining preliminary samples are employed to find the control limits. We obtain $\bar{p} = 0.037$, $UCL_p = 0.094$, and $LCL_p = 0$. All sample

TABLE 8.9   NUMBERS OF DEFECTIVE TIRES

Sample	Number of Defectives	Sample	Number of Defectives
1	6	13	4
2	3	14	2
3	2	15	12
4	3	16	2
5	1	17	4
6	5	18	3
7	2	19	6
8	4	20	7
9	0	21	4
10	6	22	3
11	4	23	5
12	2	24	7

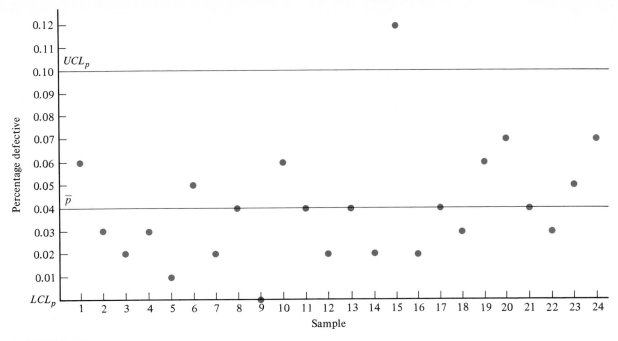

FIGURE 8.39
Trial *p* chart for tire samples

proportions are contained within the new control limits, and so the new limits are employed to construct the *p* chart.

8.10.4 c Charts

In some cases a unit is considered to be nondefective even if it possesses some defects. For instance, the wing of an airplane may have some defective rivets, and yet the entire wing may itself not be considered defective. For the purposes of quality control, our interest is in monitoring the number of defects per unit, where a unit might be defined as a single item, a collection of items, or a subassembly within an item. The appropriate control chart is known as a **c chart.**

It is assumed that the number of defects per unit satisfies a Poisson distribution with parameter λ. As is customary in quality control, we will let c (lower case) denote the random variable corresponding to the count. Since $E[c] = \mathrm{Var}\,[c] = \lambda$, employing the usual six standard deviation range yields the upper and lower control limits

$$UCL_c = \lambda + 3\sqrt{\lambda}$$

$$LCL_c = \lambda - 3\sqrt{\lambda}$$

In the case of a *p* chart, the probability of a sample proportion lying between the control limits depends upon p; analogously, in the case of a *c* chart, the probability of a count falling between the control limits depends upon λ.

Should λ not be known, it can be estimated from the data. Specifically, for the ith sample, the number of defects c_i is an unbiased estimator of λ. Thus, the desired estimator of λ is the average of the c_i,

$$\bar{c} = \frac{1}{k} \sum_{i=1}^{k} c_i$$

The center line of the c chart is determined by $\bar{c}$, and the upper and lower control limits are

Control Limits for c Chart

$$UCL_c = \bar{c} + 3\sqrt{\bar{c}}$$
$$LCL_c = \bar{c} - 3\sqrt{\bar{c}}$$

The same comments apply to setting up a c chart from preliminary data as apply to setting up a p chart. As in the case of a p chart, the lower control limit might be negative. In such an event, $LCL_c = 0$.

EXAMPLE 8.27

Suppose a plant manufactures refrigerators and each day a count is kept of the number of refrigerators that fail the finished-product inspection. To set up a c chart for the number of failures per day, 20 preliminary daily counts are made, and it is assumed that c possesses a Poisson distribution. Based on the counts given in Table 8.10,

$$\sum_{i=1}^{20} c_i = 98$$

$$\bar{c} = \frac{98}{20} = 4.90$$

$$UCL_c = 4.90 + 3\sqrt{4.90} = 11.54$$

$$LCL_c = 0$$

The corresponding c chart is illustrated in Figure 8.40. We can see that all counts lie within the control limits, and therefore we conclude that the process is in control.

TABLE 8.10 DAILY COUNTS OF REFRIGERATORS FAILING INSPECTION

Day	Count	Day	Count
1	4	11	2
2	6	12	4
3	9	13	6
4	4	14	4
5	3	15	7
6	8	16	5
7	2	17	8
8	6	18	1
9	5	19	4
10	5	20	5

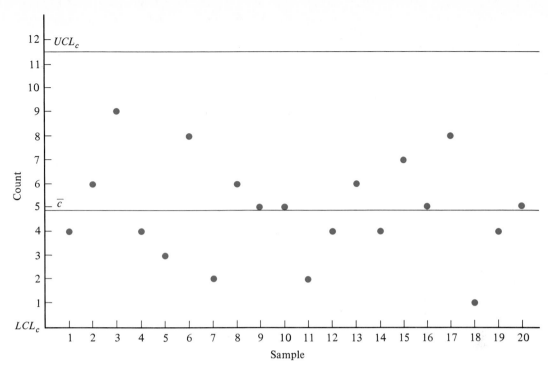

FIGURE 8.40
Trial c chart for refrigerator counts

EXERCISES

SECTION 8.1

8.1. A sample of 15 voltmeters is randomly selected from a large shipment, and each is tested to determine whether or not it meets a prespecified degree of accuracy. Based upon previous experience, the testing laboratory believes that either 5% or 20% of the voltmeters in the entire shipment will not meet the desired standard. Construct a hypothesis test to detect which of the substandard proportions applies to the shipment. Assuming the null hypothesis is $H_0: p = 0.05$, make the size of type I error as close to 0.15 as possible and find the corresponding size of type II error.

8.2. Construct a hypothesis test in Exercise 8.1 that will make α as close as possible to 0.05 and find the corresponding value of β.

8.3. Suppose the laboratory in Exercise 8.1 tests 35 voltmeters. Employ a large-sample approach to test the competing hypotheses so that α is approximately 0.10. Find the size of type II error.

8.4. Urn A contains 10 white balls and 20 black balls, whereas urn B contains 15 of each color. The labels on the urns are covered, and a subject is to select 10 balls from one of the urns. Based upon the number

of black balls selected, the subject is to decide from which urn he or she has chosen. Construct an appropriate hypothesis test with H_0 referring to urn A and with α and β as close as possible. Keep in mind that the binomial distribution is inappropriate, since the total number of balls in each urn is not very large.

8.5. The lengths of telephone calls originating at locations A_1 and A_2 have been determined to satisfy exponential distributions with parameters $b_1 = 0.25$ and $b_2 = 0.1$, respectively. A single phone call is observed and its duration is measured. Assuming our desire is to demonstrate that the point of origin is A_1, construct a hypothesis test with $\alpha = 0.10$. Find the corresponding value of β.

8.6. Redo Exercise 8.5 with $\alpha = 0.05$.

8.7. An investigator wishes to determine whether or not an intercepted coded transmission is genuine or is a fake meant to mislead. Based upon an analysis of past transmissions, it has been estimated that legitimate messages possess lengths that are normally distributed with the mean and standard deviation being 940 bits and 110 bits, respectively, and that planted bogus transmissions possess lengths that are normally distributed with mean and standard deviation 868 bits and 140 bits, respectively. Construct a hypothesis test that is applicable to the observation of a single message and that will be used to decide whether or not the message is genuine. Assume that the goal of the investigator is to prove the message is genuine. Let $\alpha = 0.08$ and find β.

SECTION 8.2

8.8. If too much torque is applied when tightening a machine bolt, the threads on the bolt or in the bolt hole can be stripped. For a particular application, the point at which thread distortion occurs is measured. Assuming 15 test bolts are measured and the sample mean of the stripping torques is employed as the test statistic, find the critical region for the hypothesis test

$$H_0:\ \mu \le 100$$
$$H_1:\ \mu > 100$$

Assume normality, $\sigma = 6.7$ pound-feet and $\alpha = 0.10$. Find $\beta(105)$, $\beta(110)$, and $\beta(115)$.

8.9. Repeat Exercise 8.8 with $\alpha = 0.05$.

8.10. The random variable X possesses a uniform distribution over the interval $[\mu - 1, \mu + 1]$. To demonstrate that $\mu > 12.4$, a single observation of X is to be taken. Construct the appropriate hypothesis test for $\alpha = 0.10$. Find $\beta(12.6)$ and $\beta(13)$.

8.11. In a given environment, CPU service time is assumed to possess an exponential distribution with mean μ. To demonstrate execution speed, it is desired to show that $\mu < 1.2$ seconds. Construct, for a single observation, an appropriate hypothesis test with $\alpha = 0.10$. Find $\beta(1)$ and $\beta(0.5)$.

8.12. Repeat Exercise 8.11 with $\alpha = 0.05$.

8.13. An experimenter wishes to demonstrate that, at 100°C, the coefficient of static friction of steel on steel when lubricated by a newly developed graphited oil is less than 0.13. Given that the measurement process is assumed to be normally distributed with standard deviation $\sigma = 0.005$ and the sample mean corresponding to a random sample of size 40 is to be employed as the test statistic, find the critical region of size $\alpha = 0.05$. Find $\beta(0.12)$.

8.14. Referring to Exercise 8.8, suppose it is desired to show that $\mu \ne 100$ pound-feet. Construct an appropriate critical region for $\alpha = 0.05$. Find $\beta(90)$, $\beta(95)$, $\beta(105)$, and $\beta(110)$.

8.15. Cartons of grommets are rated at 40 pounds. For $\alpha = 0.01$, construct a two-tailed critical region to test the null hypothesis $\mu = 40$, the sample mean corresponding to a random sample of size 10 being the test statistic. Assume the weights to be normally distributed with variance $\sigma^2 = 0.64$. Assuming an

error of 1 pound to be important to detect, find $\beta(39) = \beta(41)$. Comment.

8.16. Redo Exercise 8.15 with $\alpha = 0.10$. Again comment.

8.17. Given the sample weights

| 41.7 | 40.6 | 42.4 | 42.7 | 41.5 |
| 41.0 | 42.5 | 40.4 | 41.2 | 42.6 |

what conclusion is drawn in Exercise 8.15?

SECTION 8.3

8.18. Find $H(\mu)$ and plot the OC curve for Exercise 8.8.

8.19. Find $H(\mu)$ and plot the OC curve for Exercise 8.9.

8.20. Find $H(\mu)$ and plot the OC curve for Exercise 8.10.

8.21. Find $H(\mu)$ and plot the OC curve for Exercise 8.11.

8.22. Find $H(\mu)$ and plot the OC curve for Exercise 8.14.

8.23. Find $H(\mu)$ and plot the OC curve for Exercise 8.15.

SECTION 8.4

8.24. At a particular setting, an oxygen-acetylene welding unit is supposed to attain a temperature of 4500°F. Should the operating temperature deviate too greatly, then the unit must be serviced. Because of measurement error and the inherent imprecision of the valves, a number of independent test measurements must be taken. Experience has shown the measurements to be normally distributed with standard deviation 150°F. If 5 test measurements are to be taken, find a two-sided critical region of size $\alpha = 0.05$ to test the hypothesis $H_0:\ \mu = 4500$ using the test statistic Z. Do the five measurements

| 4755 | 4862 | 4789 | 4810 | 4790 |

result in rejection of the null hypothesis?

8.25. A certain carbon-steel alloy has a melting point of 2760°F. To detect a change in the production process, 12 samples are taken and their melting points tested. If the null hypothesis of mean melting point $\mu = 2760$ is rejected at level $\alpha = 0.10$, then it is concluded that a problem exists. Assuming the melting-point measurements to be normally distributed with a standard deviation of 24°F, find the appropriate Z-test critical region for 12 measurements. Using the OC curve, find $\beta(2750)$. What conclusion is drawn if sample measurements are as follows?

| 2732 | 2742 | 2760 | 2776 | 2720 | 2750 |
| 2702 | 2738 | 2776 | 2754 | 2779 | 2710 |

8.26. If a deviation of 300°F in Exercise 8.24 is considered to be of consequence, find the sample size that will make erroneous acceptance of the null hypothesis have probability 0.15 in the event of such a deviation. Find the corresponding acceptance critical region.

8.27. If a deviation of 40°F in Exercise 8.25 is considered meaningful, find the sample size that will make erroneous acceptance of the null hypothesis have a probability of 0.25. Find the corresponding acceptance region.

8.28. Concentrated hydrochloric acid contains 37% HCl by weight. A chemist wishes to detect a deviation from the nominal 37% value in a large shipment of bottles of acid. To do so, he or she plans to randomly select a number of bottles from the shipment. Assuming the measurements to be normally distributed with standard deviation $\sigma = 0.4$ and the two-tailed Z-test critical region to be selected so that $\alpha = 0.01$, find the sample size necessary to have $\beta(37.3) = 0.10$. What is the corresponding acceptance region? What conclusion is drawn if measurements are taken and the resulting empirical mean is $\bar{x} = 37.3$?

8.29. A lathe operator must handle jobs of various specifications, and different jobs arrive randomly. Over time it has been observed that the operator's time per job is normally distributed with standard deviation $\sigma = 4.62$ minutes. Moreover, his or her average time per job has been nominally rated at 18 minutes. It is desired to determine whether the operator's job performance has suffered due to boredom, and a hypothesis test based on the sample mean is to be used to detect reduced performance. Based on a Z test, find the critical region of size $\alpha = 0.01$. If the times

21.01 19.32 18.76 22.42 20.49
25.89 20.11 18.97 20.90

are observed, what conclusion is drawn?

8.30. A deterioration of 1.5 minutes is considered meaningful in Exercise 8.29, and it is desired to keep the probability of false acceptance of the null hypothesis down to 0.20 in the case of such a performance loss. Find the necessary sample size and the corresponding critical region.

8.31. Transmission of a digital image in a particular setting is found to take, on average, 3.45 seconds. By compressing the information in the image and then transmitting (which results in a less precise image on the other end), the transmission time can be cut. It is claimed that a particular compression algorithm yields acceptable images and results in a mean transmission time of less than 2.50 seconds. Assuming normally distributed times with standard deviation $\sigma = 0.32$ seconds, construct a critical region of size $\alpha = 0.05$ to test the speed claim. Using the OC curve, find $\beta(2.40)$. Given 15 times with a mean of $\bar{x} = 2.35$, what is the conclusion concerning the claimed transmission time?

8.32. It is important to reject the null hypothesis of Exercise 8.31 when the true mean transmission time is 2.40 seconds. What is the appropriate sample size

for $\beta(2.40) = 0.05$ if $\alpha = 0.01$? Find the corresponding critical region.

8.33. A manufacturer claims that a sealed bearing possesses a time to failure of better than 925 hours when operating at 5000 revolutions per minute. Based upon past experience it is assumed that the variance of the time-to-failure distribution is $\sigma^2 = 2500$, but it is known that the distribution is not normal. The manufacturer wishes to demonstrate the claim at the significance level $\alpha = 0.01$. It is believed crucial that detection should occur if the actual mean time to failure is 1000 hours. If a one-sided Z test is to be employed, what sample size is necessary to have $\beta(1000) = 0.10$? Find the corresponding critical region.

8.34. The top speed of a dragstrip racer in a quarter-mile race is altered by a change in tires. A tire manufacturer claims that a new make of tire will yield a top speed in excess of 185 mph. Based upon experience, the manufacturer is uncertain whether or not the standard deviation of 4.7 mph that has been previously observed will continue to be accurate when using the newly developed tire; nor is there certainty regarding normality. It is desired to find a critical region of size $\alpha = 0.05$ to test the null hypothesis H_0: $\mu \leq 185$. It is also desired that $\beta(190)$ be kept to 0.10. Assuming the conservative guess $\sigma = 6$ is employed for the standard deviation when using the OC curves, what is the necessary sample size? Construct the corresponding critical region.

8.35. Find the P value for Exercise 8.24.

8.36. Find the P value for Exercise 8.25.

8.37. Find the P value for Exercise 8.29.

8.38. Find the P value for Exercise 8.31.

SECTION 8.5

8.39. Pipe stock is automatically fed to a cutter to produce cuts of nominal length 8 feet. To test the accuracy of the equipment, 12 cuts are randomly selected and their lengths measured. Assuming the population of lengths is normally distributed, do the measurements (in feet)

8.08 8.02 8.04 8.04 8.02 8.05
8.02 8.03 8.07 8.01 8.03 8.07

yield the conclusion, at the $\alpha = 0.05$ level, that the nominal mean length should be rejected?

8.40. Suppose in Exercise 8.39 it is considered important to detect a deviation of 0.03 feet. Using the standard deviation of the test results (a risky approach), find the corresponding value of β.

8.41. Referring to Exercise 7.6, a problem with the process is indicated if the mean BTU measurement varies too greatly from 1230. Assuming the measurements to be normally distributed, do the data in Exercise

7.6 indicate a problem at the $\alpha = 0.05$ level? Assuming a 20-BTU deviation to be consequential, find the corresponding value of β by employing the sample standard deviation of the actual data.

8.42. Under the assumption of normality, employ the re-crystallization data of Exercise 7.58 to test whether or not the null hypothesis H_0: $\mu = 150$ should be rejected at the $\alpha = 0.01$ level. If $\sigma = 0.20$ is considered to be a conservative estimate of the population standard deviation, find $\beta(149.85)$.

8.43. The *mean time to repair (MTTR)* a unit is analogous to the mean time to failure (MTTF) [see Section 6.3]: it is the average time it takes to repair a unit that has been taken out of service. As an illustration, consider a robot paint-sprayer. Such a unit can fail for any number of reasons; however, let us suppose that the time to repair is normally distributed, but with unknown variance. It is claimed by the manufacturer that the MTTR is less than 3.4 hours. With a number of the units in operation, nine repair times are randomly selected and a t test is performed in an effort to demonstrate the manufacturer's claim. The following times (in hours) are observed:

$$0.8 \quad 7.4 \quad 3.8 \quad 5.5 \quad 1.3$$
$$2.7 \quad 3.5 \quad 1.1 \quad 1.8$$

Find the critical region for $\alpha = 0.05$ and determine if the manufacturer's claim is supported at that level.

8.44. Repeat Exercise 8.43 with the data

$$3.2 \quad 3.1 \quad 3.5 \quad 3.4 \quad 2.9$$
$$2.6 \quad 3.3 \quad 2.8 \quad 3.1$$

8.45. Referring to Exercise 8.43, find $\beta(3.1)$ assuming $\sigma = 2.0$ to be a conservative estimate of the standard deviation. Find $\beta(3.1)$ assuming $\sigma = 0.4$ to be a conservative estimate of the standard deviation.

8.46. Based upon the observations

$$72.5 \quad 74.2 \quad 75.1 \quad 72.1 \quad 71.8 \quad 72.5$$

can it be concluded at the $\alpha = 0.05$ level that the Rockwell hardness of a particular type of steel is in excess of 72.5? Assume hardness measurements to be normally distributed. Find the P value.

8.47. A plant manager claims that it takes more than 200 hours of on-the-job training before a typical employee in a certain job becomes self-sufficient. To test the claim, a random selection of new employees is taken and, unbeknown to the employees, the number of hours prior to each being left on his or her own is recorded. These times are

$$208 \quad 180 \quad 232 \quad 168 \quad 212$$
$$208 \quad 254 \quad 229 \quad 220 \quad 181$$

Assuming that training times are normally distributed, is the claim justified at the $\alpha = 0.05$ level? Find the P value?

8.48. Suppose an upper one-tailed t test is to be applied at level $\alpha = 0.05$ with $n = 16$. If $\bar{x} - \mu_0 = 2$, what is the maximum standard deviation that will result in rejection of the null hypothesis?

8.49. Repeat Exercise 8.48 in the case of a two-tailed test.

8.50. Write a general expression for the maximum standard deviation that will result in rejection of the null hypothesis for a one-tailed t test. Specifically, express s in terms of α, n, $\bar{x}$, and μ_0.

8.51. Referring to Exercise 8.46, if n is free to be increased and a t test is desired, what value of n is necessary to make $\beta(73) = 0.20$? Assume $\sigma = 1.8$ to be a conservative estimate of the standard deviation.

8.52. Referring to Exercise 8.39, what value of n is necessary to make the probability of not detecting a deviation of 0.05 to be 0.15? Assume 0.02 to be a conservative estimate of the standard deviation.

SECTION 8.6

8.53. To test the durability of plastic drill casings, 15 drills are randomly selected from stock and each is subjected to a series of impacts. The manufacturer believes that the proportion of casings that break under the testing procedure should be less than 10%. Construct a hypothesis test to test the counterclaim that the manufacturer's claim is not true. Find the critical region that yields a value of α as close as possible to 0.05. A 20% failure proportion is believed to be consequential. Find the probability that such a proportion will be erroneously not detected. If 15 drill casings are tested and 3 fail the impact testing, what conclusion is drawn?

8.54. Redo the critical region part of Exercise 8.53 with the intent being to demonstrate the manufacturer's claim at the 0.10 level.

8.55. A manufacturer of aircraft engines is concerned with the proportion of units in the factory that are actively in production or actively between production processes, as opposed to those sitting idle in a state of partial completion. Based upon past experience, it is believed that the proportion of units in or between production processes is 82%. To test this figure, 30 units are randomly selected by serial number and are rated as "active" or "idle." Assuming the number of units in the factory to be very large, so that the binomial distribution is applicable, construct a hypothesis test to check the null hypothesis $p = 0.82$ at level $\alpha \approx 0.10$. A percentage error of 4% is deemed serious. Find $\beta(0.78)$ and $\beta(0.86)$. If 30 units are selected and 18 are found to be active, what conclusion is drawn?

8.56. Referring to Exercise 8.55, it is more than likely that the manufacturer is concerned only if the proportion of active units drops below 82%. Construct a test to detect, at level $\alpha \approx 0.10$, too small a proportion. Find $\beta(0.78)$.

8.57. Repeat Exercise 8.56 with $\alpha \approx 0.05$.

8.58. At any given moment, some percentage of employees in a company will not be gainfully working. A supervisor believes this figure to be in excess of 20%, and wishes to construct a hypothesis test to demonstrate the belief. He or she also believes that a 25% figure represents an unacceptable level of idleness. Find n so that the probability of not detecting the latter level of idleness is 0.10 when $\alpha = 0.05$. Find the resulting critical region.

8.59. Referring to Exercise 8.56, find n so that $\beta(0.78) = 0.05$ and reconstruct the critical region using the new value of n.

8.60. Investment casting is a casting method employed for making small, complicated parts such as dental fittings. In a particular application, it is believed that quality control will find that 10% of the parts will be defective; nevertheless, the process is still considered cost effective when compared to alternative means of production. Should the 10% figure be inaccurate, either too low or too high, then the economics of production are altered and thus detection of a different defective proportion is considered significant. In particular, a defective proportion of 13% is considered especially noteworthy. Construct a two-tailed test appropriate to the given problem. Let $\alpha = 0.10$ and find n so that $\beta(0.13) = 0.10$.

8.61. Reconsider Exercise 8.60 in the event the manufacturer is interested only in detecting an increase in the proportion of defective castings.

SECTION 8.7

8.62. The random variable X counts the number of bits in coded messages emanating from a certain source. Assuming X to be approximately normally distributed, it is desired to check the null hypothesis $H_0: \sigma^2 = 160,000$ at the $\alpha = 0.10$ level. Construct the appropriate two-tailed critical region and make a decision based upon the bit counts

| 4532 | 4606 | 3511 | 4201 | 3392 | 4639 |
| 4021 | 4722 | 3470 | 3100 | 4212 | 4165 |

8.63. Referring to Exercise 8.62, it is important to detect a false null hypothesis when $\sigma = 500$. For $\alpha = 0.05$, find n so that $\beta(360,000) = 0.20$. Construct the corresponding critical region.

8.64. Referring to Exercise 8.62, it is important to detect a false null hypothesis when $\sigma = 300$. For $\alpha = 0.05$, find n so that $\beta(90,000) = 0.20$. Construct the corresponding critical region.

8.65. In Exercise 8.25 it was assumed that $\sigma = 24$. At the level $\alpha = 0.01$, is this assumption compatible with the given data?

8.66. Referring to Exercise 8.25, if the assumption $\sigma^2 = 576$ is to be checked at the $\alpha = 0.01$ level and it is required that $\beta(784) = 0.25$, then what sample size is required. Construct the corresponding critical region.

8.67. Referring to the lathe operator of Exercise 8.29, suppose the claim is made that $\sigma^2 > 9$. Is this claim supported at the $\alpha = 0.05$ level by the data? If it is desired that $\beta(16) = 0.4$, what sample size is necessary?

8.68. Referring to Exercise 8.47, can it be claimed at the $\alpha = 0.10$ level that $\sigma^2 > 15$?

8.69. Assuming $n = 22$, $\alpha = 0.05$, and $\sigma_0^2 = 8.3$, what is the minimum value of the sample variance that will result in a rejection of the null hypothesis for an upper-one-tailed test?

8.70. Run times are considered to be normally distributed in a certain batch-programming environment. Do the data (in seconds)

302 345 329 340 311 305 320 339 310 311
290 313 335 321 338 299 340 310 308 338

support the claim, at level $\alpha = 0.10$, that the variance is less than 900?

8.71. Referring to Exercise 8.70, find n so that $\beta(324) = 0.10$ at significance level $\alpha = 0.01$. Construct the corresponding critical region.

8.72. Do the data of Exercise 7.49 support, at level $\alpha = 0.05$, the claim that $\sigma^2 < 4900$?

8.73. Referring to Exercise 8.72, find n so that $\beta(1600) = 0.25$. Construct the corresponding critical region.

SECTION 8.8

8.74. The random variable X satisfies a Poisson distribution; however, we are uncertain as to whether $H_0: \lambda = 1$ or $H_1: \lambda = 2$. Use the Neymann-Pearson lemma to determine a best test based on a sample size of 2 and a significance level of $\alpha = 0.05$. Keep in mind that the discreteness of the distribution only allows an approximation to 0.05. Illustrate the form of the critical region (in terms of x_1 and x_2) graphically.

8.75. Find the form of the best test in Exercise 8.74 for an arbitrary size n. For $n = 10$, give the critical region in terms of the sample mean.

8.76. Suppose X possesses one of the following two probability mass functions:

x	0	1	2	3	4	5	6	7
$f_0(x)$	0.05	0.02	0.03	0.01	0.01	0.04	0.09	0.75
$f_1(x)$	0.44	0.06	0.09	0.01	0.03	0.14	0.03	0.20

Consider the hypothesis test

$$H_0: f_X = f_0$$
$$H_1: f_X = f_1$$

which is designed to choose between the two competing densities based upon a single observation of X. Find all critical regions of size (exactly) $\alpha = 0.05$ and choose which one among them is best.

8.77. Suppose X is normally distributed with mean 0. Use the Neyman-Pearson lemma to prove that the best critical region for testing $H_0: \sigma = \sigma_0$ against $H_1: \sigma = \sigma_1 > \sigma_0$ is of the form

$$\sum_{k=1}^{n} X_k^2 > c$$

where c is an appropriately chosen constant.

8.78. X possesses the density $f(x; \theta) = (1 + \theta)x^\theta$ for $0 < x < 1$. Using the Neyman-Pearson lemma, determine the form of the best critical region for testing $H_0: \theta = \theta_0$ against $H_1: \theta = \theta_1 < \theta_0$.

SECTION 8.9

8.79. The following windspeed measurements (in mph) are taken in a hurricane, there being no distributional assumptions concerning the underlying population.

88.1	87.3	89.2	96.5	89.7
100.3	90.1	86.3	108.9	95.6
86.3	90.1	78.8	84.3	88.6
101.8	89.3	77.6	93.2	89.2

Perform a sign test to demonstrate, at level $\alpha = 0.05$, that the population median is less than 90.0 mph. What conclusion is drawn?

8.80. A lathe is set so that the nominal diameter of steel rods is 4.000 cm. A random selection of rods yields the following diameters.

4.012 4.003 4.000 3.998 4.006 4.012
4.005 4.001 3.998 3.999 4.005 4.009
4.011 4.008 4.010 4.000

Perform a sign test to determine if the nominal diameter should be rejected at the $\alpha = 0.05$ level. What conclusion is drawn?

8.81. The nominal resistance of a wire is 0.150 ohm. Random testing of the wire stock yields the following resistance data.

0.148 0.147 0.151 0.146
0.151 0.148 0.147 0.150

Does the sign test yield the conclusion, at the $\alpha = 0.05$ level, that the median resistance is less than 0.150 ohm?

8.82. In the absence of distributional assumptions, can it be concluded from the BTU data of Exercise 7.6 that the population median exceeds 1200 BTUs? Let $\alpha = 0.01$.

8.83. In the absence of distributional assumptions, can it be concluded from the crack-advance data of Exercise 7.49 that the median crack-advance exceeds 3100 ft/sec? Let $\alpha = 0.03$.

In Exercises 8.84 and 8.85, apply the sign test to perform the hypothesis test of the cited exercise. Assuming each population is symmetric and possesses a mean, the sign test will concern the mean.

8.84. Exercise 8.25

8.85. Exercise 8.43

In Exercises 8.86 through 8.91, apply the signed-rank test in the cited exercise. In each case, assume the mean exists so that (according to the symmetry assumption necessary for the signed-rank test) the test concerns the mean of the distribution. When justified, employ the large-sample approach.

8.86. Exercise 8.79

8.87. Exercise 8.80

8.88. Exercise 8.81

8.89. Exercise 8.82

8.90. Exercise 8.83

8.91. Exercise 8.43

8.92. It is hypothesized that the median drying time for a certain type of glue is 3.1 minutes. Do the following data lead to rejection of the hypothesized value at level $\alpha = 0.05$? Use a large-sample approach.

3.0 3.2 3.4 3.1 3.2 2.8
2.9 3.4 3.6 2.9 4.0 3.0
3.8 3.7 3.0 3.7 3.2 3.1
3.4 3.9 2.9 3.6

SECTION 8.10

8.93. Sixteen-ounce soda cans are filled automatically. When the process is in statistical control, the mean content is $\mu = 16.01$ ounces, the standard deviation is $\sigma = 0.07$, and the content possesses a normal distribution. Set up the $\bar{X}$ chart. Fill in the control chart for the samples of Table 8.11 and determine at which sample the process is first seen to be out of control. Each sample consists of six cans.

8.94. When the production of household thermostats is in statistical control, the mean difference between the temperature at which the heat is switched on and the nominal setting is $\mu = 1.12°F$, the standard deviation of the difference is $\sigma = 0.08$, and the difference is normally distributed. Set up the $\bar{X}$ chart. Fill in the control chart for the samples of Table 8.12 and determine at which sample the process is first seen to be out of control. Each sample consists of four thermostats.

TABLE 8.11 MEANS OF SODA-CAN SAMPLES

Sample	$\bar{X}$	R	Sample	$\bar{X}$	R
1	16.03	0.19	11	15.96	0.20
2	16.00	0.18	12	16.08	0.22
3	15.97	0.20	13	15.83	0.24
4	15.95	0.23	14	15.97	0.22
5	16.02	0.18	15	15.84	0.20
6	16.05	0.15	16	15.95	0.22
7	16.07	0.35	17	15.93	0.19
8	16.00	0.14	18	15.93	0.17
9	15.96	0.22	19	15.85	0.17
10	16.08	0.15	20	15.94	0.34

TABLE 8.12 MEANS OF THERMOSTAT-ERROR SAMPLES

Sample	$\bar{X}$	R	Sample	$\bar{X}$	R
1	1.22	0.20	13	1.10	0.26
2	1.02	0.26	14	1.12	0.15
3	1.20	0.15	15	1.05	0.10
4	1.04	0.10	16	1.01	0.12
5	1.18	0.12	17	1.26	0.19
6	1.04	0.16	18	1.18	0.22
7	1.10	0.09	19	1.15	0.18
8	1.12	0.22	20	1.30	0.17
9	1.20	0.16	21	1.19	0.10
10	1.01	0.14	22	1.20	0.22
11	1.10	0.10	23	1.05	0.12
12	1.12	0.16	24	1.26	0.18

TABLE 8.13 MEANS AND RANGES FOR GAGE-PRESSURE SAMPLES

Sample	$\bar{X}$	R	Sample	$\bar{X}$	R
1	28.10	0.32	12	27.95	0.20
2	27.90	0.26	13	28.15	0.38
3	28.13	0.18	14	28.21	0.27
4	28.89	1.02	15	28.12	0.33
5	27.87	0.28	16	28.00	0.27
6	28.03	0.20	17	28.10	0.05
7	27.64	0.38	18	28.03	0.20
8	27.99	0.27	19	28.11	0.84
9	27.80	0.30	20	28.00	0.20
10	28.01	0.22	21	27.90	0.10
11	28.15	0.25	22	28.06	0.24

TABLE 8.14 MEANS AND RANGES FOR SAMPLE TURNAROUND TIMES

Sample	$\bar{X}$	R	Sample	$\bar{X}$	R
1	3.22	0.88	11	7.02	1.21
2	4.00	0.52	12	2.87	0.35
3	2.38	0.38	13	3.66	0.64
4	3.02	0.50	14	3.88	0.40
5	1.22	0.44	15	2.88	0.62
6	2.66	0.37	16	4.10	0.58
7	2.90	0.70	17	4.22	0.45
8	2.01	0.35	18	1.20	0.15
9	3.44	0.44	19	2.98	0.39
10	6.32	1.33	20	2.50	0.60

8.95. Assuming the mean and variance are not known in Exercise 8.93, set up the $\bar{X}$ chart using $\bar{\bar{X}}$ and $\bar{R}$. Use the trial control limit approach, treating the samples of Table 8.11 as preliminary samples.

8.96. Assuming the mean and variance are not known in Exercise 8.94, set up the $\bar{X}$ chart using $\bar{\bar{X}}$ and $\bar{R}$. Use the trial control limit approach, treating the samples of Table 8.12 as preliminary samples.

8.97. Tire-pressure gages are usually read in pounds. In a particular production process, samples of five gages are taken from different production runs and are checked on a known pressure of 28 lb. The data of Table 8.13 result from 22 samples. Set up both an $\bar{X}$ chart and an R chart, making certain that all (remaining) sample values lie within the upper and lower control limits of the respective charts.

8.98. To check the adequacy of a computer network, benchmark programs are periodically run on the network. In each test, four benchmarks are randomly selected from a collection of benchmarks which possess (essentially) the same mix of instructions and I/O calls. Based upon the mean turnaround times (in seconds) and turnaround ranges of the 20 samples of Table 8.14, set up both an $\bar{X}$ chart and an R chart. Make certain that all (remaining) sample values lie within the upper and lower control limits of the respective charts.

8.99. As bicycles leave production, each is tested to determine whether or not it is deemed satisfactory. The manufacturer accepts the fact that a small percentage will fall below an acceptable level of quality. Table 8.15 gives the numbers of unsatisfactory units in 20 preliminary samples of 40 bicycles. Use the preliminary samples to set up a p chart which applies when the process is (assumed to be) in control.

8.100. Bottles of aspirin tablets are sealed with a protective plastic strip to guard against tampering. Table 8.16 gives the number of defective strips in 20 preliminary samples of 50 bottles. Set up the p chart.

8.101. The number of daily repair reports on a subway system is modeled as a Poisson distribution. Using the preliminary sample data of Table 8.17, set up a c chart.

8.102. The daily number of absentees from a plant is modeled as a Poisson random variable. Based upon the preliminary absentee samples of Table 8.18, construct a c chart.

TABLE 8.15 NUMBERS OF DEFECTIVE BICYCLES

Sample	Number of Defectives	Sample	Number of Defectives
1	2	11	3
2	1	12	0
3	2	13	1
4	8	14	0
5	1	15	1
6	2	16	1
7	0	17	3
8	1	18	2
9	2	19	6
10	2	20	1

TABLE 8.16 NUMBERS OF DEFECTIVE SEALING STRIPS

Sample	Number of Defectives	Sample	Number of Defectives
1	1	11	3
2	1	12	2
3	2	13	0
4	8	14	1
5	1	15	2
6	0	16	0
7	1	17	3
8	2	18	2
9	1	19	2
10	1	20	0

TABLE 8.17 DAILY COUNTS OF REPAIR REPORTS

Sample	Count	Sample	Count
1	16	12	10
2	10	13	18
3	20	14	12
4	22	15	32
5	8	16	21
6	6	17	25
7	20	18	16
8	18	19	36
9	10	20	4
10	16	21	8
11	19	22	10

TABLE 8.18 DAILY ABSENTEE COUNTS

Sample	Count	Sample	Count
1	4	11	15
2	1	12	5
3	2	13	6
4	8	14	4
5	3	15	2
6	0	16	12
7	4	17	1
8	3	18	3
9	4	19	3
10	5	20	6

9

||||||||||||||||||||||||||||||||||||||||||||||||||||||||||||||||||||||||||||||||||||||||||||||||||||||||||||||||||||||||||||||||||||||||||||||||||||||

INFERENCES CONCERNING TWO PARAMETERS

Whereas Chapters 7 and 8 dealt with estimation procedures and hypothesis tests concerning a single population parameter, the present chapter concerns inferences involving two parameters, each of which comes from a distinct population. For instance, should one random variable correspond to measuring some attribute of a control population whereas another corresponds to the same attribute in a test population, it is important to decide whether or not the respective means are significantly different.

The first three sections of the chapter continue the development of hypothesis testing, the difference being that the tests concern the difference of two means, not simply a single mean. Special attention should be paid to Section 9.3, in which a particular experimental design is employed to enhance the sensitivity of the test. Having developed hypothesis tests for the difference of two means, we then discuss confidence intervals for such differences. Tests concerning two proportions follow. The next section discusses testing the equality of two variances and introduces the F distribution, which is crucial to statistical inference based upon analysis of variance. The chapter concludes with a section on distribution-free approaches.

9.1 TESTING THE DIFFERENCE OF TWO MEANS— VARIANCES KNOWN

Many diverse problems involve the comparison of two populations. These might include comparisons of the Brinell hardnesses of metals, the program execution times of two computer architectures, the life expectancies of competing makes of voltage regulators, the weights of cattle administered a particular drug versus a control group not issued the drug, and the variabilities in the completion rates of two competing production procedures. The common thread to these problems, and many others, is the comparison of corresponding parameters arising from a similar measurement process being applied to different classes of phenomena.

Since our interest is now with two populations, Definition 7.1 regarding random sampling must be suitably adapted. In the present section, and until further notice, it will be assumed that there are two random variables X and Y under consideration and that there are two random samples, $X_1, X_2, \ldots, X_{n_X}$ arising from X and $Y_1, Y_2, \ldots, Y_{n_Y}$ arising from Y. Moreover, the random samples will be assumed to be independent of each other. This means that the entire collection of sample variables is independent.

The two-tailed hypothesis test for the equality of the means μ_X and μ_Y is

$$H_0: \mu_X = \mu_Y$$

$$H_1: \mu_X \neq \mu_Y$$

Rewriting the hypotheses in terms of the difference of the means yields the logically equivalent hypothesis test

$$H_0: \mu_X - \mu_Y = 0$$

$$H_1: \mu_X - \mu_Y \neq 0$$

Since the sample means $\bar{X}$ and $\bar{Y}$ are unbiased estimators of μ_X and μ_Y, respectively, it seems reasonable to utilize the difference $\bar{X} - \bar{Y}$ as a test statistic. The following theorem is basic to this approach.

THEOREM 9.1

Suppose the random variables X and Y possess means μ_X and μ_Y and standard deviations σ_X and σ_Y, respectively. If the sample means $\bar{X}$ and $\bar{Y}$ result from independent random samples of sizes n_X and n_Y, then

$$\mu_{\bar{X}-\bar{Y}} = \mu_X - \mu_Y$$

and

$$\sigma^2_{\bar{X}-\bar{Y}} = \frac{\sigma_X^2}{n_X} + \frac{\sigma_Y^2}{n_Y}$$

Moreover, if X and Y are normally distributed, then so is $\bar{X} - \bar{Y}$.

Proof: The assertion regarding the expected value follows at once from Theorem 7.1 in conjunction with the linearity of E:

$$E[\bar{X} - \bar{Y}] = E[\bar{X}] - E[\bar{Y}] = \mu_X - \mu_Y$$

The assertion regarding the variance follows from Theorems 5.17 and 7.1:

$$\text{Var}\,[\bar{X} - \bar{Y}] = \text{Var}\,[\bar{X}] + (-1)^2\,\text{Var}\,[\bar{Y}] = \frac{\text{Var}\,[X]}{n_X} + \frac{\text{Var}\,[Y]}{n_Y}$$

Finally, the normality of $\bar{X} - \bar{Y}$ in the event that X and Y are normally distributed follows from Theorem 5.9. ∎

Supposing the random variables X and Y to be normally distributed, all conclusions of Theorem 9.1 apply, and therefore it appears reasonable to utilize the test statistic $\bar{X} - \bar{Y}$ when testing the null hypothesis H_0: $\mu_X - \mu_Y = 0$ against the alternative hypothesis H_1: $\mu_X - \mu_Y \neq 0$. Indeed, under the null hypothesis, $E[\bar{X} - \bar{Y}] = 0$. More generally, we can consider the two-tailed hypothesis test

$$H_0\text{: } \mu_X - \mu_Y = \delta_0$$

$$H_1\text{: } \mu_X - \mu_Y \neq \delta_0$$

However, rather than utilize the test statistic $\bar{X} - \bar{Y}$, the test methodology can be standardized (as it was in Section 8.4) by employing the standardized version of $\bar{X} - \bar{Y}$, namely,

$$Z = \frac{\bar{X} - \bar{Y} - \delta_0}{\sqrt{\dfrac{\sigma_X^2}{n_X} + \dfrac{\sigma_Y^2}{n_Y}}}$$

where the standardization is relative to the null hypothesis and possesses a standard normal distribution.

Intuitively, the foregoing test statistic is likely to be near 0 if the null hypothesis is true and likely to be away from 0 if it is false. Thus, for a level of significance α, the acceptance region is the interval

$$A = (-z_{\alpha/2}, z_{\alpha/2})$$

As in the case of the single-parameter Z test of Section 8.4, there exist corresponding acceptance and critical regions for upper and lower one-tailed tests. These are given in Table 9.1, together with the corresponding alternative hypotheses.

In the event that X and Y are not known to be normal, the central limit theorem allows the method to be employed for large samples. In addition, if the variances are unknown, then the sample variances S_X^2 and S_Y^2 can be used to estimate the variances σ_X^2 and σ_Y^2 in the large-sample setting.

EXAMPLE 9.1

In the design of an operating system, CPU scheduling invariably involves an analysis of queues. However, in practice it is possible that the queue discipline will not be a simple one such as first-in-first-out, nor will the arrival and service distributions necessarily fit some standard distribution such as the exponential or Erlang. Given two scheduling routines, it might be more convincing to compare them in actual practice. To make this comparison, suppose the turnaround times (waiting plus service times) of 40 randomly selected program segments are recorded for scheduling method A and 40 others are recorded for scheduling method B. A claim is made that the mean turnaround time for method A is more than 0.4 second faster than for method B. Let X and Y be the random variables describing the populations of turnaround times for methods A and B, respectively. Since $n = n_X = n_Y \geq 30$, the large-sample approach is appropriate. Thus, even if the turnaround times are not normally distributed, the Z test still applies. Also, since the testing is being done on newly designed operating systems, it is improbable that the population variances are known beforehand, and therefore the variances must be estimated by the respective empirical variances. We proceed according to our usual hypothesis testing format.

1. *Hypothesis test.* H_0: $\mu_X - \mu_Y \geq -0.4$
   $\quad\quad\quad\quad\quad\quad H_1$: $\mu_X - \mu_Y < -0.4$
2. *Level of significance.* $\alpha = 0.05$

TABLE 9.1   IMPORTANT INFORMATION RELATING TO THE Z TEST OF THE DIFFERENCE BETWEEN TWO MEANS

Null Hypothesis: H_0: $\mu_X - \mu_Y = \delta_0$

Test Statistic: $Z = \dfrac{\bar{X} - \bar{Y} - \delta_0}{\sigma_{\bar{X} - \bar{Y}}}$

where

$$\sigma_{\bar{X} - \bar{Y}}^2 = \frac{\sigma_X^2}{n_X} + \frac{\sigma_Y^2}{n_Y}$$

Alternative Hypothesis (Two-Tailed): H_1: $\mu_X - \mu_Y \neq \delta_0$

Critical Region: $C = \{z\colon z \leq -z_{\alpha/2} \text{ or } z \geq z_{\alpha/2}\}$

Values of β: $\beta(\delta) = \Phi\left(z_{\alpha/2} + \dfrac{\delta_0 - \delta}{\sigma_{\bar{X} - \bar{Y}}}\right) - \Phi\left(-z_{\alpha/2} + \dfrac{\delta_0 - \delta}{\sigma_{\bar{X} - \bar{Y}}}\right)$

Expression of n in terms of α and $\beta(\delta_1)$ when $n = n_X = n_Y$:

$$n \approx \frac{(\sigma_X^2 + \sigma_Y^2)(z_{\alpha/2} + z_\beta)^2}{(\delta_1 - \delta_0)^2}$$

Alternative Hypothesis (Upper One-Tailed): H_1: $\mu_X - \mu_Y > \delta_0$

Critical Region: $C = \{z\colon z \geq z_\alpha\}$

Values of β: $\beta(\delta) = \Phi\left(z_\alpha + \dfrac{\delta_0 - \delta}{\sigma_{\bar{X} - \bar{Y}}}\right)$

Expression of n in terms of α and $\beta(\delta_1)$ when $n = n_X = n_Y$:

$$n = \frac{(\sigma_X^2 + \sigma_Y^2)(z_\alpha + z_\beta)^2}{(\delta_1 - \delta_0)^2}$$

Alternative Hypothesis (Lower One-Tailed): H_1: $\mu_X - \mu_Y < \delta_0$

Critical Region: $\{z\colon z \leq -z_\alpha\}$

Values of β: $\beta(\delta) = 1 - \Phi\left(-z_\alpha + \dfrac{\delta_0 - \delta}{\sigma_{\bar{X} - \bar{Y}}}\right)$

Expression of n in terms of α and $\beta(\delta_1)$ when $n = n_X = n_Y$: Same as upper one-tailed test.

3. *Test statistic.* $Z = \dfrac{\bar{X} - \bar{Y} + 0.4}{\sigma_{\bar{X} - \bar{Y}}}$

4. *Sample sizes.* $n_X = n_Y = 40$

5. *Critical region.* $z \leq -z_{0.05} = -1.65$

6. *Sample value.* Suppose the 40 observations of method A yield mean $\bar{x} = 3.64$ and standard deviation $s_X = 0.53$ and the 40 observations of method B yield $\bar{y} = 4.25$ and $s_Y = 0.44$. Then

$$\sigma_{\bar{X}-\bar{Y}} \approx \sqrt{\frac{s_X^2}{n_X} + \frac{s_Y^2}{n_Y}} = \sqrt{\frac{0.28 + 0.19}{40}} = 0.11$$

and

$$z \approx \frac{\bar{x} - \bar{y} - \delta_0}{0.11} = \frac{3.64 - 4.25 + 0.40}{0.11} = -1.91$$

7. *Decision.* Since $z < -1.65$, reject the null hypothesis and conclude that scheduling method A results in a mean turnaround time of more than 0.4 second faster than method B.

Type II error concerns the probability that the test statistic falls in the acceptance region given the truth of the alternative hypothesis. In the case of an upper one-tailed test of $\mu_X - \mu_Y$ with level of significance α, for $\delta > \delta_0$,

$$\beta(\delta) = P\left(\frac{\bar{X} - \bar{Y} - \delta_0}{\sigma_{\bar{X}-\bar{Y}}} < z_\alpha; \mu_X - \mu_Y = \delta\right)$$

$$= P(\bar{X} - \bar{Y} < \delta_0 + z_\alpha \sigma_{\bar{X}-\bar{Y}}; \mu_X - \mu_Y = \delta)$$

$$= P\left(\frac{\bar{X} - \bar{Y} - \delta}{\sigma_{\bar{X}-\bar{Y}}} < z_\alpha - \frac{\delta - \delta_0}{\sigma_{\bar{X}-\bar{Y}}}; \mu_X - \mu_Y = \delta\right)$$

$$= \Phi\left(z_\alpha - \frac{\delta - \delta_0}{\sigma_{\bar{X}-\bar{Y}}}\right)$$

where the last equality follows from the fact that, under the assumption $\mu_X - \mu_Y = \delta$, the random variable $(\bar{X} - \bar{Y} - \delta)/\sigma_{\bar{X}-\bar{Y}}$ possesses a standard normal distribution. Table 9.1 provides the values of β in the cases of both one- and two-tailed tests.

EXAMPLE 9.2

For the hypothesis test of Example 9.1, the probability of wrongly accepting H_0 in the event that the mean turnaround time using scheduling method A is actually 0.5 second faster than the mean turnaround time of method B is $\beta(-0.5)$. Using the expression for β given in Table 9.1 for the lower one-tailed Z test gives

$$\beta(-0.5) \approx 1 - \Phi\left(-1.65 + \frac{-0.4 + 0.5}{0.11}\right) = 1 - \Phi(-0.74) = 0.7704$$

Now suppose the sample sizes are equal, namely, $n_X = n_Y = n$. Then

$$\sigma_{\bar{X}-\bar{Y}}^2 = \frac{\sigma_X^2 + \sigma_Y^2}{n}$$

and, for $\delta_1 > \delta_0$, in the upper one-tailed case the size of type II error is

$$\beta(\delta_1) = \Phi\left(z_\alpha - \frac{(\delta_1 - \delta_0)\sqrt{n}}{\sqrt{\sigma_X^2 + \sigma_Y^2}}\right)$$

Letting $\beta = \beta(\delta_1)$ and recognizing that $\beta = P(Z < -z_\beta)$, we obtain

$$-z_\beta = z_\alpha - \frac{(\delta_1 - \delta_0)\sqrt{n}}{\sqrt{\sigma_X^2 + \sigma_Y^2}}$$

so that

$$n = \frac{(\sigma_X^2 + \sigma_Y^2)(z_a + z_\beta)^2}{(\delta_1 - \delta_0)^2}$$

A similar computation yields the same result in the lower one-tailed case when $n = n_X = n_Y$ (see Table 9.1). In the two-tailed case, it can be shown that the approximation

$$n \approx \frac{(\sigma_X^2 + \sigma_Y^2)(z_{\alpha/2} + z_\beta)^2}{(\delta_1 - \delta_0)^2}$$

is reasonably good so long as

$$P\left(Z < -z_{\alpha/2} - \frac{|\delta_1 - \delta_0|}{\sigma_{\bar{X} - \bar{Y}}}\right)$$

is small in comparison to β. Recall that a similar approximation was applicable in the Z test of Section 8.4. These expressions for n (in the case of equal sample sizes) are summarized in Table 9.1.

As in the one-parameter case, OC curves can be useful for determining β given the sample size and, conversely, for determining the appropriate sample size given a value of β for an alternative difference. Assuming $n_X = n_Y$, the OC-curve families for testing the difference between two means of normal random variables are the same as for the Z test of Section 8.4. Specifically, Figures 8.15 and 8.16 give the OC-curve families for the two-tailed test at significance levels 0.05 and 0.01, respectively, and Figures 8.18 and 8.19 give the OC-curve families for the one-tailed tests at significance levels 0.05 and 0.01, respectively. Once again, owing to the symmetry of the test statistic, there is no distinction between upper and lower one-tailed tests. In the present case, the normalization for reading the tables is given by

$$d = \frac{|\delta - \delta_0|}{\sqrt{\sigma_X^2 + \sigma_Y^2}}$$

EXAMPLE 9.3

It is desired to test the claim that two brands of paint possess the same mean drying time. It will be assumed that drying times are normally distributed and it is necessary to keep the probability of erroneously accepting equal mean drying times at the alternative difference of 3.5 minutes to $\beta(3.5) = 0.2$. From experience it can be safely assumed that both drying-time variances are less than 8. Let X and Y be the random variables representing the two drying-time populations. Given $\alpha = 0.05$, we desire the appropriate sample size under the assumption $n_X = n_Y = n$. Using the OC-curve family of Figure 8.15 with

$$d = \frac{3.5}{\sqrt{8 + 8}} = 0.875$$

gives $n = 10$. Applying the approximation formula given in Table 9.1 yields

$$n \approx \frac{(8 + 8)(1.96 + 0.84)^2}{(3.5)^2} = 10.24$$

Note the dilemma posed by the result $n = 10$ in Example 9.3. Since the variances are unknown, a sample size of 10 will not be sufficient to employ the Z test, because the variances will have to be estimated from the samples. There are two choices available for the experimenter. He or she could make 30 observations of each drying time. This would certainly suffice to keep $\beta(3.5)$ sufficiently low; indeed, with $d = 0.875$, the OC curve of Table 8.15 corresponding to $n = 30$ gives a reading of $\beta(3.5) \approx 0$. Rather than pursue such a wasteful course, it might be better to employ the two-sample t test to be introduced in the next section.

9.2 TESTING THE DIFFERENCE OF TWO MEANS—VARIANCES UNKNOWN

In those circumstances where the variances of the normal random variables X and Y are unknown and the sample sizes are not sufficiently large to justify estimation by the sample variance, the methods of Section 9.1 are not applicable. As has been noted elsewhere, it is common for the variances to be unknown, so that variance-known procedures often depend upon large samples. In the present section, a two-sample test will be developed in the case where both random variables are normal and possess a common unknown variance. Specifically, the difference of two means will be tested under the conditions that $X_1, X_2, \ldots, X_{nx}$ and $Y_1, Y_2, \ldots, Y_{ny}$ are independent random samples arising from two normally distributed population random variables X and Y, respectively, where $\sigma_X = \sigma_Y$. Relaxation of the requirements will be discussed subsequently.

As demonstrated in Theorem 9.1, the difference $\mu_X - \mu_Y$ possesses the unbiased estimator $\bar{X} - \bar{Y}$; however, in the present context this statistic cannot be standardized to arrive at a standard normal test statistic. The situation is analogous to that encountered in Section 8.5, where the approach was to replace the variance of the population random variable by the sample variance, the result being a t test. In the present setting there is a slight difference. Letting S_X^2 and S_Y^2 denote the sample variances derived from the random samples $X_1, X_2, \ldots, X_{nx}$ and $Y_1, Y_2, \ldots, Y_{ny}$, respectively, a better estimator of the common variance can be obtained by simultaneously utilizing the data of both samples. The result is the pooled estimator defined next.

DEFINITION 9.1

Pooled Sample Variance. Suppose S_X^2 and S_Y^2 are the sample variances resulting from independent random samples of sizes n_X and n_Y arising the normal random variables X and Y, respectively, where X and Y possess the common variance σ^2. The estimator of σ^2 defined by

$$S_p^2 = \frac{n_X - 1}{n_X + n_Y - 2} S_X^2 + \frac{n_Y - 1}{n_X + n_Y - 2} S_Y^2$$

is called the pooled sample variance. The **pooled sample standard deviation** is S_p.

The pooled sample variance is an estimator of σ^2 that is formed from a weighted average of the individual sample variances. It is evident from the definition that the sample variance corresponding to the larger sample is given more weight. Should $n_X = n_Y$, then the pooled variance reduces to the unweighted average of the sample variances.

THEOREM 9.2

Under the conditions of Definition 9.1, the pooled sample variance is an unbiased estimator of the common variance.

Proof: Both S_X^2 and S_Y^2 are unbiased estimators. Thus, the linearity of the expected-value operator yields

$$E[S_p^2] = \frac{1}{n_X + n_Y - 2} [(n_X - 1)E[S_X^2] + (n_Y - 1)E[S_Y^2]]$$

$$= \frac{1}{n_X + n_Y - 2} [(n_X - 1)\sigma_X^2 + (n_Y - 1)\sigma_Y^2]$$

$$= \sigma^2$$

the last equality following from the fact that $\sigma_X = \sigma_Y = \sigma$. ∎

In Section 8.5 the test statistic for the small-sample test of the mean was obtained by replacing the standard deviation in the standardized version of the sample mean by the sample standard deviation. In the present setting, the standardized version of $\bar{X} - \bar{Y}$ is

$$\frac{\bar{X} - \bar{Y} - \mu_{\bar{X}-\bar{Y}}}{\sigma_{\bar{X}-\bar{Y}}} = \frac{\bar{X} - \bar{Y} - (\mu_X - \mu_Y)}{\sigma \sqrt{\dfrac{1}{n_X} + \dfrac{1}{n_Y}}}$$

where σ^2 is the common variance. To obtain the desired estimator, the unknown parameter σ is replaced by its pooled estimator S_p. The following theorem provides the theoretical justification for the test methodology.

THEOREM 9.3

Suppose X and Y are normally distributed random variables possessing common variance σ^2 and $X_1, X_2, \ldots, X_{n_X}$ and $Y_1, Y_2, \ldots, Y_{n_Y}$ are independent random samples arising from X and Y, respectively. Then the random variable

$$T = \frac{\bar{X} - \bar{Y} - (\mu_X - \mu_Y)}{S_p \sqrt{\dfrac{1}{n_X} + \dfrac{1}{n_Y}}}$$

possesses a t distribution with $n_X + n_Y - 2$ degrees of freedom.

Proof: Since $\sigma = \sigma_X = \sigma_Y$, Theorem 7.9 shows the random variables

$$\frac{(n_X - 1)S_X^2}{\sigma^2} \qquad \text{and} \qquad \frac{(n_Y - 1)S_Y^2}{\sigma^2}$$

possess chi-square distributions with $n_X - 1$ and $n_Y - 1$ degrees of freedom, respectively. Since the assumptions concerning the sampling procedure guarantee that these two variables are independent, Theorem 5.7 shows that their sum,

$$U = \frac{(n_X + n_Y - 2)S_p^2}{\sigma^2}$$

possesses a chi-square distribution with $n_X + n_Y - 2$ degrees of freedom. Let Z denote the standardized version of $\bar{X} - \bar{Y}$ (which was expressed in the paragraph prior to the statement of the theorem). It can be shown that U and Z are independent. Thus, according to Definition 7.16,

$$\frac{Z}{\sqrt{\dfrac{U}{n_X + n_Y - 2}}}$$

possesses a t distribution with $n_X + n_Y - 2$ degrees of freedom. Straightforward algebra shows this t variable is equal to T. ∎

Recall from Section 7.8.3 that the number of degrees of freedom is the number of freely determined deviations upon which the variance estimator is based. Writing the pooled sample variance as

$$S_p^2 = \frac{1}{n_X + n_Y - 2} \left[\sum_{i=1}^{n_X} (X_i - \bar{X})^2 + \sum_{j=1}^{n_Y} (Y_j - \bar{Y})^2 \right]$$

shows there to be $n_X - 1$ freely determined deviations $X_i - \bar{X}$ and $n_Y - 1$ freely determined deviations $Y_j - \bar{Y}$. Thus, the conclusion of the theorem is consistent with the notion of freely determined deviations.

With the form of a test statistic decided upon, so long as the conditions of Theorem 9.3 are satisfied, hypothesis tests concerning the difference of two means can be constructed in the usual t-test manner. For instance, the two-tailed test

$$H_0: \mu_X - \mu_Y = \delta_0$$
$$H_1: \mu_X - \mu_Y \neq \delta_0$$

utilizes the test statistic

$$T = \frac{\bar{X} - \bar{Y} - \delta_0}{S_p \sqrt{\dfrac{1}{n_X} + \dfrac{1}{n_Y}}}$$

For level of significance α, the acceptance region is the interval

$$(-t_{\alpha/2, n_X + n_Y - 2}, \ t_{\alpha/2, n_X + n_Y - 2})$$

Table 9.2 provides a summary of the two-sample t test, including the upper and lower one-tailed tests.

EXAMPLE 9.4

Two makes of automobile, say A and B, are being tested to see which best withstands a certain type of crash, such as a head-on collision into a retaining wall. It will be assumed that the measure of damage is based upon a scale running from 0 to 100, the scale having been devised to take into account various types of damage to the automobile. Obviously, such testing is expensive and the number of tests must be kept small. If X and Y measure the damage to cars of types A and B, respectively, so long as X and Y are (essentially) normally distributed and the variances are (essentially) the same, the two-sample t test is applicable. It is necessary to check a claim by the manufacturer of model A that model A withstands the crash better than model B.

TABLE 9.2 IMPORTANT INFORMATION RELATING TO THE t TEST OF THE DIFFERENCE BETWEEN TWO MEANS

Null Hypothesis: H_0: $\mu_X - \mu_Y = \delta_0$

Test Statistic: $T = \dfrac{\bar{X} - \bar{Y} - \delta_0}{S_p \sqrt{\dfrac{1}{n_X} + \dfrac{1}{n_Y}}}$

Alternative Hypothesis (Two-Tailed): H_1: $\mu_X - \mu_Y \neq \delta_0$

Critical Region:

$$C = \{t: t \leq -t_{\alpha/2, n_X + n_Y - 2} \text{ or } t \geq t_{\alpha/2, n_X + n_Y - 2}\}$$

Alternative Hypothesis (Upper One-Tailed): H_1: $\mu_X - \mu_Y > \delta_0$

Critical Region: $C = \{t: t \geq t_{\alpha, n_X + n_Y - 2}\}$

Alternative Hypothesis (Lower One-Tailed): H_1: $\mu_X - \mu_Y < \delta_0$

Critical Region: $C = \{t: t \leq -t_{\alpha, n_X + n_Y - 2}\}$

1. *Hypothesis test.* H_0: $\mu_X - \mu_Y \geq 0$
   $\qquad\qquad\qquad H_1$: $\mu_X - \mu_Y < 0$
2. *Level of significance.* $\alpha = 0.05$
3. *Test statistic.* $T = \dfrac{\bar{X} - \bar{Y}}{S_p \sqrt{\dfrac{1}{n_X} + \dfrac{1}{n_Y}}}$
4. *Sample sizes.* $n_X = 5$ and $n_Y = 4$.
5. *Critical region.* The test statistic possesses $n_X + n_Y - 2$ degrees of freedom. From Table A.6, the critical value is found to be $t_{0.05,7} = 1.895$. Thus,

$$C = \{t: t \leq -1.895\}$$

6. *Sample value.* Suppose the test results give the mean $\bar{x} = 43.3$ and variance $s_X^2 = 4.5$ for model A and the mean $\bar{y} = 45.2$ and variance $s_Y^2 = 3.8$ for model B. The value of the pooled variance is then

$$s_p^2 = \frac{4}{7} \times 4.5 + \frac{3}{7} \times 3.8 = 4.20$$

and the value of the test statistic is

$$t = \frac{43.3 - 45.2}{(2.05) \sqrt{\dfrac{1}{5} + \dfrac{1}{4}}} = -1.38$$

7. *Decision.* Since $t > -1.895$, the null hypothesis is not rejected. Thus, there is no legitimacy (at level of significance 0.05) to the claim that model A better withstands the type of crash under study than does model B.

When using the two-sample t test, type II error can be estimated from the appropriate OC-curve family. Moreover, given α and $\beta(\delta_1)$, for some alternative difference δ_1, OC curves can be employed to determine a suitable sample size. In both cases the restriction $n_X = n_Y$ will apply. Figures 8.21 and 8.22 apply to the two-tailed test for significance levels 0.05 and 0.01, respectively, and Figures 8.23 and 8.24 apply to the corresponding one-tailed tests. In all figures the normalized difference

$$d = \frac{|\delta - \delta_0|}{2\sigma}$$

is employed for the alternative value δ. As in the case of the one-sample t test, σ is not known and some reasonable (conservative) estimate of σ must be utilized. In the two-sample case (with $n_X = n_Y$), having been given $\beta(\delta_1)$ and having read n from the OC-curve family, the desired sample size is given by

$$n_X = n_Y = \frac{n + 1}{2}$$

Similarly, given $n_X = n_Y$, the size of type II error at an alternative difference δ can be determined from the OC-curve family by simply locating the curve parameterized by $n = 2n_X - 1$ and reading the value of the curve at d.

EXAMPLE 9.5

In Example 9.3 it was discovered that a sample size of 10 was necessary to have $\beta(3.5) = 0.2$. Subsequent to the example it was noted that a lack of knowledge regarding the variances of X and Y would force overly large samples. As in Example 9.3, suppose we believe the variances to be equal, with 8 being a conservative estimate. Then Figure 8.21, with

$$d = \frac{3.5}{2\sqrt{8}} = 0.62$$

yields $n = 23$. Thus, the required common sample size is

$$n_X = n_Y = \frac{23 + 1}{2} = 12$$

Throughout the section it has been assumed that X and Y are normally distributed and possess the same variance. As in the case of a single sample, the two-sample t test is rather robust with regard to violations of the normality assumption. Nonetheless, should either distribution be markedly nonnormal, then the test cannot be employed. There is also some degree of robustness relative to the assumption of equal variances; however, here the problem is somewhat more delicate. If the variances are suspected of being more than marginally not equal, then it might be best to forego the two-sample t-test.

A test that can be employed in the event the population random variables are normal but possess different variances is the **Smith-Satterthwaite** test. For the null hypothesis

$$H_0: \mu_X - \mu_Y = \delta_0$$

it employs the test statistic

$$T' = \frac{\bar{X} - \bar{Y} - \delta_0}{\sqrt{\dfrac{S_X^2}{n_X} + \dfrac{S_Y^2}{n_Y}}}$$

Given H_0, the distribution of T' is approximated by a t distribution with

$$\nu = \frac{\left(\dfrac{S_X^2}{n_X} + \dfrac{S_Y^2}{n_Y}\right)^2}{\dfrac{\left(\dfrac{S_X^2}{n_X}\right)^2}{n_X - 1} + \dfrac{\left(\dfrac{S_Y^2}{n_Y}\right)^2}{n_Y - 1}}$$

Since ν is not usually an integer, it must be rounded up when applying the test. To use the Smith-Satterthwaite test, proceed exactly as in the case of the two-sample t test, replacing T by T' and $n_X + n_Y - 2$ by ν. It should be recognized that T' is only approximated by a t distribution and therefore the corrresponding test lacks the precision attained by the t test (in the case where $\sigma_X = \sigma_Y$).

EXAMPLE 9.6

Various techniques may be employed to minimize the background noise in a production environment. To compare background-noise levels at two different locations, 15 random measurements will be taken at each plant site, the samples themselves being independent. X and Y will be the random variables representing the measurement populations at locations A and B, respectively. Because of changing conditions, it is very unlikely that the variances of X and Y are known. In addition, if conditions at the two sites differ, especially with respect to the types of machines in operation, then it might be risky to assume equal variances. Thus, if it is desired to show that the mean noise level at site A is more than 3 decibels greater than the mean noise level at site B, an upper-one-tailed Smith-Satterthwaite test is appropriate.

1. *Hypothesis test.* H_0: $\mu_X - \mu_Y \leq 3$
   $\qquad\qquad\qquad H_1$: $\mu_X - \mu_Y > 3$
2. *Level of significance.* $\alpha = 0.05$
3. *Test statistic.* T'
4. *Sample sizes.* $n_X = n_Y = 15$
5. *Critical region.* To obtain the critical value, it is first necessary to derive the number of degrees of freedom. Suppose the 15 observations of each random variable yield the empirical variances $s_X^2 = 2.3$ and $s_Y^2 = 4.1$. Then

$$\frac{\left(\dfrac{2.3}{15} + \dfrac{4.1}{15}\right)^2}{\dfrac{\left(\dfrac{2.3}{15}\right)^2}{14} + \dfrac{\left(\dfrac{4.1}{15}\right)^2}{14}} = 25.9$$

so that rounding up gives $\nu = 26$. The corresponding critical value is $t_{0.05,26} = 1.706$.
6. *Sample value.* Suppose the empirical means obtained from the observations are $\bar{x} = 82.4$ and $\bar{y} = 78.3$. Then

$$t' = \frac{82.4 - 78.3 - 3.0}{\sqrt{\dfrac{2.3}{15} + \dfrac{4.1}{15}}} = 1.684$$

7. *Decision.* Since $t' < 1.706$, the null hypothesis is not rejected.

9.3 TESTING THE DIFFERENCE OF MEANS FOR PAIRED OBSERVATIONS

To this point it has been assumed that X and Y are normally distributed random variables and testing has been carried out under the assumption that the random samples are independent. The latter assumption is not always appropriate. In some instances, sampling constraints prohibit the assumption of independent samples; in others, independence is not a desirable experimental design.

Suppose we wish to compare the performance of two processes, say $\mathcal{P}_X$ and $\mathcal{P}_Y$. These could be chemical, electrical, computer, or sociological processes. Data are obtained by applying each process to a number of experimental units, the result being two measurement samples. Let X and Y be the random variables describing the populations of measurements relative to processes $\mathcal{P}_X$ and $\mathcal{P}_Y$, respectively. If the processes are applied under experimental conditions that are essentially the same throughout, then the selection of independent random samples has the desired effect of not giving extra weight to any portion of either population. However, if measurements are taken under widely varying conditions, then extensive variation is introduced into both sets of measurements. If the intent is to test the difference of the means μ_X and μ_Y, then this additional within-sample variation can obscure a significant difference between the samples. A way to enhance the sensitivity of the testing procedure is to analyze the differences of **paired observations.** In this approach, pairs of measurements, one for $\mathcal{P}_X$ and one for $\mathcal{P}_Y$, are taken under similar conditions and the differences are recorded, the net effect being to focus attention on the difference of the processes rather than on the processes themselves.

As an illustration of the type of experimental design we have in mind, consider the comparison of two finishing treatments, $\mathcal{P}_X$ and $\mathcal{P}_Y$, designed to increase the longevity of steel cutting-blades. If our desire is to compare the treatments over a range of blades, then it is very possible there will be a great deal of within-sample variability in both the sample for X and the sample for Y, simply because longevity will depend not only on the treatments but also on the types of blades chosen for the samples. Because the two-sample t statistic has the pooled standard deviation in its denominator, large values of S_X and S_Y will desensitize the statistic to differences between the samples and thereby increase the likelihood of not detecting a consequential difference between the longevity-measurement means. To increase the sensitivity of the test, it might be better to randomly select n blades, apply each finishing treatment to a portion of the same blade, test the n blades to obtain pairs of measurements, and then consider the sample of differences. In this approach, both treatments are applied to each experimental unit (blade), the result being that the within-sample variability due to blade characteristics is mitigated by the design of the experiment.

To accomplish **paired-sample** testing, consider a collection of random variables $X_1, X_2, \ldots, X_n$, each identically distributed with a random variable X possessing unknown mean μ_X and unknown variance σ_X^2, and another collection of random variables $Y_1, Y_2, \ldots, Y_n$, each identically distributed with a random variable Y possessing unknown mean μ_Y and unknown variance σ_Y^2. For $k = 1, 2, \ldots, n$, the difference

$$D_k = X_k - Y_k$$

is assumed to possess a normal distribution, and the differences $D_1, D_2, \ldots, D_n$ are assumed to be independent. Since each D_k is identically distributed with the random variable $D = X - Y$,

$$\mu_{D_k} = \mu_D = \mu_X - \mu_Y$$

for $k = 1, 2, \ldots, n$; however, since X_k and Y_k are not assumed to be independent,

$$\sigma_{D_k}^2 = \sigma_{X_k}^2 + \sigma_{Y_k}^2 - 2\,\mathrm{Cov}\,[X_k, Y_k]$$

and $\mathrm{Cov}\,[X_k, Y_k]$ need not be zero.

In terms of D, a test concerning the difference of the means μ_X and μ_Y has the null hypothesis

$$H_0: \ \mu_D = \delta_0$$

and one of three alternative hypotheses, these being the two-tailed, upper one-tailed, and lower one-tailed alternatives. Under the conditions imposed upon $D_1, D_2, \ldots, D_n$, the t test of Section 9.2 applies and, in the present context, is called the **paired-sample t test.** Any decision applies directly to the difference-population random variable D. The test statistic is

$$T = \frac{\bar{D} - \delta_0}{S_D/\sqrt{n}}$$

where $\bar{D}$ is the sample mean of the differences $D_k = X_k - Y_k$, S_D is the sample standard deviation, and T possesses $n - 1$ degrees of freedom. Table 8.3 applies with μ_D in place of μ, δ_0 in place of μ_0, D in place of X, and S_D in place of S.

EXAMPLE 9.7 Any measurement device or gage must work over some range of measurements. In comparing two such devices, say A and B, the measurement variables X and Y can possess high variances resulting from a wide class of objects over which the measurement populations are considered. In comparing A and B over the range of their application, we do not wish to let the detection of a meaningful mean difference be obscured by the individual variations of X and Y. As an illustration, suppose the goal is to detect whether or not there is a significant difference between two ammeters. Eight currents of varying amperage will be measured with both ammeters, and it will be assumed that the population of all possible differences is normally distributed. We proceed by applying the paired t test to the difference-population random variable D.

1. *Hypothesis test.* $H_0: \ \mu_D = 0$
$$H_1: \ \mu_D \neq 0$$
2. *Level of significance.* $\alpha = 0.05$
3. *Test statistic.* $T = \dfrac{\bar{D}}{S_D/\sqrt{n}}$
4. *Sample sizes.* $n = 8$
5. *Critical region.* According to Table A.6, $t_{0.025,7} = 2.365$. Consequently, the critical region is

$$C = (-\infty, -2.365] \cup [2.365, \infty)$$

TABLE 9.3 CURRENT MEASUREMENTS ARISING FROM TWO AMMETERS

x_k	y_k	d_k
12.33	12.55	-0.22
2.45	2.47	-0.02
6.26	6.42	-0.16
1.84	1.85	-0.01
7.62	7.74	-0.12
10.01	10.20	-0.19
12.02	12.21	-0.19
4.11	4.16	-0.05
$\bar{x} = 7.08$	$\bar{y} = 7.20$	$\bar{d} = -0.12$
$s_X = 4.127$	$s_Y = 4.206$	$s_D = 0.083$

6. *Sample value*. Suppose the 8 pairs of amperages, listed together with their differences and relevant empirical moments in Table 9.3, are obtained by observation. Then the paired t statistic is

$$t = \frac{\bar{d}}{s_D/\sqrt{8}} = -4.089$$

7. *Decision*. Since $t < -2.365$, the null hypothesis is rejected and it is concluded that the mean of the differences is not zero.

Examination of Example 9.7 is quite revealing insofar as variational effects are concerned. By focusing on the differences, the test has not been desensitized by measurement variation resulting from widely varying conditions. Only the variation inherent in D has been taken into account. To better see the benefit of the paired t test, suppose the 8 A weights had been taken independently of the 8 B weights and suppose the same 16 observations had resulted. Assuming both X and Y to be normally distributed, the t test of Section 9.2 applies with $n_X = n_Y = n = 8$. The corresponding value of the test statistic is

$$t = \frac{\bar{x} - \bar{y}}{s_p \sqrt{\dfrac{2}{n}}}$$

Since there are $2n - 2 = 14$ degrees of freedom, the critical value is $t_{0.025,14} = 2.145$. The pooled variance is

$$s_p^2 = \frac{1}{2}(s_X^2 + s_Y^2) = 17.361$$

and

$$t = \frac{-0.12}{(4.167)\sqrt{\dfrac{1}{4}}} = -0.058$$

Thus, the unpaired t test does not yield a significant value of the test statistic. The variation in X and Y has masked the significance of the differences.

EXAMPLE 9.8

Underlying an automatic recognition system is some set of algorithms by which the input image is processed, significant features of the processed image are identified, and a decision is made as to the identification of the original image. Although the hardware can usually be assumed to be environmentally independent, that is certainly not the case with the software. The success of any recognition software depends on many external conditions, such as light (in the case of optical systems), occlusions of the target, and background texture. Consider comparing two recognition systems by testing each under ten distinct sets of conditions. In each setting, each system will be confronted with 50 scenes containing objects it is designed to recognize. The number of successful identifications at each site will be recorded. Based upon these tests, it is hoped to prove that system $\mathcal{P}_X$ outperforms system $\mathcal{P}_Y$. Because of the significant variation in both measurement populations, a paired approach is to be utilized. Although the data will be discrete, it will be assumed that the difference distribution $D = X - Y$ is closely approximated by a normal distribution, so that the paired t test can be employed.

1. *Hypothesis test.* H_0: $\mu_D \leq 0$

   H_1: $\mu_D > 0$

2. *Level of significance.* $\alpha = 0.05$

3. *Test statistic.* $T = \dfrac{\bar{D}}{S_D/\sqrt{n}}$

4. *Sample size.* $n = 10$

5. *Critical region.* Since $t_{0.05,9} = 1.833$, the critical region is

$$C = \{t: t \geq 1.833\}$$

6. *Sample value.* Based upon the information in Table 9.4,

$$t = \frac{3.7}{3.23/\sqrt{10}} = 3.618$$

7. *Decision.* Since $t > 1.833$, the alternative hypothesis is accepted, and it is concluded that system $\mathcal{P}_X$ is superior to $\mathcal{P}_Y$.

TABLE 9.4   RECOGNITION DATA FOR SYSTEMS $\mathcal{P}_X$ AND $\mathcal{P}_Y$

x_k	y_k	d_k
47	38	9
15	16	−1
42	36	6
38	36	2
25	24	1
46	39	7
35	30	5
12	12	0
42	39	3
40	35	5
$\bar{x} = 34.2$	$\bar{y} = 30.5$	$\bar{d} = 3.7$
$s_X = 12.559$	$s_Y = 9.869$	$s_D = 3.234$

Note in Example 9.8 that there is great variability of the test data from site to site. Indeed, at three sites the system successes are within one of each other. Had the same data been recorded through a random selection of sites and the unpaired t test applied, then a different decision would have been reached. Indeed, using the information in Table 9.4, the pooled variance is

$$s_p^2 = \frac{1}{2}(s_X^2 + s_Y^2) = \frac{1}{2}(157.728 + 97.397) = 127.563$$

and the unpaired t statistic is

$$t = \frac{\bar{x} - \bar{y}}{s_p\sqrt{2/10}} = \frac{3.7}{11.294/\sqrt{0.2}} = 0.733$$

For the unpaired approach, there are $20 - 2 = 18$ degrees of freedom, and the critical value is $t_{0.05,18} = 1.734$. Thus, the null hypothesis is not rejected.

To get a deeper understanding of the relative benefits of the unpaired and paired approaches, consider the unpaired and paired test statistics

$$T_1 = \frac{\bar{X} - \bar{Y}}{S_p\sqrt{2/n}}$$

and

$$T_2 = \frac{\bar{D}}{S_D/\sqrt{n}}$$

respectively, where there are n observations of each variable. For an upper one-tailed test with significance level α, rejection occurs in the unpaired test if

$$\frac{\bar{x} - \bar{y}}{s_p\sqrt{2/n}} \geq t_{\alpha,2n-2}$$

In the paired test, since $\bar{d} = \bar{x} - \bar{y}$, rejection occurs if

$$\frac{\bar{x} - \bar{y}}{s_D/\sqrt{n}} \geq t_{\alpha,n-1}$$

For any fixed α, the critical values of the t distribution decrease with increasing degrees of freedom. Thus,

$$t_{\alpha,n-1} > t_{\alpha,2n-2}$$

From the standpoint of critical values, the test statistic need not be as large in the unpaired case. However, if there is large variation within the individual measurement processes X and Y and yet at the same time X and Y are highly correlated, then S_D tends to be significantly smaller than $\sqrt{2}\, S_p$. Therefore, under such circumstances, the paired t statistic is greater than the unpaired t statistic. The result is greater sensitivity to detection on the part of the paired test.

In sum, if there is little correlation between the population random variables and these variables do not possess large variation, then it is usually better to design the experiment around independent random samples. On the other hand, if the population random variables possess large variation and are highly correlated, then it is usually better to employ paired observations. As stated, the criteria tend to represent extreme positions and therefore can serve only as general guidelines. Another problem is that we often know neither the correlation coefficient of the population random variables nor their variances.

Insofar as Examples 9.7 and 9.8 are concerned, in both cases the population variables possess significant variation. In Example 9.7 the samples also come from populations having a strong linear correlation ($\rho \approx 1$). While the correlation in Example 9.8 is more moderate ($\rho \approx 0.6$), it is still somewhat substantial.

9.4 CONFIDENCE INTERVALS FOR THE DIFFERENCE OF TWO MEANS

Based upon the statistics developed for two-sample hypothesis tests, confidence intervals can be given for the difference of two means. We begin with a confidence interval for $\mu_X - \mu_Y$ in the case where the random variables are normally distributed and the variances are known. Confidence intervals involving the unpaired and paired t statistics follow.

THEOREM 9.4

Suppose X and Y are normally distributed random variables possessing known variances σ_X^2 and σ_Y^2. If $X_1, X_2, \ldots, X_{nx}$ and $Y_1, Y_2, \ldots, Y_{ny}$ comprise independent random samples arising from X and Y, respectively, then a $(1 - \alpha)100\%$ confidence interval for $\mu_X - \mu_Y$ is given by

$$\left(\bar{x} - \bar{y} - z_{\alpha/2} \sqrt{\frac{\sigma_X^2}{n_X} + \frac{\sigma_Y^2}{n_Y}}, \ \bar{x} - \bar{y} + z_{\alpha/2} \sqrt{\frac{\sigma_X^2}{n_X} + \frac{\sigma_Y^2}{n_Y}} \right)$$

Proof: The standardization of $\bar{X} - \bar{Y}$ possesses a standard normal distribution. Thus,

$$P\left(-z_{\alpha/2} < \frac{\bar{X} - \bar{Y} - (\mu_X - \mu_Y)}{\sigma_{\bar{X} - \bar{Y}}} < z_{\alpha/2} \right) = 1 - \alpha$$

Rearrangement of the inequality gives

$$P(\bar{X} - \bar{Y} - z_{\alpha/2}\sigma_{\bar{X} - \bar{Y}} < \mu_X - \mu_Y < \bar{X} - \bar{Y} + z_{\alpha/2}\sigma_{\bar{X} - \bar{Y}}) = 1 - \alpha$$

Substitution of the expression for $\sigma_{\bar{X} - \bar{Y}}$ given in Theorem 9.1 shows the confidence interval in the statement of the theorem to be valid according to Definition 7.15. ∎

If either X or Y does not possess a normal distribution, then the central limit theorem allows the approximate use of the confidence interval of Theorem 9.4 for large samples. Moreover, if the variances are not known, then the confidence interval holds approximately in the large-sample setting with the empirical variances of the samples being used in place of the true variances.

EXAMPLE 9.9

Since $z_{0.025} = 1.96$, the data of Example 9.1 yield the approximate large-sample 95% confidence interval

$$(3.64 - 4.25 - (1.96)(0.11), \ 3.64 - 4.25 + (1.96)(0.11)) = (-0.83, -0.39)$$

When the random variables are normal and the variances are unknown, a small-sample confidence interval results from the t distribution.

If X and Y are normally distributed random variables with unknown variances and $X_1, X_2, \ldots, X_{nx}$ and $Y_1, Y_2, \ldots, Y_{ny}$ comprise independent random samples arising from X and Y, respectively, then a $(1 - \alpha)100\%$ confidence interval for $\mu_X - \mu_Y$ is given by

$$\left(\bar{x} - \bar{y} - t_{\alpha/2, nx+ny-2} \cdot s_p \sqrt{\frac{1}{n_X} + \frac{1}{n_Y}}, \bar{x} - \bar{y} + t_{\alpha/2, nx+ny-2} \cdot s_p \sqrt{\frac{1}{n_X} + \frac{1}{n_Y}} \right)$$

where s_p is the sample value of the pooled standard deviation.

Proof: The proof is similar to that of Theorem 9.4, except we employ the random variable T of Theorem 9.3. Specifically,

$$P(-t_{\alpha/2, nx+ny-2} < T < t_{\alpha/2, nx+ny-2}) = 1 - \alpha$$

Substitution of the expression defining T and rearrangement of the resulting inequality yields the probability statement whose corresponding confidence interval is the one stated in the theorem. ∎

Since $t_{0.025, 7} = 2.365$, the data of Example 9.4 yield the 95% confidence interval

$$(43.3 - 45.2 - (2.365)(2.05)\sqrt{0.45}, 43.3 - 45.2 + (2.365)(2.05)\sqrt{0.45}) = (-5.2, 1.4)$$

Theorem 9.5 holds for independent samples; the next theorem gives a confidence interval based on paired observations.

Under the conditions of paired sampling, as stated in Section 9.3, a $(1 - \alpha)100\%$ confidence interval for the difference between the means of the random variables X and Y is given by

$$\left(\bar{x} - \bar{y} - t_{\alpha/2, n-1} \cdot \frac{s_D}{\sqrt{n}}, \bar{x} - \bar{y} + t_{\alpha/2, n-1} \cdot \frac{s_D}{\sqrt{n}} \right)$$

where $D = X - Y$, n is the common sample size, and s_D is the sample value of the sample standard deviation.

Proof: Under the conditions of paired sampling,

$$T = \frac{\bar{D} - \mu_D}{S_D/\sqrt{n}}$$

possesses a t distribution with $n - 1$ degrees of freedom. Thus,

$$P\left(-t_{\alpha/2, n-1} < \frac{\bar{D} - \mu_D}{S_D/\sqrt{n}} < t_{\alpha/2, n-1} \right) = 1 - \alpha$$

Substitution of $\bar{D} = \bar{X} - \bar{Y}$ and $\mu_D = \mu_X - \mu_Y$ together with rearrangement of the inequality gives

$$P\left(\bar{X} - \bar{Y} - t_{\alpha/2,n-1} \cdot \frac{S_D}{\sqrt{n}} < \mu_X - \mu_Y < \bar{X} - \bar{Y} + t_{\alpha/2,n-1} \cdot \frac{S_D}{\sqrt{n}}\right) = 1 - \alpha$$

According to Definition 7.15, this is precisely what is meant by the confidence interval in the statement of the theorem. ∎

EXAMPLE 9.11

Since $t_{0.005,9} = 3.250$, a 99% confidence interval for the data of Example 9.8 (Table 9.4) is given by

$$\left(34.2 - 30.5 - \frac{(3.25)(3.234)}{\sqrt{10}}, 34.2 - 30.5 + \frac{(3.25)(3.234)}{\sqrt{10}}\right) = (0.4, 7.0)$$

9.5 LARGE-SAMPLE HYPOTHESIS TESTS CONCERNING TWO PROPORTIONS

Section 8.6 concerned the testing of a proportion p, which is simply the probability of a random variable X with dodomain $\{0, 1\}$ taking on the value 1. The present section considers the comparison of two proportions. Thus, there are two binary random variables X and Y, together with respective proportions p_X and p_Y, such that the probability of X being a success is

$$P(X = 1) = p_X$$

and the probability of Y being a success is

$$P(Y = 1) = p_Y$$

Corresponding to the probabilities of each of the variables being a success are the probabilities of failure, these being

$$q_X = P(X = 0) = 1 - p_X$$
$$q_Y = P(Y = 0) = 1 - p_Y$$

Given independent random samples $X_1, X_2, \ldots, X_{nx}$ and $Y_1, Y_2, \ldots, Y_{ny}$, we wish to test the equality of the proportions. Corresponding to the null hypothesis

$$H_0: p_X = p_Y$$

there are three alternative hypotheses to consider, namely,

$$H_1: p_X \neq p_Y$$
$$H_1: p_X > p_Y$$
$$H_1: p_X < p_Y$$

The null hypothesis can be rewritten as

$$H_0: p_X - p_Y = 0$$

in which case the corresponding alternative hypotheses become

$$H_1: p_X - p_Y \neq 0$$

$$H_1: p_X - p_Y > 0$$

$$H_1: p_X - p_Y < 0$$

If $\bar{X}$ and $\bar{Y}$ are the respective sample means of the samples drawn from X and Y, then, in a manner similar to that discussed in Section 7.8.4, $\hat{p}_X = \bar{X}$ and $\hat{p}_Y = \bar{Y}$ are unbiased estimators of p_X and p_Y, respectively. That is,

$$E[\hat{p}_X] = E[\bar{X}] = E[X] = p_X$$

$$E[\hat{p}_Y] = E[\bar{Y}] = E[Y] = p_Y$$

In addition, by Theorems 4.1 and 7.1,

$$\text{Var}\,[\hat{p}_X] = \text{Var}\,[\bar{X}] = \frac{\text{Var}\,[X]}{n_X} = \frac{p_X q_X}{n_X}$$

$$\text{Var}\,[\hat{p}_Y] = \text{Var}\,[\bar{Y}] = \frac{\text{Var}\,[Y]}{n_Y} = \frac{p_Y q_Y}{n_Y}$$

Owing to the linearity of the expected-value operator, $\hat{p}_X - \hat{p}_Y$ is an unbiased estimator of $p_X - p_Y$, namely,

$$E[\hat{p}_X - \hat{p}_Y] = E[\hat{p}_X] - E[\hat{p}_Y] = p_X - p_Y$$

According to Theorem 5.17, the variance of the estimator is

$$\text{Var}\,[\hat{p}_X - \hat{p}_Y] = \text{Var}\,[\hat{p}_X] + \text{Var}\,[\hat{p}_Y] = \frac{p_X q_X}{n_X} + \frac{p_Y q_Y}{n_Y}$$

Assuming n_X and n_Y to be large, the central limit theorem ensures that both $\hat{p}_X$ and $\hat{p}_Y$ are approximately normal. Consequently, their difference is approximately normal.

Under the null hypothesis, $p_X = p_Y$. Letting p denote the common proportion of the two populations and letting $q = 1 - p$, the above relations yield

$$E[\hat{p}_X - \hat{p}_Y] = 0$$

and

$$\text{Var}\,[\hat{p}_X - \hat{p}_Y] = pq \left(\frac{1}{n_X} + \frac{1}{n_Y} \right)$$

Thus, the standardized version of $\hat{p}_X - \hat{p}_Y$ is given by

$$\frac{\hat{p}_X - \hat{p}_Y}{\sqrt{pq \left(\dfrac{1}{n_X} + \dfrac{1}{n_Y} \right)}}$$

For large n, this standardized random variable possesses an approximately standard normal distribution. To use it as a test statistic, we must employ an estimate of pq. Both $\hat{p}_X$ and $\hat{p}_Y$ are unbiased estimators of p; however, a better estimator is obtained by pooling the data in a manner similar to the pooled variance of Definition 9.1. Specifically, the **pooled estimator of p** is given by the weighted average of the individual estimators:

$$\hat{p} = \frac{n_X}{n_X + n_Y}\hat{p}_X + \frac{n_Y}{n_X + n_Y}\hat{p}_Y = \frac{n_X\bar{X} + n_Y\bar{Y}}{n_X + n_Y}$$

Note that $\hat{p}$ is simply the total number of successes from both samples divided by the total number of trials; that is, $\hat{p}$ is the proportion of successes obtained by pooling the trials. Letting $\hat{q} = 1 - \hat{p}$ and then substituting these estimators into the standardized version of $\hat{p}_X - \hat{p}_Y$ gives the desired approximately standard normal test statistic

$$Z = \frac{\hat{p}_X - \hat{p}_Y}{\sqrt{\hat{p}\hat{q}\left(\dfrac{1}{n_X} + \dfrac{1}{n_Y}\right)}}$$

So long as the sample sizes are large, this test statistic can be utilized for hypothesis tests regarding the equality of proportions. The method is simply the Z test of Section 8.4. As usual, in the upper and lower one-tailed cases, the null hypothesis can take the more realistic forms $H_0: p_X - p_Y \leq 0$ and $H_0: p_X - p_Y \geq 0$, respectively.

Although we could go through an analysis of type II error and its relation to sample sizes, because the denominator of the test statistic involves an estimator, the procedure is cumbersome and will be omitted.

EXAMPLE 9.12

An *expert system* is a computer program utilizing a logical calculus to operate on a set of facts and rules, the inference rules being based upon knowledge elicited from experts. A typical application of an expert system is troubleshooting to isolate a malfunction in some complex mechanical, electrical, or physiological system. As an illustration, suppose two expert systems, A and B, have been developed to analyze cardiovascular test data and to render opinions as to the nature of a patient's condition. To compare the two systems, each is fed data relating to 40 randomly selected conditions. Let X and Y be the binary random variables corresponding to successful diagnoses on the part of expert systems A and B, respectively, and let p_X and p_Y be the respective probabilities of a success. The investigators wish to determine if, at the 0.05 level of significance, there is reason to conclude that A and B perform at different levels of effectiveness.

1. *Hypothesis test.* $H_0: p_X = p_Y$
   $H_1: p_X \neq p_Y$
2. *Level of significance.* $\alpha = 0.05$
3. *Test statistic.* $Z = \dfrac{\hat{p}_X - \hat{p}_Y}{\sqrt{\hat{p}\hat{q}\left(\dfrac{1}{n_X} + \dfrac{1}{n_Y}\right)}}$
4. *Sample sizes.* $n_X = n_Y = 40$
5. *Critical value.* $z_{0.025} = 1.96$
6. *Sample value.* Suppose system A makes the correct diagnosis in 35 cases and system B is correct in 30 cases. Under the null hypothesis the pooled estimate of p is

$$\overset{*}{p} = \frac{35 + 30}{80} = 0.8125$$

Moreover, $\bar{x} = 35/40$ and $\bar{y} = 30/40$. Therefore,

$$z = \frac{\bar{x} - \bar{y}}{\sqrt{\overset{**}{pq}\left(\dfrac{1}{40} + \dfrac{1}{40}\right)}} = \frac{0.125}{\sqrt{(0.8125)(0.1875)(0.05)}} = 1.432$$

7. *Decision.* Since $-1.96 < z < 1.96$, the null hypothesis is not rejected, and it is not concluded that the systems provide different levels of performance.

9.6 TESTING THE EQUALITY OF TWO VARIANCES

The present section concerns the comparison of the variances of two normal random variables. The null hypothesis will posit the equality of the variances and there will be the customary two-tailed and one-tailed alternatives. However, before we can consider the details of the testing procedure, it will be necessary to introduce a new distribution.

9.6.1 *F* Distribution

The testing of two variances involves a distribution that appears often in applied statistics. It arises from the ratio of two chi-square random variables. As such, it will involve two degrees-of-freedom parameters.

DEFINITION 9.2

F Distribution. Suppose W_1 and W_2 are independent chi-square random variables possessing v_1 and v_2 degrees of freedom, respectively. Then the random variable

$$F = \frac{W_1/v_1}{W_2/v_2}$$

is said to possess an *F* distribution with v_1 and v_2 degrees of freedom.

When referring to an *F* distribution, it is customary to first state the degrees of freedom associated with the numerator and then the degrees of freedom associated with the denominator. The next theorem, stated without proof, gives the density corresponding to an *F* distribution. Figure 9.1 illustrates the density.

THEOREM 9.7

If F possesses an *F* distribution with v_1 and v_2 degrees of freedom, then its density is given by

$$f_F(x) = \frac{\Gamma\left(\dfrac{v_1 + v_2}{2}\right) v_1^{v_1/2} v_2^{v_2/2}}{\Gamma\left(\dfrac{v_1}{2}\right)\Gamma\left(\dfrac{v_2}{2}\right)} \cdot \frac{x^{(v_1/2) - 1}}{(v_2 + v_1 x)^{(v_1 + v_2)/2}}$$

for $x > 0$ and $f_F(x) = 0$ otherwise.

Critical values for tests involving the *F* distribution can be obtained from Table A.7. The table gives values of the form f_{α, v_1, v_2} for $\alpha = 0.05$ and $\alpha = 0.01$, where, as illustrated in Figure 9.2,

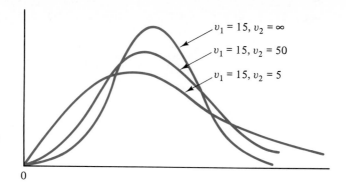

FIGURE 9.1
Various F densities
(Bowker/Lieberman,
Engineering Statistics,
2/e, © 1972, p. 121.
Reprinted by permission
of Prentice Hall Inc.,
Englewood Cliffs, New
Jersey.)

$$P(F > f_{\alpha,\nu_1,\nu_2}) = \alpha$$

For instance, $f_{0.05,8,11} = 2.95$ and $f_{0.01,10,19} = 3.43$.

For lower one-tailed tests it will be necessary to utilize critical values of the form $f_{1-\alpha,\nu_1,\nu_2}$ defined by

$$P(F < f_{1-\alpha,\nu_1,\nu_2}) = \alpha$$

and illustrated in Figure 9.3. Table A.7 gives $f_{0.05,\nu_1,\nu_2}$ but does not give $f_{0.95,\nu_1,\nu_2}$. The next theorem shows why the table is sufficient as given (at least for $\alpha = 0.05$ and $\alpha = 0.01$).

THEOREM 9.8

Under the conditions of Definition 9.2,

$$f_{1-\alpha,\nu_2,\nu_1} = \frac{1}{f_{\alpha,\nu_1,\nu_2}}$$

Proof: Let W_1, W_2, ν_1, and ν_2 be as in Definition 9.2. Then

$$1 - \alpha = 1 - P(F > f_{\alpha,\nu_1,\nu_2})$$

$$= 1 - P\left(\frac{W_1/\nu_1}{W_2/\nu_2} > f_{\alpha,\nu_1,\nu_2}\right)$$

$$= 1 - P\left(\frac{W_2/\nu_2}{W_1/\nu_1} < \frac{1}{f_{\alpha,\nu_1,\nu_2}}\right)$$

$$= P\left(\frac{W_2/\nu_2}{W_1/\nu_1} > \frac{1}{f_{\alpha,\nu_1,\nu_2}}\right)$$

FIGURE 9.2
Critical region of size α
for the upper one-tailed
F test

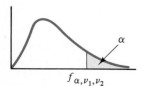

f_{α,ν_1,ν_2}

Since $(W_2/\nu_2)/(W_1/\nu_1)$ is an F distribution with ν_2 and ν_1 degrees of freedom, the definition of F-distribution critical values gives

$$1 - \alpha = P\left(\frac{W_2/\nu_2}{W_1/\nu_1} > f_{1-\alpha,\nu_2,\nu_1}\right)$$

and the theorem is proven. ∎

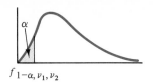

$f_{1-\alpha, \nu_1, \nu_2}$

FIGURE 9.3
Critical region of size α
for the lower one-tailed
F test

The roles of ν_1 and ν_2 can be interchanged in Theorem 9.8 to yield

$$f_{1-\alpha, \nu_1, \nu_2} = \frac{1}{f_{\alpha, \nu_2, \nu_1}}$$

As an illustration of Theorem 9.7,

$$f_{0.95, 11, 8} = \frac{1}{f_{0.05, 8, 11}} = \frac{1}{2.95} = 0.34$$

The next theorem is key to employing the F distribution to test the equality of two variances.

THEOREM 9.9

Suppose $X_1, X_2, \ldots, X_{nx}$ is a random sample of size n_X arising from a normal random variable X possessing variance σ_X^2 and $Y_1, Y_2, \ldots, Y_{ny}$ is a random sample of size n_Y arising from a normal random variable Y possessing variance σ_Y^2. In addition, suppose the random samples are independent. Then the random variable

$$F = \frac{S_X^2/\sigma_X^2}{S_Y^2/\sigma_Y^2}$$

possesses an F distribution with $n_X - 1$ and $n_Y - 1$ degrees of freedom, where S_X^2 and S_Y^2 are the sample variances corresponding to the X and Y random samples, respectively.

Proof: According to Theorem 7.9,

$$W_1 = \frac{(n_X - 1)S_X^2}{\sigma_X^2}$$

and

$$W_2 = \frac{(n_Y - 1)S_Y^2}{\sigma_Y^2}$$

possess chi-square distributions with $n_X - 1$ and $n_Y - 1$ degrees of freedom, respectively. Since the random samples are independent, so too are W_1 and W_2. According to Definition 9.2,

$$\frac{W_1/(n_X - 1)}{W_2/(n_Y - 1)}$$

which is precisely the random variable in the statement of the theorem, possesses an F distribution with $n_X - 1$ and $n_Y - 1$ degrees of freedom. ∎

9.6.2 Hypothesis Tests for Two Variances

The equality of two means was tested by relying on their difference. To test the equality of two variances, we make use of the observation that their equality can be expressed in terms of their ratio being one. Suppose X and Y are independent normal random variables with unknown means μ_X and μ_Y and unknown variances σ_X^2 and

σ_Y^2. Consider two independent random samples $X_1, X_2, \ldots, X_{n_X}$ and $Y_1, Y_2, \ldots, Y_{n_Y}$ arising from X and Y, respectively. Given the null hypothesis

$$H_0: \sigma_X^2 = \sigma_Y^2$$

the test statistic

$$F = \frac{S_X^2}{S_Y^2}$$

can be expressed as

$$F = \frac{S_X^2/\sigma_X^2}{S_Y^2/\sigma_Y^2}$$

which, according to Theorem 9.9, possesses an F distribution with $n_X - 1$ and $n_Y - 1$ degrees of freedom. In addition, since the sample variance is an unbiased estimator of the variance, if the null hypothesis is true, then the ratio of the sample variances should be near one. Thus, a plausible acceptance region for the two-tailed hypothesis test at significance level α, which has alternative hypothesis

$$H_1: \sigma_X^2 \neq \sigma_Y^2$$

is given by

$$(f_{1-\alpha/2,n_X-1,n_Y-1}, f_{\alpha/2,n_X-1,n_Y-1})$$

The corresponding critical region is

$$[0, f_{1-\alpha/2,n_X-1,n_Y-1}] \cup [f_{\alpha/2,n_X-1,n_Y-1}, \infty)$$

The upper one-tailed test has the alternative hypothesis

$$H_1: \sigma_X^2 > \sigma_Y^2$$

and the critical region

$$[f_{\alpha,n_X-1,n_Y-1}, \infty)$$

The lower one-tailed test has the alternative hypothesis

$$H_1: \sigma_X^2 < \sigma_Y^2$$

and the critical region

$$[0, f_{1-\alpha,n_X-1,n_Y-1}]$$

EXAMPLE 9.13

In Example 9.4 we applied the two-sample t test to determine whether one model of automobile held up better in a crash than another. Fundamental to the application of the test was the assumption of equal variances. Based upon the information of Example 9.4, we will now apply the F test to determine if there is sufficient reason to reject the hypothesis of equal variances at level $\alpha = 0.10$.

1. *Hypothesis test.* $H_0: \sigma_X^2 = \sigma_Y^2$
   $H_1: \sigma_X^2 \neq \sigma_Y^2$
2. *Level of significance.* $\alpha = 0.10$
3. *Test statistic.* $F = \dfrac{S_X^2}{S_Y^2}$

4. *Sample sizes.* $n_X = 5$ and $n_Y = 4$

5. *Critical region.* From Table A.7, the lower critical value is

$$f_{0.95,4,3} = \frac{1}{f_{0.05,3,4}} = \frac{1}{6.59} = 0.15$$

and the upper critical value is

$$f_{0.05,4,3} = 9.12$$

6. *Sample value.* Based upon the information of Example 9.4,

$$f = \frac{4.5}{3.8} = 1.18$$

7. *Decision.* Since $0.15 < f < 9.12$, the hypothesis of equal variances is not rejected.

An evident problem with the test of Example 9.13 is that the sample sizes are too small to avoid a large type II error. Suppose the experiment is redone with equal sample sizes of 41, an extraordinarily large number considering the expense and duration of each observation. The appropriate critical values are

$$f_{0.95,40,40} = \frac{1}{f_{0.05,40,40}} = \frac{1}{1.69} = 0.59$$

and $f_{0.05,40,40} = 1.69$. Even with a common sample size of 41 and a numerator variance that is 18% larger than the denominator variance, the F test is still not sufficiently sensitive to detect differing variances. We are led to consider the problem of β.

To examine type II error in the F test for two variances, we turn to the appropriate OC-curve families, the restriction $n_X = n_Y = n$ being required. In all cases, the curves are defined in terms of the ratio of the standard deviations,

$$\lambda = \frac{\sigma_X}{\sigma_Y}$$

For a given α and common sample size n, if $\lambda \neq 1$, then the size of type II error at the alternative ratio λ is given by the value of the OC curve corresponding to n at the value λ. The OC curves can also be used to find necessary sample sizes: given an alternative ratio λ_1 at which detection is considered important and a tolerable level of risk $\beta(\lambda_1)$ at λ_1, the necessary common sample size corresponds to the OC curve closest to the point $(\lambda_1, \beta(\lambda_1))$ when passing through the vertical line at λ_1. The two-tailed OC-curve families for $\alpha = 0.05$ and $\alpha = 0.01$ are given in Figures 9.4 and 9.5, respectively.

To use the one-tailed OC curves, it is necessary to formulate the alternative hypothesis so that it is upper one-tailed. This is always possible, since the notation X and Y can be interchanged if necessary. Specifically, name the random variables so that the alternative hypothesis takes the form $H_1: \sigma_X^2 > \sigma_Y^2$. Thus, the alternative ratio λ will always be greater than one. The one-tailed OC-curve families for $\alpha = 0.05$ and $\alpha = 0.01$ are given in Figures 9.6 and 9.7, respectively, and they are employed in the usual manner. As in the two-tailed case, these families are used under the assumption of a common sample size.

The insensitivity of the F test for testing the equality of two variances was noted

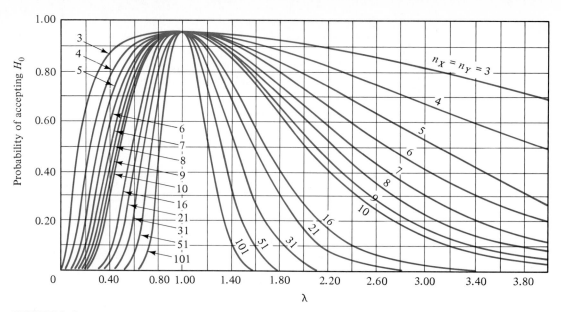

FIGURE 9.4
OC curves for the two-tailed *F* test at level $\alpha = 0.05$ (Reprinted by permission of the Institute of Mathematical Statistics.)

subsequent to Example 9.13. To further explore the difficulty, suppose it is desired to demonstrate, at significance level $\alpha = 0.05$, that a random variable X possesses a greater variance than random variable Y. As the problem is stated, the alternative hypothesis is of the appropriate form, $H_1: \sigma_X^2 > \sigma_Y^2$. In addition, suppose it is important to detect a situation in which the standard deviation of X is 50% greater than the standard deviation of Y. Then

FIGURE 9.5
OC curves for the two-tailed *F* test at level $\alpha = 0.01$ (Reprinted by permission of the Institute of Mathematical Statistics.)

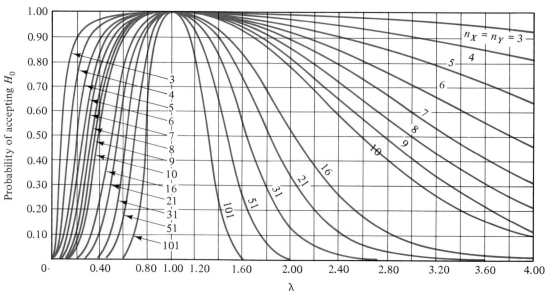

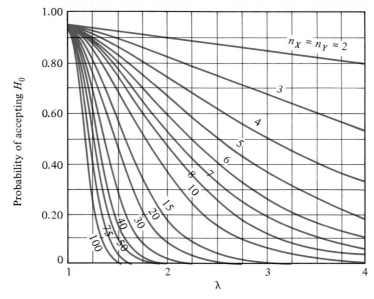

FIGURE 9.6
OC curves for the one-tailed *F* test at level $\alpha = 0.05$ (Reprinted by permission of the Institute of Mathematical Statistics.)

$$\lambda_1 = \frac{\sigma_X}{\sigma_Y} = 1.5$$

Such a requirement is certainly not stringent. Nevertheless, Figure 9.6 shows that to have $\beta(1.5) = 0.10$, it is necessary to have $n_X = n_Y = 50$.

While it might be tempting to employ the *F* test for variances prior to using the two-sample *t* test, such a course is not advised. Not only is the *F* test somewhat

FIGURE 9.7
OC curves for the one-tailed *F* test at level $\alpha = 0.01$ (Reprinted by permission of the Institute of Mathematical Statistics.)

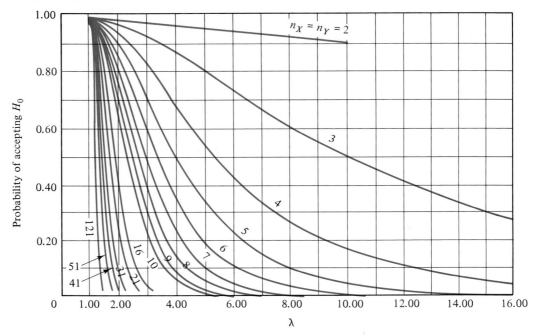

insensitive, it is also less robust with respect to the underlying assumption of normality than is the two-sample t test. Consequently, the F test might be inappropriate in settings where the two-sample t test is appropriate.

9.7 DISTRIBUTION-FREE TESTS FOR THE EQUALITY OF TWO MEANS

The two-sample t test requires the normality of the random variables and the paired t test requires the normality of the difference. The present section concerns distribution-free alternatives to the t test in the cases of paired and unpaired sampling.

9.7.1 Distribution-Free Methods in the Case of Paired Sampling

Both the sign test and the signed-rank test of Section 8.9 can be applied in the case of paired observations, the appropriate distributional restrictions applying to the difference D.

For the sign test, the appropriate null hypothesis is

$$H_0: \ \tilde{\mu}_D = \delta_0$$

where $D = X - Y$. The sign test is applied by simply counting the number of sample values for which $D_i - \delta_0 > 0$, $D_1, D_2, \ldots, D_n$ constituting the random sample of differences.

EXAMPLE 9.14

In an oil-burning furnace, fuel oil is pumped into the burner unit under pressure and then sprayed through a nozzle to become a fine mist. The mist is mixed with air and ignited by an electric spark. To compare the efficiency of two injection units, an investigator utilizes them on identical furnaces and then places 15 of each type into various homes. The homes are paired so that each pair includes two homes of similar construction, environment, and family type. The number of gallons of fuel utilized in each home for the months October through March is recorded in Table 9.5, along with the differences d_i. The investigator wishes to check

TABLE 9.5   FURNACE-FUEL DATA

x_i	y_i	d_i	$d_i - \tilde{\mu}_0$
1042	1021	21	$+1$
827	784	43	$+20$
901	891	10	-10
1214	1152	62	$+42$
1027	998	29	$+9$
684	662	22	$+2$
1020	956	64	$+44$
900	880	20	0
821	781	40	$+20$
1026	962	64	$+44$
865	842	23	$+3$
512	506	6	-14
746	740	6	-14
920	880	40	$+20$
1072	1046	26	$+6$

the claim that the mean difference is 20 gallons. Thus, the deviations of the differences from $\delta_0 = 20$ are also included in the table. The sign test is employed to check the claim. A normal approximation is employed.

1. *Hypothesis test.* H_0: $\tilde{\mu}_D = 20$
   $\qquad\qquad\qquad\quad H_1$: $\tilde{\mu}_D \neq 20$
2. *Level of significance.* $\alpha = 0.04$
3. *Test statistic.* $Z = \dfrac{V - \dfrac{n}{2}}{\sqrt{\dfrac{n}{4}}}$

4. *Sample size.* Since one of the differences is exactly 20, it must be removed from the sample. Consequently, $n = 14$.
5. *Critical value.* $z_{0.02} = 2.05$
6. *Sample value.* $z = \dfrac{11 - \dfrac{14}{2}}{\sqrt{\dfrac{14}{4}}} = 2.138$

7. *Decision.* Since $z > 2.05$, the null hypothesis is rejected.

The signed-rank test can also be employed in place of the paired t test so long as the random-variable difference $D = X - Y$ possesses a symmetric distribution. An important instance of this symmetry occurs when X possesses a density that is a translate of Y's density, regardless of whether or not X and Y are individually symmetric. Geometrically, when the densities are translates of one another, the shapes and spreads of the distributions are identical. When D is symmetric, the appropriate null hypothesis is

$$H_0\text{: } \tilde{\mu}_D = \delta_0$$

EXAMPLE 9.15

The standard measure of surface roughness is the *average roughness R_a*, which represents the arithmetic average deviation of the ordinates of profile height increments of the surface from the centerline of the surface. To compare the surface roughnesses resulting from two broaching processes, an engineer takes measurements of roughness subsequent to 20 broaching operations with each process. The sampling is done in a paired fashion, each pairing corresponding to broaching metal of the same stock. Table 9.6 gives the resulting data (in micrometers), including the differences and the rankings of the absolute differences. The desire of the engineer is to test whether there is a difference in surface roughness resulting from the two broaching processes. He or she decides to perform a signed-rank test. Since $n = 20$, a normal approximation is justified.

1. *Hypothesis test.* H_0: $\tilde{\mu}_D = 0$
   $\qquad\qquad\qquad\quad H_1$: $\tilde{\mu}_D \neq 0$
2. *Level of significance.* $\alpha = 0.05$
3. *Test statistic.* $Z = \dfrac{S - \dfrac{n(n+1)}{4}}{\sqrt{\dfrac{n(n+1)(2n+1)}{24}}}$

4. *Sample size.* $n = 20$
5. *Critical value.* $z_{0.025} = 1.96$

TABLE 9.6   SURFACE-ROUGHNESS DATA

x_i	y_i	d_i	Rank
1.77	1.56	$+0.21$	17
1.32	1.30	$+0.02$	2.5
1.83	1.85	-0.02	2.5
1.02	0.94	$+0.08$	9
1.92	1.95	-0.03	4
1.68	1.55	$+0.13$	12
2.48	2.60	-0.12	11
1.55	1.32	$+0.22$	18
0.99	0.85	$+0.14$	13.5
1.96	2.00	-0.04	5
1.63	1.56	$+0.07$	8
1.89	1.83	$+0.06$	7
0.99	0.84	$+0.15$	15
2.00	2.01	-0.01	1
1.88	1.60	$+0.28$	19
1.90	2.00	-0.10	10
0.98	0.79	$+0.19$	16
1.44	1.30	$+0.14$	13.5
2.64	2.69	-0.05	6
1.76	1.44	$+0.32$	20

6. *Sample value.* From Table 9.6, we obtain the sum of the positive-difference ranks to be $s = 170.5$. Thus,

$$z = \frac{170.5 - 105}{\sqrt{717.5}} = 2.445$$

7. *Decision.* Since $z > 1.96$, reject H_0 and assume the median of the differences is not zero.

9.7.2   Rank-Sum Test

In the absence of pairing, another distribution-free rank-type test is available, this being known as the **Wilcoxon rank-sum test** or the **Mann-Whitney test.** It, too, involves ranking to make decisions regarding a null hypothesis concerning the difference of two means. Specifically, suppose $X_1, X_2, \ldots, X_{n_X}$ and $Y_1, Y_2, \ldots, Y_{n_Y}$ are independent random samples arising from the continuous random variables X and Y, respectively. Furthermore, suppose the means of X and Y are μ_X and μ_Y, respectively, and the density of X is simply a translate of the density of Y. The rank-sum statistic W is used to test the null hypothesis

$$H_0\colon\ \mu_X - \mu_Y = \delta_0$$

against any one of the usual three alternatives.

To construct W in the case where $\delta_0 = 0$, list all the sample values in increasing order, assign ranks, and let W be the sum of the ranks corresponding to X. Regardless of the distribution of the observations,

$$W \geq 1 + 2 + \cdots + n_X = \frac{n_X(n_X + 1)}{2}$$

the minimum occurring in the event that all observations of X possess lower ranks than the least ranking observation of Y. Moreover, the maximum value of W occurs when all the observations of X possess higher ranks than the greatest ranking observation of Y. Thus,

$$
\begin{aligned}
W &\leq (n_Y + 1) + (n_Y + 2) + \cdots + (n_Y + n_X) \\
&= [1 + 2 + \cdots + (n_Y + n_X)] - (1 + 2 + \cdots + n_Y) \\
&= \frac{(n_X + n_Y)(n_X + n_Y + 1)}{2} - \frac{n_Y(n_Y + 1)}{2} \\
&= \frac{n_X(n_X + 2n_Y + 1)}{2}
\end{aligned}
$$

Under the null hypothesis H_0: $\mu_X = \mu_Y$, we would expect the sample X and Y values to distribute themselves among each other and therefore their rankings would be interspersed, the result being a middling value of W. On the other hand, falseness of the null hypothesis would lead us to expect that either the X values possess relatively higher rankings, so that W is large, or the Y values possess relatively higher rankings, so that W is small. The following theorem quantifies these considerations in terms of the mean of W.

THEOREM 9.10

Under the null hypothesis H_0: $\mu_X - \mu_Y = 0$,

$$E[W] = \frac{n_X(n_X + n_Y + 1)}{2}$$

$$\text{Var}\,[W] = \frac{n_X n_Y(n_X + n_Y + 1)}{12}$$

Proof: For $i = 1, 2, \ldots, n_X$, let R_i denote the rank of X_i. Then

$$W = R_1 + R_2 + \cdots + R_n$$

Under H_0, $\mu_X = \mu_Y$ and therefore R_i is equally likely to take on any value from 1 to $n_X + n_Y$. Consequently,

$$E[W] = \sum_{i=1}^{n_X} E[R_i] = \sum_{i=1}^{n_X} \frac{n_X + n_Y + 1}{2} = \frac{n_X(n_X + n_Y + 1)}{2}$$

The proof for the variance is somewhat complicated, because $R_1, R_2, \ldots, R_n$ do not constitute an independent collection of random variables, and it will be omitted. ∎

In the more general case, when the null hypothesis is given by H_0: $\mu_X - \mu_Y = \delta_0$, the construction of W is done in the same manner except that $X_i - \delta_0$ is used in place of X_i for $i = 1, 2, \ldots, n_X$.

For each of the three alternatives to H_0, the corresponding critical value(s) for a given significance level α can be found by employing Table A.11, in which it is required that the random variables be named so that $n_X \leq n_Y$. As in the case of the signed-rank test, the test statistic is discrete and the table values must be used to find the critical value that gives the closest approximation to the desired significance level. Once again, symmetry of the statistic allows lower one-tailed critical values to be derived from upper one-tailed critical values. The critical value w_{α, n_X, n_Y} for the upper one-tailed alternative

$$H_1: \mu_X - \mu_Y > \delta_0$$

is determined by

$$P(W \geq w_{\alpha, n_X, n_Y}; H_0) = \alpha$$

where α is chosen as close as possible to the desired level. Rejection of H_0 occurs if $W \geq w_{\alpha, n_X, n_Y}$. For the lower one-tailed alternative

$$H_1: \mu_X - \mu_Y < \delta_0$$

H_0 is rejected if

$$W \leq n_X(n_X + n_Y + 1) - w_{\alpha, n_X, n_Y}$$

Finally, for the two-tailed alternative

$$H_1: \mu_X - \mu_Y \neq \delta_0$$

H_0 is rejected if

$$W \geq w_{\alpha/2, n_X, n_Y}$$

or

$$W \leq n_X(n_X + n_Y + 1) - w_{\alpha/2, n_X, n_Y}$$

As in the signed-rank test, should a tie occur, each sample value in the tie is assigned the average of the ranks that would have been formed by the collection had they not been tied.

EXAMPLE 9.16

Fatigue failure in metal results from the repeated application of small loads, which eventually produce a crack. To test a metal for its ability to withstand fatigue, repeated stress is applied and reversed for a large number of cycles until fracture, the number of cycles being a function of the applied stress. Suppose an engineer wishes to compare two alloys with respect to fatigue, and a fatigue test is set up with a given stress factor. The number of cycles to failure for one alloy will be denoted by X and the number for the other by Y. Suppose 7 measurements pertaining to X and 8 measurements pertaining to Y are made, the data (in 10^5 cycles) being recorded in Table 9.7, along with the overall ranking of each measurement. Assuming each sample has been drawn randomly and the samples are independent, the rank-sum test is employed to compare the means of X and Y.

1. *Hypothesis test.* $H_0: \mu_X - \mu_Y = 0$
   $H_1: \mu_X - \mu_Y \neq 0$
2. *Level of significance.* Approximately 0.10
3. *Test statistic.* Rank-sum statistic W
4. *Sample size.* $n_X = 7$ and $n_Y = 8$
5. *Critical region.* Referring to Table A.11, we find that the closest value to 0.05 is 0.047. Consequently, the right critical value is $w_{0.047, 7, 8} = 71$ and the left critical value is $112 - 71 = 41$. In other words, the acceptance region is the interval [42, 70].
6. *Sample value.* Adding up the X ranks in Table 9.7 gives $w = 70$.
7. *Decision.* Since $42 \leq w \leq 70$, H_0 is accepted.

TABLE 9.7 FATIGUE DATA

x_i	Rank	y_i	Rank
2.32	7	2.30	6
2.46	13.5	2.41	11
2.38	9	2.22	5
2.51	15	2.40	10
2.20	4	2.19	3
2.46	13.5	2.14	2
2.35	8	2.12	1
		2.42	12

For large samples, in this case n_X and n_Y exceeding 8, W possesses an approximately normal distribution under H_0. Thus, the test statistic $Z = (W - \mu_W)/\sigma_W$ approximates a standard normal distribution and can be employed in the usual Z-test fashion. In the event there are ties, a modification of σ_W is necessary; however, the modification is very slight unless there are an excessive number of ties and will not be discussed.

In comparing two-sample mean-testing by means of the t test and the rank-sum procedure, the dilemma is similar to that posed in the previously discussed distribution-free tests. If the populations are normally distributed, then the t test is better; however, if they are markedly nonnormal, then the rank-sum test is superior. Since the rank-sum test is only marginally less efficient when the variables are normal and the sample sizes are large, it is preferable for large samples if there is a reasonable concern regarding normality.

EXERCISES

SECTION 9.1

9.1. To test whether it can be concluded that there is a difference in the hardness of gray cast iron resulting from two production sources, ten independent measurements of Brinell hardness are taken from the iron produced by each of the two sources. Assuming both measurement populations are normally distributed with known standard deviations $\sigma_X = \sigma_Y = 0.8$, set up the appropriate hypothesis test to check a difference in the means at the $\alpha = 0.05$ level. If the sample means yield the empirical values $\bar{x} = 160.1$ and $\bar{y} = 160.8$, what conclusion is drawn?

9.2. The lifetime of a metal-cutting tool varies according to a number of factors. A turning cutter is tested at feed rates 0.30 and 0.35 mm per revolution, the speed of the tool being held constant and the test metal being a steel of fixed hardness. Let X and Y denote the random variables corresponding to the lifetimes (in minutes) of tools tested at the feed rates 0.30 and 0.35, respectively. Given that 30 tests are independently run at feed rate 0.30, 40 tests are independently run at feed rate 0.35, the respective empirical means are $\bar{x} = 41.5$ and $\bar{y} = 40.2$ minutes, and the respective empirical standard deviations are $s_X = 3.92$ and $s_Y = 4.51$, can it be concluded at the $\alpha = 0.01$ level that the two feed rates result in different tool lives?

9.3. A sound engineer wishes to demonstrate that the frequency X of the sound emitted by one source is more than 5 hertz greater than the frequency Y of the sound emitted by another source. Independent random samples yield the values $\bar{x} = 1128$ Hz, $\bar{y} = 1121$ Hz, $s_X^2 = 12$, and $s_Y^2 = 10$. Both sample sizes are 50. Based upon the data, is the engineer's claim justified at the $\alpha = 0.01$ level of significance?

9.4. A laboratory wishes to establish that the useful life of a felt-tip pen is less than 4 minutes longer when used continuously at a specified light pressure P_x than when used continuously at a specified heavy pressure P_y. Forty pens are randomly selected and tested by machine under pressure P_x, the useful life of a test pen being determined by the number of minutes of use that expires before the track of the ink exceeds a predefined width. Independently, 40 pens are checked under pressure P_y. Given the data $\bar{x} = 26.2$ min, $\bar{y} = 23.0$ min, $s_X^2 = 4.6$, and $s_Y^2 = 5.0$, is the laboratory's claim justified at the $\alpha = 0.01$ level?

9.5. Repeat Exercise 9.4 with $s_X^2 = 1.3$, and $s_Y^2 = 3.4$.

9.6. Referring to Exercise 9.1, it is important to detect a difference of 0.3 on the Brinell scale. What is the probability that such a difference will not be detected?

9.7. Referring to Exercise 9.2, it is important to detect a difference of 1 minute. What is the probability that such a difference will not be detected?

9.8. What is the probability that the null hypothesis will not be rejected in Exercise 9.3 even if the sound from the first source has a frequency that is 6 Hz greater than the sound from the second source?

9.9. What is the probability that the alternative hypothesis will be rejected in Exercise 9.4 even if the useful life under pressure P_x is only 3 minutes longer than that under P_y?

9.10. a) Find a common sample size in Exercise 9.1 so that $\beta(0.2) = 0.15$.
   b) Repeat part (a) under the assumption that $\alpha = 0.01$ in Exercise 9.1.

9.11. a) Find a common sample size in Exercise 9.3 so that $\beta(7) = 0.10$.
   b) Repeat part (a) under the assumption that $\alpha = 0.05$ in Exercise 9.3.

9.12. a) Find a common sample size in Exercise 9.4 so that $\beta(3) = 0.20$.
   b) Repeat part (a) under the assumption that $\alpha = 0.05$ in Exercise 9.4.

SECTION 9.2

9.13. It is desired to compare the effects of unleaded and leaded gasolines on the valves of a class of automobile motors. In an effort to demonstrate that unleaded gasoline has a detrimental effect on valves, eight motors are tested with unleaded gasoline and five are tested with leaded gasoline, the data being the number of (simulated) miles run prior to a prespecified drop in cylinder compression. The data for the unleaded gasoline yield the values $\bar{x} = 47,352$ and $s_X = 6238$, whereas the data for the leaded gasoline yield the values $\bar{y} = 52,420$ and $s_Y = 5627$. Assuming both mileage measurements to be normally distributed with common variance and the samples to be independent and random, can it be concluded at the $\alpha = 0.05$ level that unleaded gasoline has a detrimental effect on valve life?

9.14. Redo Exercise 9.13 with $s_X = 4056$ and $s_Y = 3630$.

9.15. Two machines are used to fill juice containers that are nominally rated at one quart. The following data are recorded (in quarts) by measuring actual juice content in randomly selected containers:

Machine A:	1.02	1.01	1.02	1.03	1.00
	1.01				
Machine B:	1.00	0.99	0.98	0.98	1.00
	0.99				

Given that the content populations are normally distributed with equal variances, can it be concluded at the $\alpha = 0.01$ level that (on average) the machines give different quantities of juice?

9.16. Referring to Exercise 9.15, find a common sample size so that an erroneous acceptance of the null hypothesis when the actual difference of the means is 0.02 has a probability of 0.10. Assume a conservative estimate of the common variance to be 0.0002. Give the corresponding critical region.

9.17. It is desired to compare the hourly rate of an entry-level job in two fast-food chains. Eight locations for each chain are randomly selected throughout the country, the selections for each chain being independent. The following hourly rates are recorded:

Chain A:	4.25	4.75	3.80	4.50	3.90	5.00
	4.00	3.80				
Chain B:	4.60	3.65	3.80	4.00	4.80	4.00
	4.50	3.65				

Under the assumption of normality and equal variances, can it be concluded at the $\alpha = 0.05$ level that chain A pays more than chain B for the job under consideration?

9.18. Referring to Exercise 9.17 and assuming 1.10 to be a conservative estimate of the common variance, find the common sample size so that $\beta(0.10) = 0.20$. Give the corresponding critical region.

9.19. One lot of resistors is rated at 12 ohms and a second lot at 16 ohms. Independent random samples are drawn from the two lots and the following measurements are recorded:

Lot A:	12.12	12.20	12.03	12.13
	12.04			
Lot B:	16.00	16.06	15.99	16.16
	16.10	16.02		

Can it be concluded at the $\alpha = 0.05$ level that the mean resistance of lot B is at least 3.85 ohms greater than the mean resistance of A? The usual t-test assumptions apply.

9.20. Referring to Exercise 9.19, do the data justify, at the $\alpha = 0.05$ level, the conclusion that the mean resistance of lot B is less than 4.00 ohms greater than the mean resistance of lot B?

9.21. Referring to Exercise 9.19, what is the common sample size necessary to make $\beta(4.00) = 0.25$? Explain in words the meaning of $\beta(4.00)$. Use the pooled variance to estimate the common variance.

9.22. The compressive strength of concrete is affected by the time and quality of curing, as well as the type of mix and the properties of the aggregate. The following data (in psi) are obtained by independently selecting two random samples of concrete and finding the compressive strengths of the experimental units:

Cure A:	3932	4023	4203	3950	3879
	4010				
Cure B:	4398	3902	4257	4440	4262
	4321	4068			

Under the usual t-test assumptions,

a) Can it be concluded at the $\alpha = 0.05$ level that cure B results in a greater compressive strength?

b) Can it be concluded at the $\alpha = 0.05$ level that cure B results in a compressive strength that is more than 105 psi greater than the compressive strength corresponding to cure A?

c) Based upon the given data, what is the minimal value of δ_0 for which the alternative hypothesis $H_0: \mu_X - \mu_Y < -\delta_0$ is accepted?

9.23. Repeat Exercise 9.13 without the assumption of equal variances.

9.24. Repeat Exercise 9.19 without the assumption of equal variances.

SECTION 9.3

9.25. The rate of corrosion of a metal depends on a number of environmental conditions. In order to test the effects that two coatings have relative to the retardation of corrosion, metal sheets are coated over half of their surface areas with coating A and over the other half with coating B. The experimental units are then exposed to various environmental conditions and, for each, the length of time (in months) it takes to reach a prespecified degree of corrosion is recorded for the two separately coated surface areas. The following paired data are obtained:

Coating A:	7.4	15.2	12.8	20.5	10.3	8.4
Coating B:	7.9	15.9	13.3	21.5	10.8	8.8

Assuming the differences constitute a random sample from a normal population, can it be concluded at the $\alpha = 0.05$ level that coating B is superior to coating A? Find the P value.

9.26. Referring to Exercise 9.25, what would be concluded had the data resulted from independent random samples instead of pairing?

9.27. Airline B claims that its arrival times are superior to those of airline A on city-to-city flights common to both airlines. To test the claim, eight city-to-city routes are selected and a flight for each airline is randomly observed on each route. The time recorded is the time from the announced departure time to the time the plane comes to rest at the terminal. The following times (in hours) are recorded:

Airline A:	2.41	6.50	3.29	1.22	2.59
	2.81	5.35	1.78		
Airline B:	2.30	5.86	3.71	1.10	2.34
	2.24	5.00	1.95		

Assuming that the differences comprise a random sample from a normal population, is the claim of airline B justified at the $\alpha = 0.01$ level? Find the P value.

9.28. Suppose the food-chain salary data of Exercise 9.17 had been collected in the following paired manner:

Chain A:	4.25	4.75	3.80	4.50	3.90	5.00
	4.00	3.80				
Chain B:	4.00	4.60	3.65	4.50	3.80	4.80
	4.00	3.65				

What conclusion is drawn?

9.29. The composition of the air in an artificial environment is controlled by a computer chip which makes decisions based upon information received from sensors in the chamber. Tests are made with sensors purchased from companies A and B. The following data represent recorded percentages of oxygen taken in pairs according to the number of people in the chamber:

Company A:	27.6	27.9	26.3	28.5	27.9
	25.4				
Company B:	27.9	27.9	27.1	28.0	27.9
	26.9				

Assuming the conditions of the paired t test are met, can it be concluded at the $\alpha = 0.05$ level that the mean percentages are different?

9.30. Production team A is thought to be, on average, at least 7 minutes faster in completing its various tasks than team B. The following paried completion times (in minutes) are recorded:

Team A:	108.5	57.2	111.0	96.8	76.3
	168.7	98.4			
Team B:	120.4	60.0	119.5	110.2	81.3
	184.9	110.4			

Under the assumption that the conditions for applying the paired t test are satisfied, is the conjecture supported at the $\alpha = 0.10$ level of significance?

9.31. Since the paired t test actually is a single-sample t test concerning the mean of the differences, the OC-curve families corresponding to the t test of Section 8.5 apply.

a) In Exercise 9.25, what is the probability of incorrectly accepting the null hypothesis if coating B provides a month's longer protection? Use s_D to approximate σ_D.

b) In Exercise 9.29, what is the probability of erroneously rejecting the alternative hypothesis if the mean oxygen content using sensors from company A is a full percentage point higher than when using sensors from company B? Use s_D to approximate σ_D.

In Exercises 9.32 through 9.35, apply Theorem 9.4 to find a $(1 - \alpha)100\%$ confidence interval for $\mu_X - \mu_Y$ based upon the data of the cited exercise.

9.32. Exercise 9.1, $\alpha = 0.02$
9.33. Exercise 9.2, $\alpha = 0.05$
9.34. Exercise 9.3, $\alpha = 0.10$
9.35. Exercise 9.4, $\alpha = 0.05$

In Exercises 9.36 through 9.39, apply Theorem 9.5 to find a $(1 - \alpha)100\%$ confidence interval for $\mu_X - \mu_Y$ based upon the data of the cited exercise.

9.36. Exercise 9.13, $\alpha = 0.05$
9.37. Exercise 9.17, $\alpha = 0.05$
9.38. Exercise 9.19, $\alpha = 0.01$
9.39. Exercise 9.22, $\alpha = 0.05$

In Exercises 9.40 through 9.42, apply Theorem 9.6 to find a $(1 - \alpha)100\%$ confidence interval for the difference of the means based upon the cited exercise.

9.40. Exercise 9.25, $\alpha = 0.01$
9.41. Exercise 9.27, $\alpha = 0.05$
9.42. Exercise 9.28, $\alpha = 0.05$

SECTION 9.5

9.43. A computer company purchases adder ICs from two sources, A and B. Independent random samples of 200 ICs are drawn from sources A and B, the result being that 3 ICs from A do not function properly, whereas 5 from B are unsatisfactory. Can it be concluded at the level $\alpha = 0.05$ that the proportion of unsatisfactory ICs is different for source A than it is for source B?

9.44. Of 50 smoke detectors randomly selected from the stock of company A, 10 fail to give a warning at the specified threshold. Of 70 detectors randomly selected from company B, 9 fail to give a warning at the given threshold. Can it be concluded at the $\alpha = 0.01$ level that the proportion of unacceptable detectors produced by company A is greater than the proportion produced by company B?

9.45. One group of 50 randomly selected employees is led through a routine of physical exercises before beginning work and a different group of 50 is not. Based upon a collection of performance criteria, each employee is rated over a period of six months as to whether or not he or she exceeds a specified grade of performance. Of the first group, 22 employees exceed the grade; of the second group, only 16 exceed the grade. At the $\alpha = 0.05$ level, what can be concluded?

9.46. It is desired to compare the rigor of two quality inspectors. A and B, each making determinations regarding the acceptability of final bicycle assemblies.

Independent random samples of 100 are taken from the bicycles inspected by the respective inspectors. Of the 100 inspected by A, 4 are declared unsatisfactory, while of the 100 inspected by B, 8 are declared unsatisfactory. Can it be concluded at the $\alpha = 0.10$ level that the inspectors differ in their manner of inspection?

9.47. Redo Exercise 9.46 under the assumptions that the common sample size is 200, 8 of the 200 are declared unsatisfactory by inspector A, and 16 of the 200 are declared unsatisfactory by inspector B.

9.48. Redo Example 9.12 under the assumption that the common sample size is 400, A is correct 350 times, and B is correct 300 times.

9.49. Redo Example 9.12 under the assumption that A makes the correct diagnosis in all 40 cases.

SECTION 9.6

9.50. Find the following critical values:
a) $f_{0.05,7,5}$ b) $f_{0.05,5,7}$ c) $f_{0.01,9,2}$ d) $f_{0.95,10,4}$ e) $f_{0.99,5,12}$

9.51. Bores are made in a steel part either by a human being manually operating a boring machine or by a robot boring machine. The following bore depths (in centimeters) are determined from independent random samples:

Human:	4.02	3.94	4.03	4.02	3.95	4.06
	4.00					
Robot:	4.01	4.03	4.02	4.01	4.00	3.99
	4.02	4.00				

Assuming both measurement variables are normally distributed, does the F test yield, at level $\alpha = 0.05$, the conclusion that the variances are unequal?

9.52. Raw material of a certain type is received by a chemical company from two sources, A and B. The active ingredient is nominally rated at 4% by weight; however, some variability is inevitable. A number of batches are randomly selected from those delivered by the two companies, and the following active-ingredient percentages are recorded:

Source A:	4.7	3.5	4.2	4.0	4.7	3.4	3.4
	4.2						
Source B:	4.2	4.0	4.2	3.9	4.0	4.3	4.2
	4.1	4.4					

At the $\alpha = 0.05$ level, can it be concluded that the variability of the product received from source A exceeds the variability corresponding to source B? Assume the samples values to have been randomly and independently selected from normal populations.

9.53. Two rather complicated queuing disciplines for CPU service are proposed by the designers of the operating system. Rather than compare them using an analytic approach involving a mathematical model, the designers decide to compare the disciplines by placing test machines into two settings considered to be uniform in terms of the rate and duration of arriving jobs. The following waiting times (in seconds) of randomly selected jobs earmarked for the two test machines are recorded:

CPU A: 4.62   2.51   0.12   0.67   2.30   6.12
          3.01   5.55   3.00   2.42   2.61   0.52
CPU B: 2.50   2.01   3.21   1.89   1.76   2.06
          2.53   2.20   1.90   2.45   2.10   1.60

Can it be concluded at the $\alpha = 0.01$ level that the variance resulting from the CPU-A queue discipline is greater than the variance resulting from the CPU-B discipline?

Letting $\lambda_1 = \sigma_X / \sigma_Y$ in Exercises 9.54 through 9.56, find the common sample size necessary to achieve the stated value of $\beta(\lambda_1)$ in the cited exercise.

9.54. Exercise 9.51, $\beta(1.3) = 0.40$
9.55. Exercise 9.52, $\beta(1.6) = 0.10$
9.56. Exercise 9.53, $\beta(2.0) = 0.15$

SECTION 9.7

9.57. The following randomly sampled data concern the tensile strength of steel relative to the percentage of carbon, 0.1% or 0.2%.

0.1%:   34,237   29,893   31,070   24,987
         33,593   32,218
0.2%:   32,934   40,231   40,193   33,560
         35,980   34,500

Apply the rank-sum test to determine whether or not it can be concluded at the $\alpha = 0.05$ level that 0.2% carbon produces a greater tensile strength than does 0.1% carbon.

9.58. Two different lubrications, A and B, are considered for a sealed bearing. A number of bearings are randomly selected to be treated with one or the other lubrication while in production. The following data give the times to failure (in hours of continuous service) of the various test bearings.

A: 432.1   542.2   491.4   201.5   467.8
      673.1   452.0
B: 462.7   390.1   361.4   521.9   422.4
      340.8

Employing the rank-sum test, can it be concluded at level $\alpha = 0.05$ that lubrication A is superior to lubrication B?

9.59. Holes are bored by two different pieces of equipment, A and B, each hole being of nominal diameter 3/8 inch. To test the diameter equality, 25 holes are drilled with each machine and the diameters are measured. According to the rank-sum test, do the following data lead to a conclusion at level $\alpha = 0.02$ that A and B

yield different hole diameters? Use a large-sample approach.

A: 0.3762   0.3764   0.3745   0.3758   0.3759
      0.3750   0.3762   0.3750   0.3752   0.3748
      0.3759   0.3769   0.3765   0.3749   0.3759
      0.3756   0.3750   0.3750   0.3746   0.3749
      0.3765
B: 0.3750   0.3745   0.3754   0.3742   0.3751
      0.3750   0.3749   0.3739   0.3751   0.3740
      0.3740   0.3736   0.3751   0.3753   0.3742
      0.3739   0.3744   0.3740   0.3756   0.3750
      0.3735

9.60. The following data are speed readings (in mph) of randomly selected pitches thrown by baseball pitchers A and B.

A: 85.7   87.3   87.5   88.4   84.0
      87.4   90.1   85.4   83.6
B: 88.5   92.1   85.6   90.2   88.9
      89.4   87.2   87.3   88.4

Does the rank-sum test yield the conclusion, at level $\alpha = 0.05$, that pitcher B throws faster (on average) than pitcher A?

In Exercises 9.61 through 9.63, apply the rank-sum test to the cited exercise.

9.61. Exercise 9.19
9.62. Exercise 9.20
9.63. Exercise 9.25 (assume the data are not paired)

In Exercises 9.64 through 9.66, apply the sign test to perform the hypothesis test of the cited exercise. Assuming each population is symmetric and possesses a mean, the sign test will concern the mean.

9.64. Exercise 9.27
9.65. Exercise 9.29
9.66. Exercise 9.30

In Exercises 9.67 through 9.69, apply the signed-rank test in the cited exercise. In each case, asssume the mean exists so that (according to the symmetry assumption necessary for the signed-rank test) the test concerns the mean of the distribution.

9.67. Exercise 9.25
9.68. Exercise 9.29
9.69. Exercise 9.30

SUPPLEMENTAL EXERCISES FOR COMPUTER SOFTWARE

In Exercises 9.C1 through 9.C4, reperform the cited exercise employing the given data in addition to the data originally provided.

9.C1. Exercise 9.15.

A:	1.00	1.01	1.03	1.00	0.99
	1.01	1.03	1.02	1.01	
B:	0.99	0.98	0.97	1.00	0.99
	1.00	1.01	0.98	0.97	

9.C2. Exercise 9.19.

A:	12.01	12.08	12.00	12.03
	12.04	12.06		
B:	15.97	16.00	16.12	16.10
	16.03	15.98	16.06	

9.C3. Exercise 9.25.

A:	10.5	8.6	7.9	16.0	19.4
	16.9	10.1	11.1		
B:	10.9	8.8	8.5	16.8	20.0
	17.4	10.3	11.0		

9.C4. Exercise 9.30.

A:	88.0	65.3	120.6	180.4	160.3
	188.0	90.3	55.5		
B:	98.5	70.3	130.7	196.9	160.2
	197.0	99.1	62.1		

10

||||||||||||||||||||||||||||||||||||||||||||||||||||||||||||||||||||||||||||||||||||||||||||||||||||||||||||||||||||||||||||||

HYPOTHESIS TESTS CONCERNING CATEGORICAL DATA

If data are presumed to arise from a random sample governed by a particular distribution and the data are placed into categories, then the manner in which the data distribute themselves among the categories reflects the underlying distribution. Based on this heuristic principle, we conclude that hypotheses concerning the underlying distribution can be investigated by examining categorical data. Theoretically, there is a basic theorem concerning the chi-square variable that allows easy implementation of the principle in a number of settings. The first section of the chapter treats the null hypothesis that the data result from a specific random variable and the second that they result from a random variable belonging to some given family. Section 10.3 treats a quite different problem. In it, we describe a test for the independence of two variables. Here, the test concerns the degree to which the null hypothesis of independence is reflected in the distribution of data into categories based upon the two variables. The concluding section provides a test to determine whether or not several populations are homogeneous with respect to a given classification scheme.

10.1 THE χ^2 TEST FOR GOODNESS-OF-FIT

The hypotheses to be addressed here concern the manner in which observed data **fit** a hypothesized distribution. The problem is attacked by analyzing the degree to which observed frequencies match expected frequencies. More precisely, categories are defined based upon the possible outcomes of an experiment, observations are classified based upon the categories, and the category frequencies resulting from the observations are compared to the frequencies that would be expected given the truth of some null hypothesis concerning the distribution. Each category frequency is a random variable, and the general problem involves the analysis of these categorical variables.

The classification of data into a finite number of categories can be viewed as a random phenomenon in which objects are placed into cells. As such, it is described by the multinomial distribution (discussed in Example 5.3). From the present perspective, there are n independent trials (observations), and on each trial there are r cells (categories) into which the observation may be placed. The categories represent a partitioning of the observations, and, for $i = 1, 2, \ldots, r$, there exists a random variable N_i counting the number of observations (out of the n) corresponding to cell i. For any given observation, there is a probability p_i that it falls into cell i. Thus, each observation can be viewed as an outcome of a random variable X possessing codomain $\{1, 2, \ldots, r\}$, where the probability mass function of X is defined by

$$P(X = i) = p_i$$

A particular instance of this type of variable occurs in the binary case, where the codomain of the trial random variable X is $\{0, 1\}$ and the corresponding binomial variables N_0 and N_1 count the number of failures (observations of 0) and successes (observations of 1) on n Bernoulli trials.

In effect, $p_1, p_2, \ldots, p_r$ represent proportions of the X population corresponding to the r cells into which the sample values are to be placed. Questions concerning the efficacy of these probabilities can be addressed in the form of hypothesis tests. For instance, one possible null hypothesis is

$$H_0: \quad p_1 = p_{10}, \; p_2 = p_{20}, \; \ldots, \; p_r = p_{r0}$$

where the p_{i0} are the hypothesized values of the cell probabilities.

If, for $i = 1, 2, \ldots, r$, e_i is the expected frequency of cell-i observations, then

$$e_i = E[N_i] = np_i$$

Given the preceding null hypothesis, $e_i = np_{i0}$ for $i = 1, 2, \ldots, r$. If H_0 reflects the true state of nature, then we should expect the observed frequencies to be close (in some sense) to the expected frequencies. For each i, closeness can be measured by $(n_i - e_i)^2$, where n_i denotes the observed frequency (sample value) corresponding to N_i. Intuitively, it can be seen that such a measure is inadequate, because the differences between the observed and expected frequencies for cells having large expected frequencies is likely to be greater than for cells having small expected frequencies. A solution to this problem is to divide each squared difference by the expected frequency of the cell and then sum the quantities $(n_i - e_i)^2/e_i$ over all i. The next theorem, stated without proof, provides the theoretical basis for this approach. Note that the theorem is stated in terms of the random observed frequencies N_i, not the empirical observed frequencies n_i.

THEOREM 10.1

(χ^2 Test for Goodness-of-Fit). If $N_1, N_2, \ldots, N_r$ and $e_1, e_2, \ldots, e_r$ are the observed and expected frequencies corresponding to a multinomial experiment consisting of n trials, then, as n tends to infinity, the distribution of the random variable

$$W = \sum_{i=1}^{r} \frac{(N_i - e_i)^2}{e_i}$$

approaches a chi-square distribution with $r - 1$ degrees of freedom.

It has been previously noted in regard to the t distribution that degrees of freedom are interpreted in terms of freely determined parameters. In the case of the χ^2 test of the null hypothesis

$$H_0: \quad p_i = p_{i0} \quad \text{for } i = 1, 2, \ldots, r$$

only $r - 1$ of the probabilities p_i are freely determined, because once $r - 1$ of them are known, the remaining one is determined by the fact that the sum of the probabilities must be one.

When employing Theorem 10.1 with the null hypothesis specifying the values of the probabilities $p_1, p_2, \ldots, p_r$, the alternative hypothesis is given by

$$H_1: \quad p_i \neq p_{i0} \quad \text{for at least one } i$$

Since H_1 does not specify the alternative values of the p_i, it is a composite hypothesis. If W denotes the chi-square statistic as stated in the theorem, then W is small when the null hypothesis is valid and large when the null hypothesis is invalid. Thus, the critical region of size α takes the form

$$C = \{w: w \geq \chi^2_{\alpha, r-1}\}$$

Because it is the limiting form of the statistic W that possesses a chi-square distribution, one must be prudent when employing Theorem 10.1. The usual rule of thumb is to employ the χ^2 test only if $e_i \geq 5$ for all i.

EXAMPLE 10.1

One way to measure computer performance is by CPU execution time. Since there are a number of common instruction types, an appropriate way to proceed is to take a weighted average

$$\sum_{i=1}^{r} p_i t_i$$

where p_i is the probability of calling instruction I_i and t_i is the execution time of I_i. A specific set of instructions $I_1, I_2, \ldots, I_r$ in conjunction with their occurrence probabilities comprise an *instruction mix*. The execution-time rating is dependent on the accuracy of the mix. One mix in use is the *Gibson mix* given in Table 10.1, the occurrence probabilities being given in the p_i column.

Consider the problem of determining whether the Gibson mix is suitable in a particular environment. Since there are $r = 7$ classifications in the mix, the appropriate hypothesis test is one in which the null hypothesis specifies the 7 cell probabilities. The expected and observed frequencies are given in the e_i and n_i columns, respectively, of Table 10.1. The observed frequencies correspond to a sample of 200 instructions that have been randomly selected from typical programs in use. Thus, for each i, $e_i = 200p_i$. The χ^2 test is applicable with $7 - 1 = 6$ degrees of freedom.

1. *Hypothesis test.* H_0: $\quad p_1 = 0.31, \ p_2 = 0.18, \ p_3 = 0.17, \ p_4 = 0.12,$
   $\qquad\qquad\qquad p_5 = 0.07, \ p_6 = 0.04, \ p_7 = 0.11$
   $\qquad\quad H_1$: The Gibson mix does not apply.
2. *Level of significance.* $\alpha = 0.05$
3. *Test statistic.* $W \sim \chi^2_6$
4. *Sample size.* $n = 200$
5. *Critical value.* $\chi^2_{0.05,6} = 12.592$
6. *Sample value.* From the data of Table 10.1,

TABLE 10.1 DATA FOR χ^2 TEST OF GIBSON MIX

Instruction Type	Probability of Instruction Type p_i	Expected Frequency e_i	Observed Frequency n_i
Transfers to and from main memory	0.31	62	72
Indexing	0.18	36	30
Branching	0.17	34	32
Floating-point arithmetic	0.12	24	14
Fixed-point arithmetic	0.07	14	22
Shifting	0.04	8	10
Miscellaneous	0.11	22	20

$$
w = \frac{(62 - 72)^2}{62} + \frac{(36 - 30)^2}{36} + \frac{(34 - 32)^2}{34} + \frac{(24 - 14)^2}{24}
$$

$$
+ \frac{(14 - 22)^2}{14} + \frac{(8 - 10)^2}{8} + \frac{(22 - 20)^2}{22}
$$

$$
= 12.15
$$

7. *Decision.* Since $w < 12.592$, the Gibson-mix probabilities are accepted.

Keep in mind the meaning of the test. It has not demonstrated the accuracy of the Gibson mix, only that, at the 0.05 level, the sample value is not significant and the Gibson-mix probabilities are therefore not rejected.

EXAMPLE 10.2

A particularly revealing use of the χ^2 test occurs in situations where theory predicts a certain distribution of outcomes among several classes and it is desired to check the theory against actual observations. A typical example of this type of problem arises in genetics, where theory often predicts some phenotypic ratio among progeny. In particular, if two traits are involved, say $\mathcal{A}$ and $\mathcal{B}$, and each trait is characterized by a dominant and recessive gene, then there will be four phenotypes, AB, Ab, aB, and ab, where A corresponds to the dominant characteristic of trait $\mathcal{A}$ being observed, a corresponds to the recessive characteristic of trait $\mathcal{A}$ being observed, and a similar notation applies to trait $\mathcal{B}$. Depending upon the type of crossing, various phenotypic ratios among the four phenotypes are predicted. Let us assume that for a particular generation the predicted ratio is $9:3:3:1$. This means that the hypothesized probabilities are $9/16$, $3/16$, $3/16$, and $1/16$, respectively. Suppose 50 progeny are to be observed and the χ^2 test is to be employed to determine if the data differ significantly from the hypothesized ratio. If four cells are employed and e_4 is the expected number of ab phenotypes, then

$$
e_4 = np_4 = 50 \times \frac{1}{16} = 3.125 < 5
$$

Thus, the fourth (ab) class is combined with the third (aB) class before application of the test, so that all expected frequencies are at least equal to 5. Table 10.2 gives the observed frequencies and the expected frequencies. The third class is comprised of types aB and ab.

1. *Hypothesis test.* H_0: $p_1 = \dfrac{9}{16}$, $p_2 = \dfrac{3}{16}$, $p_3 = \dfrac{4}{16}$

H_1: H_0 false

TABLE 10.2 DATA FOR χ^2 TEST OF PHENOTYPIC RATIO

Phenotype	e_i	n_i
AB	28.125	20
Ab	9.375	15
aB	9.375 ⎱ 12.5	12 ⎱ 15
ab	3.125 ⎰	3 ⎰

2. *Significance level.* $\alpha = 0.05$
3. *Test statistic.* $W \sim \chi_2^2$
4. *Sample size.* $n = 50$
5. *Critical value.* $\chi_{0.05,2}^2 = 5.991$
6. *Sample value.* From the data of Table 10.2,

$$w = \frac{(28.125 - 20)^2}{28.125} + \frac{(9.375 - 15)^2}{9.375} + \frac{(12.5 - 15)^2}{12.5} = 6.22$$

7. *Decision.* Since $w > 5.991$, reject the null hypothesis.

It should be noted that the χ^2 test applies to the classification. Thus, in Example 10.2, what has been rejected is that the three classes respectively possess 9/16, 3/16, and 4/16 of the population. We did not actually test the phenotypic ratio $9:3:3:1$. To do so would have required a larger sample.

10.2 THE χ^2 TEST IN THE PRESENCE OF UNKNOWN PARAMETERS

In Section 10.1, the χ^2 statistic was employed to test the fit of a multinomial distribution when the cell probabilities are assumed to be known. In general, the cell probabilities might depend on some unknown parameters. In such a case the observed frequencies must be compared to expected frequencies based upon estimates of the unknown parameters that have been derived from the data. In these circumstances, a generalized version of the χ^2 test continues to apply.

To formulate the generalized χ^2 test, suppose the cell probabilities depend upon the parameters $\theta_1, \theta_2, \ldots, \theta_m$, none of which can be expressed in terms of the others. Then the null hypothesis takes the form

$$H_0: \quad p_i = p_{i0}(\theta_1, \theta_2, \ldots, \theta_m) \quad \text{for } i = 1, 2, \ldots, r$$

while the alternative hypothesis states that H_0 is untrue. The χ^2 test applies to this test so long as each θ_k is replaced by its maximum-likelihood estimate when computing the expected frequencies and it is recognized that the resulting χ^2 statistic possesses $r - 1 - m$ degrees of freedom. In effect, one degree of freedom is subtracted for each parameter to be estimated. In sum, if

Test Statistic for χ^2 Test in the Presence of Unknown Parameters

$$W = \sum_{i=1}^{r} \frac{[N_i - np_{i0}(\hat{\theta}_1, \hat{\theta}_2, \ldots, \hat{\theta}_m)]^2}{np_{i0}(\hat{\theta}_1, \hat{\theta}_2, \ldots, \hat{\theta}_m)}$$

where, for $k = 1, 2, \ldots, m$, $\hat{\theta}_k$ is the maximum-likelihood estimator of θ_k, then W possesses a distribution that is approximately chi-square with $r - 1 - m$ degrees of freedom. Once again, the critical region lies to the right of the critical value χ^2_{r-1-m}.

Practical use of the generalized χ^2 test is often straightforward; however, this is due to a slight misuse of the test. Strict use of the test requires the maximum-likelihood estimators be evaluated relative to the likelihood function corresponding to the multinomial density. For a single parameter θ, this means that θ must be found by maximizing the likelihood function

$$\frac{n!}{n_1! n_2! \cdots n_r!} p_{10}(\theta)^{n_1} p_{20}(\theta)^{n_2} \cdots p_{r0}(\theta)^{n_r}$$

(see Section 7.6.1 and Example 5.3). Although technically correct, such a course often leads to mathematical difficulties. Thus, we will instead find the maximum-likelihood estimators of the parameters from the sample as a whole, rather than by using the multinomial likelihood function. From the perspective of sampling, the multinomial approach treats the data in a grouped fashion; that is, the cell frequencies determine the estimates. On the other hand, the ungrouped approach considers the data as arising from a random sample $X_1, X_2, \ldots, X_n$, and the estimates are found relative to the sample. Although the resulting estimates may differ, they are usually close enough for practical purposes.

The immediate use to which we will put the generalized χ^2 test is to test the fit of a probability model. Throughout much of Chapter 7, parametric estimation was employed under the condition that the probability model was given a priori. However, recalling the discussion of Chapter 1, it is the degree to which the probability model fits the data that determines its suitability. The model should not only fit in terms of its mean, variance, or some other parameter; it should also fit the data in an overall sense. Specifically, the density should be such that the probability $P(a < X < b)$ is indicative of the proportion of sample observations falling between a and b. This is precisely the kind of condition that can be checked by the χ^2 test. If n is the sample size, then $nP(a < X < b)$ is the expected observation frequency in the interval (a, b). Moreover, if we assume that the codomain of X is partitioned into a finite number of intervals, then the χ^2 test can be employed to check the degree to which the observed frequencies match the expected frequencies. If the resulting χ^2 statistic is small, then the fit is good and the data do not give reason to change the model; however, if the χ^2 statistic is large, then there is a marked discrepancy between the expected and observed frequencies, thereby indicating a poor fit.

FIGURE 10.1

Two densities that would not be distinguished by the χ^2 test

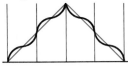

Three points should be noted. First, the class intervals determining the test must be chosen prior to the observations and should not be based upon the data. Second, once these intervals are chosen, the value of the resulting χ^2 statistic will be relative to this choice. Third, the test is sensitive to differences in the expected and observed categorization of the data, not to other features of the density. For instance, the χ^2 test will not discriminate between the two densities in Figure 10.1 because the areas under the two curves are identical over each class interval.

EXAMPLE 10.3

Suppose a supervisor hypothesizes that the number of employee absentees per day satisfies the conditions for a Poisson experiment. To test the conjecture, he or she takes a sample of 50 days over a period of six months and records the number of absentees. Since the Poisson distribution is discrete, each observation can be classified as to the number of absentees, these numbers corresponding to possible outcomes of the random variable. The possible outcomes, together with the class probabilities, expected frequencies, and observed frequencies are tab-

TABLE 10.3 DATA FOR χ^2 TEST OF POISSON DISTRIBUTION

Number of Absentees	p_{i0}	e_i	n_i
0	0.13	6.5	8
1	0.27	13.5	12
2	0.27	13.5	14
3	0.18	9.0	8
4	0.09	4.5 ⎫	3 ⎫
5	0.04	2.0 ⎬ 7.5	4 ⎬ 8
6	0.02	1.0 ⎭	1 ⎭

ulated in Table 10.3. It should be recognized that the last class is actually defined by the set of all observations greater than or equal to 6. Note, also, that the class probabilities are computed relative to the null hypothesis, which is that the population random variable X possesses a Poisson density $f(x; \lambda)$. Thus, there is one parameter to estimate from the data. According to Exercise 7.36, the maximum-likelihood estimator of λ is the sample mean. Hence,

$$\overset{*}{\lambda} = \frac{1}{50}(8 \times 0 + 12 \times 1 + 14 \times 2 + 8 \times 3 + 3 \times 4 + 4 \times 5 + 1 \times 6) = 2.04$$

It is $\overset{*}{\lambda}$ that is used to find the hypothesized class probabilities p_{i0}. Specifically, for $i = 0, 1, \ldots, 5$,

$$p_{i0} = P(X = i) = \frac{e^{-\overset{*}{\lambda}}\overset{*}{\lambda}{}^i}{i!} = \frac{e^{-2.04}(2.04)^i}{i!}$$

and, for $i = 6$,

$$p_{60} = P(X \geq 6) = 1 - \sum_{i=1}^{5} p_{i0}$$

Since the last three classes do not possess expected frequencies of at least 5, they are combined into a single class for the purposes of the χ^2 test.

1. *Hypothesis test.* H_0: X possesses a Poisson distribution.
               H_1: X does not possess a Poisson distribution.
2. *Level of significance.* $\alpha = 0.05$
3. *Test statistic.* Since there is one parameter being estimated from the data, the test statistic W is approximately chi-square with $5 - 1 - 1 = 3$ degrees of freedom.
4. *Sample size.* $n = 50$
5. *Critical value.* $\chi^2_{0.05,3} = 7.815$
6. *Sample value.*

$$w = \frac{(6.5 - 8)^2}{6.5} + \frac{(13.5 - 12)^2}{13.5} + \frac{(13.5 - 14)^2}{13.5} + \frac{(9.0 - 8)^2}{9.0} + \frac{(7.5 - 8)^2}{7.5} = 0.68$$

7. *Decision.* Since $w < 7.815$, the null hypothesis is accepted.

Notice the tightness of the fit. Not only is the value of the χ^2 statistic less than the critical value, it is very close to 0. Of course, even this tight a fit does not demonstrate conclusively that the model is accurate.

In Example 10.3 the model was discrete and thus the choice of classification cells was rather straightforward. In the continuous case the choice of classification intervals is more arbitrary, especially since the intervals must be defined independently of the sample data.

EXAMPLE 10.4

A manufacturer of light bulbs wishes to determine whether bulb life is normally distributed. A χ^2 test is planned and 100 randomly selected bulbs will be tested, their times to failure being measured in hours. The first problem is to assign observations to classes. Based upon experience, it is decided to employ the class boundaries given in Table 1.2. Suppose 100 bulbs are tested, the observations of Table 1.1 recorded, and the frequency distribution of Table 1.2 thereby obtained. Since the normal distribution possesses two parameters, μ and σ^2, these must be estimated from the data by using the maximum-likelihood estimators. In Example 7.17 these estimators were shown to be $\hat{\mu} = \bar{X}$ and

$$\hat{\sigma}^2 = \frac{1}{n} \sum_{k=1}^{n} (X_k - \bar{X})^2$$

(not S^2). Based upon the data in the frequency distribution of Table 1.2 (and using the frequency distribution for computing the sample values), we obtain

$$\overset{*}{\mu} = \bar{x} = \frac{104{,}450}{100} = 1045$$

$$\overset{*}{\sigma}{}^2 = \frac{704{,}500}{100} = 7045$$

$$\overset{*}{\sigma} = \sqrt{7045} = 84$$

where all quantities have been rounded off to the nearest integer. The class probabilities, expected frequencies, and observed frequencies are given in Table 10.4. Note that, for purposes of the χ^2 test, the upper and lower classes extend to $+\infty$ and $-\infty$, respectively. The hypothesized class probabilities are computed by using the standard normal distribution. For instance,

$$p_{10} = P(X < 850) = P\left(Z < \frac{850 - 1045}{84}\right) = 0.0102$$

and

$$p_{20} = P(850 < X < 900) = P\left(\frac{850 - 1045}{84} < Z < \frac{900 - 1045}{84}\right) = 0.0316$$

TABLE 10.4   DATA FOR χ^2 TEST OF NORMAL DISTRIBUTION

Class	X Range	Z Range	p_{i0}	e_i	n_i
	< 850	< −2.32	0.0102	1.02 ⎫	3 ⎫
1	850–900	−2.32−−1.73	0.0316	3.16 ⎬ 12.92	4 ⎬ 15
	900–950	−1.73−−1.13	0.0874	8.74 ⎭	8 ⎭
2	950–1000	−1.13−−0.54	0.1654	16.54	12
3	1000–1050	−0.54–0.06	0.2293	22.93	16
4	1050–1100	0.06–0.65	0.2183	21.83	28
5	1100–1150	0.65–1.25	0.1522	15.22	24
6	> 1150	> 1.25	0.1056	10.56	5

For this reason, the Z values corresponding to the class boundaries have been included in the table. To have all expected frequencies greater than 5, the first three classes are combined for the purposes of the χ^2 test.

1. *Hypothesis test.* H_0: X possesses a normal distribution.

   H_1: X does not possess a normal distribution.

2. *Level of significance.* $\alpha = 0.01$
3. *Test statistic.* Since there are two parameters being estimated, the test statistic W possesses a chi-square distribution with $6 - 1 - 2 = 3$ degrees of freedom.
4. *Sample size.* $n = 100$
5. *Critical value.* $\chi^2_{0.01,3} = 11.345$
6. *Sample value.*

$$
\begin{aligned}
w &= \frac{(12.92 - 15)^2}{12.92} + \frac{(16.54 - 12)^2}{16.54} + \frac{(22.93 - 16)^2}{22.93} \\
&\quad + \frac{(21.83 - 28)^2}{21.83} + \frac{(15.22 - 24)^2}{15.22} + \frac{(10.56 - 5)^2}{10.56} \\
&= 13.41
\end{aligned}
$$

7. *Decision.* Since $w > 11.35$, the null hypothesis is rejected and it is concluded that the bulb-life distribution is not normal. Given that $\alpha = 0.01$, the conclusion is quite strong. A glance at the histogram of Figure 1.1 tends to confirm the belief that the distribution is not normal—at least as indicated by the data of the present sample.

10.3 CONTINGENCY TABLES AND TESTING FOR INDEPENDENCE

Thus far, we have considered only a single classification criterion. Very often it is necessary to classify sample units according to two different criteria, say $\mathscr{C}_1$ and $\mathscr{C}_2$. A natural question arises as to whether or not the classification criteria are independent. If it is assumed that there are r cells relative to the first classification and c cells relative to the second, then the observational data can be arranged in a square array known as a **contingency table.** The table will have r rows, one for each class relating to criterion $\mathscr{C}_1$, and c columns, one for each class relating to criterion $\mathscr{C}_2$. If a sample of n units is to be taken and each unit is to be classified according to the two criteria, we will let N_{ij} denote the number of units occupying both cell i of criterion $\mathscr{C}_1$ and cell j of criterion $\mathscr{C}_2$. Thus, the contingency table will take the form given in Figure 10.2, where each n_{ij} is an observed frequency corresponding to the random variable N_{ij} and each block in the table corresponds to the ijth cell resulting from the two-way classification. In addition, if the sampling process is random, then each selection has a probability p_{ij} of being in the ijth cell, and the expected frequency of the ijth cell is

$$
e_{ij} = E[N_{ij}] = np_{ij}
$$

According to Theorem 10.1, the random variable

$$
W = \sum_{i=1}^{r} \sum_{j=1}^{c} \frac{(N_{ij} - np_{ij})^2}{np_{ij}}
$$

possesses a distribution that is approximately chi-square for sufficiently large n.

Now suppose the classification criteria are independent. If we label the cells

$\mathcal{C}_2$ Cells	$\mathcal{C}_1$ Cells			
	1	2	...	c
1	n_{11}	n_{12}	...	n_{1c}
2	n_{21}	n_{22}	...	n_{2c}
⋮	⋮	⋮		⋮
r	n_{r1}	n_{r2}	...	n_{rc}

FIGURE 10.2
Format of a contingency table

(rows) corresponding to $\mathscr{C}_1$ by 1, 2, . . . , r and the cells (columns) corresponding to $\mathscr{C}_2$ by 1, 2, . . . , c, then each classification corresponds to a random variable, say X_1 and X_2, respectively. Letting $u_i = P(X_1 = i)$ and $v_j = P(X_2 = j)$, independence of the classification criteria implies

$$p_{ij} = P(X_1 = i, X_2 = j) = P(X_1 = i)P(X_2 = j) = u_i v_j$$

for $i = 1, 2, \ldots, r$ and $j = 1, 2, \ldots, c$. Thus, the above approximately chi-square statistic takes the form

$$W = \sum_{i=1}^{r} \sum_{j=1}^{c} \frac{(N_{ij} - nu_i v_j)^2}{nu_i v_j}$$

According to the principle stated in Section 10.2, since the probabilities $u_1, u_2, \ldots, u_r$ and $v_1, v_2, \ldots, v_c$ are unknown, they must be estimated by their maximum-likelihood estimates. However, only $r - 1$ of the u_i and $c - 1$ of the v_j are freely determined, so that only $r + c - 2$ such estimates are required. Thus, W is approximately chi-square with

$$rc - 1 - (r + c - 2) = (r - 1)(c - 1)$$

degrees of freedom. Under independence, it can be shown that the maximum-likelihood estimators of u_i and v_j are

$$\hat{u}_i = \frac{M_i}{n}$$

$$\hat{v}_j = \frac{N_j}{n}$$

where

$$M_i = \sum_{j=1}^{c} N_{ij}$$

is the number of units in the ith cell for criterion $\mathscr{C}_1$ and

$$N_j = \sum_{i=1}^{r} N_{ij}$$

is the number of units in the jth cell for criterion $\mathscr{C}_2$. M_i is the random variable summing the frequencies of the ith row and N_j is the random variable summing the frequencies of the jth column. For a given set of observations, m_i and n_j are the sums of the observed frequencies in row i and column j, respectively. Thus, for $i = 1, 2, \ldots, r$ and $j = 1, 2, \ldots, c$, the maximum-likelihood estimates of u_i and v_j are

$$\overset{*}{u}_i = \frac{m_i}{n}$$

$$\overset{*}{v}_j = \frac{n_j}{n}$$

In words, the estimate of the probability of being classified in the ith cell relative to $\mathscr{C}_1$ is the number of elements in the ith cell (row) divided by the total number in the sample, and the estimate of the probability of being classified in the jth cell relative to $\mathscr{C}_2$ is the number of elements in the jth cell (column) divided by the total number in the sample.

Substituting the maximum-likelihood estimators into the statistic W above yields the test statistic

Test Statistic for χ^2 Test for Independence

$$W = \sum_{i=1}^{r} \sum_{j=1}^{c} \frac{\left(N_{ij} - \dfrac{M_i N_j}{n}\right)^2}{\dfrac{M_i N_j}{n}}$$

which possesses an approximately chi-square distribution with $(r-1)(c-1)$ degrees of freedom. When employing the χ^2 test, it is the sample value

$$w = \sum_{i=1}^{r} \sum_{j=1}^{c} \frac{(n_{ij} - e_{ij})^2}{e_{ij}}$$

that is compared to the critical value $\chi^2_{\alpha,(r-1)(c-1)}$, the expected frequencies being given by

$$e_{ij} = \frac{m_i n_j}{n}$$

In words, the expected frequency of the ijth cell is simply the product of the number of entries in the ith row times the number of entries in the jth column divided by the sample size.

EXAMPLE 10.5

The management of a company wishes to know whether or not job performance among its computer scientists is significantly dependent upon whether or not employees possess M.S. degrees or just B.S. degrees. In making the determination, a sample of 100 personnel will be randomly selected and each will be rated from 1 (highest) to 3 (lowest) relative to a standard company job-evaluation scheme. The data are recorded in the contingency table of Table 10.5, where, as is common practice, the expected frequencies have been placed in parentheses.

TABLE 10.5 CONTINGENCY TABLE FOR EDUCATIONAL LEVEL AND JOB PERFORMANCE

Degree	Performance Rating			Total
	Level 1	Level 2	Level 3	
B.S.	18 (20.4)	35 (35.36)	15 (12.24)	68
M.S.	12 (9.6)	17 (16.64)	3 (5.76)	32
Total	30	52	18	100

1. *Hypothesis test.* H_0: Classification variables independent

   H_1: Classification variables dependent
2. *Significance level.* $\alpha = 0.05$
3. *Test statistic.* Since $r = 2$ and $c = 3$, there are 2 degrees of freedom. Thus, $W \sim \chi_2^2$.
4. *Sample size.* $n = 100$
5. *Critical region.* $\chi_{0.05,2}^2 = 5.991$
6. *Sample value.* The expected frequencies are computed from the data. For instance, since $m_1 = 68$ and $n = 30$,

$$e_{11} = \frac{68 \times 30}{100} = 20.4$$

Using the data in the table, the sample value of W is

$$w = \frac{(20.4 - 18)^2}{20.4} + \frac{(35.36 - 35)^2}{35.36} + \frac{(12.24 - 15)^2}{12.24}$$
$$+ \frac{(9.6 - 12)^2}{9.6} + \frac{(16.64 - 17)^2}{16.64} + \frac{(5.76 - 3)^2}{5.76}$$
$$= 2.84$$

7. *Decision.* Since $w < 5.991$, the null hypothesis is not rejected, and thus there is no statistical proof of dependence at the 0.05 level of significance.

It is important in a setting like that of Example 10.5 to pay attention to the limitations of the conclusion. Acceptance of the null hypothesis does not preclude some variable dependency. Indeed, a look at Table 10.5 shows that the observed frequency of employees with M.S. degrees is slightly higher than expected in the highest performance category and slightly lower than expected in the lowest category. These differences, however, are not significant from the perspective of the χ^2 test; indeed, they are very slight and consequently could be attributable to chance. A second point concerns the environment in which the test has been conducted. If the employees are simply involved in coding, then the conclusion that job performance is independent of educational level is certainly plausible. It might be less plausible if the company were involved in the development of artificially intelligent systems or the design of parallel architectures.

10.4 HOMOGENEITY TESTS

Consider the problem of sampling several populations with respect to two or more categories. For instance, suppose four brands of relay switches are examined relative to the number of defective and nondefective switches. In such a case, there are four populations and two categories. If water samples are randomly collected from four different lakes and analyzed for appreciable levels of three different pollutants, then, since there are $2^3 = 8$ possible combinations of the three pollutants, there are four populations and eight categories. An important question that arises when examining apparently different populations is whether or not they are **homogeneous** with respect to the classification. Specifically, are there significantly different proportions of each population in the different categories, or is the categorical breakdown of each population essentially the same? In the case of the relay switches, it is useful to know if the proportions of defective switches corresponding to the various manufacturers are the same, or whether there is a significant difference. For the case of the water pollutants, if there is homogeneity in the manner in which the three pollutants distribute themselves, then there might be some causal relation that can be discovered.

If there are c populations and r categories, then, from a probabilistic perspective, there are c population random variables $X_1, X_2, \ldots, X_c$, each of which can take on r discrete values, say $1, 2, \ldots, r$. Thus, for $i = 1, 2, \ldots, r$ and $j = 1, 2, \ldots, c$, there is the probability

$$p_{ij} = P(X_j = i)$$

of X_j falling in category i. When an experiment is performed, there will be, for $j = 1, 2, \ldots, c$, n_j observations of X_j. If we record the number of observations of each random variable falling into each category, then the result is a contingency table of a kind similar to that discussed in Section 10.3, the differences being that the column totals, n_j, are fixed beforehand by the experimenter, and each observed frequency n_{ij} gives the number of category-i outcomes of the random variable X_j. The question of homogeneity is one concerning the equality of the probabilities (proportions) p_{ij} for fixed i. In words, are the proportions of the various random-variable observations falling into category i the same for all j? If they are, then the random variables $X_1, X_2, \ldots, X_c$ are identically distributed and represent homogeneous populations.

If, in a hypothesis test, the null hypothesis represents homogenity, then it takes the form

$$
\begin{aligned}
H_0: \quad p_{11} &= p_{12} = \cdots = p_{1c} \\
p_{21} &= p_{22} = \cdots = p_{2c} \\
\vdots \quad & \quad \vdots \qquad\qquad \vdots \\
p_{r1} &= p_{r2} = \cdots = p_{rc}
\end{aligned}
$$

Simply put, H_0 states that the proportions of the populations are constant in any given category (row). The alternative hypothesis states that there exists at least one row in which the probabilities are not all the same.

In the special case where there are just two categories, each random variable is binary, with its two possible outcomes being 0 (failure) and 1 (success). Thus, the contingency table takes the form given in Table 10.6, and, letting $p_j = P(X_j = 1)$, the null hypothesis is given by

$$H_0: p_1 = p_2 = \cdots = p_c$$

Homogeneity tests are carried out in exactly the same manner as contingency-table tests for two-way-classification for independence. The χ^2 goodness-of-fit test is employed with $(r-1)(c-1)$ degrees of freedom. The rule-of-thumb requirement that all expected frequencies must be at least 5 remains in effect.

TABLE 10.6 TWO-CATEGORY CONTINGENCY TABLE FOR HOMOGENEITY

Successes (1)	n_{11}	n_{12}	$\cdots$	n_{1c}
Failures (0)	$n_1 - n_{11}$	$n_2 - n_{12}$	$\cdots$	$n_c - n_{1c}$

EXAMPLE 10.6

Consider the relay-switch problem involving manufacturers A, B, C, and D, where 200 switches from each manufacturer are randomly selected for testing and the results of Table 10.7 are obtained. Are the populations homogeneous?

1. *Hypothesis test.* H_0: $p_1 = p_2 = p_3 = p_4$
   H_1: The proportions are not the same.
2. *Level of significance.* $\alpha = 0.01$
3. *Test statistic.* $W \sim \chi_3^2$
4. *Sample size.* $n_1 = n_2 = n_3 = n_4 = 200$
5. *Critical value.* $\chi_{0.01,3}^2 = 11.345$
6. *Sample value.* The expected frequencies are given in parentheses in the contingency table, where once again the formula is $e_{ij} = m_i n_j / n$. For instance,

$$e_{23} = \frac{770 \times 200}{800} = 192.5$$

The sample value of the approximately chi-square variable W is

$$
\begin{aligned}
w = {} & \frac{(7.5 - 6)^2}{7.5} + \frac{(7.5 - 4)^2}{7.5} + \frac{(7.5 - 4)^2}{7.5} + \frac{(7.5 - 16)^2}{7.5} \\
& + \frac{(192.5 - 194)^2}{192.5} + \frac{(192.5 - 196)^2}{192.5} + \frac{(192.5 - 196)^2}{192.5} + \frac{(192.5 - 184)^2}{192.5} \\
= {} & 13.71
\end{aligned}
$$

7. *Decision.* Because $w > 11.345$, the null hypothesis is rejected at the stringent significance level of 0.01. Thus, we can be quite confident that the proportions are not identical, or, equivalently, the brands of relay switches are not homogeneous relative to the criterion of defectiveness.

A close look at the data of Example 10.6 (Table 10.7) reveals that the rejection of the null hypothesis has resulted from the 16 defective switches in the sample of the brand D. If the χ^2 test is applied to the data and brand D is omitted, then the null hypothesis will not be rejected and it will be concluded that brands A, B, and C are homogeneous. In addition, if the samples from brands A, B, and C are grouped together and tested against the sample from brand D, then the contingency table given in Table 10.8 results. The corresponding approximately chi-square statistic W possesses 1 degree of freedom and has the sample value $w = 13.34$, a value that is

TABLE 10.7 CONTINGENCY TABLE FOR RELAY SWITCHES

Quality	Manufacturer				Total
	A	*B*	*C*	*D*	
Defective	6 (7.5)	4 (7.5)	4 (7.5)	16 (7.5)	30
Nondefective	194 (192.5)	196 (192.5)	196 (192.5)	184 (192.5)	770
Total	200	200	200	200	800

TABLE 10.8 REGROUPED CONTINGENCY TABLE

	Manufacturer		
Quality	**A–B–C**	**D**	**Total**
Defective	14 (22.5)	16 (7.5)	30
Nondefective	586 (577.5)	184 (192.5)	770
Total	600	200	800

significant even at the 0.001 level. Regrouping of the observations has demonstrated that the switches of brands A, B, and C can be treated as a single population differing significantly from the population determined by brand D.

EXAMPLE 10.7

The Pollution Standards Index (PSI) has six air-quality categories: good, moderate, unhealthful, very unhealthful, hazardous, and very hazardous. For the purposes of the present example, we will group these categories by adjacent pairs to arrive at the classification scheme: satisfactory, unhealthful, and hazardous. Suppose 150 air-quality measurements are randomly taken in different locations of city A over a period of six months and 200 are taken in city B over the same period, the results relative to the three-cell classification being given in the contingency table provided in Table 10.9. Testing the homogeneity of the data will allow us to make a determination of whether or not the cities possess significantly different air quality (relative to the given classification).

1. *Hypothesis test.* H_0: The cities are homogeneous relative to the air-quality classification, namely,
$$p_{11} = p_{12}$$
$$p_{21} = p_{22}$$
$$p_{31} = p_{32}$$
H_1: The cities are not homogeneous relative to the air-quality classification.
2. *Level of significance.* $\alpha = 0.05$
3. *Test statistic.* $W \sim \chi_2^2$
4. *Sample sizes.* $n_1 = 150$, $n_2 = 200$
5. *Critical value.* $\chi_{0.05,2}^2 = 5.991$
6. *Sample value.* The observed and expected frequencies given in Table 10.9 yield

TABLE 10.9 CONTINGENCY TABLE FOR AIR-QUALITY HOMOGENEITY TEST

	City		
Air Quality	**A**	**B**	**Total**
Satisfactory	107 (108.9)	147 (145.1)	254
Unhealthful	33 (36)	51 (48)	84
Hazardous	10 (5.1)	2 (6.9)	12
Total	150	200	350

$$w = \frac{(108.9 - 107)^2}{108.9} + \frac{(145.1 - 147)^2}{145.1} + \frac{(36 - 33)^2}{36}$$
$$+ \frac{(48 - 51)^2}{48} + \frac{(5.1 - 10)^2}{5.1} + \frac{(6.9 - 2)^2}{6.9}$$
$$= 8.68$$

7. *Decision.* Since $w > 5.991$, it is concluded that the air qualities are not homogeneous with respect to the given index.

EXERCISES

SECTION 10.1

10.1. A chemical process produces batches that are categorized according to purity into four different grades. It is thought that the percentages of batches in grades 1, 2, 3, and 4 are 20%, 45%, 25%, and 10%, respectively. Fifty batches are randomly selected and tested, their distribution into the four grades being given by

Level:	1	2	3	4
Frequency:	7	20	14	9

Do the data result in a conclusion, at the $\alpha = 0.05$ level, that the hypothesized percentages should be rejected?

10.2. A computer generates pseudorandom digits according to some routine. "Pseudorandom" means that the digit output should be (approximately) uniformly distributed among the digits 0 through 9, so that the probability of obtaining a given digit on any given outcome is 0.1. 100 randomly observed digits possess the following distribution.

Digit:	0	1	2	3	4	5
	6	7	8	9		
Frequency:	12	13	9	12	12	10
	9	12	8	3		

Perform χ^2 tests at the levels $\alpha = 0.10$, $\alpha = 0.05$, and $\alpha = 0.01$ and state your conclusions. More generally, find the P value.

10.3. There are numerous page-replacement strategies to keep the pages of a running process (program) in primary memory, the goal being to keep down the time required to bring code and data to the CPU. Consider a computer possessing three levels of memory—primary, secondary, and tertiary—each level being more remote from the CPU than the former. It is claimed that, after some period of stabilization, 60% of a program's pages are in primary memory, 35% are in secondary memory, and 5% are in tertiary memory. Fifty pages in various running programs are randomly selected, and the following memory data are recorded.

Memory:	1	2	3
Frequency:	25	16	9

Can the claim of the operating system designer be rejected at the $\alpha = 0.05$ level?

10.4. A traffic engineer, who is new to his job, reads a report dated four years earlier that the distribution of traffic leaving the city over five major arteries is 12%, 23%, 35%, 18%, and 12% for arteries A, B, C, D, and E, respectively. To check the current validity of the distribution, 200 randomly selected cars are observed leaving the city and the number leaving on each of the five arteries is recorded. The following data are obtained.

Artery:	A	B	C	D	E
Frequency:	20	39	64	36	41

Based upon the new data, does the engineer reject, at the $\alpha = 0.05$ level, the old proportions?

10.5. A company employs five machinists, each of whom produces various parts. A random selection of 100 defective parts is chosen and the number resulting from each machinist is recorded:

Machinist:	A	B	C	D	E
Frequency:	17	40	12	16	15

Can it be concluded at the $\alpha = 0.01$ level that defective parts are not uniformly distributed among the five machinests?

10.6. Referring to Exercise 10.5, discuss the implications of disregarding the 40 defectives due to machinist B and simply focusing attention on the remaining four machinists.

10.7. A particular theoretical model suggests that 95% of all offspring in a certain population should carry a certain gene, whereas 5% should not. A random sample of 100 is taken and 10% are found not to carry the gene. Is the model rejected at the $\alpha = 0.01$ level? Find the P value.

10.8. Prior to a retraining program, employees are placed into five job categories based on their various levels of skill. The result is that categories A, B, C, D, and E include 13%, 24%, 35%, 21%, and 7%, respectively, of the employees. Following retraining, 50 employees are tested and the following distribution is found.

Category:	A	B	C	D	E
Frequency:	8	17	17	5	3

Can it be concluded, at the $\alpha = 0.10$ level, that the retraining program has resulted in a significant change in performance?

SECTION 10.2

10.9. An investigator collects the gas-mileage data given in the following frequency distribution. Based upon the χ^2 test, is there reason to doubt, at the $\alpha = 0.10$ level, a conjecture that the underlying distribution is normal?

Class Mark:	30	31	32	33	34
Frequency:	15	50	100	65	20

10.10. A quality control engineer wishes to model the defective transistor count of Example 1.1 as a Poisson random variable. Given the data of Table 1.3, is such an approach rejected by the χ^2 test at level $\alpha = 0.05$?

10.11. Again referring to Table 1.3, is the supposition of a binomial distribution with $n = 4$ rejected at level $\alpha = 0.05$?

10.12. Does the χ^2 test reject, at level $\alpha = 0.20$, the supposition that the data of Table 1.6 derive from an exponential distribution?

10.13. Using the P value, is it plausible to assume that the megaflop data of Table 1.8 derive from a normal distribution?

10.14. It is conjectured that the radial distance (in yards) by which a missile misses its target possesses a uniform distribution. Twenty firings are observed and the following frequency distribution results.

Class Mark:	50	150	250	350
Frequency:	10	5	3	2

Is the uniformity conjecture refuted by the χ^2 test at level $\alpha = 0.01$? See Example 7.16 for the appropriate maximum-likelihood estimator.

10.15. It is conjectured that the sound-level readings (in decibels) from street traffic during a specific time period will possess a normal distribution. To test the hypothesis, the following frequency distribution is obtained.

Class:	<66	66–68	68–70	70–72	72–74	>74
Frequency:	4	24	35	15	8	4

Is the hypothesis sustained by the χ^2 test at the $\alpha = 0.05$ level?

10.16. Is the hypothesis that the data of Table 10.3 derive from a normal distribution refuted, at level $\alpha = 0.05$, by the χ^2 test?

SECTION 10.3

10.17. Electric motors are purchased from two manufacturers, A and B. Two hundred motors are randomly selected and subjected to continuous use under rigorous testing conditions. Depending on its time to failure, each motor is placed into a category: less than 100 hours gets category 1, between 100 and 200 hours gets category 2, between 200 and 300 hours gets category 3, and more than 300 hours gets category 4. Table 10.10 results. Is it concluded at the $\alpha = 0.10$ level that longevity classification is dependent on the manufacturer?

TABLE 10.10 CONTINGENCY TABLE FOR EXERCISE 10.17

Manufacturer	Time-to-Failure Category			
	1	2	3	4
A	8	34	42	20
B	9	50	28	9

10.18. A company makes three grades of lawn mower, A, B, and C. While it is true that the price of lawn mower A is greater than that of mowers B and C, so too is the quality. A similar comment applies to mower B's being superior and more expensive than mower C. The company wishes to determine whether or not there is a significant dependence between customer satisfaction and lawn-mower grade. It rates customer-satisfaction inquiries as (1) greatly satisfied, (2) moderately satisfied, and (3) dissatisfied. Table 10.11 results from a random customer sampling. Does the χ^2 test lead to a conclusion of dependence between grade and satisfaction level at the $\alpha = 0.05$ level of significance?

TABLE 10.11 CONTINGENCY TABLE FOR EXERCISE 10.18

Grade	Satisfaction Level		
	1	2	3
A	42	30	8
B	28	50	16
C	25	32	9

10.19. Machine bolts are categorized into three grades, 2, 5, and 8, each higher grade possessing more stringent requirements for a bolt to be declared nondefective. Table 10.12 gives the results of testing a random sample of bolts for defectives. Does the χ^2 test, at level $\alpha = 0.05$, result in a conclusion that there exists dependence between grade classification and defective–nondefective classification?

TABLE 10.12 CONTINGENCY TABLE FOR EXERCISE 10.19

Quality	Grade		
	2	5	8
Defective	12	16	24
Nondefective	208	194	150

10.20. The chi-square methodology employed throughout the chapter has depended on Theorem 10.1, which involves a limit. Hence, the statistic W is only approximated by a chi-square distribution. In the case of 2-by-2 contingency tables, this approximation can be poor for small samples. Many authors recommend the use of the **Yates' correction for continuity** in the case of 2-by-2 tables (at least when some of the expected frequencies are between 5 and 10). In the Yates' correction, the numerator terms in the summation giving the sample value of W are

$$(|n_{ij} - e_{ij}| - 0.5)^2$$

An imaging scientist wishes to determine whether the intensity of light has an effect on the ability of his or her algorithm to recognize a particular target. Considering only daytime recognition, the scientist classifies the light source as being high or low intensity and performs a series of randomly chosen recognition tests. The outcome of each test is either positive or negative, depending on whether or not the target is or is not recognized. Use the Yates' correction to perform a χ^2 test on the data of Table 10.13. Let $\alpha = 0.05$.

TABLE 10.13 CONTINGENCY TABLE FOR EXERCISE 10.20

Recognition	Light Source	
	High	Low
Positive	36	9
Negative	24	31

10.21. Repeat Exercise 10.20 without using Yates' correction.

SECTION 10.4

10.22. A paint company markets three grades, A, B, and C, of home exterior water-based paint. To test the durability of the different grades, each is applied to 30 randomly selected test houses, there being 90 test homes in all. After two years, the exterior of each house is rated as excellent, good, or poor. To allow the conclusions of the test to have wide applicability, the test houses are selected in widely varying climatic zones. The results of the rating are given in Table 10.14. Can it be concluded at the $\alpha = 0.05$ level that the paint grades are heterogenous relative to the rating system?

TABLE 10.14 CONTINGENCY TABLE FOR EXERCISE 10.22

Durability	Paint Grade		
	A	B	C
Excellent	32	18	8
Good	12	24	24
Poor	6	8	18

10.23. A company utilizes three 8-hour shifts to maintain 24-hour production. One hundred units are randomly chosen from each shift and judged defective or nondefective, the result of the testing being given in the following table.

Shift:	Day	Night	Graveyard
Defectives:	6	12	16

At the $\alpha = 0.05$ level, is it concluded that the shifts are nonhomogeneous with respect to the production of defective units?

10.24. Owing to interference, a proportion of satilite signals are too garbled to be interpreted. Three filters, A, B, and C, are tested to see if they result in differing degrees of effectiveness. Fifty messages are randomly chosen for each filter, the number of successful interpretations being given in the following table.

Filter:	A	B	C
Successes:	42	36	32

Is the hypothesis of homogeneity rejected at significance level $\alpha = 0.10$?

10.25. Reconsider Exercise 10.24 in the event the number of successes for filter C is 20.
a) Retest for homogeneity.

b) Test A and B for homogeneity while ignoring filter C.

c) Group the observations of A and B together and test the grouped data against filter C.

d) Interpret the results.

10.26. A *page fault* occurs when a running program requires a page of memory and that page is not in main memory, thus forcing a search of secondary memory and resulting in inefficient performance. (See Exercise 10.3 for a related application.) A benchmark program is run three times, employing replacement strategies A, B, and C on each succeeding run. On each run, 100 page requests are randomly observed to determine whether or not a page fault occurs, the results being tabulated in the following table.

Strategy:	A	B	C
Faults:	22	35	27

Based upon homogeneity, can it be concluded at the $\alpha = 0.01$ level that the strategies are not equivalent? Find the P value.

10.27. Cars are randomly stopped to check for safety violations. Each is classified as safe, acceptable, or unsafe. The cars are cross-classified according to longevity: new (less than 3 years), moderate (between 3 and 7 years), and old (more than 7 years). The results of the classification are given in Table 10.15. Can it be concluded at the $\alpha = 0.05$ level that longevity classes are not homogeneous with respect to safety?

TABLE 10.15 CONTINGENCY TABLE FOR EXERCISE 10.27

Safety Classification	Longevity Classification		
	New	Moderate	Old
Safe	83	73	28
Acceptable	25	35	39
Unsafe	7	12	22

10.28. Reexamine the data of Exercise 10.27 by grouping the new and moderate cars together.

11

ANALYSIS OF VARIANCE

One of the premier approaches to statistical inference involves the analysis of variation in experimental data. Put simply, the analysis-of-variance approach compares the variation attributable to the condition one is trying to detect to the variation attributable to the inherent randomness occurring within sample data. The problem addressed in the present chapter concerns the detection of differences among several means. The particulars aside, attention should be paid to the general principles behind the approach, especially the development of mathematical models corresponding to specific experimental designs. The chapter commences by explaining the basic fixed-effects, one-way-classification model and by developing the appropriate F statistic to test for equality of means in the model. Should the hypothesis of equal means be rejected, it is useful to employ some method for comparing individual pairs of means. In the second section two such multiple-comparison methods are discussed. The third section treats a different model in which the effects under study are assumed to have been randomly selected, rather than fixed as they were in the opening model. Finally, type II error is addressed.

11.1 ANALYSIS OF VARIANCE IN THE FIXED-EFFECTS, ONE-WAY-CLASSIFICATION MODEL

The hypothesis tests of Chapter 8 concerned a single parameter of a single population, whereas those of Chapter 9 concerned parameters from each of two populations. Relative to the present chapter, the testing of the difference of two means in Chapter 9 is particularly noteworthy, especially the two-tailed hypothesis test

$$H_0: \ \mu_X = \mu_Y$$

$$H_1: \ \mu_X \neq \mu_Y$$

Several approaches were taken regarding this test: a Z test where the two populations were sampled independently, a t test where the two populations were sampled independently, and a paired t test where the populations were sampled in pairs. From the perspective of comparing the effects of two processes, or treatments, the test is useful for making a decision as to whether or not the two processes have significantly different effects. For instance, are the protections afforded by two paint sealers essentially equivalent, or is there a significant difference between them? Is there a significant difference between the filtering effects of two image-processing algorithms? Do two different heat treatments result in significantly different hardnesses? Many comparison problems require testing the equality of effects; however, the tests of Chapter 9 are restricted to comparing two means. What happens when we wish to compare three sealers, or four filtering algorithms, or five heat treatments?

In the present section, we discuss a method for testing several means that depends on an analysis of the sources of variation within the experiment. In effect, the approach is to partition the variation into meaningful components, the purpose being to compare the variation attributable to differences between the populations to the variation attributable to random variation within the various populations. If the former is sufficiently large in comparison to the latter, then the procedure will result in a rejection of the null hypothesis that all populations possess the same mean.

Two points should be kept in mind. First, the analysis-of-variance methodology has far-reaching impact in statistical inference, and one should pay careful attention to the approach—not simply the particulars of its application in the present circumstances. Indeed, it will play a key role in the remainder of the text, especially in the design of experiments. Second, the methodology is model dependent. Each application depends on a suitable mathematical model. Should the experimental set-up cause the assumptions of the model to be violated to a substantial degree, then the results will not be meaningful.

11.1.1 One-Way Classification

In the present chapter it is assumed that data are obtained from r independent populations. Rigorously, there are r independent random variables $X_1, X_2, \ldots, X_r$ possessing means $\mu_1, \mu_2, \ldots, \mu_r$, respectively, and common variance σ^2. A total of r independent random samples are obtained, these being

$$X_{11}, \ X_{12}, \ \ldots, \ X_{1n_1}$$
$$X_{21}, \ X_{22}, \ \ldots, \ X_{2n_2}$$
$$\vdots \qquad \vdots \qquad \qquad \vdots$$
$$X_{r1}, \ X_{r2}, \ \ldots, \ X_{rn_r}$$

where, for $i = 1, 2, \ldots, r$, $X_{i1}, X_{i2}, \ldots, X_{in_i}$ constitute a random sample arising from X_i. For instance, $X_1, X_2, \ldots, X_r$ might be the yield points of different alloys, or the execution times of different programs written to implement the same task. If the yield points are essentially the same or the execution times only differ insignificantly, then it should be expected that the null hypothesis

$$H_0: \ \mu_1 = \mu_2 = \cdots = \mu_r$$

will be accepted if the data of the experiment are employed to find the sample value of an appropriate test statistic. Our intent is to develop test statistics based upon an

TABLE 11.1 DATA FORMAT FOR ONE-WAY-CLASSIFICATION ANALYSIS OF VARIANCE

Treatment	Observed Responses				Total	Mean
1	x_{11}	x_{12}	$\ldots$	x_{1n_1}	$T_{1.}$	$\bar{x}_{1.}$
2	x_{21}	x_{22}	$\ldots$	x_{2n_2}	$T_{2.}$	$\bar{x}_{2.}$
$\vdots$	$\vdots$	$\vdots$		$\vdots$	$\vdots$	$\vdots$
r	x_{r1}	x_{r2}	$\ldots$	x_{rn_r}	$T_{r.}$	$\bar{x}_{r.}$

analysis of variation in the experimental data. Since the r populations under study are to be examined with respect to a single criterion, the model we will be discussing leads to an analysis-of-variance technique known as **one-way-classification,** or **single-factor,** analysis of variance.

A common protocol leads naturally to the one-way-classification model. Consider an experiment in which a homogeneous collection of experimental units is randomly partitioned into subcollections $C_1, C_2, \ldots, C_r$ of sizes $n_1, n_2, \ldots, n_r$ and each subcollection is subjected to a different test treatment. The resulting r data sets can be viewed as r sample-value sets arising from r random samples. From such a perspective, the random variable X_i models the population of measurements that correspond to experimental units given treatment i. In line with this protocol, the r populations are often called **treatment populations** and the random variable X_i is called the ith **response.** This terminology is reflected in Table 11.1, which illustrates the manner in which the observational data resulting from a one-way classification are formatted. The observations corresponding to each treatment are listed in the rows of the table, with the appropriate empirical means at the ends of the rows. (The reason for the total notation "$T_{i.}$" and the mean notation "$\bar{x}_{i.}$" will shortly become clear.) It should be kept in mind that in the general model the samples can be of varying sizes; indeed, there are a variety of practical reasons why each treatment might not be applied to the same number of experimental units. Moreover, even if there happen to be an equal number of experimental units in each classification, at the completion of the experiment there might not be an equal number of valid measurements arising from each classification.

EXAMPLE 11.1

Before being used in production, raw material is often checked for purity. For instance, chemical powders are heated and the residual ash content, which represents the impurities, is weighed. The purity (in percentage) is then given by

$$\left(1 - \frac{\text{ash weight}}{\text{initial weight}}\right) 100\%$$

Suppose a pharmaceutical company wishes to compare the purities of a chemical powder that it receives from four suppliers. Various batches of the powder are randomly selected from shipments of each of the suppliers—4 from supplier 1, 5 from supplier 2, 4 from supplier 3, and 3 from supplier 4. The purity of each of the selected batches is ascertained, and the results (in percent) are presented in Table 11.2.

TABLE 11.2 ONE-WAY-CLASSIFICATION OF POWDER PURITIES

Supplier	Observations					Total	Mean
1	99.3	99.4	98.8	99.4		396.9	99.22
2	99.8	97.4	98.9	99.0	98.6	493.7	98.74
3	98.2	97.2	96.4	98.3		390.1	97.53
4	98.7	99.6	99.2			297.5	99.16

11.1.2 Fixed-Effects Model

The foregoing assumptions are rather general, and a much more restrictive linear constraint will now be imposed. While the resulting model must be viewed from an abstract mathematical perspective, it is useful to recognize its heuristic foundation in the treatment-response experimental framework. Assume for the moment that each treatment population is a subpopulation of an overall population possessing mean μ, called the **overall** or **grand** mean. Each treatment population is assumed to possess a mean μ_i (actually the mean of X_i) that differs from μ by α_i, the **effect** of treatment i. Thus,

$$\mu_i = \mu + \alpha_i$$

Further assume that the grand mean is the weighted average of the treatment means, namely,

$$\mu = \sum_{i=1}^{r} \frac{n_i \mu_i}{N}$$

where

$$N = n_1 + n_2 + \cdots + n_r$$

is the total number of observations. Since each observation is a random phenomenon, the jth observation of the ith response variable X_i takes the form

$$X_{ij} = \mu + \alpha_i + \mathrm{E}_{ij}$$

where E_{ij} is a random variable measuring the deviation of X_{ij} from the treatment mean μ_i. Note that the expected value of E_{ij} is zero:

$$E[\mathrm{E}_{ij}] = E[X_{ij}] - \mu - \alpha_i = 0$$

In addition,

$$\mathrm{Var}\,[\mathrm{E}_{ij}] = \mathrm{Var}\,[X_{ij}] = \sigma^2$$

and, owing to the independence of the random samples, for any pairs of indices (i, j) and (i', j') such that $(i, j) \neq (i', j')$,

$$\mathrm{Cov}\,[\mathrm{E}_{ij}, \mathrm{E}_{i'j'}] = 0$$

If we now impose the constraint that the random errors E_{ij} are normally distributed, then the following linear mathematical model results.

Fixed-Effects, One-Way-Classification Model

Equation:	$X_{ij} = \mu + \alpha_i + E_{ij}$
Assumptions:	1. $E_{ij} \sim N(0, \sigma)$
	2. The E_{ij} comprise an independent collection of random variables.
	3. μ is the weighted average of $\mu_1, \mu_2, \ldots, \mu_r$.

It should be noted that assumption 3 is equivalent to the condition

$$\sum_{i=1}^{r} n_i\alpha_i = 0$$

This can be seen from the identity

$$\sum_{i=1}^{r} n_i\alpha_i = \sum_{i=1}^{r} n_i(\mu_i - \mu) = \sum_{i=1}^{r} n_i\mu_i - N\mu$$

Given the model, suppose we wish to test the null hypothesis that all treatment means are equal, namely,

$$H_0: \mu_1 = \mu_2 = \cdots = \mu_r$$

Under this hypothesis,

$$\mu = \sum_{i=1}^{r} \frac{n_i\mu_i}{N} = \sum_{i=1}^{r} \frac{n_i\mu_1}{N} = \mu_1$$

Similarly, $\mu = \mu_i$ for $i = 2, 3, \ldots, r$. Thus,

$$\alpha_i = \mu_i - \mu = 0$$

for all i, and the hypothesis of equal treatment means can be phrased in terms of zero effects. Hence, it takes the form

$$H_0: \alpha_1 = \alpha_2 = \cdots = \alpha_r = 0$$

The corresponding alternative hypothesis states that not all treatment effects are equal to zero.

Implicit in the discussion thus far is that the random variables $X_1, X_2, \ldots, X_r$ represent a total of r treatment populations (of interest to the experimenter). Concomitantly, each effect α_i is modeled as a fixed constant relating to X_i and the model is known as the **fixed-effects** one-way classification. A different scenario is one in which the experimenter views the particular r treatments as a random sample of some larger collection of treatments. Then the effect α_i is a random variable depending upon which treatment has been chosen from the set of treatments. The mathematical model describing this latter case is called the **random-effects** model and will be discussed in Section 11.3.

Although the "treatment-effect-response" terminology is valuable in providing an intuitive perspective on analysis-of-variance modeling, it is the rigorous mathematical model that supports both theory and application. For instance, while it might be beneficial to conjure up the notion that the grand mean μ is the mean of some superpopulation for which $X_1, X_2, \ldots, X_r$ represent subpopulations, the model itself does not postulate the existence of such a population; rather, μ is simply a model parameter associated with the r response random variables and is defined as the weighted average of the means.

Certain random variables play key roles in analyzing the variance of the responses. For $i = 1, 2, \ldots, r$,

$$T_{i.} = X_{i1} + X_{i2} + \cdots + X_{in_i}$$

gives the (random) sum of the ith-treatment observations, and

$$\bar{X}_{i.} = \frac{T_{i.}}{n_i}$$

which is the average of the ith-treatment observations, is the sample mean associated with the random sample $X_{i1}, X_{i2}, \ldots, X_{in_i}$. In addition

$$T_{..} = \sum_{i=1}^{r} \sum_{j=1}^{n_i} X_{ij}$$

is the total sum of the observations, and we let

$$\bar{X}_{..} = \frac{T_{..}}{N}$$

denote the average of all the observations. $T_{..}$ and $\bar{X}_{..}$ are called the **grand total** and **grand average,** respectively. Although the overbar is employed for the grand average, $\bar{X}_{..}$ is not a sample mean, because the collection of response variables X_{ij} does not constitute a random sample. Nevertheless, $\bar{X}_{..}$ is a random variable, and, according to the next theorem, it is an unbiased estimator of μ.

THEOREM 11.1

Under the conditions of the fixed-effects, one-way-classification model,

$$E[\bar{X}_{..}] = \mu$$

and

$$E[\bar{X}_{i.} - \bar{X}_{..}] = \alpha_i$$

Proof: Direct substitution of the definitions in conjunction with the linearity of the expected value gives

$$E[\bar{X}_{..}] = \frac{1}{N} \sum_{i=1}^{r} \sum_{j=1}^{n_i} E[X_{ij}]$$

$$= \frac{1}{N} \sum_{i=1}^{r} \sum_{j=1}^{n_i} (\mu_i + \alpha_i)$$

$$= \frac{1}{N} \sum_{i=1}^{r} n_i(\mu_i + \alpha_i)$$

$$= \sum_{i=1}^{r} \frac{n_i}{N} \mu_i + \frac{1}{N} \sum_{i=1}^{r} n_i \alpha_i$$

$$= \mu + 0$$

and

$$E[\bar{X}_{i.} - \bar{X}_{..}] = E[\bar{X}_{i.}] - E[\bar{X}_{..}] = \mu + \alpha_i - \mu = \alpha_i \qquad \blacksquare$$

Note that in Table 11.1 we have written $T_{i.}$ to denote the observed total resulting from the ith treatment population. To be thoroughly consistent, a lower-case t should be employed; however, we have followed the common practice of employing the upper case in both the random and observed settings. Whether or not $T_{i.}$ refers to the random variable or the outcome of the random variable should be clear from the context.

Since the analysis of variation is the intent, it is customary to utilize special notation to represent various sums of squares. As noted at the outset, the analysis will be carried out by splitting the variation into meaningful components. It will be these components that yield appropriate test statistics for various hypothesis tests— in particular, the fixed-effects null hypothesis concerning zero effects. Of immediate interest are the **total sum of squares**

$$SST = \sum_{i=1}^{r} \sum_{j=1}^{n_i} (X_{ij} - \bar{X}_{..})^2$$

the **between-treatment sum of squares**

$$SSA = \sum_{i=1}^{r} n_i(\bar{X}_{i.} - \bar{X}_{..})^2$$

and the **within-treatment sum of squares**

$$SSE = \sum_{i=1}^{r} \sum_{j=1}^{n_i} (X_{ij} - \bar{X}_{i.})^2$$

SST, SSA, and SSE are random variables that measure variation in the response sample. SST measures the total variation of the responses from the grand average, SSA measures the variation of the treatment sample means from the grand average, and SSE measures the variation resulting from within-sample deviations of observations from their particular sample means. Thus, the terminology ''total,'' ''between-treatment,'' and ''within-treatment'' is appropriate. The next theorem is crucial because it shows that the total sum of squares is partitioned by the within-treatment and between-treatment sums of squares.

THEOREM 11.2

(Sum-of-Squares Identity). Under the fixed-effects, one-way-classification model,

$$SST = SSA + SSE$$

Proof: The total sum of squares can be rewritten as

$$SST = \sum_{i=1}^{r} \sum_{j=1}^{n_i} [(\bar{X}_{i.} - \bar{X}_{..}) + (X_{ij} - \bar{X}_{i.})]^2$$

$$= \sum_{i=1}^{r} \sum_{j=1}^{n_i} (\bar{X}_{i.} - \bar{X}_{..})^2$$

$$+ 2 \sum_{i=1}^{r} \sum_{j=1}^{n_i} (\bar{X}_{i.} - \bar{X}_{..})(X_{ij} - \bar{X}_{i.})$$

$$+ \sum_{i=1}^{r} \sum_{j=1}^{n_i} (X_{ij} - \bar{X}_{i.})^2$$

The middle term can be evaluated by first summing over j for each i. Each such sum over j is the sum of the deviations of the observations of X_i from the sample mean of the X_i sample. In each case these deviations must sum to zero, and therefore the middle term vanishes. Moreover, because there are no j subscripts in the first term, it reduces to SSA. Finally, the third term is simply SSE. ∎

Our intent is to detect differences among the treatment means by examining different sources of variation. Specifically, if there are differences among the means, it might be expected that the between-treatment sum of squares SSA will be larger relative to the within-treatment sum of squares SSE than if the means were identical. To devise a test statistic based upon this observation, we require the distributions of SSA and SSE. These are given in the next theorem, which is stated without proof.

THEOREM 11.3　　Under the assumptions of the fixed-effects, one-way-classification model, if the null hypothesis

$$H_0: \alpha_1 = \alpha_2 = \cdots = \alpha_r = 0$$

is true, then SSE/σ^2 and SSA/σ^2 are independent random variables possessing chi-square distributions with $N - r$ and $r - 1$ degrees of freedom, respectively.

Applying Theorems 11.2 and 11.3 in conjunction with Theorem 5.7 shows that SST/σ^2 possesses a chi-square distribution with $N - 1$ degrees of freedom. Moreover, according to Definition 9.2, an immediate corollary of Theorem 11.3 is that, under the null hypothesis,

$$F = \frac{\dfrac{SSA}{r - 1}}{\dfrac{SSE}{N - r}}$$

possesses an F distribution with $r - 1$ and $N - r$ degrees of freedom. The next theorem, which holds whether or not the null hypothesis is valid, provides the rationale for employing the preceding F distribution as a test statistic.

Under the assumptions of the fixed-effects, one-way-classification model, the expected values of the between-treatment and within-treatment sums of squares are

$$E[SSA] = (r - 1)\sigma^2 + \sum_{i=1}^{r} n_i \alpha_i^2$$

$$E[SSE] = (N - r)\sigma^2$$

Proof: According to the linear model,

$$\bar{X}_{i.} = \frac{1}{n_i} \sum_{j=1}^{n_i} \mu + \alpha_i + E_{ij} = \mu + \alpha_i + \bar{E}_{i.}$$

where, for fixed i, $\bar{E}_{i.}$ is the mean of the variables E_{ij}. Moreover, recalling the vanishing constraint on the sum of the $n_i \alpha_i$, we see that

$$\bar{X}_{..} = \frac{1}{N} \sum_{i=1}^{r} \sum_{j=1}^{n_i} \mu + \alpha_i + E_{ij}$$

$$= \frac{1}{N} \left[N\mu + \sum_{i=1}^{r} n_i \alpha_i + \sum_{i=1}^{r} \sum_{j=1}^{n_i} E_{ij} \right]$$

$$= \mu + \bar{E}_{..}$$

where $\bar{E}_{..}$ is the overall average of the E_{ij}. Consequently,

$$SSA = \sum_{i=1}^{r} n_i (\bar{X}_{i.} - \bar{X}_{..})^2 = \sum_{i=1}^{r} n_i (\alpha_i + \bar{E}_{i.} - \bar{E}_{..})^2$$

Since it has been assumed that the E_{ij} are independent with common mean 0 and variance σ^2,

$$E[\bar{E}_{i.}] = 0$$

$$E[\bar{E}_{..}] = 0$$

$$E[\bar{E}_{i.}^2] = \text{Var}\,[\bar{E}_{i.}] = \frac{\sigma^2}{n_i}$$

$$E[\bar{E}_{..}^2] = \text{Var}\,[\bar{E}_{..}] = \frac{\sigma^2}{N}$$

Using the previously derived expression for SSA yields

$$E[SSA] = \sum_{i=1}^{r} n_i E[(\alpha_i + \bar{E}_{i.} - \bar{E}_{..})^2]$$

$$= \sum_{i=1}^{r} n_i \alpha_i^2 + \sum_{i=1}^{r} n_i E[\bar{E}_{i.}^2]$$

$$+ \sum_{i=1}^{r} n_i E[\bar{E}_{..}^2] + 2 \sum_{i=1}^{r} n_i \alpha_i E[\bar{E}_{i.}]$$

$$- 2 \sum_{i=1}^{r} n_i E[\bar{E}_{i.} \bar{E}_{..}] - 2 \sum_{i=1}^{r} n_i \alpha_i E[\bar{E}_{..}]$$

which reduces to the expression given in the statement of the theorem. (Note that the third and fifth summations reduce to σ^2 and $-2\sigma^2$, respectively.)

As for the within-treatment sum of squares, employing the previously derived expression for $\bar{X}_{i.}$ and expanding the square gives

$$SSE = \sum_{i=1}^{r} \sum_{j=1}^{n_i} (X_{ij} - \bar{X}_{i.})^2$$

$$= \sum_{i=1}^{r} \sum_{j=1}^{n_i} (E_{ij} - \bar{E}_{i.})^2$$

$$= \sum_{i=1}^{r} \sum_{j=1}^{n_i} E_{ij}^2 - 2 \sum_{i=1}^{r} \bar{E}_{i.} \sum_{j=1}^{n_i} E_{ij} + \sum_{i=1}^{r} n_i \bar{E}_{i.}^2$$

$$= \sum_{i=1}^{r} \sum_{j=1}^{n_i} E_{ij}^2 - \sum_{i=1}^{r} n_i \bar{E}_{i.}^2$$

Taking the expected value yields

$$E[SSE] = \sum_{i=1}^{r} \sum_{j=1}^{n_i} E[E_{ij}^2] - \sum_{i=1}^{r} n_i E[\bar{E}_{i.}^2]$$

$$= N\sigma^2 - r\sigma^2 \qquad \blacksquare$$

Dividing SSA and SSE by the degrees of freedom corresponding to SSA/σ^2 and SSE/σ^2, respectively, gives the **mean-square** expressions

$$MSA = \frac{SSA}{r - 1}$$

$$MSE = \frac{SSE}{N - r}$$

As noted subsequent to Theorem 11.3, under the null hypothesis, MSA/MSE possesses an F distribution with $r - 1$ and $N - r$ degrees of freedom. Moreover, according to Theorem 11.4, if the null hypothesis is true, both MSA and MSE are unbiased estimators of the variance σ^2. However, also according to the theorem, if the null hypothesis is false, then MSA overestimates σ^2 by the amount

$$\frac{\sum_{i=1}^{r} n_i \alpha_i^2}{r - 1}$$

whereas MSE remains an unbiased estimator of σ^2. Thus, a critical region for H_0 is given by the upper tail of the F distribution MSA/MSE. Specifically, a critical region of size α is determined by the critical value $f_{\alpha, r-1, N-r}$, and rejection occurs when

$$F = \frac{MSA}{MSE} \geq f_{\alpha, r-1, N-r}$$

The foregoing analysis-of-variance methodology is summarized in the **ANOVA table** given in Table 11.3. Henceforth, we will employ ANOVA tables to provide the key features in analysis-of-variance procedures. In addition, ANOVA tables will be used to highlight the results obtained in specific examples.

Aside from precise mathematical results, one should appreciate the intuitive nature of analysis-of-variance reasoning. According to Theorem 11.2, the total sum of squares is equal to the sum of the within-treatment and between-treatment sums of squares. In effect, SSA measures variation that is in excess of the intrasample variations of the treatment random variables. When the treatment means are not identical, $E[SSA]$ is increased in the amount indicated in Theorem 11.4, the net effect being a positively biased estimate of σ^2 by MSA. Should the estimate be sufficiently great in comparison with the unbiased estimator MSE, then rejection of H_0 occurs.

TABLE 11.3 ANOVA TABLE FOR THE FIXED-EFFECTS, ONE-WAY-CLASSIFICATION MODEL

Source of Variation	Sum of Squares	Degrees of Freedom	Mean Square	F Test Statistic
Between treatments	SSA	$r - 1$	$MSA = \dfrac{SSA}{r - 1}$	$\dfrac{MSA}{MSE}$
Within treatments	SSE	$N - r$	$MSE = \dfrac{SSE}{N - r}$	
Total	SST	$N - 1$		

It is common to refer to SSE and MSE as the **error sum of squares** and the **error mean square,** respectively. Such terminology is appropriate to a setting in which the variation in the observations of each of the r random variables is attributed to measurement error. Specifically, if it is assumed that the observations of the ith treatment vary from their mean $\mu + \alpha_i$ only because of measurement error, or "noise," and that the underlying phenomenon being measured is deterministic, then within-treatment variation is, in fact, due to measurement variation. However, such "error" terminology can be misleading, since in many engineering applications the phenomena measured by the random variables $X_1, X_2, \ldots, X_r$ are inherently random and the fluctuations are not simply attributable to noise. Indeed, the entire treatment terminology can be misleading. Nevertheless, if one pays attention to the mathematical model, it will be seen that it is any set of random variables satisfying the model that is under study and not this or that particular experimental setting.

The next theorem provides representations with which to calculate the sum-of-squares expressions.

THEOREM 11.5

The following identities hold:

$$\text{(i) } SST = \sum_{i=1}^{r} \sum_{j=1}^{n_i} X_{ij}^2 - \frac{T_{..}^2}{N}$$

$$\text{(ii)} \quad SSA = \sum_{i=1}^{r} \frac{T_{i.}^2}{n_i} - \frac{T_{..}^2}{N}$$

$$\text{(iii)} \quad SSE = SST - SSA$$

Proof: Direct application of the definitions gives

$$SST = \sum_{i=1}^{r} \sum_{j=1}^{n_i} (X_{ij}^2 - 2X_{ij}\bar{X}_{..} + \bar{X}_{..}^2)$$

$$= \sum_{i=1}^{r} \sum_{j=1}^{n_i} X_{ij}^2 - 2\bar{X}_{..} \sum_{i=1}^{r} \sum_{j=1}^{n_i} X_{ij} + N\bar{X}_{..}^2$$

$$= \sum_{i=1}^{r} \sum_{j=1}^{n_i} X_{ij}^2 - 2\bar{X}_{..} T_{..} + N\bar{X}_{..}^2$$

$$= \sum_{i=1}^{r} \sum_{j=1}^{n_i} X_{ij}^2 - \frac{2T_{..}^2}{N} + \frac{NT_{..}^2}{N^2}$$

$$= \sum_{i=1}^{r} \sum_{j=1}^{n_i} X_{ij}^2 - \frac{T_{..}^2}{N}$$

and

$$SSA = \sum_{i=1}^{r} n_i(\bar{X}_{i.}^2 - 2\bar{X}_{..}\bar{X}_{i.} + \bar{X}_{..}^2)$$

$$= \sum_{i=1}^{r} \frac{n_i T_{i.}^2}{n_i^2} - 2\bar{X}_{..} \sum_{i=1}^{r} T_{i.} + \bar{X}_{..}^2 \sum_{i=1}^{r} n_i$$

$$= \sum_{i=1}^{r} \frac{T_{i.}^2}{n_i} - 2\frac{T_{..}}{N} T_{..} + N\bar{X}_{..}^2$$

$$= \sum_{i=1}^{r} \frac{T_{i.}^2}{n_i} - \frac{T_{..}^2}{N}$$

Part (iii) is simply a restatement of Theorem 11.2. ■

EXAMPLE 11.2

As noted in Example 11.1, the data of Table 11.2 give the percent purities of different batches of raw material obtained from four different suppliers. Under the assumptions of the fixed-effects, one-way-classification model, we will apply the analysis-of-variance methodology to perform the hypothesis test

$$H_0: \ \mu_1 = \mu_2 = \mu_3 = \mu_4$$

$H_1:$ Not all treatment means are equal.

which is equivalent to the test

$$H_0: \ \alpha_1 = \alpha_2 = \alpha_3 = \alpha_4 = 0$$

$H_1:$ Not all effects are zero.

Rejection of the null hypothesis will mean that the quality of the products is not uniform, at least insofar as quality is measured by purity; acceptance will mean that there is insufficient evidence to conclude lack of uniformity.

According to the data given in Table 11.2,

$$T_{..} = 1578.2$$

$$T_{..}^2 = 2,490,715.24$$

$$\frac{T_{..}^2}{N} = \frac{2,490,715.24}{16} = 155,669.70$$

$$\sum_{i=1}^{4} \sum_{j=1}^{n_i} x_{ij}^2 = (99.3)^2 + (99.4)^2 + \cdots + (99.2)^2 = 155,683.04$$

$$\sum_{i=1}^{4} \frac{T_{i.}^2}{n_i} = 155,676.92$$

According to Theorem 11.5,

$$SST = 155,683.04 - 155,669.70 = 13.34$$
$$SSA = 155,676.92 - 155,669.70 = 7.22$$
$$SSE = 13.34 - 7.22 = 6.12$$
$$MSA = \frac{7.22}{3} = 2.41$$
$$MSE = \frac{6.12}{12} = 0.51$$

and the sample value of F is

$$f = \frac{MSA}{MSE} = \frac{2.41}{0.51} = 4.73$$

The null hypothesis is rejected at the $\alpha = 0.05$ level, since $f_{0.05,3,12} = 3.49$ and $f > 3.49$. The entire procedure is summarized in the ANOVA table given in Table 11.4.

Now that the hypothesis of zero effects has been rejected, the effects can be estimated by means of the unbiased estimators provided by Theorem 11.1, namely,

$$\hat{\alpha}_i = \bar{X}_{i.} - \bar{X}_{..}$$

Since $\bar{x}_{..} = 1578.2/16 = 98.64$, the desired estimates are

$$\overset{*}{\alpha}_1 = \bar{x}_{1.} - \bar{x}_{..} = 99.22 - 98.64 = 0.58$$
$$\overset{*}{\alpha}_2 = \bar{x}_{2.} - \bar{x}_{..} = 98.74 - 98.64 = 0.10$$
$$\overset{*}{\alpha}_3 = \bar{x}_{3.} - \bar{x}_{..} = 97.53 - 98.64 = -1.11$$
$$\overset{*}{\alpha}_4 = \bar{x}_{4.} - \bar{x}_{..} = 99.16 - 98.64 = 0.52$$

Since the test statistic has MSE in the denominator, if there is little within-treatment variation, then it is more likely that the null hypothesis will be rejected. Intuitively, if MSE is small, then the sample means provide more precise estimation, and small differences between the sample means and the grand average are more

TABLE 11.4 ANOVA TABLE FOR EXAMPLE 11.2

Source of Variation	Sum of Squares	Degrees of Freedom	Mean Square	F Test Statistic
Treatments	7.22	3	2.41	4.73
Error	6.12	12	0.51	
Total	13.34	15		

significant. The situation is analogous to that encountered when employing the t test to test the difference of two means. As in that case, a more propitious experimental design can serve to increase the sensitivity of the experiment. Such topics will be addressed in Chapter 14.

Insofar as Example 11.2 (Table 11.2) is concerned, a chemist would most likely agree that the mean purities are not equivalent. Nevertheless, it must not be forgotten that the sensitivity of the experiment is influenced by the variation in the response observations. In particular, Theorem 11.4 states that MSE is an unbiased estimator of σ^2 and for the given data the resulting variance estimate is 0.51. Since

$$\text{Var}\,[\bar{X}_{1.}] = \text{Var}\,[\bar{X}_{3.}] = \frac{\sigma^2}{4}$$

the difference between $\bar{x}_{1.}$ and $\bar{x}_{3.}$ certainly appears significant. Had there been more within-sample variability, then MSE would have been greater and the differences in mean purity might not have been significant (from the perspective of the test statistic).

EXAMPLE 11.3

The ability of a robot to complete a task depends not only on the complexity of the task but also on the complexity of the environment in which the task is to be performed. In addition, if the task involves locating an object in the environment prior to the accomplishment of some secondary procedure, the initial positioning of the robot can cause completion times to vary. Suppose an engineer wishes to test the abilities of three robots, numbered 1, 2, and 3, to locate a box of machine parts in a room containing various objects and then to take a prespecified part from the box. Each robot is tested five times and the initial positions are randomly selected. The 15 times to completion (in seconds) are recorded in Table 11.5 and the data are used to employ a fixed-effects, one-way-classification analysis-of-variance test of the null hypothesis

$$H_0\!: \mu_1 = \mu_2 = \mu_3$$

where μ_i is the mean completion time of robot i. From the data of the table, $T_{..} = 1099$, $T_{..}^2 = 1{,}207{,}801$,

TABLE 11.5 ONE-WAY CLASSIFICATION OF ROBOT TASK TIMES

Robot	Observations					Total	Mean
1	93	56	82	104	45	380	76.0
2	42	64	112	73	100	391	78.2
3	30	55	60	98	85	328	65.6

$$\frac{T_{\cdot\cdot}^2}{N} = 80{,}520$$

$$\sum_{i=1}^{3}\sum_{j=1}^{5} x_{ij}^2 = 89{,}437$$

$$\sum_{i=1}^{3} \frac{T_{i\cdot}^2}{n_i} = 80{,}973$$

The ANOVA table of Table 11.6 gives the key information. Since $f_{0.05,2,12} = 3.89$ and $f = 0.32 < 3.89$, the null hypothesis is not rejected at level $\alpha = 0.05$. While it might be argued that the mean completion time of the third robot is quite a bit less than those of the other two, there is so much within-treatment variability that it is difficult for the test to detect lack of mean equality. Indeed, since the error mean square is 705, the *MSE* estimate of σ is $\overset{*}{\sigma} = 26.55$. Thus, the estimate of the standard deviation of $\bar{X}_{i\cdot}$ is $(26.55)/\sqrt{5} = 11.87$, so that a difference between means of 10 seconds is not significant. Perhaps a different experimental design might produce a more sensitive test. For instance, it might be better to compare times in a manner that mitigates the excessive variability introduced by random selection of initial position.

In the case of just two treatment populations, the analysis-of-variance procedure yields a test statistic for the two-tailed hypothesis test

$$H_0\colon \ \mu_1 = \mu_2$$

$$H_1\colon \ \mu_1 \neq \mu_2$$

As seen in Section 9.2, this test can be performed by using a t test. In fact, in the case of two populations it can be shown that the ANOVA F test is equivalent to the two-tailed t test for the difference of two means.

A natural question arises concerning the feasibility of employing multiple t tests to test the equality of several means. To test

$$H_0\colon \ \mu_1 = \mu_2 = \cdots = \mu_r$$

against the alternative that the means are not all equal, might we not perform two-tailed t tests on all possible pairs of means? Aside from the computational difficulty in that there are $C(r, 2)$ such pairs, a serious issue arises with respect to type I error. Even though each pairwise test might be arranged to have a type I error of size α, the probability of incorrectly rejecting H_0 is considerably greater, because H_0 would be rejected if at least one of the pairwise null hypotheses were rejected. Consequently, the test would have to be arranged so that the probability of at least one pairwise type I error is α.

TABLE 11.6 ANOVA TABLE FOR EXAMPLE 11.3

Source of Variation	Sum of Squares	Degrees of Freedom	Mean Square	F Test Statistic
Treatments	453	2	227	0.32
Error	8464	12	705	
Total	8917	14		

Concerning the robustness of the ANOVA F test, if the sample sizes are large, then the procedure is fairly robust with respect to the assumption that the treatment populations are normally distributed. Of course, should there be marked deviations from normality, the validity of the conclusion will be questionable. A more critical issue concerns the equal-variance condition. Should the sample sizes be equal, some differences in the variances can be tolerated; however, if the sizes are not equal, then the test results can be significantly skewed when the variances are not (essentially) identical.

11.2 MULTIPLE COMPARISONS

Rejection of the null hypothesis in the fixed-effects, one-way classification yields the conclusion that not all treatment means are identical, but it does not yield information regarding comparisons between pairs of means or, more generally, between subcollections of means. Tests designed to draw more specific conclusions than simple rejection are known as **multiple comparison** tests. Numerous tests have been proposed for multiple comparisons. These include the Newman-Keuls test, Scheffe's test, Tukey's test, and Duncan's test. The present section discusses the latter two. In each case the technique will be presented without supporting theory.

11.2.1 Tukey's Test

Suppose the null hypothesis is rejected in a fixed-effects, one-way model in which all treatment samples are of equal size n. Assuming there to be r treatments, the $C(r, 2)$ pairs of means can be tested for equality by the Tukey procedure. The technique involves the construction of simultaneous confidence intervals for the mean differences $\mu_i - \mu_k$, where $i \neq k$ and $i, k = 1, 2, \ldots, r$. Before giving the precise form of the confidence intervals, it is necessary to explain the manner in which the resulting collection of confidence intervals will be interpreted. Given α, for $i \neq k$, the confidence interval for $\mu_i - \mu_k$ will be of the form

$$(\bar{x}_{i.} - \bar{x}_{k.} - \kappa_\alpha, \bar{x}_{i.} - \bar{x}_{k.} + \kappa_\alpha)$$

and for $(1 - \alpha)100\%$ of all sets of sample data, all mean differences $\mu_i - \mu_k$ will be contained in their respective confidence intervals. For instance, if $\alpha = 0.05$, in only 5% of the cases in which Tukey's confidence-interval procedure is employed will there be at least one difference $\mu_i - \mu_k$ not lying within its specified confidence interval. Because α applies to the entire collection of mean differences involved in the experiment, it is termed an **experimentwise error rate.** As is always the case with confidence intervals, the precise meaning of α must be given in terms of probability statements. This will be done shortly.

Tukey's procedure utilizes a distribution known as the **studentized range distribution.** This distribution possesses two parameters, m and v. Table A.8 gives the upper one-tailed critical values for this distribution for $\alpha = 0.05$ and $\alpha = 0.01$, each critical value being of the form $Q_{\alpha, m, v}$. For $i, k = 1, 2, \ldots, r, i \neq k$, the confidence interval for $\mu_i - \mu_k$ will be of the form stated previously with

$$\kappa_\alpha = Q_{\alpha, r, N-r} \sqrt{\frac{MSE}{n}}$$

where $N - r = r(n - 1)$ is the number of degrees of freedom associated with MSE. Rigorously, the probability is $1 - \alpha$ that

$$\bar{X}_{i.} - \bar{X}_{k.} - \kappa_\alpha < \mu_i - \mu_k < \bar{X}_{i.} - \bar{X}_{k.} + \kappa_\alpha$$

for all i and k.

Insofar as pairwise mean-equality is concerned, Tukey's multiple-comparison test states that whenever the confidence interval for $\mu_i - \mu_k$ does not contain 0, then it should be concluded that μ_i and μ_k differ at the significance level α. By the form of the confidence intervals, μ_i and μ_k are judged significantly different if and only if

$$\left| \bar{x}_{i.} - \bar{x}_{k.} \right| \geq \kappa_\alpha$$

Consequently, if the values of the sample means are listed in ascending order, then those pairs whose differences are less than κ_α are assumed to result from treatment variables possessing equal means. It is common practice to write down the list of sample-mean values and to underline those concluded not to be significantly different.

EXAMPLE 11.4

Automobile hydrocarbon (and CO) emissions are controlled by elaborate emission-control systems. Such systems are affected by a myriad of factors, including operating conditions, wear on the system, and system tuning. Suppose five systems are to be compared and sample selection is randomized relative to the three conditions just cited. Each sample is to be of size four. Table 11.7 gives the observational data in hydrocarbon parts per million (ppm). First, the null hypothesis

$$H_0: \ \mu_1 = \mu_2 = \mu_3 = \mu_4 = \mu_5$$

is tested at level $\alpha = 0.05$, the key information being contained in the ANOVA table of Table 11.8. Since the sample value of the F test statistic is 18.04 and $f_{0.05,4,15} = 3.06$, mean equality is strongly rejected at the 0.05 level. In fact, $f_{0.01,4,15} = 4.89$, and the null hypothesis is also rejected at the 0.01 level.

To obtain a finer analysis of the data, we employ Tukey's multiple comparison test. According to Table A.8, the critical value at the 0.05 level is

$$Q_{0.05,5,15} = 4.37$$

Thus,

$$\kappa_{0.05} = 4.37 \times \sqrt{\frac{20.5}{4}} = 9.89$$

Reading the means from Table 11.7, listing them in ascending order, and underlining those pairs whose equality is not rejected by the test gives

$\bar{x}_{4.}$	$\bar{x}_{3.}$	$\bar{x}_{5.}$	$\bar{x}_{2.}$	$\bar{x}_{1.}$
70.0	84.5	87.0	89.5	96.0

Note that the underlining overlaps, since μ_2 is not judged significantly different than μ_3, nor is it judged significantly different than μ_1. However, 84.5 and 96.0 differ by more than 9.89, and therefore μ_3 and μ_1 are judged significantly different.

Insofar as the confidence intervals are concerned, the experimentwise 95% confidence interval for $\mu_1 - \mu_4$ is given by

$$(26 - 9.89, 26 + 9.89) = (16.11, 35.89)$$

Confidence intervals for other mean differences can be similarly constructed.

11.2.2 Duncan's Multiple Range Test

In applying Duncan's multiple range test, we will again assume that all sample sizes are equal to n. The test makes use of the **least significant studentized range,** a distribution possessing two parameters, λ and p, and whose critical values are given in Table A.9. Experimentwise error in Duncan's test will be discussed subsequent to the introduction of the procedure and the application of the test to a specific example.

Duncan's multiple range test is implemented by the following procedure:

1. List all empirical means $\bar{x}_{i\cdot}$ in ascending order.
2. Choose α and use Table A.9 with parameters $\lambda = r(n - 1)$ and p to find the **significant ranges** r_p, $p = 2, 3, \ldots, r$.
3. For $p = 2, 3, \ldots, r$, the **least significant range** is given by

$$R_p = r_p \sqrt{\frac{MSE}{n}}$$

4. In the listing of empirical sample means, suppose $\bar{x}_{i\cdot} < \bar{x}_{k\cdot}$ and there are p empirical means in the range determined by $\bar{x}_{i\cdot}$ and $\bar{x}_{k\cdot}$, which means there are p empirical means in the listing between $\bar{x}_{i\cdot}$ and $\bar{x}_{k\cdot}$, inclusively. Then μ_i and μ_k are judged significantly different if and only if

$$\bar{x}_{k\cdot} - \bar{x}_{i\cdot} > R_p$$

TABLE 11.7 ONE-WAY CLASSIFICATION OF HYDROCARBON EMISSIONS

System	Observations				Total	Mean
1	102	92	100	90	384	96.0
2	92	88	96	82	358	89.5
3	83	80	85	90	338	84.5
4	72	70	66	72	280	70.0
5	86	88	90	84	348	87.0

TABLE 11.8 ANOVA TABLE FOR EXAMPLE 11.4

Source of Variation	Sum of Squares	Degrees of Freedom	Mean Square	F Test Statistic
Treatments	1479	4	369.8	18.04
Error	308	15	20.5	
Total	1787	19		

In addition, if the equality of μ_i and μ_k is not rejected, then all population treatment means in the range determined by $\bar{x}_{i.}$ and $\bar{x}_{k.}$ are assumed to be equal. As usual, this simply means that the data do not indicate a significant departure from the hypothesis of equality.

The proviso concerning the acceptance of equality for all treatment means whose empirical means lie between empirical means not judged significantly different has practical import. Specifically, when applying the test, it is beneficial to begin by checking extreme ranges, since whenever the end means of a range are judged to be equal, ipso facto, so too are all means determined by the range. Consequently, the order of comparison can be optimized for computational efficiency.

EXAMPLE 11.5

In Example 11.4 we applied Tukey's test to the data of Table 11.7 to determine significant differences among the treatment means. To apply Duncan's multiple range test at level 0.05 to the same data, we first find, for $\lambda = 15$ and $p = 2, 3, 4,$ and 5, the significant ranges r_p from Table A.9. The least significant ranges are then found by multiplying each r_p by

$$\sqrt{\frac{MSE}{n}} = \sqrt{\frac{20.5}{4}} = 2.26$$

The following table gives the values obtained:

p	2	3	4	5
r_p	3.014	3.160	3.250	3.312
R_p	6.81	7.14	7.35	7.49

Listing the means of Table 11.7 in ascending order and proceeding "from outside in" to take advantage of the proviso concerning means within the range of two means not judged significantly different, we tabulate the following list of inequalities:

1. $96.0 - 70.0 > 7.49$ $(p = 5)$
2. $89.5 - 70.0 > 7.35$ $(p = 4)$
3. $87.0 - 70.0 > 7.14$ $(p = 3)$
4. $84.5 - 70.0 > 6.81$ $(p = 2)$
5. $96.0 - 84.5 > 7.35$ $(p = 4)$
6. $89.5 - 84.5 < 7.14$ $(p = 3)$
7. $96.0 - 87.0 > 7.14$ $(p = 3)$
8. $96.0 - 89.5 < 6.81$ $(p = 2)$

Note that 87.0 is compared to neither 84.5 nor 89.5, since they are judged not significantly different in inequality 6. The results of the test are summarized by

$\bar{x}_{4.}$	$\bar{x}_{3.}$	$\bar{x}_{5.}$	$\bar{x}_{2.}$	$\bar{x}_{1.}$
70.0	84.5	87.0	89.5	96.0

In Example 11.4, Tukey's test indicated that μ_5 and μ_1 were not significantly different; Duncan's test indicates otherwise. Note, however, that $\bar{x}_{1.} - \bar{x}_{5.} = 9.0$ is barely less than the value $\kappa_\alpha = 9.89$ of Example 11.4. Thus, μ_1 and μ_5 are not judged unequal at significance level 0.05 but would be judged so at a slightly higher level. Correspondingly, in Example 11.5, $\bar{x}_{1.} - \bar{x}_{5.}$ exceeds $R_3 = 7.14$ by a small amount. Thus, at a slightly lower level of significance, μ_1 and μ_5 would not be judged unequal. Indeed, for $\alpha = 0.01$, $r_3 = 4.347$ and $R_3 = 9.82$, a value not exceeded by $\bar{x}_{1.} - \bar{x}_{5.}$.

From the perspective of type I error, Tukey's test is conservative: it possesses an experimentwise error rate that is actually less than or equal to α. On the other hand, Duncan's test possesses an experimentwise error rate that is greater than or equal to α. Consequently, there is greater likelihood of erroneously judging two means significantly different when using Duncan's test. From a practical point of view, Duncan's test appears to be more discriminating than Tukey's test. This is borne out in Examples 11.4 and 11.5: for the data of Table 11.7, Tukey's test does not lead to the conclusion that μ_1 and μ_5 are significantly different, whereas Duncan's test does.

11.3 RANDOM-EFFECTS MODEL

Thus far, we have assumed that the random variables $X_1, X_2, \ldots, X_r$ describe a fixed set of populations of interest to the experimenter. For instance, in Example 11.2 there were four populations of purity measurements, and the conclusion applied to powders obtained from the specific suppliers. More generally, one might be interested in drawing conclusions concerning a large collection of treatments based upon an analysis of a set of sample treatments. In such a case, the random variables $X_1, X_2, \ldots, X_r$ represent a set of r populations randomly drawn from a large collection C of populations (random variables). In terms of Example 11.2, the experimenter might wish to make inferences concerning a universal supplier set, of which the four studied are but a sampling. When the treatment populations are randomly drawn from a large collection of populations, the purpose is not simply to test the equality of the specific treatment means in the analysis-of-variance study. This would lead only to an inference regarding the r population means of the study. Instead, the goal is to test the equality of the means of the random variables constituting the larger collection.

As in the fixed-effects model, the new scenario will involve r treatments, where in the present case these have been randomly selected from C, the universe of treatments. There will be r independent samples arising from the selected treatment variables, and it will be assumed that each random variable in C is normally distributed with common variance σ^2. Because of the random selection of the treatment populations, the observation model takes the form

$$X_{ij} = \mu + A_i + E_{ij}$$

$i = 1, 2, \ldots, r$ and $j = 1, 2, \ldots, n_i$, where the random variable A_i is the treatment effect for the ith selected population. A_i is random because the ith treatment has been randomly selected from C, and for this reason the model is called the **random-effects model**. The ith mean is now the random variable $\mu + A_i$ and the grand mean is once again a base value from which to view all treatment means. Specifically, we postulate the following model.

Random-Effects, One-Way-Classification Model

Equation:	$X_{ij} = \mu + A_i + E_{ij}$
Assumptions:	1. $E_{ij} \sim N(0, \sigma)$
	2. $A_i \sim N(0, \sigma_\alpha)$
	3. The E_{ij} and A_i comprise an independent collection of random variables.

An immediate consequence of assumption 3 is that

$$\text{Cov}\,[E_{ij}, E_{i'j'}] = 0$$

if $(i, j) \neq (i', j')$, and

$$\text{Cov}\,[E_{ij}, A_i] = 0$$

for all E_{ij} and A_i.

Since the means of the treatment populations are random variables, it makes no sense to define the null hypothesis in terms of the equality of these means. Instead, the random-effects model employs the hypothesis test

$$H_0: \sigma_\alpha^2 = 0$$

$$H_1: \sigma_\alpha^2 \neq 0$$

the null hypothesis being rejected in the presence of significant treatment-effect variability. In developing a suitable test statistic, we will assume equal sample sizes:

$$n_1 = n_2 = \cdots = n_r = n$$

The next theorem is fundamental. The proof is sketched to give some insight into the manner in which the relevant distributions interact. Particular attention should be paid to the fact that the theorem holds regardless of the validity of the null hypothesis.

THEOREM 11.6

Under the conditions of the random-effects model with common sample size n,

$$F = \frac{\sigma^2 MSA}{(\sigma^2 + n\sigma_\alpha^2) MSE}$$

possesses an F distribution with $r - 1$ and $N - r$ degrees of freedom.

Proof: Whether or not the null hypothesis is valid, it can be shown that SSE/σ^2 and $SSA/(\sigma^2 + n\sigma_\alpha^2)$ are independent chi-square distributions possessing $N - r$ and $r - 1$ degrees of freedom, respectively. Thus,

$$\frac{\dfrac{SSA}{(r - 1)(\sigma^2 + n\sigma_\alpha^2)}}{\dfrac{SSE}{(N - r)\sigma^2}}$$

possesses an F distribution with $r - 1$ and $N - r$ degrees of freedom. But this is precisely the random variable in the statement of the theorem. ∎

Under the null hypothesis, $\sigma_\alpha^2 = 0$ and the distribution of Theorem 11.6 reduces to

$$F = \frac{MSA}{MSE}$$

so that the test statistic is the same in both the random- and fixed-effects models. The next theorem plays an analogous role to that of Theorem 11.4 in determing the form of the critical region for the test statistic MSA/MSE.

Under the conditions of the random-effects model with common sample size n,

$$E[SSA] = (r - 1)\sigma^2 + n(r - 1)\sigma_\alpha^2$$

$$E[SSE] = (N - r)\sigma^2$$

Proof: According to the model,

$$\bar{X}_{i.} = \frac{1}{n}\sum_{j=1}^{n} \mu + A_i + E_{ij} = \mu + A_i + \bar{E}_{i.}$$

$$\bar{X}_{..} = \frac{1}{N}\sum_{i=1}^{r}\sum_{j=1}^{n} \mu + A_i + E_{ij}$$

$$= \frac{1}{N}\left[N\mu + n\sum_{i=1}^{r} A_i + \sum_{i=1}^{r}\sum_{j=1}^{n} E_{ij}\right]$$

$$= \mu + \bar{A}_. + \bar{E}_{..}$$

Consequently,

$$SSA = n\sum_{i=1}^{r} [(A_i - \bar{A}_.) + (\bar{E}_{i.} - \bar{E}_{..})]^2$$

$$= n\sum_{i=1}^{r} (A_i^2 - 2\bar{A}_. A_i + \bar{A}_.^2 + \bar{E}_{i.}^2 - 2\bar{E}_{..}\bar{E}_{i.} + \bar{E}_{..}^2)$$

$$= n\left[\sum_{i=1}^{r} A_i^2 - 2r\bar{A}_.^2 + r\bar{A}_.^2 + \sum_{i=1}^{r} \bar{E}_{i.}^2 - 2r\bar{E}_{..}^2 + r\bar{E}_{..}^2\right]$$

$$= n\sum_{i=1}^{r} A_i^2 - N\bar{A}_.^2 + n\sum_{i=1}^{r} \bar{E}_{i.}^2 - N\bar{E}_{..}^2$$

Since

$$E[A_i^2] = \text{Var}\,[A_i] = \sigma_\alpha^2$$

$$E[\bar{A}_.^2] = \frac{\sigma_\alpha^2}{r}$$

$$E[\bar{E}_{i.}^2] = \frac{\sigma^2}{n}$$

$$E[\bar{E}_{..}^2] = \frac{\sigma^2}{N}$$

applying the expected-value operator yields

$$E[SSA] = nr\sigma_\alpha^2 - \frac{N\sigma_\alpha^2}{r} + r\sigma^2 - \sigma^2$$

which reduces to the expression given in the statement of the theorem.

Now, since $X_{ij} = \mu + A_i + E_{ij}$,

$$SSE = \sum_{i=1}^{r} \sum_{j=1}^{n} (X_{ij} - \bar{X}_{i.})^2$$

$$= \sum_{i=1}^{r} \sum_{j=1}^{n} (E_{ij} - \bar{E}_{i.})^2$$

$$= \sum_{i=1}^{r} \sum_{j=1}^{n} (E_{ij}^2 - 2\bar{E}_{i.} E_{ij} + \bar{E}_{i.}^2)$$

$$= \sum_{i=1}^{r} \sum_{j=1}^{n} E_{ij}^2 - 2n \sum_{i=1}^{r} \bar{E}_{i.}^2 + n \sum_{i=1}^{r} \bar{E}_{i.}^2$$

$$= \sum_{i=1}^{r} \sum_{j=1}^{n} E_{ij}^2 - n \sum_{i=1}^{r} \bar{E}_{i.}^2$$

Hence,

$$E[SSE] = \sum_{i=1}^{r} \sum_{j=1}^{n} E[E_{ij}^2] - n \sum_{i=1}^{r} E[\bar{E}_{i.}^2]$$

$$= \sum_{i=1}^{r} \sum_{j=1}^{n} \mathrm{Var}\,[E_{ij}] - r\sigma^2$$

$$= N\sigma^2 - r\sigma^2 \qquad\qquad \blacksquare$$

It is immediate from Theorem 11.7 that under the null hypothesis both *MSA* and *MSE* are unbiased estimators of σ^2. Moreover, under the alternative hypothesis,

$$E[MSA] = \sigma^2 + n\sigma_\alpha^2$$

Thus, *MSA* overestimates σ^2 when H_0 is false and, as with fixed effects, an upper one-tailed test using the *F* statistic *MSA/MSE* is appropriate. Specifically, H_0 is rejected at significance level α if

$$F = \frac{MSA}{MSE} > f_{\alpha, r-1, N-r}$$

Both the test statistic and critical region are exactly the same as in the fixed-effects model; nevertheless, the interpretation of the test is different.

When the null hypothesis is rejected in the fixed-effects model, multiple comparisons are in order; however, such a course is not appropriate in the random-effects model. Instead, one can simply estimate the common variance of the treatment effects. From Theorem 11.7 it can be seen that

$$\hat{\sigma}_\alpha^2 = \frac{MSA - MSE}{n}$$

is an unbiased estimator of σ_α^2.

EXAMPLE 11.6

Various grades of motor oil provide varying degrees of protection for the cylinder walls and rings of an internal combustion engine. Suppose a testing laboratory is interested in determining whether or not there is a significant difference in the protection afforded by motor oils in some collection of brands and grades. An experiment is conducted in which three oils are chosen randomly from the collection under study and in which each oil is used in three different engines. The engines are run in the laboratory to simulate a given number of miles, each engine receiving periodic maintenance as suggested by its own manufacturer. The engines are selected so that each possesses the same original average cylinder compression. The test is based upon the fact that ring and cylinder-wall wear results in lower compression readings. To be certain that compression loss is not due to leaky valves, at the completion of the test mileage, new cylinder heads are installed on each engine and final compression readings, in pounds per square inch (psi), are taken. The averages of these readings are given in Table 11.9. If the random-effects null hypothesis is rejected, it will be concluded that the motor oils in the overall collection under study do not afford equal protection. Given the information in the ANOVA table of Table 11.10 and the fact that $f_{0.01,2,6} = 10.92$, the hypothesis of equal protection is rejected at the 0.01 level. An estimate of the true effect variance is given by

$$\overset{*}{\sigma}_{\alpha}^{2} = \frac{MSE - MSA}{n} = \frac{48 - 4}{3} = 14.7$$

TABLE 11.9 ONE-WAY CLASSIFICATION OF CYLINDER COMPRESSIONS

Oil	Observations			Total	Mean
1	151	156	155	462	154
2	158	159	157	474	158
3	152	150	148	450	150

TABLE 11.10 ANOVA TABLE FOR EXAMPLE 11.6

Source of Variation	Sum of Squares	Degrees of Freedom	Mean Square	F Test Statistic
Treatments	96	2	48	12
Error	24	6	4	
Total	120	8		

11.4 TYPE II ERROR IN ANALYSIS OF VARIANCE

With both fixed and random effects, the analysis-of-variance procedure leads to test statistics whose critical values are dependent upon α, the number of treatments r, and the sample sizes. We will now turn our attention to the size of type II error. As usual, the approach is to find a suitable expression for β and then to determine some criterion under which erroneous acceptance of the null hypothesis is considered a serious matter.

11.4.1 Type II Error in the Fixed-Effects Model

Restricting ourselves to a common sample size n, in the fixed-effects model the size of type II error is

$$\beta = P\left(\frac{MSA}{MSE} < f_{\alpha, r-1, N-r}; H_1\right)$$

where $N = nr$. MSA/MSE possesses an F distribution when H_0 is true, but under H_1 it possesses a **noncentral F distribution.** Like the F distribution, the noncentral F distribution has a pair of degrees-of-freedom parameters, v_1 and v_2, these being associated in the present case with MSA and MSE, respectively. However, it also possesses a third parameter λ, called the **noncentrality parameter.** In the present situation,

$$\lambda = \frac{n \sum_{i=1}^{r} \alpha_i^2}{2\sigma^2}$$

Under H_1, there is at least one nonzero treatment effect, so that $\lambda > 0$ and

$$\beta = P(F' < f_{\alpha, r-1, N-r})$$

where $F' = MSA/MSE$ is a noncentral F distribution with parameters $v_1 = r - 1$, $v_2 = N - r$, and λ.

Rather than directly utilizing F', it is common to employ OC curves constructed in terms of the parameter

$$\Phi = \sqrt{\frac{2\lambda}{r}} = \sqrt{\frac{\frac{n}{r} \sum_{i=1}^{r} \alpha_i^2}{\sigma^2}}$$

Corresponding OC-curve families are given in Figures 11.1 through 11.3 for $v_1 = 2$ to $v_1 = 4$, respectively. To employ these curves in finding β, one decides upon a value of Φ that represents a significant departure from the null hypothesis. The next example illustrates the manner in which one might arrive at a salient value of Φ. In addition, given a restriction on the size of β for a given value of Φ, it is also possible to find a suitable sample size by trial and error. This, too, will be illustrated in the next example.

EXAMPLE 11.7

Example 11.3 was concerned with detecting different mean task times among three robots. Suppose a researcher suspects that robots 1 and 2 have been programmed in such a way that their performances should be approximately the same, but that robot 3, which possesses updated guidance laws, should perform better. It is the better performance of robot 3 that he or she wishes to detect. In light of the desired detection, it is decided that β should be evaluated under the condition

$$\mu_2 = \mu_1$$

$$\mu_3 = \mu_1 - 2\sigma$$

where σ is the common standard deviation of the task times. In words, the size of type II error is desired when the true state of nature is that robots 1 and 2 possess the same mean task times and robot 3 possesses a mean task time that is two standard deviations better than the common task time of the other two robots.

Since $v_1 = r - 1 = 2$, we will have to find Φ and then utilize Figure 11.1 with

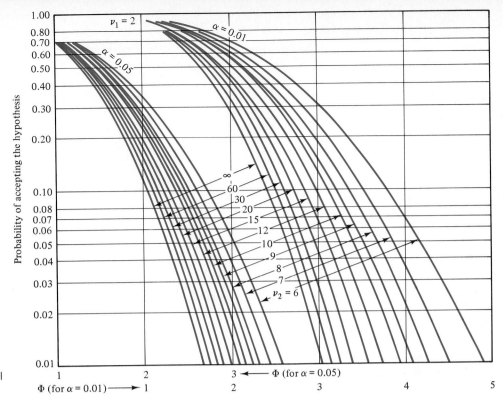

FIGURE 11.1
OC curves for fixed-effects ANOVA with $\nu_1 = 2$ (Reprinted by permission of the Institute of Mathematical Statistics.)

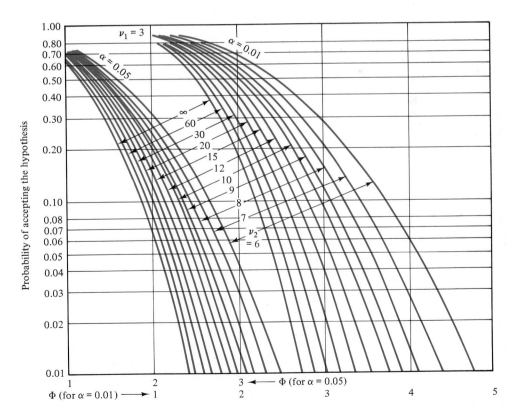

FIGURE 11.2
OC curves for fixed-effects ANOVA with $\nu_1 = 3$ (Reprinted by permission of the Institute of Mathematical Statistics.)

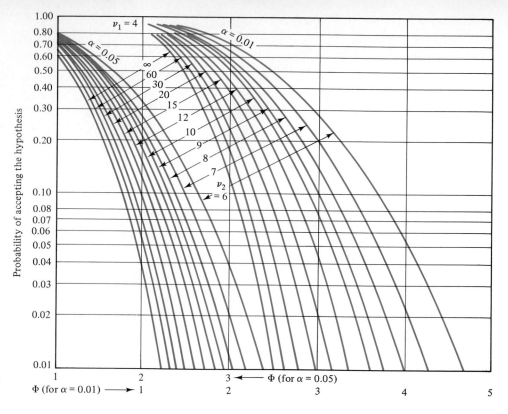

FIGURE 11.3
OC curves for fixed-effects ANOVA with $\nu_1 = 4$ (Reprinted by permission of the Institute of Mathematical Statistics.)

$\nu_2 = N - r = 12$. One of the model assumptions is that μ is the weighted average of the response means. Thus,

$$\mu = \frac{1}{3}[\mu_1 + \mu_1 + (\mu_1 - 2\sigma)] = \mu_1 - \frac{2\sigma}{3}$$

$$\alpha_1 = \alpha_2 = \mu_1 - \mu = \mu_1 - \left(\mu_1 - \frac{2\sigma}{3}\right) = \frac{2\sigma}{3}$$

$$\alpha_3 = \mu_3 - \mu = \mu_1 - 2\sigma - \left(\mu_1 - \frac{2\sigma}{3}\right) = -\frac{4\sigma}{3}$$

and

$$\Phi = \sqrt{\frac{n}{r\sigma^2}(\alpha_1^2 + \alpha_2^2 + \alpha_3^2)} = \sqrt{\frac{5}{3} \times \frac{24}{9}} = 2.11$$

Using Figure 11.1 and recalling from the original problem that $\alpha = 0.05$, we find $\beta \approx 0.18$.

Assuming the number of test robots to be fixed at $r = 3$, β can be improved by increasing the common sample size n. For instance, if it is decided that $\beta = 0.10$ is desirable, then trial and error shows that n needs to be at least 6. Indeed, if $n = 6$, then $\nu_2 = r(n - 1) = 15$, $\Phi = \sqrt{8n/9} = 2.31$, and it can be seen from Figure 11.1 that $\beta \approx 0.10$.

11.4.2 Type II Error in the Random-Effects Model

We now consider type II error in the random-effects model. For level of significance α, the probability of accepting the null hypothesis when the alternative is true, namely, when $\sigma_\alpha^2 > 0$, is given by

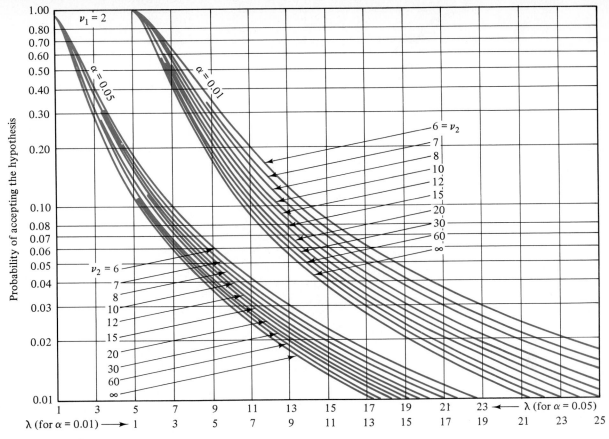

FIGURE 11.4
OC curves for random-effects ANOVA with $v_1 = 2$ (Reprinted by permission of the Institute of Mathematical Statistics.)

$$\beta = P\left(\frac{MSA}{MSE} < f_{\alpha,r-1,N-r}; \sigma_\alpha^2 > 0\right)$$

According to Theorem 11.6, if we divide both sides of the probability inequality by

$$1 + n\frac{\sigma_\alpha^2}{\sigma^2}$$

then the random variable on the left-hand side of the inequality possesses an F distribution with $r - 1$ and $N - r$ degrees of freedom. Thus,

$$\beta = P\left(F < \frac{f_{\alpha,r-1,N-r}}{1 + n\frac{\sigma_\alpha^2}{\sigma^2}}\right)$$

Letting

$$\lambda = \sqrt{1 + n\frac{\sigma_\alpha^2}{\sigma^2}}$$

shows that β decreases for increasing λ. In particular, the OC-curve families are parameterized in terms of λ, and these are given in Figures 11.4 through 11.6 for $v_1 = 2$ through $v_1 = 4$, where $v_1 = r - 1$ and $v_2 = N - r$.

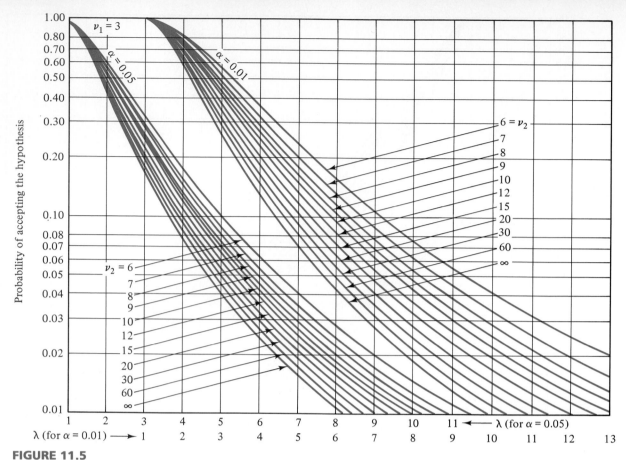

FIGURE 11.5

OC curves for random-effects ANOVA with $\nu_1 = 3$ (Reprinted by permission of the Institute of Mathematical Statistics.)

Based upon the model equation,

$$\text{Var}[X_{ij}] = \sigma_\alpha^2 + \sigma^2$$

for any observation X_{ij}. Thus, the truth of the alternative hypothesis manifests itself in increased observation variation. In employing the OC-curve parameter λ, the ratio σ_α^2/σ^2 plays the key role. The importance of detection in the presence of significant treatment-effect variability is judged in terms of the ratio of treatment-effect variance to random variance.

EXAMPLE 11.8

In the motor-oil experiment of Example 11.6, the null hypothesis was rejected at the 0.01 level. At that level, the probability of erroneously accepting the null hypothesis when

$$\frac{\sigma_\alpha^2}{\sigma^2} = 1.5$$

is found from Figure 11.4 with $\nu_1 = r - 1 = 2$, $\nu_2 = N - r = 6$, and

$$\lambda = \sqrt{1 + (1.5)n} = \sqrt{1 + 4.5} = 2.35$$

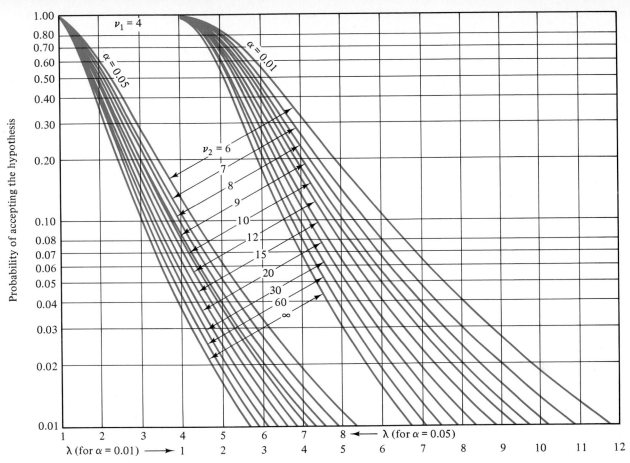

FIGURE 11.6
OC curves for random-effects ANOVA with $\nu_1 = 4$ (Reprinted by permission of the Institute of Mathematical Statistics.)

Figure 11.4 gives $\beta \approx 0.8$. At significance level $\alpha = 0.05$, the same figure gives $\beta \approx 0.55$. As usual, β can be improved by increasing n. Indeed, for $n = 8$, $\nu_2 = 21$, and at levels 0.01 and 0.05, Figure 11.4 gives $\beta \approx 0.35$ and $\beta \approx 0.25$, respectively, the appropriate parameter being

$$\lambda = \sqrt{1 + (8)(1.5)} = 3.61$$

11.5 KRUSKAL-WALLIS TEST

One of the modeling assumptions in the fixed-effects, one-way-classification model is that the error term possesses a normal distribution. The Kruskal-Wallis test is a distribution-free alternative that relaxes this restriction by considering the model

$$X_{ij} = \mu_i + E_{ij}$$

for $i = 1, 2, \ldots, r$ and $j = 1, 2, \ldots, n_i$, where the E_{ij} form a collection of independent identically distributed continuous distributions with common mean 0 and common variance σ^2. In terms of the treatment means $\mu_1, \mu_2, \ldots, \mu_r$, the null hypothesis is

$$H_0: \mu_1 = \mu_2 = \cdots = \mu_r$$

We let N denote the total number of observations in the sample.

To construct the Kruskal-Wallis test statistic K, we list the N observations in increasing order and then proceed to analyze the ranks. To that end, let R_{ij} denote the rank of X_{ij} in the ordering, let

$$R_{i.} = \sum_{j=1}^{n_i} R_{ij}$$

denote the sum of the ranks of the ith-treatment observations, and let $\bar{R}_{i.} = R_{i.}/n_i$ be the average of the ranks corresponding to the ith treatment. Under H_0, R_{ij} can attain any one of the ranks $1, 2, \ldots, N$ with equal probability. Hence,

$$E[R_{ij}] = \sum_{l=1}^{N} \frac{l}{N} = \frac{N(N+1)/2}{N} = \frac{N+1}{2}$$

and

$$E[\bar{R}_{i.}] = \frac{1}{n_i} \sum_{j=1}^{n_i} E[R_{ij}] = \frac{N+1}{2}$$

If the null hypothesis is true, it should be expected that the average treatment rankings will be nearer to $(N+1)/2$ than if the null hypothesis were false, since in the latter case some treatment rankings will cluster in the lower ranks and others will cluster in the higher ranks. The Kruskal-Wallis test statistic takes advantage of this observation by measuring, for $i = 1, 2, \ldots, r$, the deviation of $\bar{R}_{i.}$ from its mean. If these deviations are large, then the null hypothesis is rejected. Specifically, the test statistic K is defined by

$$K = \frac{12}{N(N+1)} \sum_{i=1}^{r} n_i \left(\bar{R}_{i.} - \frac{N+1}{2} \right)^2$$

$$= \frac{12}{N(N+1)} \sum_{i=1}^{r} \frac{R_{i.}^2}{n_i} - 3(N+1)$$

the latter expression being employed in computations.

The test involving K is upper one-tailed and is facilitated by the fact that, so long as the sample sizes are sufficiently large, under the null hypothesis, K possesses a distribution that is approximately chi-square with $r - 1$ degrees of freedom. The rule of thumb concerning sufficiently large samples is that, for $r = 3$, $n_i \geq 6$ for $i = 1, 2$, and 3, and for $r > 3$, $n_i \geq 5$ for $i = 1, 2, \ldots, r$. Based upon the chi-square approximation, H_0 is rejected at significance level α if $K \geq \chi^2_{\alpha, r-1}$.

EXAMPLE 11.9

To test the wattage ratings of three brands of air conditioners, each possessing a common BTU rating, an investigator randomly selects a number of each brand and randomly assigns them to various test settings. The random variables X_1, X_2, and X_3 correspond to the wattage-rating measurements of the three brands, the results (in watts) of the testing being given in Table 11.11. Also provided in the table are the overall rankings of the data. The purpose of the test is to check for equal wattage ratings.

1. *Hypothesis test.* H_0: $\mu_1 = \mu_2 = \mu_3$
   H_1: The means are not all equal.
2. *Level of significance.* $\alpha = 0.01$
3. *Test statistic.* Kruskal-Wallis statistic K
4. *Sample sizes.* $n_1 = 6$, $n_2 = 7$, $n_3 = 7$
5. *Critical value.* $\chi^2_{0.01,2} = 9.210$
6. *Sample value.* Employing the rankings, we find $R_{1.} = 56$, $R_{2.} = 93$, and $R_{3.} = 61$. Thus, the value of the Kruskal-Wallis statistic is

$$k = \frac{12}{(20)(21)}\left[\frac{(56)^2}{6} + \frac{(93)^2}{7} + \frac{(61)^2}{7}\right] - (3)(21) = 2.423$$

7. *Decision.* Since $k < 9.210$, accept H_0.

TABLE 11.11   AIR-CONDITIONER WATTAGE RATINGS

x_{1j}	Rank	x_{2j}	Rank	x_{3j}	Rank
1340	3	1352	9.5	1351	8
1352	9.5	1350	6.5	1337	2
1364	16	1369	18	1334	1
1344	4	1349	5	1356	11
1350	6.5	1474	19	1361	14
1365	17	1362	15	1359	13
		1380	20	1358	12

EXERCISES

SECTION 11.1

In Exercises 11.1 through 11.9, assume that the populations and sampling methodologies satisfy the requirements of the fixed-effects, one-way-classification model. Each exercise consists of the following parts:
a) *Find the ANOVA table.*
b) *Determine, at both $\alpha = 0.05$ and $\alpha = 0.01$, whether or not the null hypothesis of equal treatment effects is rejected, and explain the conclusion in words.*
c) *If the null hypothesis is rejected at either level, then estimate the treatment effects.*

11.1. A company produces hot-rolled steel in four separate plants. Random observations of ultimate tensile strength (in ksi) are made on steel specimens from each of the four plants, the resulting data being given in Table 11.12.

11.2. When a high-level language is compiled, the run-time efficiency depends on the compiler. A software engineer wishes to compare three compilers that have been developed for a new high-level computer

TABLE 11.12   TENSILE-STRENGTH DATA FOR EXERCISE 11.1

Plant	Observations					
A	61.2	62.0	60.9	62.1	61.8	
B	60.8	62.1	62.5	61.4	60.9	62.2
C	60.8	61.5	61.9	61.5	61.7	62.0	61.2
D	64.3	64.6	63.6	64.7	63.9	64.8	64.2

language that is to be compiled for a particular mainframe computer. Fifteen programs are randomly selected from those considered appropriate for the new language, each is coded in the new language by the same programmer, five are randomly selected for each compiler, and each is run once. The CPU times (in seconds) are presented in Table 11.13.

TABLE 11.13 CPU TIMES FOR EXERCISE 11.2

Compiler	Observations				
A	8.4	17.4	9.8	8.0	15.3
B	9.2	5.8	14.3	13.5	7.0
C	4.7	9.6	10.0	18.3	6.7

11.3. Four trial mixes of concrete are to be tested to determine whether or not there is a difference among the three insofar as compressive strength is concerned. The resulting test data (in psi) are recorded for the four mixes in Table 11.14.

TABLE 11.14. COMPRESSIVE-STRENGTH DATA FOR EXERCISE 11.3

Mix	Observations			
1	3820	4005	3693	3972
2	4052	4102	3984	4110
3	3996	3905	3880	3858
4	3880	3866	3798	3800

11.4. Four brands of grinding wheels are tested for usable life (in hours), the point at which a given wheel no longer gives reasonable service being determined by an experienced machinist. The test data are given in Table 11.15.

TABLE 11.15 USABLE LIFETIMES FOR EXERCISE 11.4

Wheel	Observations					
A	12.3	14.4	11.6	15.3	16.0	13.0
B	9.8	10.2	11.3	9.6	12.0	10.7
C	12.2	10.0	9.8	12.1	10.5	11.4

11.5. A company has four identical copiers, each to be used by a different department. Table 11.16 gives copier times (waiting plus service) in minutes for four randomly selected copying jobs in each department.

TABLE 11.16 COPIER TIMES FOR EXERCISE 11.5

Department	Observations			
1	4.22	4.45	0.21	0.42
2	2.18	0.82	1.39	10.20
3	4.90	6.28	2.10	2.53
4	1.47	0.39	5.60	0.48

11.6. Six wells are tested on three different occasions for iron content (in mg per liter), the data being given in Table 11.17.

TABLE 11.17 IRON-CONTENT DATA FOR EXERCISE 11.6

Well	Observations		
1	0.025	0.020	0.021
2	0.024	0.019	0.023
3	0.028	0.021	0.024
4	0.034	0.030	0.025
5	0.030	0.021	0.020
6	0.018	0.025	0.018

11.7. Three laboratories independently perform tests to find the specific heat of calcium. The results of the tests are given in Table 11.18.

TABLE 11.18 CALCIUM SPECIFIC-HEAT DATA FOR EXERCISE 11.7

Laboratory	Observations		
A	0.1553	0.1556	0.1551
B	0.1557	0.1568	0.1560
C	0.1566	0.1562	0.1568

11.8. Three brands of dry-cell batteries nominally rated at 1.5 volts are tested, four batteries of each brand being randomly selected. The test data (in volts) are given in Table 11.19.

TABLE 11.19 VOLTAGE DATA FOR EXERCISE 11.8

Brand	Observations			
A	1.56	1.52	1.58	1.50
B	1.50	1.54	1.49	1.53
C	1.45	1.48	1.43	1.50

11.9. It is desired to compare three sources of standard TTL (transistor-transistor logic) integrated circuits for power dissipation. The test data (in mW) are given in Table 11.20.

TABLE 11.20 POWER-DISSIPATION DATA FOR EXERCISE 11.9

Source	Observations			
A	10.12	10.08	10.09	10.11
B	10.07	10.08	10.06	10.06
C	10.04	10.07	10.03	10.06

11.10. The analysis-of-variance methodology discussed in Section 11.1 concerns a null hypothesis concerning the equality of normal population means under the assumption of equal variances. Given the normality assumption, an often-employed test concerning the null hypothesis

$$H_0: \sigma_1^2 = \sigma_2^2 = \cdots = \sigma_r^2$$

and its alternative

H_1: The variances are not all equal.

is **Bartlett's test.** Although the test can be stated for nonequal sample sizes, we restrict our attention to a common sample size n. Under this assumption, the test utilizes the statistic

$$B = \frac{[S_1^2 S_2^2 \cdots S_r^2]^{1/r}}{S_p^2}$$

where, for $i = 1, 2, \ldots, r$, S_i^2 is the sample variance for the treatment-i sample and where S_p^2 is the pooled sample-variance estimator

$$S_p^2 = \frac{1}{r} \sum_{i=1}^{r} S_i^2$$

H_0 is rejected at significance level α if $B < b_{\alpha,n,r}$, the latter being the critical value for level α, common sample size n, and number of treatments r. Critical values for $\alpha = 0.05$ are given in Table A.15.

Suppose a company employs three similar machines, A, B, and C, that utilize a revolving output spindle, whose speed is controlled by input voltage. Assuming the voltage to be fixed and the machines to have been operating sufficiently long for normal operating speed to have been achieved, four measurements (in revolutions per second) of each are taken, the resulting data being given in Table 11.21. Applying Bartlett's test with $\alpha = 0.05$, what is concluded?

TABLE 11.21 REVOLUTION DATA FOR EXERCISE 11.10

Machine	Observations			
A	12.20	12.01	12.32	11.97
B	11.97	11.92	12.00	12.02
C	12.10	12.12	12.13	12.08

11.11. Four laboratories report their findings regarding the solubility (in grams per 100 grams of water) of tartaric acid at 40°C, the test results being given in Table 11.22. Assuming the measurements of the laboratories to be independent and normally distributed, do the data lead one to conclude at the $\alpha = 0.05$ level that the variabilities of the laboratory measurement procedures are unequal?

TABLE 11.22 SOLUBILITY DATA FOR EXERCISE 11.11

Laboratory	Observations			
1	37.03	37.00	36.96	37.01
2	37.09	37.08	36.90	36.89
3	37.05	37.09	37.00	37.05
4	36.88	36.84	37.05	37.10

11.12. Apply Bartlett's test to the data of Table 11.13 at the $\alpha = 0.05$ level.

11.13. Apply Bartlett's test to the data of Table 11.19 at the $\alpha = 0.05$ level.

SECTION 11.2

Exercises 11.14 through 11.20 concern the application of Tukey's test. If the null hypothesis of the cited exercise has been rejected at the $\alpha = 0.01$ level (and, therefore, also at the $\alpha = 0.05$ level), then apply Tukey's procedure with $\alpha = 0.01$. If the null hypothesis has been rejected at the $\alpha = 0.05$ level but not at the $\alpha = 0.01$ level, then apply Tukey's procedure with $\alpha = 0.05$. Finally, if the null hypothesis has been accepted at the $\alpha = 0.05$ level, then do nothing.

11.14. Exercise 11.1
11.15. Exercise 11.3
11.16. Exercise 11.4
11.17. Exercise 11.5
11.18. Exercise 11.7
11.19. Exercise 11.8
11.20. Exercise 11.9

11.21. Using Tukey's procedure, find 95% confidence intervals for the treatment-mean differences of Exercise 11.18.

In Exercises 11.22 through 11.25, follow the instructions for Exercises 11.14 through 11.20, except employ Duncan's multiple range procedure.

11.22. Exercise 11.2
11.23. Exercise 11.6
11.24. Exercise 11.8
11.25. Exercise 11.9

SECTION 11.3

11.26. A sensor indicates when the wavelength of a light source exceeds 7000 angstroms, thereby passing into the infrared zone. A sample of five sensors are drawn from a large collection, and each is tested three times to determine at which wavelength it signals that the 7000-angstrom threshold has been exceeded. The resulting data (in angstroms) are given in Table 11.23. Construct the ANOVA table. Assuming the assumptions of the random-effects model are satisfied, is it concluded at the $\alpha = 0.05$ level that the common treatment-effect variance is greater than zero? If it is, then estimate that variance. Explain the meaning of your conclusions.

TABLE 11.23 WAVELENGTH DATA FOR EXERCISE 11.26

Sensor	Observations		
1	7010	7016	7013
2	6991	6984	6990
3	6985	6989	6990
4	7016	7010	7020
5	7017	7020	7018

11.27. Suppose the six wells discussed in Exercise 11.6 have been randomly selected from among a large number of wells in the region. Under the assumptions of the random-effects model, what do the data of Table 11.17 say about the uniformity of iron content in the region's water? Explain your answer in terms of the model, and let $\alpha = 0.05$.

11.28. Suppose the data of Table 11.20 have been derived from a random selection of three TTL ICs from among a large collection of such ICs and the taking of four power-dissipation readings for each. With $\alpha = 0.01$, is it concluded that the common variance of the treatment-effect random variables is positive? If so, then estimate that variance.

11.29. Suppose the three machines of Table 11.21 have been randomly drawn from a collection of machines. Given the assumptions of the random-effects model, what conclusion is drawn (with $\alpha = 0.05$)?

11.30. A plastics manufacturer receives a large number of batches of resin for a specific product. To test the uniformity of the batches, three are selected at random and the tensile strength of three of the parts resulting from each is tested. The results (in psi) are given in Table 11.24. Construct the ANOVA table. Is the null hypothesis of the random-effects model rejected at significance level $\alpha = 0.01$? If it is, then estimate σ_α.

TABLE 11.24 TENSILE-STRENGTH DATA FOR EXAMPLE 11.30

Batch	Observations		
A	8032	7982	8065
B	8238	8002	8306
C	7986	8376	8320

11.31. Suppose the grinding wheels of Exercise 11.4 and Table 11.15 have been randomly drawn from a large collection of grinding wheels. Based upon the random-effects model with $\alpha = 0.05$, what conclusion is reached?

SECTION 11.4

11.32. Referring to Exercise 11.2 with $\alpha = 0.05$, what is β when the true state of nature is that the mean CPU times when using compilers A and B are identical and each is σ greater than the mean CPU time when using compiler C?

11.33. Referring to Exercise 11.3 with $\alpha = 0.01$, what is β when the mean compressive strengths for mixes 1, 3, and 4 are identical and each is 2σ greater than the mean compressive strength for mix 2?

11.34. Referring to Exercise 11.4 with $\alpha = 0.01$, what is β when the mean usable lifetimes for wheels B and C are the same and each is σ greater than the mean usable lifetime for wheel A?

11.35. Referring to Exercise 11.5 with $\alpha = 0.05$, what is β when $\mu_4 = \mu_1$, $\mu_2 = \mu_1 + \sigma$, and $\mu_3 = \mu_1 + 2\sigma$?

11.36. Referring to Exercise 11.7 with $\alpha = 0.05$, what is β when $\mu_A = \mu_C + 2\sigma$ and $\mu_B = \mu_C + \sigma$? Find the necessary common sample size to make $\beta = 0.10$.

11.37. Referring to Exercise 11.26, find β when $\sigma_\alpha/\sigma = 2$. Find n so that $\beta = 0.20$.

11.38. Referring to Exercise 11.28, find β when $\sigma_\alpha/\sigma = 2.4$. Find n so that $\beta = 0.20$.

11.39. Referring to Exercise 11.29, find β when $\sigma_\alpha/\sigma = 2.2$. Find n so that $\beta = 0.10$.

In Exercises 11.40 through 11.48, apply the Kruskal-Wallis test to the data of the cited table. In all cases, state the conclusion for both α = 0.05 and α = 0.01. Because samples have been kept small to ease the computational burden, ignore the rule of thumb regarding sufficiently large samples that is stated in the text.

11.40. Table 11.12
11.41. Table 11.13
11.42. Table 11.14
11.43. Table 11.15
11.44. Table 11.16
11.45. Table 11.17
11.46. Table 11.18
11.47. Table 11.19
11.48. Table 11.20

SUPPLEMENTAL EXERCISES FOR COMPUTER SOFTWARE

In Exercises 11.C1 through 11.C3, reperform the analysis of variance in the cited exercise using the original data in conjunction with the newly provided data in the cited table.

11.C1. Exercise 11.1, Table 11.25.
11.C2. Exercise 11.4, Table 11.26.
11.C3. Exercise 11.8, Table 11.27.
11.C4. Apply a multiple comparison test to the data of Exercise 11.C2, using the additional data together with the data of Exercise 11.4.

TABLE 11.25 ADDITIONAL TENSILE-STRENGTH DATA FOR EXERCISE 11.C1

Plant	Observations						
A	61.3	62.4	62.1	60.1	59.8	61.0	
B	61.2	62.3	62.1	62.1	60.6	60.8	61.5
C	60.2	60.5	61.3	61.4	62.3	62.1	
D	63.4	64.6	64.6	64.9	63.2	64.1	63.6

TABLE 11.27 ADDITIONAL VOLTAGE DATA FOR EXERCISE 11.C3

Brand	Observations					
A	1.50	1.53	1.53	1.54	1.55	1.50
B	1.49	1.52	1.53	1.54	1.52	1.52
C	1.47	1.49	1.46	1.47	1.50	1.50

11.C5. Apply a multiple comparison test to the data of Exercise 11.C3, using the additional data together with the data of Exercise 11.8.

11.C6. Reperform the random-effects analysis of variance for the tensile-strength data of Exercise 11.30, using the additional data in Table 11.28.

TABLE 11.28 ADDITIONAL TENSILE-STRENGTH DATA FOR EXERCISE 11.C6

Batch	Observations			
A	8020	8040	8055	7998
B	8302	8322	8201	8255
C	8305	8256	8239	8144

11.C7. Redo the random-effects analysis of variance for the wavelength data of Exercise 11.26 using the additional data in Table 11.29.

TABLE 11.29 ADDITIONAL WAVELENGTH DATA FOR EXERCISE 11.C7

Sensor	Observations				
1	7010	7012	7014	7018	7010
2	7000	6988	6989	6999	7988
3	6990	6983	6988	6985	6994
4	7010	7020	7016	7017	7020
5	7024	7015	7020	7016	7018

TABLE 11.26 ADDITIONAL WHEEL-LIFETIME DATA FOR EXERCISE 11.C2

Wheel	Observations						
A	15.3	14.2	12.3	13.6	11.6	12.3	11.4
B	12.0	11.1	10.4	9.7	9.9	11.0	10.2
C	14.2	12.1	11.9	10.0	9.8	10.3	12.7

12

SIMPLE LINEAR REGRESSION

One of the oldest problems in statistics is fitting a straight line to data points in the plane. In the present chapter, it is assumed that observations of a response random variable Y are made at several points on the x axis. The data points are plotted and a straight line is fitted to the data. To make the procedure rigorous, a particular model is employed and the method of least-squares estimation is developed around the model. The principle of least-squares is to minimize (with respect to some criterion) the vertical distances of the plotted points to the fitted line. The result is the least-squares estimators for the model parameters. Using the properties of these estimators, it is possible to make inferences regarding the parameters in the underlying model.

12.1 THE SIMPLE LINEAR REGRESSION MODEL

The problem to be addressed in the present chapter involves the linear estimation (or prediction) of one variable based upon knowledge of a related variable. In the classical deterministic setting, a dependent variable y is considered to be a function $f(x)$ of an independent variable x if, given x, there exists an exact corresponding value y. In the model discussed in the present chapter, the independent variable x will continue to be deterministic but the dependent variable Y, called a **response** variable, is considered to be random. Consequently, given x, the best that can be done is to arrive at some prediction of Y.

An associated problem was discussed in Section 5.7.2. There, given a random vector (X, Y) and an observation x of X, we found the mean $E[Y|x]$ of the conditional random variable $Y|x$. Based upon the fact that the mean $E[Y|x]$ minimizes the mean-square error $E[(Y|x - c)^2]$ over all possible choices of c, $E[Y|x]$ can be considered to be the best estimate (prediction) of Y, given the observation $X = x$. Letting x vary over the codomain of X gives rise to the curve of regression (Definition 5.12). This curve gives the locus of the best (mean-square) predictions of $Y|x$.

In many practical situations, data regarding pairs of variables are obtained in such a way that one of the variables is **controllable.** The measurement process defining the controllable variable is not subject to dispersion due to randomness (except, perhaps, to a negligible extent). In many cases the controllable variable, which we will denote by x, is predetermined by the experimenter. For each x there is an observation variable, denoted by $Y|x$, which is random. Note that the notation "$Y|x$" is similar to that employed to denote a conditional random variable. In the present situation, however, the meaning is different, since x is not an outcome corresponding to one of the random variables involved in a random vector. Experimentally, the relationship between the independent (nonrandom) variable x and the dependent (random) variable Y might be examined by observing the behavior of Y at x_1, x_2, . . . , x_n, namely, by observing the random variables $Y|x_1$, $Y|x_2$, . . . , $Y|x_n$. Since, for $i = 1, 2, . . . , n$, $Y|x_i$ is a random variable, it is necessary to postulate a model in which its distributional properties are delineated.

As an illustration of the experimental setting we have in mind, consider the measurement of internal air pressure in the cabin of an airplane. If the pressure is monitored at time intervals Δx, then x is the controlled variable and the air pressure $Y|x$ is random. Experimentally, if the initial measurement is at $x = \Delta x$, then the measurement process yields observational values $y|x_1$, $y|x_2$, . . . , $y|x_n$, where n measurements are made at times $x_i = i\Delta x$. For any i, $y|x_i = y|i\Delta x$ is a sample value corresponding to the random variable $Y|x_i$. Another illustration results when the air pressure is measured at every 1000 feet as the airplane ascends to its cruising altitude. Here, $x_i = 1000i$, and, if the cruising altitude is 30,000 feet, then there are 30 random variables corresponding to 30 air-pressure measurements, namely, $Y|x_1$, . . . , $Y|x_{30}$.

Our intention is to express $Y|x$ in terms of x, at least insofar as some function of x can be employed to predict the value of $Y|x$. The conditional-expectation approach of Section 5.7.2 is not appropriate because the observations of x are not observations of a random variable; nonetheless, conditional expectation can be employed as a guide to the design of a suitable experimental model. According to Theorem 5.19, if X and Y possess a jointly normal distribution, then $Y|x$ is normally distributed for each x, and its curve of regression $\mu_{Y|x}$ is a straight line. Thus, it is certainly plausible in the present case to try to estimate the random variable $Y|x$ by a linear function of x. Keep in mind that the suitability of such an approach depends on the degree to which the actual physical phenomena conform to the resulting regression model.

The **simple linear regression** model is based upon the linear representation of the mean of $Y|x$ in terms of x. Specifically, it is postulated that there exist constants a and b such that

$$\mu_{Y|x} = a + bx$$

It is this **regression equation** that defines the **regression line** in the Cartesian plane. In terms of prediction, if a and b are known, then $Y|x$ is predicted to be its mean $a + bx$.

In practice, a and b are not known and must be estimated from observational data. To accomplish the estimation, observations will be made at n points, x_1, x_2, . . . , x_n, each x_i being known as a **regressor** or **predictor** variable. The result is a data set consisting of n points of the form (x_i, y_i), each y_i being a sample value of the random variable $Y|x_i$.

Since each observation variable can be expressed in terms of its mean and the difference from its mean, for $i = 1, 2, . . . , n$,

$$Y|x_i = a + bx_i + \mathrm{E}_i$$

where E_i is a random variable with zero expected value. Indeed,

$$E[\mathrm{E}_i] = E[Y|x_i - (a + bx_i)] = E[Y|x_i] - \mu_{Y|x_i} = 0$$

Letting $Y_i = Y|x_i$, the regression equation takes the form

$$Y_i = a + bx_i + \mathrm{E}_i$$

In terms of actual data (x_i, y_i), $i = 1, 2, \ldots, n$, the model takes the observational form

$$y_i = a + bx_i + \epsilon_i$$

where ϵ_i is the observed value of E_i.

Our goal is to find estimators $\hat{a}$ and $\hat{b}$ for a and b, respectively. The result will be an estimator equation of the form

$$\hat{\mu}_{Y|x} = \hat{a} + \hat{b}x$$

where $\hat{\mu}_{Y|x}$ is the estimator of $\mu_{Y|x}$. Given the estimation rules represented by $\hat{a}$ and $\hat{b}$, actual data will yield estimates $\overset{*}{a}$ and $\overset{*}{b}$ together with the estimated **sample (fitted) regression line**

$$\overset{*}{\mu}_{Y|x} = \overset{*}{a} + \overset{*}{b}x$$

Figure 12.1 illustrates a **scatter diagram** constructed from sample data. Note that the sample points are scattered about the theoretical and fitted regression lines (the former assumed to be known and the latter resulting from the sample points). The manner in which the sample regression line has been obtained from the sample points will be presented shortly. In Figure 12.2, the differences

$$\epsilon_i = y_i - (a + bx_i)$$

and

$$e_i = y_i - (\overset{*}{a} + \overset{*}{b}x_i)$$

have been illustrated at an arbitrary point x_i. The value e_i is known as a **residual** and measures the vertical distance from y_i to the sample regression line.

Before deriving the estimators $\hat{a}$ and $\hat{b}$, we complete the model by assuming that each E_i is normally distributed with variance σ^2 and that $\mathrm{E}_1, \mathrm{E}_2, \ldots, \mathrm{E}_n$ are independent. Thus, the simple linear regression model takes the following form.

Simple Linear Regression Model

Equation:	$Y_i = Y|x_i = a + bx_i + \mathrm{E}_i$
Assumptions:	1. $\mathrm{E}_i \sim N(0, \sigma)$
	2. The E_i comprise an independent collection of random variables.

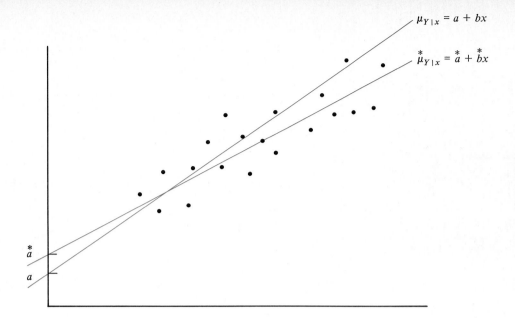

FIGURE 12.1

Scatter diagram with theoretical and sample regression lines

As a result of the first assumption, Y_i is normally distributed with mean $a + bx_i$ and variance σ^2. Note that this condition is consistent with the result of Theorem 5.19. Figure 12.3 illustrates the manner in which the distributions of the Y_i are centered along the line $a + bx$ and possess the same shape as the normal distribution with mean zero and variance σ^2.

12.2 LEAST-SQUARES ESTIMATION

The estimation problem in regression requires that the model parameters a and b be estimated from a set of sample points. Given the observed points (x_1, y_1), (x_2, y_2), . . . , (x_n, y_n), the **least-squares estimates** are selected so as to minimize the sum of

FIGURE 12.2

Illustration of e_i and ϵ_i

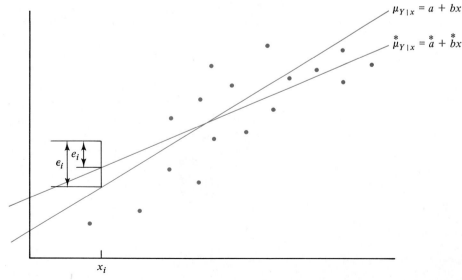

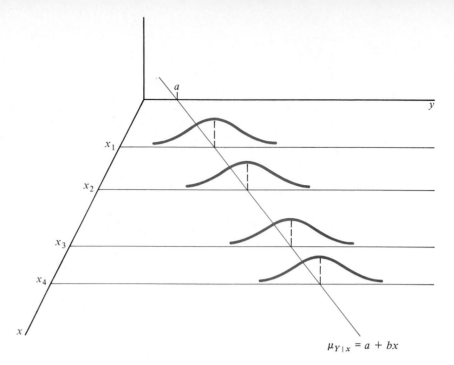

FIGURE 12.3
Distributions of Y_1, Y_2,
Y_3, . . .

x_1
x_2
x_3
x_4
a
y
x
$\mu_{Y|x} = a + bx$

the squares of the residuals. This sum is customarily called the **error sum of squares** and is given by

$$SSE = \sum_{i=1}^{n} e_i^2 = \sum_{i=1}^{n} [y_i - (\overset{*}{a} + \overset{*}{b}x_i)]^2$$

Minimization of *SSE* results in a sample regression line that **best fits** the observational data. As was the case in Chapter 11, the "error" terminology is not always appropriate. While it is true that the fluctuations in Y_i might result from measurement error, they might also result from the inherent indeterminism of the phenomenon being measured.

Before finding the least-squares estimates of a and b, we introduce some new notation that will be useful in the sequel. If X_1, X_2, . . . , X_n and Y_1, Y_2, . . . , Y_n are collections of random variables, then we define three random variables by

$$S_{XX} = \sum_{i=1}^{n} (X_i - \bar{X})^2$$

$$= \sum_{i=1}^{n} X_i^2 - \frac{1}{n} \left(\sum_{i=1}^{n} X_i \right)^2$$

$$= \sum_{i=1}^{n} X_i^2 - n\bar{X}^2$$

$$S_{YY} = \sum_{i=1}^{n} (Y_i - \bar{Y})^2$$

$$= \sum_{i=1}^{n} Y_i^2 - \frac{1}{n} \left(\sum_{i=1}^{n} Y_i \right)^2$$

$$= \sum_{i=1}^{n} Y_i^2 - n\bar{Y}^2$$

$$S_{XY} = \sum_{i=1}^{n} (X_i - \bar{X})(Y_i - \bar{Y})$$

$$= \sum_{i=1}^{n} X_i Y_i - \frac{1}{n}\left(\sum_{i=1}^{n} X_i\right)\left(\sum_{i=1}^{n} Y_i\right)$$

$$= \sum_{i=1}^{n} X_i Y_i - n\bar{X}\bar{Y}$$

The notation can be employed when either the X_i or the Y_i, or both, are not random variables but instead are sample points. If the X_i are sample points, simply replace each X_i by x_i and write S_{xx} and S_{xY}. If the Y_i are sample points, then replace each Y_i by y_i and write S_{Xy} and S_{yy}. Finally, if both the X_i and Y_i are sample points, then replace X_i by x_i, replace Y_i by y_i, and write S_{xx}, S_{yy}, and S_{xy}. Note that S_{xY} and S_{Xy} are random variables, whereas S_{xx}, S_{xy}, and S_{yy} are numbers.

THEOREM 12.1

The least-squares estimates for a and b in the simple linear regression model are

$$\overset{*}{a} = \bar{y} - \overset{*}{b}\bar{x}$$

$$\overset{*}{b} = \frac{S_{xy}}{S_{xx}}$$

where $\bar{x}$ and $\bar{y}$ are the empirical means of $x_1, x_2, \ldots, x_n$ and $y_1, y_2, \ldots, y_n$, respectively.

Proof: Treating SSE as a function of the variables $\overset{*}{a}$ and $\overset{*}{b}$, minimization is accomplished by taking partial derivatives. We obtain

$$\frac{\partial SSE}{\partial \overset{*}{a}} = -2 \sum_{i=1}^{n} (y_i - \overset{*}{a} - \overset{*}{b}x_i)$$

$$\frac{\partial SSE}{\partial \overset{*}{b}} = -2 \sum_{i=1}^{n} (y_i - \overset{*}{a} - \overset{*}{b}x_i)x_i$$

Setting these partial derivatives equal to zero yields

$$n\overset{*}{a} + \overset{*}{b} \sum_{i=1}^{n} x_i = \sum_{i=1}^{n} y_i$$

and

$$\overset{*}{a} \sum_{i=1}^{n} x_i + \overset{*}{b} \sum_{i=1}^{n} x_i^2 = \sum_{i=1}^{n} x_i y_i$$

which are known as the **normal equations.** Solving for $\overset{*}{a}$ and $\overset{*}{b}$ yields the solutions stated in the theorem. ∎

Theorem 12.1, which is stated in terms of estimates, can be correspondingly expressed in terms of estimators: the **least-squares estimators** of a and b are

TABLE 12.1 VISCOSITY–TEMPERATURE DATA FOR SAE 30 MOTOR OIL

x_i	x_i^2	y_i	y_i^2	x_iy_i
165	27,225	28.5	812.25	4702.5
170	28,900	26.1	681.21	4437.0
175	30,625	23.9	571.21	4182.5
180	32,400	22.0	484.00	3960.0
185	34,225	20.4	416.16	3774.0
190	36,100	18.5	342.25	3515.0
195	38,025	17.1	292.41	3334.5
200	40,000	15.8	249.64	3160.0
1460	267,500	172.3	3849.13	31,065.5

$$\hat{a} = \bar{Y} - \hat{b}\bar{x}$$

$$\hat{b} = \frac{S_{xY}}{S_{xx}}$$

where $\bar{Y}$ is the mean of the random variables $Y_1, Y_2, \ldots, Y_n$.

In terms of the estimators $\hat{a}$ and $\hat{b}$, the (estimator) regression line takes the form

$$\hat{\mu}_{Y|x} = \hat{a} + \hat{b}x$$

Relative to estimates resulting from a particular sample,

$$\overset{*}{\mu}_{Y|x} = \overset{*}{a} + \overset{*}{b}x$$

Since $\mu_{Y|x}$ is employed to predict future values of $Y = Y|x$, the estimator and estimate equations are often written as

$$\hat{Y} = \hat{a} + \hat{b}x$$

and

$$\overset{*}{Y} = \overset{*}{a} + \overset{*}{b}x$$

respectively. Finally, it should be recognized that, relative to the random variables $Y_1, Y_2, \ldots, Y_n$, SSE is a random variable.

EXAMPLE 12.1

The viscosity of motor oil decreases with increasing temperature. Viscosity measurements of SAE 30 motor oil are taken at 5°F increments starting at 165° and ending at 200°. The resulting data are tabulated in Table 12.1, where x denotes temperature and y denotes viscosity [in 10^{-7} (lb)(sec)/(in.)^2]. Included in the table are the summations of each column. From the data, $\bar{x} = 182.5$, $\bar{y} = 21.5375$,

$$S_{xx} = \sum_{i=1}^{8} x_i^2 - 8\bar{x}^2 = 267{,}500 - 266{,}450 = 1050$$

$$S_{xy} = \sum_{i=1}^{8} x_i y_i - 8\bar{x}\bar{y} = 31{,}065.5 - 31{,}444.75 = -379.25$$

$$\overset{*}{b} = \frac{S_{xy}}{S_{xx}} = \frac{-379.25}{1050} = -0.3612$$

$$\overset{*}{a} = \bar{y} - \overset{*}{b}\bar{x} = 21.5375 - (-0.3612)(182.5) = 87.4565$$

and the sample regression line is

$$\overset{*}{Y} = \overset{*}{\mu}_{Y|x} = 87.46 - (0.361)x$$

The degree to which the sample regression line fits the data can be seen in Figure 12.4, where the line passes through the scatter diagram formed by the data pairs.

EXAMPLE 12.2

To estimate the stress–strain curve of a particular strength of concrete, compressive stress is applied to a cylindrical concrete specimen. For increments of 0.5 ksi (kips per square inch) the strain data (in inch/inch) of Table 12.2 are recorded. Included in the table are the summations of each column. Note that the cylinder fails shortly after the stress passes 5000 psi. Recognizing that strain measurements are random at each observation stress x_i, we use the data to find the sample regression line. From the data, we find $\bar{x} = 2.75$, $\bar{y} = 0.000916$,

FIGURE 12.4
Sample regression line for Example 12.1

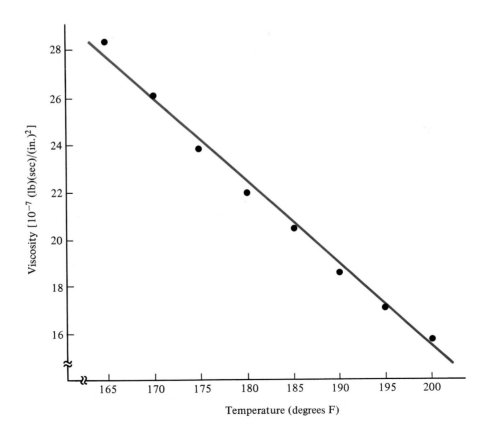

$$S_{xx} = \sum_{i=1}^{10} x_i^2 - 10\bar{x}^2 = 96.25 - 75.63 = 20.62$$

$$S_{xy} = \sum_{i=1}^{10} x_i y_i - 10\bar{x}\bar{y} = 0.03313 - 0.02519 = 0.00794$$

$$\overset{*}{b} = \frac{S_{xy}}{S_{xx}} = \frac{0.00794}{20.62} = 0.000385$$

$$\overset{*}{a} = \bar{y} - \overset{*}{b}\bar{x} = 0.000916 - (0.000385)(2.75) = -0.000143$$

and the sample regression line

$$\overset{*}{Y} = \overset{*}{\mu}_{Y|x} = -0.000143 + (0.000385)x$$

The manner in which the sample regression line fits the data can be seen in Figure 12.5.

A close look at the scatter diagram reveals that the first seven sample points are quite linear and that they begin to turn appreciably upward starting with the eighth point. This is not surprising, since stress–strain curves for concrete tend to be linear and then exhibit ever greater curvature as the yield point is approached. Based upon a visual examination of the scatter diagram or a knowledge of stress–strain curves, it might be wise to only employ the first seven points in the regression. If this is done, then we obtain $\bar{x} = 2$, $\bar{y} = 0.00061$, $S_{xx} = 7$, $S_{xy} = 0.0025$, $\overset{*}{b} = 0.000357$, $\overset{*}{a} = -0.000104$, and the fitted regression line

$$\overset{*}{Y} = \overset{*}{\mu}_{Y|x} = -0.000104 + (0.000357)x$$

which is depicted in Figure 12.6. Note how this line gives a better fit to the first seven points (and also the eighth point).

Relative to Example 12.2, two important points regarding regression can be made. First, the degree to which a straight line models the data is a function of the portion of the data considered. In practical situations, a linear relation between $\mu_{Y|x}$ and x will hold at most on some region of the x axis. For instance, stress and

TABLE 12.2 CONCRETE STRESS–STRAIN DATA

x_i	x_i^2	y_i	y_i^2	$x_i y_i$
0.5	0.25	0.00013	1.69×10^{-8}	0.00007
1.0	1.00	0.00031	9.61×10^{-8}	0.00031
1.5	2.25	0.00044	1.94×10^{-7}	0.00066
2.0	4.00	0.00060	3.60×10^{-7}	0.00120
2.5	6.25	0.00077	5.93×10^{-7}	0.00193
3.0	9.00	0.00092	8.46×10^{-7}	0.00276
3.5	12.25	0.00110	1.21×10^{-6}	0.00385
4.0	16.00	0.00131	1.72×10^{-6}	0.00524
4.5	20.25	0.00158	2.50×10^{-6}	0.00711
5.0	25.00	0.00200	4.00×10^{-6}	0.01000
27.5	96.25	0.00916	11.54×10^{-6}	0.03313

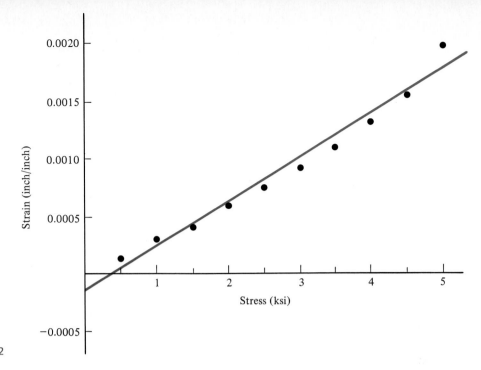

FIGURE 12.5
Sample regression line
for Example 12.2 using
all the data of Table 12.2

strain are linearly related in concrete specimens up to a point, and it is to that point that the linear model is appropriate. In general, even if all the data comprising the scatter diagram are distributed about the sample regression line, there is no reason to conclude that the linear relation will hold outside the range of observation. As a result, extrapolation along a regression line is extremely risky. In the case of concrete

FIGURE 12.6
Sample regression line
for Example 12.2 using
only the first seven data
points of Table 12.2

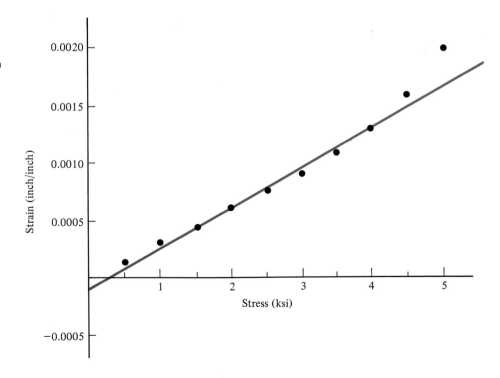

stress and strain, the cylinder will eventually collapse under increasing compressive stress.

A second point concerns linearity itself. When the observation stresses were cut to seven in Example 12.2, the regression line fit quite well. Nevertheless, even though the stress qualities of concrete make an excellent fit appear plausible in low stress ranges, an exact linear relation is still just a model to help describe phenomenal behavior. Indeed, even if a scientist believes a relation is not linear, that does not necessarily imply that a linear regression model is not suitable. Linear models are computationally and conceptually simple. Thus, they can serve as working models over ranges of the independent variable for which they provide acceptable approximations. Should a better fit be required, one can turn to a curvilinear model such as the polynomial model described in Section 13.4.

EXAMPLE 12.3

When in execution, a typical computer program spends a small proportion of its time in computation. A large amount of time is spent awaiting I/O. Consequently, there is better CPU utilization in a multiprogramming environment because programs requiring computation can access the CPU while others await I/O completion. Although an analytic account of CPU utilization requires the type of queuing analysis discussed in Chapter 6, empirical estimates of utilization efficiency can be obtained statistically.

Consider placing a set of ten benchmark programs into execution one after the other and at each step measuring the corresponding CPU utilization by finding the percentage of time the CPU is active, this measurement being taken over a fixed period of time. The data of Table 12.3 result from an experiment in which each program possesses a single utilization rate of approximately 20%. The fact that the utilization percentages increase as the number of programs increases is certainly reasonable, because with succeeding programs in execution there is less time for the queue of programs awaiting the CPU to be empty. From the data, we find $\bar{x} = 5.5$, $\bar{y} = 64.4$, $S_{xx} = 82.5$, $S_{xy} = 613$, $\overset{*}{b} = 7.43$, and $\overset{*}{a} = 23.5$. Thus, the sample regression line possesses the equation

$$\overset{*}{Y} = \overset{*}{\mu}_{Y|x} = 23.5 + (7.43)x$$

which is graphed, together with the data points, in Figure 12.7.

From the description of the experiment it is clear that the relationship between utilization and the degree of multiprogramming cannot be linear, since otherwise utilization would soon exceed 100%. In fact, in Example 13.7 we will fit a quadratic to the data points of Figure 12.7. Nonetheless, a linear model might be useful over some restricted number of programs. In particular, if we restrict the regression to seven programs, then the data of Table 12.3 yield $\bar{x} = 4$, $\bar{y} = 54.86$, $S_{xx} = 28$, $S_{xy} = 272.2$, $\overset{*}{b} = 9.71$, and $\overset{*}{a} = 16.0$, the resulting regression equation being

$$\overset{*}{Y} = \overset{*}{\mu}_{Y|x} = 16.0 + (9.71)x$$

The regression line is graphed in Figure 12.8, where it can be seen that the fit is tighter relative to the first seven points. Had linear regression only been considered over the first four points, quite a good fit would have been achieved (for those four points). The point here is twofold: (1) a linear model might serve as a first approximation even when a curvilinear model appears to be called for, and (2) a linear model might serve quite well on some restricted portion of the data even though the full data set appears to require a curvilinear fit.

In Example 12.3 the independent variable is discrete. Hence, the model applies only to integral values, even though the regression line is graphed continuously.

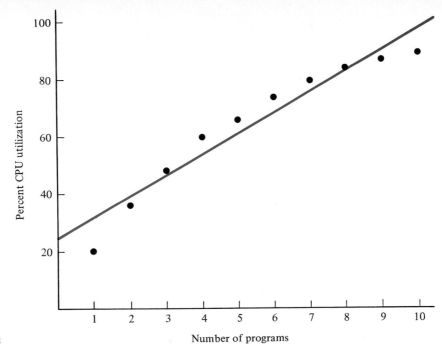

FIGURE 12.7
Sample regression line
for Example 12.3 using
all the data of Table 12.3

In the preceding examples there has been a single observation of Y for each predictor variable; however, there is no such restriction in the model. In fact, there was no provision in the theory, including Theorem 12.1, regarding the repeated use of the independent variables. For instance, rather than observe the stress–strain relation in a single concrete cylinder, one might plot the data for several cylinders and then fit the sample regression line. Such an approach would certainly help to mitigate the error due to variation among different specimens.

TABLE 12.3 CPU UTILIZATION DATA

x_i	x_i^2	y_i	y_i^2	$x_i y_i$
1	1	20	400	20
2	4	36	1296	72
3	9	48	2304	144
4	16	60	3600	240
5	25	66	4356	330
6	36	74	5476	444
7	49	80	6400	560
8	64	84	7056	672
9	81	87	7569	783
10	100	89	7921	890
55	385	644	46,378	4155

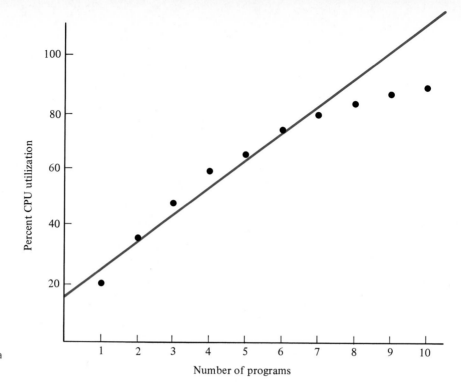

FIGURE 12.8
Sample regression line for Example 12.3 using only the first seven data points of Table 12.3

TABLE 12.4   HARDENABILITY DATA

x_i	y_i	x_i^2	y_i^2	$x_i y_i$
0	59	0	3481	0
0	58	0	3364	0
0	56	0	3136	0
10	57	100	3249	570
10	54	100	2916	540
10	52	100	2704	520
20	52	400	2704	1040
20	50	400	2500	1000
20	46	400	2116	920
30	49	900	2401	1470
30	48	900	2304	1440
30	46	900	2116	1380
40	46	1600	2116	1840
40	44	1600	1936	1760
40	42	1600	1764	1680
50	43	2500	1849	2150
50	42	2500	1764	2100
50	40	2500	1600	2000
450	884	16,500	44,020	20,410

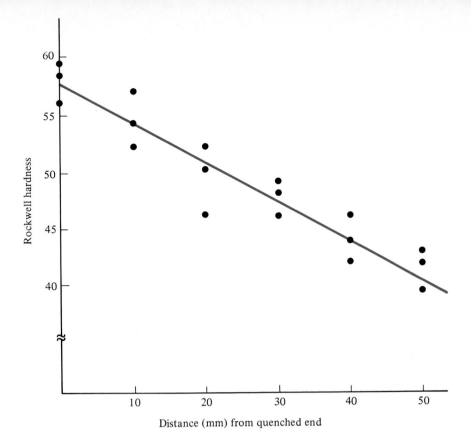

FIGURE 12.9
Sample regression line
for Example 12.4

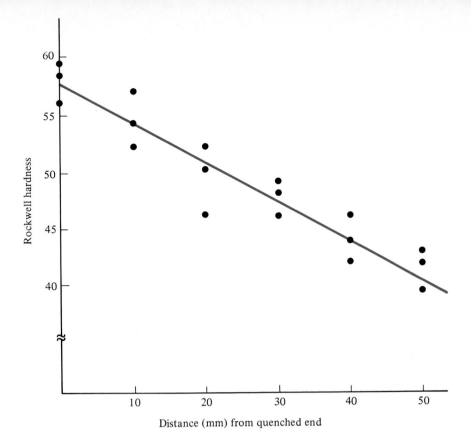

EXAMPLE 12.4

To test the hardenability of steel, a test rod is cooled at one end by quenching with water and, after the rod has completely cooled, its Rockwell hardness is tested at various distances from the quenched end. The data of Table 12.4 result from testing three rods and then measuring the hardness of each at the quenched end at increments of 10 mm from the end. The fitted regression line is depicted in Figure 12.9. According to the 18 data points, $\bar{x} = 25$, $\bar{y} = 49.111$, $S_{xx} = 5250$, $S_{xy} = -1685$, $\overset{*}{b} = -0.3210$, and $\overset{*}{a} = 57.136$, thus resulting in the sample regression line

$$\overset{*}{Y} = \overset{*}{\mu}_{Y|x} = 57.136 - (0.321)x$$

12.3 PROPERTIES OF THE LEAST-SQUARES ESTIMATORS

To make inferences concerning the least-squares estimators, a knowledge of their distributional properties is required. We now turn our attention to these properties, in all cases assuming the distributional assumptions of the simple linear regression model to be in effect, namely, that $E_1, E_2, \ldots, E_n$ are independently distributed normal random variables possessing mean zero and variance σ^2.

THEOREM 12.2

The estimator $\hat{b}$ possesses a normal distribution with mean b and variance

$$\text{Var}[\hat{b}] = \frac{\sigma^2}{S_{xx}}$$

In particular, $\hat{b}$ is an unbiased estimator of b.

Proof: According to Theorem 12.1,

$$\hat{b} = \frac{S_{xY}}{S_{xx}}$$

$$= S_{xx}^{-1} \sum_{i=1}^{n} (x_i - \bar{x})(Y_i - \bar{Y})$$

$$= S_{xx}^{-1} \left[\sum_{i=1}^{n} (x_i - \bar{x})Y_i - \bar{Y}\sum_{i=1}^{n} (x_i - \bar{x}) \right]$$

$$= S_{xx}^{-1} \sum_{i=1}^{n} (x_i - \bar{x})Y_i$$

where the last equality is obtained from the former by recognizing that the sum of the deviations of the x_i from their mean must be zero. Since each Y_i is normally distributed, the last sum is a linear combination of independent normal random variables and is therefore normal. Taking the expected value yields

$$E[\hat{b}] = S_{xx}^{-1} \sum_{i=1}^{n} (x_i - \bar{x})E[Y_i]$$

$$= S_{xx}^{-1} \sum_{i=1}^{n} (x_i - \bar{x})(a + bx_i)$$

$$= bS_{xx}^{-1} \sum_{i=1}^{n} (x_i - \bar{x})x_i$$

$$= b$$

where the second, third, and fourth equalities follow from the model equation, the sum of the deviations being zero, and the recognition that the final sum is equal to S_{xx}, respectively.

As for the variance,

$$\text{Var}\,[\hat{b}] = \text{Var}\,[S_{xx}^{-1} \sum_{i=1}^{n} (x_i - \bar{x})Y_i]$$

$$= S_{xx}^{-2} \sum_{i=1}^{n} (x_i - \bar{x})^2\,\text{Var}\,[Y_i]$$

$$= \sigma^2 S_{xx}^{-2} \sum_{i=1}^{n} (x_i - \bar{x})^2$$

$$= \sigma^2 S_{xx}^{-1}$$

where we used, in order, the expression for $\hat{b}$ derived in the first part of the proof, Theorem 5.17, the fact that Var $[Y_i] = \sigma^2$ for all i, and the definition of S_{xx}. ■

THEOREM 12.3

The estimator $\hat{a}$ possesses a normal distribution with mean a and variance

$$\text{Var}\,[\hat{a}] = \left(\frac{1}{n} + \frac{\bar{x}^2}{S_{xx}} \right) \sigma^2$$

In particular, $\hat{a}$ is an unbiased estimator of a.

Proof: Both $\bar{Y}$ and $\hat{b}$ are normal random variables. Moreover, it can be shown that they are independent. According to Theorem 12.1, $\hat{a}$ is a linear combination of $\bar{Y}$ and $\hat{b}$. Consequently, it too is normal. Furthermore,

$$
\begin{aligned}
E[\hat{a}] &= E[\bar{Y} - \bar{x}\hat{b}] \\
&= E[\bar{Y}] - \bar{x}E[\hat{b}] \\
&= E\left[\frac{1}{n}\sum_{i=1}^{n}Y_i\right] - \bar{x}E[\hat{b}] \\
&= \frac{1}{n}\sum_{i=1}^{n}E[Y_i] - \bar{x}b \\
&= \frac{1}{n}\sum_{i=1}^{n}(a + bx_i) - \bar{x}b \\
&= a + b\bar{x} - \bar{x}b \\
&= a
\end{aligned}
$$

As for the variance,

$$
\begin{aligned}
\text{Var}\,[\hat{a}] &= \text{Var}\,[\bar{Y} - \bar{x}\hat{b}] \\
&= \text{Var}\,[\bar{Y}] + \bar{x}^2\,\text{Var}\,[\hat{b}] \\
&= \frac{\sigma^2}{n} + \frac{\bar{x}^2\sigma^2}{S_{xx}}
\end{aligned}
$$

Factoring out σ^2 yields the result stated in the theorem. ∎

The final theorem to be presented in this section shows that $SSE/(n - 2)$ is an unbiased estimator of σ^2. It also gives an expression of SSE in terms of $\hat{b}$.

<table>
<tr><td>

THEOREM 12.4

</td><td>

The error sum of squares can be expressed by

$$
SSE = S_{YY} - \hat{b}S_{xY}
$$

Moreover, it has expected value

$$
E[SSE] = (n - 2)\sigma^2
$$

</td></tr>
</table>

Proof: First,

$$
\begin{aligned}
SSE &= \sum_{i=1}^{n}(Y_i - \hat{a} - \hat{b}x_i)^2 \\
&= \sum_{i=1}^{n}(Y_i - \bar{Y} + \hat{b}\bar{x} - \hat{b}x_i)^2 \\
&= \sum_{i=1}^{n}[(Y_i - \bar{Y}) - \hat{b}(x_i - \bar{x})]^2 \\
&= S_{YY} - 2\hat{b}S_{xY} + \hat{b}^2S_{xx} \\
&= S_{YY} - \hat{b}S_{xY}
\end{aligned}
$$

and the first statement is proven. Now, since $S_{xY} = \hat{b}S_{xx}$,

$$E[SSE] = E[S_{YY} - \hat{b}^2 S_{xx}]$$

$$= E\left[\sum_{i=1}^{n} Y_i^2 - n\bar{Y}^2 - \hat{b}^2 S_{xx}\right]$$

$$= \sum_{i=1}^{n} E[Y_i^2] - nE[\bar{Y}^2] - S_{xx}E[\hat{b}^2]$$

$$= \sum_{i=1}^{n} (\text{Var}\,[Y_i] + E[Y_i]^2) - n(\text{Var}\,[\bar{Y}] + E[\bar{Y}]^2) - S_{xx}(\text{Var}\,[\hat{b}] + E[\hat{b}]^2)$$

$$= n\sigma^2 + \sum_{i=1}^{n} (a + bx_i)^2 - n\left[\frac{\sigma^2}{n} + (a + b\bar{x})^2\right] - S_{xx}\left(\frac{\sigma^2}{S_{xx}} + b^2\right)$$

$$= (n - 2)\sigma^2$$

the last equality following by algebraic reduction. ∎

12.4 INFERENCES PERTAINING TO SIMPLE LINEAR REGRESSION

Given unbiased estimators of the slope b and intercept a of the model regression line, it is possible to make inferences regarding these model parameters. These include interval estimates and hypothesis tests. The inferences employ the properties of $\hat{a}$ and $\hat{b}$ discussed in the preceding section.

12.4.1 Inferences Concerning Slope

In accordance with Theorem 12.2,

$$Z = \frac{\hat{b} - b}{\dfrac{\sigma}{\sqrt{S_{xx}}}}$$

possesses a standard normal distribution. Moreover, it can be shown that SSE/σ^2 is independent of $\hat{b}$ and possesses a chi-square distribution with $n - 2$ degrees of freedom. Therefore, by Definition 7.16,

$$T_b = \frac{Z}{\sqrt{\dfrac{SSE}{\sigma^2}}} = \frac{\hat{b} - b}{\sqrt{\dfrac{MSE}{S_{xx}}}}$$

possesses a t distribution with $n - 2$ degrees of freedom, where

$$MSE = \frac{SSE}{n - 2}$$

is the **error mean square.** Using T_b, it is possible to construct a $(1 - \alpha)100\%$ confidence interval for b.

If $\overset{*}{b}$ is the estimate of b resulting from a sample of n observations, then a $(1 - \alpha)100\%$ confidence interval for b is given by

$$\left(\overset{*}{b} - t_{\alpha/2,n-2} \sqrt{\frac{MSE}{S_{xx}}}, \; \overset{*}{b} + t_{\alpha/2,n-2} \sqrt{\frac{MSE}{S_{xx}}} \right)$$

Proof: The theorem follows from rearranging the probability inequality

$$P(-t_{\alpha/2,n-2} < T_b < t_{\alpha/2,n-2}) = 1 - \alpha$$

and then applying Definition 7.15. ■

EXAMPLE 12.5

Theorem 12.5 can be employed to establish a 95% confidence interval for the slope of the regression line corresponding to the stress–strain data provided in Table 12.2 and discussed in Example 12.2. From the table,

$$S_{yy} = \sum_{i=1}^{10} y_i^2 - 10\bar{y}^2 = 0.00001154 - 0.00000839 = 0.00000315$$

Using Theorem 12.4 in conjunction with the values of $\overset{*}{b} = 0.000385$, $S_{xx} = 20.62$, and $S_{xy} = 0.00794$ found in Example 12.2 gives

$$SSE = S_{yy} - \overset{*}{b}S_{xy} = 0.0000000931$$

Hence,

$$MSE = \frac{SSE}{n-2} = 0.0000000116$$

and

$$t_{0.025,8} \sqrt{\frac{MSE}{S_{xx}}} = 2.306 \times 0.0000238 = 0.0000548$$

and, according to Theorem 12.5, a 95% confidence interval for b is given by

$$(0.000385 - 0.000055, 0.000385 + 0.000055) = (0.00033, 0.00044)$$

T_b can also be employed as the test statistic for hypothesis tests involving the null hypothesis $H_0: b = b_0$. The usual one- and two-tailed alternatives apply.

EXAMPLE 12.6

Suppose one desires to show, using the data of Table 12.2, that the slope of the model regression line of Example 12.2 exceeds 0.00037. Then the following hypothesis test applies.

1. *Hypothesis test.* $H_0: b \leq 0.00037$
   $H_1: b > 0.00037$
2. *Significance level.* $\alpha = 0.05$
3. *Test statistic.* T_b
4. *Sample size.* $n = 10$
5. *Critical region.* $t \geq 1.860$

6. *Sample value.* $t_b = \dfrac{\overset{*}{b} - 0.00037}{\sqrt{\dfrac{MSE}{S_{xx}}}} = \dfrac{0.000015}{0.0000238} = 0.630$

7. *Decision.* Accept H_0 since $t_b < 1.860$

Of particular interest is the two-tailed test

$$H_0:\ b = 0$$

$$H_1:\ b \neq 0$$

If the null hypothesis is true, then there does not exist a linear relation between x and Y. In effect, the test is used to detect whether or not there is any significant regression. Although the test can be performed using T_b, it can also be implemented by employing an analysis-of-variance approach, which we now consider.

Because it sums the squares of the residuals $Y_i - \hat{Y}_i$, SSE provides a measure of the variation between the data points and the fitted regression line. Algebraic manipulation shows that

$$S_{YY} = \sum_{i=1}^{n} (Y_i - \bar{Y})^2 = \sum_{i=1}^{n} (\hat{Y}_i - \bar{Y})^2 + \sum_{i=1}^{n} (Y_i - \hat{Y}_i)^2$$

Letting

$$SSR = \sum_{i=1}^{n} (\hat{Y}_i - \bar{Y})^2$$

yields the decomposition

$$S_{YY} = SSR + SSE$$

Looking at the terms of the partition, we see that S_{YY} measures the total variation of the observations $Y_1, Y_2, \ldots, Y_n$ about their mean, whereas SSR, called the **regression sum of squares,** is the portion of the variation due to regression and SSE is the portion due to the variation of the observations about the sample regression line. By Theorem 12.4,

$$SSR = \hat{b} S_{xY}$$

Under the null hypothesis $H_0:\ b = 0$,

$$T_b = \dfrac{\hat{b}}{\sqrt{\dfrac{MSE}{S_{xx}}}}$$

By Theorem 12.2, $\sqrt{S_{xx}}\,\hat{b}/\sigma$ possesses a standard normal distribution and, as a consequence of Example 3.38,

$$\dfrac{SSR}{\sigma^2} = \dfrac{\hat{b} S_{xY}}{\sigma^2} = \dfrac{\hat{b}^2 S_{xx}}{\sigma^2}$$

TABLE 12.5 ANOVA TABLE FOR TESTING THE SIGNIFICANCE OF REGRESSION

Source of Variation	Sum of Squares	Degrees of Freedom	Mean Square	F Test Statistic
Regression	$SSR = \hat{b}S_{xY}$	1	$MSR = \dfrac{SSR}{1}$	$\dfrac{MSR}{MSE}$
Error	$SSE = S_{YY} - \hat{b}S_{xY}$	$n - 2$	$MSE = \dfrac{SSE}{n - 2}$	
Total	S_{YY}	$n - 1$		

TABLE 12.6 DISTANCE–TEMPERATURE DATA

x_i	y_i	x_i^2	y_i^2	$x_i y_i$
0	67	0	4489	0
9	66	81	4356	594
20	63	400	3969	1260
32	65	1024	4225	2080
40	63	1600	3969	2520
50	64	2500	4096	3200
61	61	3721	3721	3721
68	63	4624	3969	4284
280	512	13,950	32,794	17,659

possesses a chi-square distribution with 1 degree of freedom. Since it can be shown that SSE/σ^2 possesses a chi-square distribution with $n - 2$ degrees of freedom and that SSR/σ^2 and SSE/σ^2 are independent, by employing Definition 9.2 and letting $MSR = SSR/1$ denote the **regression mean square,** we see that

$$F = T_b^2 = \frac{\hat{b}^2 S_{xx}}{MSE} = \frac{\hat{b}S_{xY}}{MSE} = \frac{SSR}{MSE} = \frac{MSR}{MSE}$$

possesses an F distribution with 1 and $n - 2$ degrees of freedom. Because $F = T_b^2$, the critical region for employing MSR/MSE as the test statistic is determined by the upper tail of the distribution. The entire analysis-of-variance procedure is outlined in Table 12.5.

EXAMPLE 12.7

Suppose it is postulated that there is a linear relationship between temperature (in degrees Fahrenheit) and distance (in a given direction and in miles) from a particular city. Temperature measurements are taken at a specific time and at several distances in the direction of interest from the city. The results are tabulated in Table 12.6. From the data, $\bar{x} = 35$, $\bar{y} = 64$, $S_{xx} = 4150$, $S_{yy} = 26$, $S_{xy} = -261$, $\overset{*}{b} = -0.0629$, and $\overset{*}{a} = 66.2$. The sample regression line possesses the equation

$$\overset{*}{Y} = \overset{*}{\mu}_{Y|x} = 66.2 - (0.0629)x$$

and is illustrated in Figure 12.10.

For the F test for significance of regression,

$$SSR = \overset{*}{b}S_{xy} = (-0.0629)(-261) = 16.417$$

$$SSE = S_{yy} - \overset{*}{b}S_{xy} = 26 - (-0.0629)(-261) = 9.583$$

$$MSE = \frac{9.58}{6} = 1.597$$

$$f = \frac{16.417}{1.597} = 10.280$$

Since $f_{0.05,1,6} = 5.99$, the null hypothesis is rejected at the 0.05 level, which means we conclude there is significant regression of $\mu_{Y|x}$ on x. However, $f_{0.01,1,6} = 13.75$, so that regression is not considered significant at the 0.01 level. The results of the experiment are summarized in the ANOVA table of Table 12.7.

12.4.2 Inferences Concerning the Intercept

A confidence interval for the intercept a can be constructed in the following manner. By Theorem 12.3,

$$Z = \frac{\hat{a} - a}{\sigma\left(\dfrac{1}{n} + \dfrac{\bar{x}^2}{S_{xx}}\right)^{1/2}}$$

possesses a standard normal distribution. Because $\hat{a}$ and SSE can be shown to be independent,

FIGURE 12.10
Sample regression line for Example 12.7

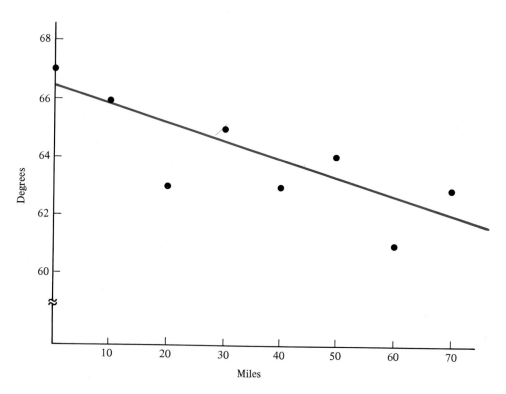

TABLE 12.7 ANOVA TABLE FOR EXAMPLE 12.7

Source of Variation	Sum of Squares	Degrees of Freedom	Mean Square	F Test Statistic
Regression	16.417	1	16.417	10.28
Error	9.583	6	1.597	
Total	26.00	7		

$$T_a = \frac{Z}{\sqrt{\dfrac{\dfrac{SSE}{\sigma^2}}{n-2}}} = \frac{\hat{a} - a}{\left[MSE\left(\dfrac{1}{n} + \dfrac{\bar{x}^2}{S_{xx}}\right)\right]^{1/2}}$$

possesses a t distribution with $n - 2$ degrees of freedom. Just as T_b leads to a confidence interval for b, T_a yields one for a. The reasoning is analogous, and we simply state the proposition.

THEOREM 12.6

If $\overset{*}{a}$ is the estimate of a resulting from a sample of n observations, then a $(1 - \alpha)100\%$ confidence interval for a is given by

$$\left(\overset{*}{a} - t_{\alpha/2,n-2}\left[MSE\left(\frac{1}{n} + \frac{\bar{x}^2}{S_{xx}}\right)\right]^{1/2}, \overset{*}{a} + t_{\alpha/2,n-2}\left[MSE\left(\frac{1}{n} + \frac{\bar{x}^2}{S_{xx}}\right)\right]^{1/2}\right)$$

EXAMPLE 12.8

Theorem 12.6 can be employed to construct a 95% confidence interval for the intercept of the regression line corresponding to the hardenability problem of Example 12.4. According to the data of Example 12.4 and Table 12.4,

$$S_{yy} = \sum_{i=1}^{18} y_i^2 - 18\bar{y}^2 = 44,020.0 - 43,414.0 = 606.0$$

$$MSE = \frac{S_{yy} - \overset{*}{b}S_{xy}}{n-2} = \frac{606.0 - 540.9}{16} = 4.07$$

$$t_{0.025,16}\left[MSE\left(\frac{1}{n} + \frac{\bar{x}^2}{S_{xx}}\right)\right]^{1/2} = (2.120)[(4.07)(0.056 + 0.119)]^{1/2} = 1.787$$

Rounding to the second decimal place gives the 95% confidence interval

$$(57.14 - 1.79, 57.14 + 1.79) = (55.35, 58.93)$$

T_a can also be employed as a test statistic for hypothesis tests involving the null hypothesis H_0: $a = a_0$, where, under the null hypothesis, the numerator of T_a is $\hat{a} - a_0$. For instance, in the upper one-tailed case, the alternative hypothesis is H_1: $a > a_0$, and the critical region consists of all t exceeding $t_{\alpha,n-2}$.

EXAMPLE 12.9

To test the claim that the intercept is less than 60 in the hardenability problem (Example 12.4), we proceed in the following manner.

1. *Hypothesis test.* H_0: $a \geq 60$

                H_1: $a < 60$
2. *Significance level.* $\alpha = 0.05$
3. *Test statistic.* T_a
4. *Sample size.* $n = 18$
5. *Critical region.* $t \leq -1.746$
6. *Sample value.* $t_a = \dfrac{\overset{*}{a} - 60.0}{0.843} = \dfrac{57.14 - 60.0}{0.843} = -3.393$
7. *Decision.* Since $t_a < -1.746$, reject H_0.

12.4.3   Interval Estimation of the Mean Response

The context of regression analysis is determined by the model equation. Although estimation depends on a finite set of x values, $x_1, x_2, \ldots, x_n$, the underlying model equation is defined for all x (though it will not be practically useful outside some restricted range of x). That equation gives the mean response of Y in terms of x, namely, $\mu_{Y|x}$. Having found estimators of a and b relative to $x_1, x_2, \ldots, x_n$, for any x we can consider the estimator of $\mu_{Y|x}$ given by

$$\hat{\mu}_{Y|x} = \hat{a} + \hat{b}x = \bar{Y} + \hat{b}(x - \bar{x})$$

The next theorem gives the important distributional properties of $\hat{\mu}_{Y|x}$.

THEOREM 12.7

The mean estimator $\hat{\mu}_{Y|x}$ possesses a normal distribution with mean $\mu_{Y|x}$ and variance

$$\text{Var}\,[\hat{\mu}_{Y|x}] = \left[\frac{1}{n} + \frac{(x - \bar{x})^2}{S_{xx}} \right] \sigma^2$$

In particular, $\hat{\mu}_{Y|x}$ is an unbiased estimator of $\mu_{Y|x}$.

Proof: The random variables $\bar{Y}$ and $\hat{b}$ are independent. Thus, since $\hat{\mu}_{Y|x}$ is a linear combination of these normal random variables, it too is normal. Moreover,

$$E[\hat{\mu}_{Y|x}] = E[\hat{a}] + xE[\hat{b}] = a + bx = \mu_{Y|x}$$

By Theorem 5.17,

$$\begin{aligned}
\text{Var}\,[\hat{\mu}_{Y|x}] &= \text{Var}\,[\bar{Y} + (x - \bar{x})\hat{b}] \\
&= \text{Var}\,[\bar{Y}] + (x - \bar{x})^2\,\text{Var}\,[\hat{b}] \\
&= \frac{\sigma^2}{n} + (x - \bar{x})^2 \frac{\sigma^2}{S_{xx}}
\end{aligned}$$

Factoring out σ^2 yields the desired result. ∎

    By proceeding in a manner analogous to the construction of the random variables T_a and T_b of Sections 12.4.1 and 12.4.2, namely, by standardizing $\hat{\mu}_{Y|x}$ and then dividing by $\sqrt{MSE}/\sigma$, we arrive at the random variable

$$T_\mu = \frac{\hat{\mu}_{Y|x} - \mu_{Y|x}}{\left[MSE\left(\dfrac{1}{n} + \dfrac{(x - \bar{x})^2}{S_{xx}} \right) \right]^{1/2}}$$

which possesses a t distribution with $n - 2$ degrees of freedom. Just as T_a and T_b yield confidence intervals for a and b, respectively, T_μ yields a confidence interval for $\mu_{Y|x}$.

THEOREM 12.8

If $\overset{*}{\mu}_{Y|x} = \overset{*}{a} + \overset{*}{b}\bar{x}$ is the estimate of $\mu_{Y|x}$ resulting from a sample of n observations, then a $(1 - \alpha)100\%$ confidence interval for $\mu_{Y|x}$ is given by

$$\left(\overset{*}{\mu}_{Y|x} - t_{\alpha/2, n-2} \left[MSE\left(\frac{1}{n} + \frac{(x - \bar{x})^2}{S_{xx}} \right) \right]^{1/2} , \right.$$

$$\left. \overset{*}{\mu}_{Y|x} + t_{\alpha/2, n-2} \left[MSE\left(\frac{1}{n} + \frac{(x - \bar{x})^2}{S_{xx}} \right) \right]^{1/2} \right)$$

Note that the confidence interval for $\mu_{Y|x}$ is of minimum width when $x = \bar{x}$ and increases in width symmetrically as x varies either below or above $\bar{x}$. Moreover, since the interval is defined at each point, by letting x vary we construct a **confidence band** about the sample regression line. This band is illustrated in Figure 12.11.

EXAMPLE 12.10

Referring to Example 12.4, Example 12.8, and Table 12.4, we will construct a 95% confidence interval for the mean Rockwell hardness at the point $x = 30$. Since $n = 18$, $MSE = 4.07$, $\bar{x} = 25$, $S_{xx} = 5250$, $\overset{*}{\mu}_{Y|30} = 47.51$, and $t_{0.025, 16} = 2.120$, Theorem 12.8 yields a 95% confidence interval with the endpoints

$$47.51 \pm (2.120) \left[4.07 \left(\frac{1}{18} + \frac{(30 - 25)^2}{5250} \right) \right]^{1/2}$$

The resulting interval is $(46.46, 48.56)$, which is of width 2.10. Note that the 95% confidence interval at $x = 35$ is wider. Specifically, its width is

$$2(2.120) \left[4.07 \left(\frac{1}{18} + \frac{(35 - 25)^2}{5250} \right) \right]^{1/2} = 2.34$$

12.4.4 A Prediction Interval for $Y|x$

For a given x, the point estimator of $Y|x$ is taken to be the estimator of the mean response at x, namely, $\hat{\mu}_{Y|x}$. Thus, we employ the estimator

$$\hat{Y}|x = \hat{\mu}_{Y|x} = \hat{a} + \hat{b}x$$

A specific set of observations yields the prediction (point estimate)

$$\overset{*}{Y}|x = \overset{*}{\mu}_{Y|x} = \overset{*}{a} + \overset{*}{b}x$$

The estimate $\overset{*}{Y}|x$ is our "best guess" concerning future occurrences of $Y|x$.

To quantify the precision of the estimator $\hat{Y}|x$, we construct a **prediction interval**

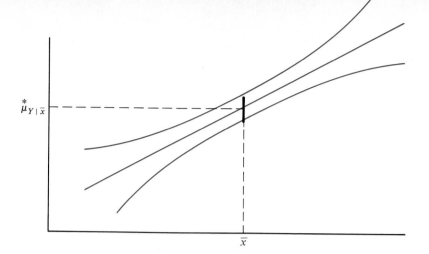

FIGURE 12.11
Confidence band for interval estimation of the mean response

for a single observation of the random variable $Y|x$. Keep in mind that $\hat{Y}|x$ is not a parameter estimator. Our concern is not with $\hat{Y}|x - \mu_{Y|x}$, the difference between the mean estimator and the mean, but with $\hat{Y}|x - Y|x$, the difference between the mean estimator and an observation of Y at x. The mean of the random variable $\hat{Y}|x - Y|x$ is

$$E[\hat{Y}|x - Y|x] = E[\hat{Y}|X] - E[Y|x] = 0$$

since both expected values are equal to $\mu_{Y|x}$. Its variance is

$$\text{Var}\,[\hat{Y}|x - Y|x] = \text{Var}\,[\hat{Y}|x] + \text{Var}\,[Y|x]$$

$$= \left[\frac{1}{n} + \frac{(x - \bar{x})^2}{S_{xx}}\right]\sigma^2 + \sigma^2$$

$$= \left[1 + \frac{1}{n} + \frac{(x - \bar{x})^2}{S_{xx}}\right]\sigma^2$$

Since $\hat{Y}|x - Y|x$ is normally distributed, the usual method of standardization and division by $\sqrt{MSE}/\sigma$ yields the random variable

$$T_x = \frac{\hat{Y}|x - Y|x}{\left[MSE\left(1 + \dfrac{1}{n} + \dfrac{(x - \bar{x})^2}{S_{xx}}\right)\right]^{1/2}}$$

which possesses a t distribution with $n - 2$ degrees of freedom. Let d denote the denominator of T_x. For a given α,

$$P(-t_{\alpha/2,n-2} < T_x < t_{\alpha/2,n-2}) = 1 - \alpha$$

Substitution of T_x and rearrangement of the inequality gives

$$P(\hat{Y}|x - t_{\alpha/2,n-2}d < Y|x < \hat{Y}|x + t_{\alpha/2,n-2}d) = 1 - \alpha$$

For a given set of observations $y_1, y_2, \ldots, y_n$, the interval

$$\left(\overset{*}{Y} \middle| x - t_{\alpha/2,n-2} \left[MSE\left(1 + \frac{1}{n} + \frac{(x - \bar{x})^2}{S_{xx}} \right) \right]^{1/2} , \right.$$

$$\left. \overset{*}{Y} \middle| x + t_{\alpha/2,n-2} \left[MSE\left(1 + \frac{1}{n} + \frac{(x - \bar{x})^2}{S_{xx}} \right) \right]^{1/2} \right)$$

is called a **$(1 - \alpha)100\%$ prediction interval** of $Y|x$, and we say that we are $(1 - \alpha)100\%$ **confident** that a given observation of $Y|x$ will fall in the interval. Be aware that it is the probability inequality that defines the meaning of the prediction interval, including the "confident" terminology, and that the meaning is not the same as for an ordinary parameter confidence interval.

It is interesting to note that, like the interval estimate for $\mu_{Y|x}$, the interval estimate for $Y|x$ produces a band about the sample regression line; however, the latter band is wider than the former. Such a phenomenon is certainly reasonable. Whereas individual responses are subject to observational variation, such variation has been "averaged out" in $\mu_{Y|x}$. Consequently, estimation of $\mu_{Y|x}$ should be more precise than estimation of $Y|x$.

EXAMPLE 12.11

In Example 12.10 we found (46.46, 48.56) to be a 95% confidence interval for the mean $\mu_{Y|30}$. The 95% prediction interval for $Y|30$ is determined by the endpoints

$$47.51 \pm (2.120)\left[4.07\left(1 + \frac{1}{18} + \frac{(30 - 25)^2}{5250} \right) \right]^{1/2}$$

which yields the prediction interval (43.11, 51.91). Note that this interval is substantially wider than the confidence interval for $\mu_{Y|30}$.

12.5 TESTING THE LINEARITY ASSUMPTION

If an investigator does not have a theoretical basis for choosing a regression model, then he or she must make a model selection based upon the distribution of sample points in a scatter diagram. In many instances a linear model is chosen, even though it is understood that it serves only as an approximation to a more complex regression relation. So long as the range of x is limited and the results are not extrapolated outside this limited range, the model can be practically beneficial. However, it may be that there is significant deviation from linearity even in a restricted range of interest. Moreover, even if the scatter diagram gives visual support to the assumption of linearity, it would certainly be advantageous to have a statistical goodness-of-fit test available in the event there is a question concerning the reasonableness of the linearity assumption. In the present section we present a goodness-of-fit test based upon an analysis of variance. As with the goodness-of-fit tests discussed in Chapter 10, a significant value of the test statistic will result in a decision to reject the null model, whereas an insignificant value will simply mean that the data do not consequentially deviate from the linearity assumption.

The test requires there to be repeated observations of Y at the x_i, with the minimum requirement being that there are at least two observations at some x_i. For $i = 1, 2, \ldots, n$, m_i will denote the number of observations at x_i and N will denote the total number of observations. For each $j = 1, 2, \ldots, m_i$, Y_{ij} denotes the jth observation of $Y|x_i$. It will be assumed that the observations $Y_{i1}, Y_{i2}, \ldots, Y_{im_i}$ constitute a random sample with sample mean $\bar{Y}_{i.}$.

The double sum

$$SSPE = \sum_{i=1}^{n} \sum_{j=1}^{m_i} (Y_{ij} - \bar{Y}_{i.})^2$$

is called the **pure error** sum of squares and provides a measure of the variability due simply to the randomness of the Y_{ij}. Meanwhile, the error sum of squares SSE measures the variation of the sample points about the sample regression line. If the distribution of the ith response variable $Y|x_i$ is centered on the regression line, as postulated by the model, then SSE will result from the natural variability of the measurement process. On the other hand, should the distribution of $Y|x_i$ not be centered on the regression line, SSE will be inflated, since the sample values Y_{ij} will distribute themselves about the mean of $Y|x_i$, which in this case differs from the model mean.

The two situations we have in mind are depicted in Figures 12.12 and 12.13. In Figure 12.12, the distributions of the random variables $Y|x_i$ are centered on the sample regression line. Hence, the sample points (indicated as dots) are clustered tightly about the line. As a consequence, SSE, which consists of terms of the form $(Y_{ij} - \hat{Y}_i)^2$, will result from the random behavior of the measurements as they fall about the centers of the distributions. The situation in Figure 12.13 is quite different. There, the distributions of the random variables $Y|x_i$ are centered on the curve $y = h(x)$. Consequently, although the sample regression line fits the data points as best as can be (in the least-squares sense), the terms in SSE are larger, due to clustering of the data points about the centers of their distributions on the curve $y = h(x)$. In this case, it is the terms of $SSPE$ that reflect pure randomness, whereas the terms of SSE reflect the differences between the distribution means and the fitted regression line.

FIGURE 12.12
Sample data compatible with the linear model

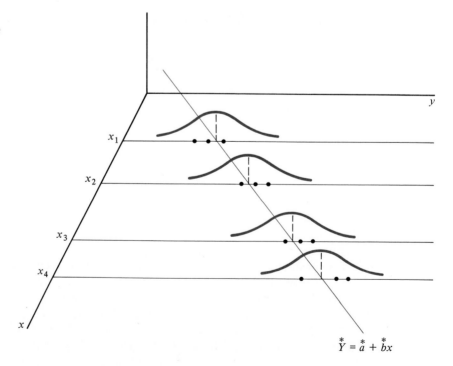

$$\overset{*}{Y} = \overset{*}{a} + \overset{*}{b}x$$

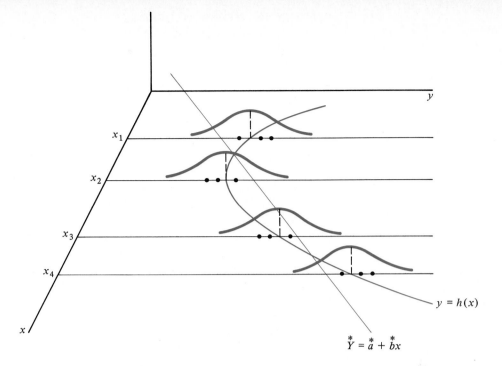

FIGURE 12.13
Sample data
incompatible with the
linear model

To quantify lack of fit, we define the **lack-of-fit** sum of squares $SSLF$ to be the difference between SSE and $SSPE$, namely,

$$SSLF = SSE - SSPE$$

Thus, SSE is decomposed into $SSLF$ and $SSPE$. Larger values of $SSLF$ with respect to $SSPE$ indicate a poorer fit of the model to the data.

It can be shown that $SSPE/\sigma^2$ and $SSLF/\sigma^2$ possess independent chi-square distributions with $N - n$ and $n - 2$ degrees of freedom, respectively. If we let

$$MSPE = \frac{SSPE}{N - n}$$

and

$$MSLF = \frac{SSLF}{n - 2}$$

be the **pure-error mean square** and **lack-of-fit mean square,** respectively, then

$$F = \frac{MSLF}{MSPE}$$

possesses an F distribution with $n - 2$ and $N - n$ degrees of freedom. F is employed as the test statistic for the upper one-tailed hypothesis test

H_0: The linear regression model is appropriate.

H_1: The linear regression model is inappropriate.

TABLE 12.8 PERFORMANCE DATA

x_i	y_{ij}	x_i^2	y_{ij}^2	$x_i y_{ij}$	$\bar{y}_{i.}$	$(y_{ij} - \bar{y}_{i.})^2$
1	16	1	256	16		2.79
1	17	1	289	17	17.67	0.45
1	20	1	400	20		5.43
2	37	4	1369	74		2.79
2	39	4	1521	78	38.67	0.11
2	40	4	1600	80		1.77
3	59	9	3481	177		2.79
3	60	9	3600	180	60.67	0.45
3	63	9	3969	189		5.43
4	75	16	5625	300		7.13
4	78	16	6084	312	77.67	0.11
4	80	16	6400	320		5.43
5	93	25	8649	465		11.09
5	96	25	9216	480	96.33	0.11
5	100	25	10,000	500		13.47
6	115	36	13,225	690		11.09
6	118	36	13,924	708	118.33	0.11
6	122	36	14,884	732		13.47
7	134	49	17,956	938		7.13
7	135	49	18,225	945	136.67	2.79
7	141	49	19,881	987		18.75
8	150	64	22,500	1200		16.00
8	152	64	23,104	1216	154.00	4.00
8	160	64	25,600	1280		36.00
108	2100	612	231,758	11,904		168.69

EXAMPLE 12.12

Consider a situation in which teams of employees (or individual employees) independently perform repetitive tasks. Such a situation might arise in assembly-line production, piece work, or with troubleshooting teams in the field. Even though each team may be expected to complete, at a minimum, some given number of tasks in an eight-hour day, the rate of performance will likely vary throughout the day. If every team worked at an identical constant rate, then the count of completed tasks throughout the day would be a straight line with slope determined by the rate. An obvious question is whether or not a linear model can be employed to describe the completion-versus-time curve. In this direction, suppose at the end of each of the eight hours, three teams are randomly observed and the total number of completions for each is tabulated, these results being given in Table 12.8 together with other relevant secondary information. From the data, $\bar{x} = 4.5$, $\bar{y} = 87.5$, $S_{xx} = 126$, $S_{yy} = 48{,}008$, $S_{xy} = 2454$, $\overset{*}{b} = 19.476$, and $\overset{*}{a} = -0.142$. Thus, the sample regression line is given by

$$\overset{*}{Y} = \overset{*}{\mu}_{Y|x} = -0.142 + (19.476)x$$

and is graphed in Figure 12.14. The F test for linear fit requires the following information:

$$SSE = S_{yy} - \overset{*}{b}S_{xy} = 213.90$$

$$SSLF = SSE - SSPE = 213.90 - 168.69 = 45.21$$

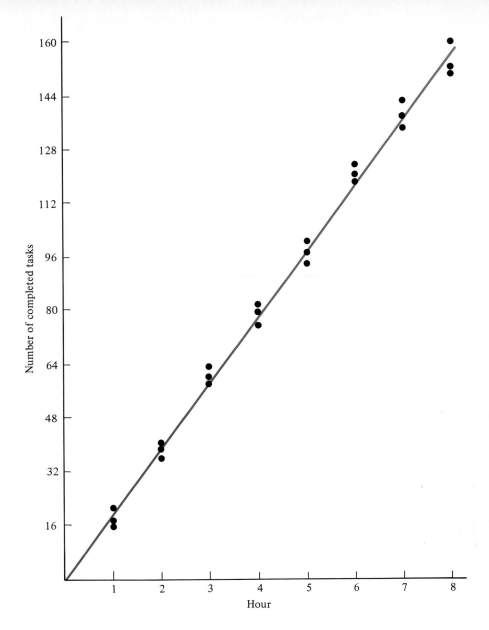

FIGURE 12.14
Sample regression line
for Example 12.12

$$MSPE = \frac{168.69}{24 - 8} = 10.543$$

$$MSLF = \frac{45.21}{6} = 7.535$$

$$f = \frac{MSLF}{MSPE} = \frac{7.535}{10.543} = 0.715$$

Since $f_{0.25,6,16} = 1.47$, the hypothesis of linearity is not even rejected at the 0.25 level. A look at Figure 12.14 shows the tight fit of the data points about the sample regression line. Table 12.9 summarizes the analysis-of-variance procedure.

TABLE 12.9 ANOVA TABLE FOR EXAMPLE 12.12

Source of Variation	Sum of Squares	Degrees of Freedom	Mean Square	F Test Statistic
Regression	47,794.10	1	47,794.10	4917.09
Error	213.90	22	9.72	
(Lack of fit)	45.21	6	7.54	0.72
(Pure error)	168.69	16	10.54	
Total	48,008	23		

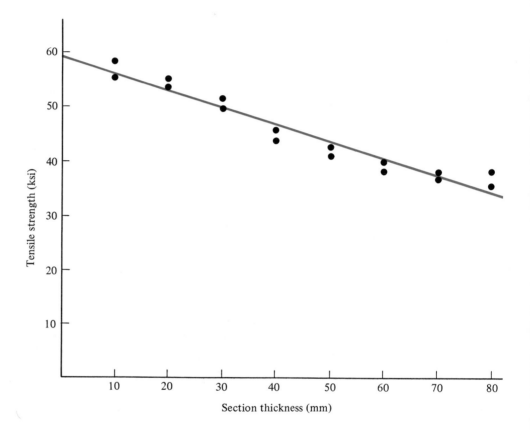

FIGURE 12.15
Sample regression line
for Example 12.13

EXAMPLE 12.13

When molten metal is poured into a mold, the casting cools inward from all bounding surfaces. Since thin sections cool more rapidly than thick ones, thin sections are likely to be stronger and finer grained than thick sections. Table 12.10 contains tensile-strength data (in ksi) for sections of various thicknesses (in mm) of a type of cast iron. Two measurements of tensile strength (y_{ij}) are taken for each thickness (x_i). The data points, together with the fitted regression line, are plotted in Figure 12.15. It appears from the figure that the linear regression hypothesis is open to question. From the data of Table 12.10, we find $\bar{x} = 45$, $\bar{y} = 45.4125$, $S_{xx} = 8400$, $S_{yy} = 874.22$, $S_{xy} = -2623$, $\overset{*}{b} = -0.3123$, and $\overset{*}{a} = 59.466$. Thus, the sample regression line in Figure 12.15 has the equation

$$\overset{*}{Y} = \overset{*}{\mu}_{Y|x} = 59.466 - (0.3123)x$$

The F test for linear fit uses the following information:

$$SSE = S_{yy} - \overset{*}{b}S_{xy} = 55.06$$

$$SSLF = SSE - SSPE = 55.06 - 11.88 = 43.18$$

$$MSPE = \frac{11.88}{8} = 1.485$$

$$MSLF = \frac{43.18}{6} = 7.197$$

$$f = \frac{MSLF}{MSPE} = \frac{7.197}{1.485} = 4.85$$

Since $f_{0.05,6,8} = 3.58$, the hypothesis of linearity is rejected at level 0.05. However, since $f_{0.01,6,8} = 6.37$, the linear regression model would not be rejected at level 0.01. In other words, $0.01 < P < 0.05$. The F test computations are summarized in the ANOVA table of Table 12.11.

TABLE 12.10 STRENGTH–THICKNESS DATA

x_i	y_i	x_i^2	y_{ij}^2	$x_i y_{ij}$	$\bar{y}_{i.}$	$(y_{ij} - \bar{y}_{i.})^2$
10	58.2	100	3387.24	582	57.0	1.44
10	55.8	100	3113.64	558		1.44
20	54.2	400	2937.64	1084	54.6	0.16
20	55.0	400	3025.00	1100		0.16
30	52.0	900	2704.00	1560	51.0	1.00
30	50.0	900	2500.00	1500		1.00
40	45.6	1600	2079.36	1824	45.0	0.36
40	44.4	1600	1971.36	1776		0.36
50	41.0	2500	1681.00	2050	41.8	0.64
50	42.6	2500	1814.76	2130		0.64
60	39.9	3600	1592.01	2394	39.0	0.81
60	38.1	3600	1451.61	2286		0.81
70	37.4	4900	1398.76	2618	37.7	0.09
70	38.0	4900	1444.00	2660		0.09
80	36.0	6400	1296.00	2880	37.2	1.44
80	38.4	6400	1474.56	3072		1.44
720	726.6	40,800	33,870.94	30,074		11.88

TABLE 12.11 ANOVA TABLE FOR EXAMPLE 12.13

Source of Variation	Sum of Squares	Degrees of Freedom	Mean Square	F Test Statistic
Regression	819.16	1	819.16	208.29
Error	55.06	14	3.93	
(Lack of fit)	43.18	6	7.20	4.85
(Pure error)	11.88	8	1.49	
Total	874.22	15		

12.6 INFERENCES CONCERNING THE CORRELATION COEFFICIENT

We now return to the bivariate setting discussed in Chapter 5, where both X and Y are random variables, and discuss inferences regarding the correlation coefficient. Recall that the correlation coefficient ρ provides a measurement of the linear relation between X and Y. A strong linear relation is indicated by $|\rho|$ near 1, whereas a weak linear relation is indicated by $|\rho|$ near 0.

12.6.1 Estimation of the Correlation Coefficient

The next definition gives the commonly employed point estimator of ρ. As will shortly be seen, in the case of jointly distributed normal variables it arises naturally by means of maximum likelihood.

DEFINITION 12.1

Sample Correlation Coefficient. Suppose

$$(X_1, Y_1), (X_2, Y_2), \ldots, (X_n, Y_n)$$

constitute a random sample corresponding to the random vector (X, Y). Then the sample correlation coefficient is the statistic

$$R = \frac{S_{XY}}{\sqrt{S_{XX}S_{YY}}}$$

FIGURE 12.16
Data indicating a bivariate distribution with ρ near $+1$

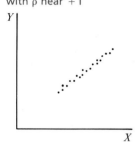

Given a set of sample values $(x_1, y_1), (x_2, y_2), \ldots, (x_n, y_n)$, the numerical estimate of ρ is

$$r = \frac{S_{xy}}{\sqrt{S_{xx}S_{yy}}}$$

The estimate r possesses properties reminiscent of those pertaining to ρ. Two of these are given in the following theorem, which, like Theorem 5.16, its companion, is stated without proof.

THEOREM 12.9

If r is the estimate of ρ corresponding to a set of sample data points, then

$$-1 \leq r \leq 1$$

Moreover, $|r| = 1$ if and only if all the sample pairs lie on a straight line.

If a set of sample data points are representative of the random vector (X, Y), then the degree of linearity exhibited by their scatter diagram should reflect the correlation coefficient of the underlying bivariate distribution. For instance, Figures 12.16 through 12.20 portray data sets arising from joint distributions with ρ near $+1$, ρ near -1, moderately positive ρ, ρ near 0 where there appears to be little relationship between the variables, and ρ near 0 where there is a tight nonlinear relationship between the variables. Should r be calculated for any of these data sets, it will likely

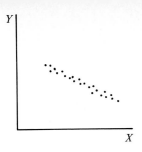

FIGURE 12.17
Data indicating a bivariate distribution with ρ near −1

FIGURE 12.18
Data indicating a bivariate distribution with ρ moderately positive

FIGURE 12.19
Data indicating a bivariate distribution with ρ near 0 and for which there is little relationship between the variables

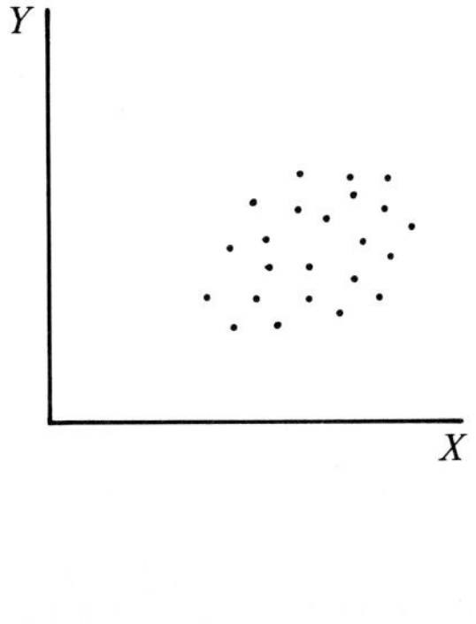

(not certainly) reflect the bivariate linear relationship between the underlying random variables.

To proceed further, we assume that X and Y possess a bivariate normal distribution, our goal being to utilize the distribution of the sample correlation coefficient to make inferences regarding ρ. Even though X and Y are jointly normally distributed, for large values of ρ the distribution of R varies greatly from normality. Nevertheless, a change of variable, called the **Fisher transformation,** transforms the distribution of R into one that is approximately normal. Specifically, it can be shown that

$$V = \frac{1}{2} \log \frac{1 + R}{1 - R}$$

possesses an approximately normal distribution with mean

$$\mu_V = \frac{1}{2} \log \frac{1 + \rho}{1 - \rho}$$

and standard deviation

$$\sigma_V = \frac{1}{\sqrt{n - 3}}$$

Obviously, the transformation requires n to be greater than 3. Furthermore, it should not be employed for small n, since in such cases the approximation to normality is not good.

Inferences regarding ρ can be made using V. Standardizing V yields

$$P\left(-z_{\alpha/2} < \frac{V - \mu_V}{\sigma_V} < z_{\alpha/2} \right) \approx 1 - \alpha$$

Rearrangement of the inequality gives

$$P(V - z_{\alpha/2}\sigma_V < \mu_V < V + z_{\alpha/2}\sigma_V) \approx 1 - \alpha$$

or, putting in the values of μ_V and σ_V,

$$P\left(V - \frac{z_{\alpha/2}}{\sqrt{n - 3}} < \frac{1}{2} \log \frac{1 + \rho}{1 - \rho} < V + \frac{z_{\alpha/2}}{\sqrt{n - 3}} \right) \approx 1 - \alpha$$

For convenience, let

$$V_1 = V - \frac{z_{\alpha/2}}{\sqrt{n - 3}}$$

$$V_2 = V + \frac{z_{\alpha/2}}{\sqrt{n - 3}}$$

Then algebraic manipulation changes the probability inequality to one involving ρ, namely,

$$P\left(\frac{e^{2V_1} - 1}{e^{2V_1} + 1} < \rho < \frac{e^{2V_2} - 1}{e^{2V_2} + 1}\right) \approx 1 - \alpha$$

According to Definition 7.15, an approximate $(1 - \alpha)100\%$ confidence interval for ρ is given by

Approximate $(1 - \alpha)100\%$ Confidence Interval for ρ

$$\boxed{\left(\frac{e^{2v_1} - 1}{e^{2v_1} + 1}, \frac{e^{2v_2} - 1}{e^{2v_2} + 1}\right)}$$

where

$$v_1 = \frac{1}{2} \log \frac{1 + r}{1 - r} - \frac{z_{\alpha/2}}{\sqrt{n - 3}}$$

$$v_2 = \frac{1}{2} \log \frac{1 + r}{1 - r} + \frac{z_{\alpha/2}}{\sqrt{n - 3}}$$

EXAMPLE 12.14

Vehicular traffic flow involves the related random variables X, which gives the vehicles per mile on the roadway, and Y, which gives the average miles per hour of the vehicles. Because of safety spacing requirements, as X increases, Y tends to decrease. Using the observations of Table 12.12, we will estimate the correlation coefficient for the joint variables X and Y, under the assumption that X and Y are (essentially) jointly normally distributed (within the range of data under discussion). From the data, $\bar{x} = 93.0$, $\bar{y} = 32.1$, $S_{xx} = 6416$, $S_{yy} = 589$, $S_{xy} = -1768$, and

$$r = \frac{-1768}{\sqrt{(6416)(589)}} = -0.909$$

which indicates a rather strong negative linear relationship. This relationship is evident in the scatter diagram of Figure 12.21.

Using r, a 95% confidence interval can be determined in a straightforward manner. Proceeding,

FIGURE 12.20
Data indicating a bivariate distribution with ρ near 0 and for which there is a tight nonlinear relationship between the variables

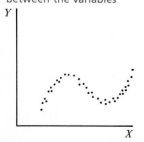

$$v = \frac{1}{2} \log \frac{1 + r}{1 - r} = \frac{1}{2} \log \frac{0.091}{1.909} = -1.522$$

$$v_1 = v - \frac{z_{0.025}}{\sqrt{7}} = -2.263$$

$$v_2 = v + \frac{z_{0.025}}{\sqrt{7}} = -0.781$$

$$e^{2v_1} = e^{-4.526} = 0.01082$$

$$e^{2v_2} = e^{-1.562} = 0.20972$$

and the desired confidence interval is

$$\left(\frac{0.01082 - 1}{0.01082 + 1}, \frac{0.20972 - 1}{0.20972 + 1}\right) = (-0.979, -0.653)$$

TABLE 12.12 TRAFFIC-FLOW DATA

x_i	y_i	x_i^2	y_i^2	x_iy_i
52	42	2704	1764	2184
106	28	11,236	784	2968
124	20	15,376	400	2480
90	36	8100	1296	3240
84	30	7056	900	2520
142	18	20,164	324	2556
105	35	11,025	1225	3675
78	40	6084	1600	3120
80	34	6400	1156	2720
69	38	4761	1444	2622
930	321	92,906	10,893	28,085

12.6.2 Hypothesis Tests Involving the Correlation Coefficient

Hypothesis tests concerning ρ can also be performed using V. If the null hypothesis is of the form H_0: $\rho = \rho_0$, where $\rho_0 \neq 1$, then the appropriate test statistic is

$$Z_\rho = \frac{V - \frac{1}{2}\log\frac{1 + \rho_0}{1 - \rho_0}}{\frac{1}{\sqrt{n - 3}}}$$

FIGURE 12.21
Scatter diagram for
Example 12.14

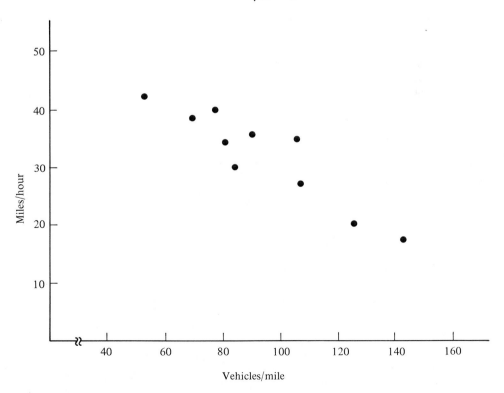

which possesses an approximately standard normal distribution when H_0 is true. One- and two-tailed critical regions are selected according to the alternative hypothesis in the usual Z-test manner.

EXAMPLE 12.15

A value of ρ less than -0.8 indicates a fairly strong negative linear correlation. Using the data of Example 12.14, we will test to see if it can be claimed that $\rho < -0.8$ for the traffic-flow variables in question.

1. *Hypothesis test.* H_0: $\rho \geq -0.8$
    H_1: $\rho < -0.8$
2. *Significance level.* $\alpha = 0.05$
3. *Test statistic.* Z_ρ
4. *Sample size.* $n = 10$
5. *Critical region.* $z \leq -1.65$
6. *Sample value.*

$$z_\rho = \frac{-1.522 - \dfrac{1}{2}\log\dfrac{1 - 0.8}{1 + 0.8}}{\dfrac{1}{\sqrt{7}}} = -1.12$$

7. *Decision.* Since $z_\rho > -1.65$, it is not concluded that $\rho < -0.8$.

Hypothesis tests concerning the null hypothesis H_0: $\rho = 0$ are especially important. Recall that jointly distributed normal random variables are independent if and only if $\rho = 0$. Thus, a rejection of H_0 and the concomitant acceptance of H_1: $\rho \neq 0$ not only determines that there is some partial linear relationship between the variables, but also that they are not independent.

Since X and Y possess a bivariate normal distribution, the theory of Section 5.8 is applicable. In particular, Theorem 5.19 applies, and for any x the conditional mean of $Y|x$ is

$$\mu_{Y|x} = \mu_Y + \rho\frac{\sigma_Y}{\sigma_X}(x - \mu_X)$$

and this straight line is the regression curve in the sense of Definition 5.12. By its definition as the conditional expectation $E[Y|x]$, $\mu_{Y|x}$ has a different interpretation in the bivariate setting than it does in the linear regression model of Section 12.1. Nevertheless, in practical applications the differences in the underlying theories tend to be ignored. The reasoning is twofold. First, the linear regression model has been constructed so as to be in agreement with the bivariate regression curve, especially in the sense that $Y|x$ is normally distributed with mean $\mu_{Y|x}$. Second, if maximum-likelihood estimation is applied to the jointly normal random vector (X, Y), then the resulting estimates of the five distribution parameters are $\hat{\mu}_X = \bar{X}$, $\hat{\mu}_Y = \bar{Y}$, $\hat{\sigma}_X = \sqrt{S_{XX}/n}$, $\hat{\sigma}_Y = \sqrt{S_{YY}/n}$, and $\hat{\rho} = R$. Thus, for a given set of sample points, the corresponding estimate of $\mu_{Y|x}$ is

$$\overset{*}{\mu}_{Y|x} = \bar{y} + r\frac{\sqrt{S_{yy}/n}}{\sqrt{S_{xx}/n}}(x - \bar{x}) = \bar{y} + \frac{S_{xy}}{S_{xx}}(x - \bar{x})$$

which is precisely the estimate derived in Theorem 12.1 for the simple linear regression model. In other words, both models lead to a similar best estimate of $\mu_{Y|x}$; however, the notions of bestness are not defined in the same manner.

In the bivariate model, it can be seen that $\rho = 0$ if and only if $\mu_{Y|x}$ is not a function of x. Hence, because of the relationship between the models, to test $\rho = 0$ against $\rho \neq 0$ is equivalent to testing $b = 0$ against $b \neq 0$. Consequently, the t test statistic

$$T_b = \frac{\hat{b}}{\sqrt{\dfrac{MSE}{S_{xx}}}}$$

can be employed as it was in Section 12.4.1, where this time the hypothesis test is

$$H_0: \rho = 0$$

$$H_1: \rho \neq 0$$

Algebraic manipulation shows that T_b can be expressed as

$$T_b = \frac{R\sqrt{n-2}}{\sqrt{1-R^2}}$$

The critical region is

$$C = \{t: t \leq -t_{\alpha/2} \text{ or } t \geq t_{\alpha/2}\}$$

EXAMPLE 12.16

Given $r = 0.5$ and a sample size of $n = 15$, one might be tempted to declare that ρ is nonzero. However, the test statistic T_b has the sample value

$$t_b = \frac{(0.5)\sqrt{13}}{\sqrt{1-(0.5)^2}} = 2.08$$

and $t_{0.025,13} = 2.16$. Thus $H_0: \rho = 0$ is not rejected in favor of $H_1: \rho \neq 0$ at significance level 0.05. In general, one must be careful when using R, because it is not a precise estimator of ρ unless n is quite large.

As defined in Definition 12.1, R is a random variable dependent on random samples corresponding to two random variables. However, if X is controllable so that it becomes a deterministic variable x and if the simple linear regression model is applicable, then R can be considered as a random variable defined in terms of the response variable Y, namely,

$$R = \frac{S_{xY}}{\sqrt{S_{xx}S_{YY}}}$$

No longer is R an estimator of ρ, since ρ is not even defined. Nevertheless, some algebraic manipulation shows that

$$R^2 = \frac{S_{YY} - SSE}{S_{YY}} = \frac{SSR}{S_{YY}}$$

Since SSR measures the variation attributable to regression and S_{YY} measures the total variation, R^2 is near 1 when most variability is attributable to regression and near 0 when most variation is random. R^2, which is called the **coefficient of determination,** measures the proportion of the total variation attributable to regression.

We close with two points concerning the bivariate normal model. First, unless the variables are normally distributed, or close to being so, the inference methods regarding the correlation coefficient are not applicable. Second, one should take care not to conclude there is a causal explanation in the event the sample correlation coefficient is near -1 or $+1$. Even if pairs of measurements display a striking linear relationship, this does not in any way guarantee there is a causal relationship between the underlying physical phenomena.

EXERCISES

SECTION 12.1

12.1. If, in the simple linear regression model, we were to assume that, for $i = 1, 2, \ldots, n$, E_i possesses a uniform distribution on $[-1, 1]$, then what would be the mean and variance for the random variables $Y|x_i$?

12.2. Repeat Exercise 12.1 with E_i possessing the density

$$f(x) = \begin{cases} -|x| + 1, & \text{if } -1 \leq x \leq 1 \\ 0, & \text{otherwise} \end{cases}$$

12.3. The estimation of a and b in Section 12.3 will require the normality and independence assumptions noted at the end of the section. Given these assumptions, what is the joint distribution of $E_1, E_2, \ldots, E_n$?

12.4. Utilizing the result of Exercise 12.3, what is the joint distribution of $Y|x_1, Y|x_2, \ldots, Y|x_n$?

SECTION 12.2

In Exercises 12.5 through 12.12, plot the given data on a scatter diagram, find the estimated regression line, and then draw the fitted line on the scatter diagram. In all cases, assume the simple linear model applies in the given range of the independent variable x.

12.5. When a metal powder is pressed and sintered (heated), the tensile strength of the finished part will depend upon the pressure. The following data give the tensile strengths (in ksi) resulting from various briquetting pressures (in ksi) for carbon-iron powder sintered at 1110°C for 30 minutes.

x_i (pressure):	200	250	300	350	400
y_i (strength):	29.3	34.2	39.0	46.2	50.5

x_i (pressure):	450	500
y_i (strength):	54.0	57.3

12.6. Redo Example 12.1 with the following data set, which includes a second set of readings.

x_i (temperature):	165	170	175	180	185

y_i (viscosity):	28.5	26.1	23.9	22.0	20.4
	28.3	26.4	24.0	21.8	20.7

x_i (temperature):	190	195	200

y_i (viscosity):	18.5	17.1	15.8
	18.5	16.9	16.0

12.7. Drying shrinkage occurs in concrete when it loses moisture by evaporation—the higher the water/cement ratio, the greater the shrinkage. The following data result from measuring the shrinkage (in 10^{-6} inches per inch) of a specific concrete element plotted against the water/cement ratio (x_i).

x_i:	0.40	0.45	0.50	0.55	0.60	0.65	0.70	0.75
y_i:	280	325	340	355	370	380	390	390

12.8. Many machine parts are subject to continuously varying levels of stress. For such parts, failures are often caused by repeated loadings at stresses far below the yield point. The *endurance limit* of a material is a point below which a part will last indefinitely without ill effects from any type of stress. The following data concern the endurance limit (in psi) for machined steel relative to the tensile strength (in psi), the latter being the independent variable. The data are derived by a special test that determines endurance limits.

x_i	y_i		
80	30.3	31.2	31.0
100	37.0	37.8	36.7
120	43.2	44.5	44.6
140	50.4	51.5	50.5
160	55.4	56.0	55.0
180	60.4	61.6	61.1
200	63.2	64.6	63.7

12.9. The *efficiency* of an assembly line is defined as the proportion of time the line is up and running. As the number of workstations in the line increases, the efficiency decreases. The following data result from measuring the line efficiencies of various lines possessing differing numbers of workstations. The lines have been chosen so that workstation breakdowns appear to occur at a somewhat uniform rate across the workstation population.

x_i:	10	15	24	28	34	40	43	50
y_i:	0.75	0.60	0.60	0.48	0.44	0.45	0.40	0.37

12.10. In developing a large computer program over time, the *failure rate* (measured in failures per second of execution time) tends to decrease as the software is being developed. The following data give numbers of failures per second relative to the number (x_i) of months of development.

x_i:	1	2	3	4	5	6	7	8
y_i:	1.5	1.3	1.7	0.6	0.7	1.2	0.5	0.7

x_i:	9	10	11
y_i:	0.3	0.2	0

Comment on the applicability of the simple linear regression model if the rates have been observed through the development of a single program rather than by selecting programs at different stages of development from a large collection of ongoing software projects of similar complexity.

12.11. In spot welding, the current has an effect on the weld strength. In a particular experiment, with all other factors being held fixed, the strength of the weld (in lb) is measured against the regressor variable current (in amps). The following data are obtained.

x_i:	4000	5000	6000	7000	8000	9000
y_i:	600	1850	2700	3650	4400	4900

12.12. The hardness of steel depends upon the carbon content of the alloy. The following data measure hardness on the Rockwell scale as a function of the percentage of carbon.

x_i	y_i		
0.2	40.4	38.3	39.3
0.3	45.3	47.2	48.0
0.4	51.3	52.4	52.2
0.5	56.4	57.0	56.9
0.6	59.2	58.6	60.0

12.13. Sometimes when a simple linear relation does not exist between the variables, one can instead find some transformation of the variables that yields a

linear relation between the transformed variables. As an example, consider the following fatigue data for magnesium AZ80. The data are the measured numbers of cycles to failure in terms of the stress (in psi) applied in the fatigue test.

x_i:	18,000	20,000	26,000	30,000	38,000
y_i:	10^8	10^7	10^6	10^5	10^4

Rather than obtain a linear regression equation for Y in terms of x, treat $U = \log_{10} Y$ as the dependent variable and obtain a regression equation of the form $U = a + bx$. Plot both the original values Y and the transformed values U in terms of x, and draw the fitted line on the latter scatter diagram.

12.14. When smoothing a surface with an abrasive, the roughness of the finished surface decreases as the abrasive grain becomes finer. The following data give measurements of surface roughness (in micrometers) in terms of the grit numbers of the grains, finer grains possessing larger grit numbers.

x_i:	24	30	36	46	54	60
y_i:	0.34	0.30	0.28	0.22	0.19	0.18

Plot the values of Y against x in a scatter diagram, determine a suitable transformation of the variables to obtain a linear scatter diagram, and then determine the regression line in terms of the transformed variables.

SECTION 12.3

12.15. For $r > 0$, derive an expression for $P(|\hat{b} - b| < r)$ using Theorem 12.2. Replace σ^2 by its unbiased estimator given in Theorem 12.4 to arrive at an approximate expression. Apply this expression to the data of Exercise 12.5 to find $P(|\hat{b} - b| < 0.1)$.

12.16. Repeat Exercise 12.15 using $\hat{a}$ and a in place of $\hat{b}$ and b.

12.17. For fixed x, find the distribution for $\hat{\mu}_{Y|x}$ under the assumptions prevailing in Section 12.3. In particular, show that $\hat{\mu}_{Y|x}$ is an unbiased estimator of $\mu_{Y|x}$.

12.18. Show that S_{XX} and S_{YY} possess chi-square distributions.

12.19. Find the distribution of S_{XY}.

12.20. Using Theorem 12.4, show that SSE/σ^2 possesses a chi-square distribution with $n - 2$ degrees of freedom.

In Exercises 12.21 through 12.26, use Theorem 12.5 to find a $(1 - \alpha)100\%$ confidence interval for b in the cited exercise.

12.21. Exercise 12.5, $\alpha = 0.05$
12.22. Exercise 12.6, $\alpha = 0.05$
12.23. Exercise 12.7, $\alpha = 0.10$
12.24. Exercise 12.9, $\alpha = 0.01$
12.25. Exercise 12.10, $\alpha = 0.01$
12.26. Exercise 12.11, $\alpha = 0.02$

12.27. Carry out the indicated details in the proof of Theorem 12.5.

In Exercises 12.28 through 12.31, use T_b to test H_0: $b = b_0$ against the alternative H_1 at the given level for the cited exercise.

12.28. Exercise 12.5, H_1: $b > 0.07$, $\alpha = 0.05$
12.29. Exercise 12.7, H_1: $b < 400$, $\alpha = 0.10$
12.30. Exercise 12.10, H_1: $b \neq 0$, $\alpha = 0.05$
12.31. Exercise 12.6, H_1: $b \neq 0$, $\alpha = 0.01$

12.32. Apply the F test to test for significance of regression at level $\alpha = 0.05$ in Exercise 12.9. Construct the ANOVA table.

12.33. Apply the F test to test for significance of regression at level $\alpha = 0.01$ in Exercise 12.11. Construct the ANOVA table.

In Exercises 12.34 through 12.37, use Theorem 12.6 to find a $(1 - \alpha)100\%$ confidence interval for a in the cited exercise.

12.34. Exercise 12.5, $\alpha = 0.05$
12.35. Exercise 12.6, $\alpha = 0.01$
12.36. Exercise 12.8, $\alpha = 0.05$
12.37. Exercise 12.10, $\alpha = 0.10$

12.38. Carry out the details for proving Theorem 12.6.
12.39. Referring to Exercise 12.5, use T_a to test the hypothesis H_0: $a \geq 15$ against the hypothesis H_1: $a < 15$ at significance level $\alpha = 0.05$.

In Exercises 12.40 through 12.43, use Theorem 12.8 to find a $(1 - \alpha)100\%$ confidence interval about $\mu_{Y|x}$ at the specified value of x for the cited exercise.

12.40. Exercise 12.5, $x = 300$, $\alpha = 0.05$
12.41. Exercise 12.9, $x = 30$, $\alpha = 0.05$
12.42. Exercise 12.10, $x = 6$, $\alpha = 0.01$
12.43. Exercise 12.12, $x = 0.4$, $\alpha = 0.05$

In Exercises 12.44 through 12.47, find a $(1 - \alpha)100\%$ prediction interval for $Y|x$ in the cited exercise.

12.44. Exercise 12.5, $x = 300$, $\alpha = 0.05$
12.45. Exercise 12.9, $x = 30$, $\alpha = 0.05$
12.46. Exercise 12.11, $x = 6500$, $\alpha = 0.01$
12.47. Exercise 12.12, $x = 0.05$, $\alpha = 0.05$

In Exercises 12.48 through 12.50, test the linearity assumption for the cited exercise at the specified significance level. In each case, construct the ANOVA table.

12.48. Exercise 12.6, $\alpha = 0.05$
12.49. Exercise 12.8, $\alpha = 0.05$
12.50. Exercise 12.12, $\alpha = 0.01$

SECTION 12.6

12.51. It is believed that weld diameter X and the shear strength Y of a weld possess a bivariate normal distribution. Fourteen welds are randomly selected, and the following data pairs are obtained, x_i being in inches and y_i in pounds.

x_i:	0.18	0.17	0.23	0.20	0.16	0.18	0.20
y_i:	1025	1012	1086	1058	1000	1020	1043

x_i:	0.22	0.20	0.25	0.16	0.22	0.17	0.19
y_i:	1079	1060	1088	1010	1075	1012	1042

Draw the scatter diagram and find the value of the sample correlation coefficient r.

12.52. At certain measurement points of a semiconductor diode, it is postulated that the reverse breakdown voltage X and the forward voltage Y possess a bivariate normal distribution. The following data pairs (in volts) result from a random sample of 12 diodes.

x_i:	98.4	103.5	98.3	100.2	102.5	96.7
y_i:	0.65	0.70	0.60	0.72	0.74	0.62

x_i:	101.4	99.8	104.6	106.2	99.6	103.4
y_i:	0.71	0.69	0.78	0.82	0.71	0.75

Draw the scatter diagram and find the value of the sample correlation coefficient r.

12.53. Referring to Exercise 12.51, find a 95% confidence interval for ρ.
12.54. Referring to Exercise 12.52, find a 90% confidence interval for ρ.
12.55. Referring to Exercise 12.51, test the null hypothesis H_0: $\rho \leq 0.4$ against the alternative hypothesis H_1: $\rho > 0.4$ at level $\alpha = 0.05$.
12.56. Referring to Exercise 12.52, test the null hypothesis H_0: $\rho \leq 0.3$ against the alternative hypothesis H_1: $\rho > 0.3$ at level $\alpha = 0.02$.
12.57. Referring to Exercise 12.51, test the null hypothesis H_0: $\rho = 0$ against the alternative hypothesis H_1: $\rho \neq 0$ at level $\alpha = 0.01$, using the statistic T_b.
12.58. Repeat Exercise 12.57 for the data of Exercise 12.52.
12.59. Estimate the regression line for the bivariate normal model of Exercise 12.51, using the data given in the exercise.

12.60. Estimate the regression line for the bivariate normal model of Exercise 12.52, using the data given in the exercise.

12.61. Find the coefficient of determination for the data of Exercise 12.7.

SUPPLEMENTAL EXERCISES FOR COMPUTER SOFTWARE

12.C1. Redo the pressure-strength regression of Exercise 12.5 using the original data in conjunction with the following additional data.

x_i:	220	270	290	320	360	390	425	475
y_i:	32.2	36.1	38.9	43.4	46.5	47.8	52.0	56.0

12.C2. Redo the pressure-strength regression of Exercise 12.5 using the original data in conjunction with the following additional data (not including the data of Exercise 12.C1).

x_i:	200	250	300	350	400	450	500
y_i:	29.0	34.6	40.1	45.8	50.7	54.0	56.7

12.C3. Redo the current-strength regression of Exercise 12.11 using the original data in conjunction with the following additional data.

x_i:	4000	5000	6000	7000	8000	9000
y_i:	750	2000	2650	3800	4550	4800
	550	1750	2650	3600	4350	4850

12.C4. Test for linearity of regression at level $\alpha = 0.05$ in Exercise 12.C2.

12.C5. Test for linearity of regression at level $\alpha = 0.05$ in Exercise 12.C3.

13

MULTIPLE LINEAR REGRESSION

Simple linear regression involves the dependency of a single random variable Y on a single independent variable. In multiple linear regression, the response variable Y is dependent on several variables, $x_1, x_2, \ldots, x_k$. The chapter commences by giving the appropriate linear model, finding the least-squares estimators of the model parameters, and discussing the properties of the estimators. Inferences are then made regarding the model parameters. Returning to the case where Y is treated as a response to a single independent variable x, we treat polynomial regression (as a particular instance of the general multiple linear model). Then we use analysis of variance to test for significance of regression in the multiple linear model. We conclude by considering the sequential selection of appropriate model variables based upon the significance of regression attributed to additional variables.

13.1 THE MULTIPLE LINEAR REGRESSION MODEL

In the simple linear regression model it is assumed that the means of the random variables $Y|x$ lie on a straight line in two-dimensional space, the underlying modeling assumption being that the response variable Y can be predicted (estimated) by considering a single independent variable x. In many scientific settings a single independent variable is inadequate to describe the mean behavior of the response variable. The situation is analogous to the deterministic multivariable setting in which a dependent variable y is described as a function of several independent variables; however, in the regression scenario, observation of the independent variables yields only a mean estimate of the response variable.

13.1.1 Description of the Model and the Normal Equations

The simple linear regression model, which expresses the response variable $Y|x$ as a linear function of x plus a random part, is naturally extended to the case in which there are k independent (controllable) variables $x_1, x_2, \ldots, x_k$. Whereas $\mu_{Y|x}$ is postulated to be a straight line in the simple linear case, in the multiple linear model the response Y depends on k variables, and its mean lies on a hyperplane in $(k + 1)$-dimensional space. Specifically, the underlying model is determined by the **multiple linear regression equation**

$$\mu_{Y|x_1,x_2,\ldots,x_k} = b_0 + b_1x_1 + b_2x_2 + \cdots + b_kx_k$$

where $b_0, b_1, \ldots, b_k$ are the model parameters. Geometrically, the point $(x_1, x_2, \ldots, x_k)$ lies in k-space and the mean is the value of the $(k + 1)$st variable. The point

$$(x_1, x_2, \ldots, x_k, \mu_{Y|x_1,x_2,\ldots,x_k})$$

lies on the **regression plane (hyperplane).** In addition, the response random variable $Y|x_1, x_2, \ldots, x_k$ is of the form

$$Y|x_1, x_2, \ldots, x_k = \mu_{Y|x_1,x_2,\ldots,x_k} + \mathrm{E}$$

where E is the random difference between the response and its mean.

Experimentally, observations of the response variable are made at n domain points, the result being a collection of sample points in $(k + 1)$-dimensional space:

$$(x_{11}, x_{21}, \ldots, x_{k1}, y_1)$$
$$(x_{12}, x_{22}, \ldots, x_{k2}, y_2)$$
$$\vdots \quad \vdots \qquad \vdots \quad \vdots$$
$$(x_{1n}, x_{2n}, \ldots, x_{kn}, y_n)$$

where, for $i = 1, 2, \ldots, n$,

$$y_i = y|x_{1i}, x_{2i}, \ldots, x_{ki}$$

denotes the sample value of the response at the observation point $(x_{1i}, x_{2i}, \ldots, x_{ki})$. The estimation problem is to find a regression plane that best fits the sample points. As in the simple linear model, "best fit" is defined in the least-squares sense.

Since the responses at the n observation points are random variables, if we assume them to be independent and identically distributed, then they comprise a random sample

$$\{Y_1, Y_2, \ldots, Y_n\} = \{Y|x_{1i}, x_{2i}, \ldots, x_{ki}: i = 1, 2, \ldots, n\}$$

where each Y_i is of the form

$$Y_i = \mu_{Y|x_{1i},x_{2i},\ldots,x_{ki}} + \mathrm{E}_i$$

The following linear model formalizes these considerations.

$$\text{Equation:} \quad Y_i = b_0 + b_1 x_{1i} + b_2 x_{2i} + \cdots + b_k x_{ki} + E_i$$

Assumptions: 1. E_i possesses mean 0 and variance σ^2.

2. The E_i constitute an independent collection of random variables.

3. E_i is normally distributed.

We have separated assumption 3 regarding normality, since it will not be utilized until the estimation theory of Section 13.3. Till then, only assumptions 1 and 2 are necessary.

In analogy to the single-variable case, given a set of observed responses y_1, $y_2, \ldots, y_n$, the **sample (fitted) regression plane** is determined by the least-squares estimates $\overset{*}{b}_0, \overset{*}{b}_1, \ldots, \overset{*}{b}_k$ of the model coefficients. Specifically, for $i = 1, 2, \ldots, n$, the ith **residual**

$$e_i = y_i - (\overset{*}{b}_0 + \overset{*}{b}_1 x_{1i} + \overset{*}{b}_2 x_{2i} + \cdots + \overset{*}{b}_k x_{ki})$$

measures the "vertical" distance from the observation to the sample regression plane, and the least-squares estimates are chosen so that the sum of the squares of the residuals,

$$SSE = \sum_{i=1}^{n} e_i^2$$

is minimized. Once again, SSE is called the **error sum of squares.** Since the residuals are defined in terms of the responses, they can be viewed as random variables. As in the single-variable case, we will follow custom and not employ distinct notation for the random and numerical sums of squares. Figure 13.1 illustrates a set of sample points in three-dimensional space arising from a set of points in the plane. Note how the sample regression plane fits the sample points, the vertical lines indicating the residuals.

In terms of the regression coefficients, SSE is a function of $k + 1$ variables and can be minimized in the usual calculus manner by finding the $k + 1$ partial derivatives $\partial SSE / \partial \overset{*}{b}_j$, for $j = 0, 1, \ldots, k$, and setting these partial derivatives equal to zero. The result will be the following $k + 1$ simultaneous **normal equations:**

Normal Equations

$$\overset{*}{b}_0 n + \overset{*}{b}_1 \sum_{i=1}^{n} x_{1i} + \overset{*}{b}_2 \sum_{i=1}^{n} x_{2i} + \cdots + \overset{*}{b}_k \sum_{i=1}^{n} x_{ki} = \sum_{i=1}^{n} y_i$$

$$\overset{*}{b}_0 \sum_{i=1}^{n} x_{1i} + \overset{*}{b}_1 \sum_{i=1}^{n} x_{1i}^2 + \overset{*}{b}_2 \sum_{i=1}^{n} x_{1i} x_{2i} + \cdots + \overset{*}{b}_k \sum_{i=1}^{n} x_{1i} x_{ki} = \sum_{i=1}^{n} x_{1i} y_i$$

$$\vdots \qquad \vdots \qquad \vdots \qquad \vdots \qquad \vdots$$

$$\overset{*}{b}_0 \sum_{i=1}^{n} x_{ki} + \overset{*}{b}_1 \sum_{i=1}^{n} x_{ki} x_{1i} + \overset{*}{b}_2 \sum_{i=1}^{n} x_{ki} x_{2i} + \cdots + \overset{*}{b}_k \sum_{i=1}^{n} x_{ki}^2 = \sum_{i=1}^{n} x_{ki} y_i$$

The normal equations can be solved for the least-squares estimates $\overset{*}{b}_0, \overset{*}{b}_1, \ldots, \overset{*}{b}_k$. Not only is the direct solution of this system rather messy, but it also tends to

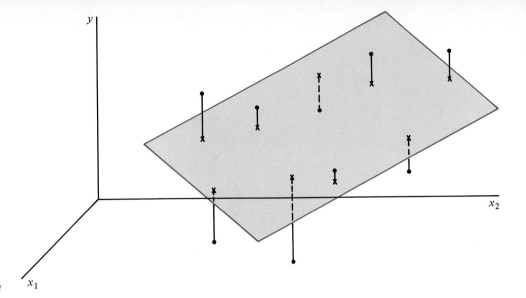

FIGURE 13.1
Sample regression plane

obscure the fundamental concepts underlying the least-squares methodology, especially insofar as the resulting least-squares estimators $\hat{b}_0, \hat{b}_1, \ldots, \hat{b}_k$ are concerned. Thus, rather than work with these equations directly, we will proceed at once to the appropriate matrix setting.

13.1.2 Matrix Representation and the Least-Squares Estimators

The observational model

$$Y_i = b_0 + b_1 x_{1i} + b_2 x_{2i} + \cdots + b_k x_{ki} + E_i$$

$i = 1, 2, \ldots, n$, constitutes a system of n linear equations in the unknowns b_0, $b_1, \ldots, b_k$:

$$Y_1 = b_0 + b_1 x_{11} + b_2 x_{21} + \cdots + b_k x_{k1} + E_1$$

$$Y_2 = b_0 + b_1 x_{12} + b_2 x_{22} + \cdots + b_k x_{k2} + E_2$$

$$\vdots$$

$$Y_n = b_0 + b_1 x_{1n} + b_2 x_{2n} + \cdots + b_k x_{kn} + E_n$$

This system can be rewritten in terms of the response random vector

$$\mathbf{Y} = \begin{pmatrix} Y_1 \\ Y_2 \\ \vdots \\ Y_n \end{pmatrix}$$

the model-parameter vector

$$\mathbf{b} = \begin{pmatrix} b_0 \\ b_1 \\ \vdots \\ b_k \end{pmatrix}$$

the error random vector

$$\mathbf{E} = \begin{pmatrix} E_1 \\ E_2 \\ \vdots \\ E_n \end{pmatrix}$$

and the $n \times (k + 1)$ **design matrix**

$$\mathbf{X} = \begin{pmatrix} 1 & x_{11} & x_{21} & \cdots & x_{k1} \\ 1 & x_{12} & x_{22} & \cdots & x_{k2} \\ \vdots & \vdots & \vdots & & \vdots \\ 1 & x_{1n} & x_{2n} & \cdots & x_{kn} \end{pmatrix}$$

The system of equations then takes the matrix form

$$\mathbf{Y} = \mathbf{Xb} + \mathbf{E}$$

For any set of observations $y_1, y_2, \ldots, y_n$, the matrix equation becomes

$$\mathbf{y} = \mathbf{Xb} + \boldsymbol{\epsilon}$$

where

$$\boldsymbol{\epsilon} = \begin{pmatrix} \epsilon_1 \\ \epsilon_2 \\ \vdots \\ \epsilon_n \end{pmatrix}$$

is the nonrandom vector of "errors."

If we now let

$$\overset{*}{\mathbf{b}} = \begin{pmatrix} \overset{*}{b}_0 \\ \overset{*}{b}_1 \\ \vdots \\ \overset{*}{b}_k \end{pmatrix}$$

denote the estimate vector corresponding to the parameter vector **b,** then a straight-forward calculation shows that the system of normal equations for the least-squares estimators given in Section 13.1.1 can be expressed in matrix format as

$$(\mathbf{X}'\mathbf{X})\overset{*}{\mathbf{b}} = \mathbf{X}'\mathbf{y}$$

where $\mathbf{X}'$ is the transpose of $\mathbf{X}$. Note that $\mathbf{X}$ is $n \times (k + 1)$, so that $\mathbf{X}'$ is $(k + 1) \times n$ and $\mathbf{X}'\mathbf{X}$ is $(k + 1) \times (k + 1)$. It is known from matrix theory that $\mathbf{X}'\mathbf{X}$ is invertible if the columns of $\mathbf{X}$ are linearly independent. Making this assumption, we at once solve the matrix representation of the normal equations to obtain

$$\overset{*}{\mathbf{b}} = (\mathbf{X}'\mathbf{X})^{-1}\mathbf{X}'\mathbf{y}$$

which provides the least-squares solution vector (assuming that $\mathbf{X}'\mathbf{X}$ is nonsingular). The next theorem formalizes these results in terms of the least-squares estimator vector

$$\hat{\mathbf{b}} = \begin{pmatrix} \hat{b}_0 \\ \hat{b}_1 \\ \vdots \\ \hat{b}_k \end{pmatrix}$$

Note that the domain points $(x_{1i}, x_{2i}, \ldots, x_{ki})$ are controllable and therefore, so long as $n \geq k + 1$, can be chosen so that the columns of $\mathbf{X}$ are linearly independent. Hence, our assumption regarding the invertibility of $\mathbf{X}'\mathbf{X}$ is a design constraint.

THEOREM 13.1

The least-squares estimator vector $\hat{\mathbf{b}}$ for the model-parameter vector $\mathbf{b}$ in the multiple linear regression model is given by

$$\hat{\mathbf{b}} = \mathbf{X}^+\mathbf{Y}$$

where

$$\mathbf{X}^+ = (\mathbf{X}'\mathbf{X})^{-1}\mathbf{X}'$$

is the **pseudoinverse** matrix for $\mathbf{X}$. For a specific observation vector $\mathbf{y}$, the least-squares estimates are determined by

$$\overset{*}{\mathbf{b}} = \mathbf{X}^+\mathbf{y}$$

EXAMPLE 13.1

The percentage yield in a chemical process depends on a number of factors, such as temperature, amount of catalyst, duration, and proportions of ingredients. Suppose the data of Table 13.1 represent the percentage output (Y), the duration (in minutes) of the process (x_1), and the percentage of catalyst (x_2). The independent variables x_1 and x_2 are controlled by the laboratory technician, and the response variable is observed. It is desired to fit a regression plane to the sample points determined by the data of Table 13.1. The design matrix is

$$\mathbf{X} = \begin{pmatrix} 1 & 8 & 0.4 \\ 1 & 8 & 0.3 \\ 1 & 7 & 0.4 \\ 1 & 7 & 0.3 \\ 1 & 6 & 0.4 \\ 1 & 6 & 0.3 \end{pmatrix}$$

Applying Theorem 13.1, we have

$$\mathbf{X}' = \begin{pmatrix} 1 & 1 & 1 & 1 & 1 & 1 \\ 8 & 8 & 7 & 7 & 6 & 6 \\ 0.4 & 0.3 & 0.4 & 0.3 & 0.4 & 0.3 \end{pmatrix}$$

$$\mathbf{X}'\mathbf{X} = \begin{pmatrix} 6 & 42 & 2.1 \\ 42 & 298 & 14.7 \\ 2.1 & 14.7 & 0.75 \end{pmatrix}$$

$$(\mathbf{X}'\mathbf{X})^{-1} = \begin{pmatrix} 20.58\overline{3} & -1.750 & -23.33\overline{3} \\ -1.750 & 0.250 & 0 \\ -23.333 & 0 & 66.66\overline{6} \end{pmatrix}$$

$$\mathbf{X}^+ = \begin{pmatrix} -2.750 & -0.41\overline{6} & -1.000 & 1.33\overline{3} & 0.750 & 3.08\overline{3} \\ 0.250 & 0.250 & 0 & 0 & -0.250 & -0.250 \\ 3.33\overline{3} & -3.33\overline{3} & 3.33\overline{3} & -3.33\overline{3} & 3.33\overline{3} & -3.33\overline{3} \end{pmatrix}$$

$$\overset{*}{\mathbf{b}} = \mathbf{X}^+\mathbf{y} = \mathbf{X}^+ \begin{pmatrix} 42.2 \\ 34.5 \\ 41.6 \\ 33.7 \\ 40.0 \\ 30.1 \end{pmatrix} = \begin{pmatrix} -4.28\overline{3} \\ 1.650 \\ 85.000 \end{pmatrix}$$

Consequently, the sample regression plane possesses the equation

$$\overset{*}{Y} = \overset{*}{\mu}_{Y|x_1,x_2} = -4.283 + 1.65x_1 + 85x_2$$

In terms of prediction, if 0.35% catalyst is employed and the reaction is allowed to run 7.5 minutes, then the predicted percentage output is

$$\overset{*}{Y} = \overset{*}{\mu}_{Y|7.5,0.35} = -4.283 + (1.65)(7.5) + (85)(0.35) = 37.842$$

TABLE 13.1 TIME–CATALYST–YIELD DATA

x_{1i}	x_{2i}	y_i
8	0.4	42.2
8	0.3	34.5
7	0.4	41.6
7	0.3	33.7
6	0.4	40.0
6	0.3	30.1

It should be clear from the preceding example that should n or k be large, the calculations become unwieldy unless performed by computer. Increasing k is particularly critical, because $\mathbf{X}'\mathbf{X}$ is $(k + 1) \times (k + 1)$ and it must be inverted. One way to simplify the computational burden is to construct the design matrix so that its columns are **orthogonal,** which simply means that the dot product of any two distinct columns is zero. For such a design matrix, $\mathbf{X}'\mathbf{X}$ will be diagonal and therefore possess a trivially obtainable inverse. There exist relatively efficient methods to achieve orthogonality in the design matrix. We will consider orthogonality in Section 13.5.

13.2 PROPERTIES OF THE LEAST-SQUARES ESTIMATORS

In the multiple linear regression model, it is best to describe the properties of the least-squares estimators in terms of the vector estimator $\hat{\mathbf{b}}$. To accomplish this end, it is necessary to define analogues of both expected value and covariance in the random-vector setting. In the succeeding definitions it will be assumed that

$$\mathbf{V} = \begin{pmatrix} V_1 \\ V_2 \\ \vdots \\ V_m \end{pmatrix}$$

is a random vector with m random-variable components.

DEFINITION 13.1

Expected Value (of a Random Vector). The expected value of the random vector $\mathbf{V}$ is the vector composed of the component expected values, namely,

$$E[\mathbf{V}] = \begin{pmatrix} E[V_1] \\ E[V_2] \\ \vdots \\ E[V_m] \end{pmatrix}$$

DEFINITION 13.2

Covariance Matrix. The covariance matrix, also called the **variance–covariance** matrix, of the random vector $\mathbf{V}$ is denoted by Cov $[\mathbf{V}]$ and is defined to be the $m \times m$ matrix whose ijth component is Cov $[V_i, V_j]$. Written out,

$$\text{Cov }[\mathbf{V}] = \begin{pmatrix} \text{Var }[V_1] & \text{Cov }[V_1, V_2] & \cdots & \text{Cov }[V_1, V_m] \\ \text{Cov }[V_2, V_1] & \text{Var }[V_2] & \cdots & \text{Cov }[V_2, V_m] \\ \vdots & \vdots & & \vdots \\ \text{Cov }[V_m, V_1] & \text{Cov }[V_m, V_2] & \cdots & \text{Var }[V_m] \end{pmatrix}$$

If $\mathbf{U}$ and $\mathbf{V}$ are random m-vectors and $\mathbf{A}$ is an $r \times m$ matrix of constants, then the linearity of E (as an operator on random variables) in conjunction with some algebraic manipulation yields

$$E[\mathbf{AV}] = \mathbf{A}E[\mathbf{V}]$$

and

$$E[\mathbf{U} + \mathbf{V}] = E[\mathbf{U}] + E[\mathbf{V}]$$

Although not quite so straightforward, it can also be shown that

$$\text{Cov }[\mathbf{AV}] = \mathbf{A}(\text{Cov }[\mathbf{V}])\mathbf{A}'$$

Returning to the multiple linear regression model, to say that $\hat{\mathbf{b}}$ is an unbiased estimator of $\mathbf{b}$ means that $E[\hat{\mathbf{b}}] = \mathbf{b}$, or, equivalently, $E[\hat{b}_j] = b_j$ for $j = 0, 1, \ldots,$ k. The next theorem states that $\hat{\mathbf{b}}$ is an unbiased estimator of $\mathbf{b}$. It also provides a useful expression for the covariance matrix of $\hat{\mathbf{b}}$.

THEOREM 13.2

In the multiple linear regression model, $E[\hat{\mathbf{b}}] = \mathbf{b}$ and

$$\text{Cov }[\hat{\mathbf{b}}] = \sigma^2(\mathbf{X}'\mathbf{X})^{-1}$$

Proof: The proof makes use of the three previously stated identities for the expected value and covariance. Keeping in mind that $E_1, E_2, \ldots, E_n$ are independent random variables with common mean 0 and common variance σ^2, letting $\mathbf{0}$ denote the zero vector, and employing the regression model equation $\mathbf{Y} = \mathbf{Xb} + \mathbf{E}$, we obtain

$$
\begin{aligned}
E[\hat{\mathbf{b}}] &= E[\mathbf{X}^+\mathbf{Y}] \\
&= \mathbf{X}^+E[\mathbf{Xb} + \mathbf{E}] \\
&= \mathbf{X}^+(\mathbf{X}E[\mathbf{b}] + E[\mathbf{E}]) \\
&= \mathbf{X}^+(\mathbf{X}E[\mathbf{b}] + \mathbf{0}) \\
&= (\mathbf{X}'\mathbf{X})^{-1}\mathbf{X}'\mathbf{Xb} \\
&= \mathbf{b}
\end{aligned}
$$

As for the covariance matrix,

$$
\text{Cov}\,[\hat{\mathbf{b}}] = \text{Cov}\,[\mathbf{X}^+\mathbf{Y}] = \mathbf{X}^+\text{Cov}\,[\mathbf{Y}](\mathbf{X}^+)'
$$

The transpose of the pseudoinverse is given by

$$
\begin{aligned}
[(\mathbf{X}'\mathbf{X})^{-1}\mathbf{X}']' &= \mathbf{X}''[(\mathbf{X}'\mathbf{X})^{-1}]' \\
&= \mathbf{X}[(\mathbf{X}'\mathbf{X})^{-1}]' \\
&= \mathbf{X}[(\mathbf{X}'\mathbf{X})']^{-1} \\
&= \mathbf{X}(\mathbf{X}'\mathbf{X})^{-1}
\end{aligned}
$$

where we have employed the matrix identities $(\mathbf{AB})' = \mathbf{B}'\mathbf{A}'$ and $(\mathbf{A}^{-1})' = (\mathbf{A}')^{-1}$. In addition, since $Y_1, Y_2, \ldots, Y_n$ are assumed to be independent,

$$
\text{Cov}\,[Y_i, Y_j] = 0
$$

for $i \neq j$, and therefore the entries off the main diagonal of $\text{Cov}\,[\mathbf{Y}]$ are zero. On the other hand, for $i = j$, $\text{Var}\,[Y_i] = \sigma^2$. Thus,

$$
\text{Cov}\,[Y] = \sigma^2\mathbf{I}
$$

where $\mathbf{I}$ is the identity matrix. Returning to the computation of $\text{Cov}\,[\hat{\mathbf{b}}]$, we have

$$
\text{Cov}\,[\hat{\mathbf{b}}] = (\mathbf{X}'\mathbf{X})^{-1}\mathbf{X}'\sigma^2\mathbf{I}\mathbf{X}(\mathbf{X}'\mathbf{X})^{-1} = \sigma^2(\mathbf{X}'\mathbf{X})^{-1} \qquad \blacksquare
$$

As in the simple linear model, variation can be analyzed in terms of the total sum of squares S_{YY}, the error sum of squares SSE, and the regression sum of squares

$$
SSR = \sum_{i=1}^{n} (\hat{Y}_i - \bar{Y})^2
$$

where the prediction $\hat{Y}_i$ is the mean estimator

$$
\hat{Y}_i = \hat{b}_0 + \hat{b}_1 x_{1i} + \hat{b}_2 x_{2i} + \cdots + \hat{b}_k x_{ki}
$$

The following theorem, stated without proof, summarizes the fundamental variational properties in the multiple linear model. It will play a key role in employing analysis of variance to test for the significance of regression in the model.

THEOREM 13.3

In the multiple linear regression model,

$$MSE = \frac{SSE}{n - k - 1}$$

is an unbiased estimator of σ^2. Moreover, the sum of squares S_{YY} can be decomposed as

$$S_{YY} = SSE + SSR$$

and the regression sum of squares is given by

$$SSR = \hat{\mathbf{b}}'(\mathbf{X}'\mathbf{Y}) - n\bar{Y}^2$$

Written out, the expression for SSR in Theorem 13.3 takes the form

$$SSR = \hat{\mathbf{b}}'(\mathbf{X}'\mathbf{Y}) - n\bar{Y}^2$$

$$= \hat{b}_0 \sum_{i=1}^{n} Y_i + \hat{b}_1 \sum_{i=1}^{n} x_{1i}Y_i + \hat{b}_2 \sum_{i=1}^{n} x_{2i}Y_i + \cdots + \hat{b}_k \sum_{i=1}^{n} x_{ki}Y_i - \frac{1}{n}\left(\sum_{i=1}^{n} Y_i\right)^2$$

In terms of estimates,

$$SSR = \overset{*}{\mathbf{b}}{}'(\mathbf{X}'\mathbf{y}) - n\bar{y}^2$$

EXAMPLE 13.2

Using the data of Example 13.1, we find

$$\overset{*}{\mathbf{b}}{}'(\mathbf{X}'\mathbf{y}) = (-4.28\bar{3} \quad 1.65 \quad 85)\begin{pmatrix} 222.1 \\ 1561.3 \\ 79.01 \end{pmatrix} = 8340.667$$

$$SSR = \overset{*}{\mathbf{b}}{}'(\mathbf{X}'\mathbf{y}) - 6\bar{y}^2 = 8340.667 - 8221.402 = 119.265$$

$$S_{yy} = \sum_{i=1}^{6} y_i^2 - 6\bar{y}^2 = 8343.350 - 8221.402 = 121.948$$

$$SSE = S_{yy} - SSR = 121.948 - 119.265 = 2.683$$

$$MSE = \frac{SSE}{6 - 2 - 1} = \frac{2.683}{3} = 0.894$$

the latter being our estimate for σ^2. If the original data of Table 13.1 are inserted into the equation for $\overset{*}{Y}$ found in Example 13.1, then the response estimates at the original points (x_{1i}, x_{2i}), $i = 1, 2, \ldots, 6$, are calculated to be 42.9167, 34.4167, 41.2667, 32.7667, 39.6167, and 31.1167, respectively. Substituting these values into

$$SSE = \sum_{i=1}^{6} (\overset{*}{Y}_i - y_i)^2$$

yields the same value as previously found.

13.3 INFERENCES PERTAINING TO MULTIPLE LINEAR REGRESSION

The present section is the multiple-regression analogue of Section 12.4. In it, we will be concerned with interval estimation of the model parameters and the mean response, as well as with the construction of a prediction interval for a single response. Hypothesis testing will also be discussed. As in the single-variable case, inference results will depend upon the assumption that the error variables E_1, E_2, . . . , E_n are normally distributed. Heretofore, it was only required that they be independent random variables possessing common mean zero and common variance σ^2. Since

$$Y_i = \mu_{Y|x_{1i}, x_{2i}, \ldots, x_{ki}} + E_i$$

the normality of E_i implies the normality of Y_i.

13.3.1 Inferences Concerning the Model Parameters

According to Theorem 13.2, $\hat{b}_j$ is an unbiased estimator of b_j for $j = 0, 1, \ldots, k$ and, according to Theorem 13.1, the $\hat{b}_j$ are determined by the matrix equation $\hat{\mathbf{b}} = (\mathbf{X}'\mathbf{X})^{-1}\mathbf{X}'\mathbf{Y}$, where $\mathbf{X}$ is the design matrix. Since the matrix $(\mathbf{X}'\mathbf{X})^{-1}$ plays an important role, we will denote its components by the notation c_{ij}. That is, we let

$$(\mathbf{X}'\mathbf{X})^{-1} = \begin{pmatrix} c_{00} & c_{01} & \cdots & c_{0k} \\ c_{10} & c_{11} & \cdots & c_{1k} \\ \vdots & \vdots & & \vdots \\ c_{k0} & c_{k1} & \cdots & c_{kk} \end{pmatrix}$$

Since the component variables of the random vector $\mathbf{Y}$ are normally distributed, the expression of $\hat{\mathbf{b}}$ as a constant matrix times $\mathbf{Y}$ means that each component estimator $\hat{b}_j$ of $\hat{\mathbf{b}}$ is a linear combination of independent normal variables and is therefore normally distributed. The variances of the $\hat{b}_j$ appear on the main diagonal of the covariance matrix of $\hat{\mathbf{b}}$ and, since Cov $[\hat{\mathbf{b}}] = \sigma^2 (\mathbf{X}'\mathbf{X})^{-1}$, the elements on the main diagonal are of the form $c_{jj}\sigma^2$. Consequently, for $j = 0, 1, \ldots, k$,

$$\text{Var } [\hat{b}_j] = c_{jj}\sigma^2$$

and hence

$$Z = \frac{\hat{b}_j - b_j}{\sigma\sqrt{c_{jj}}}$$

possesses a standard normal distribution. Since SSE/σ^2 possesses a chi-square distribution with $n - k - 1$ degrees of freedom,

$$T_j = \frac{\hat{b}_j - b_j}{\sqrt{c_{jj}MSE}}$$

possesses a t distribution with $n - k - 1$ degrees of freedom, and the next theorem follows.

THEOREM 13.4

For $j = 0, 1, \ldots, k$, a $(1 - \alpha)100\%$ confidence interval for b_j based on a collection of n observations is given by

$$(\overset{*}{b}_j - t_{\alpha/2,n-k-1}\sqrt{c_{jj}MSE}, \ \overset{*}{b}_j + t_{\alpha/2,n-k-1}\sqrt{c_{jj}MSE})$$

EXAMPLE 13.3

In Example 13.1, we found $\overset{*}{b}$ and $(\mathbf{X'X})^{-1}$ for the data of Table 13.1. In Example 13.2, MSE was found. Using this information, we can apply Theorem 13.4 to find 95% confidence intervals for the model parameters. Since $t_{0.025,3} = 3.182$, the intervals for b_0, b_1, and b_2 are

$$(-4.283 - (3.182)\sqrt{(20.583)(0.894)},$$
$$-4.283 + (3.182)\sqrt{(20.583)(0.894)}) = (-17.933, 9.367)$$
$$(1.65 - (3.182)\sqrt{(0.25)(0.894)}, \ 1.65 + (3.182)\sqrt{(0.25)(0.894)}) = (0.146, 3.154)$$

and

$$(85 - (3.182)\sqrt{(66.667)(0.894)}, \ 85 + (3.182)\sqrt{(66.667)(0.894)}) = (60.435, 109.565)$$

respectively.

For $j = 0, 1, \ldots, k$, one- and two-tailed hypothesis tests involving the null hypothesis

$$H_0\colon b_j = b_{j0}$$

can be performed with the test statistic T_j, where under H_0 the numerator of T_j is $\hat{b}_j - b_{j0}$. Of particular interest is the two-tailed test

$$H_0\colon b_j = 0$$
$$H_1\colon b_j \neq 0$$

in which the numerator of T_j is simply $\hat{b}_j$. In this test, if the null hypothesis is not rejected, then there is not convincing evidence that $b_j \neq 0$, and it is reasonable to delete the variable x_j from the model. If, however, the null hypothesis is rejected, then it is concluded that $b_j \neq 0$ and x_j cannot be deleted from the model. In effect, the test concerns the significance of regression relative to the variable x_j. A more general approach to testing the significance of regression is to apply an analysis-of-variance approach. This technique will be discussed in Section 13.5.

EXAMPLE 13.4

We test the significance of regression for the model parameter b_1 of Example 13.1.

1. *Hypothesis test.* $H_0\colon b_1 = 0$
   $H_1\colon b_1 \neq 0$
2. *Significance level.* $\alpha = 0.01$

3. *Test statistic.* T_1
4. *Sample size.* $n = 6$
5. *Critical region.* $t \leq -5.841$ or $t \geq 5.841$
6. *Sample value.*

$$t_1 = \frac{\overset{*}{b_1}}{\sqrt{c_{11} MSE}} = \frac{1.65}{\sqrt{(0.25)(0.894)}} = 3.490$$

7. *Decision.* Since $-5.841 < t_1 < 5.841$, the null hypothesis is not rejected and the regression is not considered significant (at the 0.01 level) insofar as the duration of the process is concerned.

Had we only desired 0.05 significance, then the null hypothesis would have been rejected and regression relative to the variable x_1 been considered significant. Quantitatively, the P value corresponding to the data is between 0.01 and 0.05. From a practical perspective, the significance of duration on the yield of the process is questionable.

Using T_j to test the null hypothesis H_0: $b_j = 0$ gives a test of the significance of regression relative to x_j over and above the regression attributable to the other regressor variables. For instance, in the case of two independent variables, use of T_1 might lead to rejection of H_0: $b_1 = 0$ and use of T_2 might lead to rejection of H_0: $b_2 = 0$; nevertheless, there might be significant regression in the full model. If, based on T_j, the variable x_j is deleted from the model, a new regression analysis must be performed using the remaining variables prior to testing any remaining variables.

The problem of model reduction will be discussed in depth in Section 13.5.2 in the context of analysis of variance. There it will be seen that the t test of the current section is simply a special case of an overall reduction methodology, and it is in the framework of that methodology that we discuss some of the intricacies of variable deletion.

13.3.2 Interval Estimation of the Mean and of Future Responses

Given a point $(x_{10}, x_{20}, \ldots, x_{k0})$, the sample regression plane resulting from a set of observations can be employed to form a confidence interval for the mean response at the point. According to the model, the mean response lies on the model regression plane and is given by

$$\mu_{Y|x_{10}, x_{20}, \ldots, x_{k0}} = b_0 + b_1 x_{10} + b_2 x_{20} + \cdots + b_k x_{k0}$$

which possesses the unbiased estimator

$$\hat{\mu}_{Y|x_{10}, x_{20}, \ldots, x_{k0}} = \hat{b}_0 + \hat{b}_1 x_{10} + \hat{b}_2 x_{20} + \cdots + \hat{b}_k x_{k0}$$

It can be shown that this estimator is normal. Moreover, it can be shown that it possesses variance

$$\mathrm{Var}\,[\hat{\mu}_{Y|x_{10}, x_{20}, \ldots, x_{k0}}] = \sigma^2 \mathbf{x}_0'(\mathbf{X}'\mathbf{X})^{-1}\mathbf{x}_0$$

where

$$\mathbf{x}_0 = \begin{pmatrix} 1 \\ x_{10} \\ x_{20} \\ \vdots \\ x_{k0} \end{pmatrix}$$

Division of the standardized version of the mean estimator by $\sqrt{MSE}/\sigma$ yields

$$T_\mu = \frac{\hat{\mu}_{Y|x_{10},x_{20},\ldots,x_{k0}} - \mu_{Y|x_{10},x_{20},\ldots,x_{k0}}}{\sqrt{\mathbf{x}_0'(\mathbf{X}'\mathbf{X})^{-1}\mathbf{x}_0 MSE}}$$

which possesses a t distribution with $n - k - 1$ degrees of freedom. The next theorem follows in the usual manner.

THEOREM 13.5

Based on a set of n observations, a $(1 - \alpha)100\%$ confidence interval for the mean response $\mu_{Y|x_{10},x_{20},\ldots,x_{k0}}$ is given by

$$(\overset{*}{\hat{\mu}}_{Y|x_{10},x_{20},\ldots,x_{k0}} - t_{\alpha/2,n-k-1}\sqrt{\mathbf{x}_0'(\mathbf{X}'\mathbf{X})^{-1}\mathbf{x}_0 MSE},$$
$$\overset{*}{\hat{\mu}}_{Y|x_{10},x_{20},\ldots,x_{k0}} + t_{\alpha/2,n-k-1}\sqrt{\mathbf{x}_0'(\mathbf{X}'\mathbf{X})^{-1}\mathbf{x}_0 MSE})$$

EXAMPLE 13.5

Continuing with the example pertaining to the data of Table 13.1, we will find a confidence interval for $\mu_{Y|6.5,0.35}$. Proceeding,

$$\mathbf{x}_0'(\mathbf{X}'\mathbf{X})^{-1}\mathbf{x}_0 = (1 \quad 6.5 \quad 0.35)\begin{pmatrix} 20.583 & -1.75 & -23.333 \\ -1.75 & 0.25 & 0 \\ -23.333 & 0 & 66.667 \end{pmatrix}\begin{pmatrix} 1 \\ 6.5 \\ 0.35 \end{pmatrix} = 0.129$$

$$t_{0.025,3}\sqrt{\mathbf{x}_0'(\mathbf{X}'\mathbf{X})^{-1}\mathbf{x}_0 MSE} = (3.182)\sqrt{(0.129)(0.894)} = 1.081$$

$$\overset{*}{\hat{\mu}}_{Y|6.5,0.35} = -4.283 + (1.65)(6.5) + (85)(0.35) = 36.192$$

and the 95% confidence interval is

$$(36.192 - 1.081, 36.192 + 1.081) = (35.111, 37.273)$$

Analogously to the single-variable case, for a given point $(x_{10}, x_{20}, \ldots, x_{k0})$, the estimator of Y at the point is taken to be the estimator of the mean response at the point:

$$\hat{Y}|x_{10}, x_{20}, \ldots, x_{k0} = \hat{\mu}_{Y|x_{10},x_{20},\ldots,x_{k0}}$$
$$= \hat{b}_0 + \hat{b}_1 x_{10} + \hat{b}_2 x_{20} + \cdots + \hat{b}_k x_{k0}$$

$\overset{*}{Y}|x_{10}, x_{20}, \ldots, x_{k0}$ is the predicted value pertaining to future occurrences of $Y|x_{10}, x_{20}, \ldots, x_{k0}$. As in the single-variable case, a $(1 - \alpha)100\%$ prediction interval is constructed relative to the point prediction $\overset{*}{Y}|x_{10}, x_{20}, \ldots, x_{k0}$. In the multiple regression setting, the prediction interval is given by

$$(\overset{*}{Y}|x_{10}, x_{20}, \ldots, x_{k0} - t_{\alpha/2,n-k-1}\sqrt{(1 + \mathbf{x}_0'(\mathbf{X}'\mathbf{X})^{-1}\mathbf{x}_0)MSE},$$
$$\overset{*}{Y}|x_{10}, x_{20}, \ldots, x_{k0} + t_{\alpha/2,n-k-1}\sqrt{(1 + \mathbf{x}_0'(\mathbf{X}'\mathbf{X})^{-1}\mathbf{x}_0)MSE})$$

EXAMPLE 13.6　　The 95% prediction interval at the point (6.5, 0.35) for the data of Example 13.1 can be found using the information of Example 13.5. Specifically, the interval is

$$(36.192 - (3.182)\sqrt{(1 + 0.129)(0.894)},$$

$$36.192 + (3.182)\sqrt{(1 + 0.129)(0.894)}) = (32.995, 39.389)$$

13.4　POLYNOMIAL REGRESSION

The problem of Chapter 12 was to fit a straight line to data derived from a deterministic independent variable x and a random response variable Y. Rather than model the mean response as a linear function of x, it may be more suitable to model it as a more complicated function of x. For instance, in Figure 13.2 it certainly appears that a cubic will provide a better fit than a straight line. In the present section we pay specific attention to the **polynomial regression model,** in which, for $k \geq 1$, the model equation takes the form

$$\mu_{Y|x} = b_0 + b_1x + b_2x^2 + \cdots + b_kx^k$$

and the response variable is expressed as

$$Y = b_0 + b_1x + b_2x^2 + \cdots + b_kx^k + E$$

where E is the random difference between the response and its mean. If $k = 1$, then the polynomial model reduces to the simple linear model.

If n observations are made at the points $x_1, x_2, \ldots, x_n$, then there are n equations

$$\mu_{Y|x_i} = b_0 + b_1x_i + b_2x_i^2 + \cdots + b_kx_i^k$$

$i = 1, 2, \ldots, n$. Relative to the response variables, the following polynomial regression model results.

Polynomial Regression Model

Equation:　　$Y_i = b_0 + b_1x_i + b_2x_i^2 + \cdots + b_kx_i^k + E_i$
Assumptions:　1. $E_i \sim N(0, \sigma)$
2. The E_i comprise an independent collection of random variables.

FIGURE 13.2
Polynomial regression

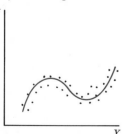

To arrive at the least-squares estimates $\overset{*}{b}_0, \overset{*}{b}_1, \ldots, \overset{*}{b}_k$, we once again minimize the error mean square, which is equal to the sum of the squares of the residuals. In this case, the residuals measure the differences between the observations and the best-fit polynomial of degree k. Specifically, for the given observations $y_1, y_2, \ldots, y_n$,

$$SSE = \sum_{i=1}^{n} [y_i - (\overset{*}{b}_0 + \overset{*}{b}_1x_i + \overset{*}{b}_2x_i^2 + \cdots + \overset{*}{b}_kx_i^k)]^2$$

is minimized for all possible choices of the model parameters. Since *SSE* is minimized by differentiating with respect to each $\overset{*}{b}_j$, setting the resulting $k + 1$ partial derivatives equal to 0, and solving, it is clear from the structure of *SSE* that the resulting least-squares estimates will be given by the same expressions that provided the least-squares estimates in the multiple linear model, except that x_i^j will be in place of x_{ji} for $j = 1, 2, \ldots, k$ and $i = 1, 2, \ldots, n$. Consequently, the desired estimates will be the components of the parameter vector

$$\overset{*}{\mathbf{b}} = \mathbf{X}^+\mathbf{y} = (\mathbf{X}'\mathbf{X})^{-1}\mathbf{X}'\mathbf{y}$$

where the design matrix is

$$\mathbf{X} = \begin{pmatrix} 1 & x_1 & x_1^2 & \cdots & x_1^k \\ 1 & x_2 & x_2^2 & \cdots & x_2^k \\ \vdots & \vdots & \vdots & & \vdots \\ 1 & x_n & x_n^2 & \cdots & x_n^k \end{pmatrix}$$

In terms of method, finding the least-squares polynomial when the response depends on a single independent variable is the same as finding a least-squares regression plane. As in the multiple linear case, it will be assumed that $\mathbf{X}'\mathbf{X}$ is nonsingular.

EXAMPLE 13.7

It was seen in Example 12.3 that a simple linear regression model gives a rough fit to the CPU-utilization data of Table 12.3, Figures 12.7 and 12.8 providing the sample regression lines computed from all 10 data points and from 7 of the data points, respectively. We will now employ the 10 data points to find the best least-squares quadratic fit. Proceeding, the design matrix is

$$\mathbf{X} = \begin{pmatrix} 1 & 1 & 1 \\ 1 & 2 & 4 \\ 1 & 3 & 9 \\ 1 & 4 & 16 \\ 1 & 5 & 25 \\ 1 & 6 & 36 \\ 1 & 7 & 49 \\ 1 & 8 & 64 \\ 1 & 9 & 81 \\ 1 & 10 & 100 \end{pmatrix}$$

$$(\mathbf{X}'\mathbf{X})^{-1} = \begin{pmatrix} 1.383333 & -0.525 & 0.041666 \\ -0.525 & 0.241287 & -0.020833 \\ -0.041666 & -0.020833 & 0.001893 \end{pmatrix}$$

$$\mathbf{X}^+ = \begin{pmatrix} 0.900000 & 0.500000 & 0.183333 & -0.050000 & -0.200000 \\ -0.304545 & -0.125757 & 0.011363 & 0.106818 & 0.160606 \\ 0.022727 & 0.007575 & -0.003787 & -0.011363 & -0.015151 \end{pmatrix}$$

$$\begin{pmatrix} -0.266666 & -0.250000 & -0.150000 & 0.033333 & 0.300000 \\ 0.172727 & 0.143181 & 0.071969 & -0.040909 & -0.195454 \\ -0.015151 & -0.011363 & -0.003787 & 0.007575 & 0.022727 \end{pmatrix}$$

and

$$\overset{*}{\mathbf{b}} = \mathbf{X}^+\mathbf{y} = \begin{pmatrix} 5.866666 \\ 16.263636 \\ -0.803030 \end{pmatrix}$$

Thus, the fitted quadratic regression line is given by

$$\overset{*}{Y} = \overset{*}{\mu}_{Y|x} = 5.867 + (16.264)x - (0.803)x^2$$

which is illustrated in Figure 13.3. The tightness of the fit relative to the sample data is evident in the figure.

From a wider perspective, the key to finding the least-squares estimates in both the multiple linear and polynomial models is that both models are linear with respect to the model parameters. Another such model would be the second-degree polynomial model in the variables x_1 and x_2, namely,

$$Y = b_0 + b_1x_1 + b_2x_2 + b_3x_1^2 + b_4x_2^2 + b_5x_1x_2 + E$$

Here, if there are n observations, then the design matrix is $n \times 6$ and the estimates of the model parameters are once again given by the matrix equation $\overset{*}{\mathbf{b}} = \mathbf{X}^+\mathbf{y}$. More generally, so long as the model equation is linear in terms of the model parameters, regression is determined by the **general linear model** equation $\mathbf{Y} = \mathbf{Xb} + \mathbf{E}$, and the parameter vector of Theorem 13.1 provides the least-squares estimates.

13.5 TESTING FOR SIGNIFICANCE OF REGRESSION BY ANALYSIS OF VARIANCE

In Section 13.3.1, a t statistic was employed to test the null hypothesis H_0: $b_j = 0$, the purpose of the test being to establish significance of regression relative to the variable x_j. In the event H_0 is accepted, x_j can be dropped from the model, since the

FIGURE 13.3
Sample regression quadratic for Example 13.7

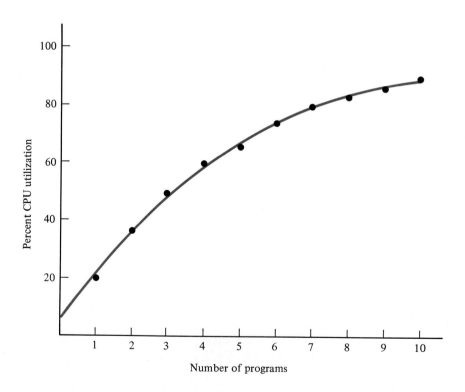

TABLE 13.2 ANOVA TABLE FOR TESTING SIGNIFICANCE OF REGRESSION

Source of Variation	Sum of Squares	Degrees of Freedom	Mean Square	F Test Statistic
Regression	$SSR = \hat{\mathbf{b}}'(\mathbf{X'Y}) - n\bar{Y}^2$	k	$MSR = \dfrac{SSR}{k}$	$F = \dfrac{MSR}{MSE}$
Error	$SSE = S_{YY} - SSR$	$n - k - 1$	$MSE = \dfrac{SSE}{n - k - 1}$	
Total	S_{YY}	$n - 1$		

data do not indicate significant regression with respect to x_j. More generally, significance of regression in the entire model, or in reduced models, can be tested by employing an analysis-of-variance approach.

13.5.1 Testing Significance of Regression in the Full Model

Significance of regression in the full model can be checked by the hypothesis test

$$H_0:\ b_1 = b_2 = \cdots = b_k = 0$$

$$H_1:\ \text{There exists at least one } j \text{ such that } b_j \neq 0.$$

Under H_0, SSR/σ^2 possesses a chi-square distribution with k degrees of freedom. Since SSE/σ^2 possesses a chi-square distribution with $n - k - 1$ degrees of freedom and the two variables are independent,

$$F = \frac{SSR/k}{SSE/(n - k - 1)} = \frac{MSR}{MSE}$$

possesses an F distribution with k and $n - k - 1$ degrees of freedom. According to Theorem 13.3, the total sum of squares S_{YY} is partitioned into the regression sum of squares SSR and the error sum of squares SSE. SSR, which consists of terms of the form $(\hat{Y}_i - \bar{Y})^2$, is a measure of the variability about the mean arising from regression. SSE, which consists of terms of the form $(\hat{Y}_i - Y_i)^2$, is a measure of the residual variability about the regression plane. Higher values of SSR relative to SSE indicate significant regression. The use of $F = MSR/MSE$ as a test statistic reflects these intuitive considerations; indeed, the test is one-tailed with rejection at level α occurring when $F \geq f_{\alpha,k,n-k-1}$. Table 13.2 gives the analysis-of-variance table for testing the significance of regression.

EXAMPLE 13.8

In Example 13.1, a sample regression plane was fit to the time–catalyst–yield data of Table 13.1. Using analysis of variance, we will test to determine the significance of regression of the entire model. The relevant hypothesis test is

$$H_0:\ b_1 = b_2 = 0$$

$$H_1:\ \text{Either } b_1 \text{ or } b_2 \text{ is nonzero.}$$

From Example 13.2, $SSR = 119.265$ and $SSE = 2.683$. Thus, the appropriate F statistic has the value

$$f = \frac{SSR/2}{SSE/3} = \frac{(119.265)/2}{(2.683)/3} = 66.678$$

which is greater than $f_{0.01,2,3} = 30.82$, so that full-model regression is considered significant at level 0.01.

13.5.2 Testing Significance of Regression in Reduced Models

Even if the null hypothesis regarding the significance of regression is rejected and it is concluded there is significant regression between the response variable and the independent variables $x_1, x_2, \ldots, x_k$, some of the variables might not be needed. Although the t statistic T_j of Section 13.3.1 can be employed to eliminate individual variables from the model, a more general analysis-of-variance procedure can be employed to test the significance of subsets of the overall set of independent variables.

Suppose the full model involves the k variables

$$x_1, x_2, \ldots, x_m, x_{m+1}, x_{m+2}, \ldots, x_k$$

and it is desired to test whether or not the reduced model involving only the m variables $x_1, x_2, \ldots, x_m$ is adequate. To arrive at an appropriate test statistic, we can employ a heuristic analysis-of-variance argument. SSR measures the variation attributable to regression in the full model. Let SSR_m denote the regression sum of squares for the reduced model. If there is significant regression attributable to the variables x_{m+1}, $x_{m+2}, \ldots, x_k$ in the presence of (over and above) $x_1, x_2, \ldots, x_m$, then one would expect SSR to be appreciably greater than SSR_m. The statistic

$$F = \frac{\dfrac{SSR - SSR_m}{k - m}}{\dfrac{SSE}{n - k - 1}}$$

possesses an F distribution with $k - m$ and $n - k - 1$ degrees of freedom and becomes significant when SSR sufficiently exceeds SSR_m. F serves as a test statistic for the hypothesis test

H_0: The reduced model is adequate.

H_1: The reduced model is inadequate.

The test is upper one-tailed.

Applying the partition $S_{YY} = SSR + SSE$ given in Theorem 13.3 and solving for SSR shows the test statistic F possesses the equivalent representation

$$F = \frac{\dfrac{SSE_m - SSE}{k - m}}{\dfrac{SSE}{n - k - 1}}$$

Complementary reasoning to that utilized in the original formulation shows the efficacy of the new form of F. Indeed, if the reduced model is adequate (H_0 true), then the reduced-model error sum of squares SSE_m, which represents the variation in the response not attributable to regression, will only slightly exceed SSE. On the other hand, if the reduced model is inadequate, we would expect SSE_m to greatly exceed SSE.

EXAMPLE 13.9

Once again referring to the duration–catalyst–yield multiple regression problem of Example 13.1, we will test to determine whether or not the data indicate that a reduced model in terms of process duration (x_1) or catalyst concentration (x_2) can be employed in place of the two-variable model. From Example 13.2, $S_{yy} = 121.948$, $SSR = 119.265$, and $SSE = 2.683$. From the data of Table 13.1,

$$S_{x_1x_1} = \sum_{i=1}^{6} x_{1i}^2 - 6\bar{x}_1^2 = 298 - 294 = 4$$

$$S_{x_2x_2} = \sum_{i=1}^{6} x_{2i}^2 - 6\bar{x}_2^2 = 0.75 - 0.735 = 0.015$$

$$S_{x_1y} = \sum_{i=1}^{6} x_{1i}y_i - 6\bar{x}_1\bar{y} = 1561.3 - 1554.7 = 6.6$$

$$S_{x_2y} = \sum_{i=1}^{6} x_{2i}y_i - 6\bar{x}_2\bar{y} = 79.01 - 77.735 = 1.275$$

$$\overset{*}{b}(1) = \frac{S_{x_1y}}{S_{x_1x_1}} = 1.65$$

$$\overset{*}{b}(2) = \frac{S_{x_2y}}{S_{x_2x_2}} = 85$$

where we have let $\overset{*}{b}(j)$ denote the slope for the sample regression line corresponding to the variable x_j. Employing similar notation for regression and error sums of squares, we obtain

$$SSE(1) = S_{yy} - \overset{*}{b}(1)S_{x_1y} = 121.948 - 10.89 = 111.058$$

$$SSE(2) = S_{yy} - \overset{*}{b}(2)S_{x_2y} = 121.948 - 108.375 = 13.573$$

$SSR(1) = 10.89$, and $SSR(2) = 108.375$. For the hypothesis test

H_0: The reduced model involving x_1 is adequate.

H_1: The reduced model involving x_1 is inadequate.

the value of the relevant F statistic is

$$f_1 = \frac{SSR - SSR(1)}{\dfrac{SSE}{3}} = \frac{119.265 - 10.89}{\dfrac{2.683}{3}} = 121.180$$

which is much greater than $f_{0.01,1,3} = 34.12$. Thus H_0 is rejected and the reduced model, which is a simple linear regression model with the variable x_1, is inadequate.

For the hypothesis test

H_0: The reduced model involving x_2 is adequate.

H_1: The reduced model involving x_2 is inadequate.

the value of the relevant F statistic is

$$f_2 = \frac{SSR - SSR(2)}{\dfrac{SSE}{3}} = \frac{119.265 - 108.375}{\dfrac{2.683}{3}} = 12.177$$

which does not exceed $f_{0.01,1,3}$. Hence, the adequacy of the reduced model is not rejected at the 0.01 level. Note, however, that $f_{0.05,1,3} = 10.13$, so adequacy of the simple linear model involving x_2 is rejected at the 0.05 level. In other words, SSR does not sufficiently exceed $SSR(2)$ to reject the reduced model at the 0.01 level of certainty, but SSR does sufficiently exceed $SSR(2)$ to reject the reduced model at the 0.05 level. The analysis of variance is summarized in Table 13.3.

Similar results regarding the significance of x_1 (relative to regression) were obtained in Examples 13.4 and 13.9. Such agreement is not accidental. In Example 13.4, the significance of x_1 in the model was tested by using the statistic T_1 to test the null hypothesis H_0: $b_1 = 0$. In Example 13.9, the significance of x_1 in the model was tested by using an F statistic to test the null hypothesis that the reduced model, absent of x_1, is adequate. In fact, the two tests are equivalent.

More generally, if the full model has k variables, testing the variable x_1 for significance of regression by using the t statistic T_1 to test the null hypothesis H_0: $b_1 = 0$ is equivalent to using the F statistic

$$F^{(1)} = \frac{SSR - SSR(2, 3, \ldots, k)}{\dfrac{SSE}{n - k - 1}}$$

to test the adequacy of the reduced model, absent of x_1, where $SSR(2, 3, \ldots, k)$ is the regression sum of squares for the reduced model having variables $x_2, x_3, \ldots, x_k$. Critical values of the t and F distributions are related by the equation

$$t^2_{\alpha/2,v} = f_{\alpha,1,v}$$

In the present situation, $|T_1| \geq t_{\alpha/2,n-k-1}$ if and only if $F^{(1)} \geq f_{\alpha,1,n-k-1}$. More generally, the significance of regression for any variable $x_j, j = 1, 2, \ldots, k$, can be equivalently

TABLE 13.3   ANOVA TABLE FOR EXAMPLES 13.8 AND 13.9

Source of Variation	Sum of Squares	Degrees of Freedom	Mean Square	F Test Statistic
Regression (x_1, x_2)	119.265	2	59.633	66.678
Error (x_1, x_2)	2.683	3	0.894	
Regression (x_1)	10.890	1	10.890	0.392
Error (x_1)	111.058	4	27.765	
Regression (x_2)	108.375	1	108.375	31.938
Error (x_2)	13.573	4	3.393	
Total	121.948	5		

TABLE 13.4 ENGINE-LIFE DATA

x_1	x_2	x_3	y
1	1	1	64,200
1	1	-1	67,600
1	-1	1	54,500
1	-1	-1	52,800
-1	1	1	58,100
-1	1	-1	56,200
-1	-1	1	49,000
-1	-1	-1	45,600

tested by using either T_j or $F^{(j)}$, the latter corresponding to the reduced model, absent of x_j.

In the next example we employ the F-statistic approach to test the adequacy of a reduced model resulting from the deletion of two variables. Not only does the example illustrate reduction of a linear model, it also demonstrates some of the advantages of using a design matrix with orthogonal columns.

EXAMPLE 13.10

Suppose an automobile engineer wishes to apply a multiple linear regression model to three factors that affect the life of an internal combustion engine, these being revolutions per minute (x_1), running temperature (x_2), and oil weight (x_3). If the engineer plans to employ each factor at only two levels, then the numerical values of the levels can be coded so that each variable takes on the values -1 and $+1$. Employing all possible combinations of the levels yields a design matrix $\mathbf{X}$ with orthogonal columns. Testing yields the data of Table 13.4, in which y is the number of simulated miles prior to the minimum cylinder compression falling below some predetermined threshold value. The design matrix is

$$
\mathbf{X} = \begin{pmatrix}
1 & 1 & 1 & 1 \\
1 & 1 & 1 & -1 \\
1 & 1 & -1 & 1 \\
1 & 1 & -1 & -1 \\
1 & -1 & 1 & 1 \\
1 & -1 & 1 & -1 \\
1 & -1 & -1 & 1 \\
1 & -1 & -1 & -1
\end{pmatrix}
$$

Applying Theorem 13.1 yields

$$
\mathbf{X}' = \begin{pmatrix}
1 & 1 & 1 & 1 & 1 & 1 & 1 & 1 \\
1 & 1 & 1 & 1 & -1 & -1 & -1 & -1 \\
1 & 1 & -1 & -1 & 1 & 1 & -1 & -1 \\
1 & -1 & 1 & -1 & 1 & -1 & 1 & -1
\end{pmatrix}
$$

$$
\mathbf{X}'\mathbf{X} = \begin{pmatrix}
8 & 0 & 0 & 0 \\
0 & 8 & 0 & 0 \\
0 & 0 & 8 & 0 \\
0 & 0 & 0 & 8
\end{pmatrix}
$$

$$(\mathbf{X}'\mathbf{X})^{-1} = \begin{pmatrix} \frac{1}{8} & 0 & 0 & 0 \\ 0 & \frac{1}{8} & 0 & 0 \\ 0 & 0 & \frac{1}{8} & 0 \\ 0 & 0 & 0 & \frac{1}{8} \end{pmatrix}$$

$$\mathbf{X}^+ = \frac{1}{8}\mathbf{X}'$$

$$\overset{*}{\mathbf{b}} = \mathbf{X}^+\mathbf{y} = \mathbf{X}^+ \begin{pmatrix} 64{,}200 \\ 67{,}600 \\ 54{,}500 \\ 52{,}800 \\ 58{,}100 \\ 56{,}200 \\ 49{,}000 \\ 45{,}600 \end{pmatrix} = \begin{pmatrix} 56{,}000 \\ 3{,}775 \\ 5{,}525 \\ 450 \end{pmatrix}$$

The sample regression plane possesses the equation

$$\overset{*}{Y} = \overset{*}{\mu}_{Y|x_1,x_2,x_3} = 56{,}000 + 3775x_1 + 5525x_2 + 450x_3$$

For analysis-of-variance purposes we find

$$\overset{*}{\mathbf{b}}'(\mathbf{X}'\mathbf{y}) = (\mathbf{X}^+\mathbf{y})'(8\mathbf{X}^+\mathbf{y}) = 8(\mathbf{X}^+\mathbf{y})'(\mathbf{X}^+\mathbf{y}) = 2.5448 \times 10^{10}$$

$$SSR = \overset{*}{\mathbf{b}}'(\mathbf{X}'\mathbf{y}) - 8\bar{y}^2 = 3.5983 \times 10^8$$

$$S_{yy} = \sum_{i=1}^{8} y_i^2 - 8\bar{y}^2 = 2.5464 \times 10^{10} - 2.5088 \times 10^{10} = 3.7590 \times 10^8$$

$$SSE = S_{yy} - SSR = 1.6070 \times 10^7$$

$$MSE = \frac{SSE}{8 - 3 - 1} = 4.0175 \times 10^6$$

$$MSR = \frac{SSR}{3} = 1.1994 \times 10^8$$

$$S_{x_1x_1} = S_{x_2x_2} = S_{x_3x_3} = 8$$

$$S_{x_1y} = \sum_{i=1}^{8} x_{1i}y_i = 30{,}200$$

$$S_{x_2y} = \sum_{i=1}^{8} x_{2i}y_i = 44{,}200$$

$$S_{x_3y} = \sum_{i=1}^{8} x_{3i}y_i = 3600$$

Since $MSR/MSE = 29.855$ and $f_{0.01,3,4} = 16.69$, significance of regression in the entire model is established at level 0.01. To determine whether or not there is significant regression relative to oil weight (x_3), we perform the hypothesis test having the null hypothesis

$$H_0: \text{The reduced model involving } x_1 \text{ and } x_2 \text{ is adequate.}$$

Once again let $\mathbf{X}$ denote the design matrix. In the reduced model it is the same as it was in the full model, except that the fourth column is omitted. We obtain

$$\mathbf{X'X} = \begin{pmatrix} 8 & 0 & 0 \\ 0 & 8 & 0 \\ 0 & 0 & 8 \end{pmatrix}$$

$$(\mathbf{X'X})^{-1} = \frac{1}{8}\begin{pmatrix} 1 & 0 & 0 \\ 0 & 1 & 0 \\ 0 & 0 & 1 \end{pmatrix}$$

$$\mathbf{X}^+ = \frac{1}{8}\mathbf{X}'$$

$$\overset{*}{\mathbf{b}} = \mathbf{X}^+\mathbf{y} = \begin{pmatrix} 56{,}000 \\ 3{,}775 \\ 5{,}525 \end{pmatrix}$$

$$SSR(1, 2) = \overset{*}{\mathbf{b}}'(\mathbf{X'y}) - 8\bar{y}^2 = 3.5821 \times 10^8$$

The relevant F test statistic is

$$\frac{\dfrac{SSR - SSR(1, 2)}{3 - 2}}{\dfrac{SSE}{8 - 3 - 1}} = \frac{1{,}620{,}000}{4{,}017{,}500} = 0.403$$

where $SSR(1, 2)$ denotes the regression sum of squares with x_1 and x_2 in the model. Since $f_{0.05,1,4} = 7.71$, the reduced model is not deemed inadequate.

Suppose we now consider removing both x_1 and x_3 from the model. Then the reduced model involves a simple linear regression with respect to x_2 and the relevant null hypothesis is

$$H_0: \text{The reduced model involving } x_2 \text{ is adequate.}$$

Using the notation of Example 13.9,

$$\overset{*}{b}(2) = \frac{S_{x_2y}}{S_{x_2x_2}} = 5525$$

$$SSE(2) = S_{yy} - \overset{*}{b}(2)S_{x_2y} = 131{,}695{,}000$$

$$SSR(2) = S_{yy} - SSE(2) = 244{,}205{,}000$$

$$f_2 = \frac{\dfrac{SSR - SSR(2)}{3 - 1}}{\dfrac{SSE}{8 - 3 - 1}} = 14.390$$

Since $f_{0.05,2,4} = 6.94$, the adequacy of the reduced model involving only x_2 is rejected at level 0.05.

In sum, dropping oil weight from the regression model is acceptable; however, dropping both revolutions per minute and oil weight is not acceptable.

An interesting aspect of Examples 13.9 and 13.10 is that in both examples the regression coefficients remaining in the model after reduction possessed the same estimates as prior to reduction. In Example 13.9, the estimates of b_1 and b_2 in the single-variable reduced models were 1.65 and 85, respectively. These were precisely the values in the sample regression equation of Example 13.1. In Example 13.10, the parameter-estimate vector for the reduced model involving x_1 and x_2 yields the estimates 3775 and 5525 for b_1 and b_2, respectively. These are the same as in the full-model. Simple linear models based on x_1, x_2, and x_3 also yield the same estimates as those given in the full model. Such invariance of the model-parameter estimates under reduction is not the rule; rather, it results from an absence of linear correlation between the regressor variables.

Given a linear regression model, there may or may not be strong linear correlation between the variables. A measure of linear correlation between the variables x_j and $x_{j'}$ is

$$r = \frac{S_{x_j x_{j'}}}{\sqrt{S_{x_j x_j} S_{x_{j'} x_{j'}}}}$$

Since x_j and $x_{j'}$ are not random variables, r is not an estimate of a bivariate correlation coefficient as it was in Section 12.6; nevertheless, it does measure correlation between controllable variables. If $r = 0$, then x_j and $x_{j'}$ are said to be **orthogonal.** Should the design matrix possess orthogonal columns, then (as in Example 13.10) the regressor variables are orthogonal. As was the case in Example 13.9, the variables can be orthogonal without the design matrix having orthogonal columns. When the regressor variables in a multiple linear regression model are mutually orthogonal, the least-squares estimates are the same in both the full and reduced models.

Let us now make the assumption that the regressor variables have been selected so that they are mutually orthogonal and each possesses mean zero, which is equivalent to the design matrix $\mathbf{X}$ having orthogonal columns. Then both $\mathbf{X'X}$ and $(\mathbf{X'X})^{-1}$ are diagonal. Specifically,

$$\mathbf{X'X} = \begin{pmatrix} n & 0 & 0 & \cdots & 0 \\ 0 & S_{x_1 x_1} & 0 & \cdots & 0 \\ 0 & 0 & S_{x_2 x_2} & \cdots & 0 \\ \vdots & \vdots & \vdots & & \vdots \\ 0 & 0 & 0 & \cdots & S_{x_k x_k} \end{pmatrix}$$

By looking at the first normal equation in the multiple linear setting, we see that $\overset{*}{b}_0 = \bar{y}$. Thus,

$$SSR = \overset{*}{\mathbf{b}}'(\mathbf{X'y}) - n\bar{y}$$

$$= (\overset{*}{b}_0 \quad \overset{*}{b}_1 \quad \cdots \quad \overset{*}{b}_k) \begin{pmatrix} n\bar{y} \\ S_{x_1 y} \\ \vdots \\ S_{x_k y} \end{pmatrix} - n\bar{y}^2$$

$$= \overset{*}{b}_1 S_{x_1 y} + \overset{*}{b}_2 S_{x_2 y} + \cdots + \overset{*}{b}_k S_{x_k y}$$

$$= SSR(1) + SSR(2) + \cdots + SSR(k)$$

TABLE 13.5 TOOL-
LIFE DATA

x_{1i}	x_{2i}	y_i
80	0.2	42
80	0.3	40
100	0.3	26
100	0.4	24
120	0.5	16
120	0.4	20

The full-model regression sum of squares is partitioned into the sum of the single-variable regression sums of squares. Consequently, the regression attributable to a newly introduced variable x_{k+1} over and above existing regression is simply the regression attributable to x_{k+1} in isolation, namely $SSR(k+1)$.

The occurrence of orthogonality relations in a multiple linear regression model depends upon judicious experimental design. Although the underlying theory regarding the least-squares estimators and the regression equation hold in the absence of orthogonality, there are advantages to an orthogonal design. Among the most important advantages is a more straightforward analysis of the increase in regression with the addition of more regressor variables.

The next example illustrates a situation in which the experiment has not been designed with orthogonality in mind.

EXAMPLE 13.11

The lifetime of a metal-cutting tool depends on a number of factors. The data of Table 13.5 have been obtained by measuring the lifetime (y) of a particular cutting tool at various speeds (x_1) and feed rates (x_2), the tests all being run on the same type of metal. The variables x_1, x_2, and y are in meters per minute, millimeters per revolution, and minutes, respectively. Based upon the data, we have

$$\mathbf{X} = \begin{pmatrix} 1 & 80 & 0.2 \\ 1 & 80 & 0.3 \\ 1 & 100 & 0.3 \\ 1 & 100 & 0.4 \\ 1 & 120 & 0.5 \\ 1 & 120 & 0.4 \end{pmatrix}$$

$$\mathbf{X'} = \begin{pmatrix} 1 & 1 & 1 & 1 & 1 & 1 \\ 80 & 80 & 100 & 100 & 120 & 120 \\ 0.2 & 0.3 & 0.3 & 0.4 & 0.5 & 0.4 \end{pmatrix}$$

$$\mathbf{X'X} = \begin{pmatrix} 6 & 600 & 2.1 \\ 600 & 61{,}600 & 218 \\ 2.1 & 218 & 0.79 \end{pmatrix}$$

$$(\mathbf{X'X})^{-1} = \begin{pmatrix} 7.91\overline{6} & -0.1125 & 10 \\ -0.1125 & 0.00229166 & -0.3\overline{3} \\ 10 & -0.3\overline{3} & 66.6\overline{6} \end{pmatrix}$$

$$\mathbf{X}^+ = \begin{pmatrix} 0.916667 & 1.916667 & -0.333333 & 0.666667 & -0.583333 & -1.583333 \\ 0.004167 & -0.029167 & 0.016667 & -0.016667 & -0.004167 & 0.029167 \\ -3.333333 & 3.333333 & -3.333333 & 3.333333 & 3.333333 & -3.333333 \end{pmatrix}$$

$$\overset{*}{\mathbf{b}} = \mathbf{X}^+\mathbf{y} = \mathbf{X}^+ \begin{pmatrix} 42 \\ 40 \\ 26 \\ 24 \\ 16 \\ 20 \end{pmatrix} = \begin{pmatrix} 81.5000 \\ -0.4417 \\ -26.6667 \end{pmatrix}$$

The sample regression plane possesses the equation

$$\overset{*}{Y} = \overset{*}{\mu}_{Y|x_1,x_2} = 81.5 - 0.4417x_1 - 26.6667x_2$$

Our goal is to examine the reduced models. From the data of Table 13.5, we find $\bar{x}_1 = 100$, $\bar{x}_2 = 0.35$, $\bar{y} = 28$, $S_{x_1x_1} = 1600$, $S_{x_2x_2} = 0.055$, $S_{yy} = 568$, $S_{x_1x_2} = 8$, $S_{x_1y} = -920$, $S_{x_2y} = -5$,

$$\overset{*}{b}(1) = \frac{S_{x_1y}}{S_{x_1x_1}} = \frac{-920}{1600} = -0.575$$

$$\overset{*}{b}(2) = \frac{S_{x_2y}}{S_{x_2x_2}} = \frac{-5}{0.055} = -90.909$$

In the full model, the estimates of b_1 and b_2 are -0.4417 and -26.6667, respectively. In the reduced models, the estimates are -0.575 and -90.909, respectively.

Turning to the significance of regression,

$$\overset{*}{\mathbf{b}}'(\mathbf{X}'\mathbf{y}) = (81.5 \quad -0.4417 \quad -26.6667) \begin{pmatrix} 168 \\ 15{,}880 \\ 53.8 \end{pmatrix} = 5243.14$$

$$SSR = \overset{*}{\mathbf{b}}'(\mathbf{X}'\mathbf{y}) - 6\bar{y}^2 = 5243.14 - 4704 = 539.14$$

$$SSE = S_{yy} - SSR = 568 - 539.14 = 28.86$$

Consequently, $MSR = SSR/2 = 269.57$, $MSE = SSE/3 = 9.62$, $MSR/MSE = 28.022$, and, since $f_{0.05,2,3} = 9.55$, significant full-model regression is supported at level 0.05.

For the reduced model with variable x_1,

$$SSR(1) = \overset{*}{b}(1)S_{x_1y} = 529$$

Thus,

$$\frac{SSR - SSR(1)}{MSE} = \frac{539.14 - 529}{9.62} = 1.054$$

Adequacy of the reduced model is not rejected at level 0.05. For the reduced model with variable x_2,

$$SSR(2) = \overset{*}{b}(2)S_{x_2y} = 454.55$$

Thus,

$$\frac{SSR - SSR(2)}{MSE} = \frac{539.14 - 454.55}{9.62} = 8.793$$

Since $f_{0.05,1,3} = 10.13$, again adequacy of the reduced model is not rejected at level 0.05. The analysis of variance is summarized in Table 13.6.

Neither variable, speed or feed rate, yielded significant regression over and above the other in Example 13.11. Examining the data of Table 13.5, we see that there appears to be linear correlation between x_1 and x_2. Indeed, based upon the computations of Example 13.11,

$$r = \frac{8}{\sqrt{1600 \times 0.055}} = 0.853$$

which is rather large. The result of such correlation is that the variables contribute redundant information in the presence of each other. This is borne out in the example, where either variable can be deleted from the model without a significant loss of regression. Alone, each is significant, but neither is significant over and above the other.

When there exists high correlation between regressor variables, we are faced with the problem of **multicollinearity.** One must be careful when interpreting results arising from experiments in which there is large correlation between regressor variables. In particular, the parameter estimates in the full model do not measure individual effects on the response variable; rather, they measure the effects in the presence of correlated regressor variables. Moreover, the additional regression attributable to a newly introduced variable may be insignificant, whereas in isolation the variable has significant effect on the response. In sum, judicious experimental design can enhance data interpretation.

Another way to look at the significance of regression involves the **coefficient of multiple determination,**

$$R^2 = \frac{SSR}{S_{YY}}$$

R^2 measures the proportion of variability explained by the regression model. $R^2 100\%$ gives the percentage of variability explained by the model. The closer R^2 is to 1, the

TABLE 13.6 ANOVA TABLE FOR EXAMPLE 13.11

Source of Variation	Sum of Squares	Degrees of Freedom	Mean Square	F Test Statistic
Regression (x_1, x_2)	539.14	2	269.57	28.025
Error (x_1, x_2)	28.86	3	9.62	
Regression (x_1)	529.00	1	529.00	54.256
Error (x_1)	39.00	4	9.75	
Regression (x_2)	454.55	1	454.55	16.026
Error (x_2)	113.45	4	28.36	
Total	568.00	5		

better the model accounts for the data. Note that the F test statistic employed in testing the significance of the full model is given by

$$F = \frac{\dfrac{SSR}{kS_{YY}}}{\dfrac{S_{YY} - SSR}{(n - k - 1)S_{YY}}} = \frac{\dfrac{R^2}{k}}{\dfrac{1 - R^2}{n - k - 1}}$$

Thus, $F = MSR/MSE$ is large when R^2 is large, and significance of regression can be judged in terms of R^2 being near 1.

Again turning our attention to testing the adequacy of a reduced model, we see that in the case of the variables $x_1, x_2, \ldots, x_m, x_{m+1}, \ldots, x_k$, the reduced set being $x_1, x_2, \ldots, x_m$, the appropriate F test statistic is given by

$$F = \frac{\dfrac{SSR - SSR_m}{(k - m)S_{YY}}}{\dfrac{S_{YY} - SSR}{(n - k - 1)S_{YY}}} = \frac{\dfrac{R^2 - R_m^2}{k - m}}{\dfrac{1 - R^2}{n - k - 1}}$$

where R_m^2 is the coefficient of multiple determination for the reduced model. Thus, the reduced model is rejected when the full-model coefficient of multiple determination significantly exceeds the reduced-model coefficient of multiple determination. The point is this: if additional independent variables increase R^2 sufficiently to put F in the critical region, then they are retained in the model; otherwise, they can be dropped.

One might be tempted to argue that extraneous x variables should be kept even if they do not increase R^2 significantly, since they might at least produce a marginally better fit, albeit at the cost of more computer time when finding the model coefficients and making estimations. Unfortunately, the cost of extraneous variables is more than simply added computation. Additional independent variables can increase the error sum of squares and thereby increase the mean square error MSE. A quick glance at the interval estimation expressions shows that these estimations lose precision as MSE increases.

13.6 SEQUENTIAL SELECTION OF VARIABLES

In many situations the independent variables are stipulated by theory, especially if there is some deterministic model of the form

$$y = b_0 + b_1 x_1 + b_2 x_2 + \cdots + b_k x_k$$

and the variability of the response results from some additive zero-mean input noise. In such a case, Y is random and a random-noise term E must be added to the model. In other situations, the variables substantially affecting the response might not be known, although there might be k variables that have been identified. Here, the problem is to select a minimal subset of the variables that accounts for observed regression. Since there are 2^k subsets of the k variables, the problem of finding an optimal collection of variables that effectively accounts for the regression can present a heavy computational burden. The dilemma is compounded by the fact that two distinct subsets of the independent variables can yield models of essentially the same adequacy. We will now present a sequential algorithm for model design based on the selection of models yielding the highest values of R^2. The procedure employs the F test to determine model adequacy at each stage.

Given a set of potential independent variables, **forward selection** proceeds recursively by checking all variables not yet incorporated into the model, incorporating that variable whose inclusion yields the greatest value of R^2 (or, equivalently, the highest value of SSR), and then checking to see if the new model shows a significant degree of regression over the prior model. Letting $x_1, x_2, \ldots, x_k$ be the potential independent variables, the algorithm can best be expressed in a number of steps.

Forward Selection Algorithm

Step 1(a). Fit the k simple regression models

$$\mu_{Y|x_1} = b_0 + b_1 x_1$$

$$\mu_{Y|x_2} = b_0 + b_2 x_2$$

$$\vdots$$

$$\mu_{Y|x_k} = b_0 + b_k x_k$$

to the data, in each case finding $\overset{*}{b}_0$ and $\overset{*}{b}_j$. Select as the first candidate variable in the model that x_j possessing the largest SSR value (largest R^2 value) and call this variable $x_{j(1)}$.

Step 1(b). By using the F test statistic, perform the hypothesis test

H_0: The reduced model is adequate.

H_1: The reduced model is inadequate.

where the reduced model is

$$\mu_Y = b_0$$

and the full model is

$$\mu_{Y|x_{j(1)}} = b_0 + b_{j(1)} x_{j(1)}$$

If the null hypothesis is accepted, then it is concluded that there does not exist significant regression relative to any of the proposed variables, and the algorithm is terminated; if the null hypothesis is rejected, then $x_{j(1)}$ is incorporated into the model and the algorithm proceeds to the next step.

Step 2(a). Fit the $k - 1$ two-variable regression models of the form

$$\mu_{Y|x_{j(1)}, x_j} = b_0 + b_{j(1)} x_{j(1)} + b_j x_j$$

to the data, where $x_j \neq x_{j(1)}$. Select as the second candidate variable in the model that x_j for which the two-variable model possesses the highest value of SSR and call this variable $x_{j(2)}$.

Step 2(b). By using the F test statistic, perform the hypothesis test

H_0: The reduced model is adequate.

H_1: The reduced model is inadequate.

where the reduced model is the model carried forward at the completion of Step 1(b) and the full model is

$$\mu_{Y|x_{j(1)}, x_{j(2)}} = b_0 + b_{j(1)}x_{j(1)} + b_{j(2)}x_{j(2)}$$

If the null hypothesis is accepted, then the algorithm is terminated, with the final regression model being the one carried forward from Step 1(b); if the null hypothesis is rejected, then $x_{j(2)}$ is incorporated into the model, and the algorithm proceeds to the next step.

Recursion [part (a)]. Assuming that m variables have been incorporated into the model, at the completion of the mth step, the model is of the form

$$\mu_{Y|x_{j(1)}, \ldots, x_{j(m)}} = b_0 + b_{j(1)}x_{j(1)} + b_{j(2)}x_{j(2)} + \cdots + b_{j(m)}x_{j(m)}$$

If $m = k$ (so that all potential variables have been included in the model), then the algorithm terminates; otherwise, proceed by fitting the $k - m$ $(m + 1)$-variable regression models of the form

$$\mu_{Y|x_{j(1)}, \ldots, x_{j(m)}, x_j} = b_0 + b_{j(1)}x_{j(1)} + \cdots + b_{j(m)}x_{j(m)} + b_j x_j$$

to the data, where, for $i = 1, 2, \ldots, m$, $x_j \neq x_{j(i)}$. Select as the $(m + 1)$st candidate variable in the model that x_j for which SSR is the greatest and call this variable $x_{j(m+1)}$.

Recursion [part (b)]. By using the F test statistic, test the adequacy of the model carried forward from the mth step against the model incorporating $x_{j(m+1)}$. If the mth-step model is adequate, then $x_{j(m+1)}$ is not incorporated into the model and the algorithm terminates, with the final model being the one carried forth from the mth step; otherwise, the model incorporating $x_{j(m+1)}$ is carried forth, and the algorithm proceeds to the next step.

Although the forward selection algorithm is computationally intensive, it is not difficult to write a program to implement it on a computer, and it (or a variant of it) is available in commercial packages. Rather than produce a numerical example, we will provide a walk-through of the algorithm to illustrate the various decision mechanisms. Subsequently, we will analyze the procedure more carefully to identify its shortcomings.

EXAMPLE 13.12

With $\alpha = 0.05$, we will construct a linear regression model by means of forward selection from among the four potential independent variables x_1, x_2, x_3, x_4. Having obtained 10 data points, we proceed according to the steps outlined in the algorithm:

Step 1(a). For $j = 1$ to 4, compute the regression sum of squares for the simple linear model employing the jth independent variable, namely,

$$SSR(j) = \frac{S_{x_jy}^2}{S_{x_jx_j}}$$

Select the index j for which $SSR(j)$ is maximum. For the sake of the example, we will assume that $j = 2$.

Step 1(b). Compute the error sum of squares $SSE(2)$ for the model involving x_2, namely,

$$SSE(2) = S_{yy} - SSR(2)$$

Let

$$F = \frac{SSR(2)}{\dfrac{SSE(2)}{n - k - 1}} = \frac{SSR(2)}{\dfrac{SSE(2)}{8}}$$

If

$$F \geq f_{0.05,1,8} = 5.32$$

then proceed to step 2(a); otherwise, terminate the algorithm and conclude there is no significant regression between Y and any of the variables x_1, x_2, x_3, and x_4. For the sake of the example, we will assume the value of F is significant and will therefore proceed to step 2(a), having already let $x_{j(1)} = x_2$.

Step 2(a). Compute the regression sums of squares corresponding to the multiple linear models for the variable x_2 in conjunction with each of the remaining variables. Obtained in the process will be the regression sums of squares, $SSR(2, 1)$, $SSR(2, 3)$, and $SSR(2, 4)$. Select the index j for which $SSR(2, j)$ is maximum and let $x_{j(2)} = x_j$. For the sake of the example, suppose $j = 4$.

Step 2(b). Compute

$$SSE(2, 4) = S_{yy} - SSR(2, 4)$$

and let

$$F = \frac{SSR(2, 4) - SSR(2)}{\dfrac{SSE(2, 4)}{7}}$$

If

$$F \geq f_{0.05,1,7} = 5.59$$

then proceed to step 3(a); otherwise, terminate the algorithm with the sample regression line

$$\hat{\mu}_{Y|x_2}^{*} = \overset{*}{b}_0 + \overset{*}{b}_2 x_2$$

For the sake of the example, we will assume the value of F to be significant and proceed to step 3(a).

Step 3(a). Compute two regression sums of squares for the multiple linear model with variables x_2 and x_4 in conjunction with each of the remaining variables, the sums being $SSR(2, 4, 1)$ and $SSR(2, 4, 3)$. Let j be the index for which $SSR(2, 4, j)$ is maximum. For the sake of the example, suppose $j = 3$.

Step 3(b). Compute

$$SSE(2, 3, 4) = S_{yy} - SSR(2, 3, 4)$$

and let

$$F = \frac{SSR(2, 3, 4) - SSR(2, 4)}{\dfrac{SSE(2, 3, 4)}{6}}$$

If

$$F \geq f_{0.05, 1, 6} = 5.99$$

then proceed to step 4(a); otherwise, terminate the algorithm with the sample regression plane

$$\hat{\mu}^*_{Y|x_2, x_4} = \overset{*}{b_0} + \overset{*}{b_2}x_2 + \overset{*}{b_4}x_4$$

For the sake of the example we will assume the value of F is not significant, and the algorithm terminates.

It should be obvious that the forward selection algorithm is not for human hands. In writing a general program, k is a variable, and the procedure of Example 13.12 must be changed to include a loop and a counter to keep track of the number of variables incorporated into the model. The F test statistic determines whether or not the routine is reentered or whether the algorithm terminates.

A flaw in the forward selection algorithm is that once a variable has been incorporated into the model, it cannot be removed at a later stage of the processing. While the selection procedure checks to determine whether or not the addition of a variable significantly increases SSR (R^2), it does not examine the possibility that the inclusion of later variables might render the inclusion of an earlier selected variable unnecessary. Specifically, owing to interrelationships among the independent variables, the incorporation of a particular variable might result in the greatest SSR at a given step, but it might be removed without significantly affecting the regression after the inclusion of further variables. A modification of the forward selection algorithm that checks for the continued significance of previously incorporated variables is known as **stepwise forward selection.** In essence, when a new variable is included in the model, an F test is performed on each reduced model obtained by dropping one of the previously incorporated variables to see if any can be dropped from the model without significantly affecting R^2.

EXAMPLE 13.13

To illustrate stepwise regression, we will modify the walk-through of Example 13.12, beginning at the conclusion of step 2(b), where x_4 has just been included in the model. In the stepwise procedure there must now be a step 2(c), in which the continued importance of x_2 in the model is checked. In effect, x_4 has been incorporated into the model because the null hypothesis in the test

H_0: The single variable x_2 is adequate.

H_1: Both x_2 and x_4 are required.

has been rejected in step 2(b). In the stepwise method, the data are employed to carry out the hypothesis test

H_0: The single variable x_4 is adequate.

H_1: Both x_2 and x_4 are required.

In the present case, the appropriate test statistic is

$$F = \frac{SSR(2, 4) - SSR(4)}{\dfrac{SSE(2, 4)}{7}}$$

If $F \geq f_{0.05,1,7} = 5.59$, then both variables are kept in the model; otherwise, x_2 is dropped and the next step is entered with only x_4 in the model. For the sake of the example, suppose both variables are kept.

At this point in Example 13.12, it was assumed that neither the inclusion of x_1 or x_3 significantly increased the value of SSR, so that the algorithm terminated with a two-variable regression in the variables x_2 and x_4. For the sake of the present example, let us assume we have found (as in Example 13.12) that $SSR(2, 4, 3) > SSR(2, 4, 1)$, but that, as measured by the appropriate F statistic, $SSR(2, 4, 3)$ is significantly greater than $SSR(2, 4)$. Thus, at the conclusion of step 3(b) the model includes the variables x_2, x_3, and x_4. Step 3(c), required in the stepwise algorithm, is to determine whether either x_2 or x_4 can be dropped from the model now that x_3 has been included. To check the importance of x_4 in the expanded model, we employ the test statistic

$$F = \frac{SSR(2, 3, 4) - SSR(2, 3)}{\dfrac{SSE(2, 3, 4)}{6}}$$

Let us assume that the value of F is significant, which means it is greater than $f_{0.05,1,6}$. Then the inclusion of x_3 in the model has not rendered x_4 superfluous. Next, the continued importance of x_2 in the model must be checked, the appropriate test statistic being

$$F = \frac{SSR(2, 3, 4) - SSR(3, 4)}{\dfrac{SSE(2, 3, 4)}{6}}$$

If F is significant, then the full model consisting of three independent variables is passed on to step 4(a); otherwise, x_2 is dropped from the model. Let us assume the latter, so that step 4(a) is entered with a two-variable model including x_3 and x_4.

Since it has already been seen that x_2 does not yield an appreciable increase in SSR when x_3 and x_4 are in the model, it need not be checked in step 4(a). Hence, the only possible inclusion in the fourth step is x_1, so that the algorithm can pass at once to step 4(b). Assuming

$$F = \frac{SSR(1, 3, 4) - SSR(3, 4)}{\dfrac{SSE(1, 3, 4)}{6}}$$

is not significant, then the algorithm terminates with the sample regression plane

$$\overset{*}{\mu}_{Y|x_3,x_4} = \overset{*}{b}_0 + \overset{*}{b}_3 x_3 + \overset{*}{b}_4 x_4$$

Even in stepwise regression, the final fitted plane might not be the one possessing maximum SSR (or, equivalently, maximum R^2) among those models possessing the same number of independent variables selected from the original collection. The reason such an anomaly can occur is that the progression of the algorithm is such that not all subcollections of the potential variables are checked. Of course, if the number of potential variables is small, as in Examples 13.12 and 13.13, then values of R^2 can be computed for all possible combinations of the variables and the desired model selected by simply analyzing models possessing maximum R^2 (maximum SSR). However, execution time grows rapidly as the number of variables under consideration increases, and therefore more efficient selection algorithms such as forward selection and stepwise forward selection are needed.

Other sequential methods are available besides forward selection. For instance, in **backward elimination** the algorithm starts with the full model, incorporating all potential independent variables, and proceeds to check sequentially for possible reductions in the model. Since the various selection procedures require computer implementation, it is best to read the literature available with whatever software package one possesses to determine the available sequential algorithms.

EXERCISES

SECTION 13.1

13.1. For the case of two independent variables x_1 and x_2, suppose the regression plane is given by

$$\mu_{Y|x_1,x_2} = 3 + 2x_1 + 5x_2$$

Find ϵ for each of the observations $(-1, 2, 12)$, $(3, 2, 17)$, and $(0, 3, 20)$.

In Exercises 13.2 through 13.7, find the sample regression plane using the least-squares estimates.

13.2. The time to complete an assembly task is to be modeled as a linear function of the complexity of the task and the time an assembler has been continuously working. Complexity (x_1) is measured on a scale from 0 to 100, work time (x_2) is in hours prior to beginning the timed assembly, and completion time (y) is in minutes. Use the following data.

x_{1i}	x_{2i}	y_i
90	8	5.6
75	4	3.8
60	0	2.0
60	8	2.4
90	4	5.2
75	0	2.9

13.3. Among the many variables that affect the performance of an automobile are the octane rating (x_1) of the fuel and the spark advance (x_2) of the ignition timing. The latter, given in degrees, measures the extent to which the fuel is ignited before a piston reaches the top of its stroke within the cylinder. The following data concern the number of seconds (y) its takes a specific automobile to accelerate from a standing position to a speed of 60 mph.

x_{1i}	x_{2i}	y_i
85	0	12.3
90	0	12.1
95	0	11.9
85	2	11.9
90	2	11.3
95	2	11.0
85	4	11.2
90	4	10.6
95	4	10.2

13.4. In Example 13.11 we will consider a multiple regression concerning the lifetime of a metal-cutting tool. The following data are obtained by measuring the lifetime (y) of a particular cutting tool at various speeds (x_1) and feed rates (x_2), the tests all being run on the same type of metal. The variables x_1, x_2, and y are in meters per minute, millimeters per revolution, and minutes, respectively.

x_{1i}	x_{2i}	y_i
90	0.2	40
90	0.3	38
90	0.4	35
90	0.4	30
120	0.2	28
120	0.3	25
120	0.4	18
120	0.4	20

13.5. The number of gallons of fuel oil used in a home depends on several factors. The following fuel-oil data are based on a sample in which the independent variables are thermostat setting (x_1), square footage (x_2) of the home, and R-value (x_3) of the insulation.

x_{1i}	x_{2i}	x_{3i}	y_i
66	2000	11	1173
70	2200	11	1560
68	3100	11	1648
70	3000	11	1845
66	1900	19	942
69	2100	19	1206
65	3300	19	1160
69	2700	19	1320

13.6. Prior to employment, a prospective employee is given two tests, one based upon grammatical skills and another based upon mathematical skills. It is postulated that there is a linear relation between scores $(x_1$ and $x_2)$ on the two tests and the score (y) on a performance test given after six months of employment. The following data are collected for seven randomly selected employees.

x_{1i}	x_{2i}	y_i
73	43	63
85	56	70
62	37	53
74	61	65
72	72	70
40	46	50
52	60	50

13.7. The data of Table 13.1 were employed in Example 13.1 to find the linear regression of yield on reaction time and catalyst percentage. Reconsider the problem using the following data, in which a third variable, reaction temperature (x_3), is measured.

x_{1i}	x_{2i}	x_{3i}	y_i
8	0.4	180	42.2
8	0.3	200	35.0
7	0.4	200	42.0
7	0.3	180	33.7
6	0.4	180	40.0
6	0.3	200	30.9

13.8. Derive the normal equations in the multiple linear regression model for the case $k = 2$.

13.9. Show that for $k = 1$, Theorem 13.1 reduces to Theorem 12.1.

13.10. Prior to Theorem 13.1, it was stated that the normal equations for the least-squares estimators can be expressed in matrix format as $(\mathbf{X'X})\mathbf{b} = \mathbf{X'y}$. Show this to be the case.

SECTION 13.2

13.11. Suppose V_1, V_2, and V_3 are jointly distributed random variables possessing a normal distribution with mean μ, an exponential distribution with parameter b, and a gamma distribution with parameters α and β, respectively. If $\mathbf{V}$ is a random vector with components V_1, V_2, and V_3, then what is $E[\mathbf{V}]$?

13.12. Suppose V_1 and V_2 are jointly distributed normal random variables with parameters μ_1, μ_2, σ_1, σ_2, and ρ. Moreover, suppose V_3 possesses a k-Erlang distribution with parameter β and V_3 is independent of both V_1 and V_2. If $\mathbf{V}$ is a random vector with components V_1, V_2, and V_3, then what are $E[\mathbf{V}]$ and Cov $[\mathbf{V}]$?

13.13. Show that if $\mathbf{U}$ and $\mathbf{V}$ are random vectors possessing m components and $\mathbf{A}$ is an $r \times m$ matrix of constants, then
a) $E[\mathbf{AV}] = \mathbf{A}E[\mathbf{V}]$
b) $E[\mathbf{U} + \mathbf{V}] = E[\mathbf{U}] + E[\mathbf{V}]$

In Exercises 13.14 through 13.19, find SSR, S_{yy}, SSE, and MSE in the cited exercise.

13.14. Exercise 13.2
13.15. Exercise 13.3
13.16. Exercise 13.4
13.17. Exercise 13.5
13.18. Exercise 13.6
13.19. Exercise 13.7

SECTION 13.3

In Exercises 13.20 through 13.23, employ Theorem 13.4 to find a $(1 - \alpha)100\%$ confidence interval for b_j in the cited exercise.

13.20. Exercise 13.3, $\alpha = 0.05$, $j = 2$
13.21. Exercise 13.4, $\alpha = 0.01$, $j = 1$
13.22. Exercise 13.6, $\alpha = 0.05$, $j = 2$
13.23. Exercise 13.7, $\alpha = 0.05$, $j = 3$

In Exercises 13.24 through 13.27, employ T_j to perform the stated hypothesis test for the cited exercise at the given level of significance.

13.24. Exercise 13.3, H_0: $b_2 \geq 0$, H_1: $b_2 < 0$, $\alpha = 0.05$
13.25. Exercise 13.5, H_0: $b_1 \leq 50$, H_1: $b_1 > 50$, $\alpha = 0.05$
13.26. Exercise 13.6, H_0: $b_2 = 0$, H_1: $b_2 \neq 0$, $\alpha = 0.05$
13.27. Exercise 13.6, H_0: $b_1 = 0$, H_1: $b_1 \neq 0$, $\alpha = 0.01$

In Exercises 13.28 through 13.31, employ Theorem 13.5 to find a $(1 - \alpha)100\%$ confidence interval for $\mu_{Y|x_{10}, x_{20}, \ldots, x_{k0}}$ in the cited exercise.

13.28. Exercise 13.2, $\mu_{Y|75,4}$, $\alpha = 0.01$
13.29. Exercise 13.3, $\mu_{Y|88,3}$, $\alpha = 0.05$
13.30. Exercise 13.4, $\mu_{Y|100,0.4}$, $\alpha = 0.05$
13.31. Exercise 13.5, $\mu_{Y|68,3000,19}$, $\alpha = 0.05$

In Exercises 13.32 through 13.35, find a $(1 - \alpha)100\%$ prediction interval at the given point for the cited exercise.

13.32. Exercise 13.2, (75,4), $\alpha = 0.01$
13.33. Exercise 13.4, (100, 0.4), $\alpha = 0.05$
13.34. Exercise 13.5, (68, 3000, 19), $\alpha = 0.05$
13.35. Exercise 13.7, (6.5, 0.35, 180), $\alpha = 0.05$

SECTION 13.4

13.36. The following fuel-consumption data have been compiled by observing randomly selected automobiles of a particular make traveling at various speeds. Fuel consumption (y_i) is in gallons per vehicle-mile and speed (x_i) is in miles per hour. Use a quadratic regression model to express fuel consumption in terms of speed. Plot the data on a scatter diagram and draw the fitted quadratic on the diagram.

x_i:	3	5	10	15	20
y_i:	0.28	0.20	0.10	0.07	0.07

13.37. Apply a quadratic least-squares fit to the cement-shrinkage data of Exercise 12.7. Plot the data on a scatter diagram and draw the estimated quadratic on the diagram. Find SSE for both the linear fit of Exercise 12.7 and the quadratic fit of the current exercise. Compare.

13.38. The cement-shrinkage data of Exercise 12.7 have been obtained from cement for which the aggregate is 80% by volume. The following data derive from cement that is only 70% aggregate by volume.

x_i: 0.40 0.45 0.50 0.55 0.60 0.65 0.70 0.75
y_i: 655 710 800 840 870 700 710 715

Plot the scatter diagram, find the best least-squares quadratic, and plot the latter on the scatter diagram.

13.39. Apply a quadratic regression to the pressure-strength data of Exercise 12.5. Draw the regression quadratic on the scatter diagram, compute SSE for both the linear regression of Exercise 12.5 and the present quadratic regression, and compare.

13.40. A linear fit was applied to the endurance-limit data of Exercise 12.8. Using only the endurance limits in the first column of the y_i data, apply a quadratic regression of y_i on x_i for the data. Redo the simple linear regression using only the data of the first y_i column. Draw both regression curves on the scatter diagram (consisting of first-column data), compute SSE for both regressions, and compare.

13.41. Referring to Exercise 13.38, apply Theorem 13.4 to find a 95% confidence interval for b_2.

13.42. Referring to Exercise 13.38, use T_2 to test the null hypothesis H_0: $b_2 = 0$ against the alternative H_1: $b_2 \neq 0$ at the $\alpha = 0.05$ level. Interpret the result of the test.

13.43. Referring to the quadratic model of Exercise 13.37, repeat the directions of Exercise 13.41. Again interpret.

13.44. Referring to the quadratic model of Exercise 13.40, repeat the directions of Exercise 13.41 with $\alpha = 0.01$.

SECTION 13.5

In Exercises 13.45 through 13.48, use analysis of variance to test for significance of regression and the adequacy of all possible reduced models at significance level α in the cited exercise. For instance, if the original model contains two independent variables, then three models must be tested (the original and two reduced). If the original model contains three independent variables, then seven models must be tested (the original and six reduced). When testing the full model, give the ANOVA table.

13.45. Exercise 13.2, $\alpha = 0.05$
13.46. Exercise 13.4, $\alpha = 0.05$
13.47. Exercise 13.6, $\alpha = 0.05$
13.48. Exercise 13.7, $\alpha = 0.01$

13.49. Find the coefficient of multiple determination for the data of Exercise 13.5.

13.50. Reconsider Exercise 13.46 under the assumption that the following coded data were obtained in Exercise 13.4.

x_{1i}	x_{2i}	y_i
-1	-1	40
-1	1	35
0	-1	30
0	1	30
1	-1	24
1	1	20

13.51. Reconsider Exercise 13.48 under the assumption that the following coded data were obtained in Exercise 13.7.

x_{1i}	x_{2i}	x_{3i}	y_i
-1	-1	-1	30.0
-1	-1	1	31.6
-1	1	-1	40.0
-1	1	1	40.2
1	-1	-1	33.4
1	-1	1	33.6
1	1	-1	42.3
1	1	1	42.0

SECTION 13.6

13.52. Apply forward selection to the variables and data of Exercise 13.6. Let $\alpha = 0.05$.

13.53. Apply forward selection to the variables and data of Exercise 13.7. Let $\alpha = 0.05$.

13.54. Apply stepwise forward selection to the variables and data of Exercise 13.7. Let $\alpha = 0.05$.

SUPPLEMENTAL EXERCISES FOR COMPUTER SOFTWARE

13.C1. Introduce a third regressor variable, years of experience (x_3), into the multiple linear model of Exercise 13.2. Perform the regression using the following newly obtained data.

x_{1i}	x_{2i}	x_{3i}	y_i
90	8	1.4	4.9
90	4	1.2	4.0
90	0	3.2	2.0
75	4	0.4	4.2
75	8	1.2	4.3
75	0	1.2	2.5
75	4	5.0	4.0
60	8	4.2	2.4
60	4	1.2	2.0
60	0	2.2	2.0
60	8	0.4	2.2

13.C2. Suppose two other regressor variables, tire pressure (x_3) and load weight (x_4), both measured in pounds, are introduced into the model of Exercise 13.3. Perform a multiple linear regression using the following new data.

x_{1i}	x_{2i}	x_{3i}	x_{4i}	y_i
85	0	28	200	12.4
90	0	25	200	12.6
87	0	28	400	12.1
90	0	25	400	13.3
88	2	28	200	11.2
93	2	25	400	11.5
85	2	28	200	12.0
90	2	25	400	12.1
85	4	28	200	11.4
90	4	25	200	10.7
95	4	28	400	10.3
88	4	25	400	11.0

13.C3. Suppose the fuel-oil experiment of Exercise 13.5 is altered so that a fourth regressor variable, average external temperature (x_4) during the months of the test, is included, x_4 being measured in degrees Fahrenheit. Perform a multiple linear regression using the following new data.

x_{1i}	x_{2i}	x_{3i}	x_{4i}	y_i
66	2000	11	34	1120
70	2400	11	28	1704
68	2400	11	36	1110
66	3000	11	33	1680
70	3200	11	28	1906
68	2500	15	30	1144
70	2200	15	38	1239
66	2900	19	30	1338
66	2100	19	34	1125
66	1800	19	35	948
68	2400	19	30	1022

13.C4. Wheel bearings are subjected to ultrarigorous operating conditions, and the following data result from measuring the time to failure (y_i), in hours, as a function of the number of revolutions per minute (x_i). Find the best-fit cubic.

x_i: 5000 5500 6000 6500 7000 7500 8000 8500
y_i: 98.7 92.8 90.1 83.2 70.8 55.4 48.4 43.3

13.C5. Worker performance (y_i) on a complex machine is rated on a scale from 0 to 100. The following data relate to performance rating as a function of the number of hours (x_i) of training. Find the best-fit cubic.

x_i: 2 4 6 8 10 12 14 16 18
y_i: 7 10 22 30 49 72 84 88 89

In Exercises 13.C6 through 13.C8, employ analysis of variance to test the full model and all reduced models of the cited exercise for significance of regression at level $\alpha = 0.05$.

13.C6. Exercise 13.C1.
13.C7. Exercise 13.C2.
13.C8. Exercise 13.C3.

In Exercises 13.C9 through 13.C11, apply a stepwise procedure (using a routine available in your software package) to arrive at a multiple linear regression model in the cited exercise.

13.C9. Exercise 13.C1.
13.C10. Exercise 13.C2.
13.C11. Exercise 13.C3.

14

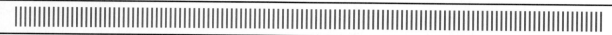

EXPERIMENTAL DESIGN

The type of information that can be extracted from experimental data depends on the mathematical model to which the data are assumed to conform. By designing an experiment along certain lines, an investigator can arrange the observations in a manner conforming to a model amenable to the desired type of analysis. The net result of such a design approach can often be the extraction of more relevant scientific information at less cost.

In conforming to a mathematical model, a measurement process must conform to the constraints imposed upon the model. These involve distributional assumptions such as normality, randomness, and (as in analysis of variance and regression) the satisfaction of equational constraints. Proper experimental design can insure reasonable conformity of the observational process with a suitable model.

The present chapter commences with a discussion of basic principles. It then moves to blocking in analysis of variance and then to two- and three-factor analysis-of-variance models. The final two sections concern special designs for multifactor experiments in which each factor possesses only two levels.

14.1 BASIC PRINCIPLES OF EXPERIMENTAL DESIGN

A prototypical example of design occurs in testing the difference of two means. If the experiment is designed so as to pair the observations, extraneous variation is diminished and the experiment (possibly) becomes more sensitive to the detection of a difference. More generally, if there is extensive within-treatment variation when performing an analysis of variance, organizing the treatment observations in **blocks** can diminish the masking effect of the variation and enhance the detection of significant differences. As will be seen in Section 14.2, blocking, like paired t testing, has the negative effect of reducing the number of degrees of freedom associated with the test statistic. When employed in appropriate circumstances, however, it does provide a beneficial design.

Two key criteria to be considered in the design of an experiment are **replication** and **randomization.** As has been seen throughout the text, repeated observation increases the precision of test statistics. Indeed, our heavy use of OC curves has been motivated by the desire to find the sample size necessary to reduce type II error to an acceptable level. In general, an experiment might require the observation of several different variables. For instance, in the fixed-effects one-way-classification analysis of variance, responses to several treatments are observed. In such cases, it is usually the replication (repeated observation) of the entire set of response variables that is useful in increasing precision, each set of responses being known as a **replicate.**

In order to mitigate the unavoidable response variation due to the nonhomogeneity of experimental units, experimental design includes judicious use of randomization. From a physical perspective, it is the application of random methods of observation that results in a measurement process satisfying the requirements of a random sample. In the treatment-response genre, complete randomization of the experimental units to which the treatments are to be applied tends to average out response differences due to variations in the experimental units.

While randomization is generally required in some form, the manner in which it is utilized can substantially affect the sensitivity of the experiment. For instance, blocking imposes a restraint on randomization that can increase experimental sensitivity to differences the experimenter wishes to detect. It is the experimenter who determines, a priori, the experimental objectives. The manner in which randomization is employed should facilitate the attainment of the objectives.

Generally, the design of an experiment should mitigate the effects of extraneous variables. The desensitizing effects of extraneous variables on one's ability to make inferences concerning variables of interest can be limited by both laboratory practice and statistical design, the latter approach being of interest herein. One of the main benefits of the analysis-of-variance technique is that it provides a framework in which to design experiments that are sensitive to desired objectives. By partitioning the variation into components based upon a particular experimental design, the investigator can often control the detrimental effect of residual variation introduced by extraneous factors. Such partitioning has been fruitfully applied in Chapters 11, 12, and 13. In the present chapter we extend the methodology with the specific intent of devising appropriate designs.

14.2 RANDOMIZED COMPLETE BLOCK DESIGN

In one-way-classification analysis of variance, measurements are taken on some number N of randomly selected experimental units. The sampling methodology is reflected in the modeling assumption that the random samples corresponding to the r treatment random variables are independent. It is due to the manner of randomization that the design is said to be **completely randomized.** Although prototypical, such a design often lacks sensitivity due to within-treatment variability. A blocking approach can sometimes be employed to increase sensitivity.

14.2.1 Fixed-Effects Model for Randomized Complete Block Design

The blocking approach is to design the experiment so that each of the r treatments is applied once under each of n settings. There are then n sets of data, each being characterized by a block of data as in Figure 14.1. Experimentally, the r treatments

Block

	1	2	$\cdots$		n
	x_{11}	x_{12}	$\cdots$		x_{1n}
	x_{21}	x_{22}	$\cdots$		x_{2n}
	$\vdots$	$\vdots$			$\vdots$
	x_{r1}	x_{r2}	$\cdots$		x_{rn}

FIGURE 14.1
Complete block design

should be randomized within each block. Probabilistically, each datum x_{ij} of Figure 14.1 is viewed as a value of a random variable X_{ij}, so that there are $N = rn$ random variables constituting the experiment:

$$X_{11}, \; X_{12}, \; \ldots, \; X_{1n}$$
$$X_{21}, \; X_{22}, \; \ldots, \; X_{2n}$$
$$\vdots \qquad \vdots \qquad \qquad \vdots$$
$$X_{r1}, \; X_{r2}, \; \ldots, \; X_{rn}$$

The overall design is called a **randomized complete block design.** It is said to be **complete** because each block contains measurements from all r treatments. Should the blocks not hold r measurements, then the design is said to be **incomplete.** If the blocks are chosen carefully, so as to reduce residual variation not arising from differing means among the treatment random variables, then better results can often be obtained. As in the case of one-way classification, if the treatments are not chosen randomly, then the design is said to be a **fixed-effects** design. Fixed effects will be studied in the present subsection; random effects will be studied subsequently.

The analysis of a randomized complete block design depends upon the introduction of certain random variables. These are

$$T_{i.} = \sum_{j=1}^{n} X_{ij}$$

the total sum of the responses corresponding to the ith treatment,

$$T_{.j} = \sum_{i=1}^{r} X_{ij}$$

the total sum of the responses corresponding to the jth block,

$$T_{..} = \sum_{i=1}^{r} \sum_{j=1}^{n} X_{ij}$$

the sum total of all responses,

$$\bar{X}_{i.} = \frac{T_{i.}}{r}$$

the average of the responses for the ith treatment,

$$\bar{X}_{.j} = \frac{T_{.j}}{n}$$

TABLE 14.1 DATA FORMAT FOR RANDOMIZED COMPLETE BLOCK DESIGN

Treatment	Block 1	Block 2	...	Block n	Treatment Total	Treatment Mean
1	x_{11}	x_{12}	...	x_{1n}	$T_{1.}$	$\bar{x}_{1.}$
2	x_{21}	x_{22}	...	x_{2n}	$T_{2.}$	$\bar{x}_{2.}$
⋮	⋮	⋮		⋮	⋮	⋮
r	x_{r1}	x_{r2}	...	x_{rn}	$T_{r.}$	$\bar{x}_{r.}$
Block Total	$T_{.1}$	$T_{.2}$	...	$T_{.n}$	$T_{..}$	
Block Mean	$\bar{x}_{.1}$	$\bar{x}_{.2}$	...	$\bar{x}_{.n}$		$\bar{x}_{..}$

the average of the responses for the jth block, and

$$\bar{X}_{..} = \frac{T_{..}}{N}$$

the overall average of the responses. The data for a randomized complete block design can be arranged as in Table 14.1, where a capital T represents an empirical total.

EXAMPLE 14.1

Very often, a numerical procedure can be accomplished by any one of a number of algorithms. Consider comparing the efficiency of three algorithms, each encoded in a high-level language, say Pascal. Each is run on four different machines. Because run times are greatly affected by choice of hardware, the experiment is blocked by using four different computers. The data (in seconds) are tabulated in Table 14.2. From the data, $\bar{x}_{3.} < \bar{x}_{1.} < \bar{x}_{2.}$. Whether or not we can draw the conclusion that algorithm 3 is superior will depend upon an analysis of variance.

The statistical model corresponding to the randomized complete block design is based on the model equation

$$X_{ij} = \mu + \alpha_i + \beta_j + E_{ij}$$

where, for $i = 1, 2, \ldots, r$ and $j = 1, 2, \ldots, n$, X_{ij} is the ijth response variable. The model parameters are the **overall (grand) mean** μ, the **effect** α_i of the ith treatment, and the **effect** β_j of the jth block. E_{ij} is the random deviation of X_{ij} from

TABLE 14.2 BLOCKED DATA FOR ALGORITHM EFFICIENCY

Algorithm	Computer 1	Computer 2	Computer 3	Computer 4	Total	Mean
1	6.42	10.51	4.00	6.20	27.13	6.78
2	6.85	11.45	4.12	6.72	29.14	7.29
3	5.60	9.50	3.94	5.66	24.70	6.18
Total	18.87	31.46	12.06	18.58	80.97	
Mean	6.29	10.49	4.02	6.19		6.75

the constant $\mu + \alpha_i + \beta_j$. As usual, the error terms E_{ij} are assumed to be independent normally distributed random variables possessing common mean 0 and common variance σ^2. Relative to the means of the random variables X_{ij}, the model equation takes the form

$$\mu_{ij} = \mu + \alpha_i + \beta_j$$

In addition to the model equation, two constraints are imposed upon the effects, namely, that the sums of the treatment effects and block effects are each zero. Note the analogy to the unblocked one-way-classification model. There, the sum of the treatment effects is assumed to be zero. The blocked model differs from the one-way-classification model in that the block effects must be taken into account. In sum, the following linear model is postulated for the fixed-effects randomized complete block design.

Model for Fixed-Effects Randomized Complete Block Design

Equation:	$X_{ij} = \mu + \alpha_i + \beta_j + E_{ij}$
Assumptions:	1. $E_{ij} \sim N(0, \sigma)$
	2. The E_{ij} comprise an independent collection.
	3. The sums of the treatment and block effects are each zero:

$$\sum_{i=1}^{r} \alpha_i = \sum_{j=1}^{n} \beta_j = 0$$

Although the model equation has been postulated in a form that clearly identifies the effects of interest, the equation can be alternatively expressed in terms of means. Let

$$\mu_{i.} = \frac{1}{n} \sum_{j=1}^{n} \mu_{ij}$$

be the average of the means of the ith-treatment random variables and

$$\mu_{.j} = \frac{1}{r} \sum_{i=1}^{r} \mu_{ij}$$

be the average of the means of the jth-block random variables. Under the constraints that the sums of the α_i and β_j must be zero,

$$\mu_{i.} = \frac{1}{n} \sum_{j=1}^{n} (\mu + \alpha_i + \beta_j)$$

$$= \frac{1}{n} \left(n\mu + n\alpha_i + \sum_{j=1}^{n} \beta_j \right)$$

$$= \mu + \alpha_i$$

Similarly,

$$\mu_{.j} = \frac{1}{r} \sum_{i=1}^{r} (\mu + \alpha_i + \beta_j) = \mu + \beta_j$$

Thus, $\alpha_i = \mu_{i.} - \mu$ measures the difference between the average of the ith-treatment means and the grand mean, and $\beta_j = \mu_{.j} - \mu$ measures the difference between the average of the jth-block means and the grand mean. In addition, using the model constraints it is straightforward to show that

$$\mu = \frac{1}{N} \sum_{i=1}^{r} \sum_{j=1}^{n} \mu_{ij}$$

Given a blocked design, there are two hypothesis tests to consider. The treatment null hypothesis and its alternative comprise the test

$$H_{0A}: \alpha_1 = \alpha_2 = \cdots = \alpha_r = 0$$

$$H_{1A}: \alpha_i \neq 0 \text{ for at least one } i$$

The block null hypothesis and its alternative comprise the test

$$H_{0B}: \beta_1 = \beta_2 = \cdots = \beta_n = 0$$

$$H_{1B}: \beta_j \neq 0 \text{ for at least one } j$$

Because of the model constraints, the treatment and block null hypotheses can be equivalently rewritten as

$$H_{0A}: \mu_{1.} = \mu_{2.} = \cdots = \mu_{r.}$$

and

$$H_{0B}: \mu_{.1} = \mu_{.2} = \cdots = \mu_{.n}$$

respectively. Since the experiment has been designed to test the equality (or lack of equality) of treatment effects, it is the treatment null hypothesis that is of interest. In deriving a test statistic for the treatment effects, we will concomitantly develop a test statistic for the block effects. However, as will be pointed out shortly, block-effect testing is problematic, owing to the manner in which the experiment has been designed.

THEOREM 14.1

Under the randomized complete block design, unbiased estimators for μ, α_i, and β_j are given by

$$\hat{\mu} = \bar{X}_{..}$$

$$\hat{\alpha}_i = \bar{X}_{i.} - \bar{X}_{..}$$

$$\hat{\beta}_j = \bar{X}_{.j} - \bar{X}_{..}$$

Proof: The proof for $\hat{\alpha}_i$ will be provided; the others are left as exercises. The model equation in conjunction with the linearity of the expected-value operator gives

$$E[\bar{X}_{i.} - \bar{X}_{..}] = \frac{1}{n} \sum_{j=1}^{n} E[X_{ij}] - \frac{1}{N} \sum_{i=1}^{r} \sum_{j=1}^{n} E[X_{ij}]$$

$$= \frac{1}{n}\left[n\mu + n\alpha_i + \sum_{j=1}^{n} \beta_j \right] - \frac{1}{N}\left[N\mu + r\sum_{j=1}^{n} \beta_j + n\sum_{i=1}^{r} \alpha_i \right]$$

According to the model constraints, the summations are all equal to zero, and the expression reduces to α_i. ∎

Since $\mu_{i.} = \mu + \alpha_i$ and $\mu_{.j} = \mu + \beta_j$, it is immediate from Theorem 14.1 that $\bar{X}_{i.}$ and $\bar{X}_{.j}$ are unbiased estimators for $\mu_{i.}$ and $\mu_{.j}$, respectively.

EXAMPLE 14.2

Assuming the experiment of Example 14.1 satisfies the conditions of a randomized complete block design, Theorem 14.1 can be applied using the data of Table 14.2. Specifically, we have the following parameter estimates:

$$\overset{*}{\mu} = \bar{x}_{..} = 6.75$$

$$\overset{*}{\alpha}_1 = \bar{x}_{1.} - \bar{x}_{..} = 6.78 - 6.75 = 0.03$$

$$\overset{*}{\alpha}_2 = \bar{x}_{2.} - \bar{x}_{..} = 7.29 - 6.75 = 0.54$$

$$\overset{*}{\alpha}_3 = \bar{x}_{3.} - \bar{x}_{..} = 6.18 - 6.75 = -0.57$$

$$\overset{*}{\beta}_1 = \bar{x}_{.1} - \bar{x}_{..} = 6.29 - 6.75 = -0.46$$

$$\overset{*}{\beta}_2 = \bar{x}_{.2} - \bar{x}_{..} = 10.49 - 6.75 = 3.74$$

$$\overset{*}{\beta}_3 = \bar{x}_{.3} - \bar{x}_{..} = 4.02 - 6.75 = -2.73$$

$$\overset{*}{\beta}_4 = \bar{x}_{.4} - \bar{x}_{..} = 6.19 - 6.75 = -0.56$$

The mean μ_{ij} of the random variable X_{ij}, which represents the run time of algorithm i on computer j, can be expressed in terms of these estimates. For instance, the mean run time of algorithm 2 on computer 3 is estimated by

$$\overset{*}{\mu}_{23} = \overset{*}{\mu} + \overset{*}{\alpha}_2 + \overset{*}{\beta}_3 = 6.75 + 0.54 - 2.73 = 4.56$$

14.2.2 Analysis of Variance in the Fixed-Effects Randomized Complete Block Design

Employing the customary analysis-of-variance approach, we define the sums of squares that will lead to the appropriate F statistics. The total sum of squares measures the total variation in the data and is given by

$$SST = \sum_{i=1}^{r} \sum_{j=1}^{n} (X_{ij} - \bar{X}_{..})^2$$

The treatments sum of squares

$$SSA = n\sum_{i=1}^{r} (\bar{X}_{i.} - \bar{X}_{..})^2$$

measures variation arising from the different treatments. The blocks sum of squares

$$SSB = r \sum_{j=1}^{n} (\bar{X}_{.j} - \bar{X}_{..})^2$$

measures variation arising from the different blocks. Finally, the error sum of squares is given by

$$SSE = \sum_{i=1}^{r} \sum_{j=1}^{n} (X_{ij} - \bar{X}_{i.} - \bar{X}_{.j} + \bar{X}_{..})^2$$

Referring to Theorem 14.1, it can be seen that

$$SSA = n \sum_{i=1}^{r} \hat{\alpha}_i^2$$

$$SSB = r \sum_{j=1}^{n} \hat{\beta}_j^2$$

As stated by the next theorem, SST can be partitioned into a sum of the other sums of squares.

THEOREM 14.2

Under the randomized complete block design,

$$SST = SSA + SSB + SSE$$

Proof: Straightforward algebra yields

$$SST = \sum_{i=1}^{r} \sum_{j=1}^{n} (X_{ij} - \bar{X}_{..})^2$$

$$= \sum_{i=1}^{r} \sum_{j=1}^{n} [(\bar{X}_{i.} - \bar{X}_{..}) + (\bar{X}_{.j} - \bar{X}_{..}) + (X_{ij} - \bar{X}_{i.} - \bar{X}_{.j} + \bar{X}_{..})]^2$$

$$= n \sum_{i=1}^{r} (\bar{X}_{i.} - \bar{X}_{..})^2 + r \sum_{j=1}^{n} (\bar{X}_{.j} - \bar{X}_{..})^2$$

$$+ \sum_{i=1}^{r} \sum_{j=1}^{n} (X_{ij} - \bar{X}_{i.} - \bar{X}_{.j} + \bar{X}_{..})^2$$

$$+ 2 \sum_{i=1}^{r} \sum_{j=1}^{n} (\bar{X}_{i.} - \bar{X}_{..})(\bar{X}_{.j} - \bar{X}_{..})$$

$$+ 2 \sum_{i=1}^{r} \sum_{j=1}^{n} (\bar{X}_{i.} - \bar{X}_{..})(X_{ij} - \bar{X}_{i.} - \bar{X}_{.j} + \bar{X}_{..})$$

$$+ 2 \sum_{i=1}^{r} \sum_{j=1}^{n} (\bar{X}_{.j} - \bar{X}_{..})(X_{ij} - \bar{X}_{i.} - \bar{X}_{.j} + \bar{X}_{..})$$

The first three terms in the expansion are SSA, SSB, and SSE, respectively. The latter three cross-product terms are all zero. For instance,

$$\sum_{i=1}^{r} \sum_{j=1}^{n} (\bar{X}_{i.} - \bar{X}_{..})(X_{ij} - \bar{X}_{i.} - \bar{X}_{.j} + \bar{X}_{..})$$

$$= \sum_{i=1}^{r} (\bar{X}_{i.} - \bar{X}_{..}) \left[\sum_{j=1}^{n} (X_{ij} - \bar{X}_{i.}) - \sum_{j=1}^{n} (\bar{X}_{.j} - \bar{X}_{..}) \right] = 0 \qquad \blacksquare$$

To construct the necessary F statistics, the distributional properties of the preceding sums of squares need to be known. Under H_{0A}, it can be shown that SST/σ^2, SSE/σ^2, and SSA/σ^2 possess chi-square distributions with $N - 1$, $(r - 1)(n - 1)$, and $r - 1$ degrees of freedom, respectively, and that SSA/σ^2 is independent of SSE/σ^2. Under H_{0B}, SST/σ^2, SSE/σ^2, and SSB/σ^2 possess chi-square distributions with $N - 1$, $(r - 1)(n - 1)$, and $n - 1$ degrees of freedom, respectively, and SSB/σ^2 is independent of SSE/σ^2. Based upon these distributional properties, we define the mean squares corresponding to the sums of squares forming the partition of SST by

$$MSA = \frac{SSA}{r - 1}$$

$$MSB = \frac{SSB}{n - 1}$$

$$MSE = \frac{SSE}{(r - 1)(n - 1)}$$

The next theorem, stated without proof, gives the expected values of the mean squares MSA, MSB, and MSE.

THEOREM 14.3

Under the fixed-effects randomized complete block design,

$$E[MSA] = \sigma^2 + \frac{n}{r - 1} \sum_{i=1}^{r} \alpha_i^2$$

$$E[MSB] = \sigma^2 + \frac{r}{n - 1} \sum_{j=1}^{n} \beta_j^2$$

$$E[MSE] = \sigma^2$$

On account of Theorem 14.3, MSA is an unbiased estimator of σ^2 under H_{0A}. Moreover, based upon the distributional properties of MSA and MSE,

$$F_A = \frac{MSA}{MSE}$$

possesses an F distribution with $r - 1$ and $(r - 1)(n - 1)$ degrees of freedom. According to Theorem 14.3, F_A is larger or smaller, depending on whether the treatment null hypothesis should be rejected or accepted. Consequently, F_A can serve as a test statistic for an upper one-tailed test of the treatment null hypothesis, the null hypothesis being rejected at significance level α if

TABLE 14.3 ANOVA TABLE FOR RANDOMIZED COMPLETE BLOCK DESIGN

Source of Variation	Sum of Squares	Degrees of Freedom	Mean Square	F Test Statistic
Treatments	SSA	$r - 1$	$MSA = \dfrac{SSA}{r - 1}$	$F_A = \dfrac{MSA}{MSE}$
Blocks	SSB	$n - 1$	$MSB = \dfrac{SSB}{n - 1}$	$F_B = \dfrac{MSB}{MSE}$
Error	SSE	$(r - 1)(n - 1)$	$MSE = \dfrac{SSE}{(r - 1)(n - 1)}$	
Total	SST	$N - 1$		

$$F_A = \frac{MSA}{MSE} \geq f_{\alpha, r-1, (r-1)(n-1)}$$

On the surface, it appears as though a parallel line of reasoning will lead to the conclusion that the block null hypothesis can be tested with the test statistic $F_B = MSB/MSE$, the critical region being defined by

$$F_B = \frac{MSB}{MSE} \geq f_{\alpha, n-1, (r-1)(n-1)}$$

Although such a test is often run, interpretation of the results is problematic, due to the manner in which the experiment has been designed. Recall that the treatments are randomized within each block; however, the blocks have not been randomly assigned to the treatments. Of course, in blocking the intent is to eliminate extraneous variability that might obscure significant differences among the treatment effects. It is block variation that is being controlled by the experimental design, and the detection of block-effect differences is not of primary interest.

The entire randomized complete block design is summarized in Table 14.3. The next theorem, whose proof is omitted, provides computationally beneficial expressions for the relevant sums of squares.

THEOREM 14.4

Under the randomized complete block design, the sums of squares are given by

$$SST = \sum_{i=1}^{r} \sum_{j=1}^{n} X_{ij}^2 - \frac{T_{..}^2}{N}$$

$$SSA = \frac{1}{N} \left[r \sum_{i=1}^{r} T_{i.}^2 - T_{..}^2 \right]$$

$$SSB = \frac{1}{N} \left[n \sum_{j=1}^{n} T_{.j}^2 - T_{..}^2 \right]$$

$$SSE = SST - SSA - SSB$$

EXAMPLE 14.3

Consider the run-time data of Table 14.2 that were discussed in Example 14.1. The computational expressions of Theorem 14.4 yield $T_{..}^2 = 6556.14$,

$$SST = (6.42)^2 + (10.51)^2 + \cdots + (5.66)^2 - \frac{6556.14}{12} = 69.0984$$

$$SSA = \frac{3[(27.13)^2 + (29.14)^2 + (24.70)^2] - 6556.14}{12} = 2.4716$$

$$SSB = \frac{4[(18.87)^2 + (31.46)^2 + (12.06)^2 + (18.58)^2] - 6556.14}{12} = 65.8112$$

$$SSE = SST - SSA - SSB = 0.8156$$

$$MSA = \frac{2.4716}{2} = 1.2358$$

$$MSB = \frac{65.8112}{3} = 21.9371$$

$$MSE = \frac{0.8156}{6} = 0.1359$$

$$f_A = \frac{1.2358}{0.1359} = 9.093$$

$$f_B = \frac{21.9371}{0.1359} = 161.421$$

The null hypothesis to be tested is

$$H_{0A}: \alpha_1 = \alpha_2 = \alpha_3 = 0$$

where α_i is the effect of algorithm i. H_{0A} is interpreted to mean that the three algorithms possess equal run times and therefore are of equal efficiency. The alternative is that the algorithms are not of equal efficiency. F_A possesses an F distribution with 2 and 6 degrees of freedom, and the appropriate critical value at significance level 0.05 is $f_{0.05,2,6} = 5.14$. Since $f_A > 5.14$, the null hypothesis is rejected, and it is concluded that the algorithms are not of equal efficiency. At significance level 0.01, the hypothesis of equal efficiency is not rejected, because $f_{0.01,2,6} = 10.92$. Thus, the P value is between 0.01 and 0.05. The analysis of variance is summarized in Table 14.4.

One might wish to use the data to demonstrate the unequal effects of the blocks (computers) by noting that $f_B > f_{0.01,3,6} = 9.78$; however, as noted previously, the blocks represent a restriction on randomness, and therefore the results of such a test can be misleading.

EXAMPLE 14.4   The state of charge of a lead-acid battery is measured by the specific gravity of the electrolyte, a specific gravity of 1.26 indicating a fully charged battery. When a hydrometer is used to test the state of charge, a reading of 1260 corresponds to a specific gravity of 1.26. Consider an investigator who wishes to test the ability of three brands of battery to maintain their states of charge over time. Three batteries of each brand will be tested on three different vehicles. Because battery wear will vary depending on the type of vehicle, the experiment is blocked by using three vehicles of each of three given makes, a block consisting of a make of vehicle. The data of Table 14.5 consist of hydrometer readings at the completion of the experiment. The analysis-of-variance computations are summarized in Table 14.6. Since

$$f_A = 37.00 > f_{0.01,2,4} = 18.00$$

the null hypothesis of zero treatment effects is rejected and it is concluded that the three brands of battery are not equal in their abilities to maintain a state of charge over a period of use.

TABLE 14.4 ANOVA TABLE FOR EXERCISE 14.3

Source of Variation	Sum of Squares	Degrees of Freedom	Mean Square	F Test Statistic
Algorithms	2.472	2	1.236	9.09
Computers	65.811	3	21.937	161.42
Error	0.816	6	0.136	
Total	69.098	11		

TABLE 14.5 BLOCKED DATA FOR BATTERY STATE OF CHARGE

Battery	Vehicle 1	Vehicle 2	Vehicle 3	Total	Mean
1	1250	1252	1244	3746	1248.67
2	1258	1258	1250	3766	1255.33
3	1248	1252	1244	3744	1248.00
Total	3756	3762	3738	11,256.00	
Mean	1252.00	1254.00	1246.00		1250.67

TABLE 14.6 ANOVA TABLE FOR EXERCISE 14.4

Source of Variation	Sum of Squares	Degrees of Freedom	Mean Square	F Test Statistic
Batteries	98.667	2	49.333	37.00
Vehicles	104.000	2	52.000	39.00
Error	5.333	4	1.333	
Total	208.000	8		

Should the hypothesis of zero treatment effects be rejected, multiple comparisons can be made among the $\bar{x}_{i.}$ as in the unblocked case by applying Duncan's multiple range test (Section 11.2.2). As in the unblocked model, the values r_p are found from Table A.9 for $p = 2, 3, \ldots, r$ and the least significant ranges are given by

$$R_p = r_p \sqrt{\frac{MSE}{n}}$$

When reading Table A.9, λ denotes the degrees of freedom associated with MSE.

EXAMPLE 14.5

The null hypothesis of zero effects was rejected at the 0.05 level in Example 14.3. Using the fact that there are 6 degrees of freedom associated with MSE, we can obtain the values of r_2 and r_3 from Table A.9. These are multiplied by $\sqrt{MSE/n} = 0.1843$ to obtain R_2 and R_3. The following table summarizes these computations.

p	2	3
r_p	3.461	3.587
R_p	0.638	0.661

Listing the means $\bar{x}_{1.}$, $\bar{x}_{2.}$, and $\bar{x}_{3.}$ in order and underlining those that are not judged significantly different under the multiple range test yields

$\bar{x}_{3.}$	$\bar{x}_{1.}$	$\bar{x}_{2.}$
6.18	6.78	7.29

We conclude that $\mu_{3.} \neq \mu_{2.}$. In sum, algorithm 3 is judged faster than algorithm 2.

Recall the dilemma in choosing between application of the t test (on randomized data) and the paired t test when testing the difference of two means. A similar dilemma arises when choosing between the completely randomized design and the block design. While blocking mitigates the detrimental effect of variation due to extraneous variables, it does cost some degrees of freedom. In the completely randomized design with treatment sample sizes equal to n, the mean-square error possesses $rn - r$ degrees of freedom, while in the block design the mean-square error possesses $(r - 1)(n - 1)$ degrees of freedom. Consequently, blocking costs $n - 1$ degrees of freedom. Although this loss can have an adverse impact on the sensitivity of detection, it will likely be more than offset by the beneficial effects of blocking if there are significant block effects. Since the loss of degrees of freedom is more crucial for small n, blocking is less likely to be beneficial for small samples, and therefore the experimenter should be more convinced of the significance of block effects before using blocking for small samples.

A second dilemma posed by the randomized complete block design has to do with the model equation itself, which implies that

$$\mu_{ij} = \mu + \alpha_i + \beta_j$$

The model postulates the **additivity** of the treatment and block effects. Specifically, for any i, j, i', and j',

$$\mu_{ij} - \mu_{ij'} = \mu_{i'j} - \mu_{i'j'}$$

In words, the difference between the ith treatment mean in block j and the ith treatment mean in block j' is equal to the difference between the i'th treatment mean in block j and the i'th treatment mean in block j'. Another way of putting the matter is that the difference of the treatment means across the various blocks remains constant. Such an assumption is certainly not always warranted. Should block j enhance the effect of treatment i and retard the effect of treatment i', while block j' retards the effect of treatment i and enhances the effect of treatment i', the additivity assumption will not hold. In the graphic of Figure 14.2, additivity holds in part (a) but not in part (b). When additivity does not hold, there is said to be **interaction** between the treatments and the blocks. In the two-way-classification design of Section 14.3, the problem of interaction will be examined more closely, and a model will be postulated that takes into account the possibility of interaction and provides the wherewithal to test for significant interaction. In the case of a randomized complete block design,

FIGURE 14.2
Interaction and noninteraction

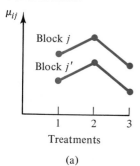

(a)

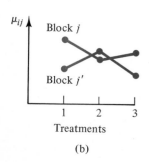

(b)

interaction cannot be readily detected and, if it exists to a considerable degree, the test for treatment effects will be detrimentally impacted. Consequently, if it is not possible to design the experiment in such a way as to make the treatment-block interaction negligible, then the block design should not be employed.

The relationship between the paired t test and the F test for the randomized complete block design goes beyond the manner in which both tests mitigate the effect of within-treatment variation on the sensitivity of the experiment. Indeed, in the case of two treatments, the tests can be employed interchangeably since $T^2 = F$ and $t^2_{\alpha/2, n-1} = f_{\alpha, 1, n-1}$.

14.2.3 Random-Effects Model for Randomized Complete Block Design

Rather than deterministically select the treatments, an investigator might select them randomly in order to arrive at a conclusion that applies to a large collection of treatments. In addition, he or she might select the blocks randomly from among a large number of conditions under which the treatments might be tested. Such an experimental design is modeled by the **random-effects** model, the model equation being

Model Equation for Random-Effects Randomized Complete Block Design

$$X_{ij} = \mu + A_i + B_j + E_{ij}$$

where A_i, B_j, and E_{ij} are independent zero-mean normally distributed random variables possessing variances σ_A^2, σ_B^2, and σ^2, respectively. The sums of squares, degrees of freedom, mean squares, and F statistics for random effects are the same as for fixed effects. There are, however, differences in the expected mean squares. In the random-effects setting, the expected values of the treatment, block, and error mean squares are

Mean-Square Expected Values in the Random-Effects Randomized Complete Block Design

$$E[MSA] = \sigma^2 + n\sigma_A^2$$
$$E[MSB] = \sigma^2 + r\sigma_B^2$$
$$E[MSE] = \sigma^2$$

The treatment hypothesis test in the random-effects design is

$$H_{0A}: \sigma_A^2 = 0$$
$$H_{1A}: \sigma_A^2 > 0$$

The block hypothesis test is

$$H_{0B}: \sigma_B^2 = 0$$
$$H_{1B}: \sigma_B^2 > 0$$

The critical regions are the same as in the fixed-effects model.

If the treatment null hypothesis is rejected, then the appropriate estimator for the treatment variance is

$$\hat{\sigma}_A^2 = \frac{MSA - MSE}{n}$$

If desired, the corresponding estimator for the block variance is

$$\hat{\sigma}_B^2 = \frac{MSB - MSE}{r}$$

EXAMPLE 14.6   In the battery–vehicle analysis of Example 14.4, it is very possible that the brands have been randomly selected from among some large collection of battery brands, so that an experimenter can determine if there are any significant differences within the collection. So as not to prejudice the experiment with respect to certain makes of vehicles, the experimenter might conceivably have randomly selected the vehicle types. Let us assume that such random choices have been made. Then the fact that F_A exceeds the critical value leads to the conclusion that $\sigma_A^2 > 0$. Practically, the conclusion means there are significant differences in the battery-brand population. The estimate of σ_A^2 is

$$\overset{*}{\sigma}_A^2 = \frac{49.333 - 1.333}{3} = 16$$

14.3   TWO-WAY CLASSIFICATION

In one-way classification, the model serves to represent a class of experiments in which a single factor affecting a collection of experimental units is under scrutiny. The treatment populations correspond to various levels of the factor of interest. A different model results from an experiment in which two factors are under consideration. For instance, suppose the output of a chemical process depends on both the temperature and the duration of the process. To compare the effects of different temperature levels, one might run a single-factor analysis of variance, the conclusion being either that temperature level is important or that it is not important. One might also run a single-factor analysis to determine the significance of various duration levels (lengths). However, due to savings of both material costs and time, it might be better to devise an experiment in which both factors, temperature and duration, can be evaluated concurrently.

14.3.1   Interaction

There is a second important reason why one might desire to test two factors in conjunction with one another. It might be that each of the factors has different effects depending on the level of the other factor. For instance, in the temperature–duration scenario, varying the temperature while maintaining a constant duration might lead one to conclusions regarding the effects of temperature, but the conclusions might apply only to the specific duration under which the experiment has been implemented. Indeed, temperature effects might vary substantially at different duration levels. In general, there might be **interaction,** or **crossing,** between the factors, thereby leading one to conclusions of very limited scope. A multifactor design is often beneficial because it facilitates testing for interaction.

The problem of interaction was briefly addressed at the conclusion of Section 14.2.2. There, the phenomenon was discussed in terms of treatment–block interaction, absence of interaction being a modeling assumption. Figure 14.2, which graphically depicts both noninteraction [part (a)] and interaction [part (b)], is immediately ap-

TABLE 14.7 TWO-WAY
CLASSIFICATION OF REACTION DATA

Temperature (degrees)	Duration (Minutes)		
	20	25	30
120	41	42	43
130	50	52	55

plicable to the two-factor setting. Simply consider the lines as corresponding to a second factor rather than to blocks.

EXAMPLE 14.7

Consider testing the effects of both temperature and duration on the output of a chemical process. The reaction is run at 120 and 130 degrees Fahrenheit and with durations of 20, 25, and 30 minutes. The percentage yields are given in Table 14.7. It appears from the data that higher temperature has a beneficial effect, regardless of the duration. In addition, it appears that duration has a negligible effect, regardless of the temperature. Figure 14.3 gives a graphic of the situation. The three lines of yield change are nearly parallel, because the increase in yield relative to temperature is essentially the same at all three duration levels. Experimentally, there does not appear to be significant interaction between the two factors in the classification. Note that the parallelism in Figure 14.2(a) represents lack of interaction as manifested by the various treatment–block means, whereas the parallelism of Figure 14.3 results from the data. Nevertheless, the underlying principle is the same.

EXAMPLE 14.8

Consider testing a vision system relative to its recognition software and the type of sensor. Suppose two different recognition algorithms, U and V, are to be tested with respect to both infrared (I) and optical (O) sensors. The system is tested in the various algorithm–sensor combinations, each combination being tested in 50 different environmental settings. The numbers of correct identifications under each combination are tabulated in Table 14.8. If one were to consider only the effects of algorithms U and V on the overall system by testing with infrared sensors, the conclusion would most likely be drawn that algorithm U is superior to algorithm V; however, the reverse conclusion would likely be drawn when testing with optical sensors. Such interaction is not surprising, since different image characteristics are revealed by the two types of sensors and therefore the performance of algorithms that utilize these characteristics for recognition is likely to be affected. This crossing effect is illustrated in Figure 14.4, where the lines of change are far from being parallel. Note that the average number of recognitions is 30 for U and 29 for V; however, a conclusion that U and V are equally effective would be misleading. Indeed, their individual effects on recognition are masked by sensor-algorithm interaction if recognition success is averaged across sensor types.

FIGURE 14.3
Apparent noninteraction between temperatures and durations

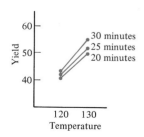

TABLE 14.8 TWO-WAY
CLASSIFICATION OF
RECOGNITION DATA

Algorithm	Sensor	
	I	O
U	42	18
V	20	38

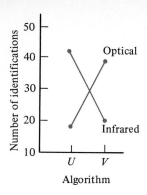

FIGURE 14.4
Apparent interaction between sensors and algorithms

In the algorithm run-time experiment discussed in Section 14.2, it was assumed that there was no interaction between the blocks and treatments. From the discussion of Example 14.8, one can see the possible difficulty with such an assumption. If it happens that the three algorithms of Example 14.1 are to be tested on computers that process programs in a somewhat similar manner, then the assumption of no interaction is likely to be valid. On the other hand, suppose one of the computers is an array processor and one of the algorithms, say algorithm 1, is so configured as to take advantage of an array architecture. It might then be very possible that algorithm 1 runs slower than the others on standard sequential processors but much faster on array processors, thereby resulting in significant treatment–block interaction. One of the advantages of the model to be discussed in the present section is that the analysis-of-variance procedure includes a test for factor interaction.

14.3.2   Fixed-Effects Two-Way-Classification Model

In general, a model that includes more than a single factor is said to be a **factorial, multifactor,** or **multiway** model. The effects of a given factor may be fixed or random, the overall model possessing fixed, random, or mixed effects. For a one-way classification, the model equation requires the effect of only a single factor in addition to the model parameter μ and the random error. In the two-way model, there must be terms for both factor effects. Furthermore, whereas the experiments of Exercises 14.7 and 14.8 were done once, in general there will be n replications of the experiment. This latter point is especially important, because with multiple replications the model can be employed to detect significant interaction. Consequently, for the ensuing discussion we will assume that $n > 1$, the upshot being the inclusion of an interaction term in the model equation.

Factors A and B will be tested with respect to a and b levels, respectively. Each replication of the experiment will consist of taking ab measurements, each indexed by a pair (i, j), $i = 1, 2, \ldots, a$ and $j = 1, 2, \ldots, b$. Given there are n replications, each observation will be of the form X_{ijk}, $k = 1, 2, \ldots, n$. The random variable X_{ijk} is called the kth **response** of the (i, j)th **treatment combination.** The $N = abn$ observations resulting from a full experiment will be tabulated as in Table 14.9.

In conjunction with the random variables X_{ijk}, we define

$$T_{ij.} = \sum_{k=1}^{n} X_{ijk}$$

which is the sum of the responses corresponding to the (i, j)th treatment combination,

$$T_{i..} = \sum_{j=1}^{b} \sum_{k=1}^{n} X_{ijk}$$

the sum of the responses corresponding to the ith level of factor A,

$$T_{.j.} = \sum_{i=1}^{a} \sum_{k=1}^{n} X_{ijk}$$

the sum of the responses corresponding to the jth level of factor B,

$$T_{...} = \sum_{i=1}^{a} \sum_{j=1}^{b} \sum_{k=1}^{n} X_{ijk}$$

TABLE 14.9 FORMAT FOR TWO-WAY CLASSIFICATION

Factor A	Factor B				Total	Mean
	1	2	...	b		
1	x_{111} x_{112} $\vdots$ x_{11n}	x_{121} x_{122} $\vdots$ x_{12n}	... ...	x_{1b1} x_{1b2} $\vdots$ x_{1bn}	$T_{1..}$	$\bar{x}_{1..}$
2	x_{211} x_{212} $\vdots$ x_{21n}	x_{221} x_{222} $\vdots$ x_{22n}	... ...	x_{2b1} x_{2b2} $\vdots$ x_{2bn}	$T_{2..}$	$\bar{x}_{2..}$
$\vdots$	$\vdots$	$\vdots$	$\vdots$	$\vdots$	$\vdots$	$\vdots$
a	x_{a11} x_{a12} $\vdots$ x_{a1n}	x_{a21} x_{a22} $\vdots$ x_{a2n}	... ...	x_{ab1} x_{ab2} $\vdots$ x_{abn}	$T_{a..}$	$\bar{x}_{a..}$
Total	$T_{.1.}$	$T_{.2.}$	...	$T_{.b.}$	$T_{...}$	
Mean	$\bar{x}_{.1.}$	$\bar{x}_{.2.}$	...	$\bar{x}_{.b.}$		$\bar{x}_{...}$

the sum of all responses,

$$\bar{X}_{ij.} = \frac{T_{ij.}}{n}$$

the response average for the (i, j)th treatment combination,

$$\bar{X}_{i..} = \frac{T_{i..}}{bn}$$

the response average for the ith level of factor A,

$$\bar{X}_{.j.} = \frac{T_{.j.}}{an}$$

the response average for the jth level of factor B, and

$$\bar{X}_{...} = \frac{T_{...}}{N}$$

the average of all responses. The level totals and their averages are included in the tabulation of the data (see Table 14.9). $T_{...}$ and $\bar{x}_{...}$ are also included. Because of the important role played by the totals in the subsequent analysis, it is useful to put them into tabular form as depicted in Table 14.10.

TABLE 14.10 FORMAT FOR TWO-WAY-CLASSIFICATION TOTALS

A	B 1	2	...	b	Total
1	$T_{11.}$	$T_{12.}$	...	$T_{1b.}$	$T_{1..}$
2	$T_{21.}$	$T_{22.}$	...	$T_{2b.}$	$T_{2..}$
⋮	⋮	⋮		⋮	⋮
a	$T_{a1.}$	$T_{a2.}$	...	$T_{ab.}$	$T_{a..}$
Total	$T_{.1.}$	$T_{.2.}$	...	$T_{.b.}$	$T_{...}$

EXAMPLE 14.9

To simultaneously test the mileage effects of three fuel mixtures and two carburetors, an investigator decides to perform a two-way analysis with three replications. Based upon such a design, 18 cars are randomly selected for the six treatment combinations, and the resulting miles-per-gallon data are given in Table 14.11. From the means of Table 14.11, it appears that fuel 3 and carburetor 2 are superior; however, the significance of the data must await an analysis of variance. Table 14.12 provides the data totals.

Taking into account the need to model the effects of both factors, the so-called **main effects,** as well as the **crossing effect,** we posit a linear model for the response

TABLE 14.11 CARBURETOR–FUEL DATA FOR TWO-WAY CLASSIFICATION

Carburetor	Fuel 1	2	3	Total	Mean
1	18.4 19.0 18.6	18.7 17.9 18.0	20.6 20.0 20.7	171.9	19.10
2	20.1 21.2 20.2	21.0 20.8 20.5	22.4 22.4 21.8	190.4	21.16
Total	117.5	116.9	127.9	362.3	
Mean	19.58	19.48	21.32		20.13

TABLE 14.12 TOTALS FOR CARBURETOR–FUEL DATA

Carburetor	Fuel 1	2	3	Total
1	56.0	54.6	61.3	171.9
2	61.5	62.3	66.6	190.4
Total	117.5	116.9	127.9	362.3

variables. For $i = 1, 2, \ldots, a$, $j = 1, 2, \ldots, b$, and $k = 1, 2, \ldots, n$, it is postulated that

$$X_{ijk} = \mu + \alpha_i + \beta_j + (\alpha\beta)_{ij} + E_{ijk}$$

where the model parameter μ is the overall mean, α_i is the effect due to level i of factor A, β_j is the effect due to level j of factor B, $(\alpha\beta)_{ij}$ is the effect of treatment interaction, and E_{ijk} is the random deviation of the responses resulting from sources not accounted for by the model. The N random variables E_{ijk} are assumed to be independent normally distributed variables possessing common mean 0 and common variance σ^2. Taking the expected value of X_{ijk} puts the model equation in terms of the response-variable means. Specifically,

$$\mu_{ijk} = \mu + \alpha_i + \beta_j + (\alpha\beta)_{ij}$$

Further constraints imposed on the model involve the effects. These require the sum of the main effects due to A to be zero, the sum of the main effects due to B to be zero, the sum of the interaction effects for any fixed level of B to be zero, and the sum of the interaction effects for any fixed level of A to be zero. The following fixed-effects two-factor linear model results.

Model for Fixed-Effects Two-Way-Classification Analysis of Variance

---

Equation: $\quad X_{ijk} = \mu + \alpha_i + \beta_j + (\alpha\beta)_{ij} + E_{ijk}$

Assumptions: 1. $E_{ijk} \sim N(0, \sigma)$

2. The E_{ijk} comprise an independent collection.

3. For both factors, the main effects sum to zero:

$$\sum_{i=1}^{a} \alpha_i = \sum_{j=1}^{b} \beta_j = 0$$

4. For $j = 1, 2, \ldots, b$,

$$\sum_{i=1}^{a} (\alpha\beta)_{ij} = 0$$

5. For $i = 1, 2, \ldots, a$,

$$\sum_{j=1}^{b} (\alpha\beta)_{ij} = 0$$

Under the constraints imposed upon the model, unbiased estimators of the model parameters are given in terms of the previously defined averages.

THEOREM 14.5

Under the fixed-effects two-way-classification model, unbiased estimators for μ, α_i, β_j, and $(\alpha\beta)_{ij}$ are given by

$$\hat{\mu} = \bar{X}_{...}$$

$$\hat{\alpha}_i = \bar{X}_{i..} - \bar{X}_{...}$$

$$\hat{\beta}_j = \bar{X}_{.j.} - \bar{X}_{...}$$

$$\widehat{(\alpha\beta)}_{ij} = \bar{X}_{ij.} - \bar{X}_{i..} - \bar{X}_{.j.} + \bar{X}_{...}$$

Proof: We will prove the result for μ, leaving the others as exercises. By the linearity of the expected-value operator and the model equation,

$$E[\bar{X}_{\ldots}] = \frac{1}{nab} \sum_{i=1}^{a} \sum_{j=1}^{b} \sum_{k=1}^{n} E[X_{ijk}]$$

$$= \frac{1}{nab} \left[nab\mu + nb \sum_{i=1}^{a} \alpha_i + na \sum_{j=1}^{b} \beta_j + n \sum_{i=1}^{a} \sum_{j=1}^{b} (\alpha\beta)_{ij} \right]$$

Due to the model assumptions regarding the sums of the effects, all the summations in the preceding expression are zero. Hence, it reduces to $nab\mu/nab = \mu$. ∎

Further insight can be gained into the model by introducing averages of the response means. Let

$$\mu_{i..} = \frac{1}{nb} \sum_{j=1}^{b} \sum_{k=1}^{n} \mu_{ijk}$$

be the average of the response means for the ith level of factor A and

$$\mu_{.j.} = \frac{1}{na} \sum_{i=1}^{a} \sum_{k=1}^{n} \mu_{ijk}$$

be the average of the response means for the jth level of factor B. Then, due to the model constraints,

$$\alpha_i = \mu_{i..} - \mu$$

$$\beta_j = \mu_{.j.} - \mu$$

The first of the preceding effect representations follows at once by expanding $\mu_{i..}$. Indeed,

$$\mu_{i..} = \frac{1}{nb} \sum_{j=1}^{b} \sum_{k=1}^{n} [\mu + \alpha_i + \beta_j + (\alpha\beta)_{ij}]$$

$$= \frac{1}{nb} \left[nb\mu + nb\alpha_i + n \sum_{j=1}^{b} \beta_j + n \sum_{j=1}^{b} (\alpha\beta)_{ij} \right]$$

Since both summations are equal to zero, the expression reduces to $\mu + \alpha_i$. The representation of β_j is demonstrated in like manner. A similar summation argument shows that

$$\mu = \frac{1}{nab} \sum_{i=1}^{a} \sum_{j=1}^{b} \sum_{k=1}^{n} \mu_{ijk}$$

To obtain a representation of $(\alpha\beta)_{ij}$, note that for each treatment combination, $X_{ij1}, X_{ij2}, \ldots, X_{ijn}$ form a collection of independent, identically distributed random variables, and they therefore constitute a random sample corresponding to the random variable describing the (i, j)th treatment combination. If $\mu_{ij.}$ denotes the mean of that random variable, then $\mu_{ijk} = \mu_{ij.}$ for $k = 1, 2, \ldots, n$. Consequently, the model of the means takes the form

$$\mu_{ij.} = \mu + \alpha_i + \beta_j + (\alpha\beta)_{ij}$$

Substitution of the previously found representations for the main effects yields the desired representation of the crossing effect:

$$(\alpha\beta)_{ij} = \mu_{ij.} - \mu_{i..} - \mu_{.j.} + \mu$$

Finally, because $\mu_{ij.} = \mu_{ijk}$ for all k,

$$\mu_{i..} = \frac{1}{b} \sum_{j=1}^{b} \mu_{ij.}$$

$$\mu_{.j.} = \frac{1}{a} \sum_{i=1}^{a} \mu_{ij.}$$

EXAMPLE 14.10

For the carburetor–fuel data of Table 14.11, Theorem 14.5 yields estimates of the model parameters. The overall mean is estimated by

$$\overset{*}{\mu} = \bar{x}_{...} = 20.13$$

and the main effects are estimated by

$$\overset{*}{\alpha}_1 = \bar{x}_{1..} - \bar{x}_{...} = 19.10 - 20.13 = -1.03$$

$$\overset{*}{\alpha}_2 = \bar{x}_{2..} - \bar{x}_{...} = 21.16 - 20.13 = 1.03$$

$$\overset{*}{\beta}_1 = \bar{x}_{.1.} - \bar{x}_{...} = 19.58 - 20.13 = -0.55$$

$$\overset{*}{\beta}_2 = \bar{x}_{.2.} - \bar{x}_{...} = 19.48 - 20.13 = -0.65$$

$$\overset{*}{\beta}_3 = \bar{x}_{.3.} - \bar{x}_{...} = 21.32 - 20.13 = 1.19$$

Interaction effects can also be estimated. For instance,

$$\begin{aligned}(\overset{*}{\alpha\beta})_{12} &= \bar{x}_{12.} - \bar{x}_{1..} - \bar{x}_{.2.} + \bar{x}_{...} \\ &= 18.20 - 19.10 - 19.48 + 20.13 = -0.25\end{aligned}$$

In the fixed-effects two-way-classification model, there are three hypothesis tests of interest. Each relates to a question of significance. First, is there any significant interaction between the factors? Here, the relevant hypothesis test is

H_{0AB}: $(\alpha\beta)_{ij} = 0$ for $i = 1, 2, \ldots, a$ and $j = 1, 2, \ldots, b$

H_{1AB}: $(\alpha\beta)_{ij} \neq 0$ for at least one index (i, j)

Second, is there any significant difference among the effects of the various levels of factor A? Since the sum of the α_i is zero, their equality is equivalent to their identically being zero. Hence, the appropriate hypothesis test for the detection of significantly different effects of factor A is

$$H_{0A}: \alpha_1 = \alpha_2 = \cdots = \alpha_a = 0$$

$$H_{1A}: \alpha_i \neq 0 \text{ for at least one } i$$

Third, is there any significant difference among the effects of the various levels of factor B? Here, the appropriate test is

$$H_{0B}: \beta_1 = \beta_2 = \cdots = \beta_b = 0$$

$$H_{1B}: \beta_j \neq 0 \text{ for at least one } j$$

As a result of the expression of the main effects in terms of $\mu_{i..}$ and $\mu_{.j.}$, the factor-A and factor-B null hypotheses can be equivalently expressed as

$$H_{0A}: \mu_{1..} = \mu_{2..} = \cdots = \mu_{a..}$$

and

$$H_{0B}: \mu_{.1.} = \mu_{.2.} = \cdots = \mu_{.b.}$$

respectively. Since significant interaction can mask main effects, it is necessary to check first for interaction before proceeding with the tests for zero main effects. Unless the hypothesis of no interaction is accepted, the main-effect tests may give misleading results. Specifically, a conclusion of insignificant main effects in the presence of strong interaction may be a consequence of masking. Should there be interaction, one-way analyses of either factor can still be performed at single levels of the other factor.

14.3.3 Analysis of Variance in the Fixed-Effects Two-Way-Classification Model

To arrive at test statistics in the two-way model, the variation must be partitioned into components. In the present setting, five sums of squares are defined.

$$SST = \sum_{i=1}^{a} \sum_{j=1}^{b} \sum_{k=1}^{n} (X_{ijk} - \bar{X}_{...})^2$$

measures the total variation in the data.

$$SSA = nb \sum_{i=1}^{a} (\bar{X}_{i..} - \bar{X}_{...})^2$$

measures variation arising from the different levels of factor A.

$$SSB = na \sum_{j=1}^{b} (\bar{X}_{.j.} - \bar{X}_{...})^2$$

measures variation arising from the different levels of factor B.

$$SSAB = n \sum_{i=1}^{a} \sum_{j=1}^{b} (\bar{X}_{ij.} - \bar{X}_{i..} - \bar{X}_{.j.} + \bar{X}_{...})^2$$

measures variation arising from treatment interaction. Finally,

$$SSE = \sum_{i=1}^{a} \sum_{j=1}^{b} \sum_{k=1}^{n} (X_{ijk} - \bar{X}_{ij.})^2$$

is the error sum of squares and measures the variation due to randomness. Referring to Theorem 14.5, we see that

$$SSA = nb \sum_{i=1}^{a} \hat{\alpha}_i^2$$

$$SSB = na \sum_{j=1}^{b} \hat{\beta}_j^2$$

$$SSAB = n \sum_{i=1}^{a} \sum_{j=1}^{b} (\widehat{\alpha\beta})_{ij}^2$$

The next theorem, whose calculational proof is omitted, provides the two-way-classification partition of SST.

Under the fixed-effects two-way classification,

$$SST = SSA + SSB + SSAB + SSE$$

To arrive at appropriate F test statistics for the three hypothesis tests under consideration, the distributional properties of the components in the partition of Theorem 14.6 need to be known, at least insofar as they pertain to the relevant null hypotheses. Under H_{0A}, SST/σ^2, SSE/σ^2, and SSA/σ^2 possess chi-square distributions with $N - 1$, $ab(n - 1)$, and $a - 1$ degrees of freedom, respectively, and SSA/σ^2 is independent of SSE/σ^2. Similar statements apply to H_{0B} and H_{0AB}, except that under H_{0B}, SSB/σ^2 possesses a chi-square distribution with $b - 1$ degrees of freedom and is independent of SSE/σ^2, and under H_{0AB}, $SSAB/\sigma^2$ possesses a chi-square distribution with $(a - 1)(b - 1)$ degrees of freedom and is independent of SSE/σ^2.

The mean squares corresponding to SSA, SSB, $SSAB$, and SSE are

$$MSA = \frac{SSA}{a - 1}$$

$$MSB = \frac{SSB}{b - 1}$$

$$MSAB = \frac{SSAB}{(a - 1)(b - 1)}$$

$$MSE = \frac{SSE}{ab(n - 1)}$$

The following theorem provides the expected values of MSA, MSB, $MSAB$, and MSE. The proof is omitted.

Under the conditions of the fixed-effects two-way classification,

$$E[MSA] = \sigma^2 + \frac{nb}{a-1} \sum_{i=1}^{a} \alpha_i^2$$

$$E[MSB] = \sigma^2 + \frac{na}{b-1} \sum_{j=1}^{b} \beta_j^2$$

$$E[MSAB] = \sigma^2 + \frac{n}{(a-1)(b-1)} \sum_{i=1}^{a} \sum_{j=1}^{b} (\alpha\beta)_{ij}^2$$

$$E[MSE] = \sigma^2$$

It is immediate from Theorem 14.7 that under H_{0A}, H_{0B}, and H_{0AB}, the respective mean squares MSA, MSB, and $MSAB$ are unbiased estimators of σ^2.

Based upon the previously discussed distributional properties, $F_A = MSA/MSE$ possesses an F distribution with $a-1$ and $ab(n-1)$ degrees of freedom. Moreover, by Theorem 14.7, we see that violation of H_{0A} results in large values of F_A. Consequently, H_{0A} is rejected for large values of F_A. More specifically, H_{0A} is rejected at significance level α if

$$F_A = \frac{MSA}{MSE} \geq f_{\alpha, a-1, ab(n-1)}$$

Similar reasoning shows that H_{0B} is rejected if

$$F_B = \frac{MSB}{MSE} \geq f_{\alpha, b-1, ab(n-1)}$$

and H_{0AB} is rejected if

$$F_{AB} = \frac{MSAB}{MSE} \geq f_{\alpha, (a-1)(b-1), ab(n-1)}$$

The entire analysis is summarized in Table 14.13.

The next proposition provides computationally beneficial expressions.

TABLE 14.13 ANOVA TABLE FOR FIXED-EFFECTS TWO-WAY CLASSIFICATION

Source of Variation	Sum of Squares	Degrees of Freedom	Mean Square	F Test Statistic
Factor A	SSA	$a-1$	$MSA = \dfrac{SSA}{a-1}$	$F_A = \dfrac{MSA}{MSE}$
Factor B	SSB	$b-1$	$MSB = \dfrac{SSB}{b-1}$	$F_B = \dfrac{MSB}{MSE}$
Interaction	$SSAB$	$(a-1)(b-1)$	$MSAB = \dfrac{SSAB}{(a-1)(b-1)}$	$F_{AB} = \dfrac{MSAB}{MSE}$
Error	SSE	$ab(n-1)$	$MSE = \dfrac{SSE}{ab(n-1)}$	
Total	SST	$N-1$		

Under the two-way classification,

$$SST = \sum_{i=1}^{a} \sum_{j=1}^{b} \sum_{k=1}^{n} X_{ijk}^2 - \frac{T_{...}^2}{N}$$

$$SSA = \frac{1}{N}\left[a \sum_{i=1}^{a} T_{i..}^2 - T_{...}^2 \right]$$

$$SSB = \frac{1}{N}\left[b \sum_{j=1}^{b} T_{.j.}^2 - T_{...}^2 \right]$$

$$SSAB = \frac{1}{N}\left[ab \sum_{i=1}^{a} \sum_{j=1}^{b} T_{ij.}^2 - a \sum_{i=1}^{a} T_{i..}^2 - b \sum_{j=1}^{b} T_{.j.}^2 + T_{...}^2 \right]$$

$$SSE = SST - SSA - SSB - SSAB$$

EXAMPLE 14.11

We are now prepared to complete the analysis-of-variance procedure for the carburetor–fuel data of Example 14.9 and Table 14.11. Applying Theorem 14.8 and using the information of Table 14.12 yields

$$SST = (18.4)^2 + (19.0)^2 + \cdots + (21.8)^2 - \frac{(362.3)^2}{18} = 7326.61 - 7292.29 = 34.32$$

$$SSA = \frac{2[(171.9)^2 + (190.4)^2] - (362.3)^2}{18} = 19.01$$

$$SSB = \frac{3[(117.5)^2 + (116.9)^2 + (127.9)^2] - (362.3)^2}{18} = 12.75$$

$$SSAB = \frac{1}{18}\{6[(56.0)^2 + (54.6)^2 + (61.3)^2 + (61.5)^2 + (62.3)^2 + (66.6)^2]$$
$$- 131{,}603.54 - 131{,}490.81 + 131{,}261.29\} = 0.59$$

and $SSE = 1.97$. The degrees of freedom specified in Table 14.13 yield the mean squares $MSA = 19.01$, $MSB = 6.375$, $MSAB = 0.295$, and $MSE = 0.164$. Thus,

$$f_{AB} = \frac{0.295}{0.164} = 1.80$$

Since F_{AB} possesses an F distribution with 2 and 12 degrees of freedom and $f_{0.05,2,12} = 3.89$, the interaction null hypothesis is not rejected at the 0.05 level, let alone at the 0.01 level. It is safe to assume that there is insignificant interaction and that $(\alpha\beta)_{ij} = 0$ for all treatment combinations. Hence, we can proceed with the analysis of the main effects. The values of the carburetor and fuel F statistics are $f_A = 115.91$ and $f_B = 38.87$, respectively. Since F_A possesses 1 and 12 degrees of freedom and $f_{0.01,1,12} = 9.33$, the hypothesis of zero carburetor effects is rejected strongly at the 0.01 level. Moreover, since F_B possesses 2 and 12 degrees of freedom and $f_{0.01,2,12} = 6.93$, the hypothesis of zero fuel effects is also rejected strongly at the same level. The analysis of variance is summarized in Table 14.14.

In Example 14.10, the estimators of Theorem 14.5 were employed to estimate the model parameters. For instance, the estimate $(\alpha\beta)_{12}^* = -0.25$ was obtained. Since the interaction null hypothesis was not rejected in Example 14.11, it may be best to simply ignore the estimation rule of Theorem 14.5 and treat all interaction effects as being 0.

TABLE 14.14 ANOVA TABLE FOR EXAMPLE 14.11

Source of Variation	Sum of Squares	Degrees of Freedom	Mean Square	F Test Statistic
Carburetors	19.01	1	19.010	115.91
Fuel	12.75	2	6.375	38.87
Interaction	0.59	2	0.295	1.80
Error	1.97	12	0.164	
Total	34.32	17		

Once the interaction null hypothesis is not rejected, if either factor null hypothesis is rejected, then Duncan's multiple range test can be applied to determine significant differences among factor levels. The test is applied to factor A with

$$R_p = r_p \sqrt{\frac{MSE}{nb}}$$

It is applied to factor B with

$$R_p = r_p \sqrt{\frac{MSE}{na}}$$

In both cases, λ is the number of degrees of freedom associated with MSE.

EXAMPLE 14.12

Since the carburetor null hypothesis was rejected in Example 14.11 and since there are only two carburetor levels, it can be concluded that $\mu_{1.} \neq \mu_{2.}$. Insofar as the fuel levels are concerned, r_2 and r_3 can be obtained from Table A.9 with $\alpha = 0.01$ and $\lambda = 12$. Using $\sqrt{MSE/na} = 0.165$, R_1 and R_2 can be obtained, the results being summarized in the following table:

p	2	3
r_p	4.320	4.504
R_p	0.713	0.743

In terms of the customary underlining approach,

$$\bar{x}_{.2} \quad \bar{x}_{.1} \quad \bar{x}_{.3}$$
$$\underline{19.48} \quad \underline{19.58} \quad 21.32$$

It is concluded that $\mu_{.3} \neq \mu_{.2}$ and $\mu_{.3} \neq \mu_{.1}$. In practical terms, fuel 3 provides better gas mileage.

14.3.4 Two-Way Classification with Random Effects

To arrive at conclusions applicable to situations in which the levels of factors A and B have been drawn from large populations of possible levels, a random-effects model is employed. The model equation is given by

$$X_{ijk} = \mu + A_i + B_j + (AB)_{ij} + E_{ijk}$$

where $i = 1, 2, \ldots, a$, $j = 1, 2, \ldots, b$, and $k = 1, 2, \ldots, n$. Since the choice of levels is random, so too are the effects. Specifically, it is assumed that A_i, B_j, $(AB)_{ij}$, and E_{ijk} are independently distributed zero-mean normal random variables possessing variances σ_A^2, σ_B^2, σ_{AB}^2, and σ^2, respectively.

The hypothesis tests for factor A, factor B, and interaction are

$$H_{0A}: \sigma_A^2 = 0$$
$$H_{1A}: \sigma_A^2 > 0$$

$$H_{0B}: \sigma_B^2 = 0$$
$$H_{1B}: \sigma_B^2 > 0$$

and

$$H_{0AB}: \sigma_{AB}^2 = 0$$
$$H_{1AB}: \sigma_{AB}^2 > 0$$

respectively.

The total, factor-A, factor-B, interaction, and error sums of squares, denoted by SST, SSA, SSB, $SSAB$, and SSE, respectively, are exactly the same as in the fixed-effects model for two-way classification. The degrees of freedom of each of the respective chi-square variables are also the same as in the fixed-effects model and therefore the mean squares MSA, MSB, $MSAB$, and MSE are defined similarly. To devise appropriate test statistics for the various tests, we must examine the expected values of the mean squares. These are different than in the fixed-effects design and are stated without proof in the next theorem.

THEOREM 14.9

Under the random-effects two-way classification,

$$E[MSA] = \sigma^2 + n\sigma_{AB}^2 + nb\sigma_A^2$$
$$E[MSB] = \sigma^2 + n\sigma_{AB}^2 + na\sigma_B^2$$
$$E[MSAB] = \sigma^2 + n\sigma_{AB}^2$$
$$E[MSE] = \sigma^2$$

To arrive at the desired test statistics, we need ratios of the mean-square expected values that are equal to 1 when the null hypotheses of interest are true. Examining Theorem 14.9 reveals that the numerator and denominator expectations of the ratio

$$F_A = \frac{MSA}{MSAB}$$

are equal when the factor-A null hypothesis is true and $E[MSA] > E[MSAB]$ when H_{0A} is false. F_A possesses an F distribution with $a - 1$ and $(a - 1)(b - 1)$ degrees of freedom, so that rejection of H_{0A} at significance level α occurs when

$$F_A \geq f_{\alpha, a-1, (a-1)(b-1)}$$

Similar reasoning shows that the appropriate test statistic for the factor-B test is given by the ratio

$$F_B = \frac{MSB}{MSAB}$$

which possesses an F distribution with $b - 1$ and $(a - 1)(b - 1)$ degrees of freedom. H_{0B} is rejected when

$$F_B \geq f_{\alpha, b-1, (a-1)(b-1)}$$

Finally, the appropriate test statistic for the interaction null hypothesis is

$$F_{AB} = \frac{MSAB}{MSE}$$

which possesses an F distribution with $(a - 1)(b - 1)$ and $ab(n - 1)$ degrees of freedom. Rejection of H_{0AB} at level α occurs when

$$F_{AB} \geq f_{\alpha, (a-1)(b-1), ab(n-1)}$$

The random-effects analysis of variance is summarized in Table 14.15. In analogy to the fixed-effects model, the interaction null hypothesis should be tested first, the factor tests being fully meaningful only if the interaction null hypothesis is not rejected. In the event the interaction null hypothesis is rejected, σ_{AB}^2 possesses the estimator

$$\widehat{\sigma}_{AB}^2 = \frac{MSAB - MSE}{n}$$

For rejection of the factor-A and factor-B null hypotheses, variance estimators are

$$\widehat{\sigma}_A^2 = \frac{MSA - MSAB}{nb}$$

$$\widehat{\sigma}_B^2 = \frac{MSB - MSAB}{na}$$

TABLE 14.15 ANOVA TABLE FOR RANDOM-EFFECTS TWO-WAY CLASSIFICATION

Source of Variation	Sum of Squares	Degrees of Freedom	Mean Square	F Test Statistic
Factor A	SSA	$a - 1$	MSA	$F_A = \dfrac{MSA}{MSAB}$
Factor B	SSB	$b - 1$	MSB	$F_B = \dfrac{MSB}{MSAB}$
Interaction	$SSAB$	$(a - 1)(b - 1)$	$MSAB$	$F_{AB} = \dfrac{MSAB}{MSE}$
Error	SSE	$ab(n - 1)$	MSE	
Total	SST	$N - 1$		

EXAMPLE 14.13

Suppose the investigator in the carburetor–fuel experiment has selected the grades of fuel from a large collection of grades and the carburetors from a large collection of carburetors, the intent being to determine if there are performance differences in the larger collections. Employing previously determined statistics, the random-effects test statistics are easily determined. Since F_{AB} is the same in both models, the interaction null hypothesis is accepted, and we can proceed to test the main effects. For these,

$$f_A = \frac{MSA}{MSAB} = \frac{19.01}{0.295} = 64.44$$

$$f_B = \frac{MSB}{MSAB} = \frac{6.375}{0.295} = 21.61$$

F_A possesses 1 and 2 degrees of freedom and $f_{0.01,1,2} = 98.50$. Thus, H_{0A} cannot be rejected at the 0.01 level. Neither can H_{0B}, since F_B possesses 2 and 2 degrees of freedom and $f_{0.01,2,2} = 99.00$. However, both null hypotheses can be rejected at the 0.05 level, since $f_{0.05,1,2} = 18.51$ and $f_{0.05,2,2} = 19.00$. Note that H_{0B} is barely rejected even at the 0.05 level. If we determine to proceed on the basis of the rejections at the 0.05 level, then estimates for the main-effect variances are

$$\overset{*}{\sigma}_A^2 = \frac{19.01 - 0.295}{9} = 2.08$$

$$\overset{*}{\sigma}_B^2 = \frac{6.375 - 0.295}{6} = 1.01$$

In the carburetor–fuel analysis, the results were different in the fixed- and random-effects cases. Intuitively, the random nature of the treatments in the random-effects case introduces greater variability and makes it harder to obtain rejection. Quantitatively, the number of degrees of freedom associated with $MSAB$ is smaller than the number associated with MSE, thereby yielding F statistics with substantially less denominator degrees of freedom in the random-effects model.

14.3.5 Two-Way Classification with Mixed Effects

A two-way classification can also be performed if levels of one of the factors, say A, are selected deterministically and levels of the other, say B, are selected randomly. In such a design, the effects of A are fixed, the effects of B are random, and the overall design is said to possess **mixed effects.** The model equation is

Model Equation for Mixed-Effects Two-Way Classification

$$\boxed{X_{ijk} = \mu + \alpha_i + B_j + (AB)_{ij} + E_{ijk}}$$

for $i = 1, 2, \ldots, a$, $j = 1, 2, \ldots, b$, and $k = 1, 2, \ldots, n$. The effects of A being fixed, the constraint is imposed on the model. In addition, B_j and E_{ijk} are independent zero-mean normally distributed random variables possessing variances σ_B^2 and σ^2, respectively. The interaction effect $(AB)_{ij}$ is random. It is assumed to possess a zero-mean normal distribution with variance $[(a-1)/a]\sigma_{AB}^2$, where the coefficient $(a-1)/a$ has been introduced to simplify the expressions for the expected mean squares. An additional constraint is placed on the interaction effects, namely, that the sum of the interaction effects over any fixed j must be zero:

$$\sum_{i=1}^{a} (AB)_{ij} = 0$$

for $j = 1, 2, \ldots, b$. As a result, for fixed j, $(AB)_{1j}, (AB)_{2j}, \ldots, (AB)_{aj}$ are not independent. In fact, it can be shown that they are negatively correlated. As for the sums of squares and the mean squares, they are defined as in the fixed- and random-effects models.

The factor-A, factor-B, and interaction hypothesis tests in the mixed-effects model are

$$H_{0A}: \alpha_1 = \alpha_2 = \cdots = \alpha_a = 0$$
$$H_{1A}: \alpha_i \neq 0 \text{ for some } i$$

$$H_{0B}: \sigma_B^2 = 0$$
$$H_{1B}: \sigma_B^2 > 0$$

and

$$H_{0AB}: \sigma_{AB}^2 = 0$$
$$H_{1AB}: \sigma_{AB}^2 > 0$$

respectively.

As usual, the appropriate F test statistics will be chosen in accordance with the expected values of the mean squares. These are

Mean-Square Expected Values in the Mixed-Effects Two-Way Classification

$$E[MSA] = \sigma^2 + n\sigma_{AB}^2 + \frac{nb}{a-1}\sum_{i=1}^{a}\alpha_i^2$$

$$E[MSB] = \sigma^2 + na\sigma_B^2$$

$$E[MSAB] = \sigma^2 + n\sigma_{AB}^2$$

$$E[MSE] = \sigma^2$$

Examination of these expected values yields the desired F statistics. For the null hypotheses H_{0A}, H_{0B}, and H_{0AB}, the F statistics are $F_A = MSA/MSAB$, $F_B = MSB/MSE$, and $F_{AB} = MSAB/MSE$, respectively. These statistics possess $a-1$ and $(a-1)(b-1)$, $b-1$ and $ab(n-1)$, and $(a-1)(b-1)$ and $ab(n-1)$ degrees of freedom, respectively. Each test is performed in the customary upper one-tailed manner. Interchanging the roles of A and B provides the analysis for A random and B fixed. Both possibilities are summarized in Table 14.16.

TABLE 14.16 ANOVA TABLE FOR MIXED-EFFECTS TWO-WAY CLASSIFICATION

Source of Variation	Sum of Squares	Degrees of Freedom	Mean Square	A Fixed, B Random F Test Statistic	B Fixed, A Random F Test Statistic
Factor A	SSA	$a - 1$	MSA	$F_A = \dfrac{MSA}{MSAB}$	$F_A = \dfrac{MSA}{MSE}$
Factor B	SSB	$b - 1$	MSB	$F_B = \dfrac{MSB}{MSE}$	$F_B = \dfrac{MSB}{MSAB}$
Interaction	SSAB	$(a - 1)(b - 1)$	MSAB	$F_{AB} = \dfrac{MSAB}{MSE}$	$F_{AB} = \dfrac{MSAB}{MSE}$
Error	SSE	$ab(n - 1)$	MSE		
Total	SST	$N - 1$			

EXAMPLE 14.14

Suppose the carburetors of Example 14.11 are chosen deterministically by the experimenter but the fuels are selected randomly. Employing information from Examples 14.11 and 14.13, the interaction null hypothesis is accepted, the value of F_A is 64.44, and the value of F_B is 38.87. Since $f_{0.01,1,2} = 98.50$ and $f_{0.05,1,2} = 18.51$, the factor-A null hypothesis is rejected at level 0.05 but accepted at level 0.01. Since $f_{0.01,2,12} = 6.93$, the factor-B null hypothesis is rejected at level 0.01.

14.3.6 Two-Way Classification with a Single Observation Per Treatment Combination

To this point, it has been assumed in the two-way model that $n > 1$. We now consider the case of fixed effects when there is only a single observation per treatment combination. In such a case, the model equation of Section 14.3.2 reduces to

$$X_{ij} = \mu + \alpha_i + \beta_j + (\alpha\beta)_{ij} + E_{ij}$$

With $n = 1$, the definition of SSE given in Section 14.3.3 reduces to zero, since each term of the sum measures the squared deviation of X_{ijk} from $\bar{X}_{ij.}$. Consequently, in the single-observation model, experimental error is not separable from two-factor interaction in the same way that it is for multiple replications.

For a single observation, Theorem 14.6 reduces to

$$SST = SSA + SSB + SSAB$$

where SSAB represents both residual and interaction variation. With $n = 1$, the expressions of Theorem 14.8 become

$$SST = \sum_{i=1}^{a} \sum_{j=1}^{b} X_{ij}^2 - \frac{T_{..}^2}{ab}$$

$$SSA = \frac{1}{ab}\left[a\sum_{i=1}^{a} T_{i.}^2 - T_{..}^2\right]$$

$$SSB = \frac{1}{ab}\left[b\sum_{j=1}^{b} T_{.j}^2 - T_{..}^2 \right]$$

$$SSAB = SST - SSA - SSB$$

SST, SSA, SSB, and $SSAB$ possess $ab - 1$, $a - 1$, $b - 1$, and $(a - 1)(b - 1)$ degrees of freedom, respectively (see Table 14.13).

With a single observation per treatment combination, the expected-value expressions of Theorem 14.7 for MSA, MSB, and $MSAB$ are modified only to the extent that $n = 1$. However, since MSE is not defined independently of $MSAB$, there is no expected mean square giving σ^2 unless either the main effects for factor A, the main effects for factor B, or the interaction effects are identically zero. In other words, one of the null hypotheses, H_{0A}, H_{0B}, or H_{0AB}, must hold. Referring to Table 14.13, the problem is evident: tests for H_{0A}, H_{0B}, and H_{0AB} utilize F statistics with the estimator (for σ^2) MSE in the denominator.

Should we assume the interaction effect to be zero, then $SSAB$ must represent residual error and can be written as SSE. Moreover, the model equation takes the form

$$X_{ij} = \mu + \alpha_i + \beta_j + E_{ij}$$

and, according to Theorem 14.7, $E[MSE] = \sigma^2$. Main effects can be tested by using the statistics $F_A = MSA/MSE$ and $F_B = MSB/MSE$. Should there actually be interaction, then misleading results might be obtained.

Tukey has developed a test to detect significant interaction in the single-observation two-factor model. We define the **nonadditivity** sum of squares SSN, which, in its computational form, is given by

$$SSN = \frac{\left[\sum_{i=1}^{a} \sum_{j=1}^{b} X_{ij} T_{i.} T_{.j} - T_{..}\left(SSA + SSB + \frac{T_{..}^2}{ab} \right) \right]^2}{ab(SSA)(SSB)}$$

Recall that interaction results in nonadditivity. Using SSN, we can partition the interaction (residual) sum of squares $SSAB$ into the nonadditivity component SSN and an error component

$$SSE = SSAB - SSN$$

$SSAB$, SSE, and SSN possess $(a - 1)(b - 1)$, $(a - 1)(b - 1) - 1$, and 1 degree of freedom, respectively. The test statistic for the hypothesis test

H_{0N}: There exists no interaction.

H_{1N}: There exists interaction.

is

$$F_N = \frac{MSN}{MSE}$$

TABLE 14.17 COMPRESSIVE-STRENGTH DATA FOR FORMULA–METHOD EXPERIMENT

Formula	Method 1	Method 2	Method 3	Total	Mean
1	3920	3860	4170	11,950	3983.3
2	4000	3930	4260	12,190	4063.3
3	3890	3900	4190	11,980	3993.3
4	4060	3990	4250	12,300	4100.0
Total	15,870	15,680	16,870	48,420	
Mean	3967.5	3920.0	4217.5		4035.0

where $MSN = SSN/1$ is the nonadditivity mean square and

$$MSE = \frac{SSE}{(a-1)(b-1)-1}$$

is the error mean square. Under the null hypothesis H_{0N}, F_N possesses an F distribution with 1 and $(a-1)(b-1)-1$ degrees of freedom. H_{0N} is rejected at level α if

$$F_N \geq f_{\alpha,1,(a-1)(b-1)-1}$$

Note the drawback of Tukey's test: rejection of the null hypothesis indicates interaction and therefore leads us not to apply the single-observation model under the assumption of no interaction; however, acceptance of the null hypothesis does not necessarily give compelling support for proceeding under the assumption of no interaction.

EXAMPLE 14.15

An investigator performs a two-factor test, with the four levels of factor A being four different cement-mix formulas and the three levels of factor B being three different methods of mixing. A single observation of compressive strength for each treatment combination is taken and the resulting data (in psi) are tabulated in Table 14.17. Although interaction is believed to be negligible, Tukey's test is applied to see if there is significant evidence to reject the hypothesis of no interaction. According to the data in the table,

$$SST = (3920)^2 + (4000)^2 + \cdots + (4250)^2 - \frac{(48,420)^2}{12} = 237,100$$

$$SSA = \frac{1}{12}\{4[(11,950)^2 + \cdots + (12,300)^2] - (48,420)^2\} = 28,300$$

$$SSB = \frac{1}{12}\{3[(15,870)^2 + (15,680)^2 + (16,870)^2] - (48,420)^2\} = 204,350$$

$$SSAB = 237,100 - 28,300 - 204,350 = 4450$$

The nonadditivity sum of squares is found in the following manner:

$$\sum_{i=1}^{4} \sum_{j=1}^{3} X_{ij}T_{i.}T_{.j} = (3920)(11{,}950)(15{,}870) + \cdots + (4250)(12{,}300)(16{,}870)$$

$$= 9{,}471{,}300{,}717{,}000$$

$$T_{..}\left(SSA + SSB + \frac{T_{..}^{2}}{ab}\right) = (48{,}420)\left[28{,}300 + 204{,}350 + \frac{(48{,}420)^{2}}{12}\right]$$

$$= 9{,}471{,}307{,}887{,}000$$

$$ab(SSA)(SSB) = (12)(28{,}300)(204{,}350) = 69{,}397{,}260{,}000$$

$$SSN = \frac{(9{,}471{,}300{,}717{,}000 - 9{,}471{,}307{,}887{,}000)^{2}}{69{,}397{,}260{,}000} = 740.79$$

Thus, the error sum of squares in Tukey's test, the corresponding error mean square, and the nonadditivity F statistic are

$$SSE = SSAB - SSN = 3709.2$$

$$MSE = \frac{3709.2}{5} = 741.84$$

$$F_{N} = \frac{MSN}{MSE} = \frac{740.79}{741.84} = 0.999$$

Since $f_{0.25,1,5} = 1.69$, the hypothesis of no interaction is not even rejected at the 0.25 level of significance.

Proceeding under the assumption of zero interaction, we employ the model for which there is no $(\alpha\beta)_{ij}$ term. Thus, we assume that $SSE = SSAB = 4450$. Then

$$MSA = \frac{28{,}300}{3} = 9433.33$$

$$MSB = \frac{204{,}350}{2} = 102{,}175$$

$$MSE = \frac{4450}{6} = 741.67$$

$$f_{A} = \frac{MSA}{MSE} = \frac{9433.33}{741.67} = 12.719$$

$$f_{B} = \frac{MSB}{MSE} = \frac{102{,}175}{741.67} = 137.76$$

F_{A} has 3 and 6 degrees of freedom and F_{B} has 2 and 6 degrees of freedom. Since $f_{0.01,3,6} = 9.78$ and $f_{0.01,2,6} = 10.92$, both the factor-A and factor-B null hypotheses are rejected at the 0.01 level, and it is concluded (at level 0.01) that both the mixing formula and the mixing method possess nonzero main effects. The ANOVA table of Table 14.18 summarizes the analysis of variance for the main-effect tests (with $SSE = SSAB$).

Under the assumption of no interaction, the single-observation two-way-classification model equation appears identical to the model equation for the randomized complete block design. Nevertheless, one should recognize that the interpretations of the models are not the same, since each involves a different set of experimental conditions. In the two-way design, ab observations are made in a completely random fashion, whereas in the block design, the observations corresponding to each replication are randomized within a block.

TABLE 14.18 ANOVA TABLE FOR MAIN-EFFECT TESTS IN EXAMPLE 14.15

Source of Variation	Sum of Squares	Degrees of Freedom	Mean Square	F Test Statistic
Formula	28,300	3	9433.33	12.72
Method	204,350	2	102,175	137.76
Error	4450	6	741.67	
Total	237,100	11		

14.4 THREE-WAY CLASSIFICATION

The concepts of the two-way-classification design can be extended to factorial experiments in which there are more than two factors. Although the computational burden grows very fast as the number of factors increases, the extension is quite straightforward for fixed effects. However, if at least one factor possesses random effects, then the composition of the test statistics is more problematic. As a result, we will restrict our attention to fixed effects and develop the analysis-of-variance procedure for the three-way classification.

It will be assumed that there are three factors A, B, and C possessing a, b, and c levels, respectively, and that there are n replications of the experiment, $n > 1$. Thus, the total number of observations is $N = nabc$. For fixed effects, the three-way-classification model is determined by the linear equation

Model Equation for Fixed-Effects Three-Way Classification

$$X_{ijkl} = \mu + \alpha_i + \beta_j + \gamma_k + (\alpha\beta)_{ij} + (\alpha\gamma)_{ik} + (\beta\gamma)_{jk} + (\alpha\beta\gamma)_{ijk} + E_{ijkl}$$

for $i = 1, 2, \ldots, a$, $j = 1, 2, \ldots, b$, $k = 1, 2, \ldots, c$, and $l = 1, 2, \ldots, n$. The model parameters α_i, β_j, and γ_k represent the main effects of the factors A, B, and C, respectively. The parameters $(\alpha\beta)_{ij}$, $(\alpha\gamma)_{ik}$, and $(\beta\gamma)_{jk}$ represent the two-factor interactions between A and B, A and C, and B and C, respectively. The parameter $(\alpha\beta\gamma)_{ijk}$ represents three-factor interaction. As usual, the E_{ijkl} constitute a collection of independent normally distributed zero-mean random variables possessing common variance σ^2. The constraint on the model parameters is that the sum over any subscript indexing the parameter must be zero. For instance,

$$\sum_{j=1}^{b} (\beta\gamma)_{jk} = \sum_{k=1}^{c} (\alpha\beta\gamma)_{ijk} = 0$$

Once again, the analysis proceeds in terms of certain averages, these involving the following summations:

$$T_{i\ldots} = \sum_{j=1}^{b} \sum_{k=1}^{c} \sum_{l=1}^{n} X_{ijkl}$$

$$T_{.j..} = \sum_{i=1}^{a} \sum_{k=1}^{c} \sum_{l=1}^{n} X_{ijkl}$$

$$T_{..k.} = \sum_{i=1}^{a} \sum_{j=1}^{b} \sum_{l=1}^{n} X_{ijkl}$$

$$T_{ij..} = \sum_{k=1}^{c} \sum_{l=1}^{n} X_{ijkl}$$

$$T_{i.k.} = \sum_{j=1}^{b} \sum_{l=1}^{n} X_{ijkl}$$

$$T_{.jk.} = \sum_{i=1}^{a} \sum_{l=1}^{n} X_{ijkl}$$

$$T_{ijk.} = \sum_{l=1}^{n} X_{ijkl}$$

$$T_{....} = \sum_{i=1}^{a} \sum_{j=1}^{b} \sum_{k=1}^{c} \sum_{l=1}^{n} X_{ijkl}$$

Based upon these sums, the following averages are defined: $\bar{X}_{i...} = T_{i...}/bcn$, $\bar{X}_{.j..} = T_{.j..}/acn$, $\bar{X}_{..k.} = T_{..k.}/abn$, $\bar{X}_{ij..} = T_{ij..}/cn$, $\bar{X}_{i.k.} = T_{i.k.}/bn$, $\bar{X}_{.jk.} = T_{.jk.}/an$, $\bar{X}_{ijk.} = T_{ijk.}/n$, and $\bar{X}_{....} = T_{....}/abcn$. Under the model constraints, unbiased estimators of the model parameters are given in terms of these averages:

Model-Parameter Estimators in the Fixed-Effects Three-Way Classification

$$\hat{\mu} = \bar{X}_{....}$$

$$\hat{\alpha}_i = \bar{X}_{i...} - \bar{X}_{....}$$

$$\widehat{(\alpha\beta)}_{ij} = \bar{X}_{ij..} - \bar{X}_{i...} - \bar{X}_{.j..} + \bar{X}_{....}$$

$$\widehat{(\alpha\beta\gamma)}_{ijk} = \bar{X}_{ijk.} - \bar{X}_{ij..} - \bar{X}_{i.k.} - \bar{X}_{.jk.} + \bar{X}_{i...} + \bar{X}_{.j..} + \bar{X}_{..k.} - \bar{X}_{....}$$

The remaining parameter estimators can be obtained by symmetry.

The total and error sums of squares are defined by

$$SST = \sum_{i=1}^{a} \sum_{j=1}^{b} \sum_{k=1}^{c} \sum_{l=1}^{n} (X_{ijkl} - \bar{X}_{....})^2$$

$$SSE = \sum_{i=1}^{a} \sum_{j=1}^{b} \sum_{k=1}^{c} \sum_{l=1}^{n} (X_{ijkl} - \bar{X}_{ijk.})^2$$

As in two-way classification, sums of squares for main and interaction effects are defined by summing the squares of the corresponding estimators over all indices, SSA by summing $\hat{\alpha}_i^2$, $SSAB$ by summing $\widehat{(\alpha\beta)}_{ij}^2$, and so on. Thus,

$$SSA = \sum_{i=1}^{a} \sum_{j=1}^{b} \sum_{k=1}^{c} \sum_{l=1}^{n} \hat{\alpha}_i^2 = nbc \sum_{i=1}^{a} (\bar{X}_{i...} - \bar{X}_{....})^2$$

$$SSAB = \sum_{i=1}^{a} \sum_{j=1}^{b} \sum_{k=1}^{c} \sum_{l=1}^{n} \widehat{(\alpha\beta)}_{ij}^2 = nc \sum_{i=1}^{a} \sum_{j=1}^{b} \widehat{(\alpha\beta)}_{ij}^2$$

$$SSABC = \sum_{i=1}^{a} \sum_{j=1}^{b} \sum_{k=1}^{c} \sum_{l=1}^{n} \widehat{(\alpha\beta\gamma)}_{ijk}^2 = n \sum_{i=1}^{a} \sum_{j=1}^{b} \sum_{k=1}^{c} \widehat{(\alpha\beta\gamma)}_{ijk}^2$$

Other sums of squares can be found in a similar manner. The fundamental sum-of-squares identity takes the form

$$SST = SSA + SSB + SSC + SSAB + SSAC + SSBC + SSABC$$

Computational expressions for the sums of squares in terms of the previously defined totals are given by

$$SST = \sum_{i=1}^{a} \sum_{j=1}^{b} \sum_{k=1}^{c} \sum_{l=1}^{n} X_{ijkl}^2 - \frac{T_{....}^2}{N}$$

$$SSA = \frac{1}{N}\left[a \sum_{i=1}^{a} T_{i...}^2 - T_{....}^2 \right]$$

$$SSB = \frac{1}{N}\left[b \sum_{j=1}^{b} T_{.j..}^2 - T_{....}^2 \right]$$

$$SSC = \frac{1}{N}\left[c \sum_{k=1}^{c} T_{..k.}^2 - T_{....}^2 \right]$$

$$SSAB = \frac{1}{cn} \sum_{i=1}^{a} \sum_{j=1}^{b} T_{ij..}^2 - \frac{T_{....}^2}{N} - SSA - SSB$$

$$SSAC = \frac{1}{bn} \sum_{i=1}^{a} \sum_{k=1}^{c} T_{i.k.}^2 - \frac{T_{....}^2}{N} - SSA - SSC$$

$$SSBC = \frac{1}{an} \sum_{j=1}^{b} \sum_{k=1}^{c} T_{.jk.}^2 - \frac{T_{....}^2}{N} - SSB - SSC$$

$$SSABC = \frac{1}{n} \sum_{i=1}^{a} \sum_{j=1}^{b} \sum_{k=1}^{c} T_{ijk.}^2 - \frac{T_{....}^2}{N}$$
$$- SSA - SSB - SSC - SSAB - SSAC - SSBC$$

$$SSE = SST - SSA - SSB - SSC - SSAB - SSAC - SSBC - SSABC$$

Based upon the distributional properties of the sums of squares, the corresponding mean squares can be found. The number of degrees of freedom corresponding to each mean square is given in Table 14.19. For instance, referring to the table, we recognize that

$$MSAB = \frac{SSAB}{(a-1)(b-1)}$$

$$MSABC = \frac{SSABC}{(a-1)(b-1)(c-1)}$$

The expected values of the mean squares MSA, $MSAB$, and $MSABC$ are

$$E[MSA] = \sigma^2 + \frac{nbc}{a-1} \sum_{i=1}^{a} \alpha_i^2$$

$$E[MSAB] = \sigma^2 + \frac{nc}{(a-1)(b-1)} \sum_{i=1}^{a} \sum_{j=1}^{b} (\alpha\beta)_{ij}^2$$

$$E[MSABC] = \sigma^2 + \frac{n}{(a-1)(b-1)(c-1)} \sum_{i=1}^{a} \sum_{j=1}^{b} \sum_{k=1}^{c} (\alpha\beta\gamma)_{ijk}^2$$

Using symmetry, expressions for $E[MSB]$ and $E[MSC]$ can be obtained from $E[MSA]$, and expressions for $E[MSAC]$ and $E[MSBC]$ can be obtained from $E[MSAB]$. As usual, $E[MSE] = \sigma^2$. Employing the expected values of the mean squares, appropriate F test statistics can be developed for the hypothesis tests that are relevant to the model. Each null hypothesis states that all parameters of a certain type are zero. For instance, the factor-A null hypothesis is

$$H_{0A}: \alpha_i = 0 \qquad \text{for all } i$$

the AB-interaction null hypothesis is

$$H_{0AB}: (\alpha\beta)_{ij} = 0 \qquad \text{for all } (i, j)$$

and the ABC-interaction null hypothesis is

$$H_{0ABC}: (\alpha\beta\gamma)_{ijk} = 0 \qquad \text{for all } (i, j, k)$$

Other null hypotheses are constructed by symmetry. The fixed-effects three-way-classification analysis is summarized in Table 14.19.

TABLE 14.19 ANOVA TABLE FOR FIXED-EFFECTS THREE-WAY CLASSIFICATION

Source of Variation	Sum of Squares	Degrees of Freedom	Mean Square	F Test Statistic
A	SSA	$a-1$	MSA	$F_A = \dfrac{MSA}{MSE}$
B	SSB	$b-1$	MSB	$F_B = \dfrac{MSB}{MSE}$
C	SSC	$c-1$	MSC	$F_C = \dfrac{MSC}{MSE}$
AB	$SSAB$	$(a-1)(b-1)$	$MSAB$	$F_{AB} = \dfrac{MSAB}{MSE}$
AC	$SSAC$	$(a-1)(c-1)$	$MSAC$	$F_{AC} = \dfrac{MSAC}{MSE}$
BC	$SSBC$	$(b-1)(c-1)$	$MSBC$	$F_{BC} = \dfrac{MSBC}{MSE}$
ABC	$SSABC$	$(a-1)(b-1)(c-1)$	$MSABC$	$F_{ABC} = \dfrac{MSABC}{MSE}$
Error	SSE	$abc(n-1)$	MSE	
Total	SST	$N-1$		

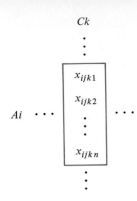

FIGURE 14.5
Data format for a three-way classification

When performing a three-way-classification analysis of variance, the hypotheses of higher-order interaction should be checked first. Just as two-factor interaction can mask the main effects in the two-way model and thereby lead to a lack of detection, three-factor interaction can have a detrimental masking effect in the three-way model. Thus, unless the ABC-interaction null hypothesis H_{0ABC} is accepted, one must be prudent when interpreting the results of the paired-interaction analysis. Similar precaution must be taken when proceeding to the main-effect tests if H_{0AB}, H_{0AC}, and H_{0BC} are not accepted.

Because of the complexity of the three-way model (and of higher-order multiway models), higher-order interaction terms are sometimes omitted from the model under the assumption that they are zero. While such a course is appealing, it is also somewhat risky unless one has strong evidence to justify it. To be on the safe side, one might better use a little more computer time and undertake the entire analysis. Once terms in the model are shown to be zero, they might then be dropped from the model in future considerations, so long as one keeps in mind the possible masking of lower-order effects by higher-order effects.

In three-way classification, convenient display of the data is important to avoid errors when entering data. To avoid confusion between levels of one factor and another, we will label the levels of factors A, B, and C by Ai, Bj, and Ck, respectively, and record data in b tables. For $j = 1, 2, \ldots, b$, table j holds data corresponding to the jth level of factor B, the rows holding data for the various levels of factor A and the columns holding data for the various levels of factor C. A typical cell of data in table j is of the form depicted in Figure 14.5. Since the totals play the key role in computation, it is convenient to also put them into tables. For $j = 1, 2, \ldots, b$, the totals T_{ijk} appear in Table 14.20, there being b such tables. The AB-combination totals $T_{ij..}$ are tabulated in the format specified by Table 14.21. The AC-combination

TABLE 14.20   FORMAT OF
REPLICATION TOTALS FOR LEVEL Bj

	$C1$	$C2$	$\ldots$	Cc
$A1$	$T_{1j1.}$	$T_{1j2.}$	$\ldots$	$T_{1jc.}$
$A2$	$T_{2j1.}$	$T_{2j2.}$	$\ldots$	$T_{2jc.}$
$\vdots$	$\vdots$	$\vdots$		$\vdots$
Aa	$T_{aj1.}$	$T_{aj2.}$	$\ldots$	$T_{ajc.}$

TABLE 14.21   FORMAT OF
AB-COMBINATION TOTALS

	$B1$	$B2$	$\ldots$	Bb	Total
$A1$	$T_{11..}$	$T_{12..}$	$\ldots$	$T_{1b..}$	$T_{1...}$
$A2$	$T_{21..}$	$T_{22..}$	$\ldots$	$T_{2b..}$	$T_{2...}$
$\vdots$	$\vdots$	$\vdots$		$\vdots$	$\vdots$
Aa	$T_{a1..}$	$T_{a2..}$	$\ldots$	$T_{ab..}$	$T_{a...}$
Total	$T_{.1..}$	$T_{.2..}$	$\ldots$	$T_{.b..}$	$T_{....}$

totals $T_{i.k.}$ and the BC-combination totals $T_{.jk.}$ are formatted similarly. Note that the row and column totals of the AB-, AC-, and BC-combination tables give sums of the form $T_{i...}$, $T_{.j..}$, and $T_{..k.}$.

EXAMPLE 14.16

Suppose an industrial engineer wishes to compare three factors affecting a particular assembly, these being the skill (A) of the personnel, the production method (B), and fatigue (C). Skill levels are rated from top down as $A1$, $A2$, and $A3$. Method complexity is rated as higher ($B1$) or lower ($B2$). Fatigue is determined by morning production ($C1$) or afternoon production ($C2$). The data correspond to a rating system devised to measure both the quantity and quality of production. Two replications are employed, and the data are tabulated in Table 14.22. Replication totals are given in Table 14.23, and combination totals are given in Tables 14.24, 14.25, and 14.26. According to the data, $T_{....} = 1700$, $T^2_{....} = 2{,}890{,}000$,

$$\frac{T^2_{....}}{N} = 120{,}416.67$$

$$SST = (92)^2 + (88)^2 + \cdots + (59)^2 - 120{,}416.67 = 2475.33$$

$$SSA = \frac{3[(667)^2 + (549)^2 + (484)^2] - 2{,}890{,}000}{24} = 2151.58$$

$$SSB = \frac{2[(875)^2 + (825)^2] - 2{,}890{,}000}{24} = 104.17$$

$$SSC = \frac{2[(864)^2 + (836)^2] - 2{,}890{,}000}{24} = 32.67$$

$$SSAB = \frac{(352)^2 + \cdots + (245)^2}{4} - 120{,}416.67 - 2151.58 - 104.17 = 116.58$$

$$SSAC = \frac{(340)^2 + \cdots + (239)^2}{4} - 120{,}416.67 - 2151.58 - 32.67 = 3.08$$

$$SSBC = \frac{(445)^2 + \cdots + (406)^2}{6} - 120{,}416.67 - 104.17 - 32.67 = 0.17$$

$$SSABC = \frac{(180)^2 + \cdots + (121)^2}{2} - 120{,}416.67 - 2151.58 - 104.17 - 32.67$$
$$- 116.58 - 3.08 - 0.17 = 1.08$$

and $SSE = 66.00$. The degrees of freedom, mean squares, and values of the F test statistics are tabulated in Table 14.27. Since F_{ABC} possesses 2 and 12 degrees of freedom and $f_{0.05,2,12} = 3.89$, H_{0ABC} is accepted and three-way interaction is assumed to be zero. F_{AC} also possesses 2 and 12 degrees of freedom and therefore AC interaction is assumed to be null. BC interaction is also concluded to be zero, the appropriate critical value being $f_{0.05,1,12} = 4.75$. On the other hand, AB interaction is quite strong, and H_{0AB} is rejected at the 0.01 level, since

$$f_{AB} = 10.60 > f_{0.01,2,12} = 6.93$$

Insofar as the main effects are concerned, factor C is not affected by higher-order interaction, and F_C possesses 1 and 12 degrees of freedom. Hence, nullity of the C effects is rejected at the 0.05 level and accepted at the 0.01 level. Even though factors A and B are affected by higher-order interaction, both F_A and F_B substantially exceed their respective critical values at the 0.01 significance level, the critical values being 6.93 and 9.33, respectively. Therefore, we conclude that neither factor A nor factor B possesses zero effects.

TABLE 14.22 THREE-WAY CLASSIFICATION OF SKILL–METHOD–FATIGUE DATA

B1

	C1	C2
A1	92	88
	88	84
A2	70	69
	74	71
A3	62	60
	59	58

B2

	C1	C2
A1	78	78
	82	77
A2	69	63
	66	67
A3	64	62
	60	59

TABLE 14.23 REPLICATION TOTALS FOR SKILL–METHOD–FATIGUE DATA

B1

	C1	C2
A1	180	172
A2	144	140
A3	121	118

B2

	C1	C2
A1	160	155
A2	135	130
A3	124	121

TABLE 14.24 *AB*-COMBINATION TOTALS FOR SKILL–METHOD–FATIGUE DATA

	B1	B2	$T_{i..}$
A1	352	315	667
A2	284	265	549
A3	239	245	484
$T_{.j..}$	875	825	1700

TABLE 14.25 *AC*-COMBINATION TOTALS FOR SKILL–METHOD–FATIGUE DATA

	C1	C2	$T_{i..}$
A1	340	327	667
A2	279	270	549
A3	245	239	484
$T_{..k.}$	864	836	1700

TABLE 14.26 *BC-*COMBINATION TOTALS FOR SKILL–METHOD–FATIGUE DATA

	C1	C2	$T_{.j.}$
B1	445	430	875
B2	419	406	825
$T_{..k.}$	864	836	1700

TABLE 14.27 ANOVA TABLE FOR EXAMPLE 14.16

Source of Variation	Sum of Squares	Degrees of Freedom	Mean Square	F Test Statistic
Skill	2151.58	2	1075.79	195.60
Method	104.17	1	104.17	18.94
Fatigue	32.67	1	32.67	5.94
AB interaction	116.58	2	58.29	10.60
AC interaction	3.08	2	1.54	0.28
BC interaction	0.17	1	0.17	0.03
ABC interaction	1.08	2	0.54	0.10
Error	66.00	12	5.50	
Total	2475.33	23		

14.5 2^k FACTORIAL DESIGNS

A special case of a multifactor design occurs when the experiment is run with each factor possessing two levels. If there are k factors, then there are 2^k treatment combinations, and each replicate of the experiment requires 2^k observations. In certain circumstances, the investigator might actually be concerned with more than two levels of each factor but to save expense decides to perform a preliminary screening experiment with just two levels of each factor. The resulting design is called a **2^k factorial design.** It is common to refer to the two levels of each factor as the **low** and **high** levels and to denote them by 0 and 1, respectively. Although these designs can be investigated as instances of general factorial designs, special methods have been developed to treat them. It will be assumed that the conditions of the completely randomized fixed-effects multiway classification are in effect.

14.5.1 The 2^2 Factorial Design

In the fixed-effects two-way classification with n replications and factors A and B, the totals take the format given in Table 14.28. When discussing the 2^2 design, it is common to let (1), a, b, and ab denote the totals of the observations corresponding to the (0, 0), (1, 0), (0, 1), and (1, 1) treatment combinations, respectively. Referring to Table 14.28, $T_{00.} = (1)$, $T_{10.} = a$, $T_{01.} = b$, and $T_{11.} = ab$. Thus, the table takes the form given in Table 14.29.

TABLE 14.28
STANDARD FORMAT
FOR TOTALS IN THE
2^2 DESIGN

	B	
A	**0**	**1**
0	$T_{00.}$	$T_{01.}$
1	$T_{10.}$	$T_{11.}$

TABLE 14.29
SPECIAL FORMAT
FOR TOTALS IN THE
2^2 DESIGN

	B	
A	**0**	**1**
0	(1)	b
1	a	ab

Based on the linear model posited in Section 14.3.2,

$$X_{ijk} = \mu + \alpha_i + \beta_j + (\alpha\beta)_{ij} + E_{ijk}$$

where, in the present circumstances, i and j are 0 or 1, and the model constraints become

$$\alpha_0 + \alpha_1 = 0$$

$$\beta_0 + \beta_1 = 0$$

$$(\alpha\beta)_{0j} + (\alpha\beta)_{1j} = 0$$

for $j = 0, 1$, and

$$(\alpha\beta)_{i0} + (\alpha\beta)_{i1} = 0$$

for $i = 0, 1$. Because of the model constraints, there are only three independent parameters for the main and crossing effects. Specifically, if α_0, β_0, and $(\alpha\beta)_{00}$ are known, then $\alpha_1 = -\alpha_0$, $\beta_1 = -\beta_0$, $(\alpha\beta)_{10} = (\alpha\beta)_{01} = -(\alpha\beta)_{00}$, and $(\alpha\beta)_{11} = (\alpha\beta)_{00}$.

According to Theorem 14.5, unbiased estimators for α_0, β_0, and $(\alpha\beta)_{00}$ are given by

$$\hat{\alpha}_0 = \bar{X}_{0..} - \bar{X}_{...}$$

$$= \frac{1}{2n}(T_{00.} + T_{01.}) - \frac{1}{4n}(T_{00.} + T_{01.} + T_{10.} + T_{11.})$$

$$= \frac{1}{4n}(T_{00.} + T_{01.} - T_{10.} - T_{11.})$$

$$= \frac{1}{4n}[(1) + b - a - ab]$$

$$\hat{\beta}_0 = \bar{X}_{.0.} - \bar{X}_{...}$$
$$= \frac{1}{2n}(T_{00.} + T_{10.}) - \frac{1}{4n}(T_{00.} + T_{01.} + T_{10.} + T_{11.})$$
$$= \frac{1}{4n}(T_{00.} + T_{10.} - T_{01.} - T_{11.})$$
$$= \frac{1}{4n}[(1) + a - b - ab]$$

$$(\widehat{\alpha\beta})_{00} = \bar{X}_{00.} - \bar{X}_{0..} - \bar{X}_{.0.} + \bar{X}_{...}$$
$$= \frac{1}{n}T_{00.} - \frac{1}{2n}(T_{00.} + T_{01.}) - \frac{1}{2n}(T_{00.} + T_{10.})$$
$$+ \frac{1}{4n}(T_{00.} + T_{01.} + T_{10.} + T_{11.})$$
$$= \frac{1}{4n}(T_{00.} - T_{01.} - T_{10.} + T_{11.})$$
$$= \frac{1}{4n}[(1) - b - a + ab]$$

According to the model constraints, $\hat{\alpha}_1 = -\hat{\alpha}_0$. Hence,

$$SSA = \sum_{i=0}^{1} \sum_{j=0}^{1} \sum_{k=1}^{n} \hat{\alpha}_i^2$$
$$= 2n[\hat{\alpha}_0^2 + (-\hat{\alpha}_0)^2]$$
$$= 4n\hat{\alpha}_0^2$$
$$= \frac{1}{4n}[ab + a - b - (1)]^2$$

Similarly, $\hat{\beta}_1 = -\hat{\beta}_0$ and

$$SSB = 2n[\hat{\beta}_0^2 + (-\hat{\beta}_0)^2]$$
$$= 4n\hat{\beta}_0^2$$
$$= \frac{1}{4n}[ab + b - a - (1)]^2$$

Finally,

$$SSAB = \sum_{i=0}^{1} \sum_{j=0}^{1} \sum_{k=1}^{n} (\widehat{\alpha\beta})_{ij}^2$$
$$= n[(\widehat{\alpha\beta})_{00}^2 + (\widehat{\alpha\beta})_{01}^2 + (\widehat{\alpha\beta})_{10}^2 + (\widehat{\alpha\beta})_{11}^2]$$
$$= n\{(\widehat{\alpha\beta})_{00}^2 + [-(\widehat{\alpha\beta})_{00}]^2 + [-(\widehat{\alpha\beta})_{00}]^2 + (\widehat{\alpha\beta})_{00}^2\}$$
$$= 4n(\widehat{\alpha\beta})_{00}^2$$
$$= \frac{1}{4n}[ab - a - b + (1)]^2$$

SSA, *SSB*, and *SSAB* each possess a single degree of freedom. *SST* takes the usual form and possesses $4n - 1$ degrees of freedom. *SSE* is computed from *SST*, *SSA*, *SSB*, and *SSAB* by

$$SSE = SST - SSA - SSB - SSAB$$

Any nontrivial linear combination of treatment totals is called a **contrast.** Examining SSA, SSB, and $SSAB$, we see that each involves a contrast, known variously as the **effect total, total effect,** or **effect contrast.** Let the A, B, and AB contrasts be defined by

$$[A] = ab + a - b - (1)$$
$$[B] = ab + b - a - (1)$$
$$[AB] = ab - a - b + (1)$$

Then the sums of squares take the form

$$SSK = \frac{[K]^2}{4n}$$

where $K = A$, B, or AB.

Having proceeded directly from the linear model for the fixed-effects two-way classification, we will now view the 2^2 design from a different perspective. Referring to Table 14.29, we see that in changing from level 0 of factor A to level 1 of factor A, the total for level 0 of factor B goes from (1) to a and the total for level 1 of factor B goes from b to ab. In the current context, the increments $[a - (1)]/n$ and $(ab - b)/n$ constitute the effects of factor A at the low and high levels of factor B, respectively. The **average effect** of factor A is

$$A = \frac{[a - (1)] + (ab - b)}{2n} = \frac{ab + a - b - (1)}{2n} = \frac{[A]}{2n}$$

where we have followed custom and denoted the average effect by the same letter as the factor. Similarly, the average effect of factor B is the average of the effects of factor B at the low and high levels of A, namely,

$$B = \frac{[b - (1)] + (ab - a)}{2n} = \frac{ab + b - a - (1)}{2n} = \frac{[B]}{2n}$$

The **average interaction effect** is defined to be the average difference between the effect of A at the high level of B and the effect of A at the low level of B, namely,

$$AB = \frac{(ab - b) - [a - (1)]}{2n} = \frac{ab - a - b + (1)}{2n} = \frac{[AB]}{2n}$$

In sum, each average effect is the corresponding contrast divided by $2n$.

Each effect contrast is a linear combination of (1), a, b, and ab whose coefficients are $+1$ and -1. Thus, the symbols $+$ and $-$ are sufficient to indicate the appropriate coefficients for each contrast. Table 14.30 gives the sign of each treatment–combination total in the contrast for each of the factorial effects in the 2^2 design. Reading down a column gives the contrast corresponding to the effect at the column head. Since these contrasts occur in the expressions giving the sums of squares and the average effects, Table 14.30 can also be useful in obtaining these. Note that the rows

TABLE 14.30 SIGNS FOR COMPUTING EFFECT CONTRASTS IN THE 2^2 DESIGN

Treatment Combination	Factorial Effect		
	A	**B**	**AB**
(1)	−	−	+
a	+	−	−
b	−	+	−
ab	+	+	+

of the table follow the order (1), a, b, ab. This ordering of the treatment combinations is called **standard order.**

Keep in mind that the average effects A, B, and AB are actually random variables. Moreover, based on the previous computations of $\hat{\alpha}_0$, $\hat{\beta}_0$, and $\widehat{(\alpha\beta)}_{00}$,

$$A = \frac{[A]}{2n} = -2\hat{\alpha}_0 = 2\hat{\alpha}_1$$

$$B = \frac{[B]}{2n} = -2\hat{\beta}_0 = 2\hat{\beta}_1$$

$$AB = \frac{[AB]}{2n} = 2\widehat{(\alpha\beta)}_{00}$$

Thus, A, B, and AB can serve as estimators for the main effects and the interaction effects. Consequently, A, B, and AB are commonly called the factor-A main effect, the factor-B main effect, and the interaction effect, respectively.

EXAMPLE 14.17

A laboratory uses ovens to heat electronic components in life testing. To test the effects of two temperatures, 600°F and 675°F, and two ovens of different brands, a 2^2 factorial experiment is run with three replications. Let A and B denote the temperature and oven factors, respectively. Moreover, let the low and high levels of A represent the two test temperatures and the low and high levels of B represent the two test ovens. Table 14.31 gives the experimental time-to-failure data in minutes. The effect contrasts, main effects, interaction effect, and effect sums of squares are

$$[A] = 541 + 528 - 642 - 617 = -190$$

$$[B] = 541 + 642 - 528 - 617 = 38$$

$$[AB] = 541 - 528 - 642 + 617 = -12$$

$$A = \frac{-190}{6} = -31.67$$

$$B = \frac{38}{6} = 6.33$$

$$AB = \frac{-12}{6} = -2.00$$

$$SSA = \frac{(-190)^2}{12} = 3008.33$$

$$SSB = \frac{(38)^2}{12} = 120.33$$

$$SSAB = \frac{(12)^2}{12} = 12.00$$

The total sum of squares is

$$SST = \sum_{i=0}^{1} \sum_{j=0}^{1} \sum_{k=1}^{3} X_{ijk}^2 - \frac{T_{...}^2}{12}$$

$$= (200)^2 + (212)^2 + \cdots + (176)^2 - \frac{(2328)^2}{12}$$

$$= 455,028 - 451,632 = 3396$$

The error sum of squares is

$$SSE = SST - SSA - SSB - SSAB$$
$$= 3396.00 - 3008.33 - 120.33 - 12.00 = 255.33$$

The analysis of variance is summarized in Table 14.32. The interaction effect is negligible and the oven effect (B) is not significant at level 0.05, since $f_{0.05,1,8} = 5.32$. However, the temperature effect (A) is significant at level 0.005, since $f_{0.005,1,8} = 14.69$.

Based on the computations of A, B, and AB, we can make estimates of the linear-model parameters α_0, β_0, and $(\alpha\beta)_{00}$:

$$\overset{*}{\alpha}_0 = \frac{-A}{2} = 15.83$$

$$\overset{*}{\beta}_0 = \frac{-B}{2} = -3.17$$

$$\overset{*}{(\alpha\beta)}_{00} = \frac{AB}{2} = -1$$

Other model-parameter estimates can be obtained from the relationships among the parameters.

14.5.2 The 2^3 Factorial Design

The 2^3 design pertains to the fixed-effects three-way classification when each of the three factors, A, B, and C, possesses two levels, 0 (low) and 1 (high). The appropriate linear model, together with model constraints, is given in Section 14.4. As in the 2^2

TABLE 14.31 TEMPERATURE–OVEN DATA

Treatment Combination	Factors		Replicates			
	Temperature (A)	Oven (B)	I	II	III	Total
(1)	0	0	200	212	205	617
a	1	0	183	175	170	528
b	0	1	210	220	212	642
ab	1	1	185	180	176	541

TABLE 14.32   ANOVA TABLE FOR EXAMPLE 14.17

Source of Variation	Sum of Squares	Degrees of Freedom	Mean Square	F Test Statistic
Temperature (A)	3008.33	1	3008.33	94.25
Oven (B)	120.33	1	120.33	3.77
AB	12.00	1	12.00	0.38
Error	255.33	8	31.92	
Total	3396.00	11		

design, it is common to denote the treatment–combination totals by letters that indicate the high-level factors in the combination. Thus, $T_{100.} = a$, $T_{010.} = b$, $T_{001.} = c$, $T_{110.} = ab$, $T_{101.} = ac$, $T_{011.} = bc$, $T_{111.} = abc$, and $T_{000.} = (1)$. It is convenient to represent treatment combinations and corresponding totals in the cube of Figure 14.6.

We could proceed as in the 2^2 design and express the model-parameter estimators of Section 14.4 in terms of the totals. For instance,

$$
\begin{aligned}
\hat{\alpha}_0 &= \bar{X}_{0\ldots} - \bar{X}_{\ldots} \\
&= \frac{1}{4n}(T_{000.} + T_{001.} + T_{010.} + T_{011.}) \\
&\quad - \frac{1}{8n}(T_{000.} + T_{001.} + T_{010.} + T_{011.} + T_{100.} + T_{101.} + T_{110.} + T_{111.}) \\
&= \frac{1}{8n}(T_{000.} + T_{001.} + T_{010.} + T_{011.} - T_{100.} - T_{101.} - T_{110.} - T_{111.}) \\
&= \frac{1}{8n}[(1) + c + b + bc - a - ac - ab - abc]
\end{aligned}
$$

We could then proceed to express the sums of squares SSA, SSB, SSC, SSAB, SSAC, SSBC, and SSABC in terms of the totals. Such a course will not be followed. Instead,

FIGURE 14.6
Treatment combinations in a 2^3 factorial design

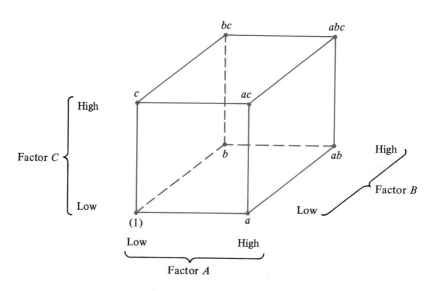

we will employ the intuitive technique discussed in the 2^2 design to find the average effects A, B, C, AB, AC, BC, and ABC. Each of these will involve a contrast in terms of the eight totals.

Referring to Figure 14.6, the effect of factor A when B and C are at the low level is $[a - (1)]/n$; the effect of factor A when B is at the high level and C is at the low level is $(ab - b)/n$; the effect of factor A when B is at the low level and C is at the high level is $(ac - c)/n$; and the effect of factor A when both B and C are at the high level is $(abc - bc)/n$. The average effect of A is therefore

$$A = \frac{a - (1) + ab - b + ac - c + abc - bc}{4n}$$

Note that the numerator of A is a contrast for which the four vertices of the right face of the cube in Figure 14.6 possess the coefficient $+1$ and the four vertices of the left face possess the coefficient -1. The average effect of B can be analyzed similarly. The result is

$$B = \frac{bc + abc + b + ab - c - ac - a - (1)}{4n}$$

where the numerator is the contrast resulting from giving the rear-face vertices the coefficient $+1$ and giving the front-face vertices the coefficient -1. The average C effect is

$$C = \frac{c + bc + abc + ac - b - ab - a - (1)}{4n}$$

The contrast in the numerator of C results from giving the top-face vertices the coefficient $+1$ and the bottom-face vertices the coefficient -1.

It is now necessary to define the interaction average effect AB. When C is at the low level, the effect of the AB interaction is the average difference in the A effect at the high and low levels of B, namely,

$$\frac{(ab - b) - [a - (1)]}{2n}$$

A similar average difference is taken for the AB effect when C is at the high level, namely,

$$\frac{(abc - bc) - (ac - c)}{2n}$$

The average AB effect is the average of the effects when C is at the high and low levels:

$$AB = \frac{ab - b - a + (1) + abc - bc - ac + c}{4n}$$

Analogous reasoning is applied to define the average effects of AC and BC, these being

$$AC = \frac{(1) - a + b - ab - c + ac - bc + abc}{4n}$$

and

$$BC = \frac{(1) + a - b - ab - c - ac + bc + abc}{4n}$$

The average ABC effect is defined as the average difference between the AB interaction effects for the high and low levels of C. These were previously found in the computation of AB. The resulting average effect is

$$ABC = \frac{(abc - bc) - (ac - c) - (ab - b) + [a - (1)]}{4n}$$

$$= \frac{abc - bc - ac + c - ab + b + a - (1)}{4n}$$

For each average effect, the numerator is a contrast, the contrasts for A, B, C, AB, AC, BC, and ABC being denoted by $[A]$, $[B]$, $[C]$, $[AB]$, $[AC]$, $[BC]$, and $[ABC]$, respectively. Each of these contrasts is a linear combination of (1), a, b, c, ab, ac, bc, and abc. The coefficients in each contrast are $+1$ and -1. Table 14.33 gives the signs for each contrast, the signs being found by reading down the appropriate column. For any treatment combination (row of the table), given the signs corresponding to the factors A, B, and C, each interaction-effect sign can be found by multiplying the signs of the factors comprising the interaction. For instance,

$$\text{sign } AB = (\text{sign } A)(\text{sign } B)$$

and

$$\text{sign } ABC = (\text{sign } A)(\text{sign } B)(\text{sign } C)$$

The row ordering (1), a, b, ab, c, ac, bc, abc is the standard ordering in the 2^3 design.

TABLE 14.33 SIGNS FOR COMPUTING EFFECT CONTRASTS IN THE 2^3 DESIGN

Treatment Combination	Factorial Effect						
	A	B	AB	C	AC	BC	ABC
(1)	$-$	$-$	$+$	$-$	$+$	$+$	$-$
a	$+$	$-$	$-$	$-$	$-$	$+$	$+$
b	$-$	$+$	$-$	$-$	$+$	$-$	$+$
ab	$+$	$+$	$+$	$-$	$-$	$-$	$-$
c	$-$	$-$	$+$	$+$	$-$	$-$	$+$
ac	$+$	$-$	$-$	$+$	$+$	$-$	$-$
bc	$-$	$+$	$-$	$+$	$-$	$+$	$-$
abc	$+$	$+$	$+$	$+$	$+$	$+$	$+$

TABLE 14.34 ALLOY DATA

Treatment Combination	Factors			Replicates		Total
	Ni (A)	Mo (B)	Ti (C)	I	II	
(1)	15	4	0.55	270	270	540
a	18	4	0.55	275	275	550
b	15	5	0.55	286	290	576
ab	18	5	0.55	293	291	584
c	15	4	0.65	270	280	550
ac	18	4	0.65	282	280	562
bc	15	5	0.65	290	286	576
abc	18	5	0.65	296	302	598

In analogy to the 2^2 design, it can be shown that the sums of squares satisfy the relation

$$SSK = \frac{[K]^2}{8n}$$

for $K = A, B, C, AB, AC, BC$, and ABC. Thus, once Table 14.33 has been used to find the effect contrasts, the relevant sums of squares can be easily found.

In the 2^2 design we saw that the random variables A, B, and AB can be expressed in terms of the estimators $\hat{\alpha}_0$, $\hat{\beta}_0$, and $(\widehat{\alpha\beta})_{00}$, respectively. A similar situation occurs in the 2^3 design. For instance, $\hat{\alpha}_0 = -[A]/8n$ in the 2^3 design and therefore

$$A = \frac{[A]}{4n} = -2\hat{\alpha}_0 = 2\hat{\alpha}_1$$

In analogy to the 2^2 design, the random variables A, B, and C are called main effects and AB, AC, BC, and ABC are called interaction effects.

EXAMPLE 14.18

Among other additives, a tool-grade steel contains nickel, molybdenum, and titanium, these to be called factors A, B, and C, respectively. While holding other additives at a constant level and maintaining a consistent cooling and reheating procedure, an engineer wishes to test the effects of different levels of nickel, molybdenum, and titanium on tensile strength. A 2^3 factorial experiment with two replications is to be employed. The high and low levels of nickel, molybdenum, and titanium are 15% and 18%, 4% and 5%, and 0.55% and 0.65%, respectively. Table 14.34 contains the experimental data (in ksi). The effect contrasts are

$$[A] = -540 + 550 - 576 + 584 - 550 + 562 - 576 + 598 = 52$$

$$[B] = -540 - 550 + 576 + 584 - 550 - 562 + 576 + 598 = 132$$

$$[AB] = 540 - 550 - 576 + 584 + 550 - 562 - 576 + 598 = 8$$

$[C] = 36$, $[AC] = 16$, $[BC] = -8$, and $[ABC] = 12$. The effects are

$$A = \frac{[A]}{4n} = \frac{52}{8} = 6.5$$

$$B = \frac{[B]}{4n} = \frac{132}{8} = 16.5$$

$$AB = \frac{[AB]}{4n} = \frac{8}{8} = 1$$

$C = 4.5$, $AC = 2$, $BC = -1$, and $ABC = 1.5$. The effect sums of squares are

$$SSA = \frac{[A]^2}{8n} = \frac{2704}{16} = 169$$

$$SSB = \frac{[B]^2}{8n} = \frac{17,424}{16} = 1089$$

$$SSAB = \frac{[AB]^2}{8n} = \frac{64}{16} = 4$$

$SSC = 81$, $SSAC = 16$, $SSBC = 4$, and $SSABC = 9$. The total sum of squares, error sum of squares, and error mean square are

$$SST = (270)^2 + (270)^2 + \cdots + (302)^2 - \frac{(4536)^2}{16}$$
$$= 1,287,416 - 1,285,956 = 1460$$

$$SSE = SST - SSA - SSB - SSAB - SSC - SSAC - SSBC - SSABC$$
$$= 1460 - 169 - 1089 - 4 - 81 - 16 - 4 - 9 = 88$$

$$MSE = \frac{88}{8} = 11$$

Table 14.35 summarizes the analysis of variance. Since $f_{0.05,1,8} = 5.32$, all main effects are significant at level 0.05. Moreover, $f_{0.01,1,8} = 11.26$, and therefore both A (nickel) and B (molybdenum) are significant at level 0.01. All interaction effects are negligible.

TABLE 14.35 ANOVA TABLE FOR EXAMPLE 14.18

Source of Variation	Sum of Squares	Degrees of Freedom	Mean Square	F Test Statistic
Nickel (A)	169	1	169	15.36
Molybdenum (B)	1089	1	1089	99.00
Titanium (C)	81	1	81	7.36
AB	4	1	4	0.36
AC	16	1	16	1.45
BC	4	1	4	0.36
ABC	9	1	9	0.82
Error	88	8	11	
Total	1460	15		

14.5.3   The General 2^k Design

The factorial procedures of Sections 14.5.1 and 14.5.2 can be generalized to k factors, the result being the 2^k design for $k = 2, 3, 4, \ldots$. The notation for treatment combinations generalizes directly, with each treatment combination being denoted by the lower-case letters corresponding to the high-level factors of the combination. For instance, a 2^6 design has the six factors A, B, C, D, E, and F, and $abdf$ denotes the treatment combination with A, B, D, and F at high levels and C and E at low levels. To find the standard order for the treatment combinations, begin with (1), introduce the factors in alphabetical order, and as each is listed for the first time, combine it (in order) with all combinations listed previously. For instance, the standard order for the 2^4 design is $(1), a, b, ab, c, ac, bc, abc, d, ad, bd, abd, cd, acd, bcd, abcd$.

To find the average effects and the sums of squares in the 2^k design, it is necessary to find the effect contrasts. Although it is possible to construct tables such as Tables 14.30 and 14.33 in designs containing more than three factors, there exists a simple computational device for finding contrasts. To find any contrast, expand the expression

$$(a \pm 1)(b \pm 1) \ldots (l \pm 1)$$

algebraically, using a minus sign in a parenthesis if the appropriate factor is included in the effect and a plus sign otherwise (the assumption being that the k factors are $A, B, \ldots, L$). When the computation is completed, replace 1 by (1). For instance, in a 2^3 design the contrast for BC is given by

$$
\begin{aligned}
[BC] &= (a + 1)(b - 1)(c - 1) \\
&= abc - ab - ac + a + bc - b - c + (1)
\end{aligned}
$$

Note that the signs of the contrast agree with those in the BC column of Table 14.33. In a 2^4 design the contrast for BCD is given by

$$
\begin{aligned}
[BCD] &= (a + 1)(b - 1)(c - 1)(d - 1) \\
&= abcd - acd + bcd - cd - abd + ad - bd + d \\
&\quad - abc + ac - bc + c + ab - a + b - (1)
\end{aligned}
$$

Having found the effect contrasts, the average effects are found by the relation

$$K = \frac{[K]}{2^{k-1}n}$$

for $K = A, B, AB, C, \ldots, AB\cdots L$. The corresponding sums of squares are given by

$$SSK = \frac{[K]^2}{2^k n}$$

For instance, in a 2^7 design, the interaction average effect corresponding to A, B, D, E, and F is

$$ABDEF = \frac{[ABDEF]}{2^6 n}$$

and the corresponding sum of squares is

$$SSABDEF = \frac{[ABDEF]^2}{2^7 n}$$

14.5.4 Yates' Method for Computing Contrasts

For large k, finding contrasts and sums of squares is computationally intensive. An algorithm for reducing the computational burden has been given by Yates. For an arbitrary k, the algorithm entails the following steps:

1. Label the rows of a table according to the treatment combinations listed in standard order. Let the 0th column of the table contain the observed totals listed in standard order. Give this column the heading "response total."
2. Form the top half of the 1st column by letting the first entry be the sum of the first pair of entries in the 0th column, the second entry be the sum of the second pair of entries in the 0th column, and so on. Form the bottom half of the 1st column by letting the first entry be the subtraction of the first entry of the 0th column from the second entry of the 0th column, the second entry be the subtraction of the third entry of the 0th column from the fourth entry of the 0th column, and so on.
3. (Recursion) For $j = 2, 3, \ldots, k$, form the jth column from the $(j - 1)$st column in the same manner that the 1st column was formed from the 0th column.
4. (Termination) The first entry of the kth column will be the grand total of the observations. The remaining entries in the kth column will be the contrasts.

Table 14.36 gives the execution of Yates' algorithm for the alloy totals of Table 14.34. The entries in column 1 are obtained in the following manner: $1090 = 540 + 550$, $1160 = 576 + 584$, $1112 = 550 + 562$, $1174 = 576 + 598$, $10 = 550 - 540$, $8 = 584 - 576$, $12 = 562 - 550$, and $22 = 598 - 576$. Proceeding recursively gives columns 2 and 3. Note that the first entry (4536) in column 3 is the response total in Example 14.18 and that the remaining entries give the effect contrasts as found in the example.

14.5.5 A Single Observation Per Treatment Combination in the 2^k Design

Even though there are only two levels for each factor in a 2^k factorial design, for large k there are quite a number of treatment combinations, the exact number being

TABLE 14.36 YATES' ALGORITHM FOR THE DATA OF TABLE 14.34

Treatment Combination	Response Total	1	2	3	Sum of Squares
(1)	540	1090	2250	4536	
a	550	1160	2286	52	169
b	576	1112	18	132	1089
ab	584	1174	34	8	4
c	550	10	70	36	81
ac	562	8	62	16	16
bc	576	12	-2	-8	4
abc	598	22	10	12	9

2^k. Therefore, it is sometimes desirable to employ a 2^k design without replication. The problem confronted in such a design is analogous to the one discussed in Section 14.3.6. There, in the case of a two-way classification, it was seen that it is not possible to estimate σ^2 unless the interaction is assumed to be null. Under the assumption of zero interaction in the single-observation two-way model, σ^2 is estimated by *MSAB* (which is then called *MSE*). In the case of a 2^k factorial experiment, there are many higher-order interactions. Should one of these be null, then its mean square can be taken as an estimate of σ^2. More generally, if a number of the higher-order interactions are presumed to be null, then σ^2 is estimated by the average of their mean squares, and it is this average that is taken as *MSE* when computing the F statistics for the remaining effects.

As a rule of thumb, the assumption of null two-factor interactions should be avoided and single-observation experiments should not be run unless $k \geq 4$. Moreover, those interactions that are assumed to be null should be selected prior to running the experiment.

As an illustration of the technique, consider a 2^5 design and suppose all interactions containing four or more factors are assumed to possess zero effects. Then it is possible to test the main effects for A, B, C, D, and E and the interaction effects for AB, AC, AD, AE, BC, BD, BE, CD, CE, DE, ABC, ABD, ABE, ACD, ACE, ADE, BCD, BCE, BDE, and CDE. The error sum of squares is taken as the sum of the sums of squares of the fourth- and fifth-order interaction sums of squares, namely,

$$SSE = SSABCD + SSABCE + SSABDE + SSACDE + SSBCDE + SSABCDE$$

(Recall that each sum of squares possesses a single degree of freedom, so that *SSE* is equivalently equal to the sum of the mean squares.) *MSE* is taken as the average of the fourth- and fifth-order interaction sums of squares: $MSE = SSE/6$. The number of higher-order interactions that are taken to be null (in this case six) is the number of degrees of freedom associated with *SSE*.

As always, there is a risk in making an a priori assumption that certain interactions possess zero effects. Indeed, if *MSE* is an average of interaction mean squares and should one or more of the interactions be significant, then *MSE* will overestimate σ^2. Since the F statistics associated with the lower-order effects each possess *MSE* in the denominator, these statistics will be reduced, and significant lower-order effects might not be detected.

EXAMPLE 14.19

Numerous factors relate to the ability of a building to withstand vibration, such as might be experienced in an earthquake. To make a preliminary analysis of four factors, investigators decide to build small test buildings and subject each to the same amount of vibration. Four factors are to be tested at two levels each. Specifically, two types of architectural design (A), of cement mix (B), of foundation (C), and of beam structure (D) are to be employed in a 2^4 factorial design with a single observation for each treatment combination. After being subjected to vibration, each of the 16 buildings will be rated on a scale from 0 to 20 to measure the degree of damage (0 being no damage). For each factor, the low level is the one deemed inferior by the investigators. The data resulting from the vibrational tests are given in the response column of Table 14.37 and Yates' method is provided in Table 14.38. The total sum of squares is

$$SST = (16)^2 + (14)^2 + \cdots + (3)^2 - \frac{(144)^2}{16} = 1546 - 1296 = 250$$

In order to make an estimation of the error, it will be assumed that third- and fourth-order interactions are negligible. Under this assumption, the error sum of squares is

$$SSE = SSABC + SSABD + SSACD + SSBCD + SSABCD = 10.5$$

and $MSE = SSE/5 = 2.1$. The results of the analysis of variance are summarized in Table 14.39. Since $f_{0.05,1,5} = 6.61$, architecture (A), cement (B), and foundation (C) are significant at the 0.05 level. Moreover, since $f_{0.01,1,5} = 16.26$, only cement (B) is significant at the 0.01 level. No two-factor interactions are significant at either level.

Note that $SSABC = 6.25$. Therefore, one might suspect that there is some ABC interaction. Indeed, $SSABC$ contributes more to SSE than do $SSABD$, $SSACD$, $SSBCD$, and $SSABCD$ together. If the assumption of no ABC interaction is substantially incorrect, then it is likely that $SSABC$ does not arise solely from randomness and that MSE overestimates σ^2. If so, then it is possible that the statistic $f_D = 5.83$ has been deflated, due to an overly large denominator, and that a significant D effect should have been detected. Note that 5.83 is not that much less than 6.61.

14.5.6 Fractional Replication

Sometimes useful information can be obtained from a factorial experiment by running only some fraction of the treatment combinations. Such designs are called **fractional factorial designs.** Although one can employ one-half, one-quarter, and, in general, $1/2^p$ designs, we restrict our attention to **one-half** designs. These are factorial designs in which only one-half of the possible treatment combinations are observed. Interpretation of the analysis of variance in fractional designs is somewhat problematic, so we will only briefly introduce the topic, leaving a thorough discussion to a course on experimental design.

Suppose an investigator wishes to study three two-level factors but only wishes to spend the resources to observe 4 of the possible 8 treatment combinations. Such an experiment is called a 2^{3-1} design. He or she can proceed by observing only the a, b, c, and abc treatment combinations. Referring to Table 14.33, we see that these are the treatment combinations that possess a plus sign in the ABC column. ABC is called the **defining relation** of the design. For each of the main and two-factor effects,

TABLE 14.37 DATA FOR VIBRATION EXPERIMENT

Treatment Combination	Architecture (A)	Cement (B)	Foundation (C)	Structure (D)	Response
(1)	0	0	0	0	16
a	1	0	0	0	14
b	0	1	0	0	8
ab	1	1	0	0	8
c	0	0	1	0	15
ac	1	0	1	0	9
bc	0	1	1	0	6
abc	1	1	1	0	3
d	0	0	0	1	14
ad	1	0	0	1	10
bd	0	1	0	1	7
abd	1	1	0	1	6
cd	0	0	1	1	10
acd	1	0	1	1	10
bcd	0	1	1	1	5
abcd	1	1	1	1	3

TABLE 14.38 YATES' METHOD FOR THE DATA OF TABLE 14.36

Treatment Combination	Response	1	2	3	4	Sum of Squares
(1)	16	30	46	79	144	
a	14	16	33	65	−18	20.25
b	8	24	37	−11	−52	169.00
ab	8	9	28	−7	6	2.25
c	15	24	−2	−29	−22	30.25
ac	9	13	−9	−23	−4	1.00
bc	6	20	−5	5	−2	0.25
abc	3	8	−2	1	−4	1.00
d	14	−2	−14	−13	−14	12.25
ad	10	0	−15	−9	4	1.00
bd	7	−6	−11	−7	6	2.25
abd	6	−3	−12	3	−4	1.00
cd	10	−4	2	−1	4	1.00
acd	10	−1	3	−1	10	6.25
bcd	5	0	3	1	0	0.00
abcd	3	−2	−2	−5	−6	2.25

TABLE 14.39 ANOVA TABLE FOR EXAMPLE 14.19

Source of Variation	Sum of Squares	Degrees of Freedom	Mean Square	F Test Statistic
Architecture (A)	20.25	1	20.25	9.64
Cement (B)	169.00	1	169.00	80.48
Foundation (C)	30.25	1	30.25	14.40
Structure (D)	12.25	1	12.25	5.83
AB	2.25	1	2.25	1.07
AC	1.00	1	1.00	0.48
AD	1.00	1	1.00	0.48
BC	0.25	1	0.25	0.12
BD	2.25	1	2.25	1.07
CD	1.00	1	1.00	0.48
Error	10.50	5	2.10	
Total	250.00	15		

a contrast involving the four treatment combinations a, b, c, and abc is defined by the columns of Table 14.33. For instance,

$$[A] = a - b - c + abc$$

and

$$[AB] = -a - b + c + abc$$

In analogy to Section 14.5.1, the contrasts $[A]$, $[B]$, and $[C]$ give main-effect estimators according to the relations $A = [A]/2$, $B = [B]/2$, and $C = [C]/2$. The contrasts $[AB]$, $[AC]$, and $[BC]$ yield the interaction-effect estimators $AB = [AB]/2$, $AC = [AC]/2$, and $[BC] = BC/2$. Note that, in terms of effects, $A = BC$, $B = AC$, and

$C = BC$. The reason for these equalities is that certain contrasts are identical, namely, $[A]$ and $[BC]$, $[B]$ and $[AC]$, and $[C]$ and $[AB]$. Any two factorial effects that possess the same contrast are called **aliases.**

Aliases are determined by employing the defining relation of the fractional design: simply multiply (modulo 2) an effect by the defining relation to obtain its alias. For instance, for the defining relation ABC in the 2^{3-1} design, the aliases of A, B, and C are

$$(A)(ABC) = A^2BC = BC$$
$$(B)(ABC) = AB^2C = AC$$
$$(C)(ABC) = ABC^2 = AB$$

Referring to Table 14.33, we see that it is also possible to employ the 2^{3-1} design by observing the treatment combinations (1), ab, ac, and bc. These result from the minus-sign entries in the ABC column of Figure 14.33. Modulo-2 multiplication of (1), ab, ac, and bc by the defining ABC relation gives the same alias structure that was obtained previously, the only difference being the responses to be observed. The fractional design associated with the plus signs of the ABC column is called the **principal fraction.**

Unless there are assumptions made regarding the negligibility of higher-order interactions, perhaps including two-factor interactions, definitive conclusions based upon analysis of variance can be elusive. The problem is that the experimental design does not provide a means for determining which of the two aliased effects are behind the response.

To illustrate the dilemma, we consider a (single replicate) 2^{4-1} design with defining relation $ABCD$. The principal fraction for such an experiment consists of the treatment combinations (1), ab, ac, ad, bc, bd, cd, and $abcd$. Contrasts exist for all effects except $ABCD$. The alias structure is found by multiplying (modulo 2) the factorial effects by $ABCD$. The following alias structure results:

$$
\begin{aligned}
A\!: \quad & (A)(ABCD) & = BCD \\
B\!: \quad & (B)(ABCD) & = ACD \\
C\!: \quad & (C)(ABCD) & = ABD \\
D\!: \quad & (D)(ABCD) & = ABC \\
AB\!: \quad & (AB)(ABCD) & = CD \\
AC\!: \quad & (AC)(ABCD) & = BD \\
AD\!: \quad & (AD)(ABCD) & = BC
\end{aligned}
$$

For any effect K, the sum of squares is given by

$$SSK = \frac{[K]^2}{2^{4-1}}$$

where $[K]$ is the contrast for K. Should SSA result in a significant F statistic, it is not possible to determine from the experiment whether A or its alias BCD is responsible for the response. Similarly, a significant value of F_{AB} might be due to CD. Analogous statements apply to B, C, D, AC, and BC. To draw conclusions based upon an analysis of variance, two assumptions need to be made. First, an estimate of σ^2 requires that two-factor interactions are presumed negligible. Second, conclusions concerning main effects require assumptions regarding the negligibility of the three-factor interactions with which they are aliased. We leave the details of these and other fractional-replication considerations to a text on experimental design.

14.6 CONFOUNDING IN THE 2^k EXPERIMENTAL DESIGN

Throughout Section 14.5 it was assumed that the treatment combinations of a replication were tested under homogeneous conditions. For instance, if a chemical process is to be tested, with three factors of the process each being varied between two levels, then the experiment is designed so that all eight treatment combinations are tested on the same batch of raw material (or at least on batches that are presumed to be homogeneous relative to the factors being tested). Should two nonhomogeneous batches be required to test the eight combinations, then batch effects must be taken into account and the methods of Section 14.5 might lead to erroneous conclusions. The root question concerns blocking. When a 2^k factorial experiment is designed in blocks, some treatment effects are **confounded** (mixed up) with block effects. The net result is that, to the experimenter, some treatment effects are indistinguishable from block effects.

14.6.1 2^k Factorial Design in Two Blocks

Consider a 2^2 design employing two blocks in which there is to be a single observation of each treatment combination. Let A and B denote the factors and (1), a, b, and ab denote the four treatment combinations. Suppose the treatment combinations are blocked according to the scheme depicted in Figure 14.7—block 1 containing (1) and ab, and block 2 containing a and b. If we assume the block effect is additive, with block 2 contributing an amount x to each response, then the observations of block 2 can be rewritten as $a + x$ and $b + x$, where a and b now represent the observations that would have occurred had all treatment combinations been applied under the conditions of block 1. Under the additivity assumption, the contrasts $[A]$, $[B]$, and $[AB]$ are

$$[A] = -(1) + (a + x) - (b + x) + ab = -(1) + a - b + ab$$

$$[B] = -(1) - (a + x) + (b + x) + ab = -(1) - a + b + ab$$

$$[AB] = (1) - (a + x) - (b + x) + ab = (1) - a - b + ab - 2x$$

(see Table 14.30). The A and B contrasts are unaffected by the blocking, and therefore the effects of A and B, $A = [A]/2$ and $B = [B]/2$, are also unaffected by the blocking. On the other hand, division of $[AB]$ by 2 shows that the interaction effect has been altered by (minus) the block effect, so that the interaction and block effects are confounded. (Keep in mind that when actually running the experiment, the order of running the treatment combinations within a block is randomized and so too is the order of running the blocks.)

Algebraically, the reason for confounding in the two-block 2^2 design is discernible from Table 14.30. Referring to Figure 14.7, we see that both treatment combinations having a $+$ sign in the AB contrast have been assigned to block 1. Thus, there is no cancellation of the block effect. Conversely, the treatment combinations having a $+$ sign in the A and B contrasts have been split between the two

FIGURE 14.7
Two-block 2^2 design with AB confounded

Block 1	Block 2
(1)	a
ab	b

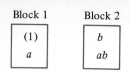

Block 1	Block 2
(1)	b
a	ab

FIGURE 14.8
Two-block 2^2 design with B confounded

blocks, thereby resulting in a cancellation of the block effect. There is no escape from the dilemma. For instance, had the blocking of Figure 14.8 been employed, then the B effect would have been confounded with the blocks, whereas the A and AB effects would have been unaltered. It is customary to confound the highest-order interaction with the blocks. Thus, in the 2^2 design with two blocks, the design of Figure 14.7 is usually employed.

A similar analysis applies to blocking a 2^3 factorial design in two blocks. Given the factors A, B, and C, suppose the treatment effects are blocked according to the scheme of Figure 14.9. Again letting x denote the effect of block 2 (under the additivity assumption), the observations of block 2 can be rewritten as $a + x$, $b + x$, $c + x$, and $abc + x$, where a, b, c, and abc represent the observations that would have occurred without the block effect. The resulting A contrast is

$$[A] = -(1) + (a + x) - (b + x) + ab - (c + x) + ac - bc + (abc + x)$$

(see Table 14.33). The x's cancel and therefore the A contrast is unaffected by the blocking. Similar statements apply to B, C, AB, AC, and BC. However, for three-factor interaction,

$$[ABC] = -(1) + (a + x) + (b + x) - ab + (c + x) - ac - bc + (abc + x)$$

and the block effect x does not cancel. Specifically, the interaction $ABC = [ABC]/4$ is increased by x. The net result is that the ABC effect is confounded with the block effect. Although a different design could leave the ABC effect unconfounded, some lower-order effect would then have to be confounded with the blocks. As noted previously, it is customary to let the highest-order effect be confounded.

Generally, in a two-block 2^k factorial design, one effect must be confounded with the blocks. The one chosen for confounding is usually the highest-order effect.

Given the confounded effect, there is an algorithmic technique for choosing the block distribution. Suppose $A_1, A_2, \ldots, A_k$ are the k factors. Then every factorial effect is of the form $A_1^{q_1} A_2^{q_2} \cdots A_k^{q_k}$, where, for $i = 1, 2, \ldots, k$, q_i is either 0 or 1. In particular, the chosen confounded effect takes the form $A_1^{x_1} A_2^{x_2} \cdots A_k^{x_k}$. Define the functional

$$Q(q_1, q_2, \ldots, q_k) = q_1 x_1 + q_2 x_2 + \cdots + q_k x_k$$

Q is called the **defining contrast.** Compute $Q = Q(q_1, q_2, \ldots, q_k)$ for each of the 2^k treatment combinations $a_1^{q_1} a_2^{q_2} \cdots a_k^{q_k}$ and evaluate Q (modulo 2) for each combination. If one block consists of all treatment combinations for which Q (modulo 2) is 0 and the other consists of all combinations for which Q (modulo 2) is 1, then the blocking will be so configured that the desired effect is confounded.

As an illustration, consider the 2^3 design with two blocks and confounded effect ABC ($A_1 A_2 A_3$). Then, for all treatment combinations, the defining contrast is

FIGURE 14.9
Two-block 2^3 design with ABC confounded

Block 1	Block 2
(1)	a
ab	b
ac	c
bc	abc

$$Q(q_1, q_2, q_3) = q_1 + q_2 + q_3$$

The values of Q for the eight treatment combinations are

$$(1): \quad Q = Q(0, 0, 0) = 0 + 0 + 0 = 0 = 0 \ (\text{mod } 2)$$

$$a: \quad Q = Q(1, 0, 0) = 1 + 0 + 0 = 1 = 1 \ (\text{mod } 2)$$

$$b: \quad Q = Q(0, 1, 0) = 0 + 1 + 0 = 1 = 1 \ (\text{mod } 2)$$

$$ab: \quad Q = Q(1, 1, 0) = 1 + 1 + 0 = 2 = 0 \,(\text{mod } 2)$$

$$c: \quad Q = Q(0, 0, 1) = 0 + 0 + 1 = 1 = 1 \,(\text{mod } 2)$$

$$ac: \quad Q = Q(1, 0, 1) = 1 + 0 + 1 = 2 = 0 \,(\text{mod } 2)$$

$$bc: \quad Q = Q(0, 1, 1) = 0 + 1 + 1 = 2 = 0 \,(\text{mod } 2)$$

$$abc: \quad Q = Q(1, 1, 1) = 1 + 1 + 1 = 3 = 1 \,(\text{mod } 2)$$

The resulting blocking scheme is the one given in Figure 14.9 (in which ABC is confounded).

The block containing (1) is called the **principal block.** The elements (treatment combinations) of the principal block form a group (in the sense of abstract algebra) with respect to modulo-2 multiplication. For instance, in the 2^4 design, if abc and ad are in the principal block, then so is

$$(abc)(ad) = a^2bcd = bcd$$

In the principal block, each element except (1) is obtainable by multiplying another two elements in the block. For instance, in the principal block of Figure 14.9,

$$(ab)(ac) = a^2bc = bc$$

$$(ab)(bc) = ab^2c = ac$$

$$(ac)(bc) = abc^2 = ab$$

The treatment combinations of the principal block can be employed to find the treatment combinations of the other block. The procedure is simply to select any treatment combination in the other block and multiply it (modulo 2) times the combinations in the principal block. The results of the multiplications are the members of the second block. For instance, in Figure 14.9, a is not in the principal block, and therefore the treatment combinations of block 2 are $(a)(1) = a$, $(a)(ab) = a^2b = b$, $(a)(ac) = a^2c = c$, and $(a)(bc) = abc$. While this approach is not particularly useful when there are only two blocks, it becomes so when there are 2^p blocks, $p > 2$.

Thus far, we have proceeded under the assumption that the experiment will be run with a single replicate. The lack of replication leaves us without an estimate of σ^2. Consequently, it is necessary to assume that certain unconfounded high-order interactions are zero and to let the error sum of squares be the sum of the sums of squares of these interactions. For instance, if there is a single replicate in a two-block 2^4 design with $ABCD$ confounded, then the analysis of variance can be carried out under the assumption that the interaction effects ABC, ABD, ACD, and BCD are zero. Under this assumption,

$$SSE = SSABC + SSABD + SSACD + SSBCD$$

SSE possesses 4 degrees of freedom and we define the error mean square to be $MSE = SSE/4$. Of course, there are the usual uncertainties in making the assumption of null interaction effects.

Let us now consider in detail the analysis of variance in the case of a two-block 2^4 design for which there is a single replicate, $ABCD$ is confounded, and the three-factor interactions are assumed to be negligible. The defining contrast is

$$Q = q_1 + q_2 + q_3 + q_4$$

and the appropriate blocking scheme is given in Figure 14.10. Among the sources of variation are A, B, C, D, AB, AC, AD, BC, BD, and CD, each of which possesses a sum of squares with one degree of freedom. There is also a sum of squares for blocks (for $ABCD$, which is confounded with blocks) that possesses a single degree of freedom. Finally, SSE possesses 4 degrees of freedom. The total number of degrees of freedom is $2^4 - 1 = 15$.

EXAMPLE 14.20

A chemical reaction requires two reactants and a catalyst. A chemist wishes to vary four factors in the reaction, these being one of the reactants (A), the catalyst (B), the duration (C), and the reaction temperature (D). Each factor is to be employed at two levels and a 2^4 factorial experiment run. Because 16 reactions are required to make a full experiment, the experiment will have to be blocked between the day and night shifts. Factor $ABCD$ will be confounded, so that the appropriate design is the one given in Figure 14.10. Block 1 represents the day shift. The yields (in kilograms) of the 16 treatment combinations are given in Table 14.40 and Yates' method is applied in Table 14.41. It is assumed that the three-factor interactions are null, so that the error sum of squares is

$$SSE = SSABC + SSABD + SSACD + SSBCD$$
$$= 0.5625 + 0.0625 + 0.0625 + 0.5625 = 1.25$$

and $MSE = 1.25/4 = 0.3125$. The usual calculation yields the total sum of squares $SST = 463.9375$. The analysis of variance is summarized in Table 14.42. Referring to the F statistics in the table, we see that reactant (A), catalyst (B), duration (C), and temperature (D) are all significant at level 0.01, since $f_{0.01,1,4} = 21.20$. In addition, the interactions AB, AD, and CD are significant at level 0.05, since $f_{0.05,1,4} = 7.71$; however, no two-factor interactions are significant at level 0.01. Finally, note the block sum of squares $SSABCD$. It certainly appears that the block effect is significant. Nevertheless, the block effect is confounded with the interaction $ABCD$, and therefore a definitive conclusion based upon the F statistic is impossible.

If k is not too large, then a two-block 2^k factorial experiment might be run with n replications, $n > 1$. An example would be a two-block 2^3 design with three replications. Figure 14.11 illustrates such a design with ABC confounded. Note that

FIGURE 14.10
Two-block 2^4 design with $ABCD$ confounded

Block 1	Block 2
(1)	a
ab	b
ac	c
ad	d
bc	abc
bd	abd
cd	acd
$abcd$	bcd

TABLE 14.40 REACTION DATA

Treatment Combination	Reactant (A)	Catalyst (B)	Duration (C)	Temperature (D)	Response
(1)	0	0	0	0	22
a	1	0	0	0	24
b	0	1	0	0	26
ab	1	1	0	0	33
c	0	0	1	0	23
ac	1	0	1	0	28
bc	0	1	1	0	28
abc	1	1	1	0	32
d	0	0	0	1	27
ad	1	0	0	1	33
bd	0	1	0	1	32
abd	1	1	0	1	38
cd	0	0	1	1	31
acd	1	0	1	1	36
bcd	0	1	1	1	34
$abcd$	1	1	1	1	42

TABLE 14.41 YATES' METHOD APPLIED TO DATA OF TABLE 14.40

Treatment Combination	Response	1	2	3	4	Sum of Squares
(1)	22	46	105	216	489	
a	24	59	111	273	43	115.5625
b	26	51	130	18	41	105.0625
ab	33	60	143	25	7	3.0625
c	23	60	9	22	19	22.5625
ac	28	70	9	19	1	0.0625
bc	28	67	12	4	−5	1.5625
abc	32	76	13	3	−3	0.5625
d	27	2	13	6	57	203.0625
ad	33	7	9	13	7	3.0625
bd	32	5	10	0	−3	0.5625
abd	38	4	9	1	−1	0.0625
cd	31	6	5	−4	7	3.0625
acd	36	6	−1	−1	1	0.0625
bcd	34	5	0	−6	3	0.5625
abcd	42	8	3	3	9	5.0625

TABLE 14.42 ANOVA TABLE FOR EXAMPLE 14.20

Source of Variation	Sum of Squares	Degrees of Freedom	Mean Square	F Test Statistic
Reactant (A)	115.56	1	115.56	369.80
Catalyst (B)	105.06	1	105.06	336.20
Duration (C)	22.56	1	22.56	72.20
Temperature (D)	203.06	1	203.06	649.80
AB	3.06	1	3.06	9.80
AC	0.06	1	0.06	0.20
AD	3.06	1	3.06	9.80
BC	1.56	1	1.56	5.00
BD	0.56	1	0.56	1.80
CD	3.06	1	3.06	9.80
Blocks (ABCD)	5.06	1	5.06	16.20
Error	1.25	4	0.31	
Total	463.94	15		

FIGURE 14.11
Three-replicate two-block 2^3 design with ABC confounded

Replicate I		Replicate II		Replicate III	
Block 1	Block 2	Block 1	Block 2	Block 1	Block 2
(1)	a	(1)	a	(1)	a
ab	b	ab	b	ab	b
ac	c	ac	c	ac	c
bc	abc	bc	abc	bc	abc

the pair of blocks is the same for each replicate. In total, there are 24 observations. Thus, there are $24 - 1 = 23$ total degrees of freedom. There is a single degree of freedom associated with each of the six unconfounded effects A, B, C, AB, AC, and BC. There are $6 - 1 = 5$ degrees of freedom associated with the six blocks (see Table 14.3). Finally, the error, which is associated with within-block variability, has its degrees of freedom obtained by subtracting the unconfounded-effect and block degrees of freedom from the total degrees of freedom. In the present case, there are $23 - 6 - 5 = 12$ degrees of freedom associated with error.

For a two-block 2^3 design with four replications, there are $2^3 \times 4 - 1 = 31$ total degrees of freedom, 6 degrees of freedom associated with unconfounded effects, 7 degrees of freedom associated with blocks, and 18 error degrees of freedom.

In the multiple-replicate case, there are sums of squares and mean squares associated with each unconfounded effect, with blocks, and with error. The unconfounded-effect sums of squares can be found in the same manner as in the unconfounded design. To find the block sum of squares, written $SSBL$, let N and m denote the number of observations and number of blocks, respectively. Then

$$SSBL = \frac{m}{N} \sum_{j=1}^{m} T_{.j}^2 - \frac{T_{..}^2}{N}$$

where $T_{.j}$ is the observation total for the jth block, and $T_{..}$ is the overall observation total. Once the sums of squares for unconfounded effects and blocks are found, SSE is found by subtracting these sums of squares from SST. The mean squares are found by dividing the sums of squares by their corresponding degrees of freedom, and F statistics are computed by dividing the mean squares by MSE.

EXAMPLE 14.21

Suppose the tensile-strength testing for the experiment of Example 14.18 is to be performed on two different pieces of test equipment and the investigator believes it prudent to design the experiment in blocks, one for each piece of equipment. Moreover, let us suppose that the data of Example 14.18 are obtained when the experiment is run. If ABC is to be confounded, then the design of Figure 14.12 results, where the response data from Example 14.18 have been written in the blocks. The sums of squares for A, B, C, AB, AC, and BC are obtained from Table 14.35, as is SST. There are 3 degrees of freedom associated with the blocks, 6 degrees of freedom associated with error,

$$SSBL = \frac{(1135)^2 + (1127)^2 + (1127)^2 + (1147)^2}{4} - \frac{(4536)^2}{16} = 67$$

$$SSE = SST - SSA - SSB - SSC - SSAB - SSAC - SSBC - SSBL = 30$$

and $MSE = 5$. The analysis of variance is summarized in Table 14.43. Since $f_{0.01,1,6} = 13.75$, all main effects are significant at level 0.01. Recall that in Example 14.18, titanium (C) was not judged significant at level 0.01. In words, had the same data been obtained by randomly assigning treatment combinations to test equipment rather than by blocking, then the effect of titanium would not have been detected at the 0.01 level. Although blocking has reduced the denominator degrees of freedom for the F statistics, this loss has apparently been more than offset by a reduction in the effect of extraneous variation.

FIGURE 14.12
Two-replicate two-block 2^3 design for Example 14.21, ABC confounded

	Replicate 1			Replicate 2		
Block 1		**Block 2**		**Block 1**		**Block 2**
(1) 270		a 275		(1) 270		a 275
ab 293		b 286		ab 291		b 290
ac 282		c 270		ac 280		c 280
bc 290		abc 296		bc 286		abc 302

TABLE 14.43 ANOVA TABLE FOR EXAMPLE 14.21

Source of Variation	Sum of Squares	Degrees of Freedom	Mean Square	F Test Statistic
Nickel (A)	169	1	169	33.80
Molybdenum (B)	1089	1	1089	217.80
Titanium (C)	81	1	81	16.20
AB	4	1	4	0.80
AC	16	1	16	3.20
BC	4	1	4	0.80
Blocks	67	3	22.33	4.47
Error	30	6	5.0	
Total	1460	15		

14.6.2 2^k Factorial Design in Four Blocks

Rather than employ two blocks, one can employ four blocks in a 2^k design. A key point concerning a four-block design is that three effects will be confounded. Moreover, given two confounded effects, a third confounded effect, called the **generalized interaction,** is automatically determined by multiplication modulo 2. For instance, in a four-block 2^5 factorial design, if the experimenter selects ABC and CDE to be confounded, then the third confounded effect must be

$$(ABC)(CDE) = ABC^2DE = ABDE$$

One must be prudent in selecting the two effects for confounding. For instance, if ABC and BCD are chosen, then the generalized interaction is $(ABC)(BCD) = AD$ (modulo 2). Thus, the choice of the three-factor interactions ABC and BCD for confounding has resulted in a confounded two-factor interaction.

Given the choice of two confounded effects, the block design is generated by two defining contrasts, Q_1 and Q_2. For instance, in a four-block 2^5 design with selected confounded effects ABC and CDE, the defining contrasts are

$$Q_1 = q_1 + q_2 + q_3$$
$$Q_2 = q_3 + q_4 + q_5$$

Each treatment combination will result in a pair of values (Q_1, Q_2). The four blocks resulting in the desired three confounded effects (the two selected plus the generalized interaction) are defined by the four ordered pairs $(0, 0)$, $(0, 1)$, $(1, 0)$, and $(1, 1)$. Specifically, all treatment combinations resulting in the pair (u, v), $u = 0, 1$ and $v = 0, 1$, are placed into the same block. For the preceding defining contrasts in the four-block 2^5 design, the appropriate blocks are those depicted in Figure 14.13. For instance, ab appears in the principal block (block 1 in the figure) because both (1) and ab possess the same defining-contrast pair, namely, $(0, 0)$. Indeed, for (1),

$$Q_1(0, 0, 0, 0, 0) = 0 + 0 + 0 = 0 = 0 \text{ (mod 2)}$$
$$Q_2(0, 0, 0, 0, 0) = 0 + 0 + 0 = 0 = 0 \text{ (mod 2)}$$

and for ab,

Block 1	Block 2	Block 3	Block 4
(1)	d	a	c
ab	e	b	ad
de	ac	cd	ae
acd	bc	ce	bd
ace	abd	ade	be
bcd	abe	bde	abc
bce	acde	abcd	cde
abde	bcde	abce	abcde
(0, 0)	(0, 1)	(1, 0)	(1, 1)

FIGURE 14.13
Four-block 2^5 design with *ABC*, *CDE*, and *ABDE* confounded

$$Q_1(1, 1, 0, 0, 0) = 1 + 1 + 0 = 2 = 0 \ (\text{mod } 2)$$

$$Q_2(1, 1, 0, 0, 0) = 0 + 0 + 0 = 0 = 0 \ (\text{mod } 2)$$

As a second illustration, *be* and *abcde* are in the same block since both possess the defining-contrast pair (1, 1). Indeed, for *be*,

$$Q_1(0, 1, 0, 0, 1) = 0 + 1 + 0 = 1 = 1 \ (\text{mod } 2)$$

$$Q_2(0, 1, 0, 0, 1) = 0 + 0 + 1 = 1 = 1 \ (\text{mod } 2)$$

and for *abcde*,

$$Q_1(1, 1, 1, 1, 1) = 1 + 1 + 1 = 3 = 1 \ (\text{mod } 2)$$

$$Q_2(1, 1, 1, 1, 1) = 1 + 1 + 1 = 3 = 1 \ (\text{mod } 2)$$

In the design of Figure 14.13, blocks 1, 2, 3, and 4 are determined by the defining-contrast pairs (0, 0), (0, 1), (1, 0), and (1, 1), respectively.

As in the two-block case, the treatment combinations of the principal block form a group under multiplication (modulo 2). For instance, in the principal block of Figure 14.13, since *ace* and *abde* are in the principal block, so too is

$$(ace)(abde) = a^2bcde^2 = bcd$$

Given the principal block, the others can be found by multiplication (modulo 2). Simply select a treatment combination not in the principal block, nor in any other block thus far formed, and multiply it by each treatment combination in the principal block. The result will be another block in the design. For instance, given that the principal block of Figure 14.13 has been found using the defining contrasts, the others can be found by multiplication (modulo 2). Suppose blocks 1 and 2 have been found. Since *a* is in neither block 1 nor block 2, a third block is found by multiplication (modulo 2): $(a)(1) = a$, $(a)(ab) = b$, $(a)(de) = ade$, $(a)(acd) = cd$, $(a)(ace) = ce$, $(a)(bcd) = abcd$, $(a)(bce) = abce$, and $(a)(abde) = bde$. Block 4 can be found by multiplying (modulo 2) *c* times each treatment combination in block 1.

Analysis of variance proceeds in the four-block design in a similar fashion to the two-block design. First consider the case of a single observation per treatment combination. In addition, consider a four-block 2^5 factorial design with *ABC*, *CDE*, and *ABDE* confounded (Figure 14.13). Since there are 32 observations, there are $32 - 1 = 31$ total degrees of freedom. Let us assume that unconfounded three-, four-, and five-factor interactions are null. Then there is a single degree of freedom associated with each of the five main effects and with each of the ten unconfounded two-factor effects. *MSE* will be computed by averaging the sums of squares for unconfounded three-, four-, and five-factor interactions. There are 13 of these, so there are 13 degrees of freedom associated with error. Finally, there are three interactions confounded with the blocks, so there are three degrees of freedom associated with blocks. Insofar as sum-of-squares computations are concerned, these are done in the usual manner (or by Yates' method). The sum of squares for blocks is then given by the sum of the sums of squares for the confounded effects. For instance, since *ABC*, *CDE*, and *ABDE* are confounded,

$$SSBL = SSABC + SSCDE + SSABDE$$

The preceding degrees-of-freedom breakdown is dependent upon the choice of confounded effects. If *BCDE* and *ABCD* are chosen for confounding, then the generalized interaction is $(BCDE)(ABCD) = AE$. If it is still assumed that three-, four-, and five-factor interactions are null, then there would again be 31 total degrees of freedom, 5 degrees of freedom associated with main effects, and 3 degrees of freedom associated with blocks. However, there would only be 9 degrees of freedom associated with unconfounded two-factor effects and there would be 14 degrees of freedom associated with error, 14 being the number of unconfounded three-, four-, and five-factor effects. Table 14.44 gives the degrees-of-freedom breakdown for the four-block 2^5 single-replicate design with *ABC*, *CDE*, and *ABDE* confounded and with *BCDE*, *ABCD*, and *AE* confounded.

If the four-block experiment is carried out with replications, then the replication analysis of variance discussed in the two-block case extends directly. For instance,

TABLE 14.44 DEGREES-OF-FREEDOM BREAKDOWN FOR THE FOUR-BLOCK 2^5 DESIGN WITH A SINGLE REPLICATE AND NULL THREE-, FOUR-, AND FIVE-FACTOR INTERACTIONS

Source of Variation	Degrees of Freedom	
	ABC, CDE, ABDE Confounded	*ABCD, BCDE, AE* Confounded
A	1	1
B	1	1
C	1	1
D	1	1
E	1	1
AB	1	1
AC	1	1
AD	1	1
AE	1	—
BC	1	1
BD	1	1
BE	1	1
CD	1	1
CE	1	1
DE	1	1
Blocks	3	3
Error	13	14
Total	31	31

TABLE 14.45 DEGREES-OF-FREEDOM BREAKDOWN FOR THE FOUR-BLOCK 2^5 DESIGN WITH TWO REPLICATES AND *ABC*, *CDE*, AND *ABDE* CONFOUNDED

Source of Variation	Degrees of Freedom
Main effects	5
Unconfounded two-factor interaction	10
Unconfounded three-factor interaction	8
Unconfounded four-factor interaction	4
Unconfounded five-factor interaction	1
Blocks	7
Error	28
Total	63

suppose the design of Figure 14.13 is altered to involve two replicates. Then there are $64 - 1 = 63$ total degrees of freedom. There are 8 blocks altogether and 7 degrees of freedom associated with blocks. *SSBL* is found by the same formula as in the two-block design. There are 28 single degree-of-freedom sums of squares associated with the unconfounded effects, each of these being found in the usual manner. Finally, by subtraction, there are $63 - 7 - 28 = 28$ degrees of freedom associated with error, *SSE* being given by

$$SSE = SST - SSBL - (SSA + SSB + \cdots + SSABCDE)$$

where, in the design of Figure 14.13, the sum within the parentheses contains all effect sums of squares except *SSABC*, *SSCDE*, and *SSABDE*. Table 14.45 gives a degrees-of-freedom breakdown in terms of degrees of freedom associated with blocks, error, main effects, and unconfounded two-, three-, four-, and five-factor interactions.

14.6.3 2^k Factorial Design in 2^p Blocks

So long as $p < k$, a 2^k factorial experiment can be designed in 2^p blocks, where each block contains 2^{k-p} treatment combinations. The total number of confounded effects will be $2^p - 1$. Of these, p are chosen by the experimenter under the restriction that none of the p chosen effects can be expressed as a product (modulo 2) of the other chosen effects. Once p effects are selected for confounding, $2^p - p - 1$ generalized interactions will result. The 2^p blocks are formed by employing p defining contrasts, $Q_1, Q_2, \ldots, Q_p$. Each block is determined by the value of a defining-contrast p-tuple $(Q_1, Q_2, \ldots, Q_p)$.

Consider the 2^5 design in $2^3 = 8$ blocks. Each block will possess $2^{5-3} = 4$ treatment combinations. Letting *ABC*, *ABDE*, and *BCD* be the three effects chosen for confounding, the remaining $2^3 - 3 - 1 = 4$ confounded effects are given by the following generalized interactions:

$$(ABC)(ABDE) = A^2B^2CDE = CDE$$

$$(ABDE)(BCD) = AB^2CD^2E = ACE$$

$$(ABC)(BCD) = AB^2C^2D = AD$$

$$(ABC)(ABDE)(BCD) = A^2B^3C^2D^2E = BE$$

Block 1	Block 2	Block 3	Block 4
(1)	*ae*	*e*	*ab*
acd	*bd*	*bc*	*de*
bce	*abc*	*abd*	*ace*
abde	*cde*	*acde*	*bcd*
(0, 0, 0)	(1, 0, 0)	(0, 1, 0)	(0, 0, 1)

Block 5	Block 6	Block 7	Block 8
a	*c*	*d*	*b*
cd	*ad*	*ac*	*ce*
bde	*be*	*abe*	*ade*
abce	*abcde*	*bcde*	*abcd*
(1, 1, 0)	(1, 0, 1)	(0, 1, 1)	(1, 1, 1)

FIGURE 14.14
Eight-block 2^5 design with *ABC*, *ABDE*, and *BCD* chosen for confounding

Note how the choice of *ABC*, *ABDE*, and *BCD* has resulted in two two-factor interactions being confounded. Prudence must be taken so as not to confound effects of major concern.

Continuing with the same problem, the defining contrasts are

$$Q_1 = q_1 + q_2 + q_3$$

$$Q_2 = q_1 + q_2 + q_4 + q_5$$

$$Q_3 = q_2 + q_3 + q_4$$

The resulting eight-block design is given in Figure 14.14. As in the two- and four-block designs, once the principal block is known, the remaining blocks can be derived by multiplication (modulo 2). For instance, since *ae* is not in the principal block (block 1), a different block contains the treatment combinations $(1)(ae) = ae$, $(acd)(ae) = cde$, $(bce)(ae) = abc$, and $(abde)(ae) = bd$. Indeed, block 2 is comprised of *ae*, *cde*, *abc*, and *bd*.

The analysis of variance in the general 2^p-block design is a direct generalization of the analyses for the two- and four-block designs. In particular, for a single replication, effect sums of squares are computed in the same manner as in the unblocked design, *SSBL* is determined by summing the sums of squares for all confounded effects, and *SSE* is determined by summing unconfounded higher-order effects that are assumed to be null.

14.6.4 Partial Confounding

In the multiple-replicate design of Figure 14.11, the three-factor interaction *ABC* is confounded with blocks in each replicate, the result being that information regarding *ABC* is masked. Rather than confound the same interaction in each replicate and thereby lose all information regarding the confounded interaction, an experiment can be redesigned so that a different interaction is confounded in each replicate. When the same interaction is confounded in a multiple-replicate 2^k factorial experiment, the design is said to be **completely confounded;** when different interactions are confounded in different replicates, the design is said to be **partially confounded.** We will restrict our attention to two-block partially confounded designs.

	Replicate I (*ABC* confounded)			Replicate II (*AC* confounded)			Replicate III (*BC* confounded)	

FIGURE 14.15

Three-replicate two-block 2^3 design with *ABC*, *AC*, and *BC* partially confounded in replicates I, II, and III, respectively

Block 1	Block 2		Block 1	Block 2		Block 1	Block 2
(1)	*a*		(1)	*a*		(1)	*b*
ab	*b*		*b*	*c*		*a*	*c*
ac	*c*		*ac*	*ab*		*bc*	*ab*
bc	*abc*		*abc*	*bc*		*abc*	*ac*

As an illustration, consider the three-replicate, two-block 2^3 design pictured in Figure 14.11 in which the interaction *ABC* is completely confounded. Figure 14.15 depicts a partially confounded 2^3 design in which *ABC*, *AC*, and *BC* are confounded in replicates I, II, and III, respectively. Because *ABC*, *AC*, and *BC* are only partially confounded, some information regarding each is obtainable from the experiment.

In a partially confounded two-block 2^k factorial design, total variation is broken down into variation arising from blocks, effects, and error. If there are r replicates, and r interactions are confounded once each, then there are $2r - 1$ degrees of freedom associated with blocks and 1 degree of freedom associated with each effect. The remaining degrees of freedom are associated with error. Sums of squares for unconfounded effects are computed in the usual manner. When using Yates' method, simply ignore the sums of squares arising for partially confounded effects. For partially confounded effects, the sums of squares are computed by the formula

$$SSK = \frac{[K]^2}{(r - 1)2^k}$$

where $r - 1$ is the number of replicates in which the effect K is not confounded and where only the data in the replicates in which K is not confounded are employed to find the contrast $[K]$. The blocks sum of squares $SSBL$ is computed in the usual manner, and SSE is obtained by subtraction.

	Replicate I (*ABC* confounded)			Replicate II (*AB* confounded)	

FIGURE 14.16

Two-replicate two-block 2^3 design for Example 14.22, *ABC* and *AB* partially confounded in replicates I and II, respectively

Block 1		Block 2			Block 1		Block 2	
(1)	270	*a*	275		(1)	270	*a*	275
ab	293	*b*	286		*c*	280	*b*	290
ac	282	*c*	270		*ab*	291	*ac*	280
bc	290	*abc*	296		*abc*	302	*bc*	286

EXAMPLE 14.22

The tensile-strength experiment of Example 14.21 involves two replications and is blocked between two pieces of test equipment according to the completely confounded design of Figure 14.12, *ABC* being the confounded interaction. Suppose it is instead decided to employ the partially confounded design of Figure 14.16, in which *ABC* is confounded in replicate I and *AB* is confounded in replicate II. Moreover, suppose the data of Figure 14.12 are again obtained, these being rewritten in Figure 14.16. Since, originally, the response totals were given in Table 14.34, the description of Yates' algorithm given in Table 14.36 is applicable, the exceptions being that the sums of squares given for *AB* and *ABC* should be ignored. Moreover, the value $SST = 1460$ found in Example 14.18 also applies, since the response totals and $T_{..} = 4536$ are unchanged. From the data of Figure 14.16,

$$SSBL = \frac{(1135)^2 + (1127)^2 + (1143)^2 + (1131)^2}{4} - \frac{(4536)^2}{16} = 35$$

Sums of squares for the partially confounded interactions are

$$SSABC = \frac{[a + b + c + abc - (1) - ab - ac - bc]^2}{1 \times 8}$$

$$= \frac{(275 + 290 + 280 + 302 - 270 - 291 - 280 - 286)^2}{8} = 50$$

$$SSAB = \frac{[ab + c + abc + (1) - a - b - ac - bc]^2}{1 \times 8}$$

$$= \frac{(293 + 270 + 296 + 270 - 275 - 286 - 282 - 290)^2}{8} = 2$$

The error sum of squares and error mean square are

$$SSE = SST - SSA - SSB - SSC - SSAB - SSAC - SSBC - SSABC - SSBL$$
$$= 1460 - 169 - 1089 - 81 - 2 - 16 - 4 - 50 - 35 = 14$$

and $MSE = 14/5 = 2.8$. The analysis of variance is summarized in Table 14.46. Of particular interest is $f_{ABC} = 17.86$, which is significant at level 0.01.

TABLE 14.46   ANOVA TABLE FOR EXAMPLE 14.22

Source of Variation	Sum of Squares	Degrees of Freedom	Mean Square	F Test Statistic
Nickel (A)	169	1	169	60.36
Molybdenum (B)	1089	1	1089	388.93
Titanium (C)	81	1	81	28.93
AB	2	1	2	0.71
AC	16	1	16	5.71
BC	4	1	4	1.43
ABC	50	1	50	17.86
Blocks	35	3	11.67	4.17
Error	14	5	2.80	
Total	1460	15		

EXERCISES

In Exercises 14.1 through 14.5, construct the ANOVA table and test for equality of treatment effects at significance levels $\alpha = 0.05$ and $\alpha = 0.01$.

14.1. A metallurgist wishes to compare the effects of three hardening methods on several alloyed steels. Because the result of hardening will depend upon the particular alloy, the decision is made to block the experiment with each block consisting of one of four alloys. Randomness is achieved within each block by randomly assigning each of three alloy specimens to one of the three hardening methods. The Rockwell-hardness data of Table 14.47 are obtained.

TABLE 14.47 BLOCKED HARDNESS DATA

Method	Alloy 1	2	3	4
1	48	52	44	58
2	47	48	41	54
3	50	55	47	60

14.2. A small electric motor can be assembled by an individual in three different manners. To decide which is most efficient, the manufacturer wishes to observe five assemblies for each assembly technique. Because assembly depends on operator skills, five operators are selected and each is treated as a block for the three assembly techniques. For each assembler, three unassembled motors are randomly assigned to the techniques and the order of assembly is randomly determined. The resulting assembly times (in minutes) are provided in Table 14.48.

TABLE 14.48 BLOCKED ASSEMBLY-TIME DATA

Technique	Assembler 1	2	3	4	5
1	34.1	30.2	37.7	35.6	30.0
2	32.7	29.1	37.8	31.5	30.0
3	31.3	28.4	37.2	31.6	29.2

14.3. An operating-system designer wishes to compare the effects of four queue disciplines for jobs awaiting CPU service. Recognizing that the efficiency of a particular discipline depends upon the computational intensity and the I/O requirements of the program population, the designer decides to block the experiment by observing each discipline in each of three distinct programming environments, the latter serving as the blocks. The data of Table 14.49 give the average observed queue lengths for each discipline in each of the three environments.

TABLE 14.49 BLOCKED QUEUE-LENGTH DATA

Discipline	Environment 1	2	3
1	5	12	8
2	4	10	6
3	4	10	7
4	4	8	5

14.4. Three laboratories have been making melting-point determinations for a class of compounds, and the question arises as to whether their methods of analysis lead to homogeneous results. Each is assigned four different compounds, the compounds serving as the blocks. Randomness is achieved within the blocks by randomly selecting one of three samples from each compound to be assigned to each laboratory. The melting-point data (in degrees centigrade) are given in Table 14.50.

TABLE 14.50 MELTING-POINT DATA

Lab	Compound 1	2	3	4
1	176.4	189.3	192.8	180.0
2	176.0	188.8	192.1	179.2
3	176.8	189.6	193.0	180.5

14.5. An automobile engineer wishes to study the effect of speed on the quantity of carbon monoxide emissions. Although attention will be fixed upon a single make of automobile traveling at different speeds, it is known that the external temperature has an effect. Thus, the tests are run at 15, 20, 25, and 30 mph, the blocks being determined by external temperatures of 20°F, 40°F, 60°F, and 80°F. Table 14.51 contains the emission test data (in grams per vehicle mile).

TABLE 14.51 CARBON MONOXIDE EMISSION DATA

Mph	Temperature (°F) 20°	40°	60°	80°
15	104	92	78	75
20	88	70	62	62
25	70	56	50	49
30	60	50	48	45

In Exercises 14.6 through 14.10, apply Theorem 14.1 to the data of the cited exercise to arrive at estimates of the model parameters μ, α_i, β_j, $\mu_{i.}$, $\mu_{.j}$, and μ_{ij} for the relevant i and j.

14.6. Exercise 14.1
14.7. Exercise 14.2
14.8. Exercise 14.3

14.9. Exercise 14.4

14.10. Exercise 14.5

In Exercises 14.11 through 14.15, if the null hypothesis of equal treatment effects is rejected at level $\alpha = 0.01$ (and, therefore, also at level $\alpha = 0.05$) in the cited exercise, apply Duncan's multiple range test using $\alpha = 0.01$. If the null hypothesis is rejected at level $\alpha = 0.05$ but not at level $\alpha = 0.01$, then apply Duncan's multiple range test at level $\alpha = 0.05$. Finally, if the null hypothesis is not rejected at either level, do nothing.

14.11. Exercise 14.1

14.12. Exercise 14.2

14.13. Exercise 14.3

14.14. Exercise 14.4

14.15. Exercise 14.5

In Exercises 14.16 through 14.20, assume the random-effects model and state the appropriate conclusion regarding σ_A^2 in the cited exercise. For instance, relative to Exercise 14.1, it might be that the experimenter has randomly selected the experimental hardening methods from a collection of hardening methods and has randomly chosen the test alloys from a collection of alloys. In all cases, let $\alpha = 0.05$. Moreover, if the null hypothesis is rejected, estimate σ_A^2.

14.16. Exercise 14.1

14.17. Exercise 14.2

14.18. Exercise 14.3

14.19. Exercise 14.4

14.20. Exercise 14.5

14.21. Use the model constraints to verify that

$$\mu = \frac{1}{N} \sum_{i=1}^{r} \sum_{j=1}^{n} \mu_{ij}$$

14.22. Show that $\alpha_1 = \cdots = \alpha_r$ if and only if $\mu_{1.} = \cdots = \mu_{r.}$.

14.23. Complete the proof of Theorem 14.1 by showing that the stated estimators $\hat{\mu}$ and $\hat{\beta}_j$ are unbiased estimators of μ and β_j, respectively.

14.24. Prove Theorem 14.3.

SECTION 14.3

In Exercises 14.25 through 14.29, construct the fixed-effects two-way-classification ANOVA table. Let $\alpha = 0.05$ and test for interaction and for zero main effects in each exercise. Interpret the results. Finally, if either factor null hypothesis is rejected, apply Duncan's multiple range test to the factor (or both factors) whose corresponding null hypothesis has been rejected.

14.25. A manager wishes to determine whether or not three office employees perform word-processing tasks at essentially equal speeds. At the same time, he or she desires to know whether completion time is affected by choice of software packages, A or B. Eigh-

teen tasks that are rated of equal difficulty are randomly assigned to operators and computers, and the data (in minutes) of Table 14.52 are obtained.

TABLE 14.52 WORD-PROCESSING TIMES

Employee	System A	System B
1	45.4	50.0
	48.3	50.9
	44.6	52.5
2	50.3	55.6
	53.2	58.0
	51.7	57.4
3	45.6	51.5
	42.1	48.3
	48.2	51.3

14.26. Example 14.7 discussed interaction in an experiment to test yield as a function of both duration and temperature. The data of Table 14.53 are derived from such an experiment. The temperatures are 140°F, 160°F, and 180°F, and the durations are 20, 30, and 40 minutes.

TABLE 14.53 YIELD DATA

Temperature	Duration 20	Duration 30	Duration 40
140	41	43	44
	40	42	44
	42	44	44
160	50	52	52
	50	51	53
	48	50	52
180	50	50	53
	49	53	55
	51	53	54

14.27. Over time, the bore of a hydraulic cylinder can become pitted and scored. A manufacturer wishes to compare two makes of cylinders and three different hydraulic fluids. Table 14.54 contains coded data, each datum representing the degree to which a given cylinder supplied with a given fluid is pitted and scored after some specified amount of usage.

TABLE 14.54 CYLINDER-WEAR DATA

Cylinder	Fluid		
	1	2	3
1	32 30 33		
	34 29 31		
2	35 31 33		
	34 30 44		

TABLE 14.56 RECOGNITION DATA

Algorithm	Filter	
	1	2
A	72 68	
	68 66	
	74 59	
B	45 52	
	50 56	
	48 57	

14.28. A shoe manufacturer wishes to compare the durabilities of three types of synthetic soles and at the same time compare the effects of people of different weights on durability. Table 14.55 gives laboratory data (in hours) that have been derived by observing wearers of different weights. Each datum gives the elapsed time until a sole reaches a prespecified state of wear. The sole types are A, B, and C, and the subjects' weights are 140, 180, and 220 pounds.

TABLE 14.55 SHOE-WEAR DATA

Sole	Weight		
	140	180	220
A	5.4	4.3	3.2
	5.0	4.8	3.6
	5.2	4.4	3.4
B	5.3	5.0	4.7
	5.4	4.9	4.5
	5.5	5.1	5.0
C	5.5	5.2	4.9
	5.8	5.4	5.0
	5.2	4.9	5.0

14.29. An image engineer wishes to determine whether or not there is a significant difference between the effects of two target-recognition algorithms, A and B. Before the application of either A or B, the image must be filtered to remove excess noise. The engineer can choose between filter F and filter G. Table 14.56 contains ratings given to the recognition system under each of the four possible algorithm-filter combinations.

In Exercises 14.30 through 14.34, apply Theorem 14.5 to find estimates of μ, α_i, β_j, and $(\alpha\beta)_{ij}$ in the cited exercise.

14.30. Exercise 14.25
14.31. Exercise 14.26

14.32. Exercise 14.27
14.33. Exercise 14.28
14.34. Exercise 14.29

In Exercises 14.35 through 14.39, assume that the levels of both factors in the cited exercise have been randomly selected from large populations of possible levels. Applying a random-effects model, find the ANOVA table. Test the random-effects interaction and factor null hypotheses at level $\alpha = 0.05$. Interpret the results. For any rejected null hypothesis, estimate the corresponding variance.

14.35. Exercise 14.25
14.36. Exercise 14.26
14.37. Exercise 14.27
14.38. Exercise 14.28
14.39. Exercise 14.29

In Exercises 14.40 through 14.44, treat the given factor of the cited exercise as fixed and the other as random. Construct the ANOVA table and perform the appropriate hypothesis tests at level $\alpha = 0.05$. Interpret the results.

14.40. Exercise 14.25, system
14.41. Exercise 14.26, duration
14.42. Exercise 14.27, cylinder
14.43. Exercise 14.28, sole
14.44. Exercise 14.29, algorithm

14.45. Complete the proof of Theorem 14.5.
14.46. Prove that, in the fixed-effects two-way classification, $\beta_j = \mu_{.j.} - \mu$.
14.47. Prove that, in the fixed-effects two-way classification,

$$\mu = \frac{1}{nab} \sum_{i=1}^{a} \sum_{j=1}^{b} \sum_{k=1}^{n} \mu_{ijk}$$

In Exercises 14.48 through 14.51, assume interaction to be negligible and apply analysis of variance using a single observation per treatment combination to the cited data. Let $\alpha = 0.05$ and test for zero main effects.

14.48. Table 14.7
14.49. First replicate of Table 14.52
14.50. First replicate of Table 14.54
14.51. First replicate of Table 14.55

14.52. Apply Tukey's test for interaction to replicate I of Table 14.53 with $\alpha = 0.05$.
14.53. Apply Tukey's test for interaction to replicate I of Table 14.55 with $\alpha = 0.01$.

SECTION 14.4

14.54. There are seven null hypotheses of interest in the three-factor model, three of which are explicitly stated in the text. Using symmetry, explicitly state the remaining four.
14.55. In the text, unbiased parameter estimators are given for μ, α_i, $(\alpha\beta)_{ij}$, and $(\alpha\beta\gamma)_{ijk}$. Using symmetry, find the remaining model-parameter estimators.
14.56. In the text, expressions are given for SSA, $SSAB$, and $SSABC$ in terms of model-parameter estimators. What are the corresponding expressions for SSB, SSC, $SSAC$, and $SSBC$?
14.57. In the text, expressions are given for $E[MSA]$, $E[MSAB]$, and $E[MSABC]$. Using symmetry, find expressions for $E[MSB]$, $E[MSC]$, $E[MSAC]$, and $E[MSBC]$.

In Exercises 14.58 through 14.60, construct the three-way-classification ANOVA table for the cited exercise. Carry out all hypothesis tests for the model and interpret the results.

14.58. In Exercise 14.26 we considered yield percentages for two factors, temperature and duration. An analysis of variance was performed using a fixed-effects two-way classification. Now consider the addition of a third factor, the percentage of catalyst. The temperature levels are 160°F and 180°F, the duration levels are 20 min and 30 min, and the catalyst levels are 0.3% or 0.4%. Table 14.57 gives the data for a fixed-effects three-way classification, there being three replications.

TABLE 14.57 YIELD DATA FOR THREE-WAY CLASSIFICATION

	20		30	
	0.3	**0.4**	**0.3**	**0.4**
160	31	33	42	43
	34	36	40	42
	35	35	41	41
180	37	39	42	48
	33	40	41	46
	35	38	40	50

14.59. In Exercise 14.28 we considered a fixed-effects two-way classification involving the factors sole type and wearer weight. In the present exercise, we include a third factor concerning the pace at which the wearer is walking. The sole and weight levels are the same as those of Exercise 14.28 and the pace levels are slow (S) and fast (F). Table 14.58 gives the elapsed times to the wear limits for each sole–weight–pace category.

TABLE 14.58 SHOE-WEAR DATA FOR THREE-WAY CLASSIFICATION

	140		**180**		**220**	
	S	**F**	**S**	**F**	**S**	**F**
A	5.6	4.8	4.9	4.0	3.8	2.8
	5.8	4.6	4.7	4.0	4.0	3.2
B	5.7	4.9	5.4	4.8	4.9	4.3
	5.3	4.9	4.6	4.4	4.5	4.3
C	6.0	5.4	6.0	5.1	5.5	5.0
	6.2	5.2	5.4	4.7	5.1	5.0

14.60. In Exercise 14.29 we considered two recognition algorithms and two filters. Suppose the images are optically created and consider the addition of a third factor, the level of light. Table 14.59 gives the data for the new fixed-effects three-way classification. The algorithm and filter levels are the same as in Exercise 14.29 and the levels of light are low (L) and high (H).

TABLE 14.59 RECOGNITION DATA FOR EXERCISE 14.60

	1		**2**	
	L	**H**	**L**	**H**
A	67	78	70	64
	63	74	70	62
	65	70	67	60
B	40	50	58	47
	48	54	60	50
	44	54	59	53

SECTION 14.5

In Exercises 14.61 through 14.64, the data from the cited table are assumed to have been derived from a 2^2 factorial

design. In each case, find the ANOVA table, draw conclusions from the F statistics at levels 0.05 and 0.01, and use the computed values of A, B, and AB to estimate the eight model parameters.

14.61. Temperatures 140° and 160° and durations 20 minutes and 30 minutes of Table 14.53.

14.62. Temperatures 160° and 180° and durations 20 minutes and 30 minutes of Table 14.53.

14.63. Cylinders 1 and 2 and fluids 1 and 2 of Table 14.54.

14.64. Algorithms *A* and *B* and filters 1 and 2 of Table 14.56.

14.65. To test the effects of motor oil, fuel mixture, and type of use on a particular type of engine, a 2^3 factorial design is employed using two replicates. Oil (*A*) is rated as low viscosity (0) or high viscosity (1), fuel (*B*) is rated as unleaded (0) or leaded (1), and usage (*C*) is rated as light (0) or heavy (1). At the completion of the tests on the 16 randomly selected engines, the condition of each is rated on a scale from 0 to 100, a higher score representing a better condition. The data of Table 14.60 result. Find the ANOVA table and draw conclusions from the *F* statistics at levels 0.05 and 0.01. Apply Yates' method.

TABLE 14.60   ENGINE-WEAR DATA FOR EXERCISE 14.65

Factor			Replicate	
A	*B*	*C*	I	II
0	0	0	70	74
0	0	1	42	50
0	1	0	68	72
0	1	1	50	48
1	0	0	82	76
1	0	1	61	58
1	1	0	88	82
1	1	1	64	60

TABLE 14.61   SHELF-LIFE DATA FOR EXERCISE 14.66

Factor			Replicate	
A	*B*	*C*	I	II
0	0	0	20	22
0	0	1	23	25
0	1	0	22	21
0	1	1	23	15
1	0	0	11	15
1	0	1	17	15
1	1	0	14	17
1	1	1	17	14

TABLE 14.62   FABRIC-QUALITY DATA FOR EXERCISE 14.68

Factor				Replicate	
A	*B*	*C*	*D*	I	II
0	0	0	0	7.4	7.8
0	0	0	1	8.0	7.8
0	0	1	0	8.5	8.2
0	0	1	1	8.8	9.0
0	1	0	0	7.3	7.2
0	1	0	1	8.0	7.5
0	1	1	0	8.0	7.6
0	1	1	1	8.6	9.0
1	0	0	0	7.0	7.5
1	0	0	1	7.5	8.3
1	0	1	0	7.8	8.0
1	0	1	1	8.0	7.8
1	1	0	0	8.0	8.4
1	1	0	1	8.7	9.0
1	1	1	0	9.0	9.2
1	1	1	1	9.0	8.5

14.66. The shelf life (in days) of a milk product is to be tested relative to storage temperature (*A*), packaging (*B*), and supplier (*C*). Temperature is rated low (0) and high (1), packaging is of types 0 and 1, and suppliers are coded 0 and 1. The resulting data for two replications are given in Table 14.61. Find the ANOVA table and draw conclusions from the *F* statistics at levels 0.05 and 0.01. Apply Yates' method.

14.67. Presuming the data of Table 14.57 to have come from a 2^3 factorial experiment, find the ANOVA table and draw conclusions from the *F* statistics at levels 0.05 and 0.01. Apply Yates' method.

14.68. The quality of a fabric is rated on a scale from 0 to 10.0. A 2^4 factorial experiment is used to test the effects of two human operators (*A*), two machines (*B*), raw material (*C*), and dye (*D*). The results from two replications are given in Table 14.62. Using analysis of variance, test for main and interaction effects at level 0.05. Apply Yates' method.

For Exercises 14.69 and 14.70, apply the single-replicate approach to 2^k factorial design to the replicate of the cited table. Find the ANOVA tables and draw relevant conclusions from the F statistics at levels $\alpha = 0.05$ and $\alpha = 0.01$. Note that our rule of thumb concerning k $\geq$ 4 is violated in Exercise 14.69.

14.69. Replicate I of Table 14.60 (assuming third-order interaction is null)

14.70. Replicate I of Table 14.62 (assuming third- and fourth-order interactions are null)

SECTION 14.6

In Exercises 14.71 through 14.76, assume the data of the cited table have come from a two-block 2^k design with the highest-level interaction effect confounded with blocks. Find the ANOVA table and draw conclusions from the relevant F statistics at levels 0.05 and 0.01.

14.71. Replicates I and II of Table 14.57
14.72. All three replicates of Table 14.57
14.73. Replicate II of Table 14.60
14.74. Both replicates of Table 14.60
14.75. Replicate II of Table 14.61
14.76. Replicate II of Table 14.62

14.77. Set up the blocks for a 2^3 design run in two blocks with *AC* confounded.

14.78. Set up the blocks for a 2^4 design run in two blocks with *ABD* confounded.

14.79. Set up the blocks for a 2^5 design run in two blocks with *ABDE* confounded.

14.80. Set up the blocks for a 2^5 design run in four blocks with *ABCDE* and *ABC* chosen for confounding.

14.81. Set up the blocks for a 2^6 design run in four blocks with *ABCD* and *ABEF* chosen for confounding.

14.82. Suppose the fabric-quality data of Table 14.62 have been derived from an experiment run in four blocks with *ABC*, *CD*, and *ABD* confounded. Perform an analysis of variance and draw conclusions at levels 0.05 and 0.01 from the relevant *F* statistics.

14.83. Repeat Exercise 14.82 under the condition that *BCD*, *ABC*, and *AD* are confounded.

In Exercises 14.84 through 14.86, suppose the data of the cited table have been derived from a partially confounded 2^k design with the stated effects confounded. In each, find the ANOVA table and draw conclusions at levels 0.05 and 0.01 based upon the relevant F statistics.

14.84. Table 14.60 with *ABC* and *AB* confounded in replicates I and II, respectively.

14.85. Table 14.61 with *ABC* and *AC* confounded in replicates I and II, respectively.

14.86. Table 14.62 with *ABCD* and *ABC* confounded in replicates I and II, respectively.

SUPPLEMENTAL EXERCISES FOR COMPUTER SOFTWARE

In Exercises 14.C1 through 14.C3, repeat the instructions of the cited exercise, employing the additional data provided in the cited table in conjunction with the original data.

14.C1. Exercise 14.1, Table 14.63.
14.C2. Exercise 14.2, Table 14.64.
14.C3. Exercise 14.26, Table 14.65.

14.C4. Repeat the instructions of Exercise 14.29 using the new data of Table 14.66.

TABLE 14.64 ADDITIONAL BLOCKED ASSEMBLY-TIME DATA FOR EXERCISE 14.C2

Technique	Assembler					
	6	**7**	**8**	**9**	**10**	**11**
1	29.3	36.0	41.4	30.0	34.5	37.8
2	28.3	36.1	35.3	29.0	32.4	36.4
3	28.3	34.2	33.3	27.3	29.5	33.4

TABLE 14.63 ADDITIONAL BLOCKED HARDNESS DATA FOR EXERCISE 14.C1

Method	Alloy					
	5	**6**	**7**	**8**	**9**	**10**
1	46	48	56	56	50	56
2	42	44	53	50	45	52
3	48	46	58	55	51	55

TABLE 14.65 ADDITIONAL YIELD DATA FOR EXERCISE 14.C3

Temperature	Duration		
	20	**30**	**40**
150	45	47	47
	44	44	50
	48	48	48
170	50	51	53
	52	49	53
	51	51	55

TABLE 14.66 RECOGNITION DATA FOR EXERCISE 14.C4

Algorithm	Filter		
	1	**2**	**3**
A	73 70 73	70 66 62	59 54 50
B	48 50 46	55 54 50	48 49 45
C	58 56 62	55 60 62	50 46 50

14.C5. Redo Exercise 14.C3 under the assumption of random effects.

14.C6. Redo Exercise 14.C4 under the assumption of random effects.

14.C7. Redo the fixed-effects three-way-classification analysis of variance of Exercise 14.58 using the additional data provided in Table 14.67.

TABLE 14.67 ADDITIONAL YIELD DATA FOR EXERCISE 14.C7

	20		**30**	
	0.3	**0.4**	**0.3**	**0.4**
170	36 35 33	37 35 36	42 40 41	45 43 47
190	36 35 38	42 39 38	41 43 42	53 50 51

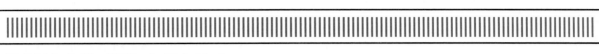

BIBLIOGRAPHY

Barlow, R. E., and F. Proschan. *Mathematical Theory of Reliability*. New York: John Wiley & Sons, Inc., 1965.

Barlow, R. E., and F. Proschan. *Statistical Theory of Reliability and Life Testing: Probability Models*. New York: Holt, Rinehart and Winston, 1975.

Baxovsky, I. *Reliability: Theory and Practice*. Englewood Cliffs, NJ: Prentice-Hall, Inc., 1964.

Bhat, U. N. *Elements of Applied Stochastic Processes*. New York: John Wiley & Sons, Inc., 1972.

Bowker, A. H., and G. J. Lieberman. *Engineering Statistics*, 2d ed. Englewood Cliffs, NJ: Prentice-Hall, Inc., 1972.

Box, G. E. P., Hunter, W. G., and J. S. Hunter. *Statistics for Experimenters*. New York: John Wiley & Sons, Inc., 1978.

Brownlee, K. A. *Statistical Theory and Methodology in Science and Engineering*, 2d ed. New York: John Wiley & Sons, Inc., 1965.

Burr, I. W. *Engineering Statistics and Quality Control*. New York: McGraw-Hill Book Company, 1953.

Canavos, G. C. *Applied Probability and Statistical Methods*. Boston: Little, Brown and Company, 1984.

Chatterjee, S., and B. Price. *Regression Analysis by Example*. New York: John Wiley & Sons, Inc., 1977.

Chernoff, H., and L. E. Moses. *Elementary Decision Theory*. New York: John Wiley & Sons, Inc., 1959.

Cochran, W. G., and G. M. Cox. *Experimental Designs*, 2d ed. New York: John Wiley & Sons, Inc., 1957.

Conover, W. J. *Practical Nonparametric Statistics*. New York: John Wiley & Sons, Inc., 1971.

Cowden, D. J. *Statistical Methods in Quality Control*. Englewood Cliffs, NJ: Prentice-Hall, Inc., 1957.

Cox, D. R., and H. D. Miller. *The Theory of Stochastic Processes*. New York: John Wiley & Sons, Inc., 1968.

Cramer, H. *Mathematical Methods of Statistics*. Princeton, NJ: Princeton University Press, 1946.

Cramer, H., and M. R. Leadbetter. *Stationary and Related Stochastic Processes*. New York: John Wiley & Sons, Inc., 1967.

Daniel, C., and F. Wood. *Fitting Equations to Data*, 2d ed. New York: John Wiley & Sons, Inc., 1980.

David, H. A. *Order Statistics*. New York: John Wiley & Sons, Inc., 1970.

DeGroot, M. H. *Optimal Statistical Decisions*. New York: McGraw-Hill Book Company, 1970.

DeGroot, M. H. *Probability and Statistics*, 2d ed. Reading, MA: Addison-Wesley Publishing Co., 1986.

Devore, J. L. *Probability and Statistics for Engineering and the Sciences*, 2d ed. Monterey, CA: Brooks/Cole Publishing Co., 1987.

Dixon, W. J., and F. J. Massey, Jr. *Introduction to Statistical Analysis*, 3d ed. New York: McGraw-Hill Book Company, 1969.

Dougherty, E. R., and C. R. Giardina. *Mathematical Methods for Artificial Intelligence and Autono-*

mous Systems. Englewood Cliffs, NJ: Prentice-Hall, Inc., 1988.

Duncan, A. J. Quality Control and Industrial Statistics, 3d ed. Homewood, Ill.: Richard D. Irwin, Inc., 1965.

Draper, N., and H. Smith. Applied Regression Analysis, 2d ed. New York: John Wiley & Sons, Inc., 1981.

Everitt, B. S. The Analysis of Contingency Tables. New York: John Wiley & Sons, Inc., 1977.

Federer, W. T. Experimental Designs, Theory and Application. New York: Macmillan Publishing Company, 1955.

Feller, W. An Introduction to Probability Theory and Its Applications, Vol. 1, 3d ed. New York: John Wiley & Sons, Inc., 1957.

Fisz, M. Probability Theory and Mathematical Statistics. New York: John Wiley & Sons, Inc., 1963.

Fraser, D. A. S. Probability and Statistics: Theory and Applications. Boston: Duxbury Press, 1976.

Freund, J. E., and R. E. Walpole. Mathematical Statistics, 4th ed. Englewood Cliffs, NJ: Prentice-Hall, Inc., 1987.

Goldberg, S. Probability—An Introduction. Englewood Cliffs, NJ: Prentice-Hall, Inc., 1960.

Grant, E., and R. Leavenworth. Statistical Quality Control, 4th ed. New York: McGraw-Hill Book Company, 1972.

Guenther, W. C. Analysis of Variance. Englewood Cliffs, NJ: Prentice-Hall, Inc., 1964.

Guttman, I., and S. S. Wilks. Introductory Engineering Statistics. New York: John Wiley & Sons, Inc., 1965.

Hahn, G. J., and S. S. Shapiro. Statistical Models in Engineering. New York: John Wiley & Sons, Inc., 1967.

Hicks, C. R. Fundamental Concepts in the Design of Experiments, 2d ed. New York: Holt, Rinehart and Winston, 1973.

Hines, W. W., and D. C. Montgomery. Probability and Statistics in Engineering and Management Science. New York: Ronald Press, 1972.

Hoel, P. G. Introduction to Mathematical Statistics, 4th ed. New York: John Wiley & Sons, Inc., 1971.

Hoel, P. G., Port, S. C., and C. J. Stone. Introduction to Probability Theory. Boston: Houghton Mifflin Co., 1971.

Hoel, P. G., Port, S. C., and C. J. Stone. Introduction to Statistical Theory. Boston: Houghton Mifflin Co., 1971.

Hoel, P. G., Port, S. C., and C. J. Stone. Introduction to Stochastic Processes. Boston: Houghton Mifflin Co., 1972.

Hogg, R. V., and A. T. Craig. Introduction to Mathematical Statistics, 4th ed. New York: Macmillan Co., 1978.

Hollander, M., and D. Wolfe. Nonparametric Statistical Methods. Boston: Houghton Mifflin Co., 1973.

Isaacson, D. L., and R. W. Madsen. Markov Chains—Theory and Applications. New York: John Wiley & Sons, Inc., 1975.

Johnson, N. L., and F. C. Leone. Statistics and Experimental Design: In Engineering and the Physical Sciences, Vols. I and II, 2d ed. New York: John Wiley & Sons, Inc., 1977.

Lehmann, E. Nonparametrics: Statistical Methods Based on Ranks. San Francisco: Holden-Day, Inc., 1981.

Lewis, T. O., and P. L. Odell. Estimation in Linear Models. Englewood Cliffs, NJ: Prentice-Hall, Inc., 1971.

Mendenhall, W. An Introduction to Linear Models and the Design and Analysis of Experiments. Belmont, CA: Wadsworth Publishing Co., 1968.

Miller, I., and J. E. Freund. Probability and Statistics for Engineers, 3d ed. Englewood Cliffs, NJ: Prentice-Hall, Inc., 1985.

Milton, J. S., and J. C. Arnold. Probability and Statistics in the Engineering and Computing Sciences. New York: McGraw-Hill Book Company, 1986.

Montgomery, D. C. Design and Analysis of Experiments. New York: John Wiley & Sons, Inc., 1976.

Montgomery, D. C., and E. A. Peck. Introduction to Linear Regression Analysis. New York: John Wiley & Sons, Inc., 1982.

Morrison, D. F. Applied Linear Statistical Methods. Englewood Cliffs, NJ: Prentice-Hall, Inc., 1983.

Morse, P. H. Queues, Inventories and Maintenance. New York: John Wiley & Sons, Inc., 1958.

Mosteller, F., and J. Tukey. Data Analysis and Regression. Reading, MA: Addison-Wesley Publishing Co., Inc., 1977.

Noether, G. E. Introduction to Statistics: A Nonparametric Approach, 2d ed. Boston: Houghton Mifflin Co., 1976.

Ott, L. An Introduction to Statistical Methods and Data Analysis. Boston: Duxbury Press, 1977.

Owen, G. Game Theory. Philadelphia: W. B. Saunders Co., 1968.

Papoulis, A. Probability, Random Variables, and Stochastic Processes, 2d ed. New York: McGraw-Hill Book Company, 1984.

Parzen, E. Modern Probability Theory and Its Applications. New York: John Wiley & Sons, Inc., 1960.

Parzen, E. Stochastic Processes. San Francisco: Holden-Day, Inc., 1962.

Rao, C. R. Linear Statistical Inference and Its Applications. New York: John Wiley & Sons, Inc., 1965.

Roberts, N. H. Mathematical Methods in Reliability

Engineering. New York: McGraw-Hill Book Company, 1964.

Ross, S. *A First Course in Probability*, 2d ed. New York: Macmillan Publishing Co., 1984.

Ruiz-Pala, E., Avila-Beloso, C., and W. W. Hines. *Waiting-Line Models*. New York: Reinhold Publishing Corporation, 1967.

Scheaffer, R. L., and J. T. McClave. *Probability and Statistics for Engineers*, 2d ed. Boston: Duxbury Press, 1986.

Soong, T. T. *Probabilistic Modeling and Analysis in Science and Engineering*. New York: John Wiley & Sons, Inc., 1981.

Tukey, T. W. *Exploratory Data Analysis*. Reading, MA: Addison-Wesley Publishing Co., Inc., 1977.

Wald, A. *Statistical Decision Functions*. New York: John Wiley & Sons, Inc., 1950.

Walpole, R. E., and R. H. Myers. *Probability and Statistics for Engineers and Scientists*, 3d ed. New York: Macmillan Publishing Co., 1985.

Weisberg, S. *Applied Linear Regression*, 2d ed. New York: John Wiley & Sons, Inc., 1985.

Weiss, L. *Statistical Decision Theory*. New York: McGraw-Hill Book Company, 1961.

Wonnacott, T. H., and R. J. Wonnacott. *Regression: A Second Course in Statistics*. New York: John Wiley & Sons, Inc., 1981.

APPENDIX

A

STATISTICAL TABLES

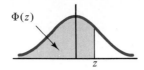

z	.00	.01	.02	.03	.04	.05	.06	.07	.08	.09
−3.4	.0003	.0003	.0003	.0003	.0003	.0003	.0003	.0003	.0003	.0002
−3.3	.0005	.0005	.0005	.0004	.0004	.0004	.0004	.0004	.0004	.0003
−3.2	.0007	.0007	.0006	.0006	.0006	.0006	.0006	.0005	.0005	.0005
−3.1	.0010	.0009	.0009	.0009	.0008	.0008	.0008	.0008	.0007	.0007
−3.0	.0013	.0013	.0013	.0012	.0012	.0011	.0011	.0011	.0010	.0010
−2.9	.0019	.0018	.0017	.0017	.0016	.0016	.0015	.0015	.0014	.0014
−2.8	.0026	.0025	.0024	.0023	.0023	.0022	.0021	.0021	.0020	.0019
−2.7	.0035	.0034	.0033	.0032	.0031	.0030	.0029	.0028	.0027	.0026
−2.6	.0047	.0045	.0044	.0043	.0041	.0040	.0039	.0038	.0037	.0036
−2.5	.0062	.0060	.0059	.0057	.0055	.0054	.0052	.0051	.0049	.0048
−2.4	.0082	.0080	.0078	.0075	.0073	.0071	.0069	.0068	.0066	.0064
−2.3	.0107	.0104	.0102	.0099	.0096	.0094	.0091	.0089	.0087	.0084
−2.2	.0139	.0136	.0132	.0129	.0125	.0122	.0119	.0116	.0113	.0110
−2.1	.0179	.0174	.0170	.0166	.0162	.0158	.0154	.0150	.0146	.0143
−2.0	.0228	.0222	.0217	.0212	.0207	.0202	.0197	.0192	.0188	.0183
−1.9	.0287	.0281	.0274	.0268	.0262	.0256	.0250	.0244	.0239	.0233
−1.8	.0359	.0352	.0344	.0336	.0329	.0322	.0314	.0307	.0301	.0294
−1.7	.0446	.0436	.0427	.0418	.0409	.0401	.0392	.0384	.0375	.0367
−1.6	.0548	.0537	.0526	.0516	.0505	.0495	.0485	.0475	.0465	.0455
−1.5	.0668	.0655	.0643	.0630	.0618	.0606	.0594	.0582	.0571	.0559
−1.4	.0808	.0793	.0778	.0764	.0749	.0735	.0722	.0708	.0694	.0681
−1.3	.0968	.0951	.0934	.0918	.0901	.0885	.0869	.0853	.0838	.0823
−1.2	.1151	.1131	.1112	.1093	.1075	.1056	.1038	.1020	.1003	.0985
−1.1	.1357	.1335	.1314	.1292	.1271	.1251	.1230	.1210	.1190	.1170
−1.0	.1587	.1562	.1539	.1515	.1492	.1469	.1446	.1423	.1401	.1379
−0.9	.1841	.1814	.1788	.1762	.1736	.1711	.1685	.1660	.1635	.1611
−0.8	.2119	.2090	.2061	.2033	.2005	.1977	.1949	.1922	.1894	.1867
−0.7	.2420	.2389	.2358	.2327	.2296	.2266	.2236	.2206	.2177	.2148
−0.6	.2743	.2709	.2676	.2643	.2611	.2578	.2546	.2514	.2483	.2451
−0.5	.3085	.3050	.3015	.2981	.2946	.2912	.2877	.2843	.2810	.2776
−0.4	.3446	.3409	.3372	.3336	.3300	.3264	.3228	.3192	.3156	.3121
−0.3	.3821	.3783	.3745	.3707	.3669	.3632	.3594	.3557	.3520	.3483
−0.2	.4207	.4168	.4129	.4090	.4052	.4013	.3974	.3936	.3897	.3859
−0.1	.4602	.4562	.4522	.4483	.4443	.4404	.4364	.4325	.4286	.4247
−0.0	.5000	.4960	.4920	.4880	.4840	.4801	.4761	.4721	.4681	.4641

TABLE A.1 (Continued)

z	.00	.01	.02	.03	.04	.05	.06	.07	.08	.09
0.0	.5000	.5040	.5080	.5120	.5160	.5199	.5239	.5279	.5319	.5359
0.1	.5398	.5438	.5478	.5517	.5557	.5596	.5636	.5675	.5714	.5753
0.2	.5793	.5832	.5871	.5910	.5948	.5987	.6026	.6064	.6103	.6141
0.3	.6179	.6217	.6255	.6293	.6331	.6368	.6406	.6443	.6480	.6517
0.4	.6554	.6591	.6628	.6664	.6700	.6736	.6772	.6808	.6844	.6879
0.5	.6915	.6950	.6985	.7019	.7054	.7088	.7123	.7157	.7190	.7224
0.6	.7257	.7291	.7324	.7357	.7389	.7422	.7454	.7486	.7517	.7549
0.7	.7580	.7611	.7642	.7673	.7704	.7734	.7764	.7794	.7823	.7852
0.8	.7881	.7910	.7939	.7967	.7995	.8023	.8051	.8078	.8106	.8133
0.9	.8159	.8186	.8212	.8238	.8264	.8289	.8315	.8340	.8365	.8389
1.0	.8413	.8438	.8461	.8485	.8508	.8531	.8554	.8577	.8599	.8621
1.1	.8643	.8665	.8686	.8708	.8729	.8749	.8770	.8790	.8810	.8830
1.2	.8849	.8869	.8888	.8907	.8925	.8944	.8962	.8980	.8997	.9015
1.3	.9032	.9049	.9066	.9082	.9099	.9115	.9131	.9147	.9162	.9177
1.4	.9192	.9207	.9222	.9236	.9251	.9265	.9278	.9292	.9306	.9319
1.5	.9332	.9345	.9357	.9370	.9382	.9394	.9406	.9418	.9429	.9441
1.6	.9452	.9463	.9474	.9484	.9495	.9505	.9515	.9525	.9535	.9545
1.7	.9554	.9564	.9573	.9582	.9591	.9599	.9608	.9616	.9625	.9633
1.8	.9641	.9649	.9656	.9664	.9671	.9678	.9686	.9693	.9699	.9706
1.9	.9713	.9719	.9726	.9732	.9738	.9744	.9750	.9756	.9761	.9767
2.0	.9772	.9778	.9783	.9788	.9793	.9798	.9803	.9808	.9812	.9817
2.1	.9821	.9826	.9830	.9834	.9838	.9842	.9846	.9850	.9854	.9857
2.2	.9861	.9864	.9868	.9871	.9875	.9878	.9881	.9884	.9887	.9890
2.3	.9893	.9896	.9898	.9901	.9904	.9906	.9909	.9911	.9913	.9916
2.4	.9918	.9920	.9922	.9925	.9927	.9929	.9931	.9932	.9934	.9936
2.5	.9938	.9940	.9941	.9943	.9945	.9946	.9948	.9949	.9951	.9952
2.6	.9953	.9955	.9956	.9957	.9959	.9960	.9961	.9962	.9963	.9964
2.7	.9965	.9966	.9967	.9968	.9969	.9970	.9971	.9972	.9973	.9974
2.8	.9974	.9975	.9976	.9977	.9977	.9978	.9979	.9979	.9980	.9981
2.9	.9981	.9982	.9982	.9983	.9984	.9984	.9985	.9985	.9986	.9986
3.0	.9987	.9987	.9987	.9988	.9988	.9989	.9989	.9989	.9990	.9990
3.1	.9990	.9991	.9991	.9991	.9992	.9992	.9992	.9992	.9993	.9993
3.2	.9993	.9993	.9994	.9994	.9994	.9994	.9994	.9995	.9995	.9995
3.3	.9995	.9995	.9995	.9996	.9996	.9996	.9996	.9996	.9996	.9997
3.4	.9997	.9997	.9997	.9997	.9997	.9997	.9997	.9997	.9997	.9998

TABLE A.2 PROBABILITY DISTRIBUTION FUNCTION FOR THE BINOMIAL DISTRIBUTION

$$B(x; n, p) = \sum_{k=0}^{x} b(k$$

n	x	.10	.20	.25	.30	.40	.50	.60	.70	.80	.90
1	0	.9000	.8000	.7500	.7000	.6000	.5000	.4000	.3000	.2000	.1000
	1	1.0000	1.0000	1.0000	1.0000	1.0000	1.0000	1.0000	1.0000	1.0000	1.0000
2	0	.8100	.6400	.5625	.4900	.3600	.2500	.1600	.0900	.0400	.0100
	1	.9900	.9600	.9375	.9100	.8400	.7500	.6400	.5100	.3600	.1900
	2	1.0000	1.0000	1.0000	1.0000	1.0000	1.0000	1.0000	1.0000	1.0000	1.0000
3	0	.7290	.5120	.4219	.3430	.2160	.1250	.0640	.0270	.0080	.0010
	1	.9720	.8960	.8438	.7840	.6480	.5000	.3520	.2160	.1040	.0280
	2	.9990	.9920	.9844	.9730	.9360	.8750	.7840	.6570	.4880	.2710
	3	1.0000	1.0000	1.0000	1.0000	1.0000	1.0000	1.0000	1.0000	1.0000	1.0000
4	0	.6561	.4096	.3164	.2401	.1296	.0625	.0256	.0081	.0016	.0001
	1	.9477	.8192	.7383	.6517	.4752	.3125	.1792	.0837	.0272	.0037
	2	.9963	.9728	.9492	.9163	.8208	.6875	.5248	.3483	.1808	.0523
	3	.9999	.9984	.9961	.9919	.9744	.9375	.8704	.7599	.5904	.3439
	4	1.0000	1.0000	1.0000	1.0000	1.0000	1.0000	1.0000	1.0000	1.0000	1.0000
5	0	.5905	.3277	.2373	.1681	.0778	.0312	.0102	.0024	.0003	.0000
	1	.9185	.7373	.6328	.5282	.3370	.1875	.0870	.0308	.0067	.0005
	2	.9914	.9421	.8965	.8369	.6826	.5000	.3174	.1631	.0579	.0086
	3	.9995	.9933	.9844	.9692	.9130	.8125	.6630	.4718	.2627	.0815
	4	1.0000	.9997	.9990	.9976	.9898	.9688	.9222	.8319	.6723	.4095
	5		1.0000	1.0000	1.0000	1.0000	1.0000	1.0000	1.0000	1.0000	1.0000
6	0	.5314	.2621	.1780	.1176	.0467	.0156	.0041	.0007	.0001	.0000
	1	.8857	.6554	.5339	.4202	.2333	.1094	.0410	.0109	.0016	.0001
	2	.9841	.9011	.8306	.7443	.5443	.3438	.1792	.0705	.0170	.0013
	3	.9987	.9830	.9624	.9295	.8208	.6563	.4557	.2557	.0989	.0158
	4	.9999	.9984	.9954	.9891	.9590	.8906	.7667	.5798	.3447	.1143
	5	1.0000	.9999	.9998	.9993	.9959	.9844	.9533	.8824	.7379	.4686
	6		1.0000	1.0000	1.0000	1.0000	1.0000	1.0000	1.0000	1.0000	1.0000
7	0	.4783	.2097	.1335	.0824	.0280	.0078	.0016	.0002	.0000	
	1	.8503	.5767	.4449	.3294	.1586	.0625	.0188	.0038	.0004	.0000
	2	.9743	.8520	.7564	.6471	.4199	.2266	.0963	.0288	.0047	.0002
	3	.9973	.9667	.9294	.8740	.7102	.5000	.2898	.1260	.0333	.0027
	4	.9998	.9953	.9871	.9712	.9037	.7734	.5801	.3529	.1480	.0257
	5	1.0000	.9996	.9987	.9962	.9812	.9375	.8414	.6706	.4233	.1497
	6		1.0000	.9999	.9998	.9984	.9922	.9720	.9176	.7903	.5217
	7			1.0000	1.0000	1.0000	1.0000	1.0000	1.0000	1.0000	1.0000

Reprinted with permission of Macmillan Publishing Company from *Probability and Statistics for Engineers and Scientists* by R. E. Walpole and R. H. Myers. Copyright © 1978 by Macmillan Publishing Company.

n	x	.10	.20	.25	.30	.40	.50	.60	.70	.80	.90
							p				
8	0	.4305	.1678	.1001	.0576	.0168	.0039	.0007	.0001	.0000	
	1	.8131	.5033	.3671	.2553	.1064	.0352	.0085	.0013	.0001	
	2	.9619	.7969	.6785	.5518	.3154	.1445	.0498	.0113	.0012	.0000
	3	.9950	.9437	.8862	.8059	.5941	.3633	.1737	.0580	.0104	.0004
	4	.9996	.9896	.9727	.9420	.8263	.6367	.4059	.1941	.0563	.0050
	5	1.0000	.9988	.9958	.9887	.9502	.8555	.6846	.4482	.2031	.0381
	6		.9991	.9996	.9987	.9915	.9648	.8936	.7447	.4967	.1869
	7		1.0000	1.0000	.9999	.9993	.9961	.9832	.9424	.8322	.5695
	8				1.0000	1.0000	1.0000	1.0000	1.0000	1.0000	1.0000
9	0	.3874	.1342	.0751	.0404	.0101	.0020	.0003	.0000		
	1	.7748	.4362	.3003	.1960	.0705	.0195	.0038	.0004	.0000	
	2	.9470	.7382	.6007	.4628	.2318	.0898	.0250	.0043	.0003	.0000
	3	.9917	.9144	.8343	.7297	.4826	.2539	.0994	.0253	.0031	.0001
	4	.9991	.9804	.9511	.9012	.7334	.5000	.2666	.0988	.0196	.0009
	5	.9999	.9969	.9900	.9747	.9006	.7461	.5174	.2703	.0856	.0083
	6	1.0000	.9997	.9987	.9957	.9750	.9102	.7682	.5372	.2618	.0530
	7		1.0000	.9999	.9996	.9962	.9805	.9295	.8040	.5638	.2252
	8			1.0000	1.0000	.9997	.9980	.9899	.9596	.8658	.6126
	9					1.0000	1.0000	1.0000	1.0000	1.0000	1.0000
10	0	.3487	.1074	.0563	.0282	.0060	.0010	.0001	.0000		
	1	.7361	.3758	.2440	.1493	.0464	.0107	.0017	.0001	.0000	
	2	.9298	.6778	.5256	.3828	.1673	.0547	.0123	.0016	.0001	
	3	.9872	.8791	.7759	.6496	.3823	.1719	.0548	.0106	.0009	.0000
	4	.9984	.9672	.9219	.8497	.6331	.3770	.1662	.0474	.0064	.0002
	5	.9999	.9936	.9803	.9527	.8338	.6230	.3669	.1503	.0328	.0016
	6	1.0000	.9991	.9965	.9894	.9452	.8281	.6177	.3504	.1209	.0128
	7		.9999	.9996	.9984	.9877	.9453	.8327	.6172	.3222	.0702
	8		1.0000	1.0000	.9999	.9983	.9893	.9536	.8507	.6242	.2639
	9				1.0000	.9999	.9990	.9940	.9718	.8926	.6513
	10					1.0000	1.0000	1.0000	1.0000	1.0000	1.0000
11	0	.3138	.0859	.0422	.0198	.0036	.0005	.0000			
	1	.6974	.3221	.1971	.1130	.0302	.0059	.0007	.0000		
	2	.9104	.6174	.4552	.3127	.1189	.0327	.0059	.0006	.0000	
	3	.9815	.8369	.7133	.5696	.2963	.1133	.0293	.0043	.0002	
	4	.9972	.9496	.8854	.7897	.5328	.2744	.0994	.0216	.0020	.0000
	5	.9997	.9883	.9657	.9218	.7535	.5000	.2465	.0782	.0117	.0003
	6	1.0000	.9980	.9924	.9784	.9006	.7256	.4672	.2103	.0504	.0028
	7		.9998	.9988	.9957	.9707	.8867	.7037	.4304	.1611	.0185
	8		1.0000	.9999	.9994	.9941	.9673	.8811	.6873	.3826	.0896
	9			1.0000	1.0000	.9993	.9941	.9698	.8870	.6779	.3026
	10					1.0000	.9995	.9964	.9802	.9141	.6862
	11						1.0000	1.0000	1.0000	1.0000	1.0000

n	x	.10	.20	.25	.30	.40	.50	.60	.70	.80	.90
								p			
12	0	.2824	.0687	.0317	.0138	.0022	.0002	.0000			
	1	.6590	.2749	.1584	.0850	.0196	.0032	.0003	.0000		
	2	.8891	.5583	.3907	.2528	.0834	.0193	.0028	.0002	.0000	
	3	.9744	.7946	.6488	.4925	.2253	.0730	.0153	.0017	.0001	
	4	.9957	.9274	.8424	.7237	.4382	.1938	.0573	.0095	.0006	.0000
	5	.9995	.9806	.9456	.8821	.6652	.3872	.1582	.0386	.0039	.0001
	6	.9999	.9961	.9857	.9614	.8418	.6128	.3348	.1178	.0194	.0005
	7	1.0000	.9994	.9972	.9905	.9427	.8062	.5618	.2763	.0726	.0043
	8		.9999	.9996	.9983	.9847	.9270	.7747	.5075	.2054	.0256
	9		1.0000	1.0000	.9998	.9972	.9807	.9166	.7472	.4417	.1109
	10				1.0000	.9997	.9968	.9804	.9150	.7251	.3410
	11					1.0000	.9998	.9978	.9862	.9313	.7176
	12						1.0000	1.0000	1.0000	1.0000	1.0000
13	0	.2542	.0550	.0238	.0097	.0013	.0001	.0000			
	1	.6213	.2336	.1267	.0637	.0126	.0017	.0001	.0000		
	2	.8661	.5017	.3326	.2025	.0579	.0112	.0013	.0001		
	3	.9658	.7473	.5843	.4206	.1686	.0461	.0078	.0007	.0000	
	4	.9935	.9009	.7940	.6543	.3530	.1334	.0321	.0040	.0002	
	5	.9991	.9700	.9198	.8346	.5744	.2905	.0977	.0182	.0012	.0000
	6	.9999	.9930	.9757	.9376	.7712	.5000	.2288	.0624	.0070	.0001
	7	1.0000	.9980	.9944	.9818	.9023	.7095	.4256	.1654	.0300	.0009
	8		.9998	.9990	.9960	.9679	.8666	.6470	.3457	.0991	.0065
	9		1.0000	.9999	.9993	.9922	.9539	.8314	.5794	.2527	.0342
	10			1.0000	.9999	.9987	.9888	.9421	.7975	.4983	.1339
	11				1.0000	.9999	.9983	.9874	.9363	.7664	.3787
	12					1.0000	.9999	.9987	.9903	.9450	.7458
	13						1.0000	1.0000	1.0000	1.0000	1.0000
14	0	.2288	.0440	.0178	.0068	.0008	.0001	.0000			
	1	.5846	.1979	.1010	.0475	.0081	.0009	.0001			
	2	.8416	.4481	.2811	.1608	.0398	.0065	.0006	.0000		
	3	.9559	.6982	.5213	.3552	.1243	.0287	.0039	.0002		
	4	.9908	.8702	.7415	.5842	.2793	.0898	.0175	.0017	.0000	
	5	.9985	.9561	.8883	.7805	.4859	.2120	.0583	.0083	.0004	
	6	.9998	.9884	.9617	.9067	.6925	.3953	.1501	.0315	.0024	.0000
	7	1.0000	.9976	.9897	.9685	.8499	.6047	.3075	.0933	.0116	.0002
	8		.9996	.9978	.9917	.9417	.7880	.5141	.2195	.0439	.0015
	9		1.0000	.9997	.9983	.9825	.9102	.7207	.4158	.1298	.0092
	10			1.0000	.9998	.9961	.9713	.8757	.6448	.3018	.0441
	11				1.0000	.9994	.9935	.9602	.8392	.5519	.1584
	12					.9999	.9991	.9919	.9525	.8021	.4154
	13					1.0000	.9999	.9992	.9932	.9560	.7712
	14						1.0000	1.0000	1.0000	1.0000	1.0000

n	x	.10	.20	.25	.30	.40	.50	.60	.70	.80	.90
							p				
15	0	.2059	.0352	.0134	.0047	.0005	.0000				
	1	.5490	.1671	.0802	.0353	.0052	.0005	.0000			
	2	.8159	.3980	.2361	.1268	.0271	.0037	.0003	.0000		
	3	.9444	.6482	.4613	.2969	.0905	.0176	.0019	.0001		
	4	.9873	.8358	.6865	.5155	.2173	.0592	.0094	.0007	.0000	
	5	.9978	.9389	.8516	.7216	.4032	.1509	.0338	.0037	.0001	
	6	.9997	.9819	.9434	.8689	.6098	.3036	.0951	.0152	.0008	
	7	1.0000	.9958	.9827	.9500	.7869	.5000	.2131	.0500	.0042	.0000
	8		.9992	.9958	.9848	.9050	.6964	.3902	.1311	.0181	.0003
	9		.9999	.9992	.9963	.9662	.8491	.5968	.2784	.0611	.0023
	10		1.0000	.9999	.9993	.9907	.9408	.7827	.4845	.1642	.0127
	11			1.0000	.9999	.9981	.9824	.9095	.7031	.3518	.0556
	12				1.0000	.9997	.9963	.9729	.8732	.6020	.1841
	13					1.0000	.9995	.9948	.9647	.8329	.4510
	14						1.0000	.9995	.9953	.9648	.7941
	15							1.0000	1.0000	1.0000	1.0000
16	0	.1853	.0281	.0100	.0033	.0003	.0000				
	1	.5147	.1407	.0635	.0261	.0033	.0003	.0000			
	2	.7892	.3518	.1971	.0994	.0183	.0021	.0001			
	3	.9316	.5981	.4050	.2459	.0651	.0106	.0009	.0000		
	4	.9830	.7982	.6302	.4499	.1666	.0384	.0049	.0003		
	5	.9967	.9183	.8103	.6598	.3288	.1051	.0191	.0016	.0000	
	6	.9995	.9733	.9204	.8247	.5272	.2272	.0583	.0071	.0002	
	7	.9999	.9930	.9729	.9256	.7161	.4018	.1423	.0257	.0015	.0000
	8	1.0000	.9985	.9925	.9743	.8577	.5982	.2839	.0744	.0070	.0001
	9		.9998	.9984	.9929	.9417	.7728	.4728	.1753	.0267	.0005
	10		1.0000	.9997	.9984	.9809	.8949	.6712	.3402	.0817	.0033
	11			1.0000	.9997	.9951	.9616	.8334	.5501	.2018	.0170
	12				1.0000	.9991	.9894	.9349	.7541	.4019	.0684
	13					.9999	.9979	.9817	.9006	.6482	.2108
	14					1.0000	.9997	.9967	.9739	.8593	.4853
	15						1.0000	.9997	.9967	.9719	.8147
	16							1.0000	1.0000	1.0000	1.0000

						p					
n	*x*	.10	.20	.25	.30	.40	.50	.60	.70	.80	.90
17	0	.1668	.0225	.0075	.0023	.0002	.0000				
	1	.4818	.1182	.0501	.0193	.0021	.0001	.0000			
	2	.7618	.3096	.1637	.0774	.0123	.0012	.0001			
	3	.9174	.5489	.3530	.2019	.0464	.0064	.0005	.0000		
	4	.9779	.7582	.5739	.3887	.1260	.0245	.0025	.0001		
	5	.9953	.8943	.7653	.5968	.2639	.0717	.0106	.0007	.0000	
	6	.9992	.9623	.8929	.7752	.4478	.1662	.0348	.0032	.0001	
	7	.9999	.9891	.9598	.8954	.6405	.3145	.0919	.0127	.0005	
	8	1.0000	.9974	.9876	.9597	.8011	.5000	.1989	.0403	.0026	.0000
	9		.9995	.9969	.9873	.9081	.6855	.3595	.1046	.0109	.0001
	10		.9999	.9994	.9968	.9652	.8338	.5522	.2248	.0377	.0008
	11		1.0000	.9999	.9993	.9894	.9283	.7361	.4032	.1057	.0047
	12			1.0000	.9999	.9975	.9755	.8740	.6113	.2418	.0221
	13				1.0000	.9995	.9936	.9536	.7981	.4511	.0826
	14					.9999	.9988	.9877	.9226	.6904	.2382
	15					1.0000	.9999	.9979	.9807	.8818	.5182
	16						1.0000	.9998	.9977	.9775	.8332
	17							1.0000	1.0000	1.0000	1,0000
18	0	.1501	.0180	.0056	.0016	.0001	.0000				
	1	.4503	.0991	.0395	.0142	.0013	.0001				
	2	.7338	.2713	.1353	.0600	.0082	.0007	.0000			
	3	.9018	.5010	.3057	.1646	.0328	.0038	.0002			
	4	.9718	.7164	.5787	.3327	.0942	.0154	.0013	.0000		
	5	.9936	.8671	.7175	.5344	.2088	.0481	.0058	.0003		
	6	.9988	.9487	.8610	.7217	.3743	.1189	.0203	.0014	.0000	
	7	.9998	.9837	.9431	.8593	.5634	.2403	.0576	.0061	.0002	
	8	1.0000	.9957	.9807	.9404	.7368	.4073	.1347	.0210	.0009	
	9		.9991	.9946	.9790	.8653	.5927	.2632	.0596	.0043	.0000
	10		.9998	.9988	.9939	.9424	.7597	.4366	.1407	.0163	.0002
	11		1.0000	.9998	.9986	.9797	.8811	.6257	.2783	.0513	.0012
	12			1.0000	.9997	.9942	.9519	.7912	.4656	.1329	.0064
	13				1.0000	.9987	.9846	.9058	.6673	.2836	.0282
	14					.9998	.9962	.9672	.8354	.4990	.0982
	15					1.0000	.9993	.9918	.9400	.7287	.2662
	16						.9999	.9987	.9858	.9009	.5497
	17						1.0000	.9999	.9984	.9820	.8499
	18							1.0000	1.0000	1.0000	1.0000

TABLE A.2 *(Continued)*

n	x	.10	.20	.25	.30	.40	.50	.60	.70	.80	.90
19	0	.1351	.0144	.0042	.0011	.0001					
	1	.4203	.0829	.0310	.0104	.0008	.0000				
	2	.7054	.2369	.1113	.0462	.0055	.0004	.0000			
	3	.8850	.4551	.2631	.1332	.0230	.0022	.0001			
	4	.9648	.6733	.4654	.2822	.0696	.0096	.0006	.0000		
	5	.9914	.8369	.6678	.4739	.1629	.0318	.0031	.0001		
	6	.9983	.9324	.8251	.6655	.3081	.0835	.0116	.0006		
	7	.9997	.9767	.9225	.8180	.4878	.1796	.0352	.0028	.0000	
	8	1.0000	.9933	.9713	.9161	.6675	.3238	.0885	.0105	.0003	
	9		.9984	.9911	.9674	.8139	.5000	.1861	.0326	.0016	
	10		.9997	.9977	.9895	.9115	.6762	.3325	.0839	.0067	.0000
	11		.9999	.9995	.9972	.9648	.8204	.5122	.1820	.0233	.0003
	12		1.0000	.9999	.9994	.9884	.9165	.6919	.3345	.0676	.0017
	13			1.0000	.9999	.9969	.9682	.8371	.5261	.1631	.0086
	14				1.0000	.9994	.9904	.9304	.7178	.3267	.0352
	15					.9999	.9978	.9770	.8668	.5449	.1150
	16					1.0000	.9996	.9945	.9538	.7631	.2946
	17						1.0000	.9992	.9896	.9171	.5797
	18							.9999	.9989	.9856	.8649
	19							1.0000	1.0000	1.0000	1.0000
20	0	.1216	.0115	.0032	.0008	.0000					
	1	.3917	.0692	.0243	.0076	.0005	.0000				
	2	.6769	.2061	.0913	.0355	.0036	.0002	.0000			
	3	.8670	.4114	.2252	.1071	.0160	.0013	.0001			
	4	.9568	.6296	.4148	.2375	.0510	.0059	.0003			
	5	.9887	.8042	.6172	.4164	.1256	.0207	.0016	.0000		
	6	.9976	.9133	.7858	.6080	.2500	.0577	.0065	.0003		
	7	.9996	.9679	.8982	.7723	.4159	.1316	.0210	.0013	.0000	
	8	.9999	.9900	.9591	.8867	.5956	.2517	.0565	.0051	.0001	
	9	1.0000	.9974	.9861	.9520	.7553	.4119	.1275	.0171	.0006	
	10		.9994	.9961	.9829	.8725	.5881	.2447	.0480	.0026	.0000
	11		.9999	.9991	.9949	.9435	.7483	.4044	.1133	.0100	.0001
	12		1.0000	.9998	.9987	.9790	.8684	.5841	.2277	.0321	.0004
	13			1.0000	.9997	.9935	.9423	.7500	.3920	.0867	.0024
	14				1.0000	.9984	.9793	.8744	.5836	.1958	.0113
	15					.9997	.9941	.9490	.7625	.3704	.0432
	16					1.0000	.9987	.9840	.8929	.5886	.1330
	17						.9998	.9964	.9645	.7939	.3231
	18						1.0000	.9995	.9924	.9308	.6083
	19							1.0000	.9992	.9885	.8784
	20								1.0000	1.0000	1.0000

TABLE A.3 PROBABILITY DISTRIBUTION FUNCTION FOR THE POISSON DISTRIBUTION

$$\Pi(x; \lambda) = \sum_{k=0}^{x} p(k; \lambda)$$

x	λ 0.1	0.2	0.3	0.4	0.5	0.6	0.7	0.8	0.9
0	0.9048	0.8187	0.7408	0.6730	0.6065	0.5488	0.4966	0.4493	0.4066
1	0.9953	0.9825	0.9631	0.9384	0.9098	0.8781	0.8442	0.8088	0.7725
2	0.9998	0.9989	0.9964	0.9921	0.9856	0.9769	0.9659	0.9526	0.9371
3	1.0000	0.9999	0.9997	0.9992	0.9982	0.9966	0.9942	0.9909	0.9865
4		1.0000	1.0000	0.9999	0.9998	0.9996	0.9992	0.9986	0.9977
5				1.0000	1.0000	1.0000	0.9999	0.9998	0.9997
6							1.0000	1.0000	1.0000

x	λ 1.0	1.5	2.0	2.5	3.0	3.5	4.0	4.5	5.0
0	0.3679	0.2231	0.1353	0.0821	0.0498	0.0302	0.0183	0.0111	0.0067
1	0.7358	0.5578	0.4060	0.2873	0.1991	0.1359	0.0916	0.0611	0.0404
2	0.9197	0.8088	0.6767	0.5438	0.4232	0.3208	0.2381	0.1736	0.1247
3	0.9810	0.9344	0.8571	0.7576	0.6472	0.5366	0.4335	0.3423	0.2650
4	0.9963	0.9814	0.9473	0.8912	0.8153	0.7254	0.6288	0.5321	0.4405
5	0.9994	0.9955	0.9834	0.9580	0.9161	0.8576	0.7851	0.7029	0.6160
6	0.9999	0.9991	0.9955	0.9858	0.9665	0.9347	0.8893	0.8311	0.7622
7	1.0000	0.9998	0.9989	0.9958	0.9881	0.9733	0.9489	0.9134	0.8666
8		1.0000	0.9998	0.9989	0.9962	0.9901	0.9786	0.9597	0.9319
9			1.0000	0.9997	0.9989	0.9967	0.9919	0.9829	0.9682
10				0.9999	0.9997	0.9990	0.9972	0.9933	0.9863
11				1.0000	0.9999	0.9997	0.9991	0.9976	0.9945
12					1.0000	0.9999	0.9997	0.9992	0.9980
13						1.0000	0.9999	0.9997	0.9993
14							1.0000	0.9999	0.9998
15								1.0000	0.9999
16									1.0000

Reprinted with permission of Macmillan Publishing Company from *Probability and Statistics for Engineers and Scientists* by R. E. Walpole and R. H. Myers. Copyright © 1978 by Macmillan Publishing Company.

TABLE A.3 *(Continued)*

					λ				
x	5.5	6.0	6.5	7.0	7.5	8.0	8.5	9.0	9.5
0	0.0041	0.0025	0.0015	0.0009	0.0006	0.0003	0.0002	0.0001	0.0001
1	0.0266	0.0174	0.0113	0.0073	0.0047	0.0030	0.0019	0.0012	0.0008
2	0.0884	0.0620	0.0430	0.0296	0.0203	0.0138	0.0093	0.0062	0.0042
3	0.2017	0.1512	0.1118	0.0818	0.0591	0.0424	0.0301	0.0212	0.0149
4	0.3575	0.2851	0.2237	0.1730	0.1321	0.0996	0.0744	0.0550	0.0403
5	0.5289	0.4457	0.3690	0.3007	0.2414	0.1912	0.1496	0.1157	0.0885
6	0.6860	0.6063	0.5265	0.4497	0.3782	0.3134	0.2562	0.2068	0.1649
7	0.8095	0.7440	0.6728	0.5987	0.5246	0.4530	0.3856	0.3239	0.2687
8	0.8944	0.8472	0.7916	0.7291	0.6620	0.5925	0.5231	0.4557	0.3918
9	0.9462	0.9161	0.8774	0.8305	0.7764	0.7166	0.6530	0.5874	0.5218
10	0.9747	0.9574	0.9332	0.9015	0.8622	0.8159	0.7634	0.7060	0.6453
11	0.9890	0.9799	0.9661	0.9466	0.9208	0.8881	0.8487	0.8030	0.7520
12	0.9955	0.9912	0.9840	0.9730	0.9573	0.9362	0.9091	0.8758	0.8364
13	0.9983	0.9964	0.9929	0.9872	0.9784	0.9658	0.9486	0.9261	0.8981
14	0.9994	0.9986	0.9970	0.9943	0.9897	0.9827	0.9726	0.9585	0.9400
15	0.9998	0.9995	0.9988	0.9976	0.9954	0.9918	0.9862	0.9780	0.9665
16	0.9999	0.9998	0.9996	0.9990	0.9980	0.9963	0.9934	0.9889	0.9823
17	1.0000	0.9999	0.9998	0.9996	0.9992	0.9984	0.9970	0.9947	0.9911
18		1.0000	0.9999	0.9999	0.9997	0.9994	0.9987	0.9976	0.9957
19			1.0000	1.0000	0.9999	0.9997	0.9995	0.9989	0.9980
20					1.0000	0.9999	0.9998	0.9996	0.9991
21						1.0000	0.9999	0.9998	0.9996
22							1.0000	0.9999	0.9999
23								1.0000	0.9999
24									1.0000

TABLE A.3 *(Continued)*

x	λ 10.0	11.0	12.0	13.0	14.0	15.0	16.0	17.0	18.0
0	0.0000	0.0000	0.0000						
1	0.0005	0.0002	0.0001	0.0000	0.0000				
2	0.0028	0.0012	0.0005	0.0002	0.0001	0.0000	0.0000		
3	0.0103	0.0049	0.0023	0.0010	0.0005	0.0002	0.0001	0.0000	0.0000
4	0.0293	0.0151	0.0076	0.0037	0.0018	0.0009	0.0004	0.0002	0.0001
5	0.0671	0.0375	0.0203	0.0107	0.0055	0.0028	0.0014	0.0007	0.0003
6	0.1301	0.0786	0.0458	0.0259	0.0142	0.0076	0.0040	0.0021	0.0010
7	0.2202	0.1432	0.0895	0.0540	0.0316	0.0180	0.0100	0.0054	0.0029
8	0.3328	0.2320	0.1550	0.0998	0.0621	0.0374	0.0220	0.0126	0.0071
9	0.4579	0.3405	0.2424	0.1658	0.1094	0.0699	0.0433	0.0261	0.0154
10	0.5830	0.4599	0.3472	0.2517	0.1757	0.1185	0.0774	0.0491	0.0304
11	0.6968	0.5793	0.4616	0.3532	0.2600	0.1848	0.1270	0.0847	0.0549
12	0.7916	0.6887	0.5760	0.4631	0.3585	0.2676	0.1931	0.1350	0.0917
13	0.8645	0.7813	0.6815	0.5730	0.4644	0.3632	0.2745	0.2009	0.1426
14	0.9165	0.8540	0.7720	0.6751	0.5704	0.4657	0.3675	0.2808	0.2081
15	0.9513	0.9074	0.8444	0.7636	0.6694	0.5681	0.4667	0.3715	0.2867
16	0.9730	0.9441	0.8987	0.8355	0.7559	0.6641	0.5660	0.4677	0.3750
17	0.9857	0.9678	0.9370	0.8905	0.8272	0.7489	0.6593	0.5640	0.4686
18	0.9928	0.9823	0.9626	0.9302	0.8826	0.8195	0.7423	0.6550	0.5622
19	0.9965	0.9907	0.9787	0.9573	0.9235	0.8752	0.8122	0.7363	0.6509
20	0.9984	0.9953	0.9884	0.9750	0.9521	0.9170	0.8682	0.8055	0.7307
21	0.9993	0.9977	0.9939	0.9859	0.9712	0.9469	0.9108	0.8615	0.7991
22	0.9997	0.9990	0.9970	0.9924	0.9833	0.9673	0.9418	0.9047	0.8551
23	0.9999	0.9995	0.9985	0.9960	0.9907	0.9805	0.9633	0.9367	0.8989
24	1.0000	0.9998	0.9993	0.9980	0.9950	0.9888	0.9777	0.9594	0.9317
25		0.9999	0.9997	0.9990	0.9974	0.9938	0.9869	0.9748	0.9554
26		1.0000	0.9999	0.9995	0.9987	0.9967	0.9925	0.9848	0.9718
27			0.9999	0.9998	0.9994	0.9983	0.9959	0.9912	0.9827
28			1.0000	0.9999	0.9997	0.9991	0.9978	0.9950	0.9897
29				1.0000	0.9999	0.9996	0.9989	0.9973	0.9941
30					0.9999	0.9998	0.9994	0.9986	0.9967
31					1.0000	0.9999	0.9997	0.9993	0.9982
32						1.0000	0.9999	0.9996	0.9990
33							0.9999	0.9998	0.9995
34							1.0000	0.9999	0.9998
35								1.0000	0.9999
36									0.9999
37									1.0000

TABLE A.4 INCOMPLETE GAMMA FUNCTION

$$\gamma(x; \alpha) = \frac{1}{\Gamma(\alpha)} \int_0^x t^{\alpha-1} e^{-t} dt$$

x	α									
	1	2	3	4	5	6	7	8	9	10
1	.632	.264	.080	.019	.004	.001	.000	.000	.000	.000
2	.865	.594	.323	.143	.053	.017	.005	.001	.000	.000
3	.950	.801	.577	.353	.185	.084	.034	.012	.004	.001
4	.982	.908	.762	.567	.371	.215	.111	.051	.021	.008
5	.993	.960	.875	.735	.560	.384	.238	.133	.068	.032
6	.998	.983	.938	.849	.715	.554	.398	.256	.153	.084
7	.999	.993	.970	.918	.827	.699	.550	.401	.271	.170
8	1.000	.997	.986	.958	.900	.809	.687	.547	.407	.283
9		.999	.994	.979	.945	.884	.793	.676	.544	.413
10		1.000	.997	.990	.971	.933	.870	.780	.667	.542
11			.999	.995	.985	.962	.921	.857	.768	.659
12			1.000	.998	.992	.980	.954	.911	.845	.758
13				.999	.996	.989	.974	.946	.900	.834
14				1.000	.998	.994	.986	.968	.938	.891
15					.999	.997	.992	.982	.963	.930

From *Probability and Statistics for Engineering and the Sciences*, 2nd Ed. by J. L. Devore. Copyright © 1987, 1982 by Wadsworth, Inc. Reprinted by permission of Brooks/Cole Publishing Company, Pacific Grove, California 93950.

TABLE A.5 CRITICAL VALUES FOR THE CHI-SQUARE DISTRIBUTION

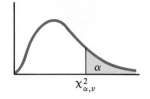

$$\chi^2_{\alpha,\nu}$$

ν	α							
	0.999	0.995	0.99	0.975	0.95	0.90	0.75	0.50
1	0.000	0.000	0.000	0.001	0.004	0.016	0.102	0.455
2	0.002	0.010	0.020	0.051	0.103	0.211	0.575	1.386
3	0.024	0.072	0.115	0.216	0.352	0.584	1.213	2.366
4	0.091	0.207	0.297	0.484	0.711	1.064	1.923	3.357
5	0.210	0.412	0.554	0.831	1.145	1.610	2.675	4.351
6	0.381	0.676	0.872	1.237	1.635	2.204	3.455	5.348
7	0.599	0.989	1.239	1.690	2.167	2.833	4.255	6.346
8	0.857	1.344	1.646	2.180	2.733	3.490	5.071	7.344
9	1.152	1.735	2.088	2.700	3.325	4.168	5.899	8.343
10	1.479	2.156	2.558	3.247	3.940	4.865	6.737	9.342
11	1.834	2.603	3.053	3.816	4.575	5.578	7.584	10.341
12	2.214	3.074	3.571	4.404	5.226	6.304	8.438	11.340
13	2.617	3.565	4.107	5.009	5.892	7.042	9.299	12.340
14	3.041	4.075	4.660	5.629	6.571	7.790	10.165	13.339
15	3.483	4.601	5.229	6.262	7.261	8.547	11.037	14.339
16	3.942	5.142	5.812	6.908	7.962	9.312	11.912	15.338
17	4.416	5.697	6.408	7.564	8.672	10.085	12.792	16.338
18	4.905	6.265	7.015	8.231	9.390	10.865	13.675	17.338
19	5.407	6.844	7.633	8.907	10.117	11.651	14.562	18.338
20	5.921	7.434	8.260	9.591	10.851	12.443	15.452	19.337
21	6.447	8.034	8.897	10.283	11.591	13.240	16.344	20.337
22	6.983	8.643	9.542	10.982	12.338	14.041	17.240	21.337
23	7.529	9.260	10.196	11.689	13.091	14.848	18.137	22.337
24	8.085	9.886	10.856	12.401	13.848	15.659	19.037	23.337
25	8.649	10.520	11.524	13.120	14.611	16.473	19.939	24.337
26	9.222	11.160	12.198	13.844	15.379	17.292	20.843	25.336
27	9.803	11.808	12.879	14.573	16.151	18.114	21.749	26.336
28	10.391	12.461	13.565	15.308	16.928	18.939	22.657	27.336
29	10.986	13.121	14.256	16.047	17.708	19.768	23.567	28.336
30	11.588	13.787	14.953	16.791	18.493	20.599	24.478	29.336
31	12.196	14.458	15.655	17.539	19.281	21.434	25.390	30.336
32	12.811	15.134	16.362	18.291	20.072	22.271	26.304	31.336
33	13.431	15.815	17.074	19.047	20.867	23.110	27.219	32.336
34	14.057	16.501	17.789	19.806	21.664	23.952	28.136	33.336
35	14.688	17.192	18.509	20.569	22.465	24.797	29.054	34.336
36	15.324	17.887	19.233	21.336	23.269	25.643	29.973	35.336
37	15.965	18.586	19.960	22.106	24.075	26.492	30.893	36.336
38	16.611	19.289	20.691	22.878	24.884	27.343	31.815	37.335
39	17.262	19.996	21.426	23.654	25.695	28.196	32.737	38.335
40	17.916	20.707	22.164	24.433	26.509	29.051	33.660	39.335
41	18.576	21.421	22.906	25.215	27.326	29.907	34.585	40.335
42	19.239	22.138	23.650	25.999	28.144	30.765	35.510	41.335
43	19.906	22.859	24.398	26.785	28.965	31.625	36.436	42.335
44	20.576	23.584	25.148	27.575	29.787	32.487	37.363	43.335
45	21.251	24.311	25.901	28.366	30.612	33.350	38.291	44.335
46	21.929	25.041	26.657	29.160	31.439	34.215	39.220	45.335
47	22.610	25.775	27.416	29.956	32.268	35.081	40.149	46.335
48	23.295	26.511	28.177	30.755	33.098	35.949	41.079	47.335
49	23.983	27.249	28.941	31.555	33.930	36.818	42.010	48.335
50	24.674	27.991	29.707	32.357	34.764	37.689	42.942	49.335
51	25.368	28.735	30.475	33.162	35.600	38.560	43.874	50.335
52	26.065	29.481	31.246	33.968	36.437	39.433	44.808	51.335
53	26.765	30.230	32.019	34.776	37.270	40.308	45.741	52.335
54	27.468	30.981	32.793	35.586	38.116	41.183	46.676	53.355
55	28.173	31.735	33.570	36.398	38.958	42.060	47.610	54.335
56	28.881	32.491	34.350	37.212	39.801	42.937	48.546	55.335
57	29.592	33.248	35.131	38.027	40.646	43.816	49.482	56.335
58	30.305	34.008	35.913	38.844	41.492	44.696	50.419	57.335
59	31.021	34.770	36.698	39.662	42.339	45.577	51.356	58.335
60	31.738	35.535	37.485	40.482	43.188	46.459	52.294	59.335

Jerrold Zar, *Biostatistical Analysis*, 2/e, © 1984, pp. 479–481. Reprinted by permission of Prentice Hall, Inc., Englewood Cliffs, New Jersey.

TABLE A.5 (Continued)

ν	0.25	0.10	0.05	0.025	0.01	0.005	0.001
				α			
1	1.323	2.706	3.841	5.024	6.635	7.879	10.828
2	2.773	4.605	5.991	7.378	9.210	10.597	13.816
3	4.108	6.251	7.815	9.348	11.345	12.838	16.266
4	5.385	7.779	9.488	11.143	13.277	14.860	18.467
5	6.626	9.236	11.070	12.833	15.086	16.750	20.515
6	7.841	10.645	12.592	14.449	16.812	18.548	22.458
7	9.037	12.017	14.067	16.013	18.475	20.278	24.322
8	10.219	13.362	15.507	17.535	20.090	21.955	26.124
9	11.389	14.684	16.919	19.023	21.666	23.589	27.877
10	12.549	15.987	18.307	20.483	23.209	25.188	29.588
11	13.701	17.275	19.675	21.920	24.725	26.757	31.264
12	14.845	18.549	21.026	23.337	26.217	28.300	32.909
13	15.984	19.812	22.362	24.736	27.688	29.819	34.528
14	17.117	21.064	23.685	26.119	29.141	31.319	36.123
15	18.245	22.307	24.996	27.488	30.578	32.801	37.697
16	19.369	23.542	26.296	28.845	32.000	34.267	39.252
17	20.489	24.769	27.587	31.191	33.409	35.718	40.790
18	21.605	25.989	28.869	31.526	34.805	37.156	42.312
19	22.718	27.204	30.144	32.852	36.191	38.582	43.820
20	23.828	28.412	31.410	34.170	37.566	39.997	45.315
21	24.935	29.615	32.671	35.479	38.932	41.401	46.797
22	26.039	30.813	33.924	36.781	40.289	42.796	48.268
23	27.141	32.007	35.172	38.076	41.638	44.181	49.728
24	28.241	33.196	36.415	39.364	42.980	45.559	51.179
25	29.339	34.382	37.652	40.646	44.314	46.928	52.020
26	30.435	35.563	38.885	41.923	45.642	48.290	54.052
27	31.528	36.741	40.113	43.195	46.963	49.645	55.476
28	32.620	37.916	41.337	44.461	48.278	50.993	56.892
29	33.711	39.087	42.557	45.722	49.588	52.336	58.301
30	34.800	40.256	43.773	46.979	50.892	53.672	59.703
31	35.887	41.422	44.985	48.232	52.191	55.003	61.098
32	36.973	42.585	46.194	49.480	53.486	56.328	62.487
33	38.058	43.745	47.400	50.725	54.776	57.648	63.870
34	39.141	44.903	48.602	51.966	56.061	58.964	65.247
35	40.223	46.059	49.802	53.203	57.342	60.275	66.619
36	41.304	47.212	50.998	54.437	58.619	61.581	67.985
37	42.383	48.363	52.192	55.668	59.893	62.883	69.346
38	43.462	49.513	53.384	56.896	61.162	64.181	70.703
39	44.539	50.660	54.572	58.120	62.428	65.476	72.055
40	45.616	51.805	55.758	59.342	63.691	66.766	73.402
41	46.692	52.949	56.942	60.561	64.950	68.053	74.745
42	47.766	54.090	58.124	61.777	66.206	69.336	76.084
43	48.840	55.230	59.304	62.990	67.459	70.616	77.419
44	49.913	56.369	60.481	64.201	68.710	71.893	78.750
45	50.985	57.505	61.656	65.410	69.957	73.166	80.077
46	52.056	58.641	62.830	66.617	71.201	74.437	81.400
47	53.127	59.774	64.001	67.821	72.443	75.704	82.720
48	54.196	60.907	65.171	69.023	73.683	76.969	84.037
49	55.265	62.038	66.339	70.222	74.919	78.231	85.351
50	56.334	63.167	67.505	71.420	76.154	79.490	86.661
51	57.401	64.295	68.669	72.616	77.386	80.747	87.968
52	58.468	65.422	69.832	73.810	78.616	82.001	89.272
53	59.534	66.548	70.993	75.002	79.843	83.253	90.573
54	60.600	67.673	72.153	76.192	81.069	84.502	91.872
55	61.665	68.796	73.311	77.380	82.292	85.749	93.168
56	62.729	69.919	74.468	78.567	83.513	86.994	94.461
57	63.793	71.040	75.624	79.752	84.733	88.236	95.751
58	64.857	72.160	76.778	80.936	85.950	89.477	97.039
59	65.919	73.279	77.931	82.117	87.166	90.715	98.324
60	66.981	74.397	79.082	83.298	88.379	91.952	99.607

TABLE A.6 CRITICAL VALUES FOR THE *t* DISTRIBUTION

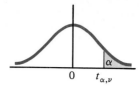

ν	α						
	.10	.05	.025	.01	.005	.001	.0005
1	3.078	6.314	12.706	31.821	63.657	318.31	636.62
2	1.886	2.920	4.303	6.965	9.925	22.326	31.598
3	1.638	2.353	3.182	4.541	5.841	10.213	12.924
4	1.533	2.132	2.776	3.747	4.604	7.173	8.610
5	1.476	2.015	2.571	3.365	4.032	5.893	6.869
6	1.440	1.943	2.447	3.143	3.707	5.208	5.959
7	1.415	1.895	2.365	2.998	3.499	4.785	5.408
8	1.397	1.860	2.306	2.896	3.355	4.501	5.041
9	1.383	1.833	2.262	2.821	3.250	4.297	4.781
10	1.372	1.812	2.228	2.764	3.169	4.144	4.587
11	1.363	1.796	2.201	2.718	3.106	4.025	4.437
12	1.356	1.782	2.179	2.681	3.055	3.930	4.318
13	1.350	1.771	2.160	2.650	3.012	3.852	4.221
14	1.345	1.761	2.145	2.624	2.977	3.787	4.140
15	1.341	1.753	2.131	2.602	2.947	3.733	4.073
16	1.337	1.746	2.120	2.583	2.921	3.686	4.015
17	1.333	1.740	2.110	2.567	2.898	3.646	3.965
18	1.330	1.734	2.101	2.552	2.878	3.610	3.922
19	1.328	1.729	2.093	2.539	2.861	3.579	3.883
20	1.325	1.725	2.086	2.528	2.845	3.552	3.850
21	1.323	1.721	2.080	2.518	2.831	3.527	3.819
22	1.321	1.717	2.074	2.508	2.819	3.505	3.792
23	1.319	1.714	2.069	2.500	2.807	3.485	3.767
24	1.318	1.711	2.064	2.492	2.797	3.467	3.745
25	1.316	1.708	2.060	2.485	2.787	3.450	3.725
26	1.315	1.706	2.056	2.479	2.779	3.435	3.707
27	1.314	1.703	2.052	2.473	2.771	3.421	3.690
28	1.313	1.701	2.048	2.467	2.763	3.408	3.674
29	1.311	1.699	2.045	2.462	2.756	3.396	3.659
30	1.310	1.697	2.042	2.457	2.750	3.385	3.646
40	1.303	1.684	2.021	2.423	2.704	3.307	3.551
60	1.296	1.671	2.000	2.390	2.660	3.232	3.460
120	1.289	1.658	1.980	2.358	2.617	3.160	3.373
∞	1.282	1.645	1.960	2.326	2.576	3.090	3.291

This figure is used by permission of the Biometrika Trust.

TABLE A.7 CRITICAL VALUES FOR THE *F* DISTRIBUTION

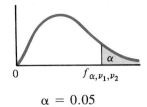

$$\alpha = 0.05$$

ν_2	ν_1								
	1	2	3	4	5	6	7	8	9
1	161.4	199.5	215.7	224.6	230.2	234.0	236.8	238.9	240.5
2	18.51	19.00	19.16	19.25	19.30	19.33	19.35	19.37	19.38
3	10.13	9.55	9.28	9.12	9.01	8.94	8.89	8.85	8.81
4	7.71	6.94	6.59	6.39	6.26	6.16	6.09	6.04	6.00
5	6.61	5.79	5.41	5.19	5.05	4.95	4.88	4.82	4.77
6	5.99	5.14	4.76	4.53	4.39	4.28	4.21	4.15	4.10
7	5.59	4.74	4.35	4.12	3.97	3.87	3.79	3.73	3.68
8	5.32	4.46	4.07	3.84	3.69	3.58	3.50	3.44	3.39
9	5.12	4.26	3.86	3.63	3.48	3.37	3.29	3.23	3.18
10	4.96	4.10	3.71	3.48	3.33	3.22	3.14	3.07	3.02
11	4.84	3.98	3.59	3.36	3.20	3.09	3.01	2.95	2.90
12	4.75	3.89	3.49	3.26	3.11	3.00	2.91	2.85	2.80
13	4.67	3.81	3.41	3.18	3.03	2.92	2.83	2.77	2.71
14	4.60	3.74	3.34	3.11	2.96	2.85	2.76	2.70	2.65
15	4.54	3.68	3.29	3.06	2.90	2.79	2.71	2.64	2.59
16	4.49	3.63	3.24	3.01	2.85	2.74	2.66	2.59	2.54
17	4.45	3.59	3.20	2.96	2.81	2.70	2.61	2.55	2.49
18	4.41	3.55	3.16	2.93	2.77	2.66	2.58	2.51	2.46
19	4.38	3.52	3.13	2.90	2.74	2.63	2.54	2.48	2.42
20	4.35	3.49	3.10	2.87	2.71	2.60	2.51	2.45	2.39
21	4.32	3.47	3.07	2.84	2.68	2.57	2.49	2.42	2.37
22	4.30	3.44	3.05	2.82	2.66	2.55	2.46	2.40	2.34
23	4.28	3.42	3.03	2.80	2.64	2.53	2.44	2.37	2.32
24	4.26	3.40	3.01	2.78	2.62	2.51	2.42	2.36	2.30
25	4.24	3.39	2.99	2.76	2.60	2.49	2.40	2.34	2.28
26	4.23	3.37	2.98	2.74	2.59	2.47	2.39	2.32	2.27
27	4.21	3.35	2.96	2.73	2.57	2.46	2.37	2.31	2.25
28	4.20	3.34	2.95	2.71	2.56	2.45	2.36	2.29	2.24
29	4.18	3.33	2.93	2.70	2.55	2.43	2.35	2.28	2.22
30	4.17	3.32	2.92	2.69	2.53	2.42	2.33	2.27	2.21
40	4.08	3.23	2.84	2.61	2.45	2.34	2.25	2.18	2.12
60	4.00	3.15	2.76	2.53	2.37	2.25	2.17	2.10	2.04
120	3.92	3.07	2.68	2.45	2.29	2.17	2.09	2.02	1.96
∞	3.84	3.00	2.60	2.37	2.21	2.10	2.01	1.94	1.88

This figure is used by permission of the Biometrika Trust.

$$\alpha = 0.05$$

ν_2	ν_1									
	10	12	15	20	24	30	40	60	120	∞
1	241.9	243.9	245.9	248.0	249.1	250.1	251.1	252.2	253.3	254.3
2	19.40	19.41	19.43	19.45	19.45	19.46	19.47	19.48	19.49	19.50
3	8.79	8.74	8.70	8.66	8.64	8.62	8.59	8.57	8.55	8.53
4	5.96	5.91	5.86	5.80	5.77	5.75	5.72	5.69	5.66	5.63
5	4.74	4.68	4.62	4.56	4.53	4.50	4.46	4.43	4.40	4.36
6	4.06	4.00	3.94	3.87	3.84	3.81	3.77	3.74	3.70	3.67
7	3.64	3.57	3.51	3.44	3.41	3.38	3.34	3.30	3.27	3.23
8	3.35	3.28	3.22	3.15	3.12	3.08	3.04	3.01	2.97	2.93
9	3.14	3.07	3.01	2.94	2.90	2.86	2.83	2.79	2.75	2.71
10	2.98	2.91	2.85	2.77	2.74	2.70	2.66	2.62	2.58	2.54
11	2.85	2.79	2.72	2.65	2.61	2.57	2.53	2.49	2.45	2.40
12	2.75	2.69	2.62	2.54	2.51	2.47	2.43	2.38	2.34	2.30
13	2.67	2.60	2.53	2.46	2.42	2.38	2.34	2.30	2.25	2.21
14	2.60	2.53	2.46	2.39	2.35	2.31	2.27	2.22	2.18	2.13
15	2.54	2.48	2.40	2.33	2.29	2.25	2.20	2.16	2.11	2.07
16	2.49	2.42	2.35	2.28	2.24	2.19	2.15	2.11	2.06	2.01
17	2.45	2.38	2.31	2.23	2.19	2.15	2.10	2.06	2.01	1.96
18	2.41	2.34	2.27	2.19	2.15	2.11	2.06	2.02	1.97	1.92
19	2.38	2.31	2.23	2.16	2.11	2.07	2.03	1.98	1.93	1.88
20	2.35	2.28	2.20	2.12	2.08	2.04	1.99	1.95	1.90	1.84
21	2.32	2.25	2.18	2.10	2.05	2.01	1.96	1.92	1.87	1.81
22	2.30	2.23	2.15	2.07	2.03	1.98	1.94	1.89	1.84	1.78
23	2.27	2.20	2.13	2.05	2.01	1.96	1.91	1.86	1.81	1.76
24	2.25	2.18	2.11	2.03	1.98	1.94	1.89	1.84	1.79	1.73
25	2.24	2.16	2.09	2.01	1.96	1.92	1.87	1.82	1.77	1.71
26	2.22	2.15	2.07	1.99	1.95	1.90	1.85	1.80	1.75	1.69
27	2.20	2.13	2.06	1.97	1.93	1.88	1.84	1.79	1.73	1.67
28	2.19	2.12	2.04	1.96	1.91	1.87	1.82	1.77	1.71	1.65
29	2.18	2.10	2.03	1.94	1.90	1.85	1.81	1.75	1.70	1.64
30	2.16	2.09	2.01	1.93	1.89	1.84	1.79	1.74	1.68	1.62
40	2.08	2.00	1.92	1.84	1.79	1.74	1.69	1.64	1.58	1.51
60	1.99	1.92	1.84	1.75	1.70	1.65	1.59	1.53	1.47	1.39
120	1.91	1.83	1.75	1.66	1.61	1.55	1.50	1.43	1.35	1.25
∞	1.83	1.75	1.67	1.57	1.52	1.46	1.39	1.32	1.22	1.00

TABLE A.7 (*Continued*)

$$\alpha = 0.01$$

ν_2	ν_1								
	1	2	3	4	5	6	7	8	9
1	4052	4999.5	5403	5625	5764	5859	5928	5981	6022
2	98.50	99.00	99.17	99.25	99.30	99.33	99.36	99.37	99.39
3	34.12	30.82	29.46	28.71	28.24	27.91	27.67	27.49	27.35
4	21.20	18.00	16.69	15.98	15.52	15.21	14.98	14.80	14.66
5	16.26	13.27	12.06	11.39	10.97	10.67	10.46	10.29	10.16
6	13.75	10.92	9.78	9.15	8.75	8.47	8.26	8.10	7.98
7	12.25	9.55	8.45	7.85	7.46	7.19	6.99	6.84	6.72
8	11.26	8.65	7.59	7.01	6.63	6.37	6.18	6.03	5.91
9	10.56	8.02	6.99	6.42	6.06	5.80	5.61	5.47	5.35
10	10.04	7.56	6.55	5.99	5.64	5.39	5.20	5.06	4.94
11	9.65	7.21	6.22	5.67	5.32	5.07	4.89	4.74	4.63
12	9.33	6.93	5.95	5.41	5.06	4.82	4.64	4.50	4.39
13	9.07	6.70	5.74	5.21	4.86	4.62	4.44	4.30	4.19
14	8.86	6.51	5.56	5.04	4.69	4.46	4.28	4.14	4.03
15	8.68	6.36	5.42	4.89	4.56	4.32	4.14	4.00	3.89
16	8.53	6.23	5.29	4.77	4.44	4.20	4.03	3.89	3.78
17	8.40	6.11	5.18	4.67	4.34	4.10	3.93	3.79	3.68
18	8.29	6.01	5.09	4.58	4.25	4.01	3.84	3.71	3.60
19	8.18	5.93	5.01	4.50	4.17	3.94	3.77	3.63	3.52
20	8.10	5.85	4.94	4.43	4.10	3.87	3.70	3.56	3.46
21	8.02	5.78	4.87	4.37	4.04	3.81	3.64	3.51	3.40
22	7.95	5.72	4.82	4.31	3.99	3.76	3.59	3.45	3.35
23	7.88	5.66	4.76	4.26	3.94	3.71	3.54	3.41	3.30
24	7.82	5.61	4.72	4.22	3.90	3.67	3.50	3.36	3.26
25	7.77	5.57	4.68	4.18	3.85	3.63	3.46	3.32	3.22
26	7.72	5.53	4.64	4.14	3.82	3.59	3.42	3.29	3.18
27	7.68	5.49	4.60	4.11	3.78	3.56	3.39	3.26	3.15
28	7.64	5.45	4.57	4.07	3.75	3.53	3.36	3.23	3.12
29	7.60	5.42	4.54	4.04	3.73	3.50	3.33	3.20	3.09
30	7.56	5.39	4.51	4.02	3.70	3.47	3.30	3.17	3.07
40	7.31	5.18	4.31	3.83	3.51	3.29	3.12	2.99	2.89
60	7.08	4.98	4.13	3.65	3.34	3.12	2.95	2.82	2.72
120	6.85	4.79	3.95	3.48	3.17	2.96	2.79	2.66	2.56
∞	6.63	4.61	3.78	3.32	3.02	2.80	2.64	2.51	2.41

$$\alpha = 0.01$$

ν_2	ν_1									
	10	12	15	20	24	30	40	60	120	∞
1	6056	6106	6157	6209	6235	6261	6287	6313	6339	6366
2	99.40	99.42	99.43	99.45	99.46	99.47	99.47	99.48	99.49	99.50
3	27.23	27.05	26.87	26.69	26.60	26.50	26.41	26.32	26.22	26.13
4	14.55	14.37	14.20	14.02	13.93	13.84	13.75	13.65	13.56	13.46
5	10.05	9.89	9.72	9.55	9.47	9.38	9.29	9.20	9.11	9.02
6	7.87	7.72	7.56	7.40	7.31	7.23	7.14	7.06	6.97	6.88
7	6.62	6.47	6.31	6.16	6.07	5.99	5.91	5.82	5.74	5.65
8	5.81	5.67	5.52	5.36	5.28	5.20	5.12	5.03	4.95	4.86
9	5.26	5.11	4.96	4.81	4.73	4.65	4.57	4.48	4.40	4.31
10	4.85	4.71	4.56	4.41	4.33	4.25	4.17	4.08	4.00	3.91
11	4.54	4.40	4.25	4.10	4.02	3.94	3.86	3.78	3.69	3.60
12	4.30	4.16	4.01	3.86	3.78	3.70	3.62	3.54	3.45	3.36
13	4.10	3.96	3.82	3.66	3.59	3.51	3.43	3.34	3.25	3.17
14	3.94	3.80	3.66	3.51	3.43	3.35	3.27	3.18	3.09	3.00
15	3.80	3.67	3.52	3.37	3.29	3.21	3.13	3.05	2.96	2.87
16	3.69	3.55	3.41	3.26	3.18	3.10	3.02	2.93	2.84	2.75
17	3.59	3.46	3.31	3.16	3.08	3.00	2.92	2.83	2.75	2.65
18	3.51	3.37	3.23	3.08	3.00	2.92	2.84	2.75	2.66	2.57
19	3.43	3.30	3.15	3.00	2.92	2.84	2.76	2.67	2.58	2.49
20	3.37	3.23	3.09	2.94	2.86	2.78	2.69	2.61	2.52	2.42
21	3.31	3.17	3.03	2.88	2.80	2.72	2.64	2.55	2.46	2.36
22	3.26	3.12	2.98	2.83	2.75	2.67	2.58	2.50	2.40	2.31
23	3.21	3.07	2.93	2.78	2.70	2.62	2.54	2.45	2.35	2.26
24	3.17	3.03	2.89	2.74	2.66	2.58	2.49	2.40	2.31	2.21
25	3.13	2.99	2.85	2.70	2.62	2.54	2.45	2.36	2.27	2.17
26	3.09	2.96	2.81	2.66	2.58	2.50	2.42	2.33	2.23	2.13
27	3.06	2.93	2.78	2.63	2.55	2.47	2.38	2.29	2.20	2.10
28	3.03	2.90	2.75	2.60	2.52	2.44	2.35	2.26	2.17	2.06
29	3.00	2.87	2.73	2.57	2.49	2.41	2.33	2.23	2.14	2.03
30	2.98	2.84	2.70	2.55	2.47	2.39	2.30	2.21	2.11	2.01
40	2.80	2.66	2.52	2.37	2.29	2.20	2.11	2.02	1.92	1.80
60	2.63	2.50	2.35	2.20	2.12	2.03	1.94	1.84	1.73	1.60
120	2.47	2.34	2.19	2.03	1.95	1.86	1.76	1.66	1.53	1.38
∞	2.32	2.18	2.04	1.88	1.79	1.70	1.59	1.47	1.32	1.00

TABLE A.8 CRITICAL VALUES FOR THE STUDENTIZED RANGE DISTRIBUTION

$$Q_{\alpha,m,\nu}$$

ν	α	2	3	4	5	6	7	8	9	10	11
5	.05	3.64	4.60	5.22	5.67	6.03	6.33	6.58	6.80	6.99	7.17
	.01	5.70	6.98	7.80	8.42	8.91	9.32	9.67	9.97	10.24	10.48
6	.05	3.46	4.34	4.90	5.30	5.63	5.90	6.12	6.32	6.49	6.65
	.01	5.24	6.33	7.03	7.56	7.97	8.32	8.61	8.87	9.10	9.30
7	.05	3.34	4.16	4.68	5.06	5.36	5.61	5.82	6.00	6.16	6.30
	.01	4.95	5.92	6.54	7.01	7.37	7.68	7.94	8.17	8.37	8.55
8	.05	3.26	4.04	4.53	4.89	5.17	5.40	5.60	5.77	5.92	6.05
	.01	4.75	5.64	6.20	6.62	6.96	7.24	7.47	7.68	7.86	8.03
9	.05	3.20	3.95	4.41	4.76	5.02	5.24	5.43	5.59	5.74	5.87
	.01	4.60	5.43	5.96	6.35	6.66	6.91	7.13	7.33	7.49	7.65
10	.05	3.15	3.88	4.33	4.65	4.91	5.12	5.30	5.46	5.60	5.72
	.01	4.48	5.27	5.77	6.14	6.43	6.67	6.87	7.05	7.21	7.36
11	.05	3.11	3.82	4.26	4.57	4.82	5.03	5.20	5.35	5.49	5.61
	.01	4.39	5.15	5.62	5.97	6.25	6.48	6.67	6.84	6.99	7.13
12	.05	3.08	3.77	4.20	4.51	4.75	4.95	5.12	5.27	5.39	5.51
	.01	4.32	5.05	5.50	5.84	6.10	6.32	6.51	6.67	6.81	6.94
13	.05	3.06	3.73	4.15	4.45	4.69	4.88	5.05	5.19	5.32	5.43
	.01	4.26	4.96	5.40	5.73	5.98	6.19	6.37	6.53	6.67	6.79
14	.05	3.03	3.70	4.11	4.41	4.64	4.83	4.99	5.13	5.25	5.36
	.01	4.21	4.89	5.32	5.63	5.88	6.08	6.26	6.41	6.54	6.66
15	.05	3.01	3.67	4.08	4.37	4.59	4.78	4.94	5.08	5.20	5.31
	.01	4.17	4.84	5.25	5.56	5.80	5.99	6.16	6.31	6.44	6.55
16	.05	3.00	3.65	4.05	4.33	4.56	4.74	4.90	5.03	5.15	5.26
	.01	4.13	4.79	5.19	5.49	5.72	5.92	6.08	6.22	6.35	6.46
17	.05	2.98	3.63	4.02	4.30	4.52	4.70	4.86	4.99	5.11	5.21
	.01	4.10	4.74	5.14	5.43	5.66	5.85	6.01	6.15	6.27	6.38
18	.05	2.97	3.61	4.00	4.28	4.49	4.67	4.82	4.96	5.07	5.17
	.01	4.07	4.70	5.09	5.38	5.60	5.79	5.94	6.08	6.20	6.31
19	.05	2.96	3.59	3.98	4.25	4.47	4.65	4.79	4.92	5.04	5.14
	.01	4.05	4.67	5.05	5.33	5.55	5.73	5.89	6.02	6.14	6.25
20	.05	2.95	3.58	3.96	4.23	4.45	4.62	4.77	4.90	5.01	5.11
	.01	4.02	4.64	5.02	5.29	5.51	5.69	5.84	5.97	6.09	6.19
24	.05	2.92	3.53	3.90	4.17	4.37	4.54	4.68	4.81	4.92	5.01
	.01	3.96	4.55	4.91	5.17	5.37	5.54	5.69	5.81	5.92	6.02
30	.05	2.89	3.49	3.85	4.10	4.30	4.46	4.60	4.72	4.82	4.92
	.01	3.89	4.45	4.80	5.05	5.24	5.40	5.54	5.65	5.76	5.85
40	.05	2.86	3.44	3.79	4.04	4.23	4.39	4.52	4.63	4.73	4.82
	.01	3.82	4.37	4.70	4.93	5.11	5.26	5.39	5.50	5.60	5.69
60	.05	2.83	3.40	3.74	3.98	4.16	4.31	4.44	4.55	4.65	4.73
	.01	3.76	4.28	4.59	4.82	4.99	5.13	5.25	5.36	5.45	5.53
120	.05	2.80	3.36	3.68	3.92	4.10	4.24	4.36	4.47	4.56	4.64
	.01	3.70	4.20	4.50	4.71	4.87	5.01	5.12	5.21	5.30	5.37
∞	.05	2.77	3.31	3.63	3.86	4.03	4.17	4.29	4.39	4.47	4.55
	.01	3.64	4.12	4.40	4.60	4.76	4.88	4.99	5.08	5.16	5.23

This figure is used by permission of the Biometrika Trust.

				m						
12	13	14	15	16	17	18	19	20	α	ν
7.32	7.47	7.60	7.72	7.83	7.93	8.03	8.12	8.21	.05	5
10.70	10.89	11.08	11.24	11.40	11.55	11.68	11.81	11.93.	.01	
6.79	6.92	7.03	7.14	7.24	7.34	7.43	7.51	7.59	.05	6
9.48	9.65	9.81	9.95	10.08	10.21	10.32	10.43	10.54	.01	
6.43	6.55	6.66	6.76	6.85	6.94	7.02	7.10	7.17	.05	7
8.71	8.86	9.00	9.12	9.24	9.35	9.46	9.55	9.65	.01	
6.18	6.29	6.39	6.48	6.57	6.65	6.73	6.80	6.87	.05	8
8.18	8.31	8.44	8.55	8.66	8.76	8.85	8.94	9.03	.01	
5.98	6.09	6.19	6.28	6.36	6.44	6.51	6.58	6.64	.05	9
7.78	7.91	8.03	8.13	8.23	8.33	8.41	8.49	8.57	.01	
5.83	5.93	6.03	6.11	6.19	6.27	6.34	6.40	6.47	.05	10
7.49	7.60	7.71	7.81	7.91	7.99	8.08	8.15	8.23	.01	
5.71	5.81	5.90	5.98	6.06	6.13	6.20	6.27	6.33	.05	11
7.25	7.36	7.46	7.56	7.65	7.73	7.81	7.88	7.95	.01	
5.61	5.71	5.80	5.88	5.95	6.02	6.09	6.15	6.21	.05	12
7.06	7.17	7.26	7.36	7.44	7.52	7.59	7.66	7.73	.01	
5.53	5.63	5.71	5.79	5.86	5.93	5.99	6.05	6.11	.05	13
6.90	7.01	7.10	7.19	7.27	7.35	7.42	7.48	7.55	.01	
5.46.	5.55	5.64	5.71	5.79	5.85	5.91	5.97	6.03	.05	14
6.77	6.87	6.96	7.05	7.13	7.20	7.27	7.33	7.39	.01	
5.40	5.49	5.57	5.65	5.72	5.78	5.85	5.90	5.96	.05	15
6.66	6.76	6.84	6.93	7.00	7.07	7.14	7.20	7.26	.01	
5.35	5.44	5.52	5.59	5.66	5.73	5.79	5.84	5.90	.05	16
6.56	6.66	6.74	6.82	6.90	6.97	7.03	7.09	7.15	.01	
5.31	5.39	5.47	5.54	5.61	5.67	5.73	5.79	5.84	.05	17
6.48	6.57	6.66	6.73	6.81	6.87	6.94	7.00	7.05	.01	
5.27	5.35	5.43	5.50	5.57	5.63	5.69	5.74	5.79	.05	18
6.41	6.50	6.58	6.65	6.73	6.79	6.85	6.91	6.97	.01	
5.23	5.31	5.39	5.46	5.53	5.59	5.65	5.70	5.75	.05	19
6.34	6.43	6.51	6.58	6.65	6.72	6.78	6.84	6.89	.01	
5.20	5.28	5.36	5.43	5.49	5.55	5.61	5.66	5.71	.05	20
6.28	6.37	6.45	6.52	6.59	6.65	6.71	6.77	6.82	.01	
5.10	5.18	5.25	5.32	5.38	5.44	5.49	5.55	5.59	.05	24
6.11	6.19	6.26	6.33	6.39	6.45	6.51	6.56	6.61	.01	
5.00	5.08	5.15	5.21	5.27	5.33	5.38	5.43	5.47	.05	30
5.93	6.01	6.08	6.14	6.20	6.26	6.31	6.36	6.41	.01	
4.90	4.98	5.04	5.11	5.16	5.22	5.27	5.31	5.36	.05	40
5.76	5.83	5.90	5.96	6.02	6.07	6.12	6.16	6.21	.01	
4.81	4.88	4.94	5.00	5.06	5.11	5.15	5.20	5.24	.05	60
5.60	5.67	5.73	5.78	5.84	5.89	5.93	5.97	6.01	.01	
4.71	4.78	4.84	4.90	4.95	5.00	5.04	5.09	5.13	.05	120
5.44	5.50	5.56	5.61	5.66	5.71	5.75	5.79	5.83	.01	
4.62	4.68	4.74	4.80	4.85	4.89	4.93	4.97	5.01	.05	∞
5.29	5.35	5.40	5.45	5.49	5.54	5.57	5.61	5.65	.01	

TABLE A.9 LEAST-SIGNIFICANT STUDENTIZED RANGES

$$\alpha = 0.05$$

					p				
λ	2	3	4	5	6	7	8	9	10
1	17.97	17.97	17.97	17.97	17.97	17.97	17.97	17.97	17.97
2	6.085	6.085	6.085	6.085	6.085	6.085	6.085	6.085	6.085
3	4.501	4.516	4.516	4.516	4.516	4.516	4.516	4.516	4.516
4	3.927	4.013	4.033	4.033	4.033	4.033	4.033	4.033	4.033
5	3.635	3.749	3.797	3.814	3.814	3.814	3.814	3.814	3.814
6	3.461	3.587	3.649	3.680	3.694	3.697	3.697	3.697	3.697
7	3.344	3.477	3.548	3.588	3.611	3.622	3.626	3.626	3.626
8	3.261	3.399	3.475	3.521	3.549	3.566	3.575	3.579	3.579
9	3.199	3.339	3.420	3.470	3.502	3.523	3.536	3.544	3.547
10	3.151	3.293	3.376	3.430	3.465	3.489	3.505	3.516	3.522
11	3.113	3.256	3.342	3.397	3.435	3.462	3.480	3.493	3.501
12	3.082	3.225	3.313	3.370	3.410	3.439	3.459	3.474	3.484
13	3.055	3.200	3.289	3.348	3.389	3.419	3.442	3.458	3.470
14	3.033	3.178	3.268	3.329	3.372	3.403	3.426	3.444	3.457
15	3.014	3.160	3.250	3.312	3.356	3.389	3.413	3.432	3.446
16	2.998	3.144	3.235	3.298	3.343	3.376	3.402	3.422	3.437
17	2.984	3.130	3.222	3.285	3.331	3.366	3.392	3.412	3.429
18	2.971	3.118	3.210	3.274	3.321	3.356	3.383	3.405	3.421
19	2.960	3.107	3.199	3.264	3.311	3.347	3.375	3.397	3.415
20	2.950	3.097	3.190	3.255	3.303	3.339	3.368	3.391	3.409
24	2.919	3.066	3.160	3.226	3.276	3.315	3.345	3.370	3.390
30	2.888	3.035	3.131	3.199	3.250	3.290	3.322	3.349	3.371
40	2.858	3.006	3.102	3.171	3.224	3.266	3.300	3.328	3.352
60	2.829	2.976	3.073	3.143	3.198	3.241	3.277	3.307	3.333
120	2.800	2.947	3.045	3.116	3.172	3.217	3.254	3.287	3.314
∞	2.772	2.918	3.017	3.089	3.146	3.193	3.232	3.265	3.294

This table has been reproduced by permission of the Biometric Society.

TABLE A.9 (*Continued*)

$$\alpha = 0.01$$

λ	p								
	2	**3**	**4**	**5**	**6**	**7**	**8**	**9**	**10**
1	90.03	90.03	90.03	90.03	90.03	90.03	90.03	90.03	90.03
2	14.04	14.04	14.04	14.04	14.04	14.04	14.04	14.04	14.04
3	8.261	8.321	8.321	8.321	8.321	8.321	8.321	8.321	8.321
4	6.512	6.677	6.740	6.756	6.756	6.756	6.756	6.756	6.756
5	5.702	5.893	5.989	6.040	6.065	6.074	6.074	6.074	6.074
6	5.243	5.439	5.549	5.614	5.655	5.680	5.694	5.701	5.703
7	4.949	5.145	5.260	5.334	5.383	5.416	5.439	5.454	5.464
8	4.746	4.939	5.057	5.135	5.189	5.227	5.256	5.276	5.291
9	4.596	4.787	4.906	4.986	5.043	5.086	5.118	5.142	5.160
10	4.482	4.671	4.790	4.871	4.931	4.975	5.010	5.037	5.058
11	4.392	4.579	4.697	4.780	4.841	4.887	4.924	4.952	4.975
12	4.320	4.504	4.622	4.706	4.767	4.815	4.852	4.883	4.907
13	4.260	4.442	4.560	4.644	4.706	4.755	4.793	4.824	4.850
14	4.210	4.391	4.508	4.591	4.654	4.704	4.743	4.775	4.802
15	4.168	4.347	4.463	4.547	4.610	4.660	4.700	4.733	4.760
16	4.131	4.309	4.425	4.509	4.572	4.622	4.663	4.696	4.724
17	4.099	4.275	4.391	4.475	4.539	4.589	4.630	4.664	4.693
18	4.071	4.246	4.362	4.445	4.509	4.560	4.601	4.635	4.664
19	4.046	4.220	4.335	4.419	4.483	4.534	4.575	4.610	4.639
20	4.024	4.197	4.312	4.395	4.459	4.510	4.552	4.587	4.617
24	3.956	4.126	4.239	4.322	4.386	4.437	4.480	4.516	4.546
30	3.889	4.056	4.168	4.250	4.314	4.366	4.409	4.445	4.477
40	3.825	3.988	4.098	4.180	4.244	4.296	4.339	4.376	4.408
60	3.762	3.922	4.031	4.111	4.174	4.226	4.270	4.307	4.340
120	3.702	3.858	3.965	4.044	4.107	4.158	4.202	4.239	4.272
∞	3.643	3.796	3.900	3.978	4.040	4.091	4.135	4.172	4.205

TABLE A.10   CRITICAL VALUES AND PROBABILITIES FOR THE SIGNED-RANK TEST

n	$S_{\alpha,n}$	α		n	$S_{\alpha,n}$	α
3	6	.125			78	.011
4	9	.125			79	.009
	10	.062			81	.005
5	13	.094		14	73	.108
	14	.062			74	.097
	15	.031			79	.052
6	17	.109			84	.025
	19	.047			89	.010
	20	.031			92	.005
	21	.016		15	83	.104
7	22	.109			84	.094
	24	.055			89	.053
	26	.023			90	.047
	28	.008			95	.024
8	28	.098			100	.011
	30	.055			101	.009
	32	.027			104	.005
	34	.012		16	93	.106
	35	.008			94	.096
	36	.004			100	.052
9	34	.102			106	.025
	37	.049			112	.011
	39	.027			113	.009
	42	.010			116	.005
	44	.004		17	104	.103
10	41	.097			105	.095
	44	.053			112	.049
	47	.024			118	.025
	50	.010			125	.010
	52	.005			129	.005
11	48	.103		18	116	.098
	52	.051			124	.049
	55	.027			131	.024
	59	.009			138	.010
	61	.005			143	.005
12	56	.102		19	128	.098
	60	.055			136	.052
	61	.046			137	.048
	64	.026			144	.025
	68	.010			152	.010
	71	.005			157	.005
13	64	.108		20	140	.101
	65	.095			150	.049
	69	.055			158	.024
	70	.047			167	.010
	74	.024			172	.005

From Dixon/Massey, *Introduction to Statistical Analysis*, 4/e, © 1984. Reprinted by permission of McGraw-Hill Publishing Company.

n_X	n_Y	w_{α, n_X, n_Y}	α	n_X	n_Y	w_{α, n_X, n_Y}	α
3	3	15	.05			40	.004
	4	17	.057		6	40	.041
		18	.029			41	.026
	5	20	.036			43	.009
		21	.018			44	.004
	6	22	.048		7	43	.053
		23	.024			45	.024
		24	.012			47	.009
	7	24	.058			48	.005
		26	.017		8	47	.047
		27	.008			49	.023
	8	27	.042			51	.009
		28	.024			52	.005
		29	.012	6	6	50	.047
		30	.006			52	.021
4	4	24	.057			54	.008
		25	.029			55	.004
		26	.014		7	54	.051
	5	27	.056			56	.026
		28	.032			58	.011
		29	.016			60	.004
		30	.008		8	58	.054
	6	30	.057			61	.021
		32	.019			63	.01
		33	.010			65	.004
		34	.005	7	7	66	.049
	7	33	.055			68	.027
		35	.021			71	.009
		36	.012			72	.006
		37	.006		8	71	.047
	8	36	.055			73	.027
		38	.024			76	.01
		40	.008			78	.005
		41	.004	8	8	84	.052
5	5	36	.048			87	.025
		37	.028			90	.01
		39	.008			92	.005

From Dixon/Massey, *Introduction to Statistical Analysis*, 4/e, © 1984. Reprinted by permission of McGraw-Hill Publishing Company.

10097	32533	76520	13586	34673	54876	80959	09117	39292	74945
37542	04805	64894	74296	24805	24037	20636	10402	00822	91665
08422	68953	19645	09303	23209	02560	15953	34764	35080	33606
99019	02529	09376	70715	38311	31165	88676	74397	04436	27659
12807	99970	80157	36147	64032	36653	98951	16877	12171	76833
66065	74717	34072	76850	36697	36170	65813	39885	11199	29170
31060	10805	45571	82406	35303	42614	86799	07439	23403	09732
85269	77602	02051	65692	68665	74818	73053	85247	18623	88579
63573	32135	05325	47048	90553	57548	28468	28709	83491	25624
73796	45753	03529	64778	35808	34282	60935	20344	35273	88435
98520	17767	14905	68607	22109	40558	60970	93433	50500	73998
11805	05431	39808	27732	50725	68248	29405	24201	52775	67851
83452	99634	06288	98083	13746	70078	18475	40610	68711	77817
88685	40200	86507	58401	36766	67951	90364	76493	29609	11062
99594	67348	87517	64969	91826	08928	93785	61368	23478	34113
65481	17674	17468	50950	58047	76974	73039	57186	40218	16544
80124	35635	17727	08015	45318	22374	21115	78253	14385	53763
74350	99817	77402	77214	43236	00210	45521	64237	96286	02655
69916	26803	66252	29148	36936	87203	76621	13990	94400	56418
09893	20505	14225	68514	46427	56788	96297	78822	54382	14598
91499	14523	68479	27686	46162	83554	94750	89923	37089	20048
80336	94598	26940	36858	70297	34135	53140	33340	42050	82341
44104	81949	85157	47954	32979	26575	57600	40881	22222	06413
12550	73742	11100	02040	12860	74697	96644	89439	28707	25815
63606	49329	16505	34484	40219	52563	43651	77082	07207	31790
61196	90446	26457	47774	51924	33729	65394	59593	42582	60527
15474	45266	95270	79953	59367	83848	82396	10118	33211	59466
94557	28573	67897	54387	54622	44431	91190	42592	92927	45973
42481	16213	97344	08721	16868	48767	03071	12059	25701	46670
23523	78317	73208	89837	68935	91416	26252	29663	05522	82562
04493	52494	75246	33824	45862	51025	61962	79335	65337	12472
00549	97654	64051	88159	96119	63896	54692	82391	23287	29529
35963	15307	26898	09354	33351	35462	77974	50024	90103	39333
59808	08391	45427	26842	83609	49700	13021	24892	78565	20106
46058	85236	01390	92286	77281	44077	93910	83647	70617	42941
32179	00597	87379	25241	05567	07007	86743	17157	85394	11838
69234	61406	20117	45204	15956	60000	18743	92423	97118	96338
19565	41430	01758	75379	40419	21585	66674	36806	84962	85207
45155	14938	19476	07246	43667	94543	59047	90033	20826	69541
94864	31994	36168	10851	34888	81553	01540	35456	05014	51176
98086	24826	45240	28404	44999	08896	39094	73407	35441	31880
33185	16232	41941	50949	89435	48581	88695	41994	37548	73043
80951	00406	96382	70774	20151	23387	25016	25298	94624	61171
79752	49140	71961	28296	69861	02591	74852	20539	00387	59579
18633	32537	98145	06571	31010	24674	05455	61427	77938	91936
74029	43902	77557	32270	97790	17119	52527	58021	80814	51748
54178	45611	80993	37143	05335	12969	56127	19255	36040	90324
11664	49883	52079	84827	59381	71539	09973	33440	88461	23356
48324	77928	31249	64710	02295	36870	32307	57546	15020	09994
69074	94138	87637	91976	35584	04401	10518	21615	01848	76938

Reprinted from Tables A15 and A16 of *A Million Random Digits with 100,000 Normal Deviates* by The RAND Corporation (New York: The Free Press, 1955). Copyright 1955 and 1983 by The RAND Corporation. Used with permission.

1.276–	1.218–	0.453–	0.350–	0.723	0.676	1.099–	0.314–	0.394–	0.633–
0.318–	0.799–	1.664–	1.391	0.382	0.733	0.653	0.219	0.681–	1.129
1.377–	1.257–	0.495	0.139–	0.854–	0.428	1.322–	0.315–	0.732–	1.348–
2.334	0.337–	1.955–	0.636–	1.318–	0.433–	0.545	0.428	0.297–	0.276
1.136–	0.642	3.436	1.667–	0.847	1.173–	0.355–	0.035	0.359	0.930
0.414	0.011–	0.666	1.132–	0.410–	1.077–	0.734	1.484	0.340–	0.789
0.494–	0.364	1.237–	0.044–	0.111–	0.210–	0.931	0.616	0.377–	0.433–
1.048	0.037	0.759	0.609	2.043–	0.290–	0.404	0.543–	0.486	0.869
0.347	2.816	0.464–	0.632–	1.614–	0.372	0.074–	0.916–	1.314	0.038–
0.637	0.563	0.107–	0.131	1.808–	1.126–	0.379	0.610	0.364–	2.626–
2.176	0.393	0.924–	1.911	1.040–	1.168–	0.485	0.076	0.769–	1.607
1.185–	0.944–	1.604–	0.185	0.258–	0.300–	0.591–	0.545–	0.018	0.485–
0.972	1.170	2.682	2.813	1.531–	0.490–	2.071	1.444	1.092–	0.478
1.210	0.294	0.248–	0.719	1.103	1.090	0.212	1.185–	0.338–	1.134–
2.647	0.777	0.450	2.247	1.151	1.676–	0.384	1.133	1.393	0.814
0.398	0.318	0.928–	2.416	0.936–	1.036	0.024	0.560–	0.203	0.871–
0.846	0.699–	0.368–	0.344	0.926–	0.797–	1.404–	1.472–	0.118–	1.456
0.654	0.955–	2.907	1.688	0.752	0.434–	0.746	0.149	0.170–	0.479–
0.522	0.231	0.619–	0.265–	0.419	0.558	0.549–	0.192	0.334–	1.373
1.288–	0.539–	0.824–	0.244	1.070–	0.010	0.482	0.469–	0.090–	1.171
1.372	1.769	1.057–	1.646	0.481	0.600–	0.592–	0.610	0.096–	1.375–
0.854	0.535–	1.607	0.428	0.615–	0.331	0.336–	1.152–	0.533	0.833–
0.148–	1.144–	0.913	0.684	1.043	0.554	0.051–	0.944–	0.440–	0.212–
1.148–	1.056–	0.635	0.328–	1.221–	0.118	2.045–	1.977–	1.133–	0.338
0.348	0.970	0.017–	1.217	0.974–	1.291–	0.399–	1.209–	0.248–	0.480
0.284	0.458	1.307	1.625–	0.629–	0.504–	0.056–	0.131–	0.048	1.879
1.016–	0.360	0.119–	2.331	1.672	1.053–	0.840	0.246–	0.237	1.312–
1.603	0.952–	0.566–	1.600	0.465	1.951	0.110	0.251	0.116	0.957–
0.190–	1.479	0.986–	1.249	1.934	0.070	1.358–	1.246–	0.959–	1.297–
0.722–	0.925	0.783	0.402–	0.619	1.826	1.272	0.945–	0.494	0.050
1.696–	1.879	0.063	0.132	0.682	0.544	0.417–	0.666–	0.104–	0.253–
2.543–	1.333–	1.987	0.668	0.360	1.927	1.183	1.211	1.765	0.035
0.359–	0.193	1.023–	0.222–	0.616–	0.060–	1.319–	0.785	0.430–	0.298–
0.248	0.088–	1.379–	0.295	0.115–	0.621–	0.618–	0.209	0.979	0.906
0.099–	1.376–	1.047	0.872–	2.200–	1.384–	1.425	0.812–	0.748	1.093–
0.463–	1.281–	2.514–	0.675	1.145	1.083	0.667–	0.223–	1.592–	1.278–
0.503	1.434	0.290	0.397	0.837–	0.973–	0.120–	1.594–	0.996–	1.244–
0.857–	0.371–	0.216–	0.148	2.106–	1.453–	0.686	0.075–	0.243–	0.170–
0.122–	1.107	1.039–	0.636–	0.860–	0.895–	1.458–	0.539–	0.159–	0.420–
1.632	0.586	0.468–	0.386–	0.354–	0.203	1.234–	2.381	0.388–	0.063–
2.072	1.445–	0.680–	0.224	0.120–	1.753	0.571–	1.223	0.126–	0.034
0.435–	0.375–	0.985–	0.585–	0.203–	0.556–	0.024	0.126	1.250	0.615–
0.876	1.227–	2.647–	0.745–	1.797	1.231–	0.547	0.634–	0.836–	0.719–
0.833	1.289	0.022–	0.431–	0.582	0.766	0.574–	1.153–	0.520	1.018–
0.891–	0.332	0.453–	1.127–	2.085	0.722–	1.508–	0.489	0.496–	0.025–
0.644	0.233–	0.153–	1.098	0.757	0.039–	0.460–	0.393	2.012	1.356
0.105	0.171–	0.110–	1.145–	0.878	0.909–	0.328–	1.021	1.613–	1.560
1.192–	1.770	0.003–	0.369	0.052	0.647	1.029	1.526	0.237	1.328–
0.042–	0.553	0.770	0.324	0.489–	0.367–	0.378	0.601	1.996–	0.738–
0.498	1.072	1.567	0.302	1.157	0.720–	1.403	0.698	0.370–	0.551–

Reprinted from Tables A15 and A16 of *A Million Random Digits with 100,000 Normal Deviates* by The RAND Corporation (New York: The Free Press, 1955). Copyright 1955 and 1983 by The RAND Corporation. Used with permission.

TABLE A.14 FACTORS FOR QUALITY CONTROL CHARTS

Number of Observations in Sample, n	Chart for Averages			Chart for Standard Deviations						Chart for Ranges						
	Factors for Control Limits			Factors for Central Line		Factors for Control Limits				Factors for Central Line			Factors for Control Limits			
	A	A_1	A_2	c_2	$1/c_2$	B_1	B_2	B_3	B_4	d_2	$1/d_2$	d_3	D_1	D_2	D_3	D_4
2	2.121	3.760	1.880	0.5642	1.7725	0	1.843	0	3.267	1.128	0.8865	0.853	0	3.686	0	3.276
3	1.732	2.394	1.023	0.7236	1.3820	0	1.858	0	2.568	1.693	0.5907	0.888	0	4.358	0	2.575
4	1.501	1.880	0.729	0.7979	1.2533	0	1.808	0	2.266	2.059	0.4857	0.880	0	4.698	0	2.282
5	1.342	1.596	0.577	0.8407	1.1894	0	1.756	0	2.089	2.326	0.4299	0.864	0	4.918	0	2.115
6	1.225	1.410	0.483	0.8686	1.1512	0.026	1.711	0.030	1.970	2.534	0.3946	0.848	0	5.078	0	2.004
7	1.134	1.277	0.419	0.8882	1.1259	0.105	1.672	0.118	1.882	2.704	0.3698	0.833	0.205	5.203	0.076	1.924
8	1.061	1.175	0.373	0.9027	1.1078	0.167	1.638	0.185	1.815	2.847	0.3512	0.820	0.387	5.307	0.136	1.864
9	1.000	1.094	0.337	0.9139	1.0942	0.219	1.609	0.239	1.761	2.970	0.3367	0.808	0.546	5.394	0.184	1.816
10	0.949	1.028	0.308	0.9227	1.0837	0.262	1.584	0.284	1.716	3.078	0.3249	0.797	0.687	5.469	0.223	1.777
11	0.905	0.973	0.285	0.9300	1.0753	0.299	1.561	0.321	1.679	3.173	0.3152	0.787	0.812	5.534	0.256	1.744
12	0.866	0.925	0.266	0.9359	1.0684	0.331	1.541	0.354	1.646	3.258	0.3069	0.778	0.924	5.592	0.284	1.719
13	0.832	0.884	0.249	0.9410	1.0627	0.359	1.523	0.382	1.618	3.336	0.2998	0.770	1.026	5.646	0.308	1.692
14	0.802	0.848	0.235	0.9453	1.0579	0.384	1.507	0.406	1.594	3.407	0.2935	0.762	1.121	5.693	0.329	1.671
15	0.775	0.816	0.223	0.9490	1.0537	0.406	1.492	0.428	1.572	3.472	0.2880	0.755	1.207	5.737	0.348	1.652
16	0.750	0.788	0.212	0.9523	1.0501	0.427	1.478	0.448	1.552	3.532	0.2831	0.749	1.285	5.779	0.364	1.636
17	0.728	0.762	0.203	0.9551	1.0470	0.445	1.465	0.466	1.534	3.588	0.2787	0.743	1.359	5.817	0.379	1.621
18	0.707	0.738	0.194	0.9576	1.0442	0.461	1.454	0.482	1.518	3.640	0.2747	0.738	1.426	5.854	0.392	1.608
19	0.688	0.717	0.187	0.9599	1.0418	0.477	1.443	0.497	1.503	3.689	0.2711	0.733	1.490	5.888	0.404	1.596
20	0.671	0.697	0.180	0.9619	1.0396	0.491	1.433	0.510	1.490	3.735	0.2677	0.729	1.548	5.922	0.414	1.586
21	0.655	0.679	0.173	0.9638	1.0376	0.504	1.424	0.523	1.477	3.778	0.2647	0.724	1.606	5.950	0.425	1.575
22	0.640	0.662	0.167	0.9655	1.0358	0.516	1.415	0.534	1.466	3.819	0.2618	0.720	1.659	5.979	0.434	1.566
23	0.626	0.647	0.162	0.9670	1.0342	0.527	1.407	0.545	1.455	3.858	0.2592	0.716	1.710	6.006	0.443	1.557
24	0.612	0.632	0.157	0.9684	1.0327	0.538	1.399	0.555	1.445	3.895	0.2567	0.712	1.759	6.031	0.452	1.548
25	0.600	0.619	0.153	0.9696	1.0313	0.548	1.392	0.565	1.435	3.931	0.2544	0.709	1.804	6.058	0.459	1.541
Over 25	$\frac{3}{\sqrt{n}}$	$\frac{3}{\sqrt{n}}$	—	—	—	++	§	++	§	—	—	—	—	—	—	—

Copyright ASTM. Reprinted with permission.

TABLE A.15 CRITICAL VALUES FOR BARTLETT'S TEST

$$b_{0.05,n,r}$$

n	2	3	4	5	6	7	8	9	10
3	.3123	.3058	.3173	.3299	*	*	*	*	*
4	.4780	.4699	.4803	.4921	.5028	.5122	.5204	.5277	.5341
5	.5845	.5762	.5850	.5952	.6045	.6126	.6197	.6260	.6315
6	.6563	.6483	.6559	.6646	.6727	.6798	.6860	.6914	.6961
7	.7075	.7000	.7065	.7142	.7213	.7275	.7329	.7376	.7418
8	.7456	.7387	.7444	.7512	.7574	.7629	.7677	.7719	.7757
9	.7751	.7686	.7737	.7798	.7854	.7903	.7946	.7984	.8017
10	.7984	.7924	.7970	.8025	.8076	.8121	.8160	.8194	.8224
11	.8175	.8118	.8160	.8210	.8257	.8298	.8333	.8365	.8392
12	.8332	.8280	.8317	.8364	.8407	.8444	.8477	.8506	.8531
13	.8465	.8415	.8450	.8493	.8533	.8568	.8598	.8625	.8648
14	.8578	.8532	.8564	.8604	.8641	.8673	.8701	.8726	.8748
15	.8676	.8632	.8662	.8699	.8734	.8764	.8790	.8814	.8834
16	.8761	.8719	.8747	.8782	.8815	.8843	.8868	.8890	.8909
17	.8836	.8796	.8823	.8856	.8886	.8913	.8936	.8957	.8975
18	.8902	.8865	.8890	.8921	.8949	.8975	.8997	.9016	.9033
19	.8961	.8926	.8949	.8979	.9006	.9030	.9051	.9069	.9086
20	.9015	.8980	.9003	.9031	.9057	.9080	.9100	.9117	.9132
21	.9063	.9030	.9051	.9078	.9103	.9124	.9143	.9160	.9175
22	.9106	.9075	.9095	.9120	.9144	.9165	.9183	.9199	.9213
23	.9146	.9116	.9135	.9159	.9182	.9202	.9219	.9235	.9248
24	.9182	.9153	.9172	.9195	.9217	.9236	.9253	.9267	.9280
25	.9216	.9187	.9205	.9228	.9249	.9267	.9283	.9297	.9309
26	.9246	.9219	.9236	.9258	.9278	.9296	.9311	.9325	.9336
27	.9275	.9249	.9265	.9286	.9305	.9322	.9337	.9350	.9361
28	.9301	.9276	.9292	.9312	.9330	.9347	.9361	.9374	.9385
29	.9326	.9301	.9316	.9336	.9354	.9370	.9383	.9396	.9406
30	.9348	.9325	.9340	.9358	.9376	.9391	.9404	.9416	.9426
40	.9513	.9495	.9506	.9520	.9533	.9545	.9555	.9564	.9572
50	.9612	.9597	.9606	.9617	.9628	.9637	.9645	.9652	.9658
60	.9677	.9665	.9672	.9681	.9690	.9698	.9705	.9710	.9716
80	.9758	.9749	.9754	.9761	.9768	.9774	.9779	.9783	.9787
100	.9807	.9799	.9804	.9809	.9815	.9819	.9823	.9827	.9830

From Dyer and Keating, "On the determination of critical values for Bartlett's test," *Journal of the American Statistical Association*, Vol. 75, 1980. Reprinted by permission of the American Statistical Association.

APPENDIX

B

||||||||||||||||||||||||||||||||||||||||||||||||||||||||||||||||||||||||||||||||||||||||

STATISTICAL SOFTWARE PACKAGES: SAS, MINITAB, SPSS

After learning procedures such as the multiple linear regression or the paired-comparisons *t* test, the student of statistics should try his or her hand at performing a fairly realistic analysis of suitable data. With the advent in recent years of the personal computer and user-friendly software, opportunities for students to gain hands-on experience are considerably greater than they were previously.

The statistical software described herein makes it possible for the statistician and student of statistics to analyze moderately large data sets. But there is another reason for becoming familiar with these computer tools: statistical analyses can be performed faster and more accurately. Owing to computational and reporting advantages, the time invested in learning to use a software package is amply rewarded. A problem, once entered into the computer, can easily be reanalyzed, saving time and increasing the likelihood that more advanced techniques will be attempted. We are not advocating that the student of engineering statistics learn all three or even two of the software packages described; rather, three are covered to give the student a choice, which should be coordinated with the institution where he or she is studying. All three packages provide the same basic statistical functions, along with good data handling and reporting capabilities. They differ in advanced functions and the method by which they process data. Also, the learning curve may be steeper for SAS, which is a larger system than the other two. Many students have already had some exposure to one of these three statistical software systems, and this may determine their choice.

B.1 OVERVIEW OF THE THREE SYSTEMS

B.1.1 Overview of MINITAB

The MINITAB system is based on a two-dimensional spreadsheet concept, in which the columns are variables and the rows are cases. Columns 1 through 3 are named

This appendix was prepared by Francis Sand, Associate Professor, Department of Mathematics and Computer Science, Fairleigh Dickinson University, Teaneck, NJ.

C1, C2, and C3, respectively, and they may be used in formulas to create new columns. For instance,

$$\text{LET C4 = 2.0*C1 - 0.04*C2^2}$$

causes the new variable C4 to be created, and in every row it will contain the value corresponding to the formula applied to the cases of C1 and C2 in that row. Thus, if the fourth row has 1 and 5 in columns 1 and 2, then it will contain $2 - 0.04*25 = 1$ in column 4 after the LET statement.

Statistical procedures are implemented by invoking their MINITAB names with a list of columns and parameters for MINITAB processing. The MINITAB statements permit generous use of inserted ''help'' words, which do nothing to the calculation but allow the user to express a concept naturally. For instance, to do a multiple linear regression of C5 on C1, C2, and C3, one can write

$$\text{REGRESS C5 on 3 variables C1 C2 and C3}$$

The lower-case words are not required in the statement, but help to clarify its meaning to the user. As a result of this command, MINITAB performs the regression analysis of C5 on C1, C2, and C3 using all the rows of the specified columns to obtain a maximal sample. The output, directed to screen and/or printer, contains the regression coefficients, their t statistics, a regression analysis of variance, and other information as demonstrated later in this appendix. MINITAB is interactive: output is obtained immediately, and mistakes can be easily corrected during a work session. The work session can be saved in a MINITAB worksheet file for later retrieval.

B.1.2 SAS Overview

SAS statistical analysis, long enjoyed by mainframe users, is now available for the IBM PC. It is a system that allows the user maximum flexibility in data processing while at the same time providing a large library of high-quality statistical procedures. SAS uses a method of naming data sets that is internal to SAS and separate from external file names (if any exist). The DATA statement in SAS permits the user to establish an SAS data set by naming it and at the same time naming the variables in it. For instance, in the following small example concerning battery lifetimes, data are entered directly into an SAS data set called SAMPLE, and, as shown in Table B.1, the variables are named TYPE and LIFE. Note that semicolons are necessary for termination of SAS statements. New variables are included in SAS data sets when defined by equations such as L100 = LIFE − 100; this statement must precede the CARDS statement. Subsequent SAS processing is done on this data set unless a new one is named. The next step in this work session might call for a ranking of batteries by life:

```
PROC RANK;
  VAR LIFE;
  RANKS RANKEDLIFE;
  OUTPUT;
```

SAS refers to the statistical procedure for ranking as PROC RANK, and VAR and RANKS following PROC identify the input and output variables, respectively. The new variable RANKEDLIFE is added to the data set SAMPLE as a result of executing the RANK procedure; however, to save it, the OUTPUT statement must be included.

To obtain a printout of the results one would insert a PRINT step:

TABLE B.1

```
DATA SAMPLE;
  INPUT TYPE LIFE;
  CARDS;
  800    120
  700     90
  701    110
  500    115
  504    115
  400     85
  401     75
  404     70
  406     80
  407     95
  ;
```

```
                      PROC PRINT;
                  TITLE RANKED BATTERY LIFE;
```

SAS will print the list of batteries, with type and life as well as the rank of each battery by life. The report will have the title "RANKED BATTERY LIFE."

SAS has numerous options with each PROC for doing statistical testing, producing output, and creating breakdowns of the data set. For instance, if desired, the printout of battery lives can be produced by type as follows:

```
                  PROC PRINT;
                  BY TYPE;
                  TITLE RANKED BATTERY LIFE BY TYPE;
```

B.1.3 SPSS Overview

The SPSS package was designed for processing survey data on mainframe computers. Today it has many features of interest to scientists and engineers and can be obtained in a PC version. The method of processing data in SPSS is similar to SAS. Each data set is defined in a data definition statement, which also names the variables and specifies the columns in which they are to be placed. For instance,

```
              DATA LIST/ X 1–5 Y 6–10 Z 11–15.
```

The newly created variables may be given descriptive labels in SPSS by a VARIABLE LABELS statement.

Many statistical procedures are available, such as the ANOVA, which performs analysis of variance on the variables indicated:

TABLE B.2

```
SPSS/PC: report format=list/
       : variables = alarms
       : price 'Dealer' 'price'
       : time(label) (12) 'Time Limits' '(entry/exit)'
       : arming(label) (13) /
       : title = 'RATINGS OF AUTO ALARM SYSTEMS – 1988'/
       : break = rating(label) (9)/
       : summary = Mean (price(2) (dollar)) 'Mean'.
```

RATINGS OF AUTO ALARM SYSTEMS – 1988

RATING	ALARM	Dealer price	Time Limits (entry/exit)	Arming/Disarming methods
Better				
	CRIMESTOPPER	399	0–28	PASSIVE/PASSIVE
	CLIFFORD III	550	30	PASSIVE/ACTIVE
Mean		$475		
Good				
	ALPINE 8101	400	0–45	PASSIVE/ACTIVE
	MAXIGUARD	370	5–30/60	PASSIVE/ACTIVE
	TECHNE UNGO	485	32	PASSIVE/ACTIVE
	CODE ALARM	325	3–30/3–60	PASSIVE/ACTIVE
Mean		$395		
Average				
	THUG BUG AVENGER	450	0–24/40	PASSIVE/ACTIVE
	PARAGON	550	13/65	PASSIVE/PASSIVE
	ASP K400-FS	455	16/45	ACTIVE/ACTIVE
Mean		$485		

All SPSS statements end in a period, and the ''/'' precedes options. The processing is entirely interactive. SPSS has extensive reporting and graphical facilities. The use of the package to produce attractive reports of statistical subjects is one of its advantages.

An example is given in Table B.2. Note the formatted appearance which the user has specified in a REPORT FORMAT statement.

Because the statistical software packages all have PC versions, except where otherwise noted, use of statistical software will be discussed in terms which are general and which apply to either mainframe or PC versions. It will not be assumed that the student is familiar with the use of software packages for engineering or scientific computation. For detailed references, the reader is strongly urged to make use of the software publishers' handbooks and user guides; we will provide only a brief introduction and some worked-out examples.

All three packages provide excellent graphics, which are, however, quite varying in the ease of access. In the case of SAS, a separate module has to be accessed for presentation-quality graphics. Since many universities and colleges may not have this module, the subject of presentation graphics will not be treated.

B.2 MINITAB

B.2.1 Interactive and Batch Modes

The MINITAB system can be used interactively or in batch mode. In the interactive mode, each command produces an immediate result with appropriate output unless it is ended with a semicolon. The purpose of the semicolon is to inform the system that subcommands are to follow. In this case, results are produced only after a period is encountered, as demonstrated in the following example.

```
REGRESS C2 on 1, C1;
PREDICT 11.
```

At the end of a work session in the interactive mode, the user of MINITAB should SAVE the worksheet (how to do this is described in Section B.2.3.) and then write the word STOP.

B.2.2 Entering Data

MINITAB has two main methods for directly entering data. The most straightforward is SET. All data are stored in tables, with each column representing a different variable and each row a different case. (Other arrangements of the data are allowed too, but we will use only the column-variable, row-case method.) Initially, columns are labeled C1, C2, C3, and so on, but the user can name them conveniently with names such as WEIGHT, TEMPERATURE, AGE. To use the SET command to enter ten values for the first variable, write

```
SET C1
1.0   2   3.5   1.5   5.7   -1.8
19   10.1   9   0.8
END
```

This will have the effect of storing the indicated values in C1 (column one). Note that two lines were used—some data had decimal points and some did not—and the spaces serve to separate values. In this example, the result will be to store the ten values in C1, and no output will be produced. The user may wish to alter or manipulate the data—for instance, using them in equations. The stored data may be referred to as C1 in subsequent work. This will be illustrated shortly. If C1 is not subsequently required, it can be erased:

$$\text{ERASE C1}$$

The more versatile command READ allows data in tabular form to be entered into several columns at once. For instance,

```
READ C1   C2    C3    C4
      1   1.0    9    0.0
      2   1.1   11    0.15
      3   1.2   13   -0.08
      4   1.0   20    0.009
END
```

causes four values to be entered into each of C1, C2, C3, and C4. Note that spaces between entries are mandatory, but otherwise the layout of the data is free-form until END is reached. Alignment helps the user but is not required. The data will be stored in the columns C1, C2, C3, and C4.

B.2.3 Using Stored Data in Equations

After storing data, the user can compute with the stored data, creating new columns in the process. The command LET followed by an equation should be used. The left side of the equation designates a new column, while the right side contains an expression. For instance,

$$\text{LET C5 = C2 / C1}$$

will create a new variable in column 5 which is the ratio of the variables C2 and C1. The equation may contain constants K1, K2, and so on, and may be fairly complicated:

$$\text{LET C6 = K1 + K2* (C2 + C3) - C4*C1}$$

To make this equation usable, K1 and K2 would have to have been previously defined in a LET command; for now we will assume that K1 = 0.5 and K2 = 10.

The LET command is also useful for correcting entries. Suppose that the fourth entry under C4 should have been 0.09 but was erroneously entered as 0.009. To correct it, write

$$\text{LET C4 (4) = 0.09}$$

This will replace the fourth entry of C4 with the correct value. Note that there is no END after LET, and that the parentheses are required only when a single value is being entered. Otherwise MINITAB operates on the entire column. At this point, if the reader has performed the example commands exactly as shown, MINITAB contains stored constants

```
K1    K2
0.5   10
```

and stored variable data in six columns (recall that the first C1 was erased):

C1	C2	C3	C4	C5	C6
1	1.0	9	0.00	1.00	100.5
2	1.1	11	0.15	0.55	121.2
3	1.2	13	-0.08	0.40	142.74
4	1.0	20	0.09	0.25	210.14

To see this table displayed on the screen or printed on paper,* use the PRINT command

PRINT C1 C2 C3 C4 C5 C6

Alternatively, because the columns are contiguous, a simpler form of the command is

PRINT C1–C6

Extensive data may be output to a file by use of the WRITE command. The filename must appear within quotes:

WRITE 'TABLE1' C1–C6

All the data shown above without the column headings will be stored in a MINITAB data file with filename 'TABLE1'. This data can later be read into another or the same worksheet with a READ command:

READ 'TABLE1' C1–C6

A worksheet may be saved (SAVE 'MYWORK') and later retrieved (RETRIEVE 'MYWORK'). Note the use of single quotes around the name of the saved worksheet. This may be done conveniently at the end of a work session, or when the user wishes to have a clear worksheet with space available for a new subject or version of the same subject. The command for clearing the worksheet is RESTART.

B.2.4 Statistics

We are now ready to do some statistics. This section will demonstrate the use of the DESCRIBE command and its subcommands. With the data previously stored in C1–C6, write

DESCRIBE C1–C6

The following statistics will be printed for each of the six columns:

N	Number of nonmissing values
NMISS	Number of missing values
MEAN	
MEDIAN	
TRMEAN	5% Trimmed mean
STDEV	Standard deviation
SEMEAN	Standard error of the mean
MAX	Maximum value
MIN	Minimum value
Q3	Third quartile
Q1	First quartile

* To obtain hardcopy output from MINITAB, issue the command PAPER; to discontinue the printout, the command NOPAPER.

The actual printout for DESCRIBE C1–C6 is shown in Table B.3. All of the subcommands can be used by themselves, either to obtain direct output:

MEAN C2

or to store the statistic as a constant:

LET K3 = STDEV (C6)
LET K4 = MEAN (C6) / K3

Note the use of an existing constant, K3, in the expression on the right side of the second of these LET statements. New statistics can easily be created in this second usage by combining two or more built-in functions:

LET K5 = (MIN (C5) + MAX (C5)) / 2

which stores the value of the midrange of the fifth variable, C5, in K5. MINITAB provides additional simple descriptive statistics such as COUNT, the number of values (whether missing or not), and SUM, the straight sum of the nonmissing values. As a result of the assignments made in the three LET examples, the stored constants in the worksheet at this point, if printed out by PRINT K1–K5, would appear as

K1	K2	K3	K4	K5
0.5	10	41.1936	0.286774	155.32

Before presenting a more advanced statistical example, we point out the use of names for variables in the MINITAB worksheet. While not required, the naming of variables makes the output easier to read. The NAME command ties columns C1, C2, and so on to user-selected names, which must always be enclosed in single quotes. The following command names the variables that will be used in the advanced statistical example:

NAME C1='HOURS' C2='RESIDUAL'

At various times from 2 to 12 hours after cleaning a swimming pool, the chlorine residual in parts per million was measured. An exponential model was thought to apply for relating residual to hours since cleaning. The following data were entered into MINITAB:

TABLE B.3

	N	MEAN	MEDIAN	TRMEAN	STDEV	SEMEAN
C1	4	2.500	2.500	2.500	1.291	0.645
C2	4	1.0750	1.0500	1.0750	0.0957	0.0479
C3	4	13.25	12.00	13.25	4.79	2.39
C4	4	0.0400	0.0450	0.0400	0.1010	0.0505
C5	4	0.550	0.475	0.550	0.324	0.162
C6	4	143.6	132.0	143.6	47.6	23.8

	MIN	MAX	Q1	Q3
C1	1.000	4.000	1.250	3.750
C2	1.0000	1.2000	1.0000	1.1750
C3	9.00	20.00	9.50	18.25
C4	-0.0800	0.1500	-0.0600	0.1350
C5	0.250	1.000	0.287	0.887
C6	100.5	210.1	105.7	193.3

```
      READ C1    C2
             2   1.8
             4   1.5
             6   1.45
             8   1.42
            10   1.38
            12   1.36
      END
```

After naming the two variables as shown above, the exponential model $Y = ae^{-bx}$ is estimated by (1) taking logarithms and (2) performing simple linear regression. After the regression results are obtained, the exponential model is restored by taking the exponential function of the estimated log regression. First create the natural logarithm of the chlorine residual using the MINITAB function LOGE:

```
LET C3 = LOGE(C2)
```

Now name this new variable:

```
NAME C3='LOGRES'
```

Then perform the simple linear regression:

```
REGRESS 'LOGRES' ON 1, 'HOURS'
```

We could have written

```
REGRESS C3 1,C1
```

the word "ON" being optional. Note that the numeral "1" is required; it tells MINITAB the number of regressor variables to be used in the equation. When this number is more than 1, the model is a multiple linear regression. The results of the regression will be displayed on the screen or printed on paper as shown in Table B.4.

The results of the regression equation can now be employed to predict the chlorine residual at other times. For example, to predict it 20 hours after cleaning, recall that the dependent variable was the logarithm of the residual, so use the ANTILOG:

```
LET K3 = ANTILOG(0.558 - 0.0239*20)
```

TABLE B.4

```
LOGRES = 0.558 - 0.0239 hours
Predictor       Coef        Stdev        t-ratio         p
Constant      0.055810    0.05246         10.64       0.000
hours        -0.023894    0.006735        -3.55       0.024

s = 0.05635      R-sq = 75.9%      R-sq(adj) = 69.9%


Analysis of Variance
SOURCE         DF         SS          MS          F         p
Regression     1       0.039964    0.039964    12.59     0.024
Error          4       0.012700    0.003175
Total          5       0.052664
```

TABLE B.5

TEST OF MU = 300.0 VS MU N.E. 300.0

	N	MEAN	STDEV	SE MEAN	T	P VALUE
LIFETIME	15	287.533	19.632	5.069	-2.46	0.028

If this constant is printed, its value will be found to be 1.20226; after 20 hours there are predicted to be 1.2 parts per million of chlorine residual in the swimming pool.

Another example illustrates the use of the t test for the mean of a normal distribution when the standard deviation is unknown. A manufacturer of batteries specifies the mean lifetime of his product to be 300 hours. A bulk buyer of these batteries records the lifetimes of a sample of 15 batteries purchased from this manufacturer and finds the sample mean to be 287.5 hours with standard deviation 19.6 hours. The user of MINITAB does not need to know the sample statistics, because TTEST will compute them. Assuming the relevant data have been entered into the worksheet and the variable has been named 'LIFETIME', TTEST can now be performed as follows:

TTEST 300 'LIFETIME'

The results will appear as shown in Table B.5. At level $\alpha = 0.05$, the buyer will conclude that the batteries are not as long-lasting as claimed.

Nonparametric tests, such as signed-rank tests, are well represented in MINITAB. For an example of these we apply the Wilcoxon rank-sum test, also known as the Mann-Whitney U test, to the data of Table 9.4 on the comparison of two recognition systems. This test is called MANN-WHITNEY in MINITAB; it tests the null hypothesis that the two samples come from populations having the same distribution. To apply this test write

MANN-WHITNEY C1 C2

assuming the recognition data are stored in columns 1 and 2. The results of this test, unlike the paired t test, do not allow us to conclude that one recognition system is superior to the other. These results are given in Table B.6.

Polynomial regression can be performed (if the degree is low) by creating extra columns in MINITAB containing the squares, cubes, and so on, and then using REGRESS to do a multiple regression. We illustrate this approach with a quadratic model for the parts per million of hydrogen present in core drillings at various distances from the base of a vacuum-cast ingot. The 'Distance' variable in C1 is first squared to create a new column C3, which is named 'SqDist':

LET C3 = C1*C1
NAME C3='SqDist'

TABLE B.6

```
MANN-WHITNEY CONFIDENCE INTERVAL AND TEST
C1            N = 10    MEDIAN =        39.00
C2            N = 10    MEDIAN =        35.50
POINT ESTIMATE FOR ETA1-ETA2 IS       4.00
95.5  PCT C.I. FOR ETA1-ETA2 IS ( -3.99,    12.00)
W =     121.5
TEST OF ETA1 = ETA2 VS. ETA1 N.E. ETA2 IS SIGNIFICANT AT
0.2265
CANNOT REJECT ALPHA AT 0.05
```

FIGURE B.1

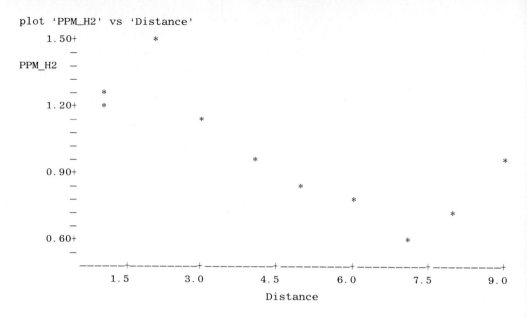

```
plot 'PPM_H2' vs 'Distance'
    1.50+              *
        -
PPM_H2  -
        -
        -     *
    1.20+     *
        -               *
        -
        -                      *                        *
    0.90+
        -                        *
        -                              *
        -                                    *
    0.60+                                          *
        -
         ------+---------+---------+---------+---------+---------+
             1.5       3.0       4.5       6.0       7.5       9.0
                               Distance
```

To make the case for the polynomial regression more convincing, we first plot 'PPM__H2' against 'Distance' as shown in Figure B.1.

The regression can now be performed as a multiple regression of 'PPM__H2' on 'Distance' and 'SqDist'. MINITAB output is shown in Table B.7.

As a final example we consider air pollution in 41 U.S. cities in the early 1970s as recorded in Federal Government statistics. The SO2 content of air in micrograms

TABLE B.7

REGRESS 'PPM_H2' on 2 variables 'Distance' and 'SqDist'

The regression equation is

PPM_H2 = 1.58 - 0.225 Distance + 0.0155 SqDist

Predictor	Coef	Stdev	t-ratio	p
Constant	1.5801	0.1658	9.53	0.000
Distance	-0.22504	0.08366	-2.69	0.031
SqDist	0.015491	0.008531	1.82	0.112

s = 0.1644   R-sq = 73.8%   R-sq(adj) = 66.3%

Analysis of Variance

SOURCE	DF	SS	MS	F	p
Regression	2	0.53162	0.26581	9.84	0.009
Error	7	0.18914	0.02702		
Total	9	0.72076			

SOURCE	DF	SEQ SS
Distance	1	0.44254
SqDist	1	0.08908

Unusual Observations

Obs.	Distance	PPM_H2	Fit	Stdev.Fit	Residual	St.Resid
2	2.00	1.5000	1.1920	0.0718	0.3080	2.08R

R denotes an obs. with a large st. resid.

TABLE B.8

```
REGRESS 'SO2' on 3 variables 'TEMP','MFG','POP70'

The regression equation is
SO2 = 58.6 - 0.593 TEMP + 0.0718 MFG - 0.0472 POP70
```

Predictor	Coef	Stdev	t-ratio	p
Constant	58.61	20.36	2.88	0.007
TEMP	-0.5927	0.3685	-1.61	0.116
MFG	0.07179	0.01598	4.49	0.000
POP70	-0.04716	0.01529	-3.08	0.004

```
s = 15.13     R-sq = 61.6%     R-sq(adj) = 58.5%
```

Analysis of Variance

SOURCE	DF	SS	MS	F	p
Regression	3	13571.2	4523.7	19.77	0.000
Error	37	8466.7	228.8		
Total	40	22037.9			

SOURCE	DF	SEQ SS
TEMP	1	4140.7
MFG	1	7254.1
POP70	1	2176.3

Unusual Observations

Obs.	TEMP	SO2	Fit	Stdev.Fit	Residual	St.Resid
11	50.6	110.00	109.81	12.55	0.19	0.02 X
30	50.4	61.00	29.13	3.31	31.87	2.16R
31	50.0	94.00	45.16	5.00	48.84	3.42R

```
R denotes an obs. with a large st. resid.
X denotes an obs. whose X value gives it large influence.
```

per cubic meter was entered into MINITAB in C1 and named 'SO2'; the regressor variables were:

C2: average annual temperature, named 'TEMP'
C3: number of manufacturing companies with 20 or more employees, named 'MFG'
C4: population (1970 census) in thousands, named 'POP70'

There are two approaches to the analysis on MINITAB: (1) specify the linear regression equation and estimate its coefficients by least squares, (2) use the stepwise method, which adds or removes variables one at a time, the ultimate equation being the result of computations. The latter approach is particularly suitable when numerous regressor variables are available. In this example we illustrate the former approach first. To specify the relationship between 'SO2' and the regressor variables, mention 'SO2' first in the REGRESS statement; specify the number, then the names of the regressor variables (or their columns). The MINITAB printout is shown in Table B.8.

For the second approach, the STEPWISE statement applies. The syntax is similar to REGRESS except that the number of regressor variables is omitted. The MINITAB printout is shown in Table B.9.

Note that the stepwise regression does not include the TEMP variable, as TEMP did not pass the default significance test for introduction of another variable. Also, the MINITAB output reveals that MFG was introduced into the equation in STEP 1, and POP70 was added to the equation in STEP 2. Subcommands FENTER = K and FREMOVE = K allow the user to modify the criteria for adding or removing variables in the equation to be estimated by STEPWISE (not shown above). Note too that R^2,

TABLE B.9

STEPWISE 'SO2' on columns C2–C4		
STEPWISE REGRESSION OF SO2 ON 3 PREDICTORS, WITH N = 41		
STEP	1	2
CONSTANT	17.62	26.42
MFG	0.0269	0.0830
T-RATIO	5.27	5.65
POP70		-0.057
T-RATIO		-4.00
S	18.2	15.4
R-SQ	41.62	58.89

the measure of regression "fit," in the two-variable stepwise equation is almost as good as R^2 in the three-variable multiple regression equation.

B.3 SAS

The advantages of SAS are that (1) its powerful data-manipulation facilities allow the statistician to handle larger and more complex data sets, and (2) SAS contains an extensive and growing library of statistical procedures. From the beginner's point of view, one must pay a price: it is more difficult to get started in statistical computing in SAS than in simpler systems like MINITAB. Nevertheless, some levels of statistical analysis in SAS are readily available to the student who is willing to spend only a short time learning to use SAS. We will illustrate SAS by means of two examples, one a paired-comparison t test and the other a multiple linear regression.

B.3.1 SAS Program Structure

Every SAS program has a number of statements telling SAS which data to use, which procedures to invoke, and which reports to print. These statements are separated by semicolons and may contain options modifying the procedures. A RUN statement is placed at the end of a program segment that the user wishes to execute. Normally, one such statement is used at the end of the program. SAS processes named data sets, which keep their names throughout a work session.

The DATA statement serves two purposes: (1) to announce that an input process is about to begin, and (2) to name the SAS data set. DATA is usually followed by an INPUT statement which names the variables to be processed and which may also contain format descriptions of data fields in the input stream. The data are introduced either by a file specification, which we shall not present here, or by means of a CARDS statement. The latter leads to direct input of data immediately following the CARDS statement. As an example of this method of data input, we show how SAS treats the input of a sample of battery lifetimes in hours. The result is an SAS data set called LIFE.

```
DATA LIFE;
  INPUT X @@;
  CARDS;
293 285 268 312 297 251 277 320 290 284
288 298 251 295 304;
```

Note (1) the use of semicolons as separators; (2) the naming of the one variable "X" in the INPUT statement; (3) the use of @@ at the end of the INPUT statement, indicating that the data are presented in a continuous stream (many to a line) rather than as a column (one to a line). To refer to the same data set later in the work session, the user only has to issue the DATA statement

```
DATA LIFE;
```

and SAS will provide the desired flow of battery lifetimes for whatever SAS procedure follows next.

B.3.2 Regression in SAS

The PC version, SAS/STAT Release 6.03, offers no less than seven alternative regression procedures: CATMOD, GLM, LIFEREG, NLIN, ORTHOREG, REG, and RSREG. Other modules of SAS contain yet more procedures which perform regression analysis on time-series data. For present purposes, we will provide a brief guide to the use of PROC REG. In SAS, all statistical procedures are called by a PROC _____ statement, with the name of the procedure substituted for _____. The procedure processes the last named data set, unless a DATA = _____ option is employed. For instance, the following segment calls for a multiple linear regression of Y on X1 and X2, which are variables associated with an existing SAS data set ABC:

```
DATA ABC;
PROC REG;
    MODEL Y = X1 X2;
RUN;
```

Note the use of the MODEL statement after PROC REG to specify the regression model, with an equals sign (=) separating the dependent from the independent variables. A standard output report results from the successful execution of PROC REG. The example which follows will show the general contents and layout of the regression output, but numerous options can be used to specify output details. The OUTPUT subcommand allows the user to create an SAS data set with the PREDICTED values and RESIDUALS in it for subsequent processing.

For our example, we use the data from Example 13.1 in the text. Percentage output (Y) of the chemical process is modeled as a linear function of the duration (X1) and percentage catalyst (X2). The data set is named CHEM1.

```
DATA CHEM1;
INPUT Y X1 X2;
CARDS;
    42.2   8   0.4
    34.5   8   0.3
    41.6   7   0.4
    33.7   7   0.3
    40.0   6   0.4
    30.1   6   0.3;
PROC REG;
    MODEL Y = X1 X2;
RUN;
```

Note that the data are organized into columns, and in this case the INPUT statement does not contain @@. After this program segment, SAS sends a report to the output section (for printing or display) which contains the regression coefficients,

PERCENT OUTPUT VS. DURATION & PERCENT CATALYST

Model: MODEL1
Dependent Variable: Y

Analysis of Variance

Source	DF	Sum of Squares	Mean Square	F Value	Prob > F
Model	2	119.265	59.6325	66.670	0.0033
Error	3	2.683	0.8944		
C Total	5	121.948			

Root MSE	0.94575	R-Square	0.9780	
Dep Mean	37.01667	Adj R-sq	0.9633	

Parameter Estimates

Variable	DF	Parameter Estimate	Standard Error	T for H0: Parameter=0	Prob > |T|
INTERCEP	1	-4.2833	4.2907	-0.998	0.3917
X1	1	1.6500	0.4728	3.489	0.0398
X2	1	85.0000	7.7220	11.007	0.0016

their t statistics, the regression analysis of variance, and so on. Table B.10 contains the output information.

Now suppose that the chemist suspects an interaction effect between duration and catalyst on the process yield. We might model this by creating a new variable $X3 = X1 * X2$ and then redoing the regression as below. But first we must create a new data set with the variable X3 in it. The OUTPUT subcommand is used to create the new data set CHEM2:

```
DATA CHEM2;
    SET CHEM1;
    X3 = X1 * X2;
    OUTPUT;
PROC REG;
    MODEL Y = X1 X2 X3 / SELECTION = STEPWISE;
RUN;
```

Note (1) the use of the SET statement to specify where the data for Y, X1, and X2 are to be found; (2) the equation defining $X3 = X1 * X2$; (3) the OUTPUT statement informing SAS to put all the variables from CHEM1 and X3 into CHEM2; (4) the SELECTION = STEPWISE option in the MODEL statement, which causes SAS to use a stepwise algorithm to enter the regressor variables one by one into the regression equation. The "/" in the MODEL statement is required to let SAS know that options are being used. As shown in Table B.11, when this stepwise regression is run, the interaction term fails the significance test for entering variables. Both X1 and X2 pass the significance test for entering variables, so the model is essentially unchanged.

B.3.3 t Test in SAS

The data set used here is the same as in Example 9.7 in the text. Two ammeters are compared by measuring electrical currents over a range of amperages. The t test is

TABLE B.11

PERCENT OUTPUT VS. DURATION & PERCENT CATALYST & INTERACTION

All variables in the model are significant at the 0.1500 level. No other variable met the 0.1500 significance level for entry into the model.

Summary of Stepwise Procedure for Dependent Variable Y

Step	Variable Entered Removed	Number In	Partial R**2	Model R**2	F	Prob > F
1	X2	1	0.8887	0.8887	31.9376	0.0048
2	X1	2	0.0893	0.9780	12.1752	0.0398

used to test whether the two devices are responding differently to the same environment. Data in amperes are input directly into an SAS data set called AMPS:

```
DATA AMPS;
    INPUT METER $ CURRENT @@;
    CARDS;
    A   12.33  B   12.55  A   2.45  B   2.47  A   6.26  B   6.42
    A   1.84   B   1.85   A   7.62  B   7.44  A  10.01  B  10.20
    A   12.02  B   12.21  A   4.11  B   4.16;
```

Note the use of the meter names A and B in the CARDS input. The $ following METER in the INPUT statement tells SAS to expect character input for this variable. Without the $, the INPUT statement would have tried to read the A or B as numeric data, which would have caused an error.

The actual t-test commands follow, beginning with PROC TTEST. The variable METER, used for pairing the data, is identified in a CLASS statement.

```
PROC TTEST;
    CLASS METER;
    VAR CURRENT;
    TITLE 'PAIRED COMPARISON OF TWO AMMETERS';
RUN;
```

Note the use of a TITLE statement to improve readability of the output. The chosen title stays in force until a new TITLE statement is encountered. The results of this t test are reported in standard output form in Table B.12.

TABLE B.12

PAIRED COMPARISONS OF TWO AMMETERS

TTEST PROCEDURE

Variable: CURRENT

METER	N	Mean	Std Dev	Std Error	Minimum	Maximum
A	8	7.0275	4.2052	1.4867	1.420	12.3300
B	8	7.1625	4.2022	1.4857	1.8500	12.5500

Variances	T	DF	Prob > |T|
Unequal	-0.0642	14.0	0.9497
Equal	-0.0642	14.0	0.9497

For H0: Variances are equal, F' = 1.00   DF = (7,7)   Prob > F' = 0.9986

SPSS was developed in the 1960s for mainframe computers as a statistical package for social scientists, who were particularly interested in its capability to cross-tabulate survey results and perform standard statistical tests on the data. A quarter-century later, this popular package is in widespread use at universities and research establishments. It was extended in the 1970s to SPSS-X: its statistical, data-manipulation, and graphics capabilities were enhanced so that it would be valuable to anyone requiring statistical analysis of data. In the 1980s the PC version made this package available to users of personal computers at home, in offices, and on the campus. It is this version which we discuss here.

B.4.1 Overview of SPSS/PC Commands

An overview of SPSS/PC commands is presented here; the reader is referred for further details to the SPSS User Guide. SPSS/PC is an interactive, command-driven system. Its commands let one manipulate and manage data, produce reports, and perform simple or complex statistical analyses. SPSS/PC performs these functions using three types of commands:

- Operation commands.
- Data definition commands.
- Procedure commands.

Operation commands affect the system rather than the data. They are used to get information about the way SPSS/PC works, or to change the way it operates:

- To get online assistance.
- To select SPSS/PC options.
- To submit SPSS/PC commands from an external file.
- To end an SPSS/PC session.

The following is a list of SPSS/PC operation commands:

Function	Commands	
Online Assistance	HELP	DISPLAY
Select Options	SET	
Submit Commands	INCLUDE	
End an SPSS/PC Session	FINISH	

Data-definition commands describe, label, and manipulate data. They are used to:

- Compute new variables.
- Change the values of existing variables.
- Identify missing values.
- Title and format output.
- Label variables and values of variables.
- Permanently or temporarily select cases to use in an analysis.

The following is a list of SPSS/PC data-definition commands:

Function	Commands
Data input	DATA LIST, BEGIN DATA, IMPORT, END DATA
Data transformations	RECORD, COUNT, COMPUTE, IF
Missing data	MISSING VALUE
Case selection	SELECT IF, SAMPLE
Weighting	PROCESS IF, WEIGHT
Output labeling	TITLE, VARIABLE LABELS
Formatting	SUBTITLE, VALUE LABELS

The DATA LIST command establishes an active system file that SPSS/PC uses throughout a work session. There can be only one DATA LIST per session, but one can create additional variables by transformations on the variables already defined. The new variables are created by the COMPUTE and COUNT commands. Variables can be labeled by the VARIABLE LABELS command. The active system file contains all the data values, labels, and dictionary information pertaining to a particular work session. To use it in a subsequent session, the user should SAVE it. In a subsequent session, it is retrieved by the GET command:

```
SAVE OUTFILE = 'MYWORK.SYS'.      <In first session>
GET FILE='MYWORK.SYS'.           <In second session>
```

Procedure commands tell SPSS/PC to produce reports, perform statistical analyses, and save data to files. For a given data set, there are procedures that:

- Print and plot the data.
- Produce descriptive statistics.
- Tabulate and analyze categorical data.
- Compare groups.
- Compute multivariate statistics.
- Calculate nonparametric statistical tests.

The following is a list of SPSS/PC procedure commands:

Function	Commands
Data display procedures	LIST, REPORT PLOT
Descriptive and categorical procedures	FREQUENCIES, DESCRIPTIVES CROSSTABS, HILOGLINEAR
Comparison of groups procedures	T-TEST, MEANS ONEWAY, ANOVA
Multivariate procedures	CORRELATION, REGRESSION FACTOR, CLUSTER
Nonparametric procedures	NPAR TESTS
Utility procedures	SORT CASES, WRITE EXPORT / IMPORT, GET / SAVE

The results of a procedure are automatically displayed on the screen after execution of the procedure. The SET command allows results to be sent to disk or printer as well. For instance, the command

```
SET PRINTER=ON/WIDTH=WIDE/LENGTH=59.
```

causes the results to be printed on wide paper with a page length of 59 lines. To send results to a disk file named 'RESULTS.LIS', use SET with DISK = ON or name a disk file by

```
SET DISK='RESULTS.LIS'.
```

After leaving SPSS/PC, one can print or edit the file with any text processor.

As an example, we consider the efficiency of three algorithms for solving a given problem as measured in run time (in seconds) on four different computers. The experiment is a completely randomized block design as described in Table 14.2 of the text. The command for entering the data into SPSS/PC is

```
DATA LIST / ALGORTM 1-2, COMPUTER 4-5, RUNTIME 7-11.
BEGIN DATA.
1   1   6.42
1   2   10.51
1   3   4.00
1   4   6.20
2   1   6.85
2   2   11.45
2   3   4.12
2   4   6.72
3   1   5.60
3   2   9.50
3   3   3.94
3   4   5.66
END DATA.

VARIABLE LABELS
ALGORTM 'ALGORITHM TESTED FOR EFFICIENCY',
COMPUTER 'CHOICES OF HARDWARE',
RUNTIME 'RUN TIME IN SECONDS'.
```

Note the use of periods as statement separators, the "/" for separating segments within a statement, and the column numbers in the DATA LIST statement to tell SPSS/PC where to locate the tabular data which follow. While the names of variables are limited to eight characters, the VARIABLE LABELS statement allows the user to label these quantities appropriately so as to produce readable reports. Each variable name in the list is followed by a description of up to 60 characters in length in single quotes.

Alternatively, one can use the DATA LIST statement with a FILE subcommand to introduce data that have been filed earlier:

```
DATA LIST FILE='SOMEDATA.DAT' / ALGORTM 1-2, COMPUTER 4-5, RUNTIME 7-11.
```

The data in the prepared file should be arranged in fixed format, and the variable names in the DATA LIST command are followed by the column numbers just as in the previous example.

The values of the variables can be examined at any time during a session by means of the LIST command. The following example shows how to list the variables ALGORTM, COMPUTER, and RUNTIME, which were previously created by a DATA LIST command.

```
LIST VARIABLES=ALGORTM, COMPUTER, RUNTIME.
```

Each procedure produces a report on its results. This report is normally displayed on the screen, but one can use the SET RESULTS= command to obtain disk files of all results as well:

```
SET RESULTS='MYREPORT.PRC'.
```

Another output command, WRITE, used with the SET command causes all the data values to be written to a disk file. In WRITE, the desired variables written out to disk are named:

```
SET RESULTS='SOMEDATA.DAT'.
WRITE VARIABLES=ALGORTM, COMPUTER, RUNTIME.
```

This will write only the data values, not the data dictionary to the file 'SOMEDATA.DAT'. SPSS/PC also has extensive reporting capabilities, which are thoroughly described in the SPSS user's manual.

B.4.2 ANOVA in SPSS/PC

We now give an example of analysis of variance using SPSS/PC. The example uses the previously discussed data on algorithm efficiency and applies ANOVA, one of the SPSS procedures for analysis of variance. To use the ANOVA in SPSS with the randomized block data of Table 14.2, one must specify the variable names:

```
ANOVA /VARIABLES= ALGORTM, COMPUTER, RUNTIME.
```

The results in standard ANOVA output form are shown in Table B.13.

TABLE B.13

SPSS/PC+

* * * ANALYSIS OF VARIANCE * * *

RUNTIME
BY   ALGORTM
COMPUTER

Source of Variation	Sum of Squares	DF	Mean Square	F	Signif of F
Main Effects	68.283	5	13.656	100.4	.00
ALGORTM	2.472	2	1.236	9.1	.02
COMPUTER	65.811	3	21.937	161.3	.00
Explained	68.283	5	13.656	100.4	.00
Residual	0.816	6	0.136		
Total	69.098	11	6.281		

12 cases were processed.
0 CASES (.0 PCT) were missing.

||||||||||||||||||||||||||||||||||||||||||||||||||||||||||||||||||||||||||||||||||||||||||||||||||||

ANSWERS TO PROBLEMS

Numerical solutions depend on the accuracy maintained during computation. Consequently, some discrepancy might exist between your solution and the one given in the text. This problem is more likely to occur later in the text where we have employed computer software to solve many of the exercises. Not only does the software provide great computational accuracy, it also employs statistical tables possessing greater accuracy than the ones given in the text. The latter difference is most relevant when interpolation is required.

CHAPTER 1

1.1, 1.6.

Class	Class Mark	FREQ	CFD	RFD	CRFD	NRFD
32.85–34.75	33.8	1	1	0.013	0.013	0.007
34.75–36.65	35.7	0	1	0.000	0.013	0.000
36.65–38.55	37.6	4	5	0.053	0.067	0.028
38.55–40.45	39.5	6	11	0.080	0.147	0.042
40.45–42.35	41.4	12	23	0.160	0.307	0.084
42.35–44.25	43.3	22	45	0.293	0.600	0.154
44.25–46.15	45.2	22	67	0.293	0.893	0.154
46.15–48.05	47.1	4	71	0.053	0.947	0.028
48.05–49.95	49.0	3	74	0.040	0.987	0.021
49.95–51.85	50.9	1	75	0.013	1.000	0.007

1.2, 1.7.

Class	Class Mark	FREQ	CFD	RFD	CRFD	NRFD
5.15–5.65	5.4	3	3	0.038	0.038	0.075
5.65–6.15	5.9	10	13	0.125	0.163	0.250
6.15–6.65	6.4	8	21	0.100	0.263	0.200
6.65–7.15	6.9	11	32	0.138	0.400	0.275
7.15–7.65	7.4	20	52	0.250	0.650	0.500
7.65–8.15	7.9	11	63	0.138	0.788	0.275
8.15–8.65	8.4	9	72	0.113	0.900	0.225
8.65–9.15	8.9	6	78	0.075	0.975	0.150
9.15–9.65	9.4	2	80	0.025	1.000	0.050

1.3, 1.8.

Class	Class Mark	FREQ	CFD	RFD	CRFD	NRFD
1.5– 8.5	5	6	6	0.086	0.086	0.012
8.5–15.5	12	8	14	0.114	0.200	0.016
15.5–22.5	19	13	27	0.186	0.386	0.027
22.5–29.5	26	21	48	0.300	0.686	0.043
29.5–36.5	33	10	58	0.143	0.829	0.020
36.5–43.5	40	11	69	0.157	0.986	0.022
43.5–50.5	47	1	70	0.014	1.000	0.002

1.4, 1.9.

Class	Class Mark	FREQ	CFD	RFD	CRFD	NRFD
0.035–0.125	0.08	4	4	0.077	0.077	0.855
0.125–0.215	0.17	16	20	0.308	0.385	3.419
0.215–0.305	0.26	26	46	0.500	0.885	5.556
0.305–0.395	0.35	6	52	0.115	1.000	1.282

1.5, 1.10.

Class	Class Mark	FREQ	CFD	RFD	CRFD	NRFD
−0.5–0.5	0	52	52	0.481	0.481	0.481
0.5–1.5	1	35	87	0.325	0.806	0.324
1.5–2.5	2	14	101	0.130	0.935	0.130
2.5–3.5	3	5	106	0.046	0.981	0.046
3.5–4.5	4	1	107	0.009	0.991	0.009
4.5–5.5	5	1	108	0.009	1.000	0.009

1.11. **a)** 0.133   **b)** 0.173   **c)** 0.093   **d)** 0.107   **1.13.** **a)** 0.386   **b)** 0.371   **c)** 0.177   **d)** 0.246

1.15. $\bar{x} = 24.7$, $\tilde{x} = 24.5$ **1.17.** $\bar{x} = 43.351$, $\tilde{x} = 43.602$ **1.19.** $\bar{x} = 0.229$, $\tilde{x} = 0.236$

1.21. $s^2 = 111.054$, $V = 42.7\%$ **1.23.** $s^2 = 8.681$, $V = 6.8\%$ **1.25.** $s^2 = 0.00505$, $V = 31.0\%$

1.27. $M.D. = 2.138$

CHAPTER 2

2.1. acceptable: a, c, d; unacceptable: b, e, f

2.3. **a)** S
  b) {2, 4, 6, . . . , 30, 3, 5, 7, 11, 13, 17, 19, 23, 29}
  c) (2, 4, 6, . . . , 30, 21, 23, 25, 27, 29}
  d) {2, 1, 3, 5, . . . , 29}
  e) {1, 9, 15, 21, 25, 27}
  f) {23, 29}
  g) {22, 24, 26, 28, 30}
  h) F
  i) {1, 3, 4, 5, 6, . . . , 30}
  j) same as i

2.5. **a)** {(1, 2), (1, 3), (1, 4), (1, 5), (1, 6), (2, 3), (2, 4), (2, 5), (2, 6), (3, 4), (3, 5), (3, 6), (4, 5), (4, 6), (5, 6)}
  b) {(1, 1), (2, 2), (3, 3), (4, 4), (5, 5), (6, 6)}
  c) {(1, 1)}

2.7. Two possibilities are $\{A, B, C\}^{20}$ and

$$\{(i, j, k): i + j + k = 20 \text{ and } 0 \le i, j, k \le 20\}$$

2.9. Let S be the union of the following sets:

$$\{s^5\}, \{us^5\}, \{s^m us^n: m, n \ge 1 \text{ and } m + n = 5\},$$
$$\{s^m us^n u: m, n \ge 0 \text{ and } m + n \le 4\}$$

2.13. a) $E \cup F \cup G$
  b) $(E \cup F \cup G)^c$
  c) $E \cap F \cap G$
  d) $E \cap F \cap G^c$
  e) $(E \cap F^c \cap G^c) \cup (E^c \cap F \cap G^c) \cup (E^c \cap F^c \cap G)$
  f) $(E \cap F \cap G^c) \cup (E \cap F^c \cap G) \cup (E^c \cap F \cap G)$
  g) $(E \cap F \cap G)^c$
  h) $E \cap (F \cup G)^c$
  i) $E \cap F$
  j) $E \cup F$

2.15. a) $\frac{2}{9}$ b) $\frac{1}{9}$ c) $\frac{1}{3}$ d) $\frac{5}{9}$

2.17. $P(E) = \frac{5}{9}, P(F) = \frac{2}{3}, P(G) = \frac{1}{3}, P(H) = \frac{1}{3}$
If the dart lands in a darkly shaded region 90 times, we might be tempted to conclude that the dart is not falling uniformly randomly across the board.

2.19. $\frac{3}{8}$ **2.21.** a) 126 b) 5550 c) 35 d) 187,460 e) 2520

2.23. Assuming order counts, the robot can pick up the items in $10! = 3,628,800$ ways. Under the given constraint, the robot can pick up the items in $(2)(6!)(4!) = 34,560$ ways.

2.25. $3!4!6! = 103,680$

2.27. There are a total of $C(12, 3) = 220$ site combinations; under the restriction, there are $4^3 = 64$ ways.

2.29. 162

2.31. a) $P(25, 4) = 303,600$
  b) $P(21, 5) = 2,441,880$
  c) $21 \times 20 \times 24 \times 23 \times 22 = 5,100,480$
  d) $5! = 120$
  e) 85,800
  f) $5 \times 4 \times 3 \times 21 \times 20 = 25,200$
  g) $5 \times 21 \times 4 \times 20 \times 3 + 21 \times 5 \times 20 \times 4 \times 19 = 184,800$

2.33. $C(7, 2)C(5, 2)/C(12, 4) = 0.4242$ **2.35.** $C(12, 7)C(5, 2)C(3, 2)C(1, 1) = 23,760$

2.37. a) $C(5, 3)/C(9, 3) = 0.1190$
  b) $C(5, 1)C(4, 2)/C(9, 3) = 0.3571$
  c) $[C(5, 2)C(4, 1) + C(5, 3)C(4, 0)]/C(9, 3) = 0.5952$

2.39. Without the restriction: $C(10, 4)C(6, 4) = 3150$
With the restriction:

$$2C(8, 3)C(5, 3) + 4C(8, 3)C(5, 4) + C(8, 4) = 2310$$

(Note that we have made the tacit assumption that the teams are differentiated with respect to some criterion, such as task.)

2.41. a) $a^6 + 6a^5b + 15a^4b^2 + 20a^3b^3 + 15a^2b^4 + 6ab^5 + b^6$
  b) $1 + 4a + 6a^2 + 4a^3 + a^4$
  c) $1 - 4a + 6a^2 - 4a^3 + a^4$
  d) $a^4 + 8a^3b + 24a^2b^2 + 32ab^3 + 16b^4$
  e) $a^4 + 2a^3b + 3a^2b^2/2 + ab^3/2 + b^4/16$

2.43. a) $P(E \cap F) = 0.08$
  b) $P(E \cup F) = 0.19$
  c) $P[(E \cup F)^c] = 0.81$
  d) $P(E^c) = 0.85$
  e) $P(E - F) = 0.07$

2.45. a) $\frac{1}{8}$ b) $\frac{1}{8}$ c) $\frac{3}{8}$ d) $\frac{3}{8}$ **2.47.** a) 0.10737 b) 0.000000102 c) 0.89263 d) 0.26844

2.49. **a)** $C(4, 2)/C(52, 2) = \frac{1}{221}$
   b) $2C(4, 2)/C(52, 2) = \frac{2}{221}$
   c) $C(12, 2)/C(52, 2) = \frac{11}{221}$
   d) $C(4, 1)C(4, 1)/C(52, 2) = \frac{8}{663}$
   e) $C(4, 1)C(48, 1)/C(52, 2) = \frac{32}{221}$
   f) $1 - C(48, 2)/C(52, 2) = \frac{33}{221}$

2.51. **a)** $\frac{4}{50}$ **b)** $\frac{27}{50}$ **c)** $\frac{3}{50}$ **d)** $\frac{16}{50}$ **e)** $\frac{32}{50}$ **f)** $\frac{12}{50}$ **2.53.** 0.14

2.57. **a)** $\frac{3}{8}$ **b)** $\frac{3}{8}$ **c)** $\frac{6}{15}$ **d)** $\frac{6}{15}$ **e)** $\frac{7}{8}$ **f)** 1 **g)** $\frac{14}{15}$ **h)** 1 **i)** 0 **2.59.** **a)** $\frac{1}{4}$ **b)** $\frac{1}{20}$ **c)** $\frac{19}{20}$ **d)** $\frac{37}{40}$

2.61. 0.9312 **2.63.** 0.06

2.65. $\dfrac{4}{10} \times \dfrac{3}{9} \times \dfrac{2}{8} = \dfrac{1}{30}$

2.67. 0.0205 **2.69.** $\frac{1}{2}$

2.71. $P(\text{actually acceptable} \mid \text{deemed unacceptable}) = 0.43114$
   $P(\text{actually unacceptable} \mid \text{deemed acceptable} = 0.00214$

2.73. $\frac{1}{2}$

2.75. Let E, F, and G denote a sum of 7, a sum of 6, and a two on the first die, respectively. Then

$$P(E \cap G) = \tfrac{1}{36} = (\tfrac{1}{6})(\tfrac{1}{6}) = P(E)P(G)$$

but

$$P(F \cap G) = \tfrac{1}{36} \neq (\tfrac{5}{36})(\tfrac{1}{6}) = P(F)P(G)$$

2.77. [for Exercise 2.60] **a)** 0.16 **b)** 0.48 **c)** 0.64
   [for Exercise 2.65] $(0.4)(0.4)(0.4) = 0.064$

2.79. **a)** 0.9999721 **b)** 0.0079721 **c)** 0.000001 **d)** 0.00002783

2.81. $P(\text{exactly one head}) = 0.0768$,
   $P(\text{at least one head}) = 0.98976$

2.85. 0.07987 **2.87.** 0.999998503

CHAPTER 3

3.1. $\Omega_X = \{2, 3, 4, \ldots, 12\}$

x	2	3	4	5	6	7	8	9	10	11	12
$P(X = x)$	$\dfrac{1}{36}$	$\dfrac{2}{36}$	$\dfrac{3}{36}$	$\dfrac{4}{36}$	$\dfrac{5}{36}$	$\dfrac{6}{36}$	$\dfrac{5}{36}$	$\dfrac{4}{36}$	$\dfrac{3}{36}$	$\dfrac{2}{36}$	$\dfrac{1}{36}$

   a) $\frac{1}{3}$ **b)** $\frac{7}{12}$ **c)** $\frac{1}{6}$ **d)** $\frac{1}{6}$

3.3. 0.1846 **3.5.** **a)** 0.135 **b)** 0.524 **c)** 0

3.9.

x	0	1	2
$f(x)$	0.55263	0.39474	0.05263

3.11. **a)** 0.18394 **b)** 0.26424 **3.13.** $c = \frac{1}{16}$, 0.909796 **3.15.** **a)** $\frac{1}{2}$ **b)** $\frac{11}{16}$

3.17.

$$F(x) = \sum_{r \leq x} C(5, r)/32$$

x	0	1	2	3	4	5
$F(x)$	$\dfrac{1}{32}$	$\dfrac{6}{32}$	$\dfrac{16}{32}$	$\dfrac{26}{32}$	$\dfrac{31}{32}$	$\dfrac{32}{32}$

The table is interpreted in the following manner: if $y < 0$, then $F(y) = 0$; if $x - 1 \leq y < x$, then $F(y) = F(x - 1)$; and if $y \geq 5$, then $F(y) = 1$.

3.19.

$$F(x) = e^{-1} \sum_{r \le x} 1/r!$$

x	0	1	2	3	$\cdots$
$F(x)$	0.3679	0.7358	0.9197	0.9810	$\cdots$

This table is to be interpreted in a manner analogous to the table of Exercise 3.17.

3.21.

$$F(x) = \begin{cases} 0, & \text{if } x < 3.8 \\ 25(x^2/2 - 3.8x + 7.22), & \text{if } 3.8 \le x < 4 \\ 0.5 - 25(x^2/2 - 4.2x + 8.8), & \text{if } 4 \le x \le 4.2 \\ 1, & \text{if } x > 4.2 \end{cases}$$

3.23. $f(x) = 4x^{-5}$ for $x > 1$ and $f(x) = 0$ for $x < 1$

3.25.

x	$-\frac{1}{2}$	0	$\frac{1}{4}$	$\frac{1}{2}$	1	$\frac{5}{4}$
$f(x)$	$\frac{1}{8}$	$\frac{1}{8}$	$\frac{1}{8}$	$\frac{1}{8}$	$\frac{1}{4}$	$\frac{1}{4}$

3.27. 0.02955

3.29. $f_Y(y) = f(\log y)$
For Exercise 3.2:

y	1	e	e^2	e^3	e^4	e^5
$f(y)$	$\dfrac{1}{32}$	$\dfrac{5}{32}$	$\dfrac{10}{32}$	$\dfrac{10}{32}$	$\dfrac{5}{32}$	$\dfrac{1}{32}$

For Exercise 3.8:

y	e	e^2	e^3	e^4	e^5	e^6	e^7
$f(y)$	0.52	0.27	0.11	0.05	0.02	0.02	0.01

3.31.

$$f_Y(y) = \frac{1}{e(\sqrt{y})!}$$

3.33.

$$f_Y(y) = \begin{cases} y^{-1/2}/5, & \text{if } 0 < y \le 4 \\ y^{-1/2}/10, & \text{if } 4 < y \le 9 \end{cases}$$

3.35. $f_Y(y) = |a|^{-1}\{f_X[(y - b)/a] + f_X[(b - y)/a]\}$

3.37. For a discrete random variable X:

y	$f(y)$
0	$F_X(0) - f_X(0)$
1	$F_X(2) - F_X(0) - f_X(2) + f_X(0)$
2	$F_X(4) - F_X(2) - f_X(4) + f_X(2)$
3	$1 - F_X(4) + f_X(4)$

For a continuous random variable, drop all f_X terms. For the specified uniform random variable:

y	0	1	2	3
$f_Y(y)$	0.2	0.2	0.2	0.4

3.39. In general, $F_Y(y) = F_X(e^y)$. For the exponential density,

$$F_Y(y) = 1 - \exp[-be^y], \text{ so that } f_Y(y) = be^y \exp[-be^y].$$

3.41. 1.88   **3.43.** 8   **3.45.** 0   **3.47.** $\frac{47}{81}$ **3.49.** a) 1   b) $4\pi^2/3$ c) $(e^{2\pi} - e^{-2\pi})/4\pi$ d) 0

3.51. $\frac{3}{8}$ **3.53.** $\mu = 2.5$, $\mu_2' = 7.5$, $\mu_3' = 25$, $\sigma^2 = 1.25$, $\sigma = 1.118$ **3.55.** $\mu = 1$, $\mu_2' = 2$, $\mu_3' = 5$, $\sigma^2 = 1$, $\sigma = 1$

3.57. $\mu = 4$, $\mu'_2 = 2401/150$, $\mu'_3 = 1602/25$, $\sigma^2 = 1/150$, $\sigma = 0.08165$

3.59. $\mu = \frac{4}{3}$, $\mu'_2 = 2$, $\mu'_3 = 4$, $\sigma^2 = \frac{2}{9}$, $\sigma = 0.4714$

3.63. **a)** For $x = 1, 2, 3, 4,$

$$f(x) = \frac{C(4, x)C(3, 4 - x)}{C(7, 4)}$$

    b) $\mu = \frac{16}{7}$, $\mu'_2 = \frac{40}{7}$
    c) $\sigma^2 = \frac{24}{49}$

3.65. For the exponential distribution, $MD(X) = 2/be$; for the uniform distribution, $MD(X) = (b - a)/4$.

3.67. For the exponential distribution, $V = 1$; for the uniform distribution, $V = 2(b - a)/\sqrt{12}(a + b)$.

3.69. **a)** 2 **b)** 0 **c)** -2.099603 **d)** 0

3.71. For the exponential distribution, $x_{0.10} = 0.1054$ and $x_{0.20} = 0.2231$; for the uniform distribution, $x_{0.10} = 0.1$ and $x_{0.20} = 0.2$.

3.73. Let $P(t) = P(|X - \mu| < t)$ and $P'(t)$ be the bound on $P(t)$ given by Chebyshev's inequality.
    a)

$$P(t) = \begin{cases} e^{-(1 - bt)} - e^{-(1 + bt)}, & \text{if } t \le 1/b \\ 1 - e^{-(1 + bt)}, & \text{if } t \ge 1/b \end{cases}$$
$$P'(t) = 1 - (tb)^{-2}$$

    b)

$$P(t) = \begin{cases} 1, & \text{if } t \ge (b - a)/2 \\ 2t/(b - a), & \text{if } t \le (b - a)/2 \end{cases}$$
$$P'(t) = 1 - (b - a)^2/12t^2$$

    c)

$$P(t) = \begin{cases} 0, & \text{if } t \le \frac{1}{2} \\ \frac{5}{8}, & \text{if } \frac{1}{2} < t \le \frac{3}{2} \\ \frac{15}{16}, & \text{if } \frac{3}{2} < t \le \frac{5}{2} \\ 1, & \text{if } t > \frac{5}{2} \end{cases}$$

$$P'(t) = 1 - \frac{5}{4}t^2$$

    d)

$$P(t) = \begin{cases} 1, & \text{if } t \ge 0.2 \\ -50[(4 + t)^2/2 - 4.2(4 + t) + 8.8], & \text{if } t \le 0.2 \end{cases}$$
$$P'(t) = 1 - 0.006575/t^2$$

    For each of the parts, the alternate form of Chebyshev's inequality results from letting $t = k\sigma$.

3.77. $M_X(t) = (\frac{1}{4} - t)^{-2}/16$ for $t < \frac{1}{4}$, $\mu = 8$, $\mu'_2 = 96$

3.79. $M_X(t) = b^2e^{at}/(b^2 - t^2)$ for $-b < t < b$, $\mu = a$, $\mu'_2 = a^2 + 2/b^2$

3.81. $M_X(t) = [(0.1)e^t + 0.9]^5$, $\mu = 0.5$, $\mu'_2 = 0.7$

CHAPTER 4

4.1. **a)** 0.8358 **b)** 0.0611 **c)** 0.0138 **d)** 0.5409   **4.3.** **a)** 0.9987 **b)** 0.3917 **c)** 14 **d)** 0.9753

4.5 P(at least 5 fail to compile) $= 0.1734$, $E[X] = 22$, Var $[X] = 2.64$

4.7. E[bit errors] $= 10$, Standard deviation $= 3.1607$ **4.9.** 0.55869

4.17. $f(0) = 0.117649$, $f(1) = 0.302526$, $f(2) = 0.324135$, $f(3) = 0.185220$, $f(4) = 0.059535$, $f(5) = 0.010206$, $f(6) = 0.000729$

4.19. $f(3) = 0.091$, $f(4) = 0.409$, $f(5) = 0.409$, $f(6) = 0.091$

4.21. Expected value $= 4.5$,
    P(no more than 3 formatted) $= 0.13132$

4.23. $E[X] = 2$, Var $[X] = 0.6667$, $P(|X - \mu| \geq 2\sigma) = 0.0476$.
Chebyshev's inequality gives an upper bound of 0.25 on the desired probability.

4.25. Expected value $= 10$, Variance $= 9.6$

4.27. The first and second desired probabilities are 0.0191 and 0.1650, respectively.

4.29. Expected value $= 141.9$, Variance $= 20,265$ **4.31.**   a) 0.07948   b) 0.98217   c) 10.00

4.33. For $p = \frac{1}{2}$, $\alpha_3 = 2.1213$ **4.35.** $E[X] = 3$, $P(X > 5) = 0.0839$

4.37.   a) 18   b) 0.2693   c) 0.2875   **4.39.** $E[X] = 15$, $P(12 < X < 20) = 0.6076$

4.41.   0.7619

4.47.   a) 0.0392   b) 0.0643   c) 0.8110   d) 0.9974
      e) 0.3259   f) 0.6210   g) 0.0021   h) 0.6050

4.51.   81.45%   **4.53.** $P(2 < X < 4) = 0.9200$, $P(|X - \mu| < 1) = 0.9931$

4.55. $P(X < 650) = 0.1151$, $P(625 < X < 675) = 0.5128$ **4.57.**   a) 64.78%   b) 103.247   c) 0.085

4.59.   2.80   **4.61.**   0.8297   **4.63.**   0.9431 (without correction), 0.9591 (with correction)

4.65. $(0.675)\sigma$ **4.71.** $P(T > 1) = 0.2705$, $P(1 < T < 2) = 0.2155$

4.73. $P(T < 2) = 0.9502$, $P(2 < T < 3) = 0.0387$ **4.75.** $P(T < 15) = 0.348$, $E[T] = 17.6$

4.77.   Time T to eighth arrival is an Erlang distribution.
      $E[T] = 1.3333$, Var $[T] = 0.2222$, $P(T < 2) = 0.911$

4.83. $P(0.2 < X < 0.8) = 0.792$, $E[X] = 0.40$ **4.85.**   0.462903

4.87. $P(X > 3) = 0.5391$, $E[X] = 3.0$ **4.91.**   0.47655

4.93.   a) 0.7506   b) 0.7520   c) 0.3931

CHAPTER 5

5.1.   a)

		\multicolumn{5}{c}{y}	
		0	1	2	3	4	Marginal
x	0	0	0	0.04762	0.09524	0.02381	0.16667
	1	0	0.07143	0.28571	0.14286	0	0.50000
	2	0.01429	0.14286	0.14286	0	0	0.30000
	3	0.00952	0.02381	0	0	0	0.03333
Marginal		0.02381	0.23810	0.47619	0.23810	0.02381	

b) For $x + y \geq 2$ and $x + y \leq 4$,

$$\frac{\binom{3}{x}\binom{5}{y}\binom{2}{4 - x - y}}{\binom{10}{4}}$$

   c) 0.96667   d) 0.66667   e) 0.68096   f) 0.59524   g) 0.69047

5.3.   a) For $x + y + u + v = 5$, $x + y + v \geq 3$, and $x + u + v \geq 1$,

$$f(x, y, u, v) = \frac{\binom{6}{x}\binom{4}{y}\binom{2}{u}\binom{8}{v}}{\binom{20}{5}}$$

b) 0.3522   c) 0.03199   d) 0.69350   e) 0.47678

5.5. To find $f_{U,V}(u, v)$, divide the values in the following table by 64:

		0	1	2	*u* 3	4	5	6
v	0	1	5	10	10	5	1	0
	1	0	1	5	10	10	5	1

Moreover, $P(U \leq V + 1) = \frac{3}{16}$.

5.7.

x	0	1	2	3	
$f_X(x)$	0.1667	0.5000	0.3000	0.0333	

y	0	1	2	3	4
$f_Y(y)$	0.0238	0.2381	0.4762	0.2381	0.0238

5.9.

$$f_X(x) = \frac{\binom{6}{x}}{\binom{20}{5}} \sum \sum \sum \binom{4}{y}\binom{2}{u}\binom{8}{v} = \frac{\binom{6}{x}\binom{14}{5-x}}{\binom{20}{5}}$$

5.11.

u	0	1	2	3	4	5	6
$f_U(u)$	$\frac{1}{64}$	$\frac{6}{64}$	$\frac{15}{64}$	$\frac{20}{64}$	$\frac{15}{64}$	$\frac{6}{64}$	$\frac{1}{64}$

v	0	1
$f_V(v)$	$\frac{1}{2}$	$\frac{1}{2}$

5.13. $f_X(0) = 0.6667, f_X(1) = 0.3333, f_Y(0) = 0.4815, f_Y(1) = 0.5185, f_W(0) = 0.4815, f_W(1) = 0.5185$

5.15. a) $\frac{1}{2}$ b) $\frac{3}{8}$ c) $\frac{9}{16}$ d) 0.58502   **5.17.** a) $\frac{1}{2}$ b) $1/\sqrt{512}$ c) $\frac{3}{4}\pi$

5.19. The value of c is 8.
a) 0.0183 b) $\frac{2}{3}$ c) 0.7476

5.21. $c = 1$

5.23.

$$f_X(x) = \begin{cases} 1, & \text{if } 0 \leq x < \frac{1}{2} \\ \frac{3}{2} - x, & \text{if } \frac{1}{2} < x \leq \frac{3}{2} \\ 0, & \text{otherwise} \end{cases}$$

$$f_Y(y) = \begin{cases} \frac{3}{2} - y, & \text{if } 0 \leq y \leq 1 \\ 0, & \text{otherwise} \end{cases}$$

5.25.

$$f_X(x) = \begin{cases} x/3, & \text{if } 0 \leq x \leq 2 \\ 2 - 2x/3, & \text{if } 2 < x \leq 3 \\ 0, & \text{otherwise} \end{cases}$$

$$f_Y(y) = \begin{cases} 1 - y/2, & \text{if } 0 \leq y \leq 2 \\ 0, & \text{otherwise} \end{cases}$$

5.27. $f_X(x) = (3x^2 + 9x/2)/17$, if $0 \leq x \leq 2; f_X(x) = 0$, otherwise
$f_Y(y) = (\frac{8}{3} + 2y)/17$, if $0 \leq y \leq 3; f_Y(y) = 0$, otherwise

5.29.

	(0, 0)	(0, 1)	(1, 0)	(1, 1)
X, Y	0.259	0.407	0.222	0.111
Y, W	0.111	0.370	0.370	0.148
X, W	0.370	0.296	0.111	0.222

5.31. Densities are zero except for nonnegative values of the variables:

$$f_{X,Y}(x, y) = 2e^{-(2x+y)}, f_{X,W}(y, w) = 6e^{-(2x+3w)},$$
$$f_{Y,W}(x, w) = 3e^{-(y+3w)}, f_X(x) = 2e^{-2x}, f_Y(y) = e^{-y},$$
$$f_W(w) = 3e^{-3w}$$

5.33.

$$F(x, y) = \begin{cases} xy, & \text{if } 0 \le x \le 2, 0 \le y \le 1 \\ 1, & \text{if } x > 2, y > 1 \\ 0, & \text{otherwise} \end{cases}$$

5.35. $F(x, y) = (1 - e^{-2x})(1 - e^{-4y})$, if $x, y \ge 0$; $F(x, y) = 0$, otherwise

5.37. The variables are dependent. In Exercise 5.2, $f(0, 0) = 0.0016$, but $f_X(0)f_Y(0) = 0.01501$.

5.39. independent   **5.41.** independent

5.43.

$$\frac{1}{72.7114} \exp\left\{-(\tfrac{1}{2})[(x - 69.3)^2/6.2 + (y - 24.4)^2/21.6]\right\}$$

5.45. For $0 \le x \le 4$ and $0 \le y \le 1$,

$$f(x, y) = \binom{4}{x}\binom{1}{y}(0.3)^{x+y}(0.7)^{5-(x+y)}$$

5.47. a)

z	2	3	4
$f(z)$	0.133	0.533	0.333

b)

z	-4	-3	-2	-1	0	1	2	3
$f(z)$	0.024	0.095	0.190	0.286	0.214	0.143	0.038	0.010

c)

z	0	1	2	3	4
$f(z)$	0.190	0.071	0.429	0.167	0.143

5.49.

z	0	1	2	3	4	5	6
$f(z)$	0.14	0.27	0.26	0.17	0.10	0.05	0.01

5.53. 0.3113   **5.55.** 0.80854

5.57. The distribution of the average is $N(90.4, 1.703)$. The variance is reduced by a factor of 2.

5.59. The distribution of the sum is $N(36.0, 0.123693)$ and the desired probability is 0.9847.

5.61. $f(2, 0) = 0.0619, f(3, 0) = 0.1048, f(4, 0) = 0.0238$
$f(2, 1) = 0.0714, f(3, 2) = 0.4286, f(4, 3) = 0.1667$
$f(4, 4) = 0.1429$

5.63. $f(0, 0) = 0.00463, f(1, -1) = 0.02778, f(2, -2) = 0.05556, f(3, -3) = 0.03704, f(4, -1) = 0.16667, f(3, 0) = 0.16667, f(2, 1) = 0.04167, f(4, 2) = 0.125, f(5, 1) = 0.25, f(6, 3) = 0.125$

5.65. If v and $v/3$ are integers, $u \ge v/3$, and $u - v/3$ is even, then

$$f(u, v) = \frac{e^{-(\lambda_1+\lambda_2)}\lambda_1^{v/3}\lambda_2^{[u-(v/3)]/2}}{(v/3)![(u - v/3)/2]!}$$

Otherwise, $f(u, v) = 0$.

5.67. For $0 < u + v < 2$ and $0 < u - v < 2$, $f(u, v) = (2 - u)/2$; otherwise, $f(u, v) = 0$.

5.69.

$$f(u, v) = (2\pi)^{-1} e^{-[v^2+(u-v^2)]/2}$$

5.71. 18.6667   **5.73.** $(n^2p^2 + np - np^2)^2$ **5.75.** 24.542

5.77. Cov $[X, Y] = -0.5$, Corr $[X, Y] = -0.7071$ **5.79.** 0.15894

5.81. Cov $[X, Y] = 0$, Corr $[X, Y] = 0$

5.83. $f(2 \mid 0) = 0.2857, f(3 \mid 0) = 0.5714, f(4 \mid 0) = 0.1429,$
$f(1 \mid 1) = 0.1429, f(2 \mid 1) = 0.5714, f(3 \mid 1) = 0.2857,$
$f(0 \mid 2) = 0.0476, f(1 \mid 2) = 0.4762, f(2 \mid 2) = 0.4762,$
$f(0 \mid 3) = 0.2857, f(1 \mid 3) = 0.7143$

5.85. For $0 \le x < 2$, $f(y \mid x) = 1/(1 - x/2)$ if $0 \le y \le 1 - x/2$, and $f(y \mid x) = 0$, otherwise.

5.87.
$$f(y \mid x) = 4e^{-4y} \text{ for } y \ge 0$$

5.89.

x	0	1	2	3
$E[Y \mid x]$	2.8572	2.1429	1.4286	0.7143

5.91. $E[Y \mid x] = 1/(2 - x)$, if $0 \le x < 2$ **5.93.** $E[Y \mid x] = \frac{1}{4}$ **5.95.** $E[X \mid y] = 1/(1 - y)$, if $0 \le y < 1$

5.97. $E[Y \mid x] = 4.2 + 0.8667(x - 2)$
$E[X \mid y] = 2.0 + 0.1846(y - 4.2)$
$$f(y \mid x) = \frac{(2\pi)^{-1/2}}{2.38} \exp\{-[y - 0.8667x - 2.4666)/2.38]^2/2\}$$
$$f(x \mid y) = \frac{(2\pi)^{-1/2}}{1.10} \exp\{-[(x - 0.1846y - 1.2247)/1.10]^2/2\}$$

5.99. a) $1119 + 5576x$
  b)
$$f(y \mid 0.184) = \frac{1}{\sqrt{2\pi}(226.2)} e^{-\frac{1}{2}\left(\frac{y - 2145}{226.2}\right)^2}$$

  c) 0.9978

CHAPTER 6

6.1.

k	$X(k)$	Probability
1	0	0.5200
	1	0.4000
	2	0.0800
2	0	0.2704
	1	0.4160
	2	0.2432
	3	0.0640
	4	0.0064
3	0	0.1406
	1	0.3245
	2	0.3145
	3	0.1638
	4	0.0484
	5	0.0077
	6	0.0005

6.3.
$$P_n(t) = \frac{e^{-5t}(5t)^n}{n!}$$

6.5. **a)**

$$P = \begin{pmatrix} \frac{1}{2} & \frac{1}{4} & \frac{1}{4} & 0 & 0 & 0 \\ \frac{1}{4} & \frac{1}{2} & 0 & \frac{1}{4} & 0 & 0 \\ \frac{1}{6} & 0 & \frac{1}{2} & \frac{1}{6} & \frac{1}{6} & 0 \\ 0 & \frac{1}{6} & \frac{1}{6} & \frac{1}{2} & 0 & \frac{1}{6} \\ 0 & 0 & \frac{1}{2} & 0 & \frac{1}{2} & 0 \\ 0 & 0 & 0 & \frac{1}{2} & 0 & \frac{1}{2} \end{pmatrix}$$

b) $(0.1539 \quad 0.1539 \quad 0.2512 \quad 0.2512 \quad 0.0949 \quad 0.0949)^T$
c) 0.1539

6.7. **a)**

$$P = \begin{pmatrix} 0.82 & 0.18 \\ 0.15 & 0.85 \end{pmatrix}$$

b) $\mathbf{p}(3) = (0.4682 \quad 0.5318)^T$
c) $(0.4545 \quad 0.5455)^T$

6.9. **a)**

$$P = \begin{pmatrix} 0.8 & 0.2 & 0 \\ 0.6 & 0.2 & 0.2 \\ 0 & 0 & 1 \end{pmatrix}$$

b) $\mathbf{p}(3) = (0.6976 \quad 0.1840 \quad 0.1184)^T$

6.11. **a)**

$$P = \begin{pmatrix} 0.95 & 0.05 \\ 0.35 & 0.65 \end{pmatrix}$$

b) $\mathbf{p}(5) = (0.8847 \quad 0.1153)^T$
c) $(0.875 \quad 0.125)^T$

6.13. **a)** The process is not Markovian because the transition probabilities depend on the last two states, not just the last one.
b) 0.58333

6.15. a) $P_0 = 0.28$, $P_1 = 0.2016$, $P_2 = 0.1452$, $P_3 = 0.1045$, $P_4 = 0.0752$
b) 0.2687   c) 0.4816   d) 0.1935   e) 2.5714   f) 9.1837   g) 1.8514   h) 0.1543   i) 0.2143

6.17. a) $P_0 = 0.25$, $P_1 = 0.1875$, $P_2 = 0.1406$, $P_3 = 0.1055$, $P_4 = 0.0791$
b) 0.3164   c) 0.4375   d) 0.2373   e) 3.0   f) 12   g) 2.25   h) 0.75   i) 1.0

6.19. The server is idle exactly 5 seconds out of every 20 seconds.   **6.21.** $R(t) = (1 + t)^{-3}I_{[0,\infty)}(t)$

6.23. $MTTF_1 = 4$, $MTTF_2 = 1$, System 1 is more reliable.   **6.25.** $h(t) = 3(t + 1)^{-1}$, for $t \geq 0$

6.27. $h_1(t) = (1 - t/8)^{-1}/8$, for $0 \leq t \leq 8$
$h_2(t) = (1 + t)^{-1}$, for $t \leq 0$

6.29. $R(t) = \exp[-4t^7/7 - 0.8t - 0.3t^2]$ **6.31.** $R(t) = e^{-(1.15)t}$, $MTTF = 0.8696$

6.33. $MTTF = 20.8333$, $P(T < 1) = 8.2 \times 10^{-5}$

6.35.

$$R(t) = 1 - [1 - \exp(-4t^7/7)][1 - \exp(-0.8t)][1 - \exp(-0.3t^2)]$$

6.37. $R(t) = e^{-0.1t}[1 - (1 - e^{-0.06t})^3]$, $MTTF = 8.685$

6.39. $R(t) = 1 - 4t^3 + 2t^4 + 2t^5 - t^6$, for $0 \leq t \leq 1$; $MTTF = 0.5905$

6.41. $MTTF = 9.375$ months, Var $[T] = 29.297$, $P(T > 6) = 0.696$ **6.43.**   1.936178

6.45. $H[X] = 2$, $H[Y] = 2.32192$, $H[Z] = 2.58496$ **6.47.**   1.85917

6.49. 0 $\longleftrightarrow$ 00000
1 $\longleftrightarrow$ 00001
2 $\longleftrightarrow$ 0001
3 $\longleftrightarrow$ 01
4 $\longleftrightarrow$ 1
5 $\longleftrightarrow$ 001
For the given encoding, $E[B] = 2.0645$.

6.51.

$$0 \longleftrightarrow 001$$
$$1 \longleftrightarrow 1$$
$$2 \longleftrightarrow 01$$
$$3 \longleftrightarrow 0001$$
$$4 \longleftrightarrow 00001$$
$$5 \longleftrightarrow 00000$$

For the given encoding, $E[B] = 2.0645$.

6.53. theoretical mean $= 3.5$ **6.55.** theoretical mean $= 6.0$ **6.57.** theoretical mean $= 27$

CHAPTER 7

7.1. For $x_1, x_2, x_3, x_4 \geq 0$,

$$f(x_1, x_2, x_3, x_4) = 3^{-4} e^{-(x_1 + x_2 + x_3 + x_4)/3}$$

otherwise, $f(x_1, x_2, x_3, x_4) = 0$.

7.3 For $x_1, x_2, \ldots, x_n = 0, 1, 2, \ldots,$

$$f(x_1, x_2, \ldots, x_n) = \frac{e^{-n\lambda}\lambda^{x_1 + x_2 + \cdots + x_n}}{x_1! x_2! \cdots x_n!}$$

7.5

$$f(x_1, x_2, \ldots, x_n) = \begin{cases} (b - a)^{-n}, & \text{if } a \leq x_1, x_2, \ldots, x_n \leq b \\ 0, & \text{otherwise} \end{cases}$$

7.7. $\hat{\mu} = 1226$

7.9.

$$f(x) = \frac{\sqrt{n}}{\sqrt{2\pi}\sigma} e^{-\frac{1}{2}\left(\frac{x - \mu}{\sigma/\sqrt{n}}\right)^2}$$

7.13. $E[\hat{\mu}] = \mu$, $\text{Var}[\hat{\mu}] = \sigma^2/2$ **7.15.** 0.9264 **7.17.** 0.9715 **7.19.** 293

7.21. 1297 **7.25.** 208 **7.27.** 147

7.29.

$$f_{Y_1}(y) = \frac{3(1560 - y)^2}{8000} I_{[1540, 1560]}(y)$$

$$f_{Y_2}(y) = \frac{3(y - 1540)(1560 - y)}{4000} I_{[1540, 1560]}(y)$$

$$f_{Y_3}(y) = \frac{3(y - 1540)^2}{8000} I_{[1540, 1560]}(y)$$

7.31.

$$f_{Y_1}(y) = \begin{cases} 0, & \text{if } y < 0 \\ 2(1 - y/2)^3, & \text{if } 0 < y < 1 \\ [(3 - y)/4]^3, & \text{if } 1 < y < 3 \\ 0, & \text{if } y > 3 \end{cases}$$

$$f_{Y_4}(y) = \begin{cases} 0, & \text{if } y < 0 \\ y^3/4 & \text{if } 0 < y < 1 \\ (y + 1)^3/64, & \text{if } 1 < y < 3 \\ 0, & \text{if } y > 3 \end{cases}$$

7.43. $\hat{\mu} = \bar{X}$ **7.45.** $\hat{p} = \bar{X}$ **7.47.** $\hat{a} = \bar{X} - 1$

7.49. 99% confidence interval is $(3169.23, 3242.20)$, 85% confidence interval is $(3185.32, 3226.11)$

7.51. 609 **7.53.** $(31,897, 33,023)$

7.55. 95% confidence interval is $(4.313, 4.657)$, 99% confidence interval is $(4.231, 4.739)$

7.57. $(3139.88, 3271.55)$

7.59. 90% confidence interval is $(0.8415, 0.9299)$, 95% confidence interval is $(0.8330, 0.9384)$

7.61. No, since $n\overset{*}{p} = 2$ is not greater than or equal to 5.   **7.63.** $(2915.0, 24,382.1)$

7.65. **a)** The mean is $\frac{1}{6}$ and the variance is $\frac{5}{468}$.

b)

$$\frac{p^{10\bar{x}+1}(1-p)^{19-10\bar{x}}}{B(10\bar{x}+2,\,20-10\bar{x})}$$

c) The mean is $(10\bar{x}+2)/22$ and the variance is $(10\bar{x}+2)(20-10\bar{x})/11{,}132$.

7.67. The posterior density is

$$f(q|x) = \frac{(2/3)^{-(x+1)}}{x!}\,q^x e^{-3q/2}$$

the mean is $2(x+1)/3$, and the standard deviation is $\dfrac{2}{3}(x+1)^{1/2}$.

7.69. For $x_1, x_2, \ldots, x_n, b > 0$,

$$f(b\,|\,x_1, x_2, \ldots, x_n) = \frac{b^n e^{-b(n\bar{x}+c)}}{\displaystyle\int_0^\infty b^n e^{-b(n\bar{x}+c)}\,db}$$

7.71.

$$\frac{\displaystyle\int_0^\infty b^{n+1} e^{-nb\bar{x}-bc}\,db}{\displaystyle\int_0^\infty b^n e^{-nb\bar{x}-bc}\,db}$$

CHAPTER 8

8.1. $H_1: p = 0.20$, $\alpha = 0.171$, $\beta = 0.167$. Reject H_0 if 2 or more fail to meet the standard.

8.3. Using a normal approximation gives critical value $c = 3.4$ for $\alpha = 0.1$. Using $c = 3.4$, $\beta = 0.064$. Reject H_0 if 4 or more fail.

8.5. $H_1: b = 0.25$. Reject H_0 if $x \leq 1.0536$; $\beta = 0.7684$. **8.7.** $H_1: \mu = 940$. Reject H_0 if $x \geq 1065$; $\beta = 0.87$.

8.9. Reject H_0 if $x \geq 102.85$; $\beta(105) = 0.1069$, $\beta(110) \approx 0$, $\beta(115) \approx 0$.

8.11. $H_1: \mu < 1.2$, the critical region is $x \leq .12643$, $\beta(1) = 0.8812$, $\beta(0.5) = 0.7766$.

8.13. $H_1: \mu < 0.13$, the critical region is $\bar{x} \leq 0.1287$, $\beta < 0.0001$.

8.15. The critical region is $\bar{x} \leq 39.35$ or $\bar{x} \geq 40.65$, and $\beta(39) = 0.084$.

8.17. Since $\bar{x} = 41.66$, reject the null hypothesis. **8.19.** $H(\mu) = \Phi[(102.85 - \mu)/1.73]$

8.21.

$$H(\mu) = \mu^{-1}\int_{0.12643}^\infty e^{-x/\mu}\,dx$$

8.23. $H(\mu) = \Phi[(40.65 - \mu)/0.253] - \Phi[(39.35 - \mu)/0.253]$

8.25. $H_1: \mu \neq 2760$, $\beta(2750) = 0.5791$. Since $z = -2.177$ and $z_{0.05} = 1.645$, reject H_0.

8.27. The necessary sample size is $n = 2$, the acceptance region remains $(-1.645, 1.645)$; however, the test statistic is now $Z = (\bar{X} - 2760)/(24/\sqrt{2})$.

8.29. $H_1: \mu > 18$, and the critical region is $z \geq 2.33$. Since $z = 1.866$, accept H_0.

8.31. $H_1: \mu < 2.50$ and $\beta(2.40) = 0.668$. Since $z = -1.815$, reject H_0.

8.33. $H_1: \mu > 925$ and to make $\beta(1000) = 0.1$, n must be 6; however, to employ the central limit theorem, we need $n = 30$.

8.35. $z = 4.490$, $p < 0.0003$ **8.37.** $z = 1.866$, $p = 0.031$

8.39. Reject $H_0: \mu = 8$, since $t_{0.025,11} = 2.201$ and $t = 6.035$.

8.41. Accept H_0, since $t_{0.025,9} = 2.262$ and $t = -1.814$; $\beta(1250) = 0.20$.

8.43. Accept H_0, since $t_{0.05,8} = 1.860$ and $t = -0.406$. **8.45.** For $\sigma = 2.0$, $\beta(3.1) \approx 0.88$; for $\sigma = 0.4$, $\beta(3.1) \approx 0.30$.

8.47. Accept $H_0: \mu \leq 200$, since $t_{0.05,9} = 1.833$ and $t = 1.092$. Since $t_{0.25,9} = 0.703$ and $t_{0.10,9} = 1.383$, $0.10 < p < 0.25$.

8.49. 3.75   **8.51.** approximately 75

8.53. H_1: $p > 0.1$, $C = \{4, 5, \ldots, 15\}$, $\alpha = 0.0556$, $\beta = 0.6482$. For 3 failures, H_0 is accepted.

8.55. Using a large-sample approach, $z = -3.136$ and H_0 is rejected; $\beta(0.78) = 0.82$, $\beta(0.86) = 0.88$.

8.57. H_1: $p < 0.82$, the critical region is $z \leq -1.645$, and $\beta(0.78) = 0.84$.

8.59. 861   **8.61.** H_1: $p > 0.1$, the critical region is $z \geq 1.28$, and $n = 738$.

8.63. $n = 25$. The critical region consists of $w \leq 12.401$ and $w \geq 39.364$.

8.65. H_1: $\sigma^2 \neq 576$. Since $w = 12.735$, accept H_0.   **8.67.** Since $w = 4.312$, accept H_0. For $\beta(16) = 0.4$, $n = 20$.

8.69. 12.9128   **8.71.** $n = 29$, $w \leq 12.461$ **8.73.** $n = 12$, $w \leq 4.575$

8.75. The critical region is defined by $10\bar{x} \geq c$. For $\alpha = 0.05$, $c = 16$ with $\alpha = P(n\bar{X} \geq 16; H_0) = 0.0487$.

8.79. actual $\alpha = 0.0577$, $C = \{0, 1, \ldots, 6\}$, $v = 8$, accept H_0 **8.81.** actual $\alpha = 0.0625$, $C = \{0, 1\}$, $v = 2$, accept H_0

8.83. actual $\alpha = 0.0287$, $C = \{11, 12, 13, 14\}$, $v = 12$, reject H_0 **8.85.** actual $\alpha = 0.0195$, $C = \{0, 1\}$, $v = 4$, accept H_0

8.87. actual $\alpha = 0.050$, critical value $= 84$, $s = 96.5$, reject H_0

8.89. actual $\alpha = 0.010$, critical value $= 50$, $s = 49.5$, accept H_0

8.91. actual $\alpha = 0.049$, critical value $= 8$, $s = 17.5$, accept H_0 **8.93.**

8.95. For the full table, $\bar{X} = 15.97$, $\bar{R} = 0.208$, $UCL = 16.07$, and $LCL = 15.87$. Remove samples 10, 12, 13, 15, and 19. Then $UCL = 16.09$, $LCL = 15.87$, and all means fall within the control limits.

8.97. For the full table, $\bar{X} = 28.05$, $\bar{R} = 0.307$, $UCL_{\bar{X}} = 28.23$, $LCL_{\bar{X}} = 27.87$, $UCL_R = 0.65$, and $LCL_R = 0$. Remove samples 4, 7, 9, and 19. Recomputation yields $UCL_{\bar{X}} = 28.18$, $LCL_{\bar{X}} = 27.90$, $UCL_R = 0.50$, and $LCL_R = 0$. Remove samples 5 and 14. Recomputation yields $UCL_{\bar{X}} = 28.17$, $LCL_{\bar{X}} = 27.91$, $UCL_R = 0.49$, and $LCL_R = 0$. Remove samples 2 and 21. The final control limits are $UCL_{\bar{X}} = 28.20$, $LCL_{\bar{X}} = 27.92$, $UCL_R = 0.50$, and $LCL_R = 0$. (Note that all control limits have been formed by rounding to second decimal place.)

8.99. For the full table, $\bar{p} = 0.04875$, $UCL_p = 0.1509$, and $LCL_p = 0$. Remove sample 4. Recomputation yields $UCL_p = 0.1346$ and $LCL_p = 0$. Remove sample 19. The final control limits are $UCL_p = 0.1216$ and $LCL_p = 0$.

8.101. For the full table, $\bar{c} = 16.2273$, $UCL_c = 28.3$, and $LCL_c = 4.1$. Remove samples 15, 19, and 20. The final control limits are $UCL_c = 26.6$ and $LCL_c = 3.4$.

CHAPTER 9

9.1. H_1: $\mu_X \neq \mu_Y$, $z = -1.957$, accept H_0 **9.3.** H_1: $\mu_X - \mu_Y > 5$, $z = 3.015$, reject H_0

9.5. H_1: $\mu_X - \mu_Y < 4$, $z = -2.334$, reject H_0 **9.7.** $\beta(1) = 0.9434$ **9.9.** $\beta(3) = 0.6119$

9.11. a) $n = 72$ b) $n = 48$ **9.13.** H_1: $\mu_X < \mu_Y$, $t = -1.476$, accept H_0

9.15. H_1: $\mu_X \neq \mu_Y$, $t = 4.443$, reject H_0 **9.17.** H_1: $\mu_X > \mu_Y$, $t = 0.552$, accept H_0

9.19. H_1: $\mu_X - \mu_Y < -3.85$, $t = -2.463$, reject H_0 **9.21.** $n_X = n_Y = 4$

9.23. $v = 10$, $t = -1.515$, accept H_0 **9.25.** H_1: $\mu_D < 0$, $t = -6.708$, reject H_0, $p < 0.005$

9.27. H_1: $\mu_D > 0$, $t = 1.439$, accept H_0, $p = 0.095$ **9.29.** H_1: $\mu_D \neq 0$, $t = -1.214$, accept H_0

9.31. a) $\beta(-1) \approx 0$ b) $\beta(1) \approx 0.22$ **9.33.** $(-0.68, 3.28)$ **9.35.** $(2.24, 4.16)$

9.37. $(-0.361, 0.611)$ **9.39.** $(-431.1, -40.7)$ **9.41.** $(-0.117, 0.479)$

9.43. H_1: $p_X \neq p_Y$, $z = -0.714$, accept H_0

9.45. Since $z = 1.236$, H_0 is accepted whether the alternative hypothesis is H_1: $p_X > p_Y$ or H_1: $p_X \neq p_Y$.

9.47. H_1: $p_X \neq p_Y$, $z = -1.684$, reject H_0 **9.49.** H_1: $p_X \neq p_Y$, $z = 3.381$, reject H_0

9.51. H_1: $\sigma_X^2 \neq \sigma_Y^2$, $f = 11.028$, reject H_0 **9.53.** H_1: $\sigma_X^2 > \sigma_Y^2$, $f = 18.913$, reject H_0

9.55. $n \approx 40$ **9.57.** H_1: $\mu_X < \mu_Y$, $w = 25$, reject H_0, $\alpha = 0.047$

9.59. H_1: $\mu_X \neq \mu_Y$, $z = 3.459$, reject H_0 **9.61.** H_1: $\mu_X - \mu_Y < -3.85$, $w = 18$, reject H_0, $\alpha = 0.041$

9.63. H_1: $\mu_X < \mu_Y$, $w = 36$, accept H_0, $\alpha = 0.047$

9.65. After removing ties, $n = 4$. Thus, the smallest nonempty two-tailed critical region is for $\alpha = 0.125$, in which case $C = \{0, 4\}$. Since $v = 1$, H_0 is accepted. If such a large value of α is unacceptable, then the test cannot be employed based on the small amount of data.

9.67. H_1: $\mu_D < 0$, $s = 0$, reject H_0, $\alpha = 0.047$ **9.69.** H_1: $\mu_D < -7$, $s = 5$, reject H_0, $\alpha = 0.109$

9.C1. $t = 5.454$, reject H_0 **9.C3.** $t = -6.698$, reject H_0

CHAPTER 10

10.1. $\chi^2 = 4.558$, accept H_0 **10.3.** $\chi^2 = 2.083$, accept H_0 **10.5.** $\chi^2 = 25.7$, reject H_0

10.7. $\chi^2 = 5.26$, reject H_0

10.9. $\chi^2 = 1.289$, accept H_0 (no reason to doubt normality)   **10.11.** $\chi^2 = 0.287$, accept H_0

10.13. $\chi^2 = 2.872$, accept H_0

10.15. The bottom two classes are combined, as are the top two. The result is $\chi^2 = 5.805$ and accept H_0.

10.17. $\chi^2 = 9.774$, reject H_0 **10.19.** $\chi^2 = 8.987$, reject H_0 **10.21.** $\chi^2 = 13.636$, reject H_0

10.23. $\chi^2 = 4.47$, accept H_0

10.25. a) $\chi^2 = 22.842$, reject H_0
 b) $\chi^2 = 2.098$, accept H_0
 c) $\chi^2 = 21.264$, reject H_0
 d) A and B equally effective, C less effective

10.27. $\chi^2 = 38.144$, reject H_0

CHAPTER 11

11.1. a)

Source	DF	SS	MS	F	p
PLANT	3	37.258	12.419	44.48	0.000
ERROR	21	5.864	0.279		
TOTAL	24	43.122			

 b) Plant is significant at level 0.01.
 c) Treatment-effect estimates for plants A, B, C, D: -0.744, -0.694, -0.830, 1.956

11.3. a)

Source	DF	SS	MS	F	p
MIX	3	115038	38346	5.08	0.017
ER-ROR	12	90622	7552		
TO-TAL	15	205659			

 b) Mix is significant at level 0.05.
 c) Treatment-effect estimates for mixes 1, 2, 3, 4: -48.8, 140.7, -11.6, -80.3

11.5. a)

Source	DF	SS	MS	F	p
COP-IER	3	11.52	3.84	0.44	0.730
ER-ROR	12	105.13	8.76		
TO-TAL	15	116.66			

b) Copier is not significant for job time.

11.7. a)

Source	DF	SS	MS	F	p
LAB	2	0.0000023	0.0000011	7.09	0.026
ER-ROR	6	0.0000010	0.0000002		
TO-TAL	8	0.0000032			

b) Laboratory is significant at level 0.05.
c) Treatment-effect estimates for laboratories A, B, C: -0.000678, 0.000156, 0.000522

11.9. a)

Source	DF	SS	MS	F	p
SOURCE	2	0.005150	0.002575	10.19	0.005
ERROR	9	0.002275	0.000253		
TOTAL	11	0.007425			

b) Source is significant at level 0.01.
c) Treatment-effect estimates for A, B, C: 0.0275, -0.005, -0.0225

11.11. H_0 not rejected, $b = 7.120$ **11.13.** H_0 not rejected, $b = 1.069$ **11.15.** $\mu_1 \neq \mu_2$ **11.17.** Do nothing.

11.19. $\mu_1 \neq \mu_3$ **11.21.** $(-0.00111, 0.0264)$, $(-0.0134, 0.0241)$, $(-0.0128, 0.0248)$ $(-0.0094, 0.0281)$

11.23. Do nothing. **11.25.** $\mu_1 \neq \mu_2$, $\mu_1 \neq \mu_3$ (at level 0.05)

11.27. $f = 2.05$; accept the null hypothesis of uniform iron content at level 0.05.

11.29. $f = 2.66$; accept the null hypothesis at level 0.05.

11.31. $f = 10.73$; the null hypothesis is rejected and the estimate of the common treatment-effects variance is 5.393.

11.33. 0.60 **11.35.** 0.40 **11.37.** $\beta = 0.13$, $n = 3$ **11.39.** $\beta = 0.09$, $n = 4$

11.41. Accept null hypothesis at level 0.05; $k = 0.740$.

11.43. Reject null hypothesis at level 0.01; $k = 9.974$.

11.45. Accept null hypothesis at level 0.05; $k = 6.848$.

11.47. Result is significant at level 0.05, but not at level 0.01; $k = 6.269$.

11.C1. Reject H_0 at level 0.01; $f = 56.03$. **11.C3.** Reject H_0 at level 0.01; $f = 15.57$.

11.C5. $\mu_1 \neq \mu_3$, $\mu_2 \neq \mu_3$ (at level 0.05)

11.C7. Reject null hypothesis at level 0.01; $f = 109.12$. The estimate of the common treatment-effects variance is 210.3.

CHAPTER 12

12.1. $a + bx_i, \frac{1}{3}$

12.3.
$$f(e_1, e_2, \ldots, e_n) = (2\pi)^{-n/2}\sigma^{-n} \exp\left[-(e_1^2 + e_2^2 + \cdots + e_n^2)/2\sigma^2\right]$$

12.5. $\overset{*}{Y} = 10.6 + 0.0965x$ **12.7.** $\overset{*}{Y} = 185 + 293x$ **12.9.** $\overset{*}{Y} = 0.777 - 0.00871x$

12.11. $\overset{*}{Y} = -2573 + 0.860x$

12.13. $\log y = 11.1 - 0.000193x$

12.15. $P(|\hat{b} - b| < r) \approx P(|Z| < r\sqrt{(n-2)S_{xx}}/SSE)$
For Exercise 12.5, $P(|\hat{b} - b| < 0.1) \approx 1$.

12.19. normal with mean $S_{xx}b$ and variance σ^2 **12.21.** (0.0839, 0.1091) **12.23.** (77.0887, 508.631)

12.25. $(-0.23597, -0.05675)$ **12.29.** reject H_0, $t = -2.647$ **12.31.** reject H_0, $t = 25.18$

12.33. reject H_0, $f = 234.97$ **12.35.** (77.588, 97.212) **12.37.** (1.327, 2.013)

12.39. reject H_0, $t = -2.485$ **12.41.** (0.476, 0.555) **12.43.** (48.517, 53.123)

12.45. (0.397, 0.634) **12.47.** (25.37, 41.41) **12.49.** reject linear model, $f = 17.843$

12.51. $r = 0.970$ **12.53.** (0.9054, 0.9907) **12.55.** reject H_0, test statistic $= 5.534$

12.57. reject H_0, $t = 13.755$ **12.59.** $y = 1043.6 + (1094.7)(x - 0.195)$ **12.61.** 0.897

12.C1. $\overset{*}{Y} = 11.1 + 0.0956x$ **12.C3.** $\overset{*}{Y} = -2512 + 0.851x$ **12.C5.** reject H_0, $f = 21.888$

CHAPTER 13

13.1. $1, -2, 2$ **13.3.** $\overset{*}{Y} = 19.0 - 0.0767x_1 - 0.358x_2$ **13.5.** $\overset{*}{Y} = -4563 + 85.2x_1 + 0.257x_2 - 34.2x_3$

13.7. $\overset{*}{Y} = -3.22 + 1.58x_1 + 82.3x_2 + 0.0037x_3$

13.11. $(\mu \quad 1/b \quad \alpha\beta)^T$

13.15. $S_{yy} = 4.0889$, $SSR = 3.963$, $SSE = 0.1256$, $MSE = 0.0209$

13.17. $S_{yy} = 633{,}094$, $SSR = 625{,}526$, $SSE = 7567$, $MSE = 1892$

13.19. $S_{yy} = 112.6$, $SSR = 110.8$, $SSE = 1.81$, $MSE = 0.91$ **13.21.** $(-0.619, -0.248)$ **13.23.** $(-0.173, 0.181)$

13.25. reject H_0, $t = 3.94$ **13.27.** reject H_0, $t = 4.71$ **13.29.** (11.0339, 11.3339) **13.31.** (1273.7, 1425.2)

13.33. (22.615, 33.810) **13.35.** (31.603, 41.347)

13.37. $\overset{*}{Y} = -102 + 1333x - 905x^2$
$SSE(\text{linear}) = 1032.1$, $SSE(\text{quadratic}) = 172.6$

13.39. $\overset{*}{Y} = -0.29 + 0.164x - 0.000097x^2$
$SSE(\text{linear}) = 8.36$, $SSE(\text{quadratic}) = 3.45$

13.41. $(-9351, -362.9)$ **13.43.** $(-1370.9, -438.68)$

13.45. Full model: $y = -4.25 + 0.100x_1 + 0.100x_2$, $R^2 = 0.979$
Reduced model (x_1): $y = -4.35 + 0.107x_1$, $R^2 = 0.925$; adequacy not rejected $(f = 7.66)$
Reduced model (x_2): $y = 2.87 + 0.194x_2$, $R^2 = 0.217$; adequacy rejected $(f = 107.719)$

13.47. Full model: $y = 17.2 + 0.482x_1 + 0.214x_2$, $R^2 = 0.889$
Reduced model (x_1): $y = 25.5 + 0.530x_1$, $R^2 = 0.812$; adequacy not rejected $(f = 2.78)$
Reduced model (x_2): $y = 39.5 + 0.385x_2$, $R^2 = 0.273$; adequacy rejected $(f = 22.19)$

13.49. 0.988

13.51. Full model: $y = 36.6 + 1.19x_1 + 4.49x_2 + 0.21x_3$
Reduced model (x_1, x_2): $y = 36.6 + 1.19x_1 + 4.49x_2$; adequacy not rejected $(f = 1.19)$
Reduced model (x_1, x_3): $y = 36.6 + 1.19x_1 + 0.21x_3$; adequacy rejected $(f = 530.4)$
Reduced model (x_2, x_3): $y = 36.6 + 4.49x_2 + 0.21x_3$; adequacy rejected $(f = 37.14)$
Reduced model (x_1): $y = 36.6 + 1.19x_1$; adequacy rejected $(f = 265.78)$
Reduced model (x_2): $y = 36.6 + 4.49x_2$; adequacy rejected $(f = 19.16)$
Reduced model (x_3): $y = 36.6 + 0.21x_3$; adequacy rejected $(f = 283.75)$

13.53. Forward selection retains only catalyst and time in the model. With catalyst alone, $R^2 = 0.896$; with catalyst and time, $R^2 = 0.984$.

13.C1. $Y = -1.88 + 0.0586x_1 + 0.178x_2 - 0.039x_3$ **13.C3.** $\overset{*}{Y} = -391 + 26.4x_1 + 0.374x_2 - 30.0x_3 - 17.5x_4$

13.C5. $\overset{*}{Y} = 18.0 - 8.40x + 1.79x^2 - 0.0612x^3$

13.C7. Leaving out load $(f = 0.31)$, tire pressure $(f = 3.78)$, or octane $(f = 3.57)$ yields a three-regressor-variable model for which adequacy is accepted. Leaving out spark $(f = 30.66)$ yields a model for which adequacy is rejected. In general, all reduced models containing spark are deemed adequate $(R^2 = 0.7565$ for spark alone) and all reduced models not including spark are deemed inadequate.

13.C9. Using MINITAB, complexity is included at the first step with $R^2 = 0.3346$ and work time is included at the second step with $R^2 = 0.6088$, the final model including two variables. Had all three variables been included, we would have had $R^2 = 0.612$.

13.C11. Using MINITAB, square feet is included at the first step with $R^2 = 0.595$, the final model including only a single regressor variable. Had all four variables been included, we would have had $R^2 = 0.775$.

CHAPTER 14

14.1 Reject H_0 at level 0.01; $f(\text{method}) = 34.1170$.

Source	DF	SS	MS
METHOD	2	60.667	30.333
ALLOY	3	284.667	94.889
ERROR	6	5.333	0.889
TOTAL	11	350.667	

14.3. Reject H_0 at level 0.05 and not at level 0.01; $f(\text{discipline}) = 7.70975$.

Source	DF	SS	MS
DISPL	3	10.917	3.639
ENVIRON	2	67.167	33.583
ERROR	6	2.833	0.472
TOTAL	11	80.917	

14.5. Reject H_0 at level 0.01; $f(\text{mph}) = 85.9436$.

Source	DF	SS	MS
MPH	3	3197.2	1065.7
TEMP	3	1285.7	428.6
ERROR	9	111.6	12.4
TOTAL	15	4594.4	

14.7. The following table gives estimates for μ_{ij}, $\mu_{i.}$, $\mu_{.j}$, and μ:

	1	2	3	4	5	All
1	34.100	30.200	37.700	35.600	30.000	33.520
2	32.700	29.100	37.800	31.500	30.000	32.220
3	31.300	28.400	37.200	31.600	29.200	31.540
ALL	32.700	29.233	37.567	32.900	29.733	32.427

14.9. The following table gives estimates for μ_{ij}, $\mu_{i.}$, $\mu_{.j}$, and μ:

	1	2	3	4	All
1	176.40	189.30	192.80	180.00	184.63
2	176.00	188.80	192.10	179.20	184.03
3	176.80	189.60	193.00	180.50	184.98
ALL	176.40	189.23	192.63	179.90	184.54

14.11. All pairs of means are judged different ($\alpha = 0.01$).   **14.13.** $\mu_{1.} \neq \mu_{2.}, \mu_{1.} \neq \mu_{4.}$ ($\alpha = 0.05$)

14.15. All pairs are judged different ($\alpha = 0.01$) except $\mu_{3.}$ and $\mu_{4.}$.

14.17. Reject H_0 at level 0.05 but not at level 0.01; $f(\text{technique}) = 5.771$. The estimate of σ_A^2 is 0.8368.

14.19. Reject H_0 at level 0.01; $f(\text{lab}) = 51.87$. The estimate of σ_A^2 is 0.2264.

14.25. Interaction is not significant and main effects are significant; $f(\text{interaction}) = 0.0055$, $f(\text{employee}) = 21.15$, $f(\text{system}) = 32.71$. By Duncan's test, $\mu_{1.} \neq \mu_{2.}$, $\mu_{3.} \neq \mu_{2.}$, and $\mu_{.1} \neq \mu_{.2}$.

Source	DF	SS	MS
EMPL	2	152.72	76.36
SYSTEM	1	118.07	118.07
INTERACTION	2	0.05	0.02
ERROR	12	43.28	3.61
TOTAL	17	314.12	

14.27. Interaction and both main effects are not significant; $f(\text{interaction}) = 0.845$, $f(\text{cylinder}) = 2.455$, $f(\text{fluid}) = 2.664$

Source	DF	SS	MS
CYLINDER	1	27.0	27.0
FLUID	2	58.5	29.3
INTERACTION	2	18.5	9.3
ERROR	6	66.0	11.0
TOTAL	11	170.0	

14.29. Interaction is significant, main effect for algorithms is significant, and main effect for filters is not significant; $f(\text{interaction}) = 13.637$, $f(\text{algorithm}) = 72.274$, $f(\text{filter}) = 0.009$. Because interaction is significant one must be prudent in interpreting the statistics. Because F_A is so great, we conclude that algorithm is significant; however, it is possible that the filter effect has been masked by interaction. Duncan's test yields $\mu_{1.} \neq \mu_{2.}$.

Source	DF	SS	MS
ALGOTRM	1	816.7	816.7
FILTER	1	0.1	0.1
INTERACTION	1	154.1	154.1
ERROR	8	90.0	11.3
TOTAL	11	1060.9	

14.31. $\overset{*}{\alpha}_1 = -5.85$, $\overset{*}{\alpha}_2 = 2.37$, $\overset{*}{\alpha}_3 = 3.48$, $\overset{*}{\beta}_1 = -1.74$, $\overset{*}{\beta}_2 = 0.15$, $\overset{*}{\beta}_3 = 1.59$, $\overset{*}{(\alpha\beta)}_{11} = 0.074$, $\overset{*}{(\alpha\beta)}_{12} = 0.185$, $\overset{*}{(\alpha\beta)}_{13} = -0.259$, $\overset{*}{(\alpha\beta)}_{21} = 0.185$, $\overset{*}{(\alpha\beta)}_{22} = -0.037$, $\overset{*}{(\alpha\beta)}_{23} = -0.148$, $\overset{*}{(\alpha\beta)}_{31} = -0.259$, $\overset{*}{(\alpha\beta)}_{32} = -0.148$, $\overset{*}{(\alpha\beta)}_{33} = 0.408$, $\overset{*}{\mu} = 48.519$

14.33. $\overset{*}{\alpha}_1 = -0.507$, $\overset{*}{\alpha}_2 = 0.170$, $\overset{*}{\alpha}_3 = 0.337$, $\overset{*}{\beta}_1 = 0.493$, $\overset{*}{\beta}_2 = 0.015$, $\overset{*}{\beta}_3 = -0.507$, $\overset{*}{(\alpha\beta)}_{11} = 0.341$, $\overset{*}{(\alpha\beta)}_{12} = 0.119$, $\overset{*}{(\alpha\beta)}_{13} = -0.459$, $\overset{*}{(\alpha\beta)}_{21} = -0.137$, $\overset{*}{(\alpha\beta)}_{22} = -0.059$, $\overset{*}{(\alpha\beta)}_{23} = 0.196$, $\overset{*}{(\alpha\beta)}_{31} = -0.204$, $\overset{*}{(\alpha\beta)}_{32} = -0.059$, $\overset{*}{(\alpha\beta)}_{33} = 0.263$, $\overset{*}{\mu} = 4.874$

14.35. Interaction is insignificant and both main effects are significant; $f(\text{interaction}) = 0.007$, $f(\text{employee}) = 3054$, $f(\text{system}) = 4723$, Var [employees] $= 12.72$, Var [system] $= 13.12$.

14.37. Interaction and both main effects are insignificant; $f(\text{interaction}) = 0.845$, $f(\text{cylinder}) = 2.90$, $f(\text{fluid}) = 3.15$.

14.39. Interaction is significant, with $f(\text{interaction}) = 13.637$; neither main effect appears significant, with $f(\text{algorithm}) = 5.30$ and $f(\text{filter}) = 0.00065$. However, the strong interaction can very possibly be masking the main effects, especially the algorithm effect, whose F statistic is almost in the critical region.

14.41. Accept the interaction null hypothesis and reject both main effect null hypotheses, with $f(\text{interaction}) = 0.2897$, $f(\text{temperature}) = 217.79$, $f(\text{duration}) = 79.88$.

14.43. The interaction null hypothesis is rejected with $f(\text{interaction}) = 8.937$. Since $f(\text{weight}) = 51.998$ is in the critical region, we conclude that weight is significant. Because of possible masking due to interaction, no conclusion is possible with respect to sole, for which $f(\text{sole}) = 4.652$.

14.49. Both factors are significant, with $f(\text{employee}) = 75.03$ and $f(\text{system}) = 196.57$.

14.51. Neither factor is significant, with $f(\text{sole}) = 3.14$ and $f(\text{weight}) = 4.52$.

14.53. Interaction is significant, with $f = 63.78$ and critical value $f_{0.01,1,3} = 34.12$.

14.55. $\hat{\beta}_j = \bar{X}_{.j.} - \bar{X}_{....}$
$\hat{\gamma}_k = \bar{X}_{..k.} - \bar{X}_{....}$
$(\widehat{\alpha\beta})_{ik} = \bar{X}_{i.k.} - \bar{X}_{i...} - \bar{X}_{..k.} + \bar{X}_{....}$
$(\widehat{\beta\gamma})_{jk} = \bar{X}_{.jk.} - \bar{X}_{.j..} - \bar{X}_{..k.} + \bar{X}_{....}$

14.57.

$$E[MSB] = \sigma^2 + \frac{nac}{b-1} \sum_{j=1}^{b} \beta_j^2$$

$$E[MSC] = \sigma^2 + \frac{nab}{c-1} \sum_{k=1}^{c} \gamma_k^2$$

$$E[MSAC] = \sigma^2 + \frac{nb}{(a-1)(c-1)} \sum_{i=1}^{a} \sum_{k=1}^{c} (\alpha\gamma)_{ik}^2$$

$$E[MSBC] = \sigma^2 + \frac{na}{(b-1)(c-1)} \sum_{j=1}^{b} \sum_{k=1}^{c} (\beta\gamma)_{jk}^2$$

14.59. Weight-sole interaction is significant and so too are all main effects; all other effects are insignificant.

Source	DF	SS	MS	E
SOLE	2	6.42	3.21	52.5
WEIGHT	2	6.01	3.00	49.2
SPEED	1	4.00	4.00	65.5
SOLE-WEIGHT	4	1.76	0.44	7.2
WEIGHT-SPEED	2	0.11	0.05	0.9
SOLE-SPEED	2	0.29	0.14	2.4
SOLE-WEIGHT-SPEED	4	0.11	0.03	0.4
ERROR	18	1.10	0.06	
TOTAL	35	19.79		

14.61. Interaction is insignificant, temperature is significant at level 0.01 and duration is significant at level 0.05; $f(\text{temperature}) = 185.26$, $f(\text{duration}) = 9.33$. Moreover, $\alpha_1 = 4.083$, $\beta_1 = 0.917$, and $(\alpha\beta)_{00} = -0.083$.

Source	DF	SS	MS
TEMP	1	200.08	200.08
DURATION	1	10.08	10.08
INTERACTION	1	0.08	0.08
ERROR	8	8.67	1.08
TOTAL	11	218.92	

14.63. Interaction and cylinder are insignificant and fluid is significant at level 0.01; $f(\text{cylinder}) = 3.57$, $f(\text{fluid}) = 32.14$. Moreover, $\alpha_1 = 0.625$, $\beta_1 = -1.875$, $(\alpha\beta)_{00} = -0.125$.

Source	DF	SS	MS
CYLINDER	1	3.125	3.125
FLUID	1	28.125	28.125
INTERACTION	1	0.125	0.125
ERROR	4	3.500	0.875
TOTAL	7	34.875	

14.65. A and C are significant at level 0.01.

Source	DF	SS	MS	F
A	1	588.06	588.06	47.76
B	1	22.56	22.56	1.83
C	1	2002.56	2002.56	162.64
AB	1	14.06	14.06	1.14
AC	1	5.06	5.06	0.41
BC	1	0.56	0.56	0.05
ABC	1	18.06	18.06	1.47
ERROR	8	98.50	12.31	
TOTAL	15	2749.00		

14.67. A, B, C, and AC are significant at level 0.01.

Source	DF	SS	MS	F
A	1	66.67	66.67	28.57
B	1	337.50	337.50	144.64
C	1	54.00	54.00	23.14
AB	1	2.67	2.67	1.14
AC	1	28.17	28.17	12.07
BC	1	0.00	0.00	0.00
ABC	1	4.17	4.17	1.79
ERROR	16	37.33	2.33	
TOTAL	23	530.50		

14.69. No significance at level 0.01.

Source	DF	SS	MS	F
A	1	528.125	528.125	25.00
B	1	28.125	28.125	1.33
C	1	1035.125	1035.125	49.00
AB	1	1.125	1.125	0.05
AC	1	0.125	0.125	0.01
BC	1	6.125	6.125	0.29
ERROR	1	21.125	21.125	
TOTAL	7	1619.875		

14.71. A, B, and C are significant at level 0.01.

Source	DF	SS	MS	F
A	1	45.56	45.56	16.20
B	1	232.56	232.56	82.69
C	1	39.06	39.06	13.89
AB	1	0.06	0.06	0.02
AC	1	10.56	10.56	3.76
BC	1	1.56	1.56	0.56
BLOCKS	1	0.56	0.56	0.20
ERROR	8	22.50	2.81	
TOTAL	15	352.40		

14.73. Two-factor interactions are grouped to obtain SSE. Factor A is significant at level 0.05, factor B is insignificant, and factor C is significant at level 0.01.

Source	DF	SS	MS	F
A	1	128	128.0	13.71
B	1	2	2.0	0.21
C	1	968	968.0	103.71
BLOCKS	1	2	2.0	0.21
ERROR	3	28	9.3	
TOTAL	7	1128		

14.75. Two-factor interactions are grouped to form *SSE*. There are no significant effects.

Source	DF	SS	MS	F
A	1	60.5	60.5	5.04
B	1	12.5	12.5	1.04
C	1	4.5	4.5	0.38
BLOCKS	1	4.5	4.5	0.38
ERROR	3	36.0	12.0	
TOTAL	7	118.0		

14.77.

Block 1	Block 2
(1)	a
b	c
ac	bc
abc	ab

14.79

Block 1	Block 2
(1)	a
ab	b
c	ac
abc	bc
ad	d
bd	abd
acd	cd
bcd	abcd
ae	e
be	abe
ace	ce
bce	abce
de	ade
abde	bde
cde	acde
abcde	bcde

14.81.

Block 1	Block 2	Block 3	Block 4
(1)	*ac*	*c*	*a*
ab	*bc*	*abc*	*b*
cd	*ad*	*d*	*acd*
abcd	*bd*	*abd*	*bcd*
ace	*e*	*ae*	*ce*
bce	*abc*	*be*	*abce*
ade	*cde*	*acde*	*de*
bde	*abcde*	*bcde*	*abde*
acf	*f*	*af*	*cf*
bcf	*abf*	*bf*	*abcf*
adf	*cdf*	*acdf*	*df*
bdf	*abcdf*	*bcdf*	*abdf*
ef	*acef*	*cef*	*aef*
abef	*bcef*	*abcef*	*bef*
cdef	*adef*	*def*	*acdef*
abcdef	*bdef*	*abdef*	*bcdef*

14.83. *AB* interaction is significant at level 0.01, as are *B*, *C*, and *D*. One must be careful not to draw any conclusion regarding *A* because of possible masking due to strong *AB* interaction.

Source	DF	SS	MS	*F*
A	1	0.28	0.28	2.61
B	1	0.98	0.98	9.09
C	1	2.88	2.88	26.70
D	1	1.36	1.36	12.62
AB	1	3.25	3.25	30.15
AC	1	0.45	0.45	4.18
BC	1	0.00	0.00	0.00
CD	1	0.10	0.10	0.94
ABD	1	0.13	0.13	1.16
ABCD	1	0.05	0.05	0.42
BLOCKS	3	0.25	0.09	0.80
ERROR	18	1.94	0.11	
TOTAL	31	11.68		

14.85. Factor *A* is significant at level 0.05.

Source	DF	SS	MS	*F*
A	1	162.56	162.56	14.37
B	1	1.56	1.56	0.14
C	1	3.06	3.06	0.27
AB	1	10.56	10.56	0.93
AC	1	0.63	0.63	0.05
BC	1	18.06	18.06	1.60
ABC	1	0.75	0.75	0.07
BLOCKS	3	0.69	0.23	0.02
ERROR	5	56.56	11.31	
TOTAL	15	254.44		

14.C1.

Source	DF	SS	MS	F
METHOD	2	132.20	66.10	51.24
ALLOY	9	590.17	65.57	50.83
ERROR	18	23.13	1.29	
TOTAL	29	745.50		

14.C3.

Source	DF	SS	MS	F
TEMPERATURE	4	580.56	145.19	89.62
DURATION	2	72.13	36.07	22.27
INTERACTION	8	9.64	1.21	0.75
ERROR	30	48.67	1.62	
TOTAL	44	711.20		

14.C5. $F_A = 119.99$, $F_B = 29.81$, $F_{AB} = 0.75$

14.C7.

Source	DF	SS	MS	F
CATALYST	1	184.08	184.08	79.6
DURATION	1	736.33	736.33	318.4
TEMPERATURE	3	141.17	47.06	20.4
CAT-DUR	1	24.08	24.08	10.4
CAT-TEMP	3	52.42	17.47	7.6
DUR-TEMP	3	2.83	0.94	0.4
CAT-DUR-TEMP	3	15.08	5.03	2.2
ERROR	32	74.00	2.31	
TOTAL	47	1230.00		

INDEX

A

Absolute value, 104
Acceptance region, 375
Active redundancy, 294
Additivity, 648 (*See also* Probability measure)
Affine transformation, 103
Alias, 694
Alternative hypothesis, 375
ANOVA table, 530
    fixed-effects one-way classification, 530
    fixed-effects three-way classification, 674
    fixed-effects two-way classification, 660
    mixed-effects two-way classification, 665
    random-effects one-way classification, 539
    random-effects two-way classification, 664
    randomized complete block design, 645
    significance of regression in the multiple linear model, 615
    significance of regression in the simple linear model, 575
Arrival rate, 277
Arrival stream, 277
Assignable cause, 441
Asymptotically unbiased, 328
Average effect, 681
Average interaction effect, 681

B

Backward elimination, 632
Bartlett's test, 553
    table of critical values, 748
Bayes estimator, 364
Bayesian estimation, 361–69
Bayes risk, 363
Bayes' theorem. *See* Conditional probability
Bernoulli trials, 136 (*See also* Binomial distribution)
Best critical region, 430
Best estimator, 364
Best test, 430
Best unbiased estimator, 335 (*See also* Minimum-variance unbiased estimator)
Best linear unbiased estimator (BLUE), 335
Beta distribution, 185–90
    coefficient of skewness, 198
    density, 163, 185
    generalized beta distribution:
        density, 189
        mean, 189
        use with PERT, 190
        variance, 189
    graph of density, 186–87
    mean, 163, 188
    moments, 188
    relationship to population proportions, 189
    variance, 163, 188
Beta function, 176
Better estimator, 334, 364

Better test, 430
Bias, 318
Biased estimator, 318
Binomial coefficients, 42
Binomial distribution, 136–44
    approximation to hypergeometric distribution, 148
    coefficient of kurtosis, 194
    coefficient of skewness, 194
    density, 137–38, 143
    graph of density, 138
    mean, 142, 143
    moment-generating function, 142
    probability distribution function, 139
    recursion formula, 194
    relationship to Bernoulli trials, 139
    simulation, 301
    sum of, 227
    table, 722
    variance, 142, 143
Binomial theorem, 42
Birthday problem, 46
Bit, 296
Blocks, 636
Brownian motion, 140

C

Cardinality, 26
Cauchy distribution, 111, 163
c chart, 451
Censored sampling, 338
Center of mass. *See* Mean
Central limit theorem, 324
Central tendency (empirical), 13–19

State probability, 270
State space, 267
State vector, 272
Stationary process, 272
Statistic, 312
Statistical control, 441
Statistical hypothesis, 374
Statistical inference, 309
Steady-state, 279
Steady-state distribution, 276
Steady-state probabilities, 279
Stochastic process, 266
   continuous-time process, 266
   discrete-time process, 266
   Poisson process, 267
Studentized range distribution, 535
   table of critical values, 739
Subjective approach, 362
Sum of squares:
   between-treatment, 526
   blocks, 643
   error, 530, 560, 600, 643, 659
   factor, 658
   lack-of-fit, 583
   nonadditivity, 668
   pure error, 582
   regression, 574
   total, 526, 658
   treatment interaction, 658
   treatments, 642
   within-treatment, 526
Sum-of-squares identity, 526
Symmetric difference, 71
Synthetic system data, 299
System failure rate, 289

T

t distribution, 353
   degrees of freedom, 353
   density, 353
   mean, 356
   table of critical values, 734
   variance, 356
Time-to-failure distribution, 283
Total effect, 681
Transition probability, 268
Transition probability matrix, 271
   one-step transition matrix, 271, 273
Treatment combination, 652
Treatment population, 522
Tukey's test for interaction, 668
Tukey's test for multiple comparisons, 535
Two-tailed test, 389
Two-way classification format, 653
   totals, 654

U

Unbiased estimator, 318
Unbiased linear estimator, 335
Uncertainty, 296 (*See also* Entropy)
Uncorrelated variables, 242
Uniform distribution:
   coefficient of kurtosis, 134
   coefficient of skewness, 133
   coefficient of variation, 133
   density, 97, 163
   graph of density, 98
   joint uniform distribution, 213
   mean, 116, 163
   mean deviation, 133
   median, 133
   moment-generating function, 126
   moments, 116
   probability distribution function, 97–98
   random values for, 745
   semi-interquartile range, 135
   simulation, 301
   variance, 120, 163

Uniformly most powerful test, 435
Upper control limit, 441
Utilization factor, 279

V

Variance: (*See also* Sample variance; Standard deviation)
   coefficient of variation (empirical), 20
   empirical, 20
   of sum, 248
   pooled sample, 467
   probability distribution, 118
   sample, 326

W

Waiting time, 280
Weibull distribution, 190–93
   density, 163, 190
   graph of density, 191
   mean, 163, 191
   moments, 191
   time-to-failure, 287
   variance, 163, 191
Wilcoxon rank-sum test. *See* Rank-sum test
Wilcoxon signed-rank test. *See* Signed-rank test

X

$\bar{X}$ chart, 441

Y

Yates' correction for continuity, 518
Yates' method for computing contrasts, 690

Critical Values for the F Distribution (Continued)

$$\alpha = 0.01$$

ν_2	ν_1								
	1	2	3	4	5	6	7	8	9
1	4052	4999.5	5403	5625	5764	5859	5928	5981	6022
2	98.50	99.00	99.17	99.25	99.30	99.33	99.36	99.37	99.39
3	34.12	30.82	29.46	28.71	28.24	27.91	27.67	27.49	27.35
4	21.20	18.00	16.69	15.98	15.52	15.21	14.98	14.80	14.66
5	16.26	13.27	12.06	11.39	10.97	10.67	10.46	10.29	10.16
6	13.75	10.92	9.78	9.15	8.75	8.47	8.26	8.10	7.98
7	12.25	9.55	8.45	7.85	7.46	7.19	6.99	6.84	6.72
8	11.26	8.65	7.59	7.01	6.63	6.37	6.18	6.03	5.91
9	10.56	8.02	6.99	6.42	6.06	5.80	5.61	5.47	5.35
10	10.04	7.56	6.55	5.99	5.64	5.39	5.20	5.06	4.94
11	9.65	7.21	6.22	5.67	5.32	5.07	4.89	4.74	4.63
12	9.33	6.93	5.95	5.41	5.06	4.82	4.64	4.50	4.39
13	9.07	6.70	5.74	5.21	4.86	4.62	4.44	4.30	4.19
14	8.86	6.51	5.56	5.04	4.69	4.46	4.28	4.14	4.03
15	8.68	6.36	5.42	4.89	4.56	4.32	4.14	4.00	3.89
16	8.53	6.23	5.29	4.77	4.44	4.20	4.03	3.89	3.78
17	8.40	6.11	5.18	4.67	4.34	4.10	3.93	3.79	3.68
18	8.29	6.01	5.09	4.58	4.25	4.01	3.84	3.71	3.60
19	8.18	5.93	5.01	4.50	4.17	3.94	3.77	3.63	3.52
20	8.10	5.85	4.94	4.43	4.10	3.87	3.70	3.56	3.46
21	8.02	5.78	4.87	4.37	4.04	3.81	3.64	3.51	3.40
22	7.95	5.72	4.82	4.31	3.99	3.76	3.59	3.45	3.35
23	7.88	5.66	4.76	4.26	3.94	3.71	3.54	3.41	3.30
24	7.82	5.61	4.72	4.22	3.90	3.67	3.50	3.36	3.26
25	7.77	5.57	4.68	4.18	3.85	3.63	3.46	3.32	3.22
26	7.72	5.53	4.64	4.14	3.82	3.59	3.42	3.29	3.18
27	7.68	5.49	4.60	4.11	3.78	3.56	3.39	3.26	3.15
28	7.64	5.45	4.57	4.07	3.75	3.53	3.36	3.23	3.12
29	7.60	5.42	4.54	4.04	3.73	3.50	3.33	3.20	3.09
30	7.56	5.39	4.51	4.02	3.70	3.47	3.30	3.17	3.07
40	7.31	5.18	4.31	3.83	3.51	3.29	3.12	2.99	2.89
60	7.08	4.98	4.13	3.65	3.34	3.12	2.95	2.82	2.72
120	6.85	4.79	3.95	3.48	3.17	2.96	2.79	2.66	2.56
∞	6.63	4.61	3.78	3.32	3.02	2.80	2.64	2.51	2.41